Metal Toxicology Handbook

Metal Toxicology Handbook

Edited by
Debasis Bagchi and Manashi Bagchi

CRC Press
Taylor & Francis Group
Boca Raton London New York

CRC Press is an imprint of the
Taylor & Francis Group, an **informa** business

Library of Congress Cataloging-in-Publication Data
Names: Bagchi, Debasis, 1954- editor. | Bagchi, Manashi, editor.
Title: Metal toxicology handbook / edited by Debasis Bagchi, Manashi Bagchi.
Description: First edition. | Boca Raton : Taylor & Francis, 2020. |
Includes bibliographical references and index.
Identifiers: LCCN 2020029060 (print) | LCCN 2020029061 (ebook) |
ISBN 9781138345249 (hardback) | ISBN 9780429438004 (ebook)
Subjects: LCSH: Metals—Toxicology. | Organs (Anatomy)
Classification: LCC RA1231.M52 M483 2020 (print) | LCC RA1231.M52 (ebook) |
DDC 615.9/253—dc23
LC record available at https://lccn.loc.gov/2020029060
LC ebook record available at https://lccn.loc.gov/2020029061

ISBN: 978-1-138-34524-9 (hbk)
ISBN: 978-0-429-43800-4 (ebk)

Typeset in Times
by codeMantra

Dedicated to our only loving daughter, Deepanjali Bagchi, who is the prime source of our spontaneous inspiration and motivation.

Debasis Bagchi and Manashi Bagchi

Contents

Part I
Introduction: Metals, Metalloids, Redox Biology, and Neurodegeneration

Amit Madeshiya, Pradipta Banerjee, Suman Santra, Nandini Ghosh,
Sayantani Karmakar, Debasis Bagchi, Sashwati Roy, and Amitava Das

Abhai Kumar, Smita Singh, and Rameshwar Nath Chaurasia

Part II
Pathology of Metal Toxicity

Odete Mendes and Chidozie Amuzie

Yu Xu, Ning Wang, and Yibin Feng

Part III
Mechanisms of Restoring Metabolic Homeostasis

Bernard W. Downs, Manashi Bagchi, Bruce S. Morrison, Jeffrey Galvin,
Steve Kushner, and Debasis Bagchi

Ashfaque Hossain, Muhammad Manjurul Karim, Tania Akter Jhuma, and
Godfred A. Menezes

Preface

Heavy metals and metalloids, singly or in combination, induce their toxic manifestations either through acute or chronic pathology. Especially, long-term chronic exposure to diverse heavy metals and metalloids to humans and animals can lead to numerous physical, muscular, neurological, nephrological, and diverse degenerative diseases and dysfunctions including multiple sclerosis, muscular dystrophy, Parkinson's and Alzheimer's diseases, cardiovascular disorders, and several others. Especially, chronic long-term exposure to selected heavy metals causes genetic impairment, mutation, and loss of genomic integrity, and disrupts hormones leading to endocrine and reproductive system impairment causing birth defects, teratogenicity, and diverse types of deadly cancers. The major mechanisms of heavy metals and metalloids include oxidative stress mediated through enhanced production of oxygen-free radicals and degradation of biological macromolecules including lipids, proteins, nucleic acids, DNA single-strand breaks, and DNA fragmentation, which leads to several dysfunctions such as cardiac dysfunctions, neurotoxicity, nephrotoxicity, and carcinogenesis. Heavy metals such as lead, mercury, arsenic, cadmium, thallium, and hexavalent chromium are well recognized for their enormous toxicity. It is important to note that these metals are listed in the World Health Organization's ten most noxious chemicals of major public health concern. The immediate vital signs of acute heavy metal exposure include nausea, vomiting, diarrhea, and acute abdominal pain. Mercury has been identified as the most toxic heavy metal, and mercury poisoning is known as acrodynia or pink disease. Similarly, lead, another toxic heavy metal, was an integral part of painting.

Over several centuries of research and outstanding inventions, scientists unveiled that although it has been identified that several metals are toxic, it is now been identified that dose is the driving force of toxic manifestations. In the recent past, it has been established that zinc in appropriate dose boosts immunity and reproductive health, mitigates oxidative stress and oxidative DNA damage, and facilitates connective tissue repair as well as shortens the duration of cold and flu. According to Ancient Ayurveda, consumption of drinking water in copper cups boosts overall health and immunity, helps in digestion, kills bacteria, mitigates inflammation, heals wounds, improves joint and thyroid health, and slows down aging. Currently, these claims have been established by researchers. Similarly, selected metals, namely, trivalent chromium, vanadium, boron, and silicon, have exhibited potential health benefits provided that these minerals are used in the right doses, as per recommendation.

In this academic textbook, we exhibit the toxicity, safety, and proper human utilization in a collection of 28 chapters from eminent researchers in the field, and the chapters are divided into seven major sections including a commentary titled "A Treatise on Metal Toxicology" from the editors' desk. The initial introduction section provides a broad overview of metals, metalloids, redox biology, and neurodegeneration. This section consists of two chapters: the first chapter reviews the intricate features on the roles of metals, metalloids, redox biology, and neurodegeneration, while the second chapter highlights the role of metals in neurodegeneration.

The second section discusses the pathology of metal toxicity in two chapters. The first chapter focuses on the pathological manifestations and mechanisms of metal toxicity, and the second chapter elaborates the incidences on common metal contamination and toxicological manifestation in the quality control of Chinese medicines. The third section highlights the mechanism and salient features of restoring metabolic homeostasis in two dedicated chapters. The first chapter elaborates the development and utilization of a novel prodosomed-electrolyte and phytochemical formulation technology in restoring and attenuating metabolic homeostasis, and the second chapter elaborates the intricate aspects on the functional roles of heavy metals on gut microbiome. The fourth section demonstrates the aspects of radionuclides toxicity. An extensive review by Dr. Fco. Javier Guillén Gerada on radionuclides toxicity is presented in this section.

The fifth section discusses the benefits of metals in four individual chapters. The first chapter provides an update on bioavailability of silicon and health, the second and third chapters by the eminent professors Dr. John B. Vincent and Dr. Harry G. Preuss demonstrate the health beneficial effects of trivalent chromium, and the fourth chapter discusses the beneficial effects of vanadium was demonstrated by another eminent researcher Dr. Satinath Mukhopadhyay and his brilliant associate.

The sixth section titled "Toxic Manifestations by Diverse Heavy Metals and Metalloids" provides 16 chapters by distinguished researchers in their respective fields. The first four chapters on copper, cadmium, and cobalt were crafted by distinguished academicians and well-renowned researchers. The fifth and sixth chapters discuss the toxicological mechanism and manifestation by iron. The seventh chapter discusses an extensive interesting chapter of vanadium safety and toxicity and emphasized the importance of maintaining the right balance and highlighted the dose-dependent beneficial and adverse effects. The eighth and ninth chapters highlight the diverse aspects of lead toxicity, and the tenth chapter elaborates on the salient features of arsenic toxicity vividly. Toxic manifestation by mercury is discussed in the eleventh chapter. An overview of tungsten toxicity is provided in the twelfth chapter by Dr. Jennifer L. Freeman, an eminent researcher from Purdue University. The thirteenth chapter by Dr. Mohd. Kamran Khan and his eminent team members from Selcuk University, Turkey, discusses the salient features of boron toxicity, and the fourteenth chapter sphere-headed by Dr. Marvin A. Soriano-Ursúa and his esteemed colleagues provides a more in-depth analysis of the intricate features on boron toxicity. The fifteenth and sixteenth chapters on lithium and nickel toxicity, respectively, were crafted by an Australian scientist Dr. Tanveer, who emphasized the salient features of lithium and nickel toxicity.

The editors wrote a commentary titled "A Treatise on Metal Toxicity" and summarize a vivid scenario of metal toxicity, safety, and its consequences.

Our sincere gratitude and thanks to all our eminent researchers, contributors, as well as helpful CRC Press/Taylor & Francis editorial team members including Stephen Zollo, Randy Brehm, and Laura Piedrahita for their continued support, cooperation, and assistance.

Debasis Bagchi and Manashi Bagchi

Editors

Debasis Bagchi, PhD, MACN, CNS, MAIChE, received his PhD in medicinal chemistry in 1982. He is the Director of Scientific Affairs, VNI, Inc., Lederach, Pennsylvania; an Adjunct Faculty at Texas Southern University, Houston, Texas and a past Faculty in the Department of Pharmacological and Pharmaceutical Sciences at the University of Houston College of Pharmacy, Houston, Texas. He served as the Chief Scientific Officer at Cepham, Inc., Belmont, New Jersey, from June 2013 through December 2018. He was the Senior Vice President of Research and Development of InterHealth Nutraceuticals, Inc., Benicia, California, from 1998 until February 2011, and then as the Director of Innovation and Clinical Affairs of Iovate Health Sciences, Oakville, Ontario, Canada, until June 2013.

Dr Bagchi received the Master of American College of Nutrition Award in October 2010. He is the Past Chairman of International Society of Nutraceuticals and Functional Foods (ISNFF), Past President of American College of Nutrition, Clearwater, Florida, and Past Chair of the Nutraceuticals and Functional Foods Division of Institute of Food Technologists (IFT), Chicago, Illinois. He is serving as a Distinguished Advisor at the Japanese Institute for Health Food Standards (JIHFS), Tokyo, Japan. He is a Member of the Study Section and Peer Review Committee of the National Institutes of Health (NIH), Bethesda, Maryland. He has published 372 papers in peer-reviewed journals, has written 41 books, and has 20 patents. He is also a Member of the Society of Toxicology, Member of the New York Academy of Sciences, Fellow of the Nutrition Research Academy, and Member of the TCE stakeholder Committee of the Wright-Patterson Air Force Base, Ohio. He is the Associate Editor of the *Journal of Functional Foods, Journal of the American College of Nutrition,* and *Archives of Medical and Biomedical Research,* and he also serves as an Editorial Board Member of numerous peer-reviewed journals, including *Antioxidants and Redox Signaling, Cancer Letters, Food and Nutrition Research, Toxicology Mechanisms and Methods, The Original Internist,* and other peer-reviewed journals.

Manashi Bagchi, PhD, FACN, received her PhD in chemistry in 1984. She is currently the Chief Scientific Officer of Dr. Herbs LLC, Concord, California. She is also a Consultant for VNI, Lederach, Pennsylvania. She served as an Associate Professor at the Creighton University School of Pharmacy and Allied Health Profession, Omaha, Nebraska, from September 1990 until August 1999. Later, she served as the Director of Research at InterHealth Nutraceuticals, Benicia, California, from September 1999 to July 2009. She is a Member of the Study Section and Peer Review Committee of the National Institutes of Health (NIH), Bethesda, Maryland. Her research interests include free radicals, human diseases, toxicology, carcinogenesis, anti-aging and anti-inflammatory pathophysiology, mechanistic aspects of cytoprotection by antioxidants and chemoprotectants, regulatory pathways in obesity and gene expression, diabetes, arthritis, and efficacy and safety of natural botanical products and dietary supplements. She is a Member of the Society of Toxicology (Reston, Virginia), the New York Academy of Sciences (New York, New York), and the Institutes of Food Technologists (Chicago, Illinois). She is a Fellow and currently Board Member of the

American College of Nutrition (Clearwater, Florida). She has published 225 papers in peer-reviewed journals and 2 books, *Genomics, Proteomics and Metabolomics in Nutraceuticals and Functional Foods* and *Bio-Nanotechnology: A Revolution in Food, Biomedical and Health Sciences*, from Wiley-Blackwell. She has delivered invited lectures at various national and international scientific conferences and organized workshops and group discussion sessions. She serves as an Editorial Board Member of the *Journal of the American College of Nutrition,* and is a reviewer of many peer-reviewed journals. She received funding from various institutions and agencies, including the U.S. Air Force Office of Scientific Research, the National Institute of Aging, the Nebraska State Department of Health, and the Cancer Society of Nebraska.

Both Dr. D. Bagchi and Dr. M. Bagchi have extensively researched diverse heavy metals and metalloids, including cadmium, chromium(III), chromium(VI), arsenic, nickel, platinum, zinc, vanadium, manganese, magnesium, bismuth, cobalt, calcium, potassium, iron, and silver since 1993 to demonstrate the diverse molecular mechanisms of toxicity as well as to explore the beneficial effects of selected metals on human health. They have published approximately 75 peer-reviewed publications and book chapters as author or co-author. Their research studies have been supported by the Biomedical Research Support Grant (BRSG), the U.S. Air Force Office of Scientific Research (AFOSR), the National Institutes of Health (NIH), the U.S. Army Medical Research Institute of Infectious Diseases (USAMRIID), the Nebraska State Department of Health, and the Procter and Gamble Company (P&G).

Contributors

Antonio Abad-García
Departamentos de Fisiología y Bioquímica
and
Sección de Estudios de Posgrado e
 Investigación
Escuela Superior de Medicina
Instituto Politécnico Nacional
Mexico City, Mexico

Ahmed Alshrief
Department of Pharmacy Practice
Manipal College of Pharmaceutical Sciences
Manipal Academy of Higher Education
Manipal, India

Chidozie Amuzie
Janssen Pharmaceuticals
Beerse, Belgium

Debasis Bagchi
Department of Pharmacological and
 Pharmaceutical Sciences
Texas Southern University
Houston, Texas, USA
and
Research & Development
Victory Nutrition International, Inc.
Lederach, Pennsylvania, USA

Manashi Bagchi
Research & Development
Dr. Herbs LLC
Concord, California, USA

Arnab Bandyopadhyay
Structural Biology and Bioinformatics Division
CSIR–Indian Institute of Chemical Biology
Kolkata, India

Pradipta Banerjee
Department of Biochemistry and Plant
 Physiology
MS Swaminathan School of Agriculture
Centurion University of Technology and
 Management
Parlakhemundi, India

Samudra Prosad Banik
Department of Microbiology
Maulana Azad College
Kolkata, India

Mónica Barrón-González
Departamentos de Fisiología y Bioquímica
and
Sección de Estudios de Posgrado e
 Investigación
Escuela Superior de Medicina
Instituto Politécnico Nacional
Mexico City, Mexico

Mohiuddin Ahmed Bhuiyan
Department of Pharmacy
University of Asia Pacific
Dhaka, Bangladesh

Rameshwar Nath Chaurasia
Department of Neurology
Institute of Medical Sciences
Banaras Hindu University
Varanasi, India

Debbie C. Crans
Department of Chemistry
and
Cell and Molecular Program
Colorado State University
Fort Collins, Colorado, USA

Ahana Das
Department of Microbiology
Maulana Azad College
Kolkata, India

Amitava Das
Department of Surgery
IU Health Comprehensive Wound Center
Indiana Center for Regenerative Medicine
 and Engineering
Indiana University School of Medicine
Indianapolis, Indiana, USA

Biddut Deb Nath
Department of Physiotherapy
Centre for the Rehabilitation of the
    Paralysed (CRP)
Dhaka, Bangladesh

Dipti Debnath
Department of Microbiology
Immunology and Genetics
University of North Texas Health
    Science Center
Fort Worth, Texas, USA

Bernard W. Downs
Department of Research and
    Development
Victory Nutrition International, Inc.
Lederach, Pennsylvania, USA

Dibyendu Dutta
Department of Health and Family
    Welfare
Darjeeling District Hospital
Government of West Bengal
Darjeeling, India

Eunice D. Farfán-García
Departamentos de Fisiología y Bioquímica
and
Sección de Estudios de Posgrado e
    Investigación
Escuela Superior de Medicina
Instituto Politécnico Nacional
Mexico City, Mexico

Yibin Feng
Li Ka Shing Faculty of Medicine
School of Chinese Medicine
The University of Hong Kong
Hong Kong, China

Jennifer L. Freeman
School of Health Sciences
Purdue University
West Lafayette, Indiana, USA

Jeffrey Galvin
Vitality Medical Wellness Institute, PLLC
Charlotte, North Carolina, USA

Fco. Javier Guillén Gerada
Laruex, Department of Applied Physics
Faculty of Veterinary Sciences
University of Extremadura
Cáceres, Spain

Sait Gezgin
Faculty of Agriculture
Department of Soil Science and Plant Nutrition
Selcuk University
Konya, Turkey

Nandini Ghosh
Department of Surgery
IU Health Comprehensive Wound Center
Indiana Center for Regenerative Medicine and
    Engineering
Indiana University School of Medicine
Indianapolis, Indiana, USA

Rituparna Ghosh
Department of Physiology
Bhairab Ganguly College
Kolkata, India

Mehmet Hamurcu
Faculty of Agriculture
Department of Soil Science and Plant Nutrition
Selcuk University
Konya, Turkey

Md. Akil Hossain
Veterinary Drugs and Biologics Division
Animal and Plant Quarantine Agency
Gyeongsangbuk-do, Republic of Korea

Ashfaque Hossain
Department of Medical Microbiology and
    Immunology
RAK Medical and Health Sciences University
Ras Al Khaimah, United Arab Emirates

Noman Jahangir
Department of Trauma & Orthopaedics
Leeds Children Hospital
West Timperly, UK

Tania Akter Jhuma
Department of Microbiology
University of Dhaka
Dhaka, Bangladesh

Muhammad Manjurul Karim
Department of Microbiology
University of Dhaka
Dhaka, Bangladesh

Sayantani Karmakar
International Institute of Innovation and
    Technology (IIIT)
Kolkata, India

Mohd. Kamran Khan
Faculty of Agriculture
Department of Soil Science and
    Plant Nutrition
Selcuk University
Konya, Turkey

Abhai Kumar
Department of Neurology
Institute of Medical Sciences
Banaras Hindu University
Varanasi, India

Shilia Jacob Kurian
Department of Pharmacy Practice
Manipal College of Pharmaceutical Sciences
Manipal Academy of Higher Education
Manipal, India

Steve Kushner
ALM Research and Development
Oldsmar, Florida, USA

Judith A. MacGregor
Toxicology Consulting Services
Bonita Springs, Florida, USA

Amit Madeshiya
Department of Surgery
IU Health Comprehensive Wound Center
Indiana Center for Regenerative Medicine and
    Engineering
Indiana University School of Medicine
Indianapolis, Indiana, USA

Sayantan Maitra
Department of Health and Family Welfare
Institute of Pharmacy
Government of West Bengal
Jalpaiguri, India

Rajib Majumder
Department of Biotechnology
Adamas University
Jagannathpur, India

Keith Robert Martin
School of Health Studies
Center for Nutraceutical and Dietary
    Supplement Research
University of Memphis
Memphis, Tennessee, USA

Odete Mendes
Safety Assessent
Charles River Laboratories
Wilmington, Massachusetts, USA

Godfred A. Menezes
Department of Medical Microbiology
    and Immunology
RAK Medical and Health Sciences University
Ras Al Khaimah, United Arab Emirates

José Ángel Corbacho Merino
Laruex, Department of Applied Physics
Faculty of Veterinary Science
University of Extremadura
Cáceres, Spain

Sonal Sekhar Miraj
Department of Pharmacy Practice
Manipal College of Pharmaceutical Sciences
Manipal Academy of Higher Education
Manipal, India

Bruce S. Morrison
Morrison Family and Sports Medicine
Huntingdon Valley, Pennsylvania, USA

Bidisha Mukherjee
Department of Endocrinology and Metabolism
Institute of Postgraduate Medical Education
    and Research
Kolkata, India

Satinath Mukhopadhayay
Department of Endocrinology and Metabolism
Institute of Postgraduate Medical Education
    and Research
Kolkata, India

Sreedharan Nair
Department of Pharmacy Practice
Manipal College of Pharmaceutical Sciences
Manipal Academy of Higher Education
Manipal, India

Sreejayan Nair
School of Pharmacy
University of Wyoming
Laramie, Wyoming, USA

Anamika Pandey
Faculty of Agriculture
Department of Soil Science and
    Plant Nutrition
Selcuk University
Konya, Turkey

Rokeya Pervin
Foot and Mouth Disease Division
Animal and Plant Quarantine Agency
Gyeongsangbuk-do, Republic of Korea

Kenneth R. Phelps
Stratton Veteran Affairs Medical Center and
    Albany Medical Collage
Albany, New York, USA

Pooja Gopal Poojari
Department of Pharmacy Practice
Manipal College of Pharmaceutical
    Sciences
Manipal Academy of Higher Education
Manipal, India

Kahoana Postal
Department of Chemistry
Colorado State University
Fort Collins, Colorado, USA
and
Department of Chemistry
Universidade Federal do Paraná
Curitiba, Paraná, Brazil

Harry G. Preuss
Department of Biochemistry
Georgetown University Medical Center
Washington, District of Columbia, USA

Mahadev Rao
Department of Pharmacy Practice
Manipal College of Pharmaceutical
    Sciences
Manipal Academy of Higher Education
Manipal, India

Diana Rodríguez-Vera
Departamentos de Fisiología y Bioquímica
and
Sección de Estudios de Posgrado
    e Investigación
Escuela Superior de Medicina
Instituto Politécnico Nacional
Mexico City, Mexico

Sashwati Roy
Department of Surgery
IU Health Comprehensive Wound Center
Indiana Center for Regenerative Medicine
    and Engineering
Indiana University School of Medicine
Indianapolis, Indiana, USA

Marufa Rumman
Departments of Oncology and Pathology
Karmanos Cancer Institute
Wayne State University School of Medicine
Detroit, Michigan, USA

Zobia Saeed
Department of Public Health
Royal Liverpool University Hospital
Manchester, UK

Suman Santra
Department of Surgery
IU Health Comprehensive Wound Center
Indiana Center for Regenerative Medicine
    and Engineering
Indiana University School of Medicine
Indianapolis, Indiana, USA

Soisungwan Satarug
Kidney Disease Research Center,
    Translational Research Institute
The University of Queensland
Brisbane, Australia

Julia J. Segura-Uribe
Departamentos de Fisiología y Bioquímica
and
Sección de Estudios de Posgrado e
Investigación
Escuela Superior de Medicina
Instituto Politécnico Nacional
Mexico City, Mexico

Smita Singh
Department of Geriatric Medicine
Institute of Medical Sciences
Banaras Hindu University
Varanasi, India

Marvin A. Soriano-Ursúa
Departamentos de Fisiología y Bioquímica
and
Sección de Estudios de Posgrado e
Investigación
Escuela Superior de Medicina
Instituto Politécnico Nacional
Mexico City, Mexico

Ayyaz Sultan
Department of Cardiology
Royal Albert Edward Infirmary
Wrightington, Wigan and Leigh NHS
Foundation Trust Hospitals
Wigan, UK

Mohsin Tanveer
Stress Physiology Research Group
School of Land and Food
and
Tasmania Institute of Agriculture
University of Tasmania
Hobart, Tasmania, Australia

Girish Thunga P
Department of Pharmacy Practice
Manipal College of Pharmaceutical Sciences
Manipal Academy of Higher Education
Manipal, India

Md. Hafiz Uddin
Departments of Oncology and Pathology
Karmanos Cancer Institute
Wayne State University School of Medicine
Detroit, Michigan, USA

Muhammad Umar
Department of Trauma & Orthopaedics
Royal Liverpool University Hospital
Manchester, UK

John B. Vincent
Department of Chemistry and Biochemistry
The University of Alabama
Tuscaloosa, Alabama, USA

Lei Wang
State Key Laboratory of Desert and Oasis
Ecology
Xinjiang Institute of Ecology and Geography
Chinese Academy of Sciences
Urumqi, PR China

Ning Wang
Li Ka Shing Faculty of Medicine
University of Hong Kong
School of Chinese Medicine
Hong Kong, China

Ola Wasel
School of Health Sciences
Purdue University
West Lafayette, Indiana, USA

Yu Xu
Li Ka Shing Faculty of Medicine
University of Hong Kong
School of Chinese Medicine
Hong Kong, China

Fatma Gokmen Yilmaz
Department of Soil Science and Plant Nutrition
Selcuk University
Konya, Turkey

PART I

Introduction
Metals, Metalloids, Redox Biology, and Neurodegeneration

CHAPTER 1

Role of Metals and Metalloids in Redox Biology[1]

Amit Madeshiya
Indiana University School of Medicine

Pradipta Banerjee
Centurion University of Technology & Management

Suman Santra and Nandini Ghosh
Indiana University School of Medicine

Sayantani Karmakar
International Institute of Innovation and Technology (IIIT)

Debasis Bagchi
Texas Southern University
Victory Nutrition International, Inc.

Sashwati Roy and Amitava Das
Indiana University School of Medicine

CONTENTS

[1] Amit Madeshiya and Pradipta Banerjee have contributed equally.

1.1 INTRODUCTION

Oxidation-reduction (redox) reactions play a critical role in biological systems [1]. These involve the basic functions of life ranging from respiration to metabolism. Alterations in these redox reactions may lead to changes in physiological processes and promote progression of various diseases that can even prove fatal to the body [2]. An imbalance in homeostasis of reactive oxygen species (ROS) is the main cause of several diseases [3,4]. Basically there are two types of ROS: (a) primary ROS, generated via metabolic process or after oxygen activation by physical irradiation, and (b) secondary ROS, generated by primary ROS via interacting with enzyme/metal-catalyzed reaction [5]. Cellular antioxidant defense systems damage signaling pathways and biomolecules when overwhelmed with high levels of free radical production. Imbalance between free radicals and antioxidants in the human body is termed as oxidative stress, which takes place as a result of increased ROS production and decreased elimination [6]. Oxidative stress is frequently implicated in a number of biochemical physiological and pathological reactions and pathways [6]. Pathological processes such as cardiovascular dysfunction, diabetes, atherosclerosis, inflammation, and apoptosis occur during oxidative stress [6].

In human physiology, metals and metalloids play pivotal roles as active molecules that participate in several physiological processes including enzyme–substrate reaction, metal transporter system, and redox signaling pathway. Most of the metals are required by the human body at optimum levels since high concentration of metals is toxic for the body [7]. Heavy metals such as lead and mercury have been proven to be fatal to human health when ingested through food. These heavy metals can access the human body through numerous routes including skin, respiration, and contaminated water or food. Some of these metals react with other constituents of the body such as oxygen and chloride. The reaction of these metals inside the body eventually produces ROS that causes oxidative stress which ultimately leads to impaired kidney function, neurological diseases, endocrine diseases, and different types of cancers [8].

High concentration of iron in the body gives rise to free radicals that overpower the cellular antioxidant defense mechanisms, degrade biomolecules, and dysregulate cell signaling pathways [9]. Copper has the potential to induce oxidative stress either by catalyzing ROS formation through a Fenton-like reaction or by significantly decreasing the glutathione levels [10]. Chromium is considered as an occupational carcinogen that not only targets the lungs but also leads to adverse health conditions including gastrointestinal symptoms, hypotension, hepatic and renal failures, and sometimes stomach tumors [11]. The trivalent forms of the metalloid arsenic (As^{3+}) are the most toxic and react with the thiol groups of proteins leading to neurological disorders [12]. Hallmarks of chronic exposure to arsenic include skin lesions, peripheral neuropathy, and anemia. Zinc deficiency is associated with poor diet and related to increased oxidative damage that results in increased lipid, protein, and DNA oxidation [13]. Cadmium enters the human body through the lungs and skin, and accumulates in the intestine and kidneys [14]. Cadmium-induced testicular damage and necrosis have been well documented. Lead damages cellular components via increased oxidative stress through direct ROS generation and *via* depletion of the cellular antioxidant pool [10]. Roles of some important metals and metalloids in redox biology and their implications in physiology and pathological states will be discussed in this chapter.

1.2 IRON

Iron is one of the most essential trace elements of the earth which exhibits biological activities from bacteria to mammals. It has a wide range of oxidation states, i.e., -2 to $+6$, but biologically active oxidation states are $+2$ and $+3$. Fe^{+2} is soluble in mostly all biological fluids. Iron acts as the major metal component in many proteins (hemoglobin) and enzymes, and also plays important

$$\text{Fenton reaction: } Fe^{2+} + H_2O_2 \rightarrow Fe^{3+} + OH + OH^-$$

$$\text{Reduction of } Fe^{3+} \text{ by superoxide: } Fe^{3+} + O^{2-} \rightarrow Fe^{2+} + O_2$$

$$\text{Haber-Weiss reaction: } O^{2-} + H_2O_2 \rightarrow O_2 + OH + OH^-$$

Figure 1.1 Redox reactions involving iron.

roles in growth, development, normal cellular function, and enzymatic actions. Oxygen, sulfur, and nitrogen serve as the major biological ligands of iron [15]. In mammalian system, four classes of iron-containing proteins are there, namely, iron–sulfur enzymes (flavoproteins, hemeflavoproteins), heme proteins (hemoglobin, myoglobin, cytochromes), proteins for storage and transport (transferrin, lactoferrin, ferritin, hemosiderin), and other iron-containing or activated enzymes (sulfur, non-heme enzymes) [16,17]. Fe-S clusters act as complex protein cofactors that are bound to cysteinyl sulfur in the active sites of proteins [18]. Naturally occurring Fe-S clusters having low potential (-300 mV) undergo redox transition from (4Fe-4S) to (4Fe-4S)$^{+2}$ [19]. In the electron transport chain, iron interconverts between Fe^{+2} (reduced) to Fe^{+3} (oxidized) states.

Optimum levels of iron are necessary to maintain homeostasis, as deficiency or a high amount of iron leads to human diseases. Frey and Reed reported that in normal physiological condition, iron metabolism depends on (a) the actions of hormone, hepcidin, and iron exporter protein, ferroportin, and (b) iron regulatory proteins that bind iron-responsive elements [20]. When there is an iron overload, hydroxyl ions are produced which leads to oxidative stress in iron-sensitive tissues. Generation of ROS by iron-mediated pathway is an acute pathophysiological condition that may result in cell death in various organisms [21]. Iron-mediated cell death is often termed as ferroptosis.

Cells in the redox state are usually dependent on iron (and copper) redox couple and are maintained within strict physiological limits. Rate of iron absorption in the proximal intestine and rate of iron released are prevented by homeostatic mechanism. Unused cellular iron by other ferroproteins accumulates in ferretin, and its iron-binding capacity is limited. Hemochromatosis is a typical condition where patients suffer from iron overload causing severe organ damage. Interestingly, free iron can generate damaging reactive free radicals via the Fenton reaction [6] (Figure 1.1). Free iron has deleterious effects. When an organism is overloaded with iron, the Fenton reaction plays a significant role in vivo. The superoxide radicals generated participate in the Haber–Weiss reaction. It is a combination of the Fenton reaction and the reduction of Fe(III) by superoxide (Figure 1.1).

Free radical attacks permanently modify genetic materials and are involved in mutagenesis, aging, and carcinogenesis. Free-radical-mediated DNA damage is observed in various cancer tissues. Double bonds of DNA bases are added by the hydroxyl radical generated during the catalytic action of iron(II) during the Fenton reaction. Till date, over 100 oxidative products of DNA have been identified [9]. One of the acclaimed oxidative products of DNA is 8-hydroxyguanine (8-OH-G), and its presence is reported in human urine. These oxidized DNA products are carcinogenic and serve as biomarkers for carcinogenesis and oxidative stress [9]. Asbestos containing around 30% by weight of iron is associated with elevated risk of mutagenesis and carcinogenesis in occupational exposure of workers. Iron chelation is considered as a suitable approach for cancer prevention. High levels of iron in the system can govern the risk of coronary disease and myocardial infarction. High level of ROS is related to hypertension-mediated cardiovascular disease, metabolic syndrome, and obesity in which iron plays a crucial role [6].

1.3 LEAD

Lead is a persistent heavy metal with unusual chemical and physical properties which are used in various industrial applications [10]. Lead is used as a radiation shield in various industries. Lead is usually toxic to both plants and animals. Persistent nature of lead causes its accumulation in the environment such as water, soil, and dust [10]. Toxic effects of lead can be diagnosed through its elevated level in the blood. The blood serum level of leads 10 µg/dL or higher is considered to be toxic and responsible for a variety of disorders. Like iron, two independent pathways are involved in free radical-induced damage by lead. The first induces direct ROS formation including singlet oxygen, hydrogen peroxides, and hydroperoxides. The second mechanism involves depletion of the cellular antioxidant pool. This interrelation between these two mechanisms not only causes an increase in ROS level but also depletes antioxidant pool. Lead inhibits two specific enzymes, namely, glutathione reductase and delta-aminolevulinic acid dehydrogenase (ALAD) [10].

1.4 MERCURY

Mercury (Hg) is a universal environmental pollutant found ubiquitously, and it provokes potential toxic effects on human beings. Hg(II) binds with dissolved organic matters in marine environment such as humic acid. Elemental mercury Hg(0) is commonly found in water and sediments. Hg(II) can be efficiently reduced to Hg(0) in the presence of 0.2 mg/L reduced humic acid, while production of Hg(0) is inhibited by complex formation as humic acid concentration increases [22]. Mercury is generally known to bio-accumulate and bio-magnify as methylmercury (MeHg). The bio-accumulation of MeHg by consumption of big predatory marine fishes is toxic for human beings, and it can directly target the central nervous system (CNS). The exact mechanisms of MeHg-induced intoxication are not clear. MeHg is considered as an electrophile which favorably reacts with nucleophilic groups such as thiols and selenols of proteins and low-molecular-weight molecules. This type of interaction induces oxidative stress, which may cause potential damages by interrupting the functionality of different lipids, nucleic acids, and protein molecules, which leads to modulation of various signal transduction pathways [23]. It is reported that MeHg induces glutathione (GSH, reduced form) depletion [24] and inhibits the function of at least two thioredoxin forms (Trx1 and TrxR) by binding to the thiol groups [25].

Interestingly, it is hypothesized that MeHg can effectively induce oxidative stress and acts as an electrophilic agent, that can activate and upregulate Nrf2, a transcription factor which helps in regulation of antioxidant-mediated defense mechanism. This postulate is supported by several *in vitro* [26–29] and *in vivo* [30] research findings which shows an upregulation in the Nrf2-related gene expression as well as an upregulation in the Nrf2 nuclear translocation, after exposures to MeHg.

1.5 CADMIUM

Cadmium (Cd) occurs naturally in earth's crust and is usually present in combination with other compounds. Since the past two centuries, the industrial activities resulted in the heavy discharge of Cd into the environment. Nowadays, the exposure of Cd increased up to a significant higher level and listed as one of the 126 priority pollutants by the United States Environmental Protection Agency. In humans, Cd exposure takes place primarily via consumption of contaminated water and food, or tobacco [31]. As the Cd is a non-degradable pollutant and bio-accumulate in food chain, the risk of exposure to it is increasing steadily. Acute exposure of Cd leads to pulmonary edema and respiratory tract irritation, while chronic exposure may result in renal dysfunction, anemia, osteoporosis, and even bone fractures [32]. Cd is carcinogenic for a number of tissues [33] and is classified as a human carcinogen [34].

Though Cd is a bivalent cation and cannot produce free radical directly, multiple studies have shown the increased ROS production after its exposure [35–40]. In several studies, it has also been reported that Cd induces the oxidative stress in different organs [41,42] that finally results in the physiological damage of these organs [43–45]. The effect of Cd-induced oxidative stress among the animal and plant cells and tissues are described in several reviews [41,42,46,47]. Cd shows higher affinity with thiols, and among them, most of the thiols are antioxidants such as glutathione (GSH) which is abundantly present in cells and act as the major target of free Cd ions. Therefore, the depleted reduced GSH pool induced by free Cd ions causes the redox imbalance and finally results in the oxidative environment [48]. There are several reports available explaining the Cd-induced ROS production among the different organelles and tissues. Hepatocytes exposed to Cd have displayed Nicotinamide adenine dinucleotide phosphate (reduced) (NADPH)-dependent ROS production [49].

Cd is believed to upsurge the free Fe concentration, by its replacement in several proteins which, in turn, escalates the free redox-active metals, thereby increasing the formation of •OH radicals [50,51] *via* the Fenton reaction. The oxidized metal is reduced by the Haber–Weiss reaction with superoxide radicals $\left(O_2^{-\circ}\right)$ as a substrate [52]. A different study suggested that the chronic exposure of Cd may lead to upregulation of NO_{x4} gene expression in the mice kidneys [53]. However, the specific role of NO_{x4} in Cd toxicity is not fully understood but is believed to be interlinked to the generation of free radicals for signal transduction to activate the antioxidative defense system or adaptive immunity. The ROS production could activate protective signaling [54].

1.6 CHROMIUM

Similarly to Cd, Cr is also used in many industries such as batteries and metal plantings, and thus released to the environment [55]. The highest level of Cr is found in water especially in developing countries. A number of studies have reported the association of Cr toxicity with nephrotoxicity, hepatotoxicity, oxidative stress, and dysfunctional endocrine function [56]. Evidences suggested that the ROS may play a significant role in initiating the cellular injury which can finally result in development of cancer [57,58]. One of the key mechanisms of metal-induced ROS production is the "Fenton-type reaction," and in this category, Cr(III), (V), and (IV) play most important roles especially to generate the •OH [59]. Another type of reaction is the "Haber–Weiss reaction" which also involves Cr(III), (IV), (V), and (VI) to generate the •OH [60,61]. Apart from the roles in these types of reactions, Cr can also directly interact with several cellular molecules to form free radicals. For example, cysteine or penicillamine can react with Cr(VI) to generate corresponding thiol radicals [62].

Furthermore, it has also been observed that the metals along with ROS or metal-induced ROS collectively affect the number of genes at expressional or transcriptional level. It induces the Src kinase, MAPKs, and some transcriptional factors like NF-κβ, AP-1, p53, etc. Src, a non-receptor tyrosine kinase, is stated to be triggered by Cr(III) and ROS [63]. Similarly, Cr also affects the MAPK signaling pathway [64]. This MAPK activation results in the cell cycle arrest, DNA damage, and altered expression of many genes [65]. Evidences suggested that the Cr toxicity can induce the apoptosis in the BEAS cells and activated JNK and p38 CL3 human lung cancer cell lines [66]. Recently, authors have reported that Cr is capable to activate or affect the activation of NF-κβ [67]. NF-κβ is the prototype of redox-sensitive transcription factor; it can also be activated by several stimuli like cytokines, MAPK signaling, and ROS. In the Jurkat cells, it was observed that Cr(VI) can cause the induction of NF-κβ [68]. In line of this evidence, Ye et al. suggested that the reduction of Cr(IV) to the lower oxidation state absolutely required for the activation of NF-κβ. However, the activation process can be enhanced by superoxide dismutase and decreased by metal chelator and •OH radical scavengers. The observation suggested that the •OH radicals play a key role in the activation of NF-κβ [31,68].

1.7 MANGANESE

Manganese has multiple oxidation states ranging from Mn(−III) to Mn(+VII). The leading oxidation states of Mn in biological systems are Mn(II) and Mn(III), which have higher positive potential [69]. Tovmasyan et al. [70] reported that Mn porphyrin-based redox modulators (MnPs) can act as pro-oxidants, resulting in mild oxidative stress, and it triggers the organism to deploy its own endogenous antioxidative defenses. Mn plays vital roles in cell death mechanisms through regulating manganese superoxide dismutase (SOD2) that contains MnO through caspases. Occupational exposures to Mn cause "manganism," a neurotoxic disorder [71]. Mn is an important component of the redox interface between an organism and its environment, and it plays a role in apoptosis. In current scenario, there is only a little understanding about natural exposure of Mn to mitochondrial redox homeostasis, cell survival, and cell death signaling pathways. Mn may compete or substitute several other biologically active metals such as Fe, Cd, Mg, or Ca, and thereby can alter their biological functions [72–75].

1.8 COPPER

Copper exists either in cuprous (+1) or in cupric (+II) form in living organisms. It acts as an important cofactor in many enzymatic reactions essentially redox reactions. In electron transport chain, copper participates as a key player. Oxidative stress by copper occurs by two mechanisms [10]:

  i. Copper can directly induce ROS generation by Fenton-like reactions.
  ii. Glutathione levels in the body are significantly decreased by elevated level of copper [76].

Additionally, the atherosclerosis is also found to be linked with copper, and the most profound evidence is probably the interaction of copper and homocysteine generating free radicals.

1.9 ZINC

The redox inert metal zinc does not directly participate in oxidation-reduction reaction. However, in some cases, zinc acts as antioxidants. Over supplementation of iron or copper, poor dietary zinc intake, and excess dietary phytate intake cause zinc deficiency [10]. Loss of appetite, delay in wound healing, poor immune function, and dermatitis are few common symptoms of zinc deficiency [10]. Zinc has important roles in the development of immune system and also helps in maintaining insulin and blood glucose concentration. Oxidation of DNA, protein, and lipids is common due to oxidative damage during zinc deficiency. ROS formation caused by zinc deficiency is associated with lipid peroxidation in lung damage, formation of conjugated dienes and liver microsomes, lipoprotein oxidation, and galactosamine-induced hepatitis in rats [10].

1.10 ARSENIC

Arsenic is a colorless and odorless metalloid. It may exist in oxidation states of +5 and +3. Arsenic has the ability to form both inorganic and organic compounds in the human body as well as in the environment. The presence of arsenic compounds in food, water, or air is a huge threat to human health [10]. Arsenite [As(III)] and arsenate [As(V)] are the inorganic forms of arsenic that can be methylated to form monomethylarsonic acid (MMA) or dimethylarsenic acid (DMA). GSH mediates the breakdown of inorganic arsenic by reduction of two electrons followed by oxidative methylation to form pentavalent organic arsenic [10].

Arsenic may increase the inflammation by inducing oxidative stress *via* cycling between the oxidation states of metals or interacting with the antioxidants. The As-induced ROS mainly includes the superoxide anion $\left(O_2^{-\circ}\right)$, hydroxyl radical ($^{\bullet}OH$), hydrogen peroxide ($H_2O_2$), singlet oxygen ($^1O_2$), and peroxyl radicals [77]. Yamanaka et al. were the first group who demonstrated the arsenic-induced free radical formation [78,79]. The generation of $\left(O_2^{-\circ}\right)$ and H_2O_2 in the arsenic exposed cell lines including human vascular smooth muscle cells [80], human–hamster hybrid cells [81], and vascular endothelial cells [82] has been reported. Interestingly, other reports have observed the induction of H_2O_2 among the arsenic exposed HEL30 [83], NB4 [84], and CHOK1 [85] cell lines. Further, it was also observed that antioxidant enzyme catalase and superoxide dismutase significantly suppress the arsenic-induced ROS [86]. Thus, there are enough evidences clearly supporting the role of arsenic in the ROS production in the cellular system.

1.11 CONCLUDING REMARKS

A redox reaction is an integrated oxidation-reduction chemical reaction, which basically involves the transfer of electrons between two chemical species, basically gaining an electron (oxidation) and losing an electron (reduction) resulting in a transition in which the oxidation states of atoms are changed. In living physiological and pathological processes, redox reactions play a vital role, and impaired homeostasis of oxygen free radicals plays a leading role in multiple diseases, disorders, and dysfunctions. Earlier, we have discussed the intricate mechanistic aspects of selected metals and metalloids, which influence the redox pathways and modulate several pathophysiological roles. This chapter discussed diverse metals and metalloids including iron, lead, mercury, cadmium, chromium, zinc, and arsenic and their participation in diverse redox reactions in diverse pathophysiological processes leading to an array of degenerative diseases and dysfunctions. Intricate mechanistic aspects were discussed extensively.

REFERENCES

1. Schafer, F.Q. and G.R. Buettner, Redox environment of the cell as viewed through the redox state of the glutathione disulfide/glutathione couple. *Free Radic Biol Med*, 2001. **30**(11): pp. 1191–1212.
2. Pham-Huy, L.A., H. He, and C. Pham-Huy, Free radicals, antioxidants in disease and health. *Int J Biomed Sci*, 2008. **4**(2): pp. 89–96.
3. Dan Dunn, J., et al., Reactive oxygen species and mitochondria: A nexus of cellular homeostasis. *Redox Biol*, 2015. **6**: pp. 472–485.
4. Gorlach, A., et al., Reactive oxygen species, nutrition, hypoxia and diseases: Problems solved? *Redox Biol*, 2015. **6**: pp. 372–385.
5. Valko, M., H. Morris, and M.T. Cronin, Metals, toxicity and oxidative stress. Curr Med Chem, 2005. **12**(10): pp. 1161–1208.
6. Gudjoncik, A., et al., Iron, oxidative stress, and redox signaling in the cardiovascular system. *Mol Nutr Food Res*, 2014. **58**(8): pp. 1721–1738.
7. Zhu, F., et al., Assessment of heavy metals in some wild edible mushrooms collected from Yunnan Province, China. *Environ Monit Assess*, 2011. **179**(1–4): pp. 191–199.
8. Phaniendra, A., D.B. Jestadi, and L. Periyasamy, Free radicals: Properties, sources, targets, and their implication in various diseases. *Indian J Clin Biochem*, 2015. **30**(1): pp. 11–26.
9. Imam, M.U., et al., Antioxidants mediate both iron homeostasis and oxidative stress. *Nutrients*, 2017. **9**(7): pp. 1–19.
10. Jomova, K. and M. Valko, Advances in metal-induced oxidative stress and human disease. *Toxicology*, 2011. **283**(2–3): pp. 65–87.
11. Wise, J.P., Sr., S.S. Wise, and J.E. Little, The cytotoxicity and genotoxicity of particulate and soluble hexavalent chromium in human lung cells. *Mutat Res*, 2002. **517**(1–2): pp. 221–229.
12. Jomova, K., et al., Arsenic: Toxicity, oxidative stress and human disease. *J Appl Toxicol*, 2011. **31**(2): pp. 95–107.

13. Oteiza, P.I. and G.G. Mackenzie, Zinc, oxidant-triggered cell signaling, and human health. *Mol Aspects Med*, 2005. **26**(4–5): pp. 245–255.

14. Maret, W. and J.M. Moulis, The bioinorganic chemistry of cadmium in the context of its toxicity. *Met Ions Life Sci*, 2013. **11**: pp. 1–29.

15. Kell, D.B., Iron behaving badly: Inappropriate iron chelation as a major contributor to the aetiology of vascular and other progressive inflammatory and degenerative diseases. *BMC Med Genomics*, 2009. **2**: p. 2.

16. Andrews, N.C. and P.J. Schmidt, Iron homeostasis. *Annu Rev Physiol*, 2007. **69**: pp. 69–85.

17. Andrews, N.C., Forging a field: the golden age of iron biology. *Blood*, 2008. **112**(2): pp. 219–230.

18. Meyer, J., Iron-sulfur protein folds, iron-sulfur chemistry, and evolution. *J Biol Inorg Chem*, 2008. **13**(2): pp. 157–170.

19. Outten, F.W. and E.C. Theil, Iron-based redox switches in biology. *Antioxid Redox Signal*, 2009. **11**(5): pp. 1029–1046.

20. Frey, P.A. and G.H. Reed, The ubiquity of iron. *ACS Chem Biol*, 2012. **7**(9): pp. 1477–1481.

21. Dixon, S.J. and B.R. Stockwell, The role of iron and reactive oxygen species in cell death. *Nat Chem Biol*, 2014. **10**(1): pp. 9–17.

22. Gu, B., et al., Mercury reduction and complexation by natural organic matter in anoxic environments. *Proc Natl Acad Sci USA*, 2011. **108**(4): pp. 1479–1483.

23. Antunes Dos Santos, A., et al., Oxidative Stress in Methylmercury-Induced Cell Toxicity. *Toxics*, 2018. **6**(3): pp. 1–15.

24. Sarafian, T. and M.A. Verity, Oxidative mechanisms underlying methyl mercury neurotoxicity. *Int J Dev Neurosci*, 1991. **9**(2): pp. 147–153.

25. Carvalho, C.M., et al., Inhibition of the human thioredoxin system. A molecular mechanism of mercury toxicity. *J Biol Chem*, 2008. **283**(18): pp. 11913–11923.

26. Ni, M., et al., Comparative study on the response of rat primary astrocytes and microglia to methylmercury toxicity. *Glia*, 2011. **59**(5): pp. 810–820.

27. Wang, L., et al., Methylmercury toxicity and Nrf2-dependent detoxification in astrocytes. *Toxicol Sci*, 2009. **107**(1): pp. 135–143.

28. Culbreth, M., Z. Zhang, and M. Aschner, Methylmercury augments Nrf2 activity by downregulation of the Src family kinase Fyn. *Neurotoxicology*, 2017. **62**: pp. 200–206.

29. Toyama, T., et al., Cytoprotective role of Nrf2/Keap1 system in methylmercury toxicity. *Biochem Biophys Res Commun*, 2007. **363**(3): pp. 645–650.

30. Feng, S., et al., Sulforaphane prevents methylmercury-induced oxidative damage and excitotoxicity through activation of the Nrf2-ARE pathway. *Mol Neurobiol*, 2017. **54**(1): pp. 375–391.

31. Vangronsveld, J., F. Van Assche, and H. Clijsters, Reclamation of a bare industrial area contaminated by non-ferrous metals: In situ metal immobilization and revegetation. *Environ Pollut*, 1995. **87**(1): pp. 51–59.

32. Goering, P., M. Waalkes, and C. Klaassen, Toxicology of cadmium. *Toxicol Metals*, 1995, Springer. **115**: pp. 189–214.

33. Waalkes, M.P., Cadmium carcinogenesis in review. *J Inorg Biochem*, 2000. **79**(1–4): pp. 241–244.

34. International Agency for Research on Cancer, *IARC Monographs on the Evaluation of the Carcinogenic Risks to Humans: Beryllium, Cadmium, Mercury, and Exposures in the Glass Manufacturing Industry*. Vol. 58. 1993. Lyon, France: World Health Organization.

35. Hassoun, E.A. and S.J. Stohs, Cadmium-induced production of superoxide anion and nitric oxide, DNA single strand breaks and lactate dehydrogenase leakage in J774A.1 cell cultures. *Toxicology*, 1996. **112**(3): pp. 219–226.

36. Hart, B.A., et al., Characterization of cadmium-induced apoptosis in rat lung epithelial cells: Evidence for the participation of oxidant stress. *Toxicology*, 1999. **133**(1): pp. 43–58.

37. Szuster-Ciesielska, A., et al., The inhibitory effect of zinc on cadmium-induced cell apoptosis and reactive oxygen species (ROS) production in cell cultures. *Toxicology*, 2000. **145**(2–3): pp. 159–171.

38. Thévenod, F., et al., Up-regulation of multidrug resistance P-glycoprotein via nuclear factor-κB activation protects kidney proximal tubule cells from cadmium-and reactive oxygen species-induced apoptosis. *J Biol Chem*, 2000. **275**(3): pp. 1887–1896.

39. Galan, A., et al., The role of intracellular oxidation in death induction (apoptosis and necrosis) in human promonocytic cells treated with stress inducers (cadmium, heat, X-rays). *Eur J Cell Biol*, 2001. **80**(4): pp. 312–320.

40. Wang, Y., et al., Cadmium inhibits the electron transfer chain and induces reactive oxygen species. *Free Radic Biol Med*, 2004. **36**(11): pp. 1434–1443.

41. Bertin, G. and D. Averbeck, Cadmium: Cellular effects, modifications of biomolecules, modulation of DNA repair and genotoxic consequences (a review). *Biochimie*, 2006. **88**(11): pp. 1549–1559.

42. Thévenod, F., Cadmium and cellular signaling cascades: To be or not to be? *Toxicol Appl Pharmacol*, 2009. **238**(3): pp. 221–239.

43. Jarup, L., et al., Health effects of cadmium exposure: A review of the literature and a risk estimate. *Scand J Work Environ Health*, 1998. **24**(Suppl 1): pp. 1–51.

44. Nawrot, T.S., et al., Cadmium-related mortality and long-term secular trends in the cadmium body burden of an environmentally exposed population. *Environ Health Persp*, 2008. **116**(12): pp. 1620–1628.

45. Järup, L. and A. Åkesson, Current status of cadmium as an environmental health problem. *Toxicol Appl Pharmacol*, 2009. **238**(3): pp. 201–208.

46. Waisberg, M., et al., Molecular and cellular mechanisms of cadmium carcinogenesis. *Toxicology*, 2003. **192**(2–3): pp. 95–117.

47. Cuypers, A., K. Smeets, and J. Vangronsveld, Heavy metal stress in plants. In *Plant Stress Biology: From Genomics to Systems Biology*, Hirt, H. (ed.) 2009: pp. 161–178, Hoboken, NJ: John Wiley & Sons.

48. Lopez, E., et al., Cadmium induces reactive oxygen species generation and lipid peroxidation in cortical neurons in culture. *Free Radic Biol Med*, 2006. **40**(6): pp. 940–951.

49. Fotakis, G., et al., Cadmium chloride-induced DNA and lysosomal damage in a hepatoma cell line. *Toxicol In Vitro*, 2005. **19**(4): pp. 481–489.

50. Casalino, E., C. Sblano, and C. Landriscina, Enzyme activity alteration by cadmium administration to rats: the possibility of iron involvement in lipid peroxidation. *Archiv Biochem Biophys*, 1997. **346**(2): pp. 171–179.

51. Dorta, D.J., et al., A proposed sequence of events for cadmium-induced mitochondrial impairment. *J Inorg Biochem*, 2003. **97**(3): pp. 251–257.

52. Winterbourn, C.C., Comparison of superoxide with other reducing agents in the biological production of hydroxyl radicals. *Biochem J*, 1979. **182**(2): p. 625.

53. Thijssen, S., et al., Low cadmium exposure triggers a biphasic oxidative stress response in mice kidneys. *Toxicology*, 2007. **236**(1–2): pp. 29–41.

54. Souza, V., et al., NADPH oxidase and ERK1/2 are involved in cadmium induced-STAT3 activation in HepG2 cells. *Toxicol Lett*, 2009. **187**(3): pp. 180–186.

55. Bagchi, D., et al., Cadmium- and chromium-induced oxidative stress, DNA damage, and apoptotic cell death in cultured human chronic myelogenous leukemic K562 cells, promyelocytic leukemic HL-60 cells, and normal human peripheral blood mononuclear cells. *J Biochem Mol Toxicol*, 2000. **14**(1): pp. 33–41.

56. Jin, Y., et al., Embryonic exposure to cadmium (II) and chromium (VI) induce behavioral alterations, oxidative stress and immunotoxicity in zebrafish (Danio rerio). *Neurotoxicol Teratol*, 2015. **48**: pp. 9–17.

57. Wang, S. and X. Shi, Molecular mechanisms of metal toxicity and carcinogenesis. *Mol Cell Biochem*, 2001. **222**(1–2): pp. 3–9.

58. Leonard, S.S., J.J. Bower, and X. Shi, Metal-induced toxicity, carcinogenesis, mechanisms and cellular responses. *Mol Cell Biochem*, 2004. **255**(1–2): pp. 3–10.

59. Shi, X., N.S. Dalal, and K.S. Kasprzak, Generation of free radicals from model lipid hydroperoxides and H_2O_2 by Co(II) in the presence of cysteinyl and histidyl chelators. *Chem Res Toxicol*, 1993. **6**(3): pp. 277–283.

60. Shi, X.L. and N.S. Dalal, The role of superoxide radical in chromium(VI)-generated hydroxyl radical: the Cr(VI) Haber-Weiss cycle. *Arch Biochem Biophys*, 1992. **292**(1): pp. 323–327.

61. Shi, X. and N.S. Dalal, Vanadate-mediated hydroxyl radical generation from superoxide radical in the presence of NADH: Haber-Weiss vs Fenton mechanism. *Arch Biochem Biophys*, 1993. **307**(2): pp. 336–341.

62. Shi, X., et al., Chromate-mediated free radical generation from cysteine, penicillamine, hydrogen peroxide, and lipid hydroperoxides. *Biochim Biophys Acta*, 1994. **1226**(1): pp. 65–72.

63. Jones, R.J., V.G. Brunton, and M.C. Frame, Adhesion-linked kinases in cancer; emphasis on src, focal adhesion kinase and PI 3-kinase. *Eur J Cancer*, 2000. **36**(13 Spec No): pp. 1595–1606.

64. Alam, J., et al., Mechanism of heme oxygenase-1 gene activation by cadmium in MCF-7 mammary epithelial cells. Role of p38 kinase and Nrf2 transcription factor. *J Biol Chem*, 2000. **275**(36): pp. 27694–27702.

65. Samet, J.M., et al., Activation of MAPKs in human bronchial epithelial cells exposed to metals. *Am J Physiol*, 1998. **275**(3): pp. L551–L558.

66. Iryo, Y., et al., Involvement of the extracellular signal-regulated protein kinase (ERK) pathway in the induction of apoptosis by cadmium chloride in CCRF-CEM cells. *Biochem Pharmacol*, 2000. **60**(12): pp. 1875–1882.

67. Ding, M. and X. Shi, Molecular mechanisms of Cr(VI)-induced carcinogenesis. *Oxygen/Nitrogen Radicals*, 2002. **37**: pp. 293–300.

68. Ye, J., et al., Chromium(VI)-induced nuclear factor-kappa B activation in intact cells via free radical reactions. *Carcinogenesis*, 1995. **16**(10): pp. 2401–2405.

69. Chen, J.Y., et al., Differential cytotoxicity of Mn(II) and Mn(III): Special reference to mitochondrial [Fe-S] containing enzymes. *Toxicol Appl Pharmacol*, 2001. **175**(2): pp. 160–168.

70. Tovmasyan, A., et al., Design, mechanism of action, bioavailability and therapeutic effects of mn porphyrin-based redox modulators. *Med Princ Pract*, 2013. **22**(2): pp. 103–130.

71. Smith, M.R., et al., Redox dynamics of manganese as a mitochondrial life-death switch. *Biochem Biophys Res Commun*, 2017. **482**(3): pp. 388–398.

72. Tjalve, H., et al., Uptake of manganese and cadmium from the nasal mucosa into the central nervous system via olfactory pathways in rats. *Pharmacol Toxicol*, 1996. **79**(6): pp. 347–356.

73. Roth, J.A., Homeostatic and toxic mechanisms regulating manganese uptake, retention, and elimination. *Biol Res*, 2006. **39**(1): pp. 45–57.

74. Gunter, T.E., et al., Manganese transport via the transferrin mechanism. *Neurotoxicology*, 2013. **34**: pp. 118–127.

75. Gavin, C.E., K.K. Gunter, and T.E. Gunter, Manganese and calcium transport in mitochondria: Implications for manganese toxicity. *Neurotoxicology*, 1999. **20**(2–3): pp. 445–453.

76. Kardos, J., et al., Copper signalling: causes and consequences. *Cell Commun Signal*, 2018. **16**(1): p. 71.

77. Halliwell, B. and M. Whiteman, Measuring reactive species and oxidative damage in vivo and in cell culture: How should you do it and what do the results mean? *Br J Pharmacol*, 2004. **142**(2): pp. 231–255.

78. Yamanaka, K., et al., Mutagenicity of dimethylated metabolites of inorganic arsenics. *Chem Pharm Bull (Tokyo)*, 1989. **37**(10): pp. 2753–2756.

79. Yamanaka, K., et al., Induction of DNA damage by dimethylarsine, a metabolite of inorganic arsenics, is for the major part likely due to its peroxyl radical. *Biochem Biophys Res Commun*, 1990. **168**(1): pp. 58–64.

80. Lynn, S., et al., NADH oxidase activation is involved in arsenite-induced oxidative DNA damage in human vascular smooth muscle cells. *Circ Res*, 2000. **86**(5): pp. 514–549.

81. Liu, S.X., et al., Induction of oxyradicals by arsenic: Implication for mechanism of genotoxicity. *Proc Natl Acad Sci USA*, 2001. **98**(4): pp. 1643–1648.

82. Barchowsky, A., et al., Stimulation of reactive oxygen, but not reactive nitrogen species, in vascular endothelial cells exposed to low levels of arsenite. *Free Radic Biol Med*, 1999. **27**(11–12): pp. 1405–1412.

83. Corsini, E., et al., Sodium arsenate induces overproduction of interleukin-1alpha in murine keratinocytes: Role of mitochondria. *J Invest Dermatol*, 1999. **113**(5): pp. 760–765.

84. Jing, Y., et al., Arsenic trioxide selectively induces acute promyelocytic leukemia cell apoptosis via a hydrogen peroxide-dependent pathway. *Blood*, 1999. **94**(6): pp. 2102–2111.

85. Wang, T.S., et al., Arsenite induces apoptosis in Chinese hamster ovary cells by generation of reactive oxygen species. *J Cell Physiol*, 1996. **169**(2): pp. 256–268.

86. Nordenson, I. and L. Beckman, Is the genotoxic effect of arsenic mediated by oxygen free radicals? *Hum Hered*, 1991. **41**(1): pp. 71–73.

CHAPTER 2

Role of Metals in Neurodegeneration

Abhai Kumar, Smita Singh, and Rameshwar Nath Chaurasia
Banaras Hindu University

CONTENTS

2.1 INTRODUCTION

Metals can be divided into two groups: essential and non-essential metals. Essential metals include chromium, cobalt, copper (Cu), iron (Fe), lithium, magnesium, manganese (Mn), nickel, selenium, and zinc (Zn). These metals are involved in a variety of physiological processes such as electron transport, oxygen transportation, protein modification and neurotransmitter synthesis, redox reactions, immune response, cell adhesion, protein, and carbohydrate metabolism (1,2). The generation of free radicals and oxidative stress leads to neurodegeneration; free radicals are strongly associated with redox-active metals. The redox state of the cell is largely linked to iron and copper redox couple and is maintained within physiological limits. The intracellular concentration of free iron is kept negligible through iron-regulating system during normal state; however, during stress condition, excessive superoxide acts as an oxidant of [4Fe-4S] cluster containing enzymes releasing free iron from iron-containing molecules (3). The released iron could participate in the

Fenton reaction generating highly reactive hydroxyl radical which causes cellular death leading to neurodegeneration (3).

The roles of metals are important for plants and animals, but their requirement is very minimal; hence, excessive ingestion of metal leads to accumulation in various organs, including brain. Elevated level of metals induces various cellular abnormalities such as mitochondrial dysfunction, DNA fragmentation, protein misfolding, endoplasmic reticulum (ER) stress, autophagy, dysregulation, and activation of apoptosis (4). The effect might result in alteration in neurotransmission which leads to neurodegeneration and could manifest as cognitive problems, movement disorders, and learning and memory dysfunction. The metal-induced neurotoxicity causes multiple diseases in human such as Alzheimer's disease (AD), amyotrophic lateral sclerosis (ALS), autism spectrum disorders (ASDs), Guillain–Barre disease (GBD), Gulf War syndrome (GWS), Huntington's disease (HD), Parkinson's disease (PD), and Wilson's disease (WD). This chapter discusses in detail the effects of essential and non-essential metals on neurotoxicity and neurological diseases.

2.2 ESSENTIAL METALS

2.2.1 Copper

Copper (Cu) is an essential trace element and transition metal required for physiological activities in mammals. The role of Cu is very significant as it acts as a cofactor of various enzymes such as cytochrome c oxidase and superoxide dismutase (SODs) which play important role in electron transport, oxygen transportation, protein modification, and neurotransmitter synthesis (2). The increase in Cu level results in the generation of reactive oxygen species (ROS), DNA damage, and mitochondrial dysfunction (2). The association of copper with the occurrence of AD, ALS, HD, PD, WD, and Prion diseases in human is well established (3,5). Copper induces aggregation of amyloid precursor protein, amyloid peptide, and found to be associated with onset of AD disease. Further, it interacts with alpha-synuclein and promotes its self-aggregation, resulting in PD (6,7). The mutation in Cu-Zn (SOD) might result in oxidative stress which can also lead to motor neuron degeneration in patients with ALS (7). The level of Cu is significantly higher in HD patients as compared to healthy individuals (8). The Prion disease caused by abnormal protein isoform mainly PrP[Sc] has high binding capacity with Cu which makes them resistant from proteasomal degradation leading to neurodegeneration (2). The mixture of Cu and Ag induces the affinity of Cu with normal cellular isoform protein (PrP[c]) in Prion diseases (9). However, inverse correlation is observed with Cu and Mn in patients with Prion disease (10). Reports are available that Cu did not have a significant role in the formation of PrP[Sc] in Prion disease; therefore, intake of copper in diet decreases the onset of Prion disease (11). Cu has both potential risk and beneficial effect on neurons; therefore, improved methods for measurement of copper trafficking in brain to evaluate potential risk and development of novel therapeutics are needed for Cu-induced neurotoxicity and cognitive dysfunction.

2.2.2 Iron

Iron (Fe) plays important roles in oxygen transport and cellular respiration as it acts as a cofactor for a variety of enzymes and proteins mainly hemoglobin. The intake of iron inside the body occurs through food consumption, and the accumulation of Fe inside the body and brain occurs due to disruption in metabolism and homeostasis. Hemolysis leads to breakdown of red blood cells and immature blood-brain barrier (BBB), which result in accumulation of Fe causing increased ROS generation, lipid peroxidation, protein oxidation, DNA damage, dopamine autoxidation, and mitochondrial fragmentation (12). The imbalance in iron homeostasis leads to a variety of neurological disorders such as AD, PD, HD, ALS, and neurodegeneration with brain iron accumulation (NBIA) (13). The accumulation of iron in brain induces the aggregation of alpha-synuclein which leads to the onset of PD (14). The accumulation of Fe induces aberrant aggregation of beta-amyloid and toxicity, which is the hallmark of AD (15). The Huntingtin protein (htt) which has a central role in HD pathology also regulates Fe homeostasis. The mutated SOD-1 gene in ALS patients and accumulation of Fe in the motor cortex causes damage through Fe-induced oxidative stress. The genetic studies conducted for different mutated genes in patients with NBIA depicted that majority of mutation were not related to Fe homeostasis but rather to autophagy, mitochondria metabolism, and lipid metabolism. The Fe accumulation acts as downstream factor for disease pathology in previously present genetic mutation (16). These genes have undefined Fe regulatory functions which will define whether Fe accumulation is the cause of disease or vice versa.

2.2.3 Manganese

Manganese (Mn) is the trace element and nutrition necessary for biological process in the human body. Mn acts as a cofactor for metalloproteins, such as MnSOD and arginase, and plays an important role in functions of glutamine synthetase, hydrolases, and lyases (17). The primary route of exposure is through consumption of food but exposure to high level occurs through inhalation in occupational hazard industries such as welding and mining (18). The neurological defect is detected during the critical stage of child development having exposure to soy-based infant formula containing high level of Mn. The chronic exposure of Mn has a debilitating neurological effect, although overexposure of Mn leads to PD known as manganism (18). Manganism is characterized by tremors, lethargy, and speech impediment with occasional accompaniment of psychosis (18). In manganism, the level of Mn is elevated in dopaminergic neuron of substantia nigra which leads to motor deficits observed in manganism (19). The elevated level of Mn induces increased level of transcription for ER-related genes, ROS production, mitochondrial dysfunction, autophagy, altered acetylcholinesterase (AChE) activity, changes in cyclic AMP (cAMP) signaling, iron dyshomeostasis, and dysfunctional astrocytic activity (19). The exposure of Mn in neurons leads to increase in marker of programmed cell death, increase in internucleosomal DNA cleavage, and activation of JNK, p38, and pro-apoptotic effector caspase-3(20). The dopaminergic SH-SY5Y exposed to Mn increases ER stress response proteins, including ER chaperone GRP94 and pro-apoptotic GADD154/CHOP protein as well as phosphorylated elF2 (eukaryotic translation initiation factor 2) increased significantly (21). The decreased expression of autophagy-related protein Beclin1and activation of mammalian target of rapamycin (mTOR/p70 ribosomal S6 protein kinase (p70s6k) in Mn-treated dopaminergic neurons leads to neurodegeneration (21). The manganese transporter SLC30A10 protein is abundantly expressed in liver and basal ganglia; the loss-of-function mutants result in accumulation of Mn which results in dystonia, Parkinsonism, and hyper manganesemia (22). Mn can also catalyze autoxidation of dopamine whose toxic metabolite can destroy dopaminergic cells suggesting similar pathology between PD and manganism (23). The short-term Mn exposure prevents from Prion proteins and decreases the incidence of Prion diseases, although long-term exposure promotes stabilization and aggregation of infectious protein (24). Mn is an essential metal, although it's homeostatic and signaling pathway is still not delineated, further investigation which might provide insight into Mn regulation, manganism, and other forms of Parkinsonism.

2.2.4 Zinc

Zinc (Zn) is an essential trace metal (second most abundant transition metal after Fe) required for human and other living organisms. The antioxidant response and regulation of gene transcription are mediated by more than 300 enzymes and metalloproteins which consist of Zn as a cofactor (25). Zn concentration is high in testes, liver, kidney, and brain, further Zn deficiency affects childhood mental and physical development as well as learning abilities (26). Zn higher concentration in body decreases the absorption of Cu and Fe, promoting ROS production in the mitochondria, disrupts activities of metabolic enzymes, and activates apoptotic process (4). The imbalance in Zn homeostasis has been associated with AD, brain trauma, cerebral ischemia, epilepsy, and vascular dementia (VaD). At lower concentration, Zn suppresses fibrillar aggregation of amyloid protein preventing from neurodegeneration but at higher concentration, it induces amyloid fibrillar aggregation leading to neurodegeneration (27). The increase in concentration of Zn with glutamate in synaptic cleft during ischemic condition signifies its role in ischemic-induced neuronal death (28). The dosage of Zn and its activities at different stages of children's and adults should be elucidated to differentiate neurotoxic and beneficial concentration of Zn on brain development.

2.3 NON-ESSENTIAL METALS

2.3.1 Aluminum

Aluminum (Al) is the third most abundant element and the most abundant metal in the earth's crust. Al is extensively used in a variety of human usable products such ascans, cookware, cars, food preservatives, and vaccine adjuvant (29). Aluminum is highly reactive with carbon and oxygen making it toxic to living organism. The intake of Al from dietary intake and environmental exposure is cleared by kidney; however, Al salt in vaccine adjuvant remains biologically available and accumulates in nervous system (29). Al affects more than 200 biological reactions which cause negative effect on central nervous system and brain development, such as axonal transport, neurotransmitter synthesis, synaptic transmission, phosphorylation or de-phosphorylation of proteins, protein degradation, gene expression, peroxidation, and inflammatory responses (30). The conformational and topological changes occur in DNA due to binding of histone–DNA complex with Al; further, it decreases the expression of neurofilament (NF) and tubulin, altered expression of genes of NF, amyloid precursor protein (APP), neuron-specific enolase, decreased expression of transfer in receptor, altered expression of RNA polymerase 1, altered expression of oxidative stress marker genes (SOD1), glutathione reductase, and altered expression of β-APP secretase (31). Al increases the activity of protein kinase C and cytoskeleton proteins, while inhibiting the activity of protein phosphatase inducing non-enzymatic phosphorylation of tau; further, Al can also cause abnormal accumulation of tau proteins in neuroblastoma cells, neurofibrillary degeneration in vivo and accumulation of amyloid β protein (Aβ) in vivo (32). Al has been associated with AD, ALS, ASD, GBD, multiple sclerosis, and GWS in human (29). A recent report suggested that Al exposure increases the risk of AD by 70% although exposure to Al-adjuvant vaccines in children increases the incidences of ASD. The level of Al is significantly high in the hair, blood, and urine of children with autism. Further, Al also reduces long-term memory and cognition, and increases anxiety and neuronal death in motor cortex and spinal cord due to mitochondrial dysfunction and oxidative stress (33). The role of genetic background and increasing susceptibility for Al-induced neurotoxicity is not clear, although studies needed to establish the relationship between genetic factors and aluminum-induced neurotoxicity.

2.3.2 Arsenic

Arsenic (As) is a toxic metalloid and well-known carcinogen which affects 200 million people worldwide. As is found in wood preservatives and also known for contamination of ground water. Early exposure to As causes lower brain weight and reduction in glia and neurons (34). Research suggests that As exposure induces mitochondrial oxidative stress, imbalance of intracellular Ca^{+2}, disruption in ATP production, altered membrane potential, changes in cytoskeletal morphology, and neuronal cell death among other effect (34). As exposure leads to AD and ALS as it was reported that dimethyl arsenic acid a metabolite of As has been shown to increase β-amyloid level in AD (34). As exposure decreases the translocation of NF and is linked to decreased NF content at peripheral nerves, which providing understanding of aberrant NF distribution in ALS (35).

2.3.3 Cadmium

Cadmium (Cd) is known carcinogen and non-essential transition heavy metal, which enters peripheral and central neurons through olfactory bulb or nasal mucosa which damages permeability of BBB (36). The exposure to Cd is high in miners, welders, smokers, and workers in battery production; the chronic exposure to Cd induces oxidative stress and suppresses gene expression, DNA damage repair, and apoptosis (37). Chronic exposure to Cd severely interferes with the normal function of nervous system, more in infants and children as compared with adult (37). Cd is etiological factor of neurodegenerative diseases such as PD and AD (38,39). Studies reported that Cd accelerates aggregation of tau peptide leading to AD and symptoms such as headache, megrim, olfactory dysfunction, slowing of motor dysfunction decreased equilibrium, learning and ability, and PD-like symptom (38). Cd exposure in zebra fish results in decreased brain size and unclear brain subdivision in mid-hindbrain region; Cd also decreases morphological changes in rat cortical neurons (axon and dendrites) and inhibited neurite outgrowth in PC12 cells (40). The role of Cd as carcinogen is well established; although very less work has been done to estimate the neurotoxic effect and mechanism of action for neuronal damage.

2.3.4 Lead

Lead (Pb) is a toxic heavy metal and ubiquitously present in the environment. The Pb enters inside human body through inhalation and oral ingestion. Pb is excreted through urine, bile, and rest bind with red blood cell, which accumulates in bone (41). The Pb exposure leads to oxidative stress, mitochondrial dysfunction, changes to golgi apparatus, and increased glial filament in astrocytes. It also disrupts Ca^{+2} homeostasis, interferes with phosphorylation of PKC, and decreases nitric oxide production (41). The main target for Pb-induced neurotoxicity is the nervous system, and children are very sensitive to Pb intoxication. Hippocampus is the primary region for Pb accumulation, although it also accumulates in other brain regions and has been reported to affect intelligence, memory, executive functions, attention, speed processing, language, emotions, visuospatial, and motor skills (42). Pb exposure results in decreased intellectual ability in dose dependent manner, impaired verbal concept formation, grammatical reasoning difficulty, poor command following (43). The methods to measure the level of Pb exposure and effective treatment after Pb exposure should be future strategy of research.

2.3.5 Methylmercury

Methylmercury (MeHg) is a xenobiotic toxic organic metal compound derived from inorganic mercury (Hg). The source of release of mercury in environment is through anthropogenic sources such as industrial waste, coal mining, natural sources such as volcanoes and forest fires that release back Hg into the atmosphere (44). Hg, which exudates in water, is readily methylated in MeHg by sulfate-reducing bacteria and a variety of other anerobic bacteria (45). MeHg has a high affinity for sulfur and can cross BBB by binding onto thiol groups of proteins; it can also bind to cysteine, allowing for the possibility of amino acid transporter (45). MeHg can accumulate in brain and was also responsible for epidemics at Minamata Bay and Iraq; children were presented with a variety of central nervous system (CNS) disorders, including ataxia, paralysis, retardation, dysarthria, dysesthesia, and cerebral palsy (46). Studies conducted with the families having high level of

MeHg exposure found that their brain have accumulation of inorganic Hg; furthermore, cerebral and occipital lobe atrophy leads to vision and motor issue (46). The exposure of MeHg affects various cellular processes such as dopamine metabolism, neural stem cell differentiation and causes mitochondrial dysfunction by release of ROS and intracellular calcium influx (47). Further, MeHg exposure induces amyloid deposition in the hippocampus and decreases cerebrospinal fluid as hallmark of AD (48).The prevention strategies are more prevalent than treatment of MeHg exposure, government agencies has their own advisories for prevention of their respective population. The major source of entry of MeHg is through sea food, which has many confounding beneficial effect, this limits the understanding of MeHg neurotoxicity on human.

2.3.6 Thallium

Thallium (Tl) is a naturally occurring trace element and extremely toxic heavy metal, sparsely found in the earth's crust (49). The anthropogenic sources are industrial processing of cements, non-industrial means such as rodenticide, and consumption of food from contaminated soil. Tl enters inside the body through inhalation and skin (49). Tl exposure leads to non-neurological symptoms such asalopecia, hepatic dysfunction, gastroenteritis and neurological disorder includes polyneuropathy, cranial nerve deficits, paresthesia, and loss of sensation, ataxia, and psychosis (49). Tl poisoning results in inhibition of an enzyme acetylcholinesterase (AChE), which breakdowns acetylcholine responsible for neurotransmission that might cause peripheral neuropathy (50). Tl concentration is lower in brain than other parts of the body, and the highest concentration is present in hypothalamus as reported earlier (51). Tl interferes with K^+-dependent process because of the similarity in basic nature of univalent ions, which affects ATP generation; therefore, at cellular level, high level of Tl decreases ATP production, increases in ROS formation, glutathione oxidation, decreases in dopamine and serotonin level (49). The risk of exposure to general public is comparatively low, and no treatment is available due to lack of understanding in the signaling pathway of Tl. Complicated symptom results in misdiagnosis with Tl-induced neurotoxicity; therefore, basic mechanism for Tl-induced neurotoxicity needs to be elucidated.

2.4 NEUROTOXICITY INDUCED BY METAL MIXTURE

The environment constitutes a mixture of metals which has significant effect on the homeostasis of other metals. The changes in homeostasis of one metal have significant effect on the other metals (such as Mn, Fe, Cd, Cu, and Zn) that are transported by shared transporters or controlled by overlapping signaling pathways. The combined neurotoxic effect of Pb and As tends to increase the probability of intellectual disability when compared to exposure of single metal (52). The combined exposure of Pb and Cd affects mental and psychomotor development in children (52). The combined exposure of Pb and Mn in children below 2 years of age results deficits in cognition and language development as compared with single metal exposure. Pb exposure was associated with lower IQ scores among children with high blood Mn level (53). In contrast to Pb and As synergistic effect, Pb along with Mn, Hg, or Cd exposure tends to work in antagonistic effect, the MeHg and Pb combined exposure on cognitive deficit was less additive (54). The co-exposure of two metals might increase its retention and redistribution of individual metals. Co-exposure of Se and Hg increased retention of both metals and resulted in redistribution of Hg in blood and organ (55). The intraperitoneal administration of Pb and oral intake of Mn resulted in alteration in motor activity, learning ability, biogenic amine level, and Pb level in brain, which were more severe than in rats exposed either to Mn or Pb alone (56). The combined exposure to Pb and Mn results in decreased brain weight as compared with individual exposure (56). The monaminergic system is influenced by As/Pb interaction in mice; further, the enhanced Pb accumulation and reduced As accumulation in brain decrease norepinephrine level in hippocampus and increased serotonin level in midbrain and frontal cortex when compared with single metal exposure (57). The human exposure to metal mixture in current environmental condition is very evident; to understand synergistic and antagonistic effects of metal mixture on health requires further studies.

2.5 METAL-INDUCED NEUROTOXICITY AND TREATMENT

The metal poisoning occurrence in patients needs to transfer the patient from exposing environment followed by gastrointestinal decontamination. The Mn concentration decreases rapidly in blood of children with cholestasis receiving parental nutrition after discontinuation of Mn supplementation (58). The patients with chronic metal exposure which accumulates in the nervous system, bone, and other tissue rather than plasma need better therapeutic approach to prevent from metal-induced chronic damage. Currently, chelation therapy is a common treatment to remove additional metals in the body for both chronic and acute metal poisoning. Chelators include British antilewisite (BAL), succimer (DMSA), Prussian blue, calcium disodium edetate ($CaNa_2EDTA$), and D-Penicillamine (Cuprimine). The chelators have the negative effects on the body, such as headache, fatigue, renal failure, nasal congestion, gastrointestinal side effects, and life-threatening hypocalcemia; the concentration and dosage of these chelators are the major concern when performing chelation therapy. The exporter of specific metal such as SLC30A10 which is newly identified Mn transporter facilitate in the efflux of excessive intracellular Mn in a patient with dystonia, parkinsonism, hypermanganesemia, and manganism (59). Future studies to identify metal-specific transporters are in great need to design therapeutics to treat metal-induced toxicity.

2.6 CONCLUSIONS AND FUTURE DIRECTIONS

Metals play an important role in our daily life as they act as cofactors for various enzymes involved in cellular activity. The acute and chronic exposure of metal results in their bioaccumulation in various organs mainly, brain, kidney, and bones. The accumulation of metal in brain leads to neurotoxicity which causes diseases such as AD, ALS, autism, and PD. The metal accumulated in nervous system leads to oxidative stress, mitochondrial dysfunction, and protein misfolding, which causes neuronal death, since the process of neuronal regeneration is very slow, therefore, injured neurons have to expend greater energy to synthesize neurotransmitters and maintain homeostasis. The increase in life span of general population, the chances for exposure to greater level of metals for individuals and increase risk of occurrence of neurological diseases. Currently, there is a growing demand in investigating metal-induced neurotoxicity. Future studies need to focus more on the joint effect of metal mixture exposure, identifying specific transporters of each metal as well as developing target-specific therapeutics for patients with metal poisoning.

REFERENCES

1. Banci L: *Metallomics and the Cell*. Dordrecht, the Netherlands: Springer; 2013; 12.
2. Desai V, Kaler SG: Role of copper in human neurological disorders. *Am J Clin Nutr.* 2008; 88(3): 855S–858S.
3. Liochev SI, Fridovich I: The role of O_2 in the production of HO: In vitro and in vivo. *Free Radic Biol Med.* 1994; 16:29–33.
4. Wright RO, Baccarelli A: Metals and neurotoxicology. *J Nutr.* 2007; 137(12): 2809–2813.

5. Strausak D, Mercer JF, Dieter HH, Stremmel W, Multhaup G: Copper in disorders with neurological symptoms: Alzheimer's, Menkes, and Wilson diseases. *Brain Res Bull.* 2001; 55(2): 175–185.

6. Roos PM, Vesterberg O, Nordberg M: Metals in motor neuron diseases. *Exp Biol Med (Maywood).* 2006; 231(9): 1481–1487.

7. Rasia RM, Bertoncini CW, Marsh D, Hoyer W, Cherny D, Zweckstetter M, Griesinger C, Jovin TM, Fernández CO: Structural characterization of copper(II) binding to alpha-synuclein: Insights into the bioinorganic chemistry of Parkinson's disease. *Proc Natl Acad Sci USA.* 2005; 102(12): 4294–4299.

8. Fox JH, Kama JA, Lieberman G, Chopra R, Dorsey K, Chopra V, Volitakis I, Cherny RA, Bush AI, Hersch S: Mechanisms of copper ion mediated Huntington's disease progression. *PLoS One.* 2007; 2(3): e334.

9. Valensin D, Padula EM, Hecel A, Luczkowski M, Kozlowski H: Specific binding modes of Cu(I) and Ag(I) with neurotoxic domain of the human prion protein. *J Inorg Biochem.* 2016; 155: 26–35.

10. Thackray AM, Knight R, Haswell SJ, Bujdoso R, Brown DR: Metal imbalance and compromised antioxidant function are early changes in prion disease. *Biochem J.* 2002; 362(1): 253–258.

11. Quaglio E, Chiesa R, Harris DA: Copper converts the cellular prion protein into a protease-resistant species that is distinct from the scrapie isoform. *J Biol Chem.* 2001; 276(14): 11432–11438.

12. Nohl H, Kozlov AV, Gille L, Staniek K: Cell respiration and formation of reactive oxygen species: Facts and artifacts. *Bio chem Soc Trans.* 2003; 31(Pt 6): 1308–1311.

13. Salvador GA, Uranga RM, Giusto NM: Iron and mechanisms of neurotoxicity. *Int J Alzheimers Dis.* 2010; 2011: 720658.

14. Pyatigorskaya N, Sharman M, Corvol JC, Valabregue R, Yahia-Cherif L, Poupon F, Cormier-Dequaire F, Siebner H, Klebe S, Vidailhet M, Brice A, Lehéricy S: High nigral iron deposition in LRRK2 and Parkin mutation carriers using R2* relaxometry. *Mov Disord.* 2015; 30(8): 1077–1084.

15. Ayton S, Lei P, Bush AI: Metallostasis in Alzheimer's disease. *Free Radic Biol Med.* 2013; 62: 76–89.

16. Meyer E, Kurian MA, Hayflick SJ: Neurodegeneration with brain iron accumulation: Genetic diversity and pathophysiological mechanisms. *Annu Rev Genomics Hum Genet.* 2015; 16: 257–279.

17. Horning KJ, Caito SW, Tipps KG, Bowman AB, Aschner M: Manganese is essential for neuronal health. *Annu Rev Nutr.* 2015; 35: 71–108.

18. Karki P, Smith K, Johnson J Jr, Aschner M, Lee E: Role of transcription factor yin yang 1 in manganese-induced reduction of astrocytic glutamate transporters: Putative mechanism for manganese-induced neurotoxicity. *Neurochem Int.* 2015; 88: 53–59.

19. Robison G, Sullivan B, Cannon JR, Pushkar Y: Identification of dopaminergic neurons of the substantia nigra pars compacta as a target of manganese accumulation. *Metallomics.* 2015; 7(5): 748–755.

20. Roth JA, Horbinski C, Higgins D, Lein P, Garrick MD: Mechanisms of manganese-induced rat pheochromocytoma (PC12) cell death and cell differentiation. *Neurotoxicology.* 2002; 23(2): 147–157.

21. Zhang J, Cao R, Cai T, Aschner M, Zhao F, Yao T, Chen Y, Cao Z, Luo W, Chen J: The role of autophagy dysregulation in manganese-induced dopaminergic neurodegeneration. *Neurotox Res.* 2013; 24(4): 478–490.

22. Tuschl K, Clayton PT, Gospe SM Jr, Gulab S, Ibrahim S, Singhi P, Aulakh R, Ribeiro RT, Barsottini OG, Zaki MS, Del Rosario ML, Dyack S, Price V, Rideout A, Gordon K, Wevers RA, Chong WK, Mills PB: Syndrome of hepatic cirrhosis, dystonia, polycythemia, and hypermanganesemia caused by mutations in SLC30A10, a manganese transporter in man. *Am J Hum Genet.* 2012; 90(3): 457–466.

23. Kwakye GF, Paoliello MM, Mukhopadhyay S, Bowman AB, Aschner M: Manganese-induced parkinsonism and Parkinson's disease: Shared and distinguishable features. *Int J Environ Res Public Health.* 2015; 12(7): 7519–7540.

24. Choi CJ, Anantharam V, Saetveit NJ, Houk RS, Kanthasamy A, Kanthasamy AG: Normal cellular prion protein protects against manganese-induced oxidative stress and apoptotic cell death. *Toxicol Sci.* 2007; 98(2): 495–509.

25. Hambidge M: Human zinc deficiency. *J Nutr.* 2000; 130(5S Suppl): 1344S–1349S.

26. Prasad AS: Impact of the discovery of human zinc deficiency on health. *J Am Coll Nutr.* 2009; 28(3): 257–265.

27. Cuajungco MP, Fagét KY: Zinc takes the center stage: Its paradoxical role in Alzheimer's disease. *Brain Res Brain Res Rev.* 2003; 41(1): 44–56.

28. Weiss JH, Sensi SL, Koh JY: Zn²⁺: A novel ionic mediator of neural injury in brain disease. *Trends Pharmacol Sci.* 2000; 21(10): 395–401.

29. Shaw CA, Tomljenovic L: Aluminum in the central nervous system (CNS): Toxicity in humans and animals, vaccine adjuvants, and autoimmunity. *Immunol Res.* 2013; 56(2–3): 304–316.

30. Kawahara M, Kato-Negishi M: Link between aluminum and the pathogenesis of Alzheimer's disease: The integration of the aluminum and amyloid cascade hypotheses. *Int J Alzheimers Dis.* 2011; 2011: 276–393.

31. Muma NA, Singer SM: Aluminum-induced neuropathology: Transient changes in microtubule-associated proteins. *Neurotoxicol Teratol.* 1996; 18: 679–690.

32. Guy SP, Jones D, Mann DM, Itzhaki RF: Human neuroblastoma cells treated with aluminium express an epitope associated with Alzheimer's disease neurofibrillary tangles. *NeurosciLett.* 1991; 121:166–168.

33. De Marchi U, Mancon M, Battaglia V, Ceccon S, Cardellini P, Toninello A: Influence of reactive oxygen species production by monoamine oxidase activity on aluminum-induced mitochondrial permeability transition. *Cell Mol Life Sci.* 2004; 61(19–20): 2664–2671.

34. Prakash C, Soni M, Kumar V: Mitochondrial oxidative stress and dysfunction in arsenic neurotoxicity: A review. *J Appl Toxicol.* 2016; 36(2): 179–188.

35. DeFuria J, Shea TB: Arsenic inhibits neurofilament transport and induces perikaryal accumulation of phosphorylated neurofilaments: Roles of JNK and GSK-3beta. *Brain Res.* 2007; 1181: 74–82.

36. Luevano J, Damodaran C: A review of molecular events of cadmium-induced carcinogenesis. *J Environ Pathol Toxicol Oncol.* 2014; 33(3): 183–194.

37. Wang B, Du Y: Cadmium and its neurotoxic effects. *Oxid Med Cell Longev.* 2013; 2013: 8034.

38. Okuda B, Iwamoto Y, Tachibana H, Sugita M: Parkinsonism after acute cadmium poisoning. *Clin Neurol Neurosurg.* 1997; 99(4): 263–265.

39. Jiang LF, Yao TM, Zhu ZL, Wang C, Ji LN: Impacts of Cd(II) on the conformation and self-aggregation of Alzheimer's tau fragment corresponding to the third repeat of microtubule-binding domain. *Biochim Biophys Acta.* 2007; 1774(11): 1414–1421.

40. López E, Figueroa S, Oset-Gasque MJ, González MP: Apoptosis and necrosis: Two distinct events induced by cadmium in cortical neurons in culture. *Br J Pharmacol.* 2003; 138(5): 901–911.

41. Selvín-Testa A, Capani F, Loidl CF, Pecci-Saavedra J: The nitric oxide synthase expression of rat cortical and hippocampal neurons changes after early lead exposure. *Neurosci Lett.* 1997; 236(2): 75–78.

42. Mason LH, Harp JP, Han DY: Pb neurotoxicity: Neuropsychological effects of lead toxicity. *Biomed Res Int.* 2014; 2014: 840547.

43. Campbell TF, Needleman HL, Riess JA, Tobin MJ: Bone lead levels and language processing performance. *Dev Neuropsychol.* 2000; 18(2): 171–186.

44. Li WC, Tse HF: Health risk and significance of mercury in the environment. *Environ Sci Pollut Res Int.* 2015; 22(1): 192–201.

45. Parks JM, Johs A, Podar M, Bridou R, Hurt RA Jr, Smith SD, Tomanicek SJ, Qian Y, Brown SD, Brandt CC, Palumbo AV, Smith JC, Wall JD, Elias DA, Liang L: The genetic basis for bacterial mercury methylation. *Science.* 2013; 339(6125): 1332–1335.

46. Davis LE, Kornfeld M, Mooney HS, Fiedler KJ, Haaland KY, Orrison WW, Cernichiari E, Clarkson TW: Methyl mercury poisoning: Long-term clinical, radiological, toxicological, and pathological studies of an affected family. *Ann Neurol.* 1994; 35(6): 680–688.

47. Yuntao F, Chenjia G, Panpan Z, Wenjun Z, Suhua W, Guangwei X, Haifeng S, Jian L, Wanxin P, Yun F, Cai J, Aschner M, Rongzhu L: Role of autophagy in methyl mercury-induced neurotoxicity in rat primary astrocytes. *Arch Toxicol.* 2016; 90(2): 333–345.

48. Kim DK, Park JD, Choi BS: Mercury-induced amyloid-beta (Aβ) accumulation in the brain is mediated by disruption of Aβ transport. *J Toxicol Sci.* 2014; 39(4): 625–635.

49. Galván-Arzate S, Santamaría A: Thallium toxicity. *ToxicolLett.* 1998; 99(1):1–13.

50. Repetto G, Sanz P, Repetto M: In vitro effects of thallium on mouse neuroblastoma cells. *Toxicol In Vitro.* 1994; 8(4): 609–611.

51. Ibrahim D, Froberg B, Wolf A, Rusyniak DE: Heavy metal poisoning: clinical presentations and pathophysiology. *Clin Lab Med.* 2006; 26(1): 67–97.

52. Sanders AP, Claus Henn B, Wright RO: Perinatal and childhood exposure to cadmium, manganese, and metal mixtures and effects on cognition and behavior: A review of recent literature. *Curr Environ Health Rep.* 2015; 2(3): 284–294.

53. Lin CC, Chen YC, Su FC, Lin CM, Liao HF, Hwang YH, Hsieh WS, Jeng SF, Su YN, Chen PC: In utero exposure to environmental lead and manganese and neurodevelopment at 2 years of age. *Environ Res.* 2013; 123:52–57.

54. Yorifuji T, Debes F, Weihe P, Grandjean P: Prenatal exposure to lead and cognitive deficit in 7- and 14-year-old children in the presence of concomitant exposure to similar molar concentration of methyl-mercury. *Neurotoxicol Teratol.* 2011; 33(2): 205–211.

55. Kalia K, Murthy RC, Chandra SV: Tissue disposition of 54Mn in lead pretreated rats. *Ind Health.* 1984; 22(1): 49–52.

56. Chandra SV, Murthy RC, Saxena DK, Lal B: Effects of pre- and postnatal combined exposure to Pb and Mn on brain development in rats. *Ind Health.*1983; 21(4): 273–279.

57. Mejía JJ, Díaz-Barriga F, Calderón J, Ríos C, Jiménez-Capdeville ME: Effects of lead-arsenic combined exposure on central monoaminergic systems. *NeurotoxicolTeratol.* 1997; 19(6):489–497.

58. Jang DH, Hoffman RS: Heavy metal chelation in neurotoxic exposures. *NeurolClin.* 2011; 29(3): 607–622.

59. Leyva-Illades D, Chen P, Zogzas CE, Hutchens S, Mercado JM, Swaim CD, Morrisett RA, Bowman AB, Aschner M, Mukhopadhyay S: SLC30A10 is a cell surface-localized manganese efflux transporter, and parkinsonism-causing mutations block its intracellular trafficking and efflux activity. *J Neurosci.* 2014; 34(42): 14079–14095.

PART II

Pathology of Metal Toxicity

CHAPTER 3

Pathological Manifestations and Mechanisms of Metal Toxicity

Odete Mendes
Charles River Laboratories

Chidozie Amuzie
Janssen Pharmaceuticals

CONTENTS

3.1 GENERAL INTRODUCTION

Environmental exposure to metals, associated with ever-increasing levels of metal contaminants, can occur via the digestive system (mainly through what we eat and drink), from the atmosphere or exposure to contaminated water. Inside the body, there are numerous organs and systems that are impacted by their presence. Metals can be widely distributed and accumulated in organs including the brain, heart, liver, kidney, and brain, as well as in the bone. Metals can also frequently accumulate in intercellular space. Here we aim to briefly summarize what are some of the body systems (including nervous, immune, and reproductive) and organs most frequently affected by metal toxicity: the liver and kidney. We describe the pathologic manifestations of metal presence and some of the molecular pathways associated with metal-related toxicological effects.

3.2 CENTRAL AND PERIPHERAL NERVOUS SYSTEM

3.2.1 Introduction

The nervous system, both peripheral and central, is frequently affected by metal toxicity. Common heavy metals that are neurotoxicants include Aluminum, Arsenic, Lead, Manganese, Mercury, and Cadmium. Main pathological effects of heavy metals in the central nervous system (CNS) consist of neuronal loss that may be associated with specific sub-anatomical regions depending on the mechanism of toxicity of specific metal and is often accompanied by microglial and/or astrocyte reactivity. In the peripheral nervous system, Wallerian degeneration is the most common effect of metal toxicity. The most frequent mechanism of toxicity common to many metals is the formation of reactive oxygen species (ROS) and induction of oxidative stress. Additionally, specific metals may affect specific neurotransmitters and their receptors or activate multiple proinflammatory or apoptotic signal transmission pathways.

3.2.2 Aluminum

Human exposure to Aluminum (Al) occurs mainly from food and drinking water, with possible occupational exposure related to mining and welding. Al is neurotoxic to both experimental animals and humans with effects in cognition, behavior, and motor function associated with accumulation of neurofibrillary tangles and loss of synapses and dendrites in multiple brain subregions (Curtis, 2013).

Rats exposed to Aluminum chloride exhibit evidence of neurodegeneration and neuronal cell death in the hippocampus (Liaquat et al., 2019). These lesions are characterized by the presence of pyknotic neurons and neuropil vacuolation. Neurodegeneration and pyknotic nuclei can be observed in the CA1, CA2, and dentate gyrus of rats exposed to $AlCl_3$ (Liaquat et al., 2019). Chronic feeding of Al to rodents will lead to accumulation of the metal in the CA1 hippocampal pyramidal cells. When immunostained to show microtubules, increased thickness and distortion of microtubules can be observed, as well as increased dystrophic neurites. Al accumulation in the rat hippocampus also causes decreases in synapse density (Walton, 2009).

In humans, the brain can accumulate Aluminum (Al) over time. This accumulation can lead to the loss of cholinergic neurons in the cerebral cortex due to effects on cholinergic neurotransmission. Cholinergic neurons are especially susceptible to Al neurotoxicity due to variations in levels of acetylcholinesterase and choline transferase (Klotz et al., 2017). The hippocampus is a preferential area for Al accumulation; here, it can modify calcium signaling pathways and impact neuronal plasticity and memory. Aluminum accumulates in characteristic lesions of Alzheimer's disease (AD) (Klotz et al., 2017) including senile plaques and neurofibrillary tangles. There has been speculation that Al may be associated with AD because elevated levels of Al have been found in brains affected with the disease and there is a lesion overlap (Curtis, 2013). Al salts have been reported to cause increased levels of amyloid precursor protein, and biological effects of colloidal Al may resemble those of the toxic 25–35 β-amyloid fragment leading to both formation of ROS, with both an increase in lipid peroxidation and neuronal nitric oxide synthase activation (nNOS) causing oxidative stress and aggregation of β-amyloid (Bondy, 2010) (*Handbook of Clinical Neurology (Occupational Neurology)*), 2015). However, other neurodegenerative diseases have also been associated with Al accumulation where it has been shown to play a role in both cognitive dysfunction, and learning and memory impairment that are key symptoms of neurodegeneration.

One possible mechanism of toxicity for Al is that it can cause a decline in cyclic adenosine monophosphate (cAMP) response element-binding protein (CREB) phosphorylation and downregulation of transcription and protein expression in CREB target genes such as brain-derived neurotrophic factor (BDNF) and related levels of target of rapamycin complex 1 (TORC) and NAD-dependent protein deacetylase sirtuin-1 (SIRT1) that are associated with aberrant hippocampus neuronal ultrastructure (Yan et al., 2017). Al and Mercury (Hg) have also been shown to contribute to inflammation in the CNS that can lead to both degeneration and neuronal loss. The increase in proinflammatory signaling has been reported to aim the genetic apparatus including transcription adenosine triphosphate (ATP) generation. Al also impacts the regulation (up) of nuclear factor kappa-light-chain-enhancer of activated B cells (NF-ƙB) and upregulation of microRNAs and subsequent downregulation of neuronal brain specific genes (Alexandrov, Pogue, & Lukiw, 2018). Also part of the proinflammatory properties of this metal is the ability to increase glial reactivity, macrophage activation, and to produce proinflammatory cytokines, such as tumor necrosis factor alpha (TNF-α) and interleukin-1 alpha (IL-1α), and acute phase proteins (Bondy, 2010).

3.2.3 Arsenic

Arsenic (As) exposure is either occupational (from manufacturing herbicides and pesticides or the smelting industry) or environmental (from contaminated water either directly or indirectly, via eating seafood or ingestion). Arsenic is a very toxic metalloid in both its organic and inorganic forms, of which As III is considered the most toxic. It is considered highly toxic to the skin causing vesicle formation and ocular corneal damage that may lead to blindness (C. Li, Srivastava, & Athar, 2016).

Acute As poisoning can lead to sensory loss with the presence of Wallerian axonal degeneration that can be reversible if exposure ceases (Curtis, 2013). In the nervous system, As toxicity impacts nerve conducting velocity and can lead to degeneration of myelin and axons, associated with peripheral neuropathy. Toxicity may also be associated with vascular events that can lead to cerebral edema and/or petechia formation (*Haschek and Rousseaux's Handbook of Toxicologic Pathology*, 2013). As is known to cause astrocyte apoptosis, reduced astrocytic process, and reduced expression of astrocytic glial fibrillary acidic protein (GFAP), important for the maintenance of the cellular cytoskeleton. Additionally, direct impact on neuronal viability (Flora, Mittal, Pachauri, & Dwivedi, 2012; Ma, Sasoh, Kawanishi, Sugiura, & Piao, 2010; Piao et al., 2005), synaptic vesicles, and neurite growth has been reported impacting the hippocampus (Tolins, Ruchirawat, & Landrigan, 2014). With chronic exposure, peripheral neuropathy can arise characterized as a dying-back axonopathy with the presence of demyelination (Curtis, 2013). A correlation between As exposure and cognitive dysfunction (Tyler & Allan, 2014) has also been reported.

The mechanisms of As toxicity include the formation of ROS and effects of the cytoskeleton such as the production of neurofilaments (*Handbook of Clinical Neurology (Occupational Neurology)*, 2015).

Increased immunohistochemical staining for 8-hydroxy-2′-deoxyguanosine (8-OHdG), an oxidative DNA product, can be observed after exposure to As in both the cerebral and cerebellar cortex (Piao et al., 2005) 8-OHdG immunoreactivity can be observed in the cerebral cortex and cerebellar cortex, particularly in the Purkinje cells and granular cells (Piao et al., 2005). ROS formation is associated with lipid peroxidation that, in turn, causes mitochondria damage that may lead to neurodegeneration. Peroxidative membrane damage is also linked to degenerative processes. As can also cause decreased levels of antioxidant glutathione (GSH) and decreased activity of the antioxidant enzyme glutathione peroxidase that further increases oxidative damage that, in turn, impacts both neuronal development and viability (Tolins et al., 2014).

Arsenic has also been shown to activate the Nrf2-Keap1 (nuclear factor erythroid 2-related factor 2-Kelch-like ECH-associated protein) pathway through a p62-dependent mechanism. Nrf2 is a transcription factor activated in response to oxidative stress that is responsible for detoxification and establishment of redox balance; however, in the case of As, it is still unclear if this alternative way to activate the pathway may be associated with a non-protective mechanism that can potentiate toxicity and/or carcinogenicity (Lau, Whitman, Jaramillo, & Zhang, 2013).

Additionally, As and its metabolites have toxic effects in enzymes involved in the cellular energy pathways and as deoxyribonucleic acid (DNA) repair and synthesis.

As neurotoxicity has also been associated with activation of p38 and c-Jun N-terminal kinase (JNK3) protein of the mitogen-activated protein kinase (MAPK) pathway, as well as caspase activation that impacted cerebellar neuronal viability/apoptosis. In the sciatic nerve, As III causes a reduction of neurofilament protein L (low molecular weight) NF-L expression; this causes disruption and destabilization of the cytoskeleton (Mochizuki, 2019).

3.2.4 Lead

Environmental lead (Pb) contamination remains a major problem for individuals living in urban or industrial areas, with most of the Pb exposure originating from food and water. Additionally, one of the primary sources of Pb exposure in children is lead-containing household paint (Curtis, 2013).

Pb crosses the blood-brain barrier (BBB) and can accumulate in the brain's gray matter (Austin, Freeman, & Guilarte, 2016).

Children can develop lead encephalopathy, characterized by severe brain edema due to extravasation of the capillaries associated with loss of neuronal cells and gliosis. Prominence of cerebellar and cerebral capillaries with endothelial cell swelling and subsequent necrosis causing cerebral edema can be observed (Curtis, 2013). Gliosis and neuronal cell loss are more prominent in the middle to deep layers of the brain cortex. Degeneration and necrosis of Purkinje cells (Abubakar et al., 2019) may also be present (Abubakar et al., 2019; *Haschek and Rousseaux's Handbook of Toxicologic Pathology*, 2013).

Adults with occupational exposure can develop a peripheral neuropathy (also known as "lead palsy") that has been associated demyelination of the nerve fibers and segmental nerve degeneration (*Handbook of Clinical Neurology (Occupational Neurology)*, 2015). Cumulative exposure may also be associated with an impact in neurobehavior (Curtis, 2013).

Mechanism of Pb toxicity include competition with calcium (Ca^{++}) ions, affinity to protein sulfhydryl groups, and interference with Zinc (Zn) and Gamma aminobutyric acid (GABA) production in the CNS. Low levels of Pb appear capable of replacing calcium to act as a second messenger to activate multiple cellular pathways. The protein kinase C pathway has been shown to be sensitive to this effect and is associated with neurotransmitter excitotoxicity or decreased neurotransmission, leading to altered cognitive function or altered behavior patterns. Lead also appears to preferentially accumulate in endothelial cells, impacting normal BBB activity and brain prefusion both fundamental to brain development and function (Goldstein, 1990). Trace Pb amounts and/or cumulative Pb accumulation have been associated with low intelligence quotient (IQ) and severe learning or behavioral problems (Grandjean & Herz, 2015). Additional mechanisms of toxicity also include production of ROS and alteration of glutamate and aspartate neurotransmission pathways (*Handbook of Clinical Neurology (Occupational Neurology)*, 2015).

As far as neurotransmitter release, Pb is shown to impair vesicular release and to reduce the number of fast-releasing sites in the hippocampal neurons (CA1 and CA3 mossy fibers) and causes a reduction of the size of postsynaptic densities in CA3 dendrites (Guariglia, Stansfield, McGlothan, & Guilarte, 2016). Additionally, Lead exposure has been reported to impact the expression of α-amino-3-hydroxy-5-methyl-4-isoxazolepropionic acid (AMPA) glutamate receptor 2 expression that can be associated with excitotoxicity (Ishida, Kotake, Sanoh, & Ohta, 2017). Alteration of neurotransmitter release profiles and of their receptors has also been documented in other cell types such as cerebrovascular endothelia cells, astrocyte, and microglia.

Other mechanisms of Pb neurotoxicity also include cytotoxicity, apoptosis (Sanders, Liu, Buchner, & Tchounwou, 2009), and inflammation. In terms of inflammation, Pb has been associated with increased gene expression of interleukin-6 (IL-6) and transforming growth factor beta 1 (TGF-β1) with higher expression in the frontal cortex, cerebellum, striatum, hippocampus, and substantia nigra. A possible mechanism for such upregulation may be associated with activator protein 1 (AP-1) or selective promoter factor 1 (SP1) regulatory elements and impacting the Janus kinase (JAK) pathway. Other impacted proinflammatory elements are cyclooxygenase-2 (COX-2), nitric oxide synthase 2 (NOS2), NF-ƙB, toll-like receptor 4 (TRL4) (Chibowska et al., 2016), and extracellular signal-regulated protein kinase (ERK) 1/2 (Ishida et al., 2017).

Additionally, both high and low levels of Pb have been shown to cause adverse effects in developing organisms and to impact in utero neurocognitive development (Sanders et al., 2009), with younger individuals being more sensitive to toxic effects of lead than juveniles or adults (Rader, Peeler, & Mahaffey, 1981).

3.2.5 Manganese

Manganese (Mn) is essential for brain development and brain function; however, it is also associated with disease in the CNS including Parkinson's (PD) and AD. Food is the main source of Mn.

Additional occupational exposure may occur in the mining, smelting, and manufacturing industries, as well as from pesticide applications for agricultural use. The brain is the most sensitive organ to Mn toxicity, where it impacts the release of dopamine and dopaminergic neurons (Curtis, 2013).

Perhaps the best-known Mn-related disease, Manganism is a neurological disorder similar to PD occurring when excess Mn accumulates in the basal ganglia, specifically in the globus pallidus, subthalamic nucleus, substantia nigra, and striatum. However, the substantia nigra is largely spared in Manganism that is mostly linked to degeneration of GABAnergic neurons in the globus pallidus, where it accumulates and impacts the nigrostriatal pathway system. Mn toxicity has been associated with the loss of neurons in the globus pallidus and the putamen; the caudate nucleus and the subthalamic nucleus can also be affected (Cordova et al., 2012). Mn toxicity can also affect other regions of the brain including the cerebellum, red nucleus, pons, cortex, thalamus, and the anterior horn of the spinal cord. Mn-induced disease differs from PD by the absence of Lewy bodies (Kwakye, Paoliello, Mukhopadhyay, Bowman, & Aschner, 2015). Initially there is a deregulation in the dopaminergic system, with potential effects in D2 dopamine receptors. As the disease progresses, however, catecholamine levels decrease due to loss of striatonigral dopaminergic neurons that lead to the development of PD-like symptoms (*Handbook of Developmental Neurotoxicology*, 1998).

The balance and threshold at which Mn changes from being a critical component of physiological processes and becomes part of the pathophysiological processes are delicate and may change with age and state of neurodevelopment. One example of this delicate balance in the role of Mn is its relevance in the activity of sodium oxide dismutase 2 (SOD2) that can be fundamental to the response of oxidative stress but when askew is a critical element of neurotoxic and neurodegenerative processes (Pfalzer & Bowman, 2017).

Mn at neurotoxic levels also affects the levels of acetylcholine (ACh) and thus learning, memory, and locomotion. Mn can have the ability to alter choline acetyltransferase (ChaT) activity, which can have an effect in anti-inflammatory reactions and the production of inflammatory cytokines such as IL6, TNF-α (Peres et al., 2016).

Other mechanisms of toxicity of Mn include the interference of astrocyte glutamate transport or altering the expression of N-methyl-D-aspartate receptor (NMDA), leading to excitotoxicity (E. Lee, Karki, Johnson, Hong, & Aschner, 2017). Mn can also increase the phosphorylation of ERK1/2 and AKT (protein kinase B) as well as ROS production and caspase activity, and altered mitochondrial respiratory chain activities (Cordova et al., 2012).

Mn affects the norepinephrine signaling. GABA in the stratum can also be decreased as well as levels of glutamate and glutamine synthetases (*Handbook of Clinical Neurology (Occupational Neurology)*, 2015). Mn has been suggested to regulate glutamine synthetase activity that may impact glutamate trafficking and glutaminergic signaling by astrocytes in the globus pallidus (Kwakye et al., 2015). Astrocytes are the main reservoir for Mn in the brain. Accumulated Mn within the mitochondria can impact ATP generating pathways and apoptosis pathways, leading to cytotoxicity. Additionally, Mn can impact the synthesis of glutathione synthetase (GSS) and the production of antioxidants leading to an imbalance of the ROS (Szpetnar, Luchowska-Kocot, Boguszewska-Czubara, & Kurzepa, 2016).

Mn may also affect the uptake of dopamine, norepinephrine, serotonin, glutamate, and GABA (*Neurotoxicology: Approaches and Methods*, 1995). Mn also has been reported to have a role in the development of actual PD, where excessive MN intake has been directly associated with dopaminergic cell death.

Mitochondrial accumulation is a very important cellular step for Mn toxicity and impacts movement of other mitochondrial relevant ions. Mn^{2+} inhibits Ca^{2+} efflux over both the (sodium) Na^+-dependent and Na^+-independent mechanisms, and that Mn^{2+} does not share the Na^+-dependent Ca^{2+} efflux mechanism (Gavin, Gunter, & Gunter, 1990). Mitochondrial accumulation results in the generation of ROS and may also interfere with Ca^{2+}-activated ATP production. Imbalances of both Mn^{2+} and Ca^{2+} signaling in striatal astrocytes may be further associated with release of cytochrome

C (cyt C) that may activate caspase-dependent apoptosis regulated by protein kinase C (PKCδ) that, in turn, regulates protein phosphatase 2A (PP2A) that is known to regulate the activity of tyrosine hydroxylase (TH) responsible for dopamine synthesis (Tarale et al., 2016).

Mn research suggests that part of the mechanism for Mn neurotoxicity may be related to neuro-inflammatory responses including striatal modulation of phosphatidylinositol 3 kinase (PI3K); activation of MAPK and/or Janus kinase (JAK)-signal transducer and activator of transcription (STAT) 3, and cyclooxygenase-2 (COX 2) signaling (R. G. Lucchini, Aschner, Landrigan, & Cranmer, 2018). Glial activation is characteristic of brain Mn toxicity and associated with activation on inducible nitric oxide synthase (iNOS) (Harischandra et al., 2019).

Accumulation of Mn in the developing brain may be associated with changes in neurological structure such as the basal ganglia, impacting the dopaminergic pathways to the frontal lobes responsible for coordinating higher-level cognitive and executive functions. This can lead to lower IQ (R. G. Lucchini et al., 2018), reduced memory function, attention deficit hyperactivity disorder (ADHD), and affect other functions (R. Lucchini et al., 2017).

3.2.6 Mercury

Fish consumption is the principal source of exposure to methylmercury. Additionally, occupational exposure to Mercury can also occur in the chlor-alkali industry or as part of medicinal exposure to dental Mercury amalgams or drug preparations containing Mercury. Toxicity to the nervous system associated with mercurial compounds occurs mostly with methylmercury (MeHg), causing neuronal necrosis, lysis and phagocytosis, and replacement with glial cells that in acute cases is accompanied by cerebral edema (Curtis, 2013). Increased expression of endothelial cell markers as well as hyperpermeability evidence can be observed with subacute exposure to methylmercury (Takahashi et al., 2017) impacting the BBB and causing vascular disturbances (Figure 3.1).

Lesions associated in MeHg toxicity include atrophy of the folia and sulci of the lateral lobes of the cerebellum most evident in the granular cells with relative preservation of adjacent Purkinje cells (Antunes Dos Santos et al., 2016) that may degenerate in chronic cases. Proliferation of the Bergman's glia is commonly observed, but except for some thinning of the myelin and gliosis the white matter may not be affected. Also very sensitive to Mercury toxicity are the sensory neurons of the dorsal spinal ganglia (Chang, 1990). Neuronal necrosis can occur in the hypothalamus, thalamus,

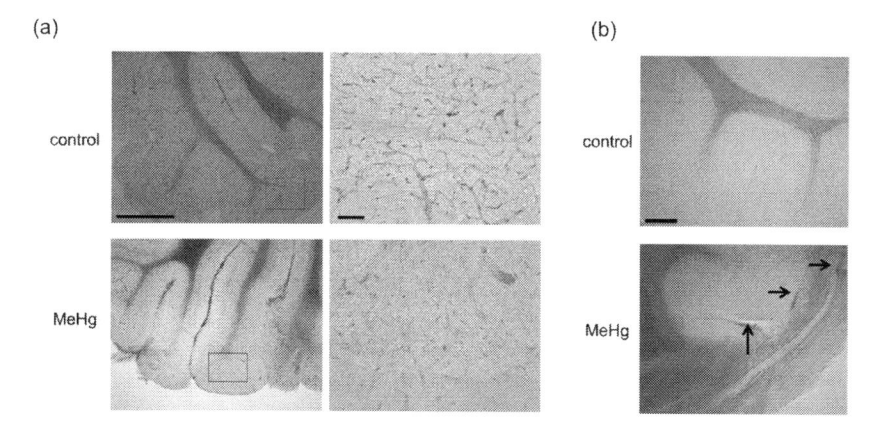

Figure 3.1 Methylmercury damages the blood-brain barrier. Exposure to methylmercury decreases the expression of endothelial cell markers, in (a) rat brain specimens that were stained using an antibody against rat endothelial cell antigen-1 (RECA-1). Scale bars, 100 and 10 µm (left and right panel, respectively) and (b) increasedd IgG extravasation. Scale bar, 25 µm (Takahashi et al., 2017).

midbrain, and basal ganglia (*Handbook of Clinical Neurology (Occupational Neurology)*, 2015). With chronicity, atrophy of the brain and hydrops can be observed characterized by gross atrophy of the cerebellar folia particularly in the lateral lobes and vermices (Chang, 1990). Other Mercury compounds (like aryl mercury) can also be associated with Wallerian degeneration and demyelination of sensory nerve fibers, dorsal root ganglia, and peripheral nerves (*Haschek and Rousseaux's Handbook of Toxicologic Pathology*, 2013). Additionally, Hg is associated with neuronopathy, affecting the cell body followed by secondary nerve fiber degeneration (dying-forward), tertiary myelin loss, and target cell atrophy or death (trans-synaptic degeneration). These neuronopathies are irreversible because neurons are terminally differentiated and can be associated with degeneration and death of astrocytes (El-Fawal, Waterman, De Feo, & Shamy, 1999). Other pathological manifestations of Mercury toxicity include neuronal spongiosis and local inflammatory cell infiltration (Abdel-Salam et al., 2018). It is also reported that inorganic Mercury accumulates in the locus ceruleus, which plays a role in cognition and neuroinflammation and potentially in associated diseases or conditions such as, aging, AD, and PD (Pamphlett, Bishop, Kum Jew, & Doble, 2018).

Methylmercury (MeHg) toxicity is mediated via glutathione that reacts with MeHg to form the MeHg–glutathione complex, an essential step in Mercury detoxification. Imbalance in the detoxification process, when, glutathione S-transferase (GST) catalyzes the MeHg–glutathione complex conjugation reaction, may lead to injury (Abdel-Salam et al., 2018). Binding to glutathione also leads to ROS defense imbalance in oxidative stress that ultimately causes injury to cells (Patrick, 2002). Alterations of glutathione peroxidase (GPx) and/or glutathione reductase (GR) have been associated with methylmercury (MeHg) exposure, leading to an increase in susceptibility to injury due to oxidants. Additionally, MeHg interacts with biological molecules such as proteins, leading to intracellular disruption and production of ROS (Unoki et al., 2018). Mg also binds to thiol groups and SH (sulfhydryl) residues of cysteine (Naganuma, Furuchi, Miura, Hwang, & Kuge, 2002) forming Mercury sulfhydryl bonds binding thus to molecules such as albumin, glutathione, and cysteine (Farina, Rocha, & Aschner, 2011).

Hg effects are thought to be associated with decreased numbers of muscarinic receptors (*Haschek and Rousseaux's Handbook of Toxicologic Pathology*, 2013). Mercury accumulates in many different cell types in the CNS with special prevalence in astrocytes where it impacts glutamate and aspartate signaling leading to excitotoxicity (El-Fawal et al., 1999). Also associated with the mechanism of Hg neurotoxicity is the increase in immunoglobulins following exposure. Synapses are the main site for Hg neurotoxicity as Hg blocks pre-synaptic Na and CA^{2+} channels and interferes with protein kinase C and calmodulin (El-Fawal et al., 1999). Another mechanism of MeHg-induced neurotoxicity is glutamate dyshomeostasis. *In vivo* MeHg inhibits glutamate uptake into synaptic vesicles (Farina et al., 2011).

Methylmercury (MeHg) is also a potent neurotoxicant to the developing brain (Kern et al., 2017), due to its persistence within the CNS. Hg compounds can cross the placenta and cause cerebellar and cerebral deformities in a developing fetus resulting in impairment of movement, visuospatial perception, and speech; or a disease called Fetal Minamata Disease (FMD). Mechanisms of neurodevelopmental Hg toxicity include alteration of neurotransmission via glutamate and Ca^{2+} signaling; mitochondrial alterations that lead to ROS production; alterations of neuronal cytoskeleton and perturbations of cell differentiation (Antunes Dos Santos et al., 2016).

3.2.7 Cadmium

Cadmium (Cd) is one of the seven most toxic heavy metals among environmental pollutants. It frequently enters systemic circulation following lung or intestinal absorption. It enters the neuron via voltage-gated calcium channels and can cause numerous disturbances both in the peripheral and CNS. These include motor dysfunction and peripheral neuropathy, mental and learning disabilities, and it has also been associated with neurodegenerative diseases. The ability of Cd to cross the BBB, as well as ependymal and pial surfaces is limited in adult individuals making young and developing subjects more sensitive to its toxicity and related histopathological damage in the cerebral and cerebellar cortices. Cd has also been linked to diseases where there is BBB impairment such as AD and PD (Branca, Morucci, & Pacini, 2018). Cd tends to accumulate in the choroid plexus where concentration is greater than in the cerebrospinal fluid (CSF) or other portions of the brain. Mechanisms of protection of the choroid plexus include high levels of metal binding ligand such as cystine and higher activity of superoxide dismutase and catalase. Effects in the choroid plexus include deterioration of the plexus structure characterized by loss of microvilli, rupture of the cells apical surface, and increased number of membranal blebs. Additionally, there can be an increase in cell debris in the ventricular lumen and increase in cytoplasmic vacuoles and lysosomes with presence if nuclear irregularity. It is also being reported that Cd injury has been associated with micro-vessel injury. Administration of Cd to Sprague Dawley rats resulted in necrosis of the rostral ventrolateral medulla (RVLM), the brain stem site that maintains blood pressure and sympathetic vasomotor tone. This lesion was characterized by shrunken neurons that lost cytoplasmic and nuclear detail and nuclear fragmentation, karyorrhexis (S. M. Chen et al., 2019).

Cd can also significantly increase the levels of lipid peroxidation in the parietal cortex, striatum, and cerebellum in a mechanism that is related to increased ROS and alteration of mitochondrial function associated with increased levels of COX, activation of MAPK, and mTOR (mammalian target of rapamycin). Additionally, Cd can be transported to the olfactory bulbs by the olfactory neurons causing olfactory dysfunction (B. Wang & Du, 2013; H. Wang, Zhang, Abel, Storm, & Xia, 2018). Once it enters the cell, cytotoxicity can be mediated by inhibition of DNA repair and production of ROS leading to oxidative stress. Intracellular Cd regulates Ca^{2+} signaling via the 1,4,5-trisphosphate (IP_3) pathway and ryanodine, and it increases intracellular Ca^{2+} concentrations by increased promoting calcium efflux from the endoplasmic reticulum (Branca et al., 2018). Oxidative stress is one of the main components of Cd pathology, with altered glutathione levels and catalase enzyme activity (Menon, Chang, & Kim, 2016). Exposure to Cd, associated with decreased endothelial cell viability, was related to increased production of ROS. This causes a decrease in rat brain phospholipids (PLs) due to the activation of phospholipase A2 enzymes (sPLA2, cPLA2) leading to the release of arachidonic acid, increased COX2 and generation of proinflammatory cytokines including IL1, IL6, TNF-α, iNOS, and interferon gamma (INF-γ). Additionally, increases in pro-apoptotic molecules, such as Bcl-2-associated X protein (Bax), and decreases in anti-apoptosis B-cell lymphoma 2 (Bcl2) were also observed with Cd exposure (Sivaprakasam, Vijayakumar, Arul, & Nachiappan, 2016). The mechanism of Cd-induced neuronal apoptosis has been associated with the Fas/FasL (Fas-Ligand) system involving the mitochondrial apoptotic pathway with increased expression of Fas, Fas-L, Fas-associated death domain (FADD) and cleaved caspase 8 (Yuan et al., 2018).

In the developing brain, injury can occur in the cortical neurons, causing neuronal loss and morphological changes in axons and dendrites. Cd is also associated with decreased enzymatic activity of superoxide dismutase (Gupta, Gupta, & Chandra, 1991). Also impacted are motor neurons of the spinal cord ventral horns. Cd increases serotonin sensibility in the CNS, inhibits the release of acetylcholine, blocks synaptic transmission of peripheral cholinergic and adrenergic synapses, decreases brain acetylcholinesterase (AChE) concentration and activity impacting the brain cholinergic mechanisms (Carageorgiou et al., 2004), and it also causes a decrease in depolarization-evoked exocytotic release of glutamate (B. Wang & Du, 2013).

3.3 IMMUNE SYSTEM

3.3.1 Introduction

Heavy metal toxicity to the immune system can have multiple manifestations. Heavy metals are considered immunosuppressive and can be ranked according to their immunosuppressive potential as Mercury, Copper, Manganese, Cobalt, Lead, Cadmium, Chromium (Krzystyniak, Tryphonas, &

Fournier, 1995). Systemic exposure to heavy metals such as As, Cd, Pb, Hg, and Ni can impair the immune response, decrease host resistance to infectious agents, and impact response to neoplasia. Metal toxicity can also induce immune-stimulation, also known as immune modulation. For instance Cd, Pb, and Hg that have been shown to have immunosuppressive properties at high concentrations may have the opposite effect at lower concentrations (Descotes & Vial, 1994). The dose-response curve of heavy metal toxicants, such as As, Cd, Cr, and Pb has been described as "hormesis" where a low dose may be immune stimulatory, a mid dose has no effect and a high dose could cause immune suppression (*Handbook on the Toxicology of Metals*).

Also, important to consider is sensitization to certain heavy metals, such as beryllium, that can be translated into immediate-type hypersensitivity response without an increase in IgE or IgG. Metals such as nickel and beryllium can also cause type IV delayed response mediated by activated macrophages and T lymphocytes. Finally, exposure to some metals, such as Gold salts and Mercury-containing compounds, has been related to interstitial immune complex nephritis with deposition of immune complexes that is similar in pathogenesis to an autoimmune disease (Krzystyniak et al., 1995). Sensitizing metals the cause allergic contact dermatitis include Nickel, Chromium, Cobalt, Platinum, Mercury, Tungsten, Vanadium, Copper, Zinc, Palladium, Tin, Iodine, Lithium, and Gold. Sensitizing metals that cause asthma and rhinitis include Nickel, Chromium, Cobalt, Platinum, Mercury, and Vanadium (Di Gioacchino et al., 2007).

Special consideration should be given when exposure to metals occurs at the nanosize level, as they may be considered as "non-self" and may initiate a process called "opsonization" (Gatto & Bardi, 2018). There is a high likelihood for the uptake of engineered nanomaterials (ENMs) by antigen-presenting cells (APC) impacting interactions with lymphocytes and other types of cells and leading to immunomodulation and/or autoimmune-like disease. TiO_2 and carbon nanotubules can also impact neutrophils, dendritic cells, B and T cells, natural killer (NK) cells, and mast cells (Smith, Brown, Zamboni, & Walker, 2014). Here, we aim to further describe immunologic effects associated with specific metals.

3.3.2 Arsenic

3.3.2.1 Immunosuppression

Arsenic (As) can affect the cell-mediated immune response. It can suppress T cell-dependent antigen stimulation with decreases in cluster of differentiation (CD)45RA+ CD69+ (naïve activated) T cells and alteration of the CD4+/CD8+ ratio and TNF-α levels. Additionally, it has shown to increase the risk of infectious disease (Attreed, Navas-Acien, & Heaney, 2017). T-lymphocyte activation is significantly decreased, with decreases in the proliferation of Th1/Th2 cells and decreases in secretion of IL-2, IL-4, IL-5, IL-10, IFN-γ, and TNF-α as well as downregulation of F-actin and CD54 adhesion molecules (Dangleben, Skibola, & Smith, 2013). The imbalance in Th1/Th2 is regulated by increased activation of NF-κB (Y. Wang et al., 2017). In humans, arsenic exposure has also been associated with an increased percentage of Th cells, and dose-dependent changes in monocytes, NK cells, and an increase in circulating TH17 cells (Lauer et al., 2019) (Figure 3.2).

Decreases in IgG have also been reported with As exposure and are ascribed to immunosuppressive-like effects in addition to direct effects in hematopoietic cells causing erythrocytopenia and leucocytopenia (Patra, Bandyopadhyay, Bandyopadhyay, & Mandal, 2013). Associated with these immune effects is the evidence of increased production of ROS and potentially (peroxide) O^2 and alterations in basal nitric oxide (NO) and iNOS and increased apoptosis (Dangleben et al., 2013).

In ethylnitrosourea (ENU) and other carcinogenic studies, where As was used as the cancer promotor, As causes an imbalance of cytokines profiles, alterations of immunoglobulin secretion profiles, T cell response, and apoptosis of immune cells including CD4 cells activation via TNF R1

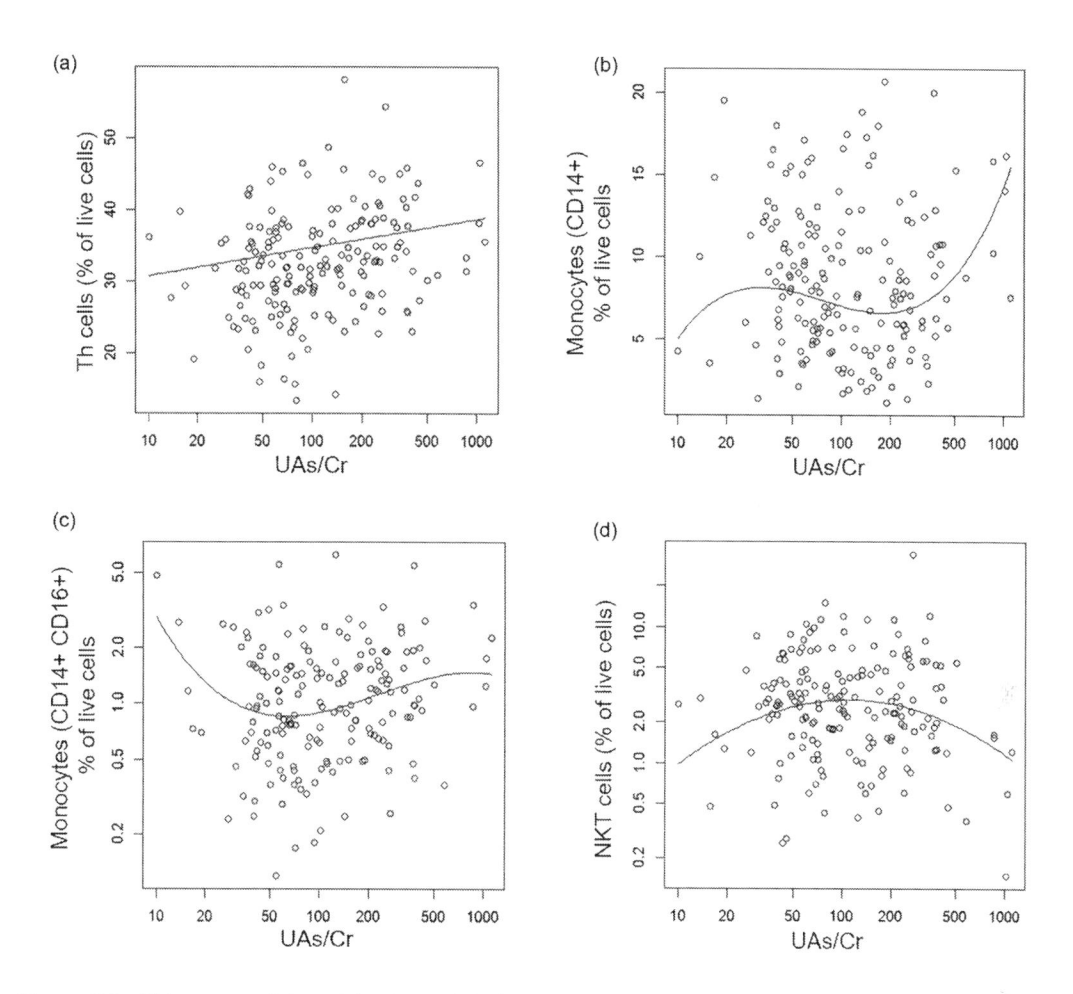

Figure 3.2 Urinary arsenic/creatinine levels (UAs/Cr) are associated with the expression of immune cell markers including total Th cells (A), CD14 (B)-positive monocytes and a subset on monocytes positive for both CD14 and CD16 (C) and natural killer cells (E). (Copyright © 2019 Lauer et al. This is an open access article distributed under the terms of the Creative Commons Attribution License, which permits unrestricted use, distribution, and reproduction in any medium, provided the original author and source are credited.)

(Acharya et al., 2010; C. H. Lee, Liao, & Yu, 2011). Apoptosis of human mononuclear cells has also been associated with As exposure in vitro, with B lymphocytes and monocytes being significantly affected (de la Fuente et al., 2002). Furthermore, As may dysregulate macrophage functions via activation of transcription factor 4 (ATF4) causing a significant downregulation of CD 11b phagocytic functions that can also be associated with impact in apoptotic cell death (Srivastava et al., 2016).

Low doses of As can dramatically suppress autoimmunity in animal disease models impacting Treg cells, the NF-κB, and caspase-3 pathways (*Handbook on the Toxicology of Metals*).

3.3.2.2 *Immunomodulation*

Arsenic can also affect the humoral immune response. It impacts the transplacental immunoglobulin transport. Additionally, increases in serum immunoglobulin (Ig)A, IgG, and IgE were observed with human as exposure. Increases in Granulocyte-Macrophage Colony Stimulating Factor (GM-CSF) secretion could also be associated with chronic inflammatory stimulation (Patra et al., 2013).

3.3.3 Cadmium

3.3.3.1 Immunosuppression

Cadmium (Cd) has been found to induce immunosuppression following intraperitoneal and oral acute or chronic injections (Burchiel et al., 1987). Cd can accumulate in circulating white blood cells, especially lymphocytes. In these and other hematopoietic cells, there is activation of metallothionein-IIA (MT-IIA) and heat shock protein 70 (HSP 70) as well as modulation of Fas-receptor-induced apoptosis causing immune-modulation of T cells response (Tsangaris, Botsonis, Politis, & Tzortzatou-Stathopoulou, 2004).

Administration of cadmium to BALB/c mice resulted in inhibition of cell proliferative responses and increase in Cd-related apoptosis with thymic cortical cell depletion and increase in splenic red pulp with diminished white pulp, effects in both organs are thought to be associated with oxidative stress (Pathak & Khandelwal, 2007). Cd can cause time-dependent apoptosis in the thymocytes and splenocytes isolated from mice treated with Cd (1.8 mg/kg) intra peritoneally for 18, 24, 48, and 72 h (Pathak & Khandelwal, 2007).

While toxicity has been reported in adults, the developing thymus also appears to a target for Cd toxicity affecting thymocyte development in the newborn. Additionally, alteration of splenocyte cytokine profile with low production of interferon gamma and reduced IL2 production has also been observed, as well as effect in the Treg population. CD4+ CD25+ FoxP3+ (characteristic of Natural regulatory T cells-nTreg) (Holaskova, Elliott, Hanson, Schafer, & Barnett, 2012).

In pulmonary disease, cadmium impacts macrophage response to invading pathogens via inhibition of the IKK (IκB kinase) complex (composed of two catalytic subunits (IKKα and IKKβ) and a regulatory subunit (IKKγ)) in the NF-κB signaling pathway (Cox et al., 2016), Additionally, Cd modulates an array of cytokines, including IL-1α (Odewumi et al., 2016).

Production of ROS with generation of free radicals has long been associated with Cd mechanisms of toxicity, including immunotoxicity, and is frequently associated with activation of redox-sensitive transcription factors that include NF-κB, AP-1, and Nrf2 (Liu, Qu, & Kadiiska, 2009). In mice, response to T cell-driven antigens and macrophage activation is decreased with Cd as well as an impediment of host response due to effects in B cell proliferation (Blakley, 1985).

3.3.3.2 Immunomodulation

Cadmium-induced immune enhanced effects have also been described and immune potentiation has been observed with Cd exposure via interference with normal processes of lymphocyte activation and differentiation. Cadmium affects bone marrow immature hematopoietic cell development with an impact on B cell development, causing substantial shifts in cell size distribution and mean cellular volume. Additionally, impacts in B cell population in other lymphoid organs, such as the spleen, have also been documented (Stelzer & Pazdernik, 1983).

3.3.4 Mercury

3.3.4.1 Immunosuppression

Methylmercury is a well-known immunotoxicant. Mercury impacts the immune system defense and surveillance mechanisms (Zefferino, Piccoli, Ricciardi, Scrima, & Capitanio, 2017). The developing immune system is sensitive to the effects of methylmercury. Maternal exposure can cause newborns to have decreases in CD4+ CD45RA+ cells and IgM levels. Similarly, in rodents, MeHg can cause suppression of humoral immunity and reduction in NK cell activity. Toxicity to the hematopoietic bone marrow cell lines; decreases in CD3CD8 splenocytes, decreases in the CD4/CD8 ratio, and decreases in T cell-dependent KHL response, as well as decreased lymphoproliferative response were observed in the post-natal period after exposure (Tonk et al., 2010).

3.3.4.2 Immunomodulation

Hg has immunomodulatory effects by stimulating T cell proliferation and, in some species, increasing B cells with upregulation of CD71 and CD23 markers and increases in IL4 and INF-γ (*Handbook on the Toxicology of Metals*).

3.3.4.3 Sensitization and Autoimmunity

Type II hypersensitivity reactions, with development of basement membrane antibodies in humans and rats, and type III reactions with subsequent deposition of immune complexes have been observed after Hg exposure. Additionally, Hg has been associated with Type IV allergic contact dermatitis. Metals such as Hg, (also Gold, Au) can form strong bonds with protein chains, may oxidize or cause protein denaturation, or can alter the protein profile and the way these are presented to the immune system as antibodies. These mechanisms may lead to the display unrecognized parts of peptides (*cryptic epitopes*) that are then presented to T cells instead of the immunodominant epitopes and because the T cells that are not tolerant to these cryptic epitopes they react and generate an autoimmune response. Hg also interacts with major autoantigen fibrillarin, potentially causing an autoimmune-like response (*Handbook on the Toxicology of Metals*).

3.3.5 Beryllium

3.3.5.1 Sensitization

Exposure to Beryllium (Be) can cause berylliosis that is associated with a mutation of the gene for human leucocyte antigen (HLA)-DPB1, impacting the presentation of Be to T cells. This causes a type IV hypersensitive that manifests itself as non-necrotizing granulomatous inflammation, frequently in the lung, that can be associated with proliferation of T cells, and cytokine increases (INF-γ and TNF-α) (*Handbook on the Toxicology of Metals*). Berylliosis can also be observed in other organs and resembles sarcoidosis. Characteristic formation of multinucleated giant cells that may contain submicron particles of Be is diagnostic for this disease usually along with occupational exposure to Be (Butnor et al., 2003).

3.3.6 Nickel

3.3.6.1 Immunosuppression

Nickel (Ni) has been reported as toxic to the intestinal mucosa of broilers, where it causes a significant decrease in IgA, IgG, and IgM that is associated with a decrease in IG-producing B cells (Wu et al., 2014b). Also associated with Ni exposure, TLR2-2, and toll-like receptor 4 (TLR4) mRNA expression levels in the cecum and cecal tonsil were decreased, implying a potential effect in TLR transduction (Wu et al., 2014a).

Mechanisms of toxicity include activation of apoptotic pathways via Fas abd c-myc (Guo, Chen, et al., 2015) and upregulation of inflammatory response via NF-κB and impacting interleukin-2 (IL-2), interleukin-4 (IL-4), and interleukin-13 (IL-13) (Guo, Deng, et al., 2015).

3.3.6.2 Sensitization

Nickel (Ni) is responsible for cell-mediated delayed hypersensitivity. Type IV hypersensitive is the most common type of sensitization effect of Ni, that it is considered the most common skin sensitizers. In a study of patients with metal hypersensitivity that also included a cohort of healthy patients, Nickel was found to be the metal to cause the highest levels of sensitivity response both in patients with known metal hypersensitivity and in control patients (Stejskal et al., 1999).

Type I hypersensitivity reactions such as asthma have also been reported with Ni exposure. It reacts with innate immune receptor TLR4 and induces NF-κB leading to a proinflammatory cytokine response (*Handbook on the Toxicology of Metals*).

3.3.7 Lead

3.3.7.1 Immunosuppression

One of the immune toxicant effects of Lead (Pb) is the increase of susceptibility to infection that has been observed in rats and mice. This can be due to many factors including the capacity of Pb to bind to antibodies causing a decrease in circulating antibodies. Lead can also suppress interferon response increasing susceptibility to viral diseases (Damstra, 1977).

In general, Pb appears to have a negative effect on the immune system even at low concentrations. Lead (Pb) induces a proinflammatory mechanism of toxicity (Metryka et al., 2018). It also increases the expression of class II molecules on B cells and interferes with interactions between T cells and macrophages, impacting their response to oxidative stress. Pb can also modify T cell response, leading to a Th2 shift affecting dendritic cells and their functions, and it also affects the production of cytokines such as IFN gamma (*Handbook on the Toxicology of Metals*). Pb may have effects Th1 and Th2 response affecting numbers of CD4 helper cells and serum concentration of IL-4 via activation of cAMP-dependent kinase and anti-CD3/CD28-stimulated naive T cells. The cytokines affected by Pb include IL-12, IL-1B, and IL-6. Both Th1 and Th2 cytokines are impacted by Pb that also impacts Th17 cells (Fenga, Gangemi, Di Salvatore, Falzone, & Libra, 2017). Pb suppresses delayed-type hypersensitivity (DTH) response to pathogens (Jane Kasten-Jolly, 2019). Lead causes downregulation of TGF-β and can disturb tolerance to oral antigens. Lead exposure can negatively impact expression in the absolute numbers of CD4 cells as well as CD8 and CD16 (Mishra, 2009). In the developing immune system, Lead causes a marked shift in the T helper function from Th1 to Th2 that causes increased levels of IgE (Curtis, 2013).

3.3.7.2 Autoimmunity

Pb enhances autoimmune manifestations in animal models prone to this type of diseases (*Handbook on the Toxicology of Metals*). Exposure to Pb can lead to the production of autoantibodies against neural proteins, including myelin basic protein and GFAP that can either lead to or aggravate brain diseases such as AD (Mishra, 2009).

3.4 REPRODUCTIVE SYSTEM

3.4.1 Introduction

Heavy metals have been associated with reproductive effects in both males and females. Effects in males are mainly seen in both quantitative and qualitative sperm parameters including sperm motility and morphology. In females, disruption of the female reproductive cycle or effects in fetal development has been reported. Here, we briefly summarize examples of reproductive system metal effects.

3.4.2 Lead

Collective evidence supports a role of Lead in decreasing sperm quality and/or number, both with experimental and natural exposure across species. Lead exposure in young (52-day-old) Wistar rats, with significant increases in circulating Lead levels, was associated with decreases in intratesticular sperm counts, intratesticular testosterone, and decreases in ventral prostate weights (Sokol, Madding, & Swerdloff, 1985). Lead-related decreases in sperm count and/or morphological quality observed in mice were associated with decreased structural DNA integrity and chromatin damage in spermatozoa (C. Li et al., 2018). There are few reports of Lead-related effects on female reproductive system and the available studies reported decreased fertility in Wistar rats following very high (1.5 mg/kg) exposure to lead acetate (Uchewa & Ezugworie, 2019). This decreased fertility was associated with edema and necrosis of the uterus and decreased uterine organ weight. Another study with a similar design in rats did not indicate any significant progesterone or gross pathological changes but reported inflammatory changes (thickening and damage of epithelial cells) in the endometrium (Nakade et al., 2015).

An epidemiological study from a Ukrainian city reported that high Lead exposures (above the WHO guidelines) were associated with changes in sperm number and sperm morphology in infertile men (Onul, Biletska, Stus, & Polion, 2018); another report from a Nigerian city linked Lead exposure, in concert with Cadmium, to oligospermia and azoospermia (Famurewa & Ugwuja, 2017). A third study, conducted in Croatia, reported Lead exposure-associated increases in percentages of pathologic sperm, short sperm and immature sperm, as well as changes in testosterone, estradiol, and prolactin A (Telisman, Colak, Pizent, Jurasovic, & Cvitkovic, 2007). A study in peripubertal girls suggested a role of high Lead exposure in decreasing Inhibin B, a marker of ovarian follicular development (Gollenberg, Hediger, Lee, Himes, & Louis, 2010), while another study associated increased Lead exposure with preeclampsia (Vigeh et al., 2006). The mechanism of Lead-related effects on the reproductive system support a central role through the hypothalamus–pituitary–testicular axis and a local role predominately related to changes in oxidant/antioxidant balance as major factors for toxicity development (Gandhi et al., 2017). The main central mechanism of Lead-related effects is most relevant during pubertal development stages, while the local oxidant effects are active at most stages of reproductive life. Several studies that explored antioxidant therapies in cases of Lead-related testicular toxicities in laboratory animals have reported amelioration of Lead-related reproductive effects (Anjum, Madhu, Reddy, & Reddy, 2017; Hassan, El-Neweshy, Hassan, & Noreldin, 2019; Kumar et al., 2013; Ommati et al., 2019) suggesting that the mechanism of damage to spermatozoa could be related to oxidant injury.

3.4.3 Cadmium

The reproductive effects of Cadmium in Wistar rats are, like Lead, mostly observed on testes and mimic some of the sperm quality effects already described for Lead such as decreased sperm count, spermatocyte DNA damage, and oxidant-related injury associated with increased testicular lactate dehydrogenase (Kumar et al., 2013; Nava-Hernandez et al., 2009). As in rats, increased seminal Cd or Lead in men, have been associated with increased spermatocyte DNA damage and decline in sperm quality (Pant et al., 2014). Furthermore, exposures to Lead and Cadmium may co-occur in some populations making attribution of reproductive effects difficult in epidemiologic studies. Epidemiologic studies of young (20–45-year-old) men (Famurewa & Ugwuja, 2017) and peripubertal (6–11-year-old) girls (Gollenberg et al., 2010) have linked Cadmium exposures (seminal and plasma) alone or together with Lead, respectively, to increased azoospermia and oligospermia or decreased Inhibin B, a marker of ovarian follicular development. In male rats experimentally exposed to Cadmium during gestation and/or early lactation, there is also a Cadmium-related decrease in the luminal size of seminiferous tubules (Corpas & Antonio, 1998).

3.4.4 Vanadium

Inhalation exposure to toxic levels of Vanadium pentoxide in mice resulted in testicular toxicity (Fortoul et al., 2007) characterized by necrosis of spermatogonia, spermatocytes, and Sertoli cells. Oral exposure of Vanadium in the form of Sodium Metavanadate did not result in testicular toxicity, although epididymal weights were decreased at the highest dose. Exposure of pregnant rats and mice to Vanadium indicates accumulation within fetal membranes but not the fetus itself (Altamirano-Lozano, Alvarez-Barrera, Mateos-Nava, Fortoul, & Rodriguez-Mercado, 2014) suggesting that likelihood for Vanadium-related direct effects on the developing fetus is low. However, epidemiological studies have associated increased exposure to Vanadium with premature rupture of membranes in pregnant women (Jin et al., 2018) and possible changes in semen quality (DNA integrity of spermatozoa) in men (Y. X. Wang et al., 2018). Taken together, the evidence for direct testicular and/or developments injury indicates that the toxicity may be dependent on exposure route, form of Vanadium exposure and exposure concentration.

3.4.5 Gallium and Indium

Gallium and Indium Arsenide are used in the semiconductor industry and have been implicated in reproductive toxicity. They reportedly caused male reproductive toxicity in rodents (Omura, Hirata, et al., 1996; Omura, Tanaka, et al., 1996). The effects in rats and mice consist of decreased sperm count and increased morphological abnormalities, especially within the epididymis, while the effects in Hamsters were limited to Gallium Arsenide that included testicular spermatid retention and subsequent reduction in sperm numbers within epididymis. Although Gallium Arsenide showed testicular effects by inhalation exposure, subcutaneous exposure of Gallium Nitrate was not associated with any testicular toxicity (Colomina, Llobet, Sirvent, Domingo, & Corbella, 1993) suggesting that Gallium may not be the direct toxicant. Gallium Arsenide also caused significant lung injury and anemia in rodents that have testicular toxicity, and these other organ system injuries can cause hypoxia, which has been proposed as an indirect cause of Gallium Arsenide-related testicular toxicity (Bomhard, Cohen, Gelbke, & Williams, 2012). To date, no epidemiology studies have implicated Gallium or Indium directly in male or female reproductive effects. Therefore, the reproductive effects of Gallium/Indium Arsenide in rodents, whether direct or indirect, are a hazard that requires appropriate risk assessment in applicable exposure scenarios.

3.5 LIVER

3.5.1 Introduction

The liver has an important role in metal metabolism, biotransformation, and elimination; therefore it is fundamental in the response to metal toxicity and potential for metal bioaccumulation. The liver is also a critical organ for metal detoxification, predominately through the induction of metalloproteins such as metallothionein (MT), transferrin, and ceruloplasmin (Wolff, Lee, Abouhamed, & Thevenod, 2008). Disruption of hepatic detoxification processes or excessive metal intake leads invariably to the development of liver injury. Acute liver lesions, frequently observed as hepatocyte necrosis, are usually the result of exposure to high metal doses in a short period of time and can, non-uncommonly, be associated with death. Chronic liver injury is a continuum of ongoing inflammatory insult that is accompanied by increased connective tissue (fibrosis) that, depending on severity and persistence of metal exposure, can lead to cirrhosis and in some cases can ultimately be associated with liver cancer.

3.5.2 Arsenic

Acute Arsenic poisoning can cause death within 24 h; however, sublethal losses may lead to liver injury within 24–48 h post ingestion. Acute As lesions are usually characterized by hepatic necrosis with marked elevation of serum aminotransferase levels. Liver injury can also occur with chronic As exposure in both humans and animal models. Chronic exposure to environmental toxicants, including Arsenic, has been associated with fatty liver disease and liver fibrosis observed even in non-obese individuals (Bambino et al., 2018). As exposure that can also cause hepatomegaly, icterus, ascites (*Haschek and Rousseaux's Handbook of Toxicologic Pathology*, 2013), and hepato-portal sclerosis (Liu & Waalkes, 2008). Liver lesions can progress into hepatic chronic inflammation (cirrhosis) or cancer development (J. Li et al., 2015) (Curtis, 2013). Inorganic Arsenic is metabolized in the liver via Arsenic 3 methyltransferase (AS3MT), and this reaction produces ROS as well as DNA methylation. One other mechanism of As toxicity is the potential to cause protein misfolding (Bambino et al., 2018). Although the exact pathways involved in As-induced hepatotoxicity have not been completely elucidated, during the formation of ROS, nuclear factor erythroid 2-related factor 2 (Nrf2), that is a molecule involved in the response to oxidative stress, is upregulated and could induce the expression of heme oxygenase-1 (HO-1) that, in turn, can have a role in the attenuation of As-induced oxidative injury in the liver (Zhang et al., 2017). After six weeks of As exposure Nrf2-/- mice liver lesions were more severe, and these mice were thought to be more sensitive to alterations on DNA methylation, oxidative damage, and apoptosis. After exposure, an increase in Nrf2 expression, as well as downstream genes, was observed in mice, and thus it is considered an important aspect if the early antioxidant response to As exposure (J. Li et al., 2015). Exposure to As has also been associated with modification in the expression of microRNAs (miRNA), targeting (for instance) Glutamate Cysteine Ligase (GCL) messenger RNA (mRNAs), and the disruption of these may lead to modifications in the *in vivo* antioxidative response ultimately impacting lesions outcomes in the liver (Ren et al., 2015).

There is growing evidence that the liver could be one of the primary targets for As carcinogenesis. In addition to altered DNA methylation and DNA oxidative damage, other possible mechanisms of carcinogenesis include impaired DNA damage repair, stimulation of hyperproliferation, apoptotic tolerance, and aberrant estrogenic signaling (Liu & Waalkes, 2008).

3.5.3 Cadmium

Cadmium has a half-life of about 15–30 years and the liver is one of the main organs where Cd accumulates and can cause hepatic injury (Neathery & Miller, 1975). In the liver, Cd binds to MT forming the Cadmium–MT complex that can be eliminated via the biliary or urinary systems. Hepatic MT is fundamental to the liver response to Cd exposure. Transgenic MT mice are more resistant to Cd-related injury as opposite to MT null mice that are more sensitive (Klaassen & Liu, 1998). Excessive Cd can cause acute hepatic necrosis (*Haschek and Rousseaux's Handbook of Toxicologic Pathology*, 2013). In rodent models of disease chronic Cd exposure causes hepatic non-specific chronic inflammation, granulomatous inflammation, apoptosis, and/or hepatocyte regeneration. In humans, exposure to Cd has been associated with necroinflammation in men and women, and nonalcoholic fatty liver disease and non-alcoholic steatohepatitis in men, both characterized by the presence of hepatic steatosis with or without increased levels of liver enzymes, respectively (Hyder et al., 2013). Low levels of Cd exposure can result in liver function changes that include liver mitochondrial protein oxidation related to fatty acid metabolism that, together with increased oxidative stress (Skipper, Sims, Yedjou, & Tchounwou, 2016), may have a role in pathways, such as c-Jun N-terminal kinase (JNK), associated with nonalcoholic fatty liver disease (NAFLD) (Go et al., 2015). As with other metals, Cd toxicity and carcinogenic potential have been associated with pathways activated with oxidative stress. Cd has been reported to generate hydrogen peroxide, superoxide anion, and hydroxyl radicals. In acute toxicity, these can activate pathways such as NF-κB and Nrf2. With chronic exposure, Cd tolerance may occur that can lead to less ROS-related signal that together for aberrant gene expression may play a role in the Cd toxicity and carcinogenesis (Liu et al., 2009).

3.5.4 Copper

Most commonly Copper (Cu) toxicity occurs due to excessive oral intake and is characterized by gastrointestinal discomfort with abdominal pain, nausea, and diarrhea. Ingestion of large amounts of Cu may lead to hepatic necrosis that in severe cases may lead to death (Tchounwou, Newsome, Williams, & Glass, 2008). Dietary exposure to large amounts of copper, in the sulfate form, might result in hepatocellular necrosis as well as gastrointestinal symptoms (Curtis, 2013). In cases of both Copper- and/or Iron-related hepatocellular necrosis, there may be fine granular intracytoplasmic pigments in hepatocytes that can be stained with various histochemical stains to confirm Copper or Iron as the initiator of the insult.

Chronic toxicity of Cu has been characterized in Wilson's disease, where there is an impairment of Cu metabolism, due to mutations of the ATP7B gene, that leads to the accumulation of Cu in multiple organs including the liver. Individuals with this disease have difficulty eliminating Cu via the biliary system, causing accumulation of Cu and reduced incorporation of Cu in ceruloplasmin that lead to accumulation primarily in the liver but also in the brain. This accumulation can lead to chronic liver inflammation/cirrhosis, acute liver failure, and/or hepatocellular carcinoma (Patil, Sheth, Krishnamurthy, & Devarbhavi, 2013). Another disease where Copper exposure can lead to hepatic dysfunction is Indian Childhood cirrhosis in which sensitivity to Cu toxicity may have an autosomal recessive component (Curtis, 2013).

One of the mechanisms of Cu toxicity is the alteration of antioxidant enzyme activity with reductions in glutathione (GSH), catalase, and superoxide dismutase (SOD) and increases in lipid peroxidation products caused by Cu nanoparticles (Anreddy, 2018). Cu also affects lipid and glucose metabolism, and it has been associated with NAFLD by causing oxidative stress, low-level inflammation and mitochondria dysfunction; and impacting micronutrient balance of, for example, Iron and fructose (Aigner, Weiss, & Datz, 2015). In chronic exposure, carcinogenicity may arise from the significant induction of stress genes related to proteotoxic effects, for example, heat shock protein 70kD (HSP 70), inflammatory/oxidative imbalance (*c-fos*) and growth arrest and DNA damage (p53) (Tchounwou et al., 2008).

3.5.5 Iron

Iron (Fe) hepatic overload in adults can have multiple etiologies, including hereditary hemochromatosis (when there is abnormal Fe absorption in the intestinal tract), excessive Iron intake, and repeated blood transfusions. Increased accumulation of Fe in any parenchymal organs is called hemosiderosis. When it becomes excessive and causes parenchymal lesions, such as increase fibrous connective tissue, it is called hemochromatosis. Hepatic hemochromatosis has been associated with increased risk of hepatocellular carcinoma (Anderson & Frazer, 2017; Curtis, 2013). However, hepatic Iron overload is also frequently observed with chronic liver disease regardless of its etiology. The {Rubino, 2015 #256} major route of Iron entry into the hepatocyte is via receptor-mediated endocytosis of transferrin by transferrin receptor-1 (TFR1). After internalization, the Iron is released into the lysosomal compartment and can be delivered into other organelles. Iron can cycle between two stable oxidation states, ferrous Iron (iron(II) or Fe^{2+}) and ferric Iron (Iron(III) or Fe^{3+}) (Sangkhae & Nemeth, 2017). When in excess, Fe is oxidized into the ferric form and can be sequestered within the ferritin (Bloomer & Brown, 2019). Increased levels of Fe can accelerate the Fenton reaction in which the ferrous form of Iron interacts with H_2O_2 and generates highly reactive hydroxyl radical ($Fe^{2+} + H_2O_2 \rightarrow Fe^{3+} + {}^{\cdot}OH + OH^- $ & $ Fe3^+ + H_2O_2 \rightarrow Fe^{2+} + {}^{\cdot}OOH + H^+$). These ROS can lead to hepatocellular injury by disruptively reacting with biological structures such as unsaturated lipids in cellular membranes, DNA and RNA components, and other essential chemical groups in catalytic and signaling proteins (Rubino, 2015). Prolonged injury causes chronic stimulation of hepatic stellate cells (HSC) that become sources of transforming growth factor-beta (TGF-β) that, in turn, stimulates further proliferation of HSC and their differentiation into myofibroblasts producers of extracellular matrix that can lead to a fibrotic state. This state can progress into cirrhosis and ultimately cancer (Mehta, Farnaud, & Sharp, 2019). Iron has also been implicated with inflammatory pathways such as NF-ƙB in liver Kupffer cells that can increase inflammatory cytokines like, tumor necrosis factor-alpha (TNF-α) and interleukin-6 (IL-6) (Bloomer & Brown, 2019). Increased Fe associated with increased intestinal absorption has also been reported as a possible factor in the etiology of nonalcoholic fatty liver disease (NAFLD) (Malik, Wilting, Ramadori, & Naz, 2017). A key hepatic hormone that regulates Iron systemic levels is hepcidin (HAMP or HEPC). Disturbances in HEPC are associated with many Iron overload disorders including chronic inflammatory diseases and cancer (Sangkhae & Nemeth, 2017).

3.6 KIDNEY

3.6.1 Introduction

Some metals, usually acquired through environmental and/or dietary exposure, are toxic to the kidney in humans and animals. The toxic forms of the metal may be protein bound (Cadmium–MT) and generally affect renal tubular epithelium, partially because these epithelia reabsorb the metals from glomerular filtrate and have high cytoplasmic concentrations of the metals. Four major metal nephrotoxicants are discussed below with respect to the mechanism of renal toxicity and exposure scenarios, when available. Other toxicants such as Chromium, Platinum, Bismuth, and Gallium will not be discussed in detail as they generally share some of the same site-specific mechanisms of injury described for the metals below.

3.6.2 Cadmium

Cadmium exposure occurs through inhalation and food intake. Additionally, smoking, mining, and Nickel–Cadmium battery manufacturing are also exposure scenarios. For most of human population exposure is dietary, through foods that accumulate Cadmium such as shellfish and rice. Daily intake is about 10–30 µg/day of which 10% is retained in the body with the kidney storing half of the body burden of Cadmium (Curtis, 2013) and may result in renal injury at renal cortical Cadmium concentrations of 50 µg/g (Klaassen, Liu, & Choudhuri, 1999).

Cadmium exposure is associated with an increase in MT synthesis and subsequent binding of this protein to Cadmium to form a Cadmium–metallothionein complex (Cd-MT), the ultimate nephrotoxicant (Nordberg, Goyer, & Nordberg, 1975). Cd-MT, like other circulating metalloproteins, filters through the glomeruli and is endocytosed in the proximal tubules through megallin/cubilin and ADP-ribosylation factor GTPases (Wolff, Lee, Abouhamed, & Thevenod, 2008). Cadmium produces segment-specific pathology in the S1 and S2 segments of the proximal tubule, and this injury is associated with increases in urinary excretion of glucose, amino acids, calcium, and cellular enzymes (Curtis, 2013) and with chronic exposure may progress to an interstitial nephritis.

3.6.3 Mercury

Environmental exposure to organic, elemental, and salt forms of Mercury occurs in humans and animals. In the body, Mercury is oxidized within red blood cells and other tissues to the inorganic Mercury, the major form in tissues (Curtis, 2013). Inorganic Mercury binds to sulfhydryl groups in proteins and is present in the glomerular filtrates in conjugates of low-molecular-weight proteins such as glutathione. They accumulate in the kidney through tubular reabsorption and initiate injury to the S3 segment of the proximal tubule (Curtis, 2013). Mercury uptake in the kidney involves organic anion transporters (OAT 1 and 3) and mechanism of injury is thought to involve binding and inactivation of sulfhydryl groups within proteins such as sodium/potassium ATPases with subsequent dysfunction of organellar or cellular transport functions (Rana, Tangpong, & Rahman, 2018). With progressive injury, tubular reabsorption may be impaired resulting in decreased solute reabsorption, glycosuria, and proteinuria.

3.6.4 Lead

People are exposed to Lead through various leaded products (paints, batteries, gasoline) that, in the last two decades, have been mostly banned or strictly controlled in most countries. Current Lead exposure is most likely to occur through food and/or water.

Exposure to Lead causes pathologic changes in the proximal tubule that are believed to be a result of oxidative stress induced by Lead, modulation of intracellular calcium homeostasis, and/or effects on the respiratory chain of mitochondria (Rana et al., 2018). Rats exposed to Pb can develop renal toxicity characterized by cytomegaly/karyomegaly in the proximal tubular epithelial cells (at 5 ppm Lead acetate in drinking water in males; 25 ppm in females), in addition to nuclear inclusion body formation and increased numbers of Iron-positive granules (Fowler, Kimmel, Woods, McConnell, & Grant, 1980). Nuclear inclusion bodies are a diagnostic feature of Lead toxicity in kidney, and energy-dispersive X-ray microanalysis has demonstrated the highest intracellular Lead concentrations in these inclusion bodies (Fowler et al., 1980). Microscopically, inclusions bodies are homogeneous, and eosinophilic and are now known to represent a Pb protein complex where MT surrounds Lead that in a nontoxic state (Waalkes, Liu, Goyer, & Diwan, 2004). MT-null mice do not form inclusion bodies after exposure to Pb and are more hypersensitive to Lead-induced nephropathy and to Lead-induced kidney damage (Waalkes et al., 2004). Chronic and progressive Lead nephrotoxicity is also associated with interstitial fibrosis, loss of nephrons, azotemia, and renal failure, in addition to increased incidence of gout (Curtis, 2013).

3.6.5 Uranium

Some epidemiological studies indicate renal toxicity in some cases of accidental or occupational Uranium exposure while others do not show renal effects (Keith et al., 2013). However, experimental studies in animals through different routes indicate that more soluble forms of Uranium can cause renal toxicity in different species (J. Chen, Meyerhof, & Tracy, 2004; Diamond, Morrow, Panner, Gelein, & Baggs, 1989; Keith et al., 2013). Very high concentrations of soluble forms of Uranium have been associated with renal tubular damage in experimental settings. The main site of toxicity is in the proximal tubule, where damage to tubular epithelial cells result in a dysfunction of tubular resorption and lead to consequent glucosuria and proteinuria (Keith et al., 2013).

3.7 SUMMARY AND CONCLUSIONS

The pathological effects of metal exposure are well described in toxicology literature. The understanding of the mechanisms behind these lesions is continuing to evolve as new tools for the study of the molecules involved in disease establishment and progression are constantly evolving.

Albeit modest, our goal in this summary of metal-induced pathology is to provide a brief insight on biochemical and molecular mechanisms associated with pathophysiological metal-induced processes.

REFERENCES

Abdel-Salam, A. M., Al Hemaid, W. A., Afifi, A. A., Othman, A. I., Farrag, A. R. H., & Zeitoun, M. M. (2018). Consolidating probiotic with dandelion, coriander and date palm seeds extracts against mercury neurotoxicity and for maintaining normal testosterone levels in male rats. *Toxicol Rep*, 5, 1069–1077. doi:10.1016/j.toxrep.2018.10.013.

Abubakar, K., Muhammad Mailafiya, M., Danmaigoro, A., Musa Chiroma, S., Abdul Rahim, E. B., & Abu Bakar M. Z. (2019). Curcumin attenuates lead-induced cerebellar toxicity in rats via chelating activity and inhibition of oxidative stress. *Biomolecules*, 9(9). doi:10.3390/biom9090453.

Acharya, S., Chaudhuri, S., Chatterjee, S., Kumar, P., Begum, Z., Dasgupta, S., . . . Chaudhuri, S. (2010). Immunological profile of arsenic toxicity: A hint towards arsenic-induced carcinogenesis. *Asian Pac J Cancer Prev*, 11(2), 479–490.

Aigner, E., Weiss, G., & Datz, C. (2015). Dysregulation of iron and copper homeostasis in nonalcoholic fatty liver. *World J Hepatol*, 7(2), 177–188. doi:10.4254/wjh.v7.i2.177.

Alexandrov, P. N., Pogue, A. I., & Lukiw, W. J. (2018). Synergism in aluminum and mercury neurotoxicity. *Integr Food Nutr Metab*, 5(3). doi:10.15761/ifnm.1000214.

Altamirano-Lozano, M. A., Alvarez-Barrera, L., Mateos-Nava, R. A., Fortoul, T. I., & Rodriguez-Mercado, J. J. (2014). Potential for genotoxic and reprotoxic effects of vanadium compounds due to occupational and environmental exposures: An article based on a presentation at the 8th International Symposium on Vanadium Chemistry, Biological Chemistry, and Toxicology, Washington DC, August 15–18, 2012. *J Immunotoxicol*, 11(1), 19–27. doi:10.3109/1547691X.2013.791734.

Anderson, G. J., & Frazer, D. M. (2017). Current understanding of iron homeostasis. *Am J Clin Nutr*, 106 (Suppl 6), 1559s–1566s. doi:10.3945/ajcn.117.155804.

Anjum, M. R., Madhu, P., Reddy, K. P., & Reddy, P. S. (2017). The protective effects of zinc in lead-induced testicular and epididymal toxicity in Wistar rats. *Toxicol Ind Health*, 33(3), 265–276. doi:10.1177/0748233716637543.

Anreddy, R. N. R. (2018). Copper oxide nanoparticles induces oxidative stress and liver toxicity in rats following oral exposure. *Toxicol Rep*, 5, 903–904. doi:10.1016/j.toxrep.2018.08.022.

Antunes Dos Santos, A., Appel Hort, M., Culbreth, M., Lopez-Granero, C., Farina, M., Rocha, J. B., & Aschner, M. (2016). Methylmercury and brain development: A review of recent literature. *J Trace Elem Med Biol*, 38, 99–107. doi:10.1016/j.jtemb.2016.03.001.

Attreed, S. E., Navas-Acien, A., & Heaney, C. D. (2017). Arsenic and immune response to infection during pregnancy and early life. *Curr Environ Health Rep*, 4(2), 229–243. doi:10.1007/s40572-017-0141-4.

Austin, R. N., Freeman, J. L., & Guilarte, T. R. (2016). Neurochemistry of lead and manganese. *Metallomics*, 8(6), 561–562. doi:10.1039/c6mt90017h.

Bambino, K., Zhang, C., Austin, C., Amarasiriwardena, C., Arora, M., Chu, J., & Sadler, K. C. (2018). Inorganic arsenic causes fatty liver and interacts with ethanol to cause alcoholic liver disease in zebrafish. *Dis Model Mech*, 11(2). doi:10.1242/dmm.031575.

Blakley, B. R. (1985). The effect of cadmium chloride on the immune response in mice. *Can J Comp Med*, 49(1), 104–108.

Bloomer, S. A., & Brown, K. E. (2019). Iron-Induced Liver Injury: A Critical Reappraisal. *Int J Mol Sci*, 20(9). doi:10.3390/ijms20092132.

Bomhard, E. M., Cohen, S. M., Gelbke, H. P., & Williams, G. M. (2012). Evaluation of the male reproductive toxicity of gallium arsenide. *Regul Toxicol Pharmacol*, 64(1), 77–86. doi:10.1016/j.yrtph.2012.06.005.

Bondy, S. C. (2010). The neurotoxicity of environmental aluminum is still an issue. *Neurotoxicology*, 31(5), 575–581. doi:10.1016/j.neuro.2010.05.009.

Branca, J. J. V., Morucci, G., & Pacini, A. (2018). Cadmium-induced neurotoxicity: Still much ado. *Neural Regen Res*, 13(11), 1879–1882. doi:10.4103/1673-5374.239434.

Burchiel, S. W., Hadley, W. M., Cameron, C. L., Fincher, R. H., Lim, T. W., Elias, L., & Stewart, C. C. (1987). Analysis of heavy metal immunotoxicity by multiparameter flow cytometry: correlation of flow cytometry and immune function data in B6CF1 mice. *Int J Immunopharmacol*, 9(5), 597–610. doi:10.1016/0192-0561(87)90127-5.

Butnor, K. J., Sporn, T. A., Ingram, P., Gunasegaram, S., Pinto, J. F., & Roggli, V. L. (2003). Beryllium detection in human lung tissue using electron probe X-ray microanalysis. *Mod Pathol, 16*(11), 1171–1177. doi:10.1097/01.Mp.0000094090.90571.Ed.

Carageorgiou, H., Tzotzes, V., Pantos, C., Mourouzis, C., Zarros, A., & Tsakiris, S. (2004). In vivo and in vitro effects of cadmium on adult rat brain total antioxidant status, acetylcholinesterase, (Na+, K+)-ATPase and Mg2+-ATPase activities: protection by L-cysteine. *Basic Clin Pharmacol Toxicol, 94*(3), 112–118.

Curtis, D. K. (2013). *Casarett & Doull's Toxicology: The Basic Science of Poisons.* (2013). (8 ed. Vol. 1). New York: McGraw Hill Education.

Chang, L. W. (1990). The neurotoxicology and pathology of organomercury, organolead, and organotin. *J Toxicol Sci, 15*(Suppl 4), 125–151. doi:10.2131/jts.15.supplementiv_125.

Chen, J., Meyerhof, D. P., & Tracy, B. L. (2004). Model results of kidney burdens from uranium intakes. *Health Phys, 86*(1), 3–11. doi:10.1097/00004032-200401000-00003.

Chen, S. M., Phuagkhaopong, S., Fang, C., Wu, J. C. C., Huang, Y. H., Vivithanaporn, P., . . . Tsai, C. Y. (2019). Dose-dependent acute circulatory fates elicited by cadmium are mediated by differential engagements of cardiovascular regulatory mechanisms in brain. *Front Physiol, 10*, 772. doi:10.3389/fphys.2019.00772.

Chibowska, K., Baranowska-Bosiacka, I., Falkowska, A., Gutowska, I., Goschorska, M., & Chlubek, D. (2016). Effect of lead (Pb) on inflammatory processes in the brain. *Int J Mol Sci, 17*(12). doi:10.3390/ijms17122140.

Colomina, M. T., Llobet, J. M., Sirvent, J. J., Domingo, J. L., & Corbella, J. (1993). Evaluation of the reproductive toxicity of gallium nitrate in mice. *Food Chem Toxicol, 31*(11), 847–851. doi:10.1016/0278-6915(93)90223-l.

Cordova, F. M., Aguiar, A. S., Jr., Peres, T. V., Lopes, M. W., Goncalves, F. M., Remor, A. P., . . . Leal, R. B. (2012). In vivo manganese exposure modulates Erk, Akt and Darpp-32 in the striatum of developing rats, and impairs their motor function. *PLoS One, 7*(3), e33057. doi:10.1371/journal.pone.0033057.

Corpas, I., & Antonio, M. T. (1998). Study of alterations produced by cadmium and cadmium/lead administration during gestational and early lactation periods in the reproductive organs of the rat. *Ecotoxicol Environ Saf, 41*(2), 180–188. doi:10.1006/eesa.1998.1690.

Cox, J. N., Rahman, M. A., Bao, S., Liu, M., Wheeler, S. E., & Knoell, D. L. (2016). Cadmium attenuates the macrophage response to LPS through inhibition of the NF-kappaB pathway. *Am J Physiol Lung Cell Mol Physiol, 311*(4), L754–L765. doi:10.1152/ajplung.00022.2016.

Damstra, T. (1977). Toxicological properties of lead. *Environ Health Perspect, 19*, 297–307. doi:10.1289/ehp.7719297.

Dangleben, N. L., Skibola, C. F., & Smith, M. T. (2013). Arsenic immunotoxicity: A review. *Environ Health, 12*(1), 73. doi:10.1186/1476-069x-12-73.

de la Fuente, H., Portales-Perez, D., Baranda, L., Diaz-Barriga, F., Saavedra-Alanis, V., Layseca, E., & Gonzalez-Amaro, R. (2002). Effect of arsenic, cadmium and lead on the induction of apoptosis of normal human mononuclear cells. *Clin Exp Immunol, 129*(1), 69–77. doi:10.1046/j.1365-2249.2002.01885.x.

Descotes, J., & Vial, T. (1994). Immunotoxic effects of xenobiotics in humans: A review of current evidence. *Toxicol In Vitro, 8*(5), 963–966. doi:10.1016/0887-2333(94)90227-5.

Di Gioacchino, M., Verna, N., Di Giampaolo, L., Di Claudio, F., Turi, M. C., Perrone, A., . . . Boscolo, P. (2007). Immunotoxicity and sensitizing capacity of metal compounds depend on speciation. *Int J Immunopathol Pharmacol, 20*(2 Suppl 2), 15–22. doi:10.1177/03946320070200s204.

Diamond, G. L., Morrow, P. E., Panner, B. J., Gelein, R. M., & Baggs, R. B. (1989). Reversible uranyl fluoride nephrotoxicity in the Long Evans rat. *Fundam Appl Toxicol, 13*(1), 65–78. doi:10.1016/0272-0590(89)90307-2.

El-Fawal, H. A., Waterman, S. J., De Feo, A., & Shamy, M. Y. (1999). Neuroimmunotoxicology: Humoral assessment of neurotoxicity and autoimmune mechanisms. *Environ Health Perspect, 107* (Suppl 5), 767–775. doi:10.1289/ehp.99107s5767.

Famurewa, A. C., & Ugwuja, E. I. (2017). Association of blood and seminal plasma cadmium and lead levels with semen quality in non-occupationally exposed infertile men in Abakaliki, South East Nigeria. *J Family Reprod Health, 11*(2), 97–103.

Farina, M., Rocha, J. B., & Aschner, M. (2011). Mechanisms of methylmercury-induced neurotoxicity: Evidence from experimental studies. *Life Sci, 89*(15–16), 555–563. doi:10.1016/j.lfs.2011.05.019.

Fenga, C., Gangemi, S., Di Salvatore, V., Falzone, L., & Libra, M. (2017). Immunological effects of occupational exposure to lead (Review). *Mol Med Rep, 15*(5), 3355–3360. doi:10.3892/mmr.2017.6381.

Flora, S. J., Mittal, M., Pachauri, V., & Dwivedi, N. (2012). A possible mechanism for combined arsenic and fluoride induced cellular and DNA damage in mice. *Metallomics, 4*(1), 78–90. doi:10.1039/c1mt00118c.

Fortoul, T. I., Bizarro-Nevares, P., Acevedo-Nava, S., Pinon-Zarate, G., Rodriguez-Lara, V., Colin-Barenque, L., . . . Saldivar-Osorio, L. (2007). Ultrastructural findings in murine seminiferous tubules as a consequence of subchronic vanadium pentoxide inhalation. *Reprod Toxicol, 23*(4), 588–592. doi:10.1016/j.reprotox.2007.03.004.

Fowler, B. A., Kimmel, C. A., Woods, J. S., McConnell, E. E., & Grant, L. D. (1980). Chronic low-level lead toxicity in the rat. III. An integrated assessment of long-term toxicity with special reference to the kidney. *Toxicol Appl Pharmacol, 56*(1), 59–77. doi:10.1016/0041-008x(80)90131-3.

Gandhi, J., Hernandez, R. J., Chen, A., Smith, N. L., Sheynkin, Y. R., Joshi, G., & Khan, S. A. (2017). Impaired hypothalamic-pituitary-testicular axis activity, spermatogenesis, and sperm function promote infertility in males with lead poisoning. *Zygote, 25*(2), 103–110. doi:10.1017/S0967199417000028.

Gatto, F., & Bardi, G. (2018). Metallic nanoparticles: General research approaches to immunological characterization. *Nanomaterials (Basel), 8*(10). doi:10.3390/nano8100753.

Gavin, C. E., Gunter, K. K., & Gunter, T. E. (1990). Manganese and calcium efflux kinetics in brain mitochondria. Relevance to manganese toxicity. *Biochem J, 266*(2), 329–334. doi:10.1042/bj2660329.

Go, Y. M., Sutliff, R. L., Chandler, J. D., Khalidur, R., Kang, B. Y., Anania, F. A., . . . Jones, D. P. (2015). Low-dose cadmium causes metabolic and genetic dysregulation associated with fatty liver disease in mice. *Toxicol Sci, 147*(2), 524–534. doi:10.1093/toxsci/kfv149.

Goldstein, G. W. (1990). Lead poisoning and brain cell function. *Environ Health Perspect, 89*, 91–94. doi:10.1289/ehp.908991.

Gollenberg, A. L., Hediger, M. L., Lee, P. A., Himes, J. H., & Louis, G. M. (2010). Association between lead and cadmium and reproductive hormones in peripubertal U.S. girls. *Environ Health Perspect, 118*(12), 1782–1787. doi:10.1289/ehp.1001943.

Grandjean, P., & Herz, K. T. (2015). Trace elements as paradigms of developmental neurotoxicants: Lead, methylmercury and arsenic. *J Trace Elem Med Biol, 31*, 130–134. doi:10.1016/j.jtemb.2014.07.023.

Guariglia, S. R., Stansfield, K. H., McGlothan, J., & Guilarte, T. R. (2016). Chronic early life lead (Pb(2+)) exposure alters presynaptic vesicle pools in hippocampal synapses. *BMC Pharmacol Toxicol, 17*(1), 56. doi:10.1186/s40360-016-0098-1.

Guo, H., Chen, L., Cui, H., Peng, X., Fang, J., Zuo, Z., . . . Wu, B. (2015). Research advances on pathways of nickel-induced apoptosis. *Int J Mol Sci, 17*(1). doi:10.3390/ijms17010010.

Guo, H., Deng, H., Cui, H., Peng, X., Fang, J., Zuo, Z., . . . Chen, K. (2015). Nickel chloride (NiCl$_2$)-caused inflammatory responses via activation of NF-kappaB pathway and reduction of anti-inflammatory mediator expression in the kidney. *Oncotarget, 6*(30), 28607–28620. doi:10.18632/oncotarget.5759.

Gupta, A., Gupta, A., & Chandra, S. V. (1991). Gestational cadmium exposure and brain development: A biochemical study. *Ind Health, 29*(2), 65–71. doi:10.2486/indhealth.29.65.

Handbook of Clinical Neurology (Occupational Neurology). (2015). (F. B. a. D. S. Michael J. Aminoff Ed. Vol. 131). Waltman, MA: Elsevier.

Handbook of Developmental Neurotoxicology. (1998). San Diego, CA: Academic Press, Inc.

Handbook on the Toxicology of Metals. (2007). (Vol. 1). San Diego, CA: Academic Press.

Harischandra, D. S., Ghaisas, S., Zenitsky, G., Jin, H., Kanthasamy, A., Anantharam, V., & Kanthasamy, A. G. (2019). Manganese-induced neurotoxicity: New insights into the triad of protein misfolding, mitochondrial impairment, and neuroinflammation. *Front Neurosci, 13*, 654. doi:10.3389/fnins.2019.00654.

Haschek and Rousseaux's Handbook of Toxicologic Pathology (2013). (3rd ed.). San Diego, CA: Academic Press, Inc.

Hassan, E., El-Neweshy, M., Hassan, M., & Noreldin, A. (2019). Thymoquinone attenuates testicular and spermotoxicity following subchronic lead exposure in male rats: Possible mechanisms are involved. *Life Sci, 230*, 132–140. doi:10.1016/j.lfs.2019.05.067.

Holaskova, I., Elliott, M., Hanson, M. L., Schafer, R., & Barnett, J. B. (2012). Prenatal cadmium exposure produces persistent changes to thymus and spleen cell phenotypic repertoire as well as the acquired immune response. *Toxicol Appl Pharmacol, 265*(2), 181–189. doi:10.1016/j.taap.2012.10.009.

Hyder, O., Chung, M., Cosgrove, D., Herman, J. M., Li, Z., Firoozmand, A., . . . Pawlik, T. M. (2013). Cadmium exposure and liver disease among US adults. *J Gastrointest Surg, 17*(7), 1265–1273. doi:10.1007/s11605-013-2210-9.

Ishida, K., Kotake, Y., Sanoh, S., & Ohta, S. (2017). Lead-induced ERK activation is mediated by GluR2 non-containing AMPA receptor in cortical neurons. *Biol Pharm Bull*, *40*(3), 303–309. doi:10.1248/bpb. b16-00784.

Jane Kasten-Jolly, D. A. L. (2019). *Advances in Neurotoxicology*, (Vol. 3). Academic Press; Elsevier.

Jin, S., Xia, W., Jiang, Y., Sun, X., Huang, S., Zhang, B., . . . Li, Y. (2018). Urinary vanadium concentration in relation to premature rupture of membranes: A birth cohort study. *Chemosphere*, *210*, 1035–1041. doi:10.1016/j.chemosphere.2018.07.110.

Keith, S., Faroon, O., Roney, N., Scinicariello, F., Wilbur, S., Ingerman, L., . . . Diamond, G. (2013). *Toxicological Profile for Uranium*. Atlanta, GA: ATSDR.

Kern, J. K., Geier, D. A., Homme, K. G., King, P. G., Bjorklund, G., Chirumbolo, S., & Geier, M. R. (2017). Developmental neurotoxicants and the vulnerable male brain: A systematic review of suspected neurotoxicants that disproportionally affect males. *Acta Neurobiol Exp (Wars)*, *77*(4), 269–296.

Klaassen, C. D., & Liu, J. (1998). Induction of metallothionein as an adaptive mechanism affecting the magnitude and progression of toxicological injury. *Environ Health Perspect*, *106* (Suppl 1), 297–300. doi:10.1289/ehp.98106s1297.

Klaassen, C. D., Liu, J., & Choudhuri, S. (1999). Metallothionein: An intracellular protein to protect against cadmium toxicity. *Annu Rev Pharmacol Toxicol*, *39*, 267–294. doi:10.1146/annurev. pharmtox.39.1.267.

Klotz, K., Weistenhofer, W., Neff, F., Hartwig, A., van Thriel, C., & Drexler, H. (2017). The health effects of aluminum exposure. *Dtsch Arztebl Int*, *114*(39), 653–659. doi:10.3238/arztebl.2017.0653.

Krzystyniak, K., Tryphonas, H., & Fournier, M. (1995). Approaches to the evaluation of chemical-induced immunotoxicity. *Environ Health Perspect*, *103* (Suppl 9), 17–22. doi:10.1289/ehp.95103s917.

Kumar, B. A., Reddy, A. G., Kumar, P. R., Reddy, Y. R., Rao, T. M., & Haritha, C. (2013). Protective role of N-Acetyl L-Cysteine against reproductive toxicity due to interaction of lead and cadmium in male Wistar rats. *J Nat Sci Biol Med*, *4*(2), 414–419. doi:10.4103/0976-9668.117021.

Kwakye, G. F., Paoliello, M. M., Mukhopadhyay, S., Bowman, A. B., & Aschner, M. (2015). Manganese-induced Parkinsonism and Parkinson's disease: Shared and distinguishable features. *Int J Environ Res Public Health*, *12*(7), 7519–7540. doi:10.3390/ijerph120707519.

Lau, A., Whitman, S. A., Jaramillo, M. C., & Zhang, D. D. (2013). Arsenic-mediated activation of the Nrf2-Keap1 antioxidant pathway. *J Biochem Mol Toxicol*, *27*(2), 99–105. doi:10.1002/jbt.21463.

Lauer, F. T., Parvez, F., Factor-Litvak, P., Liu, X., Santella, R. M., Islam, T., ... Burchiel, S. W. (2019). Changes in human peripheral blood mononuclear cell (HPBMC) populations and T-cell subsets associated with arsenic and polycyclic aromatic hydrocarbon exposures in a Bangladesh cohort. *PLoS One*, *14*(7), e0220451. doi:10.1371/journal.pone.0220451.

Lee, C. H., Liao, W. T., & Yu, H. S. (2011). Aberrant immune responses in arsenical skin cancers. *Kaohsiung J Med Sci*, *27*(9), 396–401. doi:10.1016/j.kjms.2011.05.007.

Lee, E., Karki, P., Johnson, J., Jr., Hong, P., & Aschner, M. (2017). Manganese control of glutamate transporters' gene expression. *Adv Neurobiol*, *16*, 1–12. doi:10.1007/978-3-319-55769-4_1.

Li, C., Srivastava, R. K., & Athar, M. (2016). Biological and environmental hazards associated with exposure to chemical warfare agents: Arsenicals. *Ann N Y Acad Sci*, *1378*(1), 143–157. doi:10.1111/nyas.13214.

Li, C., Zhao, K., Zhang, H., Liu, L., Xiong, F., Wang, K., & Chen, B. (2018). Lead exposure reduces sperm quality and DNA integrity in mice. *Environ Toxicol*, *33*(5), 594–602. doi:10.1002/tox.22545.

Li, J., Duan, X., Dong, D., Zhang, Y., Li, W., Zhao, L., . . . Li, B. (2015). Hepatic and nephric NRF2 pathway up-regulation, an early antioxidant response, in acute arsenic-exposed mice. *Int J Environ Res Public Health*, *12*(10), 12628–12642. doi:10.3390/ijerph121012628.

Liaquat, L., Sadir, S., Batool, Z., Tabassum, S., Shahzad, S., Afzal, A., & Haider, S. (2019). Acute aluminum chloride toxicity revisited: Study on DNA damage and histopathological, biochemical and neurochemical alterations in rat brain. *Life Sci*, *217*, 202–211. doi:10.1016/j.lfs.2018.12.009.

Liu, J., Qu, W., & Kadiiska, M. B. (2009). Role of oxidative stress in cadmium toxicity and carcinogenesis. *Toxicol Appl Pharmacol*, *238*(3), 209–214. doi:10.1016/j.taap.2009.01.029.

Liu, J., & Waalkes, M. P. (2008). Liver is a target of arsenic carcinogenesis. *Toxicol Sci*, *105*(1), 24–32. doi:10.1093/toxsci/kfn120.

Lucchini, R., Placidi, D., Cagna, G., Fedrighi, C., Oppini, M., Peli, M., & Zoni, S. (2017). Manganese and developmental neurotoxicity. *Adv Neurobiol*, *18*, 13–34. doi:10.1007/978-3-319-60189-2_2.

Lucchini, R. G., Aschner, M., Landrigan, P. J., & Cranmer, J. M. (2018). Neurotoxicity of manganese: Indications for future research and public health intervention from the Manganese 2016 conference. *Neurotoxicology, 64*, 1–4. doi:10.1016/j.neuro.2018.01.002.

Ma, N., Sasoh, M., Kawanishi, S., Sugiura, H., & Piao, F. (2010). Protection effect of taurine on nitrosative stress in the mice brain with chronic exposure to arsenic. *J Biomed Sci, 17* (Suppl 1), S7. doi:10.1186/1423-0127-17-s1-s7.

Malik, I. A., Wilting, J., Ramadori, G., & Naz, N. (2017). Reabsorption of iron into acutely damaged rat liver: A role for ferritins. *World J Gastroenterol, 23*(41), 7347–7358. doi:10.3748/wjg.v23.i41.7347.

Mehta, K. J., Farnaud, S. J., & Sharp, P. A. (2019). Iron and liver fibrosis: Mechanistic and clinical aspects. *World J Gastroenterol, 25*(5), 521–538. doi:10.3748/wjg.v25.i5.521.

Menon, A. V., Chang, J., & Kim, J. (2016). Mechanisms of divalent metal toxicity in affective disorders. *Toxicology, 339*, 58–72. doi:10.1016/j.tox.2015.11.001.

Metryka, E., Chibowska, K., Gutowska, I., Falkowska, A., Kupnicka, P., Barczak, K., . . . Baranowska-Bosiacka, I. (2018). Lead (Pb) exposure enhances expression of factors associated with inflammation. *Int J Mol Sci, 19*(6). doi:10.3390/ijms19061813.

Mishra, K. P. (2009). Lead exposure and its impact on immune system: A review. *Toxicol In Vitro, 23*(6), 969–972. doi:10.1016/j.tiv.2009.06.014.

Mochizuki, H. (2019). Arsenic neurotoxicity in humans. *Int J Mol Sci, 20*(14). doi:10.3390/ijms20143418.

Naganuma, A., Furuchi, T., Miura, N., Hwang, G. W., & Kuge, S. (2002). Investigation of intracellular factors involved in methylmercury toxicity. *Tohoku J Exp Med, 196*(2), 65–70. doi:10.1620/tjem.196.65.

Nakade, U. P., Garg, S. K., Sharma, A., Choudhury, S., Yadav, R. S., Gupta, K., & Sood, N. (2015). Lead-induced adverse effects on the reproductive system of rats with particular reference to histopathological changes in uterus. *Indian J Pharmacol, 47*(1), 22–26. doi:10.4103/0253-7613.150317.

Nava-Hernandez, M. P., Hauad-Marroquin, L. A., Bassol-Mayagoitia, S., Garcia-Arenas, G., Mercado-Hernandez, R., Echavarri-Guzman, M. A., & Cerda-Flores, R. M. (2009). Lead-, cadmium-, and arsenic-induced DNA damage in rat germinal cells. *DNA Cell Biol, 28*(5), 241–248. doi:10.1089/dna.2009.0860.

Neathery, M. W., & Miller, W. J. (1975). Metabolism and toxicity of cadmium, mercury, and lead in animals: A review. *J Dairy Sci, 58*(12), 1767–1781. doi:10.3168/jds.S0022-0302(75)84785-0.

Neurotoxicology: Approaches and Methods (1995). (J. Louis Chang and W. Slikker Eds.). San Diego, CA: Academic Press, Inc.

Nordberg, G. F., Goyer, R., & Nordberg, M. (1975). Comparative toxicity of cadmium-metallothionein and cadmium chloride on mouse kidney. *Arch Pathol, 99*(4), 192–197.

Odewumi, C. O., Latinwo, L. M., Ruden, M. L., Badisa, V. L., Fils-Aime, S., & Badisa, R. B. (2016). Modulation of cytokines and chemokines expression by NAC in cadmium chloride treated human lung cells. *Environ Toxicol, 31*(11), 1612–1619. doi:10.1002/tox.22165.

Ommati, M. M., Jamshidzadeh, A., Heidari, R., Sun, Z., Zamiri, M. J., Khodaei, F., . . . Shirazi Yeganeh, B. (2019). Carnosine and histidine supplementation blunt lead-induced reproductive toxicity through antioxidative and mitochondria-dependent mechanisms. *Biol Trace Elem Res, 187*(1), 151–162. doi:10.1007/s12011-018-1358-2.

Omura, M., Hirata, M., Tanaka, A., Zhao, M., Makita, Y., Inoue, N., . . . Ishinishi, N. (1996). Testicular toxicity evaluation of arsenic-containing binary compound semiconductors, gallium arsenide and indium arsenide, in hamsters. *Toxicol Lett, 89*(2), 123–129. doi:10.1016/s0378-4274(96)03796-4.

Omura, M., Tanaka, A., Hirata, M., Zhao, M., Makita, Y., Inoue, N., . . . Ishinishi, N. (1996). Testicular toxicity of gallium arsenide, indium arsenide, and arsenic oxide in rats by repetitive intratracheal instillation. *Fundam Appl Toxicol, 32*(1), 72–78. doi:10.1006/faat.1996.0108.

Onul, N. M., Biletska, E. M., Stus, V. P., & Polion, M. Y. (2018). The role of lead in the etiopathogenesis of male fertility reduction. *Wiad Lek, 71*(6), 1155–1160.

Pamphlett, R., Bishop, D. P., Kum Jew, S., & Doble, P. A. (2018). Age-related accumulation of toxic metals in the human locus ceruleus. *PLoS One, 13*(9), e0203627. doi:10.1371/journal.pone.0203627.

Pant, N., Kumar, G., Upadhyay, A. D., Patel, D. K., Gupta, Y. K., & Chaturvedi, P. K. (2014). Reproductive toxicity of lead, cadmium, and phthalate exposure in men. *Environ Sci Pollut Res Int, 21*(18), 11066–11074. doi:10.1007/s11356-014-2986-5.

Pathak, N., & Khandelwal, S. (2007). Role of oxidative stress and apoptosis in cadmium induced thymic atrophy and splenomegaly in mice. *Toxicol Lett, 169*(2), 95–108. doi:10.1016/j.toxlet.2006.12.009.

Patil, M., Sheth, K. A., Krishnamurthy, A. C., & Devarbhavi, H. (2013). A review and current perspective on Wilson disease. *J Clin Exp Hepatol, 3*(4), 321–336. doi:10.1016/j.jceh.2013.06.002.

Patra, P. H., Bandyopadhyay, S., Bandyopadhyay, M. C., & Mandal, T. K. (2013). Immunotoxic and genotoxic potential of arsenic and its chemical species in goats. *Toxicol Int, 20*(1), 6–10. doi:10.4103/0971-6580.111533.

Patrick, L. (2002). Mercury toxicity and antioxidants: Part 1: Role of glutathione and alpha-lipoic acid in the treatment of mercury toxicity. *Altern Med Rev, 7*(6), 456–471.

Peres, T. V., Schettinger, M. R., Chen, P., Carvalho, F., Avila, D. S., Bowman, A. B., & Aschner, M. (2016). Manganese-induced neurotoxicity: A review of its behavioral consequences and neuroprotective strategies. *BMC Pharmacol Toxicol, 17*(1), 57. doi:10.1186/s40360-016-0099-0.

Pfalzer, A. C., & Bowman, A. B. (2017). Relationships between essential manganese biology and manganese toxicity in neurological disease. *Curr Environ Health Rep, 4*(2), 223–228. doi:10.1007/s40572-017-0136-1.

Piao, F., Ma, N., Hiraku, Y., Murata, M., Oikawa, S., Cheng, F., . . . Yokoyama, K. (2005). Oxidative DNA damage in relation to neurotoxicity in the brain of mice exposed to arsenic at environmentally relevant levels. *J Occup Health, 47*(5), 445–449. doi:10.1539/joh.47.445.

Rader, J. I., Peeler, J. T., & Mahaffey, K. R. (1981). Comparative toxicity and tissue distribution of lead acetate in weanling and adult rats. *Environ Health Perspect, 42*, 187–195. doi:10.1289/ehp.8142187.

Rana, M. N., Tangpong, J., & Rahman, M. M. (2018). Toxicodynamics of Lead, Cadmium, Mercury and Arsenic- induced kidney toxicity and treatment strategy: A mini review. *Toxicol Rep, 5*, 704–713. doi:10.1016/j.toxrep.2018.05.012.

Ren, X., Gaile, D. P., Gong, Z., Qiu, W., Ge, Y., Zhang, C., . . . Wu, H. (2015). Arsenic responsive microRNAs in vivo and their potential involvement in arsenic-induced oxidative stress. *Toxicol Appl Pharmacol, 283*(3), 198–209. doi:10.1016/j.taap.2015.01.014.

Rubino, F. M. (2015). Toxicity of glutathione binding metals: A review of targets and mechanisms. *Toxics, 3*(1), 20–62. Published 2015 Jan 26. doi:10.3390/toxics3010020.

Sanders, T., Liu, Y., Buchner, V., & Tchounwou, P. B. (2009). Neurotoxic effects and biomarkers of lead exposure: A review. *Rev Environ Health, 24*(1), 15–45.

Sangkhae, V., & Nemeth, E. (2017). Regulation of the iron homeostatic hormone hepcidin. *Adv Nutr, 8*(1), 126–136. doi:10.3945/an.116.013961.

Sivaprakasam, C., Vijayakumar, R., Arul, M., & Nachiappan, V. (2016). Alteration of mitochondrial phospholipid due to the PLA2 activation in rat brains under cadmium toxicity. *Toxicol Res (Camb), 5*(6), 1680–1687. doi:10.1039/c6tx00201c.

Skipper, A., Sims, J. N., Yedjou, C. G., & Tchounwou, P. B. (2016). Cadmium chloride induces DNA damage and apoptosis of human liver carcinoma cells via oxidative stress. *Int J Environ Res Public Health, 13*(1). doi:10.3390/ijerph13010088.

Smith, M. J., Brown, J. M., Zamboni, W. C., & Walker, N. J. (2014). From immunotoxicity to nanotherapy: The effects of nanomaterials on the immune system. *Toxicol Sci, 138*(2), 249–255. doi:10.1093/toxsci/kfu005.

Sokol, R. Z., Madding, C. E., & Swerdloff, R. S. (1985). Lead toxicity and the hypothalamic-pituitary-testicular axis. *Biol Reprod, 33*(3), 722–728. doi:10.1095/biolreprod33.3.722.

Srivastava, R. K., Li, C., Wang, Y., Weng, Z., Elmets, C. A., Harrod, K. S., . . . Athar, M. (2016). Activating transcription factor 4 underlies the pathogenesis of arsenic trioxide-mediated impairment of macrophage innate immune functions. *Toxicol Appl Pharmacol, 308*, 46–58. doi:10.1016/j.taap.2016.07.015.

Stejskal, V. D., Danersund, A., Lindvall, A., Hudecek, R., Nordman, V., Yaqob, A., . . . Lindh, U. (1999). Metal-specific lymphocytes: Biomarkers of sensitivity in man. *Neuro Endocrinol Lett, 20*(5), 289–298.

Stelzer, K. J., & Pazdernik, T. L. (1983). Cadmium-induced immunotoxicity. *Int J Immunopharmacol, 5*(6), 541–548. doi:10.1016/0192–0561(83)90047-4.

Szpetnar, M., Luchowska-Kocot, D., Boguszewska-Czubara, A., & Kurzepa, J. (2016). The influence of manganese and glutamine intake on antioxidants and neurotransmitter amino acids levels in rats' brain. *Neurochem Res, 41*(8), 2129–2139. doi:10.1007/s11064-016-1928-7.

Takahashi, T., Fujimura, M., Koyama, M., Kanazawa, M., Usuki, F., Nishizawa, M., & Shimohata, T. (2017). Methylmercury causes blood-brain barrier damage in rats via upregulation of vascular endothelial growth factor expression. *PLoS One, 12*(1), e0170623. doi:10.1371/journal.pone.0170623.

Tarale, P., Chakrabarti, T., Sivanesan, S., Naoghare, P., Bafana, A., & Krishnamurthi, K. (2016). Potential role of epigenetic mechanism in manganese induced neurotoxicity. *Biomed Res Int, 2016,* 2548792. doi:10.1155/2016/2548792.

Tchounwou, P. B., Newsome, C., Williams, J., & Glass, K. (2008). Copper-induced cytotoxicity and transcriptional activation of stress genes in human liver carcinoma (HepG(2)) cells. *Met Ions Biol Med, 10,* 285–290.

Telisman, S., Colak, B., Pizent, A., Jurasovic, J., & Cvitkovic, P. (2007). Reproductive toxicity of low-level lead exposure in men. *Environ Res, 105*(2), 256–266. doi:10.1016/j.envres.2007.05.011.

Tolins, M., Ruchirawat, M., & Landrigan, P. (2014). The developmental neurotoxicity of arsenic: Cognitive and behavioral consequences of early life exposure. *Ann Glob Health, 80*(4), 303–314. doi:10.1016/j.aogh.2014.09.005.

Tonk, E. C., de Groot, D. M., Penninks, A. H., Waalkens-Berendsen, I. D., Wolterbeek, A. P., Slob, W., . . . van Loveren, H. (2010). Developmental immunotoxicity of methylmercury: The relative sensitivity of developmental and immune parameters. *Toxicol Sci, 117*(2), 325–335. doi:10.1093/toxsci/kfq223.

Tsangaris, G. T., Botsonis, A., Politis, I., & Tzortzatou-Stathopoulou, F. (2004). Cadmium induces Fas downregulation in a human immature T-cell line. *Cancer Genomics Proteomics, 1*(1), 77–86.

Tyler, C. R., & Allan, A. M. (2014). The effects of arsenic exposure on neurological and cognitive dysfunction in human and rodent studies: A review. *Curr Environ Health Rep, 1,* 132–147. doi:10.1007/s40572-014-0012-1.

Uchewa, O. O., & Ezugworie, O. J. (2019). Countering the effects of lead as an environmental toxicant on the microanatomy of female reproductive system of adult wistar rats using aqueous extract of Ficus vogelii. *J Trace Elem Med Biol, 52,* 192–198. doi:10.1016/j.jtemb.2018.12.016.

Unoki, T., Akiyama, M., Kumagai, Y., Goncalves, F. M., Farina, M., da Rocha, J. B. T., & Aschner, M. (2018). Molecular pathways associated with methylmercury-induced Nrf2 modulation. *Front Genet, 9,* 373. doi:10.3389/fgene.2018.00373.

Vigeh, M., Yokoyama, K., Ramezanzadeh, F., Dahaghin, M., Sakai, T., Morita, Y., . . . Kobayashi, Y. (2006). Lead and other trace metals in preeclampsia: a case-control study in Tehran, Iran. *Environ Res, 100*(2), 268–275. doi:10.1016/j.envres.2005.05.005.

Waalkes, M. P., Liu, J., Goyer, R. A., & Diwan, B. A. (2004). Metallothionein-I/II double knockout mice are hypersensitive to lead-induced kidney carcinogenesis: Role of inclusion body formation. *Cancer Res, 64*(21), 7766–7772. doi:10.1158/0008-5472.CAN-04-2220.

Walton, J. R. (2009). Brain lesions comprised of aluminum-rich cells that lack microtubules may be associated with the cognitive deficit of Alzheimer's disease. *Neurotoxicology, 30*(6), 1059–1069. doi:10.1016/j.neuro.2009.06.010.

Wang, B., & Du, Y. (2013). Cadmium and its neurotoxic effects. *Oxid Med Cell Longev, 2013,* 898034. doi:10.1155/2013/898034.

Wang, H., Zhang, L., Abel, G. M., Storm, D. R., & Xia, Z. (2018). Cadmium exposure impairs cognition and olfactory memory in male C57BL/6 mice. *Toxicol Sci, 161*(1), 87–102. doi:10.1093/toxsci/kfx202.

Wang, Y., Zhao, H., Shao, Y., Liu, J., Li, J., & Xing, M. (2017). Copper or/and arsenic induce oxidative stress-cascaded, nuclear factor kappa B-dependent inflammation and immune imbalance, trigging heat shock response in the kidney of chicken. *Oncotarget, 8*(58), 98103–98116. doi:10.18632/oncotarget.21463.

Wang, Y. X., Chen, H. G., Li, X. D., Chen, Y. J., Liu, C., Feng, W., . . . Lu, W. Q. (2018). Concentrations of vanadium in urine and seminal plasma in relation to semen quality parameters, spermatozoa DNA damage and serum hormone levels. *Sci Total Environ, 645,* 441–448. doi:10.1016/j.scitotenv.2018.07.137.

Wolff, N. A., Lee, W. K., Abouhamed, M., & Thevenod, F. (2008). Role of ARF6 in internalization of metal-binding proteins, metallothionein and transferrin, and cadmium-metallothionein toxicity in kidney proximal tubule cells. *Toxicol Appl Pharmacol, 230*(1), 78–85. doi:10.1016/j.taap.2008.02.008.

Wu, B., Cui, H., Peng, X., Fang, J., Zuo, Z., Deng, J., & Huang, J. (2014a). Analysis of the toll-like receptor 2-2 (TLR2-2) and TLR4 mRNA expression in the intestinal mucosal immunity of broilers fed on diets supplemented with nickel chloride. *Int J Environ Res Public Health, 11*(1), 657–670. doi:10.3390/ijerph110100657.

Wu, B., Cui, H., Peng, X., Fang, J., Zuo, Z., Deng, J., & Huang, J. (2014b). Toxicological effects of nickel chloride on IgA+ B Cells and sIgA, IgA, IgG, IgM in the intestinal mucosal immunity in broilers. *Int J Environ Res Public Health, 11*(8), 8175–8192. doi:10.3390/ijerph110808175.

Yan, D., Jin, C., Cao, Y., Wang, L., Lu, X., Yang, J., . . . Cai, Y. (2017). Effects of aluminium on long-term memory in rats and on SIRT1 mediating the transcription of CREB-dependent gene in hippocampus. *Basic Clin Pharmacol Toxicol, 121*(4), 342–352. doi:10.1111/bcpt.12798.

Yuan, Y., Zhang, Y., Zhao, S., Chen, J., Yang, J., Wang, T., . . . Liu, Z. (2018). Cadmium-induced apoptosis in neuronal cells is mediated by Fas/FasL-mediated mitochondrial apoptotic signaling pathway. *Sci Rep, 8*(1), 8837. doi:10.1038/s41598-018-27106-9.

Zefferino, R., Piccoli, C., Ricciardi, N., Scrima, R., & Capitanio, N. (2017). Possible mechanisms of mercury toxicity and cancer promotion: Involvement of gap junction intercellular communications and inflammatory cytokines. *Oxid Med Cell Longev, 2017*, 7028583. doi:10.1155/2017/7028583.

Zhang, Y., Wei, Z., Liu, W., Wang, J., He, X., Huang, H., . . . Yang, Z. (2017). Melatonin protects against arsenic trioxide-induced liver injury by the upregulation of Nrf2 expression through the activation of PI3K/AKT pathway. *Oncotarget, 8*(3), 3773–3780. doi:10.18632/oncotarget.13931.

CHAPTER 4

An Overview of Heavy Metal Research in Traditional Chinese Medicine

Yu Xu, Ning Wang, and Yibin Feng*
The University of Hong Kong

CONTENTS

4.1 INTRODUCTION

The heavy metal defined as any non-biologically degradable metal or metalloid present in the environment has evoked global attention, mainly because that metal contamination can cause toxicological risk to human [1–3]. With the increases of widespread environmental contamination and the variation of environmental pollution, metal residues are also prevalent and can contaminate traditional herb medicine, evoking severe side effects such as symptoms of chronic toxicity, renal failure, and liver damage [4]. Traditional Chinese medicine (TCM) originated from natural substances (e.g., plant, mineral, animal) as the preferred selection in China, Japan, and South Africa, for their self-treatment to protect against the development of some diseases [5]. Nowadays, The World Health Organization (WHO) has pointed out that about 65%–80% of the global population depends on TCM for primary healthcare. The high demand for TCM in developing countries has notably contributed to their perception that TCM is natural sources, thus has fewer side effects compared to chemical drugs as well as TCM are usually treated as dietary components and are beneficial for the human body. However, TCM can take up heavy metals from the environment [6], and it is easy to be contaminated with toxic heavy metal [7]. Moreover, heavy metal residues will persist in the TCM even after pretreatment, causing heavy metal poisoning [8,9]. Therefore, it is necessary for us to analyze heavy metal in order to ensure their dosage safety and make the related standard or regulations that limit their content to evaluate the biosafety of TCM.

* corresponding author

4.2 RESEARCH PROGRESS OF HEAVY METAL IN TCM

The inductively coupled plasma optical emission spectrometry (ICP-OES) detection of heavy metal of 44 herb plants collected from three different herb medicine markets in Istanbul/Turkey has shown a slight difference (above permissible limits) in their concentrations so it has confirmed that heavy metal contamination remained in herb plants and herb genetic variations could take effects on the accumulation of various metal elements at different contents. While the accumulated toxic effects they produce in humans might contribute to mutagenic actions. In general, it was expected that the aforementioned medicinal plants do not have any serious risk to human health [10]. For one thing, it has been reported by The International Agency for Research on Cancer (IARC) that antimony (Sb), arsenic (As), beryllium (Be), cadmium (Cd), cobalt (Co), chromium (Cr), nickel (Ni), lead (Pb), and vanadium (V) are identified as potentially carcinogenic to humans, since these toxic heavy metals may lead to DNA damage. Considering the potential intake of a possible toxic dose of heavy metals during taking TCM, it has been emphasized that the establishment of a probable monitor system for such carcinogenic metals in TCM is essential and useful for a human to protect against excessive toxic heavy metal exposures [11,12]. For the other thing, the mineral contained in formulations also has taken beneficial effects on vital organs and promoted the general well-being of the human body; these beneficial effects of heavy metal cannot be overlooked. For example, it has been indicated that manganese has played an essential action in *Rhodobryum ontariense* tea traditionally used for hypertension and some other heart disorders [13].

Moreover, the heavy metal content in an edible herb plant of Epimedii Folium used in China has been analyzed in different four species (i.e., *Epimedium brevicornu*, *Epimedium pubescens*, *Epimedium sagittatum*, and *Epimedium wushanense*) and various part of the plant including roots, leaves, and stems. It has been proved that there was a significant variation of heavy metals among various parts, different species, and cultivated and wild-growing plants. Cu and Pb were found in four different species, and Cd and As were not detected both in wild-growing and cultivated plants, while the level of Cu and Pb in cultivated plants were significantly lower than those in wild-growing plants. Moreover, in different organs, the level of Cd was unable to be detected, and the distribution of Cu, As, and Pb was specificity, and the apparent high concentrations of Cu and Pb were detected in leaves and roots, respectively. Owing to the variation of heavy metal content in TCM, it was not surprising to detect the excessive level of Cu, Cd, As, and Pb in several samples of Epimedii Folium collected from Chinese markets, the levels of which were beyond the nationally acceptable limits [14]. The content of heavy metal that various TCMs took up from the environment was different and was influenced by many genetic, molecular, physiological, and ecological traits [15]. Different from heavy metal found in raw herbs, some of which were considered as heavy metal hyperaccumulators, heavy metals might be intentionally added into Chinese patent medicine, such as arsenic, mercury, and lead. It was found out that none of the heavy metal from the plant origin in mother tinctures plant exceeded the permissible heavy metal limits, but substantial heavy metal concentrations of cadmium (Cd), lead (Pb), and mercury (Hg) were able to be determined in the fungus section of *Amanita muscaria* from the mother tincture. So, it was suggested that a risk-based approach to assess heavy metal for permanent control should be focused on the accumulation of heavy metals for organisms, including fungi [16].

4.2.1 Analytical Methods for Heavy Metal Detection in TCM

With the wide application of TCM throughout the world, more and more researches have paid attention to monitoring the level of heavy metal contaminants. Because several heavy metals can take cumulative effect on the living organisms, their high toxicity to human could be taken up even in low concentrations. The first step to monitor the heavy metals in TCM is to establish simple and sensitive analytical systems for the quality control of TCM in order to ensure TCM usage safety. Considering the specificity of heavy metals in TCM that some of them are incapable of keeping with the detection limits of the conventional analytical techniques, different speciation and analysis methods for trace amounts of heavy metals in various TCM such as herb plants and Chinese patent medicine have been established [17,18]. For example, Campos et al. have detected the lead level in herb plant by an online pre-concentration flow injection-flame atomic absorption spectroscopy (FAAS) determination [19] that can reach the ppb (One part per billion) level. These rapid and sensitive methodologies included ICP-OES [10,20] or inductively coupled plasma mass spectrometry (ICP-MS) [21], graphite furnace atomic absorption spectroscopy (GFAAS) [22,23], X-ray fluorescence spectrometry, anodic stripping voltammetry, and high-performance liquid chromatography(HPLC) [20,24]. Generally, the commonly applied analytical method for detection is GFAAS [20,25,26], which was also known as electrothermal atomic absorption spectrometry (ETAAS). GFAAS has been considered as the most extensively applied analytical method for Pb and Cd detection in TCM [27], this techniques treated the samples by depositing them into small graphite or pyrolytic carbon-coated graphite tube, and the samples can be heated to vaporize and atomize in order to reduce the analysis error. Moreover, the establishment and optimal of suitable measurement conditions and experimental parameters can eliminate the difficulties made from the interference iron from TCM. The most effective hydride generation method for the detection of antimony (Sb) and arsenic (As) can eliminate the interference from complex concomitant metal ion by adding the masking agent and maintaining consistent with a valence of analytes [28]. The common analytical methods for the determination of arsenic (As) in TCM include colorimetric/absorbance measurement and luminescence-based methods, electrochemical methods, inductively coupled plasma-atomic emission spectrometry (ICP-AES). The Hydride generation atomic fluorescence spectrometry (HG-AFS) is the most widely used ideal detection technique for hydride-forming heavy metals (mainly As, Se, and Sb) and Hg [29] due to its advantages of high selectivity, sensitivity, and little interference [30]. The elemental analysis of ICP-OES and FAAS detection was conducted in the medicinal plants for diabetes treatment in Eritrea such as *Aloe camper*, *Lepidium sativum*, *Meriandra dianthera*, *Nigella sativa*, and *Brassica nigra* [31]. The various forms of TCM, including raw herb and decoction pieces, made the analytical system complex. Also, different pretreatment process influences the content of heavy metals in TCM. Taking an example of Bupleuri Radix, ICP-AES determination has been established for simultaneously detecting a total of 45 paired original medicines, decoction pieces, and vinegar-processed after microwave digestion. It was confirmed that decoction treatment could decrease the heavy metals intake and the element transfer ratio can reach under 50% and decrease in order with Hg < Mn <Co < As. In order to establish the specified metal analysis for the risk assessment of Lian Hua Qing Wen capsule, the levels of As, Cd, and Pb was 0.38, 0.07, and 1.60 mg/kg, respectively, which were less than the acceptable dose according to the official regulation in China, named "import and export of medicinal plants and preparation of green industry state standards" [32]. It is suggested that heavy metal detection for TCM is a long-term and necessary regular monitor to ensure drug usage safety [33]. These analytical methods also have been applied for the detection of toxic heavy metals in the blood sample. Some toxic (Pb, Cd, and As) and essential metals (Zn, Cu, Cr, Ni, and Co) in the blood samples collected from asthmatic patients in Karachi of Pakistan have been analyzed using atomic absorption spectrophotometer for monitoring the quality of TCM [34]. Recent toxic heavy metal speciation and analysis methods are listed in Table 4.1.

4.2.2 External and Internal Reasons for Heavy Metal Contamination of TCM

It is no doubt that the heavy metal toxicity of the herbs is associated with their growing environmental contamination, so their origins, as well as collection location, should be considered as a cause for heavy metal contamination [40]. It has been documented that herb cultivation in soils with a high level of heavy metals is a critical mechanism of heavy metal contamination. The frequencies

Table 4.1   The Common Analytical Techniques for Five Toxic Metals Detection [35]

Analyte	Analytical Method
Hg	Graphite furnace atomic absorption spectrometry Hydride-generation atomic fluorescence spectrometry
Pb	Hydride-generation atomic fluorescence spectrometry Inductively coupled plasma atomic emission spectroscopy [36]
Cd	Inductively coupled plasma atomic emission spectroscopy [37] Graphite furnace atomic absorption spectrometry
Cr	Inductively coupled plasma atomic emission spectroscopy Graphite furnace atomic absorption spectrometry
As	Graphite furnace atomic absorption spectrometry [38] Hydride-generation atomic fluorescence spectrometry [39]

of mineral accumulation from contaminated soil or atmosphere in the environment are different in different herb medicine located in different countries. By assessing the heavy metal (Cd, Cu, Cr, Ni, Pb, and Zn) changes in soils near heavily trafficked roads in Korea and the different heavy metal content in both unwashed and washed foliar, leaves, as well as the deposition of traffic-related particles, has been shown that the growing location has taken a vital role in herb medicine heavy metal uptake. The herb samples collected far away from heavily trafficked roads have shown that they accumulated more levels of Ni and Cr, which might support the claim regarding the role of environmental issues in the influence of heavy metal in TCM [41]. Moreover, even the same species of herb plants absorbed different levels of heavy metals owing to a different geographical area, while different contents of heavy metals also showed in different herb species growing in the same collection location. Ten different medicinal herbs planted in five different geographical locations were collected and the results of five different heavy metals such as Pb, As, Cr, Hg, Cd have indicated that the influence of collection location on the heavy metal content of herb medicine [42]. According to Moran's I analysis, the level of As and Hg may not be influenced by growing location. It has been reported that Cr seems to have clusters of high values in the Northeast of China, but Pb and Cd made clusters of high values in the Southwest of China. [43]. These research studies generally indicated that the various collection locations varied the mineral levels and influenced the heavy metal toxicity of medicinal herbs. It was suggested that the medicinal herb should be collected from areas without heavy metals contamination. With exposure to a high heavy metal-contaminated environment, the effects on the growth and metabolism of the medicine herbs are determined by many physiological, molecular, genetic, and ecological traits [15]. So it is necessary that metallic element concentrations in medicinal herb should be monitored before their therapeutic application for pharmaceutical purposes [44].

More research has investigated the connection between the heavy metal available concentrations of medicinal herb and their level in soil. The levels of Cd, Pb, and Zn in contaminated soil exceeded the maximum limit of acceptable level and were found to make correlation with these in medicinal plants, but there was no relationship with the content of Cu, Mn, and Fe. Moreover, four different medicinal herb species, including *Urtica dioica*, *Plantago lanceolata*, *Hypericum perforatum*, and *Achillea millefolium* accumulated in heavy metals at different degrees. So it is suggested that herb plants should only be allowed to cultivate and to collect from soil safety controlled land that is monitored via a regular basis [12]. The potential environment risk index (RI) and the heavy metals' pollution indexes (HMPIs) have pointed out the cultivation status for medicinal plants. For example, the average contents of heavy metals (As, Cd, Cr, Cu, Pb, and Hg) in cultivation soil of herb medicine (Panax notoginseng) in Yunnan Province of China were 57.1, 0.4, 102.4, 35.1, 61.6, and 0.3 mg/kg, respectively. The HMPI can be ranked in the order of Pb < Cr < Cu < Hg < Cd < As. The different degrees of potential environmental risk in soil were divided into slight, middle, strong, very strong, and extremely strong with the proportion values of 5.41%, 21.62%, 35.14%, 10.81%, and 27.03%, respectively. The RI pointed out that there is about 30% of soil sites above the level of strong and extremely strong. According to the assessment of heavy metals of the root of Panax notoginseng and its cultivation soil in Yunnan, it can estimate the ecological risk of cultivation and make beneficial effects for the planting area selection [45]. It is surprising that higher levels of contaminated heavy metals such as As, Cd, Pb, and Cr were detected in wild-grown samples compared with cultivated samples. Without any apparent reason for these differences, there are several possible explanations can be considered that the wild Chinese medicine herb growth near the industrial or traffic road area that are sources of heavy metals or some heavy metals from the cultivated soil have been absorbed by the previous crop [43]. However, we can conclude that there is a higher contamination risk for medicine herb if their species are collected from industrial or traffic road sites than that from natural areas.

Different medicinal plants have different capacity to absorb different amounts of heavy metals from soil selectively; it can be considered as the internal cause of heavy metal contamination in TCM. For example, it was well known that even trace amount of Cd makes side effects on human health and the physiology of individual organisms [46], but *Clematis gouriana*, commonly used for the treatment of skin diseases, absorbed extremely high levels of Cd. The soil over covered metalliferous waste heaps or growth substrates contaminated by heavy metal has been demonstrated to influence the accumulation of heavy metal and selected secondary metabolite concentrations in *Echium vulgare L* species. The increase of allantoin, 4-hydroxybenzoic acid, and shikonin in the roots and shoots of three species has been proven to be contributed by the contaminated environmental conditions [47]. Therefore, the alteration of the metabolites composition of herb caused by heavy metal contamination can seriously influence the efficacy and quality of herb products. Besides this main reason for cultivation, the obvious reason is manufacturing and circulation. For specific treatment purpose, minerals that contain toxic heavy metals are deliberately added into some TCMs. For example, Xiong Huang (Realgar, As_4S_4) and Zhu Sha (Cinnabar, HgS) known as minerals containing toxic metals (Hg and As) are commonly applied in certain Chinese medicine prescriptions. There were about 53 kinds of Chinese medicine preparations containing Xiong Huang, and Zhu Sha has been recorded in the China Pharmacopoeia [48]. Such deliberate addition was not strictly defined as "contamination," because according to ancient concepts, these additive heavy metal has certain beneficial effects on human health, but these notions are not generally accepted by the modern science. If there is strong evidence supported that their benefits outweigh the risks, heavy metals maybe have a specific place in TCM. For instance, it has been confirmed that Realgar can take beneficial effects on leukemia, but no functional mechanism has been reported [49].

4.3 INTERNATIONAL STANDARDS REGARDING HEAVY METALS AND THEIR SAFETY LIMIT IN TCM

Recently, traditional medicines have been extensively examined by laboratories over the world for their metal element content in order to control their quality. Mercury (Hg), arsenic (Cd), and lead (Pb) can impose serious health risk to humans even at very low concentrations. Nearly, every nation has relevant official regulations on permissible limits of heavy metal content. The official organizations from different countries have treated these metals as the most commonly detected toxic metals in TCM; others contain cadmium, copper, and thallium [50]. The rigorous application of Good Agricultural and Collection Practices (GACP) and Good Manufacturing Practice (GMP) approved by WHO member states and European Union is an important measure to improve the quality of TCM during their cultivation and manufacturing process. The WHO has published the main principle guidelines on GACP and GMP for pharmaceutical products in 2003. WHO guidelines for assessing quality of herbal medicines with reference to contaminants and residues have been published as a new guideline in 2007 since GACP guidelines were formulated by WHO for medicinal herb in 2003. Other countries or regions such as the European Union, Japan, and China have issued national and regional GACP and GMP guidelines for good agriculture and collection practices of traditional medicine, which assured that the cultivation environment conditions including the irrigation water and soil reach the required limits from harmful heavy metals and toxicological substances [51]. Moreover, the permissible limits for the occurrence of certain heavy metals including As, Hg, Pb, and Cd have been standardized in herbal drugs by the US Food and Drug Administration (FDA) and the WHO. But not all heavy metals have the safety limits standardized by WHO because many other metals are still potential dietary micronutrients. The application and distribution of TCM products have been regulated by many countries such as the Australian Therapeutic Goods Administration (TGA), TGA has made regulation for TCM products if they are cultivated in Australia, and the assessment and license for the commercial importation of TCM are also required by TGA. Meanwhile, some other countries also have adopted guidelines for the heavy metals in TCM products. It is a challenge for the regulation producers that need to make effective strategies to reduce or eliminate the heavy metal risk. However, it is critical for human health. Lastly, the Good Supply Practices (GSP) should be approved by the market of herbal medicine products. Herbal products sold in the European Union should follow the simplified procedure introduced by European Directive 2004/24/EC and are needed to be manufactured by the requirements for marketing authorization. In Canada, Australia, the United States, and Europe, the respective national regulatory framework has been contributed to the medicine product license holders to ensure the quality of raw herb medicine. The sponsors of the final medicinal herbal formulation need to place detailed measures and requirements for approval of medicine ingredients through the supply chain back to the source and manufacture of the material. Most of the international firms have tried to follow vendor audit guides and strict GMP protocols to qualify herb origins and manufacture [52]. China has certified about 63 national GACP bases where herb cultivation specialized fields were adherent from 2004 to 2009, as well as local GACP bases with a large area of planting at about 11,000 km². Although the quality of Chinese herbal medicine has been improved after GACP implementation, there are several disadvantages, including the normalized management, lack of scientific standard operating procedures, the shortage of well-trained farmers, as well as the continued problems caused by other relevant experts, need to be solved. It would be an effective method to control the toxic heavy metal level in TCM by legal enforcement of GACP, GMP guidelines implementation [53], and the flow diagram of TCM circulation for the detection of heavy metal was shown in Figure 4.1 [49,53].

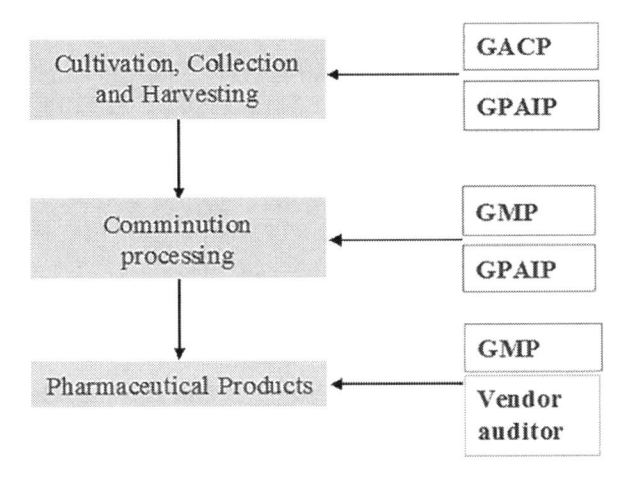

Figure 4.1 The value chain for the control of heavy metal in the traditional circulation of TCM.

Table 4.2   The Limit Control of Heavy Metal in TCM in Some Countries and Regions [59–62]

	Pb	As	Cd	Hg	Cu
China	5.0–10	2.0–10	0.3–5	0.1–1	20
US	5	2	0.3	0.2	-
England	5	5	1	0.2	-
Germany	5	5	0.2	0.1	-
EU	0.12–1.5	-	0.5–1	0.5–1	-
Japan	20	5	-	-	-
Malaysia	10	5	0.3	0.5	10
Vietnam	10	4	1	0.5	-
WHO	5	-	0.2–0.3	0.1	-
ISO	10	4	2	3	-

Numerous studies have disclosed that herbal medicines from different regions, including Africa, Europe, and South America have been determined with high contents of lead and some toxic metals [50]. Seven kinds of traditional herbs consumed in the United Arab Emirates. Copper was determined by AAS analysis method. Unfortunately, the unsafe levels of heavy metals have been detected in the analyzed herbs, which mostly exceeded WHO acceptable limits [54]. The high level of Cd in 79 samples of herbal medicine in Italy has also been detected (up to 0.75 mg/L) and in ginseng purchased from the United States, Europe, and Asia [55]. Even worse, there are still some TCM products that have no precise regulation to control their quality, for instance, Hispanic herbal products sold in Yerberias stores throughout the southwest are not regulated by any official agency [56]. Some Chinese patent medicines containing the heavy metal that exceeds the acceptable level have also been reported, for example, Hg and As were present in local herbal and Ayurvedic liquid preparations far beyond the accepted limits of official regulations for health drugs while other metals such as manganese, zinc, and iron were under the therapeutic limits [57]. Although the acceptable levels of all heavy metal in TCM are not entirely standardized, most countries or regions have made a draft for the permissible limits for the toxic heavy metals as shown in Table 4.2, such as the European Pharmacopoeia limited the levels of Cd, As, Pb, and Hg. Moreover, the United States has made the standard for the daily dietary allowance of essential metals and the acceptable levels of heavy metals, and the Food and Agriculture Organization have also proposed permissible levels of toxic substances that can be ingested every week [58].

4.4 DISCUSSION AND CONCLUSION

In this chapter, we pointed out that heavy metal contamination of medicinal herbal products occurs during cultivation, cross-contamination during their process as well as their deliberate adulteration as therapeutic reagents. It is evident that heavy metal contamination during cultivation in traditional medicine was influenced by the surrounding environments including industrial or agricultural pollution, although this clustering effect on the heavy metal level evoked by environment pollution needs further examination [43]. Control and abatement of environmental contamination can be considered as a part of practical efforts on the improvement of the quality and standardization of traditional medicine. It has been reported that several heavy metals contained in TCM sometimes occur in acceptable levels such as the higher level of Cd was observed compared with published standards in some TCM, but previous research on Chinese herb medicine does not detect this metal, so further studies on Cd levels in TCM are warranted [43]. There are few limits for essential dietary minerals, or acceptable levels of all heavy metals in TCM have not yet been standardized by official entities. Therefore, medicinal herb from contaminated cultivation areas should be banned and discouraged if possible [44]. The analytical methods for heavy metal detection have been successfully established, so it is not difficult to examine the content of heavy metals in TCM samples. However, the detailed pre-treatment process and analytical conditions are time-consuming; profession is needed to make regulation about the heavy metal in TCM, especially, the herbal companies should establish the related guidelines following the national regulations to guarantee the quality of their products [53]. Given that heavy metals in TCM perceive long-term toxicity, the application of TCM shall be under critical observation. The causes of the toxicities that were contributed to the chemical and heavy metal contents in TCM have been confirmed, but the possible adverse health effects resulting from TCM consumption have not been quantified accidental toxicity is a possibility [48].

REFERENCES

1. Hsu, M.J.; Selvaraj, K.; Agoramoorthy, G. Taiwan's industrial heavy metal pollution threatens terrestrial biota. *Environ Pollut* **2006**, *143*, 327–334, doi:10.1016/j.envpol.2005.11.023.
2. Blicharska, E.; Komsta, Ł.; Kocjan, R.; Gumieniczek, A.; Robak, A. A preliminary study on the effect of mineralization parameters on determination of metals in Viscum album species. *Open Chemistry* **2010**, *8*, doi:10.2478/s11532-009-0133-9.
3. El-Ansary, A.; Bjorklund, G.; Tinkov, A.A.; Skalny, A.V.; Al Dera, H. Relationship between selenium, lead, and mercury in red blood cells of Saudi autistic children. *Metab Brain Dis* **2017**, *32*, 1073–1080, doi:10.1007/s11011-017-9996-1.
4. Andrew, A.S.; Warren, A.J.; Barchowsky, A.; Temple, K.A.; Klei, L.; Soucy, N.V.; O'Hara, K.A.; Hamilton, J.W. Genomic and proteomic profiling of responses to toxic metals in human lung cells. *Environ Health Perspect* **2003**, *111*, 825–835, doi:10.1289/ehp.111-1241504.
5. Meena, A.K.; Bansal, P.; Kumar, S.; Rao, M.M.; Garg, V.K. Estimation of heavy metals in commonly used medicinal plants: A market basket survey. *Environ Monit Assess* **2010**, *170*, 657–660, doi:10.1007/s10661-009-1264-3.
6. Patel, A.; Patra, D.D. Influence of heavy metal rich tannery sludge on soil enzymes vis-a-vis growth of Tagetes minuta, an essential oil bearing crop. *Chemosphere* **2014**, *112*, 323–332, doi:10.1016/j.chemosphere.2014.04.063.
7. Purev, M.; Kim, Y.J.; Kim, M.K.; Pulla, R.K.; Yang, D.C. Isolation of a novel catalase (Cat1) gene from Panax ginseng and analysis of the response of this gene to various stresses. *Plant Physiol Biochem* **2010**, *48*, 451–460, doi:10.1016/j.plaphy.2010.02.005.
8. Ting, A.; Chow, Y.; Tan, W. Microbial and heavy metal contamination in commonly consumed traditional Chinese herbal medicines. *J Tradit Chin Med* **2013**, *33*, 119–124.
9. Li, Z.; Wu, L.; Sun, S.; Gao, J.; Zhang, H.; Zhang, Z.; Wang, Z. Disinfection and removal performance for *Escherichia coli*, toxic heavy metals and arsenic by wood vinegar-modified zeolite. *Ecotoxicol Environ Saf* **2019**, *174*, 129–136, doi:10.1016/j.ecoenv.2019.01.124.
10. Ozyigit, I.I.; Yalcin, B.; Turan, S.; Saracoglu, I.A.; Karadeniz, S.; Yalcin, I.E.; Demir, G. Investigation of heavy metal level and mineral nutrient status in widely used medicinal plants' leaves in turkey: Insights into health implications. *Biol Trace Elem Res* **2018**, *182*, 387–406, doi:10.1007/s12011-017-1070-7.
11. Shazia Jabeen, M.T.S.; Khan, S.; Hayat, M.Q. Determination of major and trace elements in ten important folk therapeutic plants of Haripurbasin Pakistan. *J Med Plants Res* **2010**, *4*, 8.
12. Muhammad Ashraf, M.Q.H.; Mumtaz, A.S. A study on elemental contents of medicinally important species of Artemisia L. (Asteraceae) found in Pakistan. *J Med Plants Res* **2010**, *4*, 8, doi:10.5897/JMPR10.460.
13. Pejin, B.; Kien-Thai, Y.; Stanimirovic, B.; Vuckovic, G.; Belic, D.; Sabovljevic, M. Heavy metal content of a medicinal moss tea for hypertension. *Nat Prod Res* **2012**, *26*, 2239–2242, doi:10.1080/14786419.2011.648190.
14. Yang, X.H.; Zhang, H.F.; Niu, L.L.; Wang, Y.; Lai, J.H. Contents of heavy metals in chinese edible herbs: Evidence from a case study of epimedii folium. *Biol Trace Elem Res* **2018**, *182*, 159–168, doi:10.1007/s12011-017-1075-2.

15. Dhir, B.; Sharmila, P.; Pardha Saradhi, P.; Nasim, S.A. Physiological and antioxidant responses of Salvinia natans exposed to chromium-rich wastewater. *Ecotoxicol Environ Saf* **2009**, *72*, 1790–1797, doi:10.1016/j.ecoenv.2009.03.015.

16. Busch, J.; Werner, W.; Huwer, A. Study of the risk of heavy metal transfer to homoeopathic mother tinctures. *Pharmeur Bio Sci Notes* **2012**, *2012*, 55–71.

17. Wanbang, X.; Liwei, Y.; Jinfeng, L. Analysis techniques of heavy metals ions in traditional Chinese medicine. *Mod Sci Instrum* **2012**, 25–27.

18. Yuan, X.; Chapman, R.L.; Wu, Z. Analytical methods for heavy metals in herbal medicines. *Phytochem Anal* **2011**, *22*, 189–198, doi:10.1002/pca.1287.

19. Campos, M.M.; Tonuci, H.; Silva, S.M.; de, S.A.B.; de Carvalho, D.; Kronka, E.A.; Pereira, A.M.; Bertoni, B.W.; de, C.F.S.; Miranda, C.E. Determination of lead content in medicinal plants by pre-concentration flow injection analysis-flame atomic absorption spectrometry. *Phytochem Anal* **2009**, *20*, 445–449, doi:10.1002/pca.1145.

20. Gomez, M.R.; Cerutti, S.; Sombra, L.L.; Silva, M.F.; Martínez, L.D. Determination of heavy metals for the quality control in argentinian herbal medicines by ETAAS and ICP-OES. *Food Chem Toxicol* **2007**, *45*, 1060–1064, doi:10.1016/j.fct.2006.12.013.

21. Filipiak-Szok, A.; Kurzawa, M.; Szlyk, E. Determination of toxic metals by ICP-MS in Asiatic and European medicinal plants and dietary supplements. *J Trace Elem Med Biol* **2015**, *30*, 54–58, doi:10.1016/j.jtemb.2014.10.008.

22. Mihaljev, Z.; Zivkov-Balos, M.; Cupic, Z.; Jaksic, S. Levels of some microelements and essential heavy metals in herbal teas in Serbia. *Acta Pol Pharm* **2014**, *71*, 385–391.

23. He, J.H.; Cheng, Y.Y.; Yang, T.; Zou, H.Y.; Huang, C.Z. Functional preserving carbon dots-based fluorescent probe for mercury (II) ions sensing in herbal medicines via coordination and electron transfer. *Anal Chim Acta* **2018**, *1035*, 203–210, doi:10.1016/j.aca.2018.06.053.

24. Bai, G.Y.; Wei, C.; Ouyang, L.; Xie, Q.; Liu, Y.Q.; Wang, J.; Wang, J.Y. Determination of 5 heavy metals in Chinese traditional medicines and extraction liquid containing mineral materials. *Guang Pu Xue Yu Guang Pu Fen Xi* **2011**, *31*, 256–259.

25. Liu, Y.; Wu, J.; Wei, W.; Xu, R. Simultaneous determination of heavy metal pollution in commercial traditional Chinese medicines in China. *J Nat Med* **2013**, *67*, 887–893, doi:10.1007/s11418-012-0656-9.

26. Akram, S.; Najam, R.; Rizwani, G.H.; Abbas, S.A. Determination of heavy metal contents by atomic absorption spectroscopy (AAS) in some medicinal plants from Pakistani and Malaysian origin. *Pak J Pharm Sci* **2015**, *28*, 1781–1787.

27. Hina, B.; Rizwani, G.H.; Shareef, H.; Ahmed, M. Atomic absorption spectroscopic analysis of some Pakistani herbal medicinal products used in respiratory tract infections. *Pak J Pharm Sci* **2012**, *25*, 247–253.

28. Fan Jiawen, L.S.; Dui, W.; Weisong, X.; Chanting, H. Research on the metal detection in Traditional Chinese Medincine. *Stud Trace Elements Health* **2018**, 35, 38–39.

29. Sánchez-Rodas, D.; Corns, W.T.; Chen, B.; Stockwell, P.B. Atomic Fluorescence Spectrometry: A suitable detection technique in speciation studies for arsenic, selenium, antimony and mercury. *J Anal Atomic Spectr* **2010**, *25*, 933, doi:10.1039/b917755h.

30. Ma, J.; Sengupta, M.K.; Yuan, D.; Dasgupta, P.K. Speciation and detection of arsenic in aqueous samples: a review of recent progress in non-atomic spectrometric methods. *Anal Chim Acta* **2014**, *831*, 1–23, doi:10.1016/j.aca.2014.04.029.

31. Sium, M.; Kareru, P.; Keriko, J.; Girmay, B.; Medhanie, G.; Debretsion, S. Profile of trace elements in selected medicinal plants used for the treatment of diabetes in eritrea. *Sci World J* **2016**, *2016*, 2752836, doi:10.1155/2016/2752836.

32. Lin, L.X.; Li, S.X.; Zheng, F.Y. Application of in vitro bionic digestion and biomembrane extraction for metal speciation analysis, bioavailability and risk assessment in lianhua qingwen capsule. *Zhongguo Zhong Yao Za Zhi* **2014**, *39*, 2330–2335.

33. Li, K.; Luo, J.; Ding, T.; Dou, X.; Hu, Y.; Zhang, X.; Yang, M. Multielements determination and metal transfer investigation in herb medicine Bupleuri Radix by inductively coupled plasma-mass spectrometry. *Food Sci Nutr* **2018**, *6*, 2005–2014, doi:10.1002/fsn3.701.

34. Hina, B.; Rizwani, G.H.; Zahid, H. Hematological screening of heavy metals among patients of asthma using medicinal herbs in Karachi, Pakistan. *Pak J Pharm Sci* **2014**, *27*, 1899–1904.

35. Yang, L.; Fang, L.X.; Hong, L.Q.; Hua, D.G. Overview on adverse reaction of Traditional Chinese Meicine with heavy meatal and As. *Chin J Hosp Phar* **2008**, *4*, 301–304.

36. Pavlova, D.; Karadjova, I. Toxic element profiles in selected medicinal plants growing on serpentines in Bulgaria. *Biol Trace Elem Res* **2013**, *156*, 288–297, doi:10.1007/s12011-013-9848-8.

37. Kalny, P.; Wyderska, S.; Fijalek, Z.; Wroczynski, P. Determination of selected elements in different pharmaceutical forms of some Polish herbal medicinal products. *Acta Pol Pharm* **2012**, *69*, 279–283.

38. Okatch, H.; Ngwenya, B.; Raletamo, K.M.; Andrae-Marobela, K. Determination of potentially toxic heavy metals in traditionally used medicinal plants for HIV/AIDS opportunistic infections in Ngamiland District in Northern Botswana. *Anal Chim Acta* **2012**, *730*, 42–48, doi:10.1016/j.aca.2011.11.067.

39. Zhu, Z.; Liu, J.; Zhang, S.; Na, X.; Zhang, X. Evaluation of a hydride generation-atomic fluorescence system for the determination of arsenic using a dielectric barrier discharge atomizer. *Anal Chim Acta* **2008**, *607*, 136–141, doi:10.1016/j.aca.2007.11.041.

40. Awodele, O.; Popoola, T.D.; Amadi, K.C.; Coker, H.A.; Akintonwa, A. Traditional medicinal plants in Nigeria—remedies or risks. *J Ethnopharmacol* **2013**, *150*, 614–618, doi:10.1016/j.jep.2013.09.015.

41. Kim, H.S.; Kim, K.R.; Kim, W.I.; Owens, G.; Kim, K.H. Influence of road proximity on the concentrations of heavy metals in Korean Urban agricultural soils and crops. *Arch Environ Contam Toxicol* **2017**, *72*, 260–268, doi:10.1007/s00244-016-0344-y.

42. Annan, K.; Dickson, R.A.; Amponsah, I.K.; Nooni, I.K. The heavy metal contents of some selected medicinal plants sampled from different geographical locations. *Pharmacognosy Res* **2013**, *5*, 103–108, doi:10.4103/0974-8490.110539.

43. Harris, E.S.; Cao, S.; Littlefield, B.A.; Craycroft, J.A.; Scholten, R.; Kaptchuk, T.; Fu, Y.; Wang, W.; Liu, Y.; Chen, H., et al. Heavy metal and pesticide content in commonly prescribed individual raw Chinese Herbal Medicines. *Sci Total Environ* **2011**, *409*, 4297–4305, doi:10.1016/j.scitotenv.2011.07.032.

44. Sarma, H.; Deka, S.; Deka, H.; Saikia, R.R. Accumulation of heavy metals in selected medicinal plants. *Rev Environ Contam Toxicol* **2011**, *214*, 63–86, doi:10.1007/978-1-4614-0668-6_4.

45. Ou, X.; Wang, L.; Guo, L.; Cui, X.; Liu, D.; Yang, Y. Soil-plant metal relations in panax notoginseng: An ecosystem health risk assessment. *Int J Environ Res Public Health* **2016**, *13*, doi:10.3390/ijerph13111089.

46. Meena, A.K.; Bansal, P.; Kumar, S.; Rao, M.M.; Garg, V.K. Estimation of heavy metals in commonly used medicinal plants: a market basket survey. *Environ Monit Assess* **2009**, *170*, 657–660, doi:10.1007/s10661-009-1264-3.

47. Dresler, S.; Rutkowska, E.; Bednarek, W.; Stanislawski, G.; Kubrak, T.; Bogucka-Kocka, A.; Wojcik, M. Selected secondary metabolites in *Echium vulgare* L. populations from nonmetalliferous and metalliferous areas. *Phytochemistry* **2017**, *133*, 4–14, doi:10.1016/j.phytochem.2016.11.001.

48. Cooper, K.; Noller, B.; Connell, D.; Yu, J.; Sadler, R.; Olszowy, H.; Golding, G.; Tinggi, U.; Moore, M.R.; Myers, S. Public health risks from heavy metals and metalloids present in traditional Chinese medicines. *J Toxicol Environ Health A* **2007**, *70*, 1694–1699, doi:10.1080/15287390701434885.

49. Zhang, J.; Wider, B.; Shang, H.; Li, X.; Ernst, E. Quality of herbal medicines: Challenges and solutions. *Complement Ther Med* **2012**, *20*, 100–106, doi:10.1016/j.ctim.2011.09.004.

50. Jarup, L. Hazards of heavy metal contamination. *Br Med Bull* **2003**, *68*, 167–182.

51. Sahoo, N.; Manchikanti, P.; Dey, S. Herbal drugs: Standards and regulation. *Fitoterapia* **2010**, *81*, 462–471, doi:10.1016/j.fitote.2010.02.001.

52. Dong, W.; Guan.H.F., Liu, X.-q. Research progress on techniques of removing heavy metals and pesticides from traditional Chinese medicine. *J Shenyang Pharm Univ* **2009**, *26*, 152–156.

53. Govindaraghavan, S.; Sucher, N.J. Quality assessment of medicinal herbs and their extracts: Criteria and prerequisites for consistent safety and efficacy of herbal medicines. *Epilepsy Behav* **2015**, *52*, 363–371, doi:10.1016/j.yebeh.2015.03.004.

54. Dghaim, R.; Al Khatib, S.; Rasool, H.; Ali Khan, M. Determination of heavy metals concentration in traditional herbs commonly consumed in the United Arab Emirates. *J Environ Public Health* **2015**, *2015*, 973878, doi:10.1155/2015/973878.

55. Ting, A.; Chow, Y.; Tan, W. Microbial and heavy metal contamination in commonly consumed traditional Chinese herbal medicines. *J Tradit Chin Med* **2013**, *33*, 119–124, doi:10.1016/s0254-6272(13)60112-0.

56. Levine, M.; Mihalic, J.; Ruha, A.M.; French, R.N.; Brooks, D.E. Heavy metal contaminants in yerberia shop products. *J Med Toxicol* **2013**, *9*, 21–24, doi:10.1007/s13181-012-0231-5.

57. Hajra, B.; Qayum, I.; Orakzai, S.; Hussain, F.; Faryal, U.; Aurangzeb. Evaluation of toxic heavy metals in ayurvedic syrups sold in local markets of Hazara, Pakistan. *J Ayub Med Coll Abbottabad* **2015**, *27*, 183–186.

58. Domingo, J.L. Health risks of dietary exposure to perfluorinated compounds. *Environ Int* **2012**, *40*, 187–195, doi:10.1016/j.envint.2011.08.001.

59. Shaoping, H.W.Z.J.L. Strategies and current status on limit control of heavy metals in traditional Chinese medicine. *Chin J Pharm Anal* **2007,** 27, 1849–1853.

60. Chang, H. Analysis on excessive heavy metal problem of Chinese materia medica exports to EU from perspective of UK ban. *Chin Tradit Herbal Drugs* **2016**, *47*, 1820–1824.

61. Jiang, Z.J.; Zhang, H.M.; Yu Z.B.; Li. D.Y. International Market Standards for Export of Chinese Medicinal Materials. *Modern Chin Med* **2018**, 20, 217–238.

62. Lian-hua, Z.; Yin-hui, Y.; Yi-chen, H.; Shi-hai, Y.; Hong-yu, J.; Jian-he, W.; Mei-hua, Y. Current situation analysis and countermeasures on contamination of heavy metal in traditional Chinese medicinal materials in China. *Chin Tradit Herbal Drugs* **2014**, *45*, 1199–1206.

PART III

Mechanisms of Restoring Metabolic Homeostasis

CHAPTER 5

Development and Utilization of a Novel Prodosomed-Electrolyte and Phytochemical Formulation Technology to Restore Metabolic Homeostasis

Bernard W. Downs
Victory Nutrition International, Inc.

Manashi Bagchi
Dr. Herbs LLC.

Bruce S. Morrison
Morrison Family and Sports Medicine

Jeffrey Galvin
Vitality Medical Wellness Institute

Steve Kushner
ALM R&D

Debasis Bagchi
Texas Southern University
Victory Nutrition International, Inc.

CONTENTS

5.1 INTRODUCTION

Regular ingestion of high-quality nutrients is vital to all aspects of health.[1] Concurrently, a vast portfolio of nutrients is required in a uniform fashion to ensure the body has adequate tools with which to properly function. It is extremely difficult, as shown even in some early research, to get adequate nutrition through diet alone. As such, researchers recommended the intake of at least a multivitamin and mineral supplement.[2] To this end, nutritional supplements have become an extremely important tool to augment the diet for acquiring these nutrients.

Perhaps most important, the digestive system must be able to disintegrate and dissolve foodstuffs into readily usable micronutritional particles that can be absorbed and utilized at a cellular level to ensure both restoration and maintenance of optimal health.[3] If essential nutrients are available, the body can synthesize the arsenal of non-essential vitamins, amino acids, and fatty acids. However, the body is unable to synthesize any minerals.[3,4] Minerals must be consumed and absorbed in order to be utilized for biological needs.[3] Minerals make an extremely important contribution to achieving and maintaining health through the role of electrolytes.[3,4]

Electrolytes are the ionized or ionizable constituents of a living cell, blood, and other body fluids.[4–6] Electrolyte concentrations are critically regulated within rather narrow physiologic ranges essential to human health.[7–10] The major electrolytes in intercellular fluids are sodium (Na^+), chloride (Cl^-), potassium (K^+), calcium (Ca^{2+}), phosphate $\left(PO_4^{3-}\right)$, and magnesium (Mg^{2+}) and are crucial for the basic physiological functions, including action-dependent functions, cardiac rhythm control, muscle contraction, and energy storage, among many others.[5,6] In human physiology, electrolytes and fluids are mandatory for the proper functioning of the cells, tissues, and organs.[5,8,9,11] Electrolytes are widely distributed in the blood, body fluids, tissues of the body and are important cofactors for the proper function of many crucial enzymes involved in DNA and protein synthesis and energy metabolism.[5–7] Electrolytes are eliminated through urine, sweat, feces, skin sloughing, and menstruation.[5,6] It is obvious that electrolytes present in the blood, body fluids, and tissues contribute to the wellness of physiological health.[7,8] Electrolyte levels are generally determined in blood tests.[1] Electrolyte levels must stay within a reasonably narrow range to maintain optimal pH levels or adverse health effects can occur.[5,6,11,12] Therefore, adequate electrolyte replacement is crucial in order to keep pace with the rate of excretion and maintain optimal amounts of electrolytes for healthy biological functions.[3–9]

Basically, electrolytes are comprised of electrically charged micronutrients and mineral ions that help the human body perform its regular function including (a) production of energy; (b) muscle performance, including contraction and expansion; (c) maintain the ideal life-support properties of fluids and their homeostatic equilibrium in the body; and (d) maintain the ideal pH (acid/base) balance in the various tissue compartments of the body.[5–7]

From the perspective of biochemistry, an ideal electrolyte produces an electrical conducting media when dissolved in water, a polar solvent. Following dissolution, it produces cations and anions that should disperse uniformly.[5,6]

In addition, it is advisable to consume foods rich in electrolytes, including spinach, turkey, potatoes, beans, avocados, oranges, soybeans (edamame), strawberries, and bananas to name a few.[5,6,9]

5.2 SOURCES OF DIETARY ELECTROLYTES

A healthy diet is important to maintain the ideal physiological electrolyte balance. Ideal food sources of electrolytes are fruits, vegetables, dairy, nuts, and seeds.[5–7] Table 5.1 demonstrates some dietary sources of electrolytes in food. It is worthwhile to mention that carbon dioxide (CO_2) and bicarbonate (HCO_3) are naturally produced from the intake of air and water and act as electrolytes for homeostatic acid and alkaline buffering in the plasma (i.e. extracellular fluids).[5–8] So, they are not considered here as being required to be consumed from a regular diet.

Table 5.1 Sources of Physiological Electrolytes in Foods[5,6]

Dietary Electrolytes	Foods
Sodium	Pickled foods, cheese, and table salt
Chloride	Table salt
Potassium	Fruits and vegetables such as bananas, avocado, and sweet potato
Magnesium	Seeds and nuts
Calcium	Dairy products, fortified dairy alternatives, and green leafy vegetables

5.3 ELECTROLYTE BALANCE AND IMBALANCE: PHYSIOLOGICAL CONSEQUENCES

Several essential electrolytes including sodium, potassium, chloride, calcium, magnesium, phosphate are physiologically present in a human body, which regulates proper nerve and muscle functions, maintains acid–base balance in the plasma and hydration.[5,6,8,13] Properly balanced electrolytes have several vital physiological functions including[5,6]:

a. Maintain the fluid levels in blood plasma, as well as in the whole body
b. Maintain the blood pH in slightly alkaline range between 7.35 and 7.45
c. Blood clotting
d. Optimize heart functioning (beating of the heart)
e. Enable muscle contraction and expansion
f. Facilitate transmission of nerve signals and cell-to-cell communication from heart, muscle, nerve cells to other vital cells
g. Participate as a constituent/catalyst in the synthesis of new tissues
h. Act as cofactors for important enzymes involved in DNA, protein synthesis, and energy metabolism. It is very important to note that, except for sodium, it is evident that the imperfect diet consumed by most Americans is insufficient in supplying adequate amounts of electrolytes for optimal physiological functions.

As already mentioned, excretion of electrolytes mandates replacement to maintain physiological homeostatic properties.[5–7] If electrolyte intake is insufficient, the body will repartition (steal) mineral ions from non-critical biological compartments (i.e. saliva) and structural substances (i.e. muscles, organs, and bones) to supply electrolytes for critical functions for life support, such as pH homeostasis in the blood and intracellular compartments.[5,8] In the blood, maintaining the ideal pH is crucial to ensure the effective and efficient utilization of oxygen and water. In multiple physiological conditions, extreme shifts in blood electrolyte levels, becoming too high or low, induce an electrolyte imbalance. Both low and high electrolyte levels can have a harmful effect on human health and can be fatal in selected cases.[6,7] During vomiting, diarrhea, or excessive sweating (such as from extreme physical exertion), severe dehydration can occur.[5,8] It is advisable to consume both water and electrolytes, especially sodium, chloride, and potassium during and following excessive sweating.[13,14] However, loss of electrolytes from sweat varies from person to person. Moreover, long periods of exercise, particularly in a hot climate, can cause significant electrolyte loss.[15] Endurance athletes, who are exercising for longer hours or exercising in extreme heat, should consider electrolyte-enriched drinks to replace their electrolyte losses.[15,16] An electrolyte imbalance may occur due to various dysfunctions and physiological conditions including:

1. Persistent diarrhea or vomiting
2. Excessive sweating and high fever
3. Not consuming enough foods or water
4. Chronic degenerative diseases

5. Emphysema or chronic respiratory problems
6. Metabolic alkalosis (a physiological condition of higher than normal blood pH usually induced as an extreme alkaline buffering response to anaerobic/hypoxic conditions)
7. Intake of pharmaceutical drugs including steroids, antibiotics, diuretics, or laxatives
8. Dehydration caused by desert climate of high temperature, vomiting, or diarrhea.

It is very important to mention that sometimes an electrolyte imbalance may not result in obvious observable symptoms. As for example, low potassium, a condition known as hypokalemia[17] content, may not exhibit symptoms. But it may drastically affect the glycogen levels, a major source of energy for the muscles in the body or induce abnormal cardiac rhythms.[16,18,19] This can lead to several other problems including eating disorders, kidney disease, muscle weakness, spasms, cramps, respiratory problems, and paralysis.[5,7] Severe burns can also cause electrolyte imbalances.[19,20]

Mild electrolyte disturbances don't seem to cause any major problems.[5–7] However, severe imbalances can cause several symptoms,[4,5] including:

1. Headaches
2. Rapid or irregular heartbeat
3. Numbness and tingling
4. Muscle weakness and cramps, twitches or "charley horses"
5. Confusion
6. Convulsions
7. Predispose an anaerobic tissue environment conducive for infections and inflammation, etc.

5.4 KEY PHYSIOLOGICAL FUNCTIONS OF DIVERSE ELECTROLYTES

Electrolytes are crucial for optimal brain and muscle functions, and they also support the body's internal physiological ecology for maintaining hydration and regulation of internal pH in a variety of body tissues.[5,6,21] A mild alkaline pH (aerobic metabolism) is necessary to optimize the ability for cells to effectively use oxygen and water.[22]

5.4.1 Nervous System

Nerve cells perform physiological cell-to-cell communications mediated through electrical signals throughout the body, which are known as electrical impulses.[20] These impulses are generated by changes of the electrical charge of nerve cell membranes.[19] Movement of the electrolyte sodium across the nerve cell membrane is involved in causing these changes. During this physiological process, a set of chain reactions sets off moving more sodium ions across the nerve cell membrane.[20]

5.4.2 Muscle Functioning

Calcium plays a major role in muscle functioning, especially in muscle contraction.[15] It permits muscle fibers to slide together as well as move over each other as the muscle shortens and contracts.[15,23] In addition, magnesium is essential in this process so that the muscle fibers can slide outward and muscles can relax and expand after contraction.[15,23]

5.4.3 Hydration Status

For the optimal hydration in the body, adequate water must be present both inside and outside non-bone cells in the body.[24,25] Particularly, sodium maintains fluid balance through osmosis, which will remarkably reduce the potential for dehydration.[24,25]

5.4.4 pH Homeostasis

Acid is the concentration of hydrogen ions [H+], and alkaline is the concentration of hydroxyl ions [OH-].[26-31] Neutral pH is where acid and alkaline combine in the form of H_2O, which is why an inability to effectively utilize oxygen also reduces the ability to effectively hydrate.[27-30] Aerobic cell respiration utilizes oxygen to "oxidize" or burn glucose (i.e. "aerobic glycolysis") to generate cellular ATP at about 38 ATP molecules per glucose molecule.[27-31] The ability to use oxygen is crucial in this biochemical physiology, which requires a mild alkaline pH. A reduction or loss in the ability to use oxygen induces anaerobic glycolysis, or a mechanical breakdown of glucose, resulting in about 2 ATP molecules per glucose molecule, with a concomitant increase in oxygen free radical generation.[27,29-31] Moreover, in the anaerobic scenario, cells will require significantly greater amounts of glucose substrate to generate more ATP. Blood should maintain a physiological pH between 7.35 and 7.45, with the ideal blood pH being 7.4. A pH shift of greater than 0.3 of a point in either direction, i.e. <7.1 or >7.7, can have severe, devastating or even lethal health consequences. To stay healthy, your body needs to maintain its blood pH as close to 7.4 as possible.[27-31] To achieve this, pH buffers in the tissues are utilized, both acid and alkaline. Keep in mind, that one very important function of hemoglobin is to carry and deliver oxygen to cells. Therefore, a mild alkaline pH is crucial to maintain the integrity of red blood cell oxygen-carrying capabilities. When electrolyte ions are becoming low and insufficient to maintain the requisite pH in cells, cells will expend histidine from the heme iron-bound protein of hemoglobin for alkaline buffering.[28,30] This histidine expenditure is what causes the depletion of heme iron and its reassignment/repartitioning to and accumulation in other tissues of the body. This scenario results in the appearance of iron deficiency anemia (IDA) but is more accurately termed "chronic anemia syndrome" (CAS) and is caused by a deficiency in alkaline buffers.[22,30,31] But the first step in this sequela of pathophysiological events is a deficiency in electrolyte ions.[30,31]

Thus, a right balance of electrolytes is essential for maintaining an optimal blood pH level that enables effective oxygen delivery and cellular utilization, which also reduces the production of oxygen free radicals.[22,28,31]

5.5 CONSTITUENTS FOR AN IDEAL ELECTROLYTE FORMULATION AND ITS FUNCTION

Sodium: Electrolytes carry an electrical charge specifically when it is dissolved in body fluids including blood. Accordingly, sodium is mainly located in blood and in the fluid around the cells.[32,33] The key roles of sodium are to maintain normal nerve and muscle functions and maintain the homeostatic properties of body fluids.[33-35] Food and drink are the major dietary source of sodium intake, while sweat and urine facilitate sodium loss from the human body. Sodium regulates the amount of fluid in blood and around cells.[32-36] Sodium is moved across cell membranes by the sodium–potassium pump, an enzyme also known as sodium/potassium ATPase. Moreover, the kidney acts as a prime regulator of good health by regulating hydroxyl ion exchange (onto CO_2 to make HCO_3 [bicarbonate] and off HCO_3 to make CO_2) to maintain the ideal blood pH.[32-37] In addition, the kidneys maintain a consistent level of sodium balance in the human body by regulating urinary excretion of sodium from the body. However, when sodium balance gets disrupted due to pathophysiological influences then the concentration of sodium in the blood may be too low (hyponatremia) or too high (hypernatremia).[32-37]

It is important to mention that a human physiological system continually and routinely tracks the blood volume and sodium concentration. When, the balance gets disrupted and both values become too high, physiological sensors in the cardiovascular system and kidneys identify the upsurge.[33-37] Then, those physiological sensors stimulate the kidneys to enhance sodium excretion, returning blood volume to normal.

Concurrently, when blood volume or sodium concentration goes down, the physiological sensors trigger the mechanism to increase blood volume.[33–36] Two distinct pathways have been elucidated[32–35]:

a. The adrenal glands are stimulated by the kidneys to secrete more hormone aldosterone, and aldosterone, in turn, directs the kidney to excrete potassium and retain sodium.[32–36] With the increased level of sodium, less urine is generated, and eventually blood volume goes up.
b. Vasopressin, an anti-diuretic hormone secreted by the pituitary gland, effectively helps the kidneys to conserve water.[33–37]

It is important to note that it is quite challenging to maintain fluid and sodium balance in aged individuals.[32,33]

An older population has a reduced ability to maintain sodium and fluid balance for multiple reasons[35–37]:

i. Older people sense thirst less quickly and less intensely, as well as they might not drink enough fluids when required.[33–36] Thus, decreased thirst plays a vital role in older population.
ii. Advancing age significantly disrupts excretory physiology. Kidneys in older population may not reclaim water and electrolytes from the urine, and thus, more water and electrolytes may be excreted in the urine.[33–36]
iii. It is important to note that in an older population, 45% of body weight is fluid, while in younger population, 60% of the body weight is fluid. This signifies that while fever, fasting, or not drinking enough may cause a slight loss of fluid and sodium in the general population, it will cause serious consequence in older population.[33,36]
iv. Older people may have a disability that impairs their ability to get water when they need it. This inability to get water when needed can cause even more serious problems.[32]
v. The influence of drugs including higher intake of high blood pressure medicine, diabetes mellitus, or compromised cardiovascular, vascular medicines, which are generally taken more by an older population, can induce a greater excretion of fluids and, in turn, induce detrimental effects of fluid loss.[21,22]

All the above conditions can result in a considerable fluid loss, which is exacerbated by the reduced consumption of adequate water. These conditions can induce a high sodium level in blood, a condition known as hypernatremia and/or dehydration, which overburdens kidney function.[32,33] This is more common in an older population, which can result in confusion, metabolic dysfunction, coma, or death.[32–37]

Chloride: Edible salt is the prime source of chlorides in human body. Following ingestion, it is absorbed in the intestine, and extra chloride is excreted in the urine. Chloride plays an important role in helping the body to maintain a normal balance of fluids, while it is an important indicator for a wide array of clinical conditions.[37–41] Chloride is generally tested in blood, sweat, urine, and feces. After sodium, chloride is an important and abundant electrolyte in the serum, which plays a prime role in the regulation of body fluids, proper blood volume, electrolyte balance, blood pressure, preservation of electrical neutrality and pH (acid-base homeostasis) of body fluids.[38,41] It also maintains the fluid homeostasis inside and outside the cells. Moreover, it is intricately linked to multiple physiological conditions. Abnormal chloride levels alone indicate a serious metabolic disorder including metabolic acidosis or alkalosis.[38–41] Irregularities in chloride channel expression and function can lead to multiple diseases and disorders in diverse organs. A chloride test can be performed to assess the level of chloride in the blood or urine, as well as in the sweat.[38–41]

Increased or reduced chloride levels can lead to detrimental or even disastrous consequences including[38,41]:

a. An excessive increase in chloride level, a condition known as hyperchloremia, leads to diarrhea, certain kidney diseases, and in the hyperactivity of the parathyroid glands, which can demineralize the skeleton.[38–41]

b. Decreased chloride (hypochloremia): On the other hand, chloride is normally lost in the urine, sweat, and other excretions. Massive loss of chloride can occur during heavy sweating and vomiting, as well as in the diseases of the adrenal glands and kidney diseases.[37,38,41]

Bicarbonate: In blood and other fluids, bicarbonate ions $\left(HCO_3^-\right)$ act as alkaline buffers (homeostatically juxtaposed to CO_2) to maintain pH balance in extracellular blood plasma, for example.[42–48] Generally, serum bicarbonate levels lie between 22 and 30 mmol/L. However, when acidity is increased by ingested foods, medications, chronic disease, or dehydration, bicarbonate levels can increase to maintain the homeostatic equilibrium of an alkaline blood pH of 7.4.[42–48] The lungs regulate oxygen and carbon dioxide exchange and the kidneys regulate carbon dioxide and bicarbonate exchange in the kidney tubules.[44,45,48] An excessive and prolonged anaerobic/acidic metabolic burden, exceeding alkaline pH buffering demands, is a significant reason for the most prevalent causes of mortality in chronic diseases, i.e. pneumonia, infections (sepsis), and/or kidney failure.[43,45–48]

Potassium: Multiple essential functions are accomplished by potassium. Potassium primarily helps to metabolize sugar to glycogen to provide energy for daily tasks.[37,49] This includes the regulation of nerve impulses and muscle contractions, controls the flow of fluids and nutrients in and out of the cells, and helps to maintain the blood pressure, in part, by counteracting with sodium.[49,50] Potassium and sodium exchange in cells are regulated by the electrogenic transmembrane ATPase enzyme, also known as the sodium–potassium pump.[49] For every ATP molecule the pump uses, three sodium ions are exported, and two potassium ions are imported. Therefore, there is a net export of a single positive charge per pump cycle. Ideal blood potassium level should lie in between 3.5 and 5 mmol/L.[37,49,50]

Wholegrain bread, wheat bran and granola, peanut butter, apricots, bananas, melon, mango, oranges, pears, potatoes, tomatoes, tomato sauces, parsnips, spinach, broccoli, carrots, milk, and yogurt are potassium-rich foods (250 mg/serving). In contrast, white bread, rice, apples, berries, grapes, pears, peaches, asparagus, green beans, carrots, cabbage, cauliflower, corn, eggplant, poultry, tuna, and eggs are examples of low-potassium foods.[6,49]

The kidneys maintain blood potassium content within the normal range. However, patients with kidney disease have from an impaired ability to a total inability to efficiently eliminate excess potassium.[17,49,50] Conditions of low potassium are known as hypokalemia.[17] Mild hypokalemia exhibits no clear signs and symptoms, however, moderate to extreme hypokalemia causes muscle weakness, fatigue, cardiovascular arrhythmias, and in the most severe cases can cause a fatal heart attack.[17,49,50]

On the other hand, certain medications including angiotensin-converting enzyme (ACE) inhibitors such as enalapril, captopril, and lisinopril, and angiotensin receptor blockers such as irbesartan and valsartan, recommended for high blood pressure and cardiovascular disorders, can elevate blood potassium levels.[17,49,50] Thus, patients with diabetes and cardiovascular distress are at increased risk of potassium overload, disruption of intracellular pH buffering capabilities and ATP production due to intake of these medications.[49,50]

It is very important to note that the sodium:potassium ratio in the diet is more important than the level of either mineral alone.[49,50] The Reference Daily Intake (RDI) for potassium is approximately three times higher than for calcium, making it a very important nutrient.[17,37,49,50]

Magnesium: It is well known that over 300 enzymes require magnesium ions for their catalytic action.[12,14,51] In addition, magnesium is essential for the formation of bones and teeth, as well as for normal nerve and muscle functioning.[51,52] Moreover, magnesium is required for the metabolism of calcium and potassium and is involved in a variety of numerous metabolic activities including relaxation of smooth muscles especially those surrounding the bronchial tubes in the lungs, skeletal muscle contraction, and excitation of neurons in the brain, along with a variety of other metabolic activities.[12,15,51–53]

Magnesium is another electrolyte that carries an electric charge following dissolution in body fluids such as blood. But most of the magnesium remains uncharged and bound to proteins or stored in the bone. It is interesting to note that bone contains about half of the body's magnesium, while blood contains very little.[51–53] Magnesium in the blood largely depends on the foods we eat and is mostly excreted in the urine and stool. Very little magnesium gets stored in the body. Magnesium levels in the body are closely linked with sodium, potassium, and calcium metabolism, and are regulated by the kidney.[14,51–53] Magnesium enters the body through the diet; and the amount of magnesium that is absorbed is a result of feedback mechanisms that depend upon the concentration of magnesium in the body. Too little magnesium stimulates absorption from the intestine, while too much decreases absorption. The level of magnesium in the bloodstream can be either very high, a condition known as hypermagnesemia, or very low, termed as hypomagnesemia.[12,14,54]

Hypomagnesemia can occur for multiple reasons. Dietary deficiencies reduce the amount of absorbed magnesium but increase the absorbable percentage of consumed magnesium. Certain diseases and medications reduce the ability of the intestine to absorb magnesium or increase magnesium excretion.[51–53] Common causes of low magnesium include overindulgence of alcohol and its associated malnutrition and dehydration due to chronic diarrhea and use of medications like diuretics used to control high blood pressure.[51,52,54] It is interesting to note that 50% of the ICU patients may become magnesium deficient. Symptoms include arrhythmias, muscles with weakness and cramps, nausea, vomiting, breathing difficulties, and nervous system dysfunctions potentially causing confusion, hallucinations, and seizures.[51–53]

Hypermagnesemia often occurs in patients with kidney dysfunctions, in which the excretion of magnesium is restricted. In these patients, a high intake of dietary magnesium either from diet or magnesium-containing medications may cause an increased magnesium level.[12,14] However, absorption and excretion of magnesium are linked to other electrolytes, and numerous diseases are associated with high magnesium levels including diabetic ketoacidosis, adrenal insufficiency, and hyperparathyroidism.[14,51] In addition, hypermagnesemia is often associated with hypocalcemia and hyperkalemia.[12,14,52]

Calcium: Like other electrolytes, calcium carries an electrical charge when dissolved in blood.[21,23] However, most of the body's calcium remains unchanged. Calcium is essential for the formation of strong bones and teeth, as well as for skeletal muscle contraction and stabilizing blood pressure.[21,23,54] Calcium imbalances cause hypercalcemia or hypocalcemia.[21,23,56–58]

Too much calcium in the blood causes hypercalcemia which can lead to kidney diseases, hyperparathyroidism, tuberculosis, sarcoidosis, and lung and breast cancers.[21,23,37] Excessive use of antacids, calcium, and vitamin D supplements (although vitamin D is the subject of some dosage controversy), and diverse medications including lithium, theophylline, or certain water pills can cause hypercalcemia.[56,57] On the other hand, lack of adequate calcium in the bloodstream causes hypocalcemia. Hypocalcemia induces kidney failure, hypoparathyroidism, pancreatitis, prostate cancer, and malabsorption.[21,23,57] Furthermore, use of medications including heparin, osteoporosis drugs, and antiepileptic drugs can cause hypocalcemia.[21,23]

Vitamin C: Vitamin C is a potent scavenger of free radicals and acts as an antioxidant with potential anti-inflammatory and immune-enhancing properties.[56] Vitamin C inhibits cellular oxidative damage, as well as participates in various metabolic reactions including the synthesis of collagen, a vital protein instrumental for the formation of connective tissue, cartilage, bone, skin, and tendons.[55] Vitamin C offers an interesting synergy with carnitine, which modulates the transportation of fat for its metabolism.[55,60] Furthermore, vitamin C forms elastin and connective tissues. Collagen, carnitine, elastin, and connective tissues in the body provide elasticity, resilience, strength, and stability.[55,59,60] In turn, both collagen and carnitine require and utilize vitamin C, respectively, for diverse biochemical functions and health benefits. Research demonstrated that administration in moderate to high doses of vitamin C may protect against atherosclerosis and hypertension.[59,60] Vitamin C administration over a period of 30 days produced a significant reduction in total cholesterol, very low-density lipoprotein, and low-density lipoprotein, but no changes were observed in triglycerides and high-density lipoprotein levels.[55,60] However, overdose of vitamin C may cause digestive distress and kidney stones.[55]

Sources of vitamin C include fruits and vegetables including bell pepper, orange, grapefruit, kiwi, broccoli, strawberry, Brussels sprouts, tomato, cantaloupe, cabbage, cauliflower, potato, spinach and green peas, and a very limited amount in cooked meats.[59,60] Adult men need 90 mg of vitamin C/day, while adult females require 75 mg/day.[55] Cigarette smokers need an additional 35 mg/day.[55]

Phosphorous: Phosphorous is an important element in human body, which is necessary for the formation of bones, teeth, and energy production.[56,57] In fact, phosphate is instrumental for cellular energy, formation of cell membranes, and DNA replication. Phosphorous combines with oxygen to form phosphate, which acts as an electrolyte. Foods including milk, egg yolks, chocolate, and soft drinks are the best source of dietary phosphates.[56,57] It is excreted in the urine and stool. Symptoms of phosphorous deficiency include anxiety, fatigue, irregular breathing, irritability, joint stiffness, numbness, weakness, and irregular changes in body weight.[56,57]

Bone contains 85% of the body's phosphate, while the remaining 15% is located inside the cells which is involved in energy production.[56,57] However, a high level of phosphate causes hyperphosphatemia, while low phosphate level causes hypophosphatemia. Hypophosphatemia, in turn, may be either acute or chronic.[57,61]

Hyperphosphatemia, i.e. abnormally high serum phosphate levels, may result from increased phosphate intake, decreased phosphate excretion, or a disorder that shifts intracellular phosphate to extracellular space.[61] Hyperphosphatemia can occur with overdose of oral phosphate administration or random use of enemas containing phosphate. Other causes include hypocalcemia, diabetic ketoacidosis, crush injuries, nontraumatic rhabdomyolysis, systemic infections, and tumor lysis syndrome.[56,57] However, hyperphosphatemia can be spurious in cases of hyperproteinemia, dyslipidemia, hemolysis, or hyperbilirubinemia. It may also lead to morbidity.[56,57]

On the other hand, symptoms of hypophosphatemia include muscle weakness, followed by stupor, coma, and death. During acute hypophosphatemia, phosphate level in blood suddenly goes dangerously low, while during chronic hypophosphatemia, the bones weaken, resulting in bone pain and fractures. People may become weaker and lose their appetite.[56,57] The body uses large amounts of phosphate to recover from certain disorders which can cause acute hypophosphatemia.[61] Acute hypophosphatemia can also occur in people recovering from physiological conditions such as starvation or severe undernutrition, diabetic ketoacidosis, severe alcoholism, and severe burns.[57] A sudden drop in phosphate level causes abnormal heart rhythm and even death. In chronic hypophosphatemia, low phosphate level over an extended period can be caused due to excretion of excessive phosphates from the body.[61] Symptoms of hypophosphatemia include hyperparathyroidism and chronic diarrhea. Use of diuretics or aluminum-enriched antacids for a longer period or use of large amounts of theophylline can also aggravate hypophosphatemia.[61] It is interesting to note that people who survived concentration camps ultimately died because their already low phosphate level suddenly dropped when these individuals started consuming a normal diet, a phenomenon termed "refeeding syndrome."[61]

Phytochemical/Phytopharmaceutical Complex: Phytopharmaceuticals are defined as botanical medicines whose efficacy is determined by one or more plant substances or active ingredients. Phytochemicals include prebiotics and probiotics, as well as several chemical compounds such as polyphenols and derivatives, carotenoids, and thiosulfates. The largest group of these comprises polyphenols, which can be subclassified into four main groups: flavonoids (including eight subgroups), phenolic acids (such as curcumin), stilbenoids (such as resveratrol), and lignans.[62]

Water Extracts of Fossilized Plants and Apple Extract (Commercially Available as ElavATP®): A proprietary, clinically researched combination of a water extract of fossilized plants and apple extract, especially rich in bioflavonoids, carbon, magnesium, nitrogen, oxygen, sulfur, and a magnesium-ATP complex that is utilized to enhance cellular ATP levels and enhance electrolyte, nitrogen, and oxygen utilization.[63,64]

Broad Spectrum Whole Fruit and Vegetable Concentrates and Extracts (Commercially Available as Spectra®): Provides broad spectrum botanical concentrates and extracts rich in phytosaccharides; polyphenols, including bioflavonoids; and organic acids that supports efficient cell-to-cell communications; cellular oxygen utilization; glycoprotein synthesis; and antioxidant activity, among others, within the body.[65]

Bioflavonoids: Structurally diverse flavonoids, such as quercetin, are important plant components that strengthen connective tissue; serve as potent antioxidants, anti-allergic, anti-inflammatory agents, immune modulators,[66] and cadioprotectants[67]; boost nitric oxide level,[68–70] and contribute hydroxyl groups for numerous biological needs including the authors' notion of alkaline buffering as needed. In addition to all the electrical contributions made by electrolytes cited above, when needed, electrolytes can be utilized to fund metalloprotein synthesis.[71] For example, the synthesis of collagen protein utilizes the synergistic combination of electrolytes, saccharides, and bioflavonoids.

5.6 REFERENCE DAILY INTAKE (RDI)

In food and dietary supplement industries, RDI is used in nutrition labeling to exhibit the daily value of foods, whereas the Recommended Dietary Allowances (RDAs) are a set of nutrition recommendations which involve both in the Dietary Reference Intake (DRI) and the RDIs used for nutritional labeling.[19] Table 5.2 demonstrates the new RDI values[19] of the electrolytes used in a formulation.

Proprietary Prodosomed Electrolyte Technology: A novel state-of-the-art electrolyte formulation was designed by our team to demonstrate unique and sustained beneficial effects. This formulation is enriched with sodium, chloride, potassium, magnesium, calcium, vitamin C, and phosphorous, as well as a proprietary, clinically researched combination of a water extract of fossilized plants and apple extract (commercially available as ElavATP®), especially rich in bioflavonoids, carbon, magnesium, nitrogen, oxygen, sulfur, and a magnesium-ATP complex. This complex is utilized to enhance cellular ATP levels and enhance electrolyte, nitrogen, and oxygen utilization. In addition, a complex of broad-spectrum whole fruit and vegetable concentrates (commercially available as Spectra®) is included. All the active ingredients in this formula are Prodosomed to facilitate rapid and prolonged simultaneous absorption for optimal synergistic effects.

The manufacturing was performed using a non-Genetically Modified Organism (non-GMO) lecithin containing a minimum of 85% phosphatidylcholine, which was impregnated and saturated using solar-dried electrolytes from sea water to ensure the availability of free ions. Importantly, the

Table 5.2 Reference Daily Intake (RDI)[19] of the Recommended Electrolytes

Electrolyte Constituents	Male (Age 19–30 years)	Female (Age 19–30 years)
Sodium	1.5 g	1.5 g
Chloride	2.3 g	2.3 g
Potassium	4.7 g	4.7 g
Magnesium	400 mg	310 mg
Calcium	1,300 mg	1,300 mg
Vitamin C	90 mg	75 mg
Phosphorous	700 mg	700 mg
Spectra enriched in bioflavonoids	Not Applicable	Not Applicable
ElevATP	Not Applicable	Not Applicable

Table 5.3 Constituents in PerformLyte Formulation

Ingredients	Amount (mg)
Magnesium (derived from sea water)	60
Chloride (from sea water)	156
Potassium (from citrate)	150
Sodium (from sodium ascorbate)	40
Calcium (from tricalcium phosphate)	136
Vitamin C (from sodium ascorbate)	310
Phosphorous (from tricalcium phosphate)	56
Polyphenol-rich botanical extracts and concentrates (Spectra®)	100
Proprietary blend of ancient peat and apple extracts (ElevATP®)	150

Complete range of solar-dried electrolytes from sea water.

full spectrum of naturally occurring electrolytes from sea water provides the same proportions of electrolytes that exist in human blood.

The primary electrolyte formulation is based on prevailing research and includes botanical phytonutrients that are blended in a proprietary high-speed dry-blending process with the electrolyte-enhanced phospholipids that ensures optimal bio-encapsulation to achieve a novel energetically enhanced SK713 SLP (Prodosome®) technology, commercially available as PerformLyte™.

Institutional Review Board (IRB)-approved clinical studies have been conducted focusing on the SK713 SLP proprietary delivery technology to assess its ability to improve absorption and utilization of nutrients along with beneficial outcomes to the body. The research demonstrated improvements in absorption, based on the rapidity of beneficial effects on important blood properties, hemoglobin and neutrophil restoration, and several other biomarkers.[22,72,73]

Table 5.3 exhibits the formulation of this unique electrolyte formulation.

5.7 CONCLUSION

This state-of-the-art formulation was engineered by anticipating human body requirements, available laboratory findings, and metabolic response to structurally diverse nutritional needs while remaining strictly within RDI guidelines. This novel formulation was innovated after many years of investigative research by assessing the safety evaluation, therapeutic benefits, and stability of this state-of-the-art formulation. Basically, the daily intake and excretion of ions including sodium, chloride, potassium, calcium, phosphorous, magnesium, and vitamin C, supported by the Spectra, rich in bioflavonoids, demonstrated to boost physiological nitric oxide and ElevATP, a supplement shown to boost ATP level in the blood, were used to maintain blood pH appropriately. Further research is in progress to unveil novel mechanistic aspects of human performance and function.

REFERENCES

1. Lorenzo-Lopez L, Maseda A, De Labra C, Regueiro-Folgueira L, Rodriguez-Villamil JL, Millan-Calenti JC. Nutritional determinants of frailty in older adults: A systematic review. *BMC Geriatr.* 2017; 17: 108, published online 2017 May 15. doi: 10.1186/s12877-017-0496-2.
2. Fletcher RH, Fairfield KM. Vitamins for chronic disease prevention in adults: Clinical applications. *JAMA.* 2002; 287: 3127–3129.
3. Edelman S, Leibman J. Anatomy of body water and electrolytes. *Am J Med.* 1959; 27(2): 256–277.
4. Healthline. Should You Supplement Your Diet with Electrolytes? https://www.healthline.com/nutrition/electrolytes#supplements (accessed Jan 24, 2020).

5. Subcommittee on the Tenth Edition of the RDAs, Food and Nutrition Board Commission on Life Sciences National Research Council. Water and Electrolytes (Chapter 11), in: *Recommend Dietary Allowances.* 10th Edition; Natural Academies National Academy Press: Washington, DC, 1989; pp. 247–261 https://www.ncbi.nlm.nih.gov/books/NBK234935/ (accessed Jan 24, 2020).

6. Committee on Diet and Health Food and Nutrition Board Commission on Life Sciences National Research Council. Electrolytes (Chapter 15), in: *Diet and Health: Implications for Chronic Disease Risk.* ISBN-10: 0-309-03994-0. National Research Council (US) Committee on Diet and Health, National Academies Academy Press: Washington, DC, 1989, pp. 413–430. https://www.ncbi.nlm.nih.gov/books/NBK218743/pdf/Bookshelf_NBK218743.pdf

7. Sqambato F, Sqambato E, De Santo NG. Jacques Loeb (1859–1924) and his forgotten contributions to electrolyte and acid-base physiology in the organism as a whole. *G Ital Nefrol.* 2016; 33(Suppl 66): 33.S66.28.

8. Loeb J. The role of salts in the preservation of life. *Science.* 1911; 34(881): 653–665.

9. Loeb G. The origin of the conception of physiologically balanced salt solutions. *J Biol Chem.* 1918; XXXIV: 503–504.

10. Loeb J. Chemische Konstitution und physiologische Wirksamkeit der Säuren. *Biochem Zeitschr.* 1909; 15: 254–271.

11. Loeb J. Is the antagonistic action of salt due to oppositely charged ions? *J Biol Chem.* 1914; XIX: 431–443.

12. Loeb J. Weber's law and antagonistic salt action. *Proc Natl Acad Sci USA.* 1915; 1(8): 439–444.

13. Preuss HG, Memon S, Dadgar A, Gongwei J. Effect of high sugar diets on renal fluid, electrolyte and mineral handling in rats: relationship to blood pressure. *J Am Coll Nutr.* 1994; 13(1): 73–82.

14. Blaine J, Chonchol M, Levi M. Renal control of calcium, phosphate, and magnesium homeostasis. *Clin J Am Soc Nephrol.* 2015; 10(7): 1257–1272.

15. Stecker RA, Harty PS, Jagim AR, Candow DG, Kerksick CM. Timing of ergogenic aids and macronutrients on muscle and exercise performance. *J Int Soc Sports Nutr.* 2019; 16(1): 37. doi: 10.1186/s12970-019-0304-9.

16. Iqbal S, Klammer N, Ekmekcioglu C. The effect of electrolytes on blood pressure: A brief summary of meta-analyses. *Nutrients.* 2019; 11(6). pii: E1362. doi: 10.3390/nu11061362.

17. Kardalas E, Paschou SA, Anagnostis P, Muscogiuri G, Siasos G, Vryonidou A. Hypokalemia: a clinical update. *Endocr Connect.* 2018; 7(4): R135–R146. doi: 10.1530/EC-18-0109. Epub 2018 Mar 14.

18. Ahuja M, Chung WY, Lin WY, McNally BA, Muallem S. Ca2+ signaling in exocrine cells. *Cold Spring Harb Perspect Biol.* 2019. pii: a035279. doi: 10.1101/cshperspect.a035279.

19. Wikipedia. Reference Daily Intake. https://en.wikipedia.org/wiki/Reference_Daily_Intake (accessed Jan 24, 2020).

20. Riggs JE. Neurologic manifestations of fluid and electrolyte disturbances. *Neurol Clin.* 1989; 7(3): 509–523.

21. Barstow C, Braun M. Electrolyes: Calcium disorders. *FP Essent.* 2017 Aug; 459: 29–34.

22. Corbier JR, Downs BW, Kushner S, Aloisio T, Bagchi D, Bagchi M. VMP35 MNC, a novel iron-free supplement, enhances cytoprotection against anemia in human subjects: A novel hypothesis. *Food Nutr Res.* 2019, 63: 3410. doi: 10.29219/fnr.v63.3410.

23. Ebashi S, Endo M. Calcium and muscle contraction. *Prog Biophys Mol Biol.* 1968; 18: 123–166, IN9-IN12, 167–183.

24. Armstrong LE, Johnson EC. Water intake, water balance, and the elusive daily water requirement. *Nutrients.* 2018; 10(12). pii: E1928. doi: 10.3390/nu10121928.

25. Orru S, Imperlini E, Nigro E, Alfieri A, Cevenini A, Polito R, Daniele A, Buono P, Mancini A. Role of functional beverages on sport performance and recovery. *Nutrients.* 2018; 10(10). pii: E1470. doi: 10.3390/nu10101470.

26. Hamm LL, Nakhoul N, Hering-Smith KS. Acid-base homeostasis. *Clin J Am Soc Nephrol.* 2015; 10(12): 2232–2242. doi: 10.2215/ CJN.07400715.

27. Reddy P, Mooradian AD. Clinical utility of anion gap in deciphering acid-base disorders. *Int J Clin Pract.* 2009; 63(10): 1516–1525. doi: 10.1111/j.1742–1241.2009.02000.x.

28. Thomas CP, Hamawi K. What is the role of acidemia and alkalemia in the pathogenesis of metabolic acidosis? https://www.medscape.com/answers/242975-154551/what-is-the-role-of-acidemia-and-alkalemia-in-the-pathogenesis-of-metabolic-acidosis [cited 20 December 2018].

29. Hopkins E, Sharma S. Physiology, acid base balance. *Stat Pearls*. https://www.ncbi.nlm.nih.gov/books/ NBK507807/ [cited 20 December 2018].

30. Phypers B, Pierce T. Lactate physiology in health and disease. *Contin Educ Anaesth Crit Care Pain*. 2006; 6(3): 128–132. doi: 10.1093/bjaceaccp/mkl018.

31. Preuss HG. Fundamentals of clinical acid-base evaluation. *Clin Lab Med*. 1993; 13(1): 103–116.

32. Stanhewicz AE, Kenney WL. Determinants of water and sodium intake and output. *Nutr Rev*. 2015; 73 (Suppl 2): 73–82. doi: 10.1093/nutrit/nuv033.

33. Canaud B, Kooman J, Selby NM, Taal M, Francis S, Kopperschmidt P, Malerhofer A, Kotanko P, Titze J. Sodium and water handling during hemodialysis: New pathophysiologic insights and management approaches for improving outcomes in end-stage kidney disease. *Kidney Int*. 2019; 95(2): 296–309. doi: 10.1016/j.kint.2018.09.024.

34. Verbrugge FH. Utility of urine biomarkers and electrolyte for the management of heart failure. *Curr Heart Fail Rep*. 2019; 16(6): 240–249. doi: 10.1007/s11897-019-00444-z.

35. Müller DN, Wilck N, Haase S, Kleinewietfeld M, Linker RA. Sodium in the microenvironment regulates immune responses and tissue homeostasis. *Nat Rev Immunol*. 2019; 19(4): 243–254. doi: 10.1038/s41577-018-0113-4.

36. Braun MM, Mahowald M. Electrolytes: Sodium disorders. *FP Essent*. 2017; 459: 11–20.

37. Harris L, Braun M. Electrolytes: Oral electrolytes solutions. *FP Essent*. 2017; 459: 35–38.

38. Seifter JL, Chang HY. Extracellular acid-base balance and ion transport between body fluid compartments. *Physiology (Bethesda)*. 2017; 32(5): 367–379. doi: 10.1152/physiol.00007.2017.

39. Nagami GT. Hyperchloremia - Why and how. *Nefrologia*. 2016; 36(4): 347–353. doi: 10.1016/j.nefro.2016.04.001.

40. Rottgen TS, Nickerson AJ, Rajendran VM. Calcium activated Cl-channel: Insights on the molecular identity in epithelial tissues. *Int J Mol Sci*. 2018; 19(5). pii: E1432. doi: 10.3390/ijms19051432.

41. Berend K, van Hulstelin LH, Gans RO. Chloride: The queen of electrolytes. *Eur J Intern Med*. 2012; 23(3): 203–211. doi: 10.1016/j.ejim.2011.11.013.

42. Beaume J, Braconnier A, Dolley-Hitze T, Bertocchio JP. Bicarbonate: From physiology to treatment for all clinicians. *Nephrol Ther*. 2018; 14(1): 13–23. doi: 10.1016/j.nephro.2017.02.014. Epub 2017 Nov 15.

43. Wiggins SV, Steegborn C, Levin LR, Buck J. Pharmacological modulation of the $CO_2/HCO_3^-/pH$-, calcium-, and ATP-sensing soluble adenylyl cyclase. *Pharmacol Ther*. 2018; 190: 173–186. doi: 10.1016/j.pharmthera.2018.05.008. Epub 2018 May 26.

44. Amaral Silva D, Al-Gousous J, Davies NM, Bou Chacra N, Webster GK, Lipka E, Amidon G, Löbenberg R. Simulated, biorelevant, clinically relevant or physiologically relevant dissolution media: The hidden role of bicarbonate buffer. *Eur J Pharm Biopharm*. 2019; 142: 8–19. doi: 10.1016/j.ejpb.2019.06.006.

45. Berend K, de Vries AP, Gans RO. Physiological approach to assessment of acid-base disturbances. *N Engl J Med*. 2014; 371(15): 1434–1445. doi: 10.1056/NEJMra1003327.

46. Seifter JL. Integration of acid-base and electrolyte disorders. *N Engl J Med*. 2014; 371(19): 1821–1831. doi: 10.1056/NEJMra1215672.

47. Rein JL, Coca SG. "I don't get no respect": The role of chloride in acute kidney injury. *Am J Physiol Renal Physiol*. 2019; 316(3): F587–F605. doi: 10.1152/ajprenal.00130.2018. Epub 2018 Dec 12.

48. van Niekerk J, Kersten R, Beuers U. Role of bile acids and the biliary HCO_3^- Umbrella in the pathogenesis of primary biliary cholangitis. *Clin Liver Dis*. 2018; 22(3): 457–479. doi: 10.1016/j.cld.2018.03.013. Epub 2018 May 19.

49. Weir MR, Espallit R. Clinical perspectives on the rationale for potassium supplementation. *Postgrad Med*. 2015; 127(5): 539–548. doi: 10.1080/00325481.2015.1045814.

50. Elliott TL, Braun M. Electrolytes: Potassium disorders. *FP Essent*. 2017; 4(59): 21–28.

51. Ahmed F, Mohammed A. Magnesium: The forgotten electrolyte – Review on hypomagnesemia. *Med Sci (Basel)*. 2019; 7(4). pii: E56. doi: 10.3390/medsci7040056.

52. Glasdam SM, Glasdam S, Peters GH. The importance of magnesium in the human body: A systematic literature review. *Adv Clin Chem*. 2016; 73: 169–193.

53. Elin RJ, Magnesium: The fifth but forgotten electrolyte. *Am J Clin Pathol* 1994; 102: 616–622.

54. Gelli R, Ridi F, Baglioni P. The importance of being amorphous: Calcium and magnesium phosphates in human body. *Adv Colloid Interface Sci*. 2019; 269: 219–235. doi: 10.1016/j.cis.2019.04.011. Epub 2019 Apr 27.

55. Thompson DJ, Heintz JF, Phillips PH. Effect of magnesium, fluoride, and ascorbic acid on metabolism of connective tissue. *J Nutr.* 1964; 84(1): 27–30.

56. Terry J. The other electrolytes: Magnesium, calcium, and phosphorous. *J Intraven Nurs.* 1991; 14(3): 167–176.

57. Madiraca J, Hoch C. Electrolyte series. Calcium and Phosphorous. *Nurs 2019 Crit Care* 2018; 13(2): 24–31. doi: 10.1097/01.CCN.0000529938.18274.f4. www.nursingcriticalcare.com.

58. Beloosesky Y, Grinblat J, Weiss A, Grosman B, Gafter U, Chagnac A. Electrolyte disorders following oral sodium phosphate administration for bowel cleansing in elderly patients. *Arch Intern Med.* 2003; 163(7): 803–808.

59. Meng QH, Irwin WC, Visvanathan K, Vitamin C and aberrant electrolyte results. *Clin Chem Lab Med.* 2005; 43(4): 454–456.

60. Miller MJ. Injuries to athletes: Evaluation of ascorbic acid and water-soluble citrus bioflavonoids in the prophylaxis of injuries in athletes. *Med Times* 1960; 88: 313–316.

61. Yu ASL, Stubbs JR. Evaluation and Treatment of Hypophosphatemia. https://www.uptodate.com/contents/evaluation-and-treatment-of-hypophosphatemia (accessed Jan 26 2020).

62. Carrera-Quintanar L, López Roa RI, Quintero-Fabián S, Sánchez-Sánchez MA, Vizmanos B, Ortuño-Sahagún D. Phytochemicals that influence gut microbiota as prophylactics and for the treatment of obesity and inflammatory diseases. *Mediators Inflamm.* 2018; 2018: 9734845. doi: 10.1155/2018/9734845. eCollection 2018. Review.

63. Joy JM, Vogel RM, Moon JR, Falcone PH, Mosman MM, Kim MP. Twelve weeks supplementation with an extended-release caffeine and ATP-enhancing supplement may improve body composition without affecting hematology in resistance-trained men. *J Int Soc Sports Nutr.* 2016; 13: 25. doi: 10.1186/s12970-016-0136-9.

64. Joy JM, Vogel RM, Moon JR, Falcone PH, Mosman MM, Pietrzkowski Z, Reyes T, Kim MP. Ancient peat and apple extracts supplementation may improve strength and power adaptations in resistance trained men. *BMC Complement Altern Med.* 2016; 16: 224. Published online 2016 Jul 18. doi: 10.1186/s12906-016-1222-x.

65. Joy JM, Falcone PH, Vogel RM, Mosman MM, Kim MP, Moon JR. Supplementation with a proprietary blend of ancient peat and apple extract may improve body composition without affecting hematology in resistance-trained men. *Appl Physiol Nutr Metabol.* 2015, 40(11): 1171–1177. doi: 10.1139/apnm-2015-0241.

66. Rasines-Perea Z, Teisseedre PL. Grape polyphenols' effects in human cardiovascular diseases and diabetes. *Molecules* 2017; 22(1). pii: E68. doi: 10.3390/molecules22010068.

67. Micek J, Jurikova T, Skrovankova S, Sochor J. Quercetin and its anti-allergic immune response. *Molecules.* 2016; 21(5). pii: E623. doi: 10.3390/molecules21050623.

68. Duarte J, Francisco V, Perez-Vizcaino F. Modulation of nitric oxide by flavonoids. *Food Funct.* 2014; 5(8): 1653–1668. doi: 10.1039/c4fo00144c.

69. Bondonno CP, Croft KD, Ward N, Considine MJ, Hodgson JM. Dietary flavonoids and nitrate: effects on nitric oxide and vascular function. *Nutr Rev.* 2015; 73(4): 216–245. doi: 10.1093/nutrit/nuu014.

70. Huk I, Brovkovych V, Nanobash VJ, Weigel G, Neumayer C, Partyka L, Patton S, Malinski T. Bioflavonoid quercetin scavenges superoxide and increases nitric oxide concentration in ischaemia-reperfusion injury: An experimental study. *Br J Surg.* 1998; 85(8): 1080–1085. doi: 10.1046/j.1365-2168.1998.00787.x.

71. Prabhulkar S, Tian H, Wang X, Zhu J-J, Li C-Z. Engineered proteins: Redox properties and their applications. *Antiox Redox Signal.* 2012; 17(12): 1796–1822. doi: 10.1089/ars.2011.4001.

72. Wang Q, Guo R, Nair S, Smith D, Bisha B, Nair AS, Nair R, Downs BW, Kushner S, Bagchi M. Safety and efficacy of N-SORB®, a proprietary KD120 MEC metabolically activated enzyme formulation: A randomized, double-blind, placebo-controlled study. *J Am Coll Nutr.* 2019; 38(7): 577–585. doi: 10.1080/07315724.2019.1586591.

73. Downs BW, Kushner SW, Bagchi M, Swaroop A, Bagchi D. Safety and efficacy of a novel KD120 MEC multi-enzyme complex (N-SORB®) in human volunteers. *FASEB J.* 2017; 31: lb312.

CHAPTER 6

Gut Microbiome and Heavy Metals

Ashfaque Hossain
RAK Medical and Health Sciences University

Muhammad Manjurul Karim and Tania Akter Jhuma
University of Dhaka

Godfred A. Menezes
RAK Medical and Health Sciences University

CONTENTS

6.1 TERMINOLOGY

Microbiota: Microorganisms such as bacteria, fungi, archaea, protozoa, and viruses harbored by normal, healthy individuals.

Microbiome: Total microbial genomes associated with the human body. Currently, however, microbiome is being used to denote two different concepts: (a) The collective microbial genome and (b) the sum of all microscopic life forms, viz., microbes within an environment i.e. host-inhabiting microbes and their genes and genomes.

Virome: Total viruses (both prokaryotic and eukaryotic) present in and on our body.

Pan-genome: The collection of genes found across all members of a species.

Metagenomics: Genomic analysis of DNA extracted directly from clinical or environmental samples; it can assess community diversity (through 16S rRNA metagenomics) or fuller complexity (through whole genome shotgun metagenomics).

Next-Generation Sequencing (NGS): Advanced DNA sequencing technology that enables rapid sequencing of DNA fragment of large sizes (in excess of 1 million base pairs); even the entire human genome (3 billion bases) can be sequenced in 1 day.

Metabolome: Total metabolites found in a tissue or a sample. This profile represents a snapshot in time of what chemicals are present in the sample.

Dysbiosis: Transition of a healthy microbiome (eubiotic microbiome) to an unhealthy microbiome; imbalance in the intestinal microbiota, usually representing a disease state or health condition.

Eubiosis: Transition of an unhealthy, dysbiotic microbiome to healthy microbiome, opposite of dysbiosis.

Enterotype: Classification of human gut microbiome according to its characteristics. Microbiome of enterotype-1, enterotype-2, and enterotype-3 represents eubiotic, dysbiotic, and mixed-type microbiome, respectively (Arumugam et al., 2011).

SCFA: Short-chain fatty acids; acetic acid, propionic acid, and butyric acid (Figure 6.4).

BCAA: Branched-chain amino acids; valine, isoleucine, and leucine (Figure 6.4).

Xenobiotics: This term is used to describe the presence of two types of chemical substances in an organism (a) which are not naturally produced by that organism and (b) which are present in much higher concentration than usual.

Fecal Microbial Transplantation (FMT): It is a procedure that involves transfer of fecal material from a healthy person with eubiotic microbiome (presumably contain gut health bacteria) to the gastrointestinal tract (GIT) of a person with dysbiotic microbiome with the aim of inducing eubiosis in the recipient.

Heavy Metals: Naturally occurring elements with high atomic number and five times higher density than that of water are described as heavy metals (HMs).

Probiotics: Living microorganism (bacteria and yeast) that can inhabit the gut and contribute to the health of the host.

Prebiotics: Food ingredients which are non-digestible by humans but promote the growth of beneficial microorganisms (probiotics) in the intestines.

Gut Remediation: Binding and elimination of HMs from gut through feces by probiotics.

Germ-Free Animals: Animals that have no microbiome i.e. no microorganisms living in or on them. Such animals are raised in sterile conditions and provided with sterile food. Immediately after birth, various methods are employed control their exposure to viral, bacterial or parasitic agents. Such animals are crucial in microbiome studies.

Gnotobiotic Animals: Animals in which only known microorganisms are present. Such animals are developed by introducing known, specific organisms to germ-free animals which are very useful in determining the role of different members of microbiome singly and in combination.

Firmicutes/Bacteroidetes Ratio: Firmicutes and Bacteroidetes are the two major phyla of bacteria present in the human gut microbiome and together they represent 90% of the total microbiome. Increase in Firmicutes/Bacteroidetes ratio indicates transition of gut microbiome toward dysbiosis and has been linked with a variety of disease conditions.

Akkermansia muciniphila: A mucin-degrading bacterial member of gut microbiome; it is considered as an indicator bacterium of eubiotic microbiome.

6.2 GUT MICROBIOME AND ITS COMPOSITION

A microbiome is the collection of genomes from all the microorganisms found in a particular environment. Humans, plants, and other animals all have microbiomes; these can be generalized to their entire organism, or broken down into specific microbiomes for different locations on them, the GIT or gut of human, for example. Microbiota, sometimes used interchangeably, usually refers to all microorganisms that are found in an ecosystem which all play a role in maintaining the environment's stability. The human gut, representing one of the largest interfaces (250–$400\,m^2$), is a pool of microorganisms dominated by bacteria with minor contribution from archaea, fungi, and protozoa. The dynamic population of microorganisms in the GIT, termed as "gut microbiome" has developed a complex and mutually beneficial relationship with the host over thousands of years (Thursby and Juge, 2017; Assefa and Köhler, 2020). Approximately 35,000 species in our gut microbiome comprises of six bacterial phyla including Firmicutes (>70%), Bacteroidetes (>30%), Proteobacteria (<5%), Actinobacteria (<2%), Fusobacteria, and Verrucomicrobia (<1%) (Belizário et al., 2018), where the first two phyla, Firmicutes and Bacteroidetes that dominate others, represent 90% of the total bacterial pool (Claus et al., 2016). Microbiota composition in GIT demonstrates the physiological condition of that area which is influenced by numerous nutritional, chemical, and immunological gradients (Thursby and Juge, 2017). Figure 6.1 demonstrates diversity of bacteria in GIT depending on its different pH conditions. Quantitatively, the bacterial density in stomach (10^1–10^3 CFU/mL) steadily increases in jejunum/ileum (<10^5 CFU/mL) down to the colon (10^9–10^{12} CFU/mL), where they reach the highest cell density, comprising 99% of total gastrointestinal (GI) population (Belizário et al., 2018). Overall, the gut microbiome is a distinct "organ" of around 1–$2\,kg$ with 100 trillion individual microorganisms and its overall genetic potentiality is as high as 150 times of human genome (Tilg and Kaser, 2011; Rosenfeld, 2017).

6.3 FACTORS MODULATING GUT MICROBIOME

The microbial diversity takes shape during the first hour of birth, and the microbiota becomes quite stable in an adult as the immune system matures (Thursby and Juge, 2017). Therefore, the reorientation of gut microbiota is accomplished by periodic replacement of old phyla with new bacterial phyla (Ley et al., 2006; Thursby and Juge, 2017). Since intestinal microbiota performs multiple protective and metabolic functions, their significance in maintaining wellbeing of the host is getting appreciated. Different chemical, dietary, and immunological gradients along the gut influence the density and composition of microbiota (Hossain et al., 2015; Thursby and Juge, 2017). The combination of various factors, including age, illness, genotype, mode of delivery during birth, feeding methods, host immune system, geographical location, surgery, smoking,

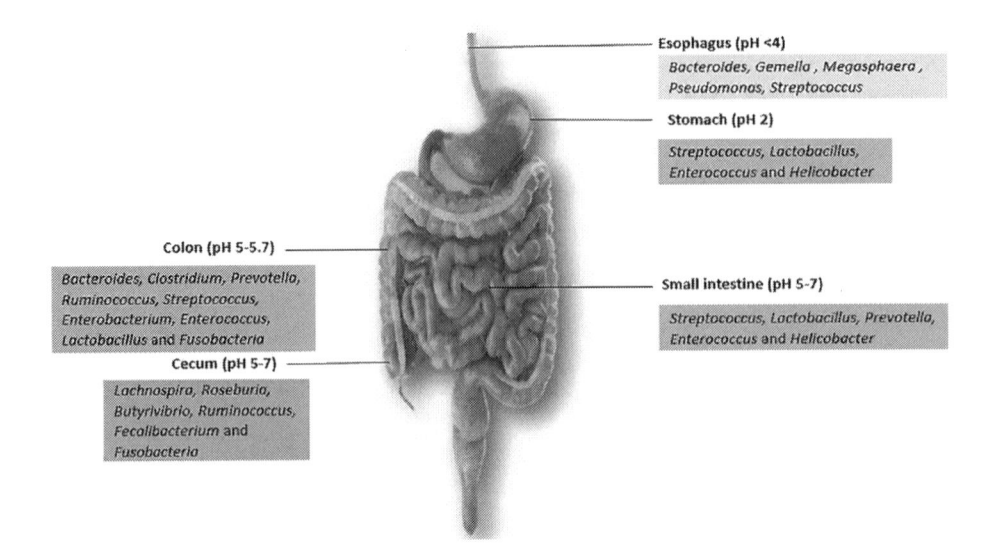

Figure 6.1   Microbiota in gastrointestinal tract (GIT) surviving to different pH conditions

depression, living arrangements (urban, rural), antibiotic treatment, mutation/lateral gene transfer, dietary composition, lifestyle, social interactions, and environmental exposure of various HMs (e.g., As, Hg, Cr, Cd, Pb, etc.), xenobiotic compounds, halogenated compounds (fluorides), metalloids (As) may cause chaotic shifts in intestinal microbiota (Figure 6.2), resulting in the

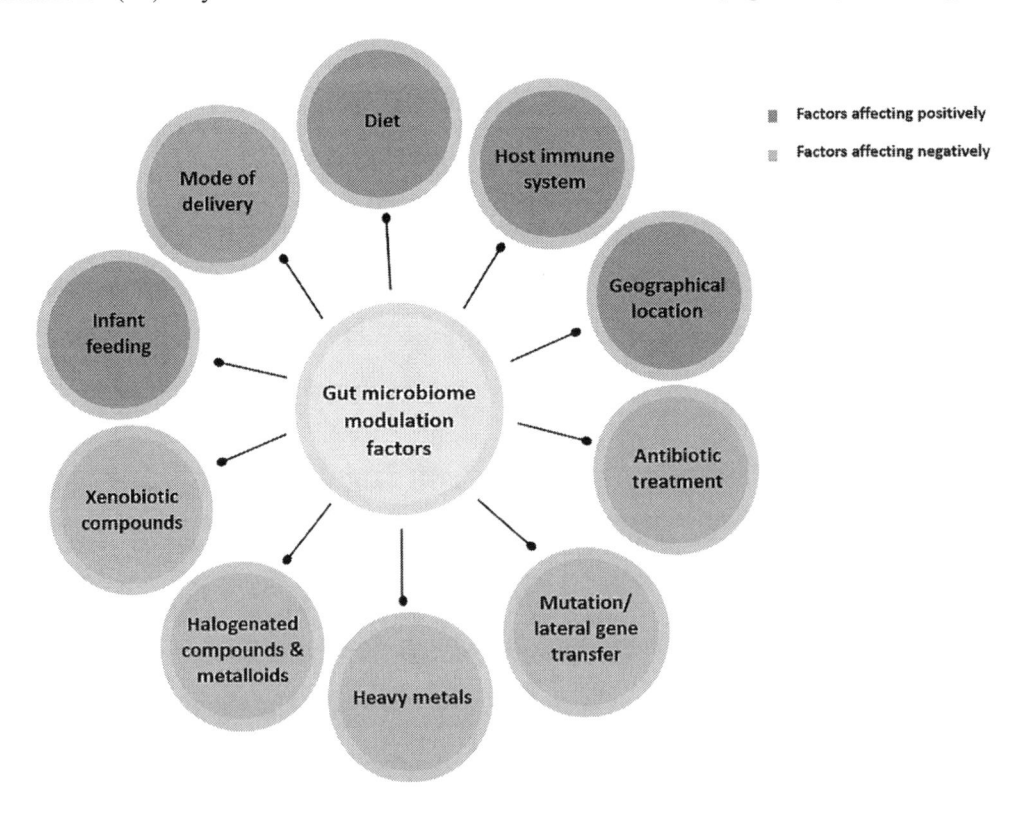

Figure 6.2   Factors regulating gut microbiome positively and negatively

proliferation of specific bacterial phyla (Tilg and Kaser, 2011; Claus et al., 2016; Rosenfeld, 2017; Thursby and Juge, 2017; Hasan and Yang, 2019). In this review, concern with HM exposure notably arsenic, mercury, chromium, cadmium, and lead on gut microbiota is discussed.

6.4 DYSBIOSIS AND EUBIOSIS

Dysbiosis and eubiosis are two very common terms in microbiome studies. Dysbiosis represents unhealthy microbiome in contrast to eubiosis that denotes healthy microbiome (Terminology; Figure 6.3) (Schippa and Conte, 2014; Hooks and O'Malleya, 2017). The most important characteristic of a healthy and balanced microbiome (eubiotic microbiome) is that it is flexible to withstand intrusion of different abiotic (diet, environmental factors) and biotic (microorganisms) stresses (Iebba et al., 2016). A healthy microbiome is essential for promoting health, and its perturbation (induction of dysbiotic state) leads to various disease states. A variety of microbiome-derived and host-derived factors are engaged in constantly maintaining the healthy state (eubiotic state) of the microbiome.

Intestinal immune system plays an integral part in maintaining the eubiotic state, and microbiome-derived products also stimulate the host immune system (example SCFA stimulates production of host antimicrobial peptide cathelicidin-3). In addition, microbiome product bacteriocin acts to selectively eliminate certain types of bacteria (Iebba et al., 2016). The gut microbiota plays a key role by symbiotic and mutualistic interactions and contributes to immune development including energy production, development of immune cells, food digestion, and homeostasis of epithelial cells, thereby maintaining a healthy GIT (Lu et al., 2015). However, when an imbalance or disruption arises between commensal and pathogenic bacteria in the intestine, owing to the ecological disturbance, dysbiosis ensues (Spor et al., 2011), resulting in innate immunity dysfunction (Belizário et al., 2018). Many diseases such as diabetes, obesity, cardiovascular diseases, hypertension, allergies, cancer, and inflammatory bowel disease have been related to specific bacterial dysbiosis (Claus et al., 2016; Li et al., 2017; Richardson et al., 2018).

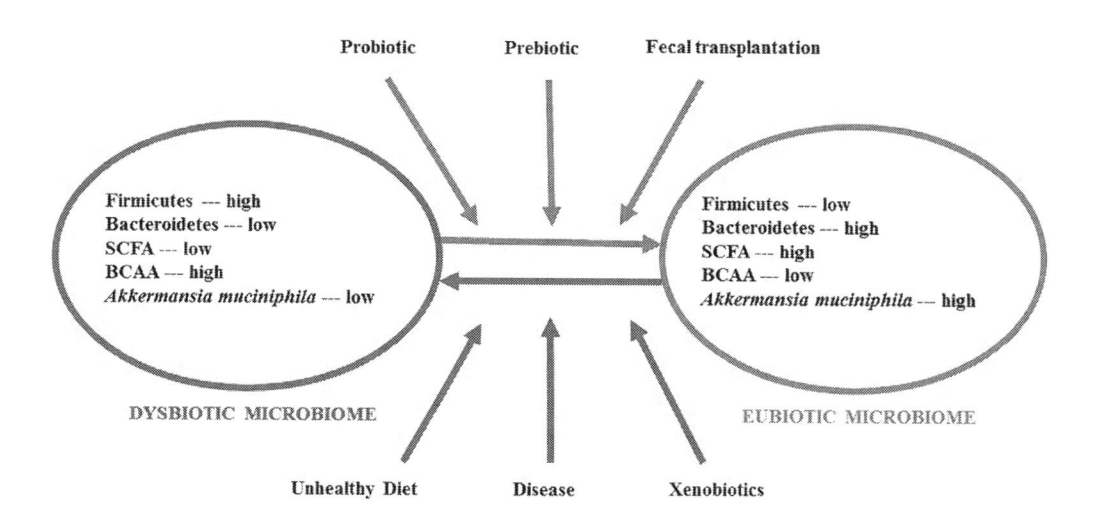

Figure 6.3 Factors influencing directional transition between dysbiotic and eubiotic microbiomes

6.5 SHORT-CHAIN FATTY ACIDS (SCFAs) AND BRANCHED-CHAIN AMINO ACIDS (BCAAs)

One of the major functions of gut microbiome that is beneficial to the host is fermentation of non-digestible carbohydrates such as resistant starches, cellulose, hemicellulose, gums, and pectins, resulting in the generation of SCFAs (acetic acid, propionic acid, butyric acid; Figure 6.4), which serve as energy source to colonic epithelial cells, which is characteristic of microbiome of type-1 enterotype (Arumugam et al., 2011; Terminology). In addition, SCFAs stimulate differentiation and proliferation of intestinal epithelial cells (Cummings and Macfarlane, 1991; Sharma and Devi, 2014; Hossain et al., 2015). SCFAs are considered as markers of eubiotic microbiome. Although proteins constitute an essential component of healthy diet, excessive consumption of protein and fat diet (for example, Western diet) leads to the development of dysbiotic microbiome (type-2 enterotype; Arumugam et al., 2011; Terminology) (Bifari et al., 2017; Hussain et al., 2019), especially in overweight individuals, which is characterized by excess production of BCAAs (valine, leucine, and isoleucine, Figure 6.4). Excess BCAA has been linked with insulin resistance and development of type-2 diabetes (Ley et al., 2006; Lynch and Adams, 2014; Karusheva et al., 2019). Enrichment of proteolytic bacterial population leads to enhanced degradation of gut mucin, and hence increased gut permeability and resultant inflammation due to translocation of lipopolysaccharide (LPS) and other bacterial products (Hossain et al., 2016; Hussain et al., 2019).

6.6 ENTEROTYPES

An enterotype is a classification of human gut microbiome which changes rapidly according to interventions and not constant for any individuals. Microbiome of enterotype 1 is present in people whose major calorie source is complex carbohydrate-rich diet (eubiotic microbiome; dominant bacteria is *Bacteroides*). Microbiome of enterotype 2 is present in people whose major energy source is protein and fat (dysbiotic microbiome; dominant bacteria - *Prevotella*). Microbiome of enterotype 3 exhibits mixed characteristics of both type-1 and type-2 enterotypes. (dominant bacteria, *Ruminococcus*). (Arumugam et al., 2011; Terminology).

6.7 *AKKERMANSIA MUCINIPHILA*

Abundance of the gut microbiome member *Akkermansia muciniphila* is considered as a marker organism of healthy microbiome. *A. muciniphila* is strictly anaerobic, non-motile, non-spore-forming, Gram-negative bacteria. It can grow only on mucin, utilizing it as the sole carbon and

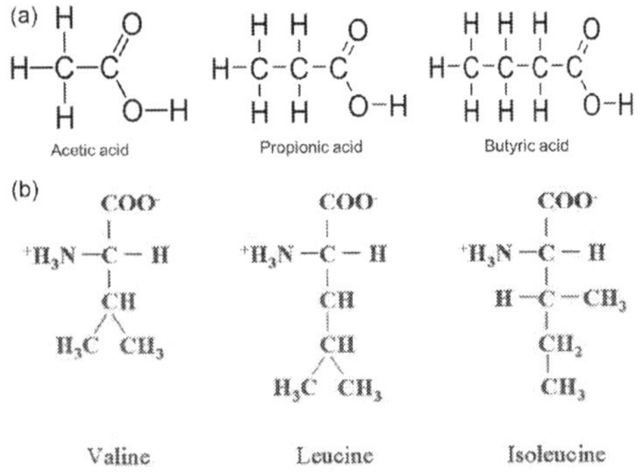

Figure 6.4　(a) Short-chain fatty acids (SCFA) and (b) branched-chain amino acids (BCAA)

nitrogen source (Derrien, 2004; de vos, 2017). It is present in the gut in the range of 3 to 5% in healthy (eubiotic) microbiome. Numerous studies have been carried on the probiotic characteristics of this microbiome member and it has linked to a variety of diseases and health conditions (generalized inflammation, type 2 diabetes, obesity, ulcerative colitis, and many others) in a positive way i.e. decrease in abundance increased disease burden (Everard et al., 2013; Caesar et al., 2015; Earley et al., 2019). In preclinical studies, *A. muciniphila* boosted response to immunotherapy for cancer. Considering the numerous ways, it can bring health benefits to humans; this microbiome member has been described as the next-generation beneficial microbe (Naito et al., 2018).

6.8 VIROME

Viruses inhabit every ecosystem on the earth including human body. Viruses are the most abundant biological entity on earth with an estimated number of 10^{31} (Suttle, 2007; Rohwer et al., 2009). Virome represents the total viruses present in and on our body (Güemes et al., 2016). Virome contains diverse collection of viruses infecting our body cells, bacteriophages infecting members of gut microbiome, non-pathogenic and pathogenic RNA and DNA viruses, which infect eukaryotic cells, prophages, giant viruses, and also several types of plant viruses (Raghavendra and Pullaiah, 2018). Microbiome including virome which affects the physiology of the host and can cause a variety of acute and chronic diseases. As existence and multiplication of viruses are interconnected with other living cells, virome can't be studied alone; a pan-genomic approach is needed. So, virome, as an intrinsic component of healthy gut microbiome, is being analyzed using NGS in context with bacteria and the host genome for gaining useful insights into the ever-evolving inter-kingdom interactions (Wylie et al., 2012; Garmaeva et al., 2019). As gut virome infects members of gut microbiome, it indirectly impacts our health and disease states by modulating the structure and function of the gut microbiome (Mukhopadhya et al., 2019).

6.9 EFFECTS OF HEAVY METALS ON GUT MICROBIOME

Due to anthropogenic activities, rapid industrialization, and urbanization, hazardous toxic pollutants are generated with concomitant release of HMs into the atmosphere, water, and soil (McConnell and Edwards, 2008; Thevenon et al., 2011; Monachese et al., 2012; Feng et al., 2019). Lead, chromium, mercury, arsenic, and cadmium have been shown to be the most common HM contaminants in our food, soil, air, and water ecosystems (Figure 6.5).

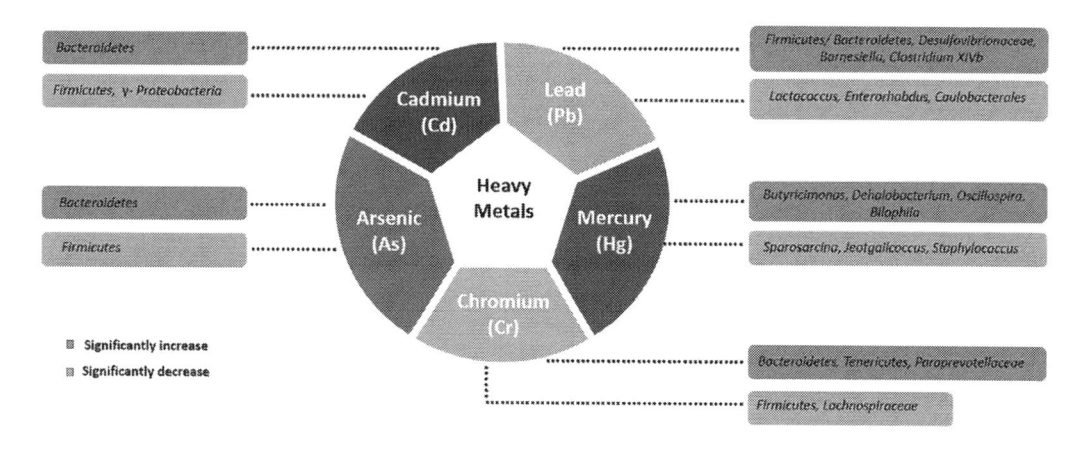

Figure 6.5 Effects of heavy metals on the composition of gut microbiota

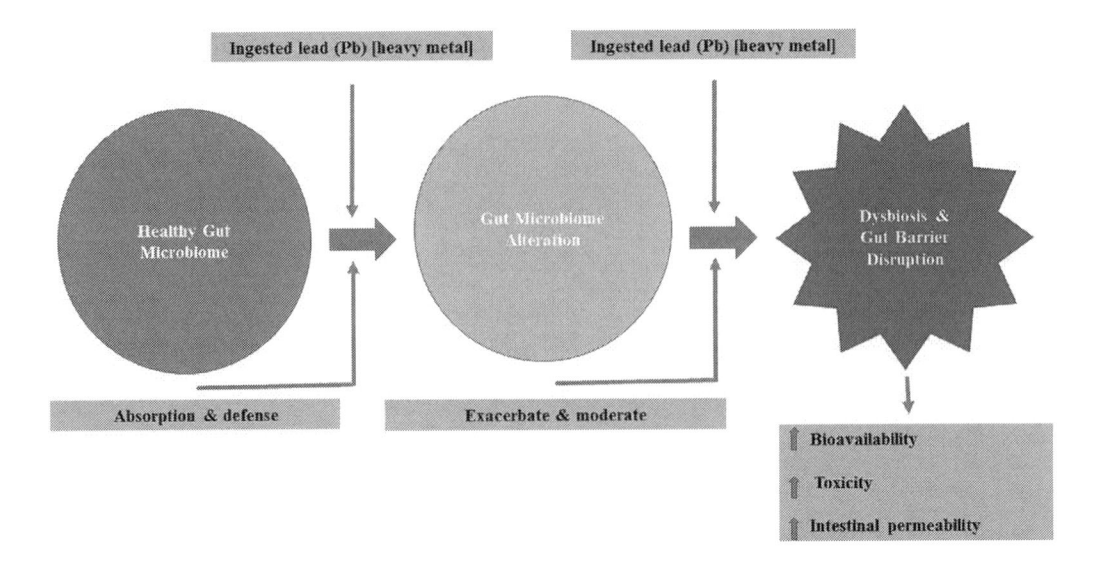

Figure 6.6 Microbiome-heavy metal interactions

Human gut can absorb about 40–60% of ingested metals across the intestinal barrier. Since the gut microbiota inhabits in a niche at the external environment and host epithelium interfaces, its composition is predominantly governed by exposure to HMs, making them an excellent biomarker for characterizing the local environment (Richardson et al., 2018). The notable factors affecting microbiome composition are intestinal location (e.g., stomach, jejunum, ileum, cecum, or colon; lumen vs. mucus layer), microenvironmental conditions (e.g., pH, oxygen availability, and redox potential) as well as the abundance of susceptible/resistant strains, and overall diversity and metabolic activity of the local microbial community (Assefa and Köhler, 2020). In mice model, exposure to Cd, As, and Cr (VI) decreases the ratio of *Firmicutes* to *Bacteroidetes* significantly in gut microbiome (Breton et al., 2013; Lu et al., 2014; Zhang et al., 2015; Wu et al., 2017; Feng et al., 2019). In case of mercury (2 mg/kg), Gram-positive bacteria *Sporosarcina*, *Jeotgailcoccus*, and *Staphylococcus* were significantly decreased, while Gram-negative bacteria *Butyricimonas*, *Dehalobacterium*, *Oscillospira*, and *Bilophila* were increased in the cecum of female mice significantly (Ruan et al., 2019) (Figure 6.5). Overall, the influence of HMs on gut microbiota is dose dependent. In the system, HMs may cause adverse effects including inflammation, oxidative stress, tissue damage, and gastrointestinal disorder (Lemire et al., 2013; Feng et al., 2019; Assefa and Köhler, 2020) (Figure 6.6).

6.10 MERCURY

Regarded as one of the most toxic HMs, mercury (Hg) is biomagnified in the aquatic food web as a result of industrial chemical processing or discarded electrical products, leading to high risk to human exposure (Lloyd et al., 2016; Bridges et al., 2018; Ruan et al., 2019). In the environment, there are three different forms of Hg: elemental Hg (Hg^0), inorganic mercury (Hg^+ and Hg^{2+}), and organic (such as methyl, ethyl, and phenyl) mercury (Monachese et al., 2012; Panel and Chain, 2012), mentioned hereafter in order of toxicity from lower to higher. Although all forms of Hg have adverse effects, methylmercury (MeHg) is the most common form of organic mercury as it bioaccumulates and biomagnifies in the food chain (Bridges et al., 2018; Ruan et al., 2019; Assefa and Köhler, 2020). The GIT absorbs about 95% of MeHg, which has biological half-life of 39–70 days depending on body burden, during when it can travel to reach brain and can affect central nervous system

(CNS) leading to neurotoxicity in animals and human (Li et al., 2017, 2019). Organic mercury is preferentially absorbed in the intestinal epithelium of fish, especially in large shark and tunas, hence posing a threat to the seafood lovers (Monachese et al., 2012). The European Commission fixed the maximum permissible levels of mercury in some fishery products at 0.5 mg/kg, while the limit was 1 mg/kg for muscle meat of fish, such as angler fish, Atlantic catfish, bonito, eel, grenadier, marlin, megrim, etc. (EC, 2006). Another metabolite, dimethyl mercury [$(CH_3)_2Hg$] appears to exert elevated toxic effects, a few ml of which is enough to cause death. Such ingested mercury can cause various digestive disorders, such as suppression of the production of trypsin, chymotrypsin, and pepsin. Other symptoms include abdominal pain, indigestion, inflammatory bowel disease, ulcers, and bloody diarrhea (Rice et al., 2014).

6.10.1 Microbiome Modulation by Mercury

The mercury-induced disruption of the gut flora was first reported in monkeys where the increase of both the mercury- and antibiotic-resistant bacteria was evidenced (Summers et al., 1993). Another study on *Porcellio scaber* (an isopod) also confirmed the presence of Hg-resistant bacteria because of exposure to Hg in the gut produced complete elimination of *Bacteriodetes* associated with decrease in *Actinobacteria*, *Betaproteobacteria*, and *Alphaproteobacteria* levels (Lapanje et al., 2010). Such an alteration of gut flora leads to disturbances of several host system such as CNS (Li et al., 2019) through gut-microbiome-brain axis (Figure 6.7) (Rosenfeld, 2017; Kim and Shin, 2018). Gut microbiome shift causes an increase in "gut leakiness" which plays a vital role in many gut-microbiome-brain comorbidity disorders (Rosenfeld, 2017).

Gut microbiota regulates the brain activity by forming the microbiota-gut-brain (MGB) axis, composed of the CNS, the enteric nervous system, and the digestive system. The microbiota establishes a bidirectional interaction with major parts of the CNS that are supposed to function on the hypothalamic–pituitary–adrenal (HPA) axis and the vagus nerve by producing bacterial metabolites via tryptophan metabolism (Rhee et al., 2009). Multiple stress such as anxiety, depression

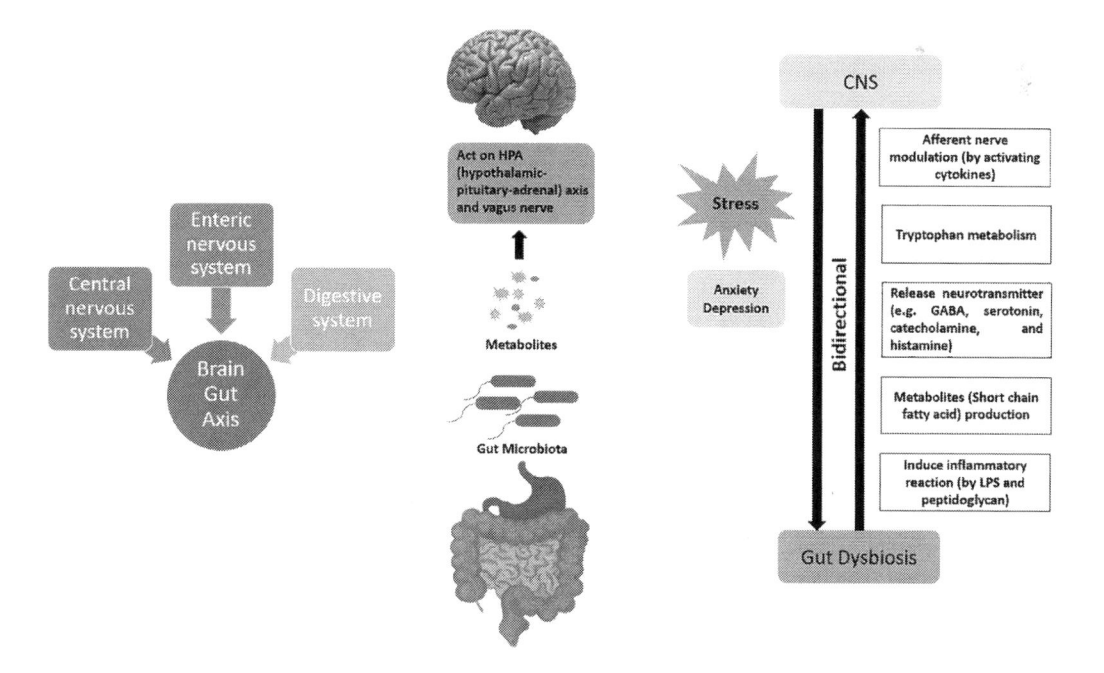

Figure 6.7 Microbiota-gut-brain (MGB) axis

influence gut microbiome and cause gut dysbiosis that further affect CNS by inducing inflammatory reaction, metabolites production, afferent nerve modulation, and release of neurotransmitter (Carabotti et al., 2015) (Figure 6.7).

6.11 ARSENIC

Arsenic (As), a colorless and tasteless metalloid element, is the most widespread environmental toxicant in the world (Monachese et al., 2012; Coryell et al., 2019). It is ubiquitous in the environment, originated primarily from the natural weathering of the earth's crust (Coryell et al., 2019). In the environment, it is available either in trivalent (As^{3+}) or pentavalent (As^{5+}) forms in both organic and inorganic compounds. While inorganic As is more toxic than organic one (Claus et al., 2016), the toxicity of arsenite (As^{3+}) happens to be higher than that of arsenate (As^{5+}) (Gokulan et al., 2018).

Arsenic exposure is naturally connected to water, soil, food, airborne particulates, etc., having a concentration of 1–2 µg/L present in natural waters. Nonetheless, concentrations can be significantly elevated in waters, particularly groundwater, where there are sulfide mineral deposits and sedimentary deposits derived from volcanic rocks (WHO, 2017). Consequently, it becomes a major health concern as it causes substantial risk of disease in millions of people worldwide, especially in South and East Asia (Ahsan et al., 2000; Monachese et al., 2012; Lu et al., 2014; Dong et al., 2017). The regulatory standard for the arsenic level in drinking water is set at 10 µg/L or 10 ppb (USEPA, 2002; WHO, 2017).

Arsenic contamination via food, water, soil, and dust, and inhalation of atmospheric arsenic causes the development of different acute and chronic toxicity resulting in various illnesses (Monachese et al., 2012; Coryell et al., 2019). The acute toxicity includes bloody urine, GI discomfort, diarrhea, headaches, vomiting, convulsions, coma, and death; whereas the chronic toxicity causes skin lesions, diabetes, blisters, blackfoot disease, organ failure/damage, metabolic dysregulation, diabetes mellitus, cardiovascular disease, pregnancy complications, and neurological symptoms (Lu et al., 2014; Richardson et al., 2018; Coryell et al., 2019). The International Agency for Research on Cancer (IARC) has listed arsenic as group I carcinogens which causes various cancers such as lung, skin, bladder, etc. (Richardson et al., 2018; Coryell et al., 2019). Interestingly, perturbation of gut microbiome is believed to be associated with arsenic-mediated human illnesses (Monachese et al., 2012).

6.11.1 Perturbation of the Gut Microbiome by Arsenic

Since the primary route of human exposure to arsenic in our body is through ingestion of contaminated drinking water or food, intestinal microbiota are the most vulnerable to arsenic-mediated toxicity and were found to play a role in arsenic-related diseases (Monachese et al., 2012; Dong et al., 2017; Roggenbeck et al., 2019). In fact, composition of the gut microbial community is determined by arsenic dose and duration of exposure (Coryell et al., 2019), which was demonstrated in both lab animals and humans.

The normal murine colon microbiota is a complex microbial community dominated primarily by Firmicutes (e.g., *Clostridium, Coprococcus, Ruminococcus, Lactobacillus*) and Bacteroidetes (e.g., *Bacteroides, Parabacteroides*) but also harbors populations of *Deferribacter, Acinetobacter, Enterobacter*, and *Bifidobacterium*. A significant reduction of Firmicutes was shown following an exposure of arsenic through drinking water in mice (Lu et al., 2014, 2015; Richardson et al., 2018). Other studies, however, found increased abundance of *Firmicutes* associated with decreased population of *Bacteroidetes* (Guo et al., 2014). In human, arsenic exposure in US infants showed a positive association of *Firmicutes* phylum and a negative association of *Bacteroides* and *Bifidobacterium* (Hoen et al., 2018). In Bangladeshi children (4–6 years of age), high levels of arsenic in-home drinking water were associated with a greater abundance of Gammaproteobacteria in the microbiome, more specifically, members of the *Enterobacteriaceae* family (Dong et al., 2017). However, in the US infants, there was a negative interaction between *Enterobacteriaceae* and urinary arsenic (Hoen et al., 2018).

6.11.2 Influence of Probiotics and Prebiotics on Gut Microbiome Dysbiosis-Induced by Arsenic and Mercury

Probiotic in food is a safe and inexpensive way to fight exposure to toxic HMs. Both prebiotic and probiotic preserve intestinal barrier integrity by promoting commensal bacterial growth. *Bifidobacterium* is an important probiotic which is essential for intestinal microbial homeostasis, gut barrier, and LPS reduction (Li et al., 2017; Belizário et al., 2018). Other probiotic especially lactic acid bacteria can contribute to the elimination of HM, thanks to their high affinity for HM. Yogurt containing *Lactobacillus rhamnosus* provided beneficial effects on pregnant women and children by reducing bioaccumulation of mercury and arsenic (Gokulan et al., 2018). Study on Tanzanian pregnant women showed probiotic yogurt consumption had a protective effect against arsenic (Bisanz et al., 2014). In vitro, *Lactobacillus acidophilus* and *Lactobacillus crispatus* DSM20584 were able to remove arsenic (50–1,000 ppb) from contaminated water within 4h of exposure in a concentration-dependent manner (Monachese et al., 2012).

Prebiotics have a protective effect against HMs. Dietary supplementation of galactooligosaccharide (GOS) can be considered as a potentially protective prebiotic that can benefit the host by regulating the microenvironment of the gut to maintain gut microbiota homeostasis (Zhai et al., 2019). Prebiotic such as oligofructose has a protective role against arsenic-induced alteration in the gastrointestinal microbiome (Starr, 2011). The combination of prebiotic and probiotic could therefore play important roles in controlling the modulation of gut microbiome by HMs.

6.12 CHROMIUM

Chromium (Cr) occurs in the environment primarily in two valence states, trivalent chromium (Cr III) and hexavalent chromium (Cr VI). Exposure may occur from natural or industrial sources of chromium. Chromium III is much less toxic than chromium VI, which is considered as a potent carcinogen. Humans are exposed to chromium either via ingestion of chromium-containing food (non-occupational exposure) or via inhalation (occupational exposure). Presence of Cr(VI) in the range of 1–10ppm is safe as it is efficiently converted to less toxic form by the gut microbiome (Sun et al., 2015). The body can detoxify some amount of chromium VI to chromium III. The respiratory tract is the major target organ for chromium VI toxicity, for acute (short-term) and chronic (long-term) inhalation exposures (EPA, 1998).

6.12.1 Chromium and Human Health

Although HMs are generally regarded as toxicants causing detrimental effects on human health, there are exceptions. Chromium, which is regarded as an environmental toxicant, has some health benefits at low levels. Cr(III) compounds are used as micronutrients and nutritional supplements and have been demonstrated to exhibit a significant number of health benefits in animals and humans. It facilities the action of the hormone insulin; patients with Cr deficiency may develop diabetes, which can be corrected by Cr supplementations. Cr is used also as a supplement for appetite reduction in weight loss efforts (Balik et al., 2007). In rat model, experimental chromium deficiency has been shown to lead to high serum cholesterol levels and to the formation of atherosclerotic plaques (Swaroop et al., 2019).

6.12.2 Effects of Chromium on Gut Microbiome

Gut microbiome is remarkably capable of coping with exposure to chromium, either acute or chronic, and this contributes significantly to offering protection against ingested load (Upreti et al., 2008). Use of germ-free animals and gnotobiotic animals provided useful data on effect of HMs on microbiome, in general, and also on the role of specific microbiome member (Richardson et al., 2018). In an interesting study, gut microbiome member *Lactobacillus plantarum* strain TW1-1 was demonstrated to be capable of carrying out "gut remediation" i.e. binding and removal of chromium from gut. In a mice model, the authors showed that this strain was able to bind chromium and promoted its excretion in the feces. This excretion was accompanied by reduction in Cr accumulation in different tissues. Metagenomic studies by 16S rRNA gene sequence analysis and NGS of DNA obtained from the stool samples showed that this probiotic strain was capable of shifting Cr-induced dysbiotic microbiome towards eubiosis i.e. 62% (49 out 67) of the operational taxonomic units altered by Cr were restored by *L. plantarum* strain TW1-1 (Wu et al., 2017). Finding of this study shows that 'selective bacteria' have the ability to carry out the tasks of "gut remediation" and also restoration of eubiotic microbiome.

6.12.3 Reduction of Cr(VI) to Cr(III) by Gut Microbiome

Gut microbiome can efficiently reduce Cr(VI) to Cr(III), thus converting a highly toxic form to a relatively less toxic form. Microbiome members engage diverse mechanism to carryout reduction of Cr(VI) to Cr(III). High amounts of glutathione (GSH), synthesized by certain gut microbiome members can efficiently reduce Cr(VI); moreover, bacterial enzymes, nitroreductases can also reduce Cr(VI) to Cr(III) (Upreti et al., 2008; Wu et al., 2017). The enhanced toxicity of Cr(VI) is attributed to its high solubility and rapid permeability which makes it capable of rapidly interacting with the cellular macromolecules. On the other hand, Cr(III) is less soluble and is less efficient in permeating cell membranes. Reduction of Cr(VI) to Cr(III) by gut microbiome thus enables sequestration and subsequent expulsion from the body with feces (Wu et al., 2017). Fecal bacterial isolates sequestrated 3.8 ± 1.7 µg Cr(VI)/10^9 bacteria, which amounts to elimination of 11–24 mg Cr(VI) daily with feces (Upreti et al., 2008; Sun et al., 2015).

6.13 CADMIUM

Cadmium (Cd) is present in high concentration in air, soil, and water, and it is listed in the top ten environmental HM pollutants of concern by environmental health agencies (ATSDR, 2015; Skolarczyk et al., 2018). It is established as an environmental pollutant that negatively impacts human health.

6.13.1 Cadmium and Human Health

Cadmium can enter our body system through consumption of eating cadmium-contaminated food, drinking cadmium-contaminated water, and by breathing cadmium-contaminated air. However, exposure via food and drink is not of great concern. Traces of cadmium generally present in tobacco plants, and cigarette smoke is the source of common exposure. Smokers have about twice as much cadmium in their bodies as nonsmokers (Telisman et al., 1997). Many harmful health effects can result from exposure to cadmium. It has a long half-life, in the range of 10–30 years in the human kidney and low, chronic levels have been linked with kidney dysfunction. Other health conditions associated with Cd exposure are deregulation of blood pressure, diabetes, and osteoporosis (Messner and Bernhard, 2010; Jarup and Akesson, 2009; Johri et al., 2010). Consumption of food and drinks containing high levels of Cd can induce vomiting and diarrhea. Inhalation of high doses of Cd can cause irritation and damage to the lungs and can lead to death. However, this is a rare scenario and chronic low dose exposure, which is more common is of the greatest concern to us (Liu et al., 2019).

6.13.2 Effects of Cadmium on Gut Microbiome

Studies have shown that Cd exposure can induce dysbiotic state by altering both the structure and function of gut microbiome in animal models exposed to sublethal dose when administered by drinking water. Cd induces decrease in the number and diversity of the members of the gut microbiota as revealed by 16S rRNA gene sequencing analysis. Moreover, a significant decrease in *Bacteroidetes* (marker of eubiotic microbiome) and probiotic bacterial species such as *Lactobacillus* spp. and *Bifidobacter* spp. was noted. Firmicutes (marker of dysbiotic microbiome), however, remained unaltered (Liu et al., 2014). Cd-induced dysbiotic state of microbiome exhibits typical characteristics of a dysbiotic microbiome as induced by other agents such as high-fat diet. Increase in gut permeability results in increased serum LPS levels leading to transcriptional upregulation of LPS-inducible genes encoding for inflammatory cytokine culminating in a dysbiotic state (Fazeli et al., 2011; Breton et al., 2013; Zhang, 2015). Moreover, Cd exposure also induced changes in gut metabolome as significant reduction in SCFA production was noted (Liu et al., 2014). Probiotics such as *L. plantarum, L. rhamnosus, Propionibacterium freudenreichii,* and *Shermanii* have the potential to detoxify lead and cadmium after binding to them (Feng et al., 2019). In addition, Cd exposure also induced phenotypic changes in the gut microbiota with regard to resistance to the metal. Overall, gram-positive bacteria were more sensitive to cadmium compared to the gram-negative bacteria possibly because of their differential capacity to uptake the metal (Fazeli et al., 2011).

6.14 LEAD

Lead (Pb) has been of varied use both in ancient and modern civilizations. It has been increasingly used in the twentieth century (Lester, 2020). This has led to extensive human exposure to Pb-induced toxicity that is quite prominent in HM poisoning, further complicated by its ability to stay in the environment for a prolonged period (WHO, 2015), thereby making it a public health concern worldwide (Gomez *et al.,* 2019).

6.14.1 Lead and Human Health

The toxicity of lead has been well recognized in a multiple systems and organs (Bisanz et al., 2014) because of its high bioavailability in the GIT and permeability through the blood brain barrier (BBB). Exposure to Pb can impair different physiological functions in the body including, reproductive system, digestive system, nervous system, and respiratory system. Acute Pb poisoning with high mortality rate could involve compromised red blood cell (RBC) functions, functional damage to peripheral nervous system (PNS) or central nervous system (CNS) and/or prominent renal dysfunction (Grant, 2008; Lester, 2020).

Pb is thought to have no physiological role (Lester, 2020). A concentration of <10 µg/dl in blood, Pb could cause hearing loss (NTP, 2012), reduced IQ scores, social anomalies, and greater hazard of Attention Deficit Hyperactivity Disorder (ADHD) (Naicker *et al.,* 2012; WHO, 2015). At the level ≤10 µg/dl, it can end with impaired neurological (Lanphear et al., 2005) and physical (Kaji and Nishi, 2006) growth in children, with adverse concerns in adulthood. At ≤30 µg/dl, Pb upsets Vitamin D metabolism and calcium levels, resulting in improper growth and bone expansion (WHO, 2015). Gastro-intestinal (GIT) anomalies are likely to occur at blood Pb levels ≥50 µg/ dl (WHO, 2010). Acute blood Pb levels of 70-100 µg/dl can produce severe renal defects. The symptoms are usually reversible; however, prolonged Pb exposures may result in irreversible chronic effects. At very high blood lead concentration (≥100 µg/dl), symptoms such as convulsions, muscular paralysis coma and eventually death can ensue.

6.14.2 Lead and Gut Microbiome

Pb led negligible but precise alterations in gut bacterial commensal communities. The alterations were at both family and genus levels. There was particularly small numbers of *Lachnospiraceae* and great numbers of *Lactobacillaceae* and *Erysipelotrichaceacae* (primarily due to fluctuations in *Turicibacter* spp.) (Breton et al., 2013).Recently, at multi-omics research demonstrated that Pb exposure changed the gut microbiota development with parallel effect on several metabolic pathways, such as detoxification and oxidative stress resulting in gut microbiome toxicity (Figure 6.6) (Gao et al., 2017; Assefa and Köhler, 2020). Another study was performed for influence of up to 8 weeks of oral Pb ingestion on the conformation of the murine intestinal microbiome. The alterations were at both family and genus levels. There was particularly small numbers of *Lachnospiraceae* and great numbers of *Lactobacillaceae* and *Erysipelotrichaceacae* (primarily due to fluctuations in *Turicibacter spp)* (Breton et al., 2013).

6.14.3 Role of Probiotics in Lead Toxicity

The elevated blood Pb levels were associated with increases in Succinivibrionaceae and Gammaproteobacteria with comparative great quantity in stool. Locally produced nutritious probiotic food denotes an inexpensive way to overcome toxic metal exposures in developing countries, particularly in proximity to mining cites (Bisanz et al., 2014). Lactobacilli species can play the role of detoxification. Both living and dead *L. plantarum* CCFM8661 offered a significant protective effect against Pb-induced toxicity in mice (Tian et al., 2012).

6.15 CONCLUSIONS AND FUTURE PERSPECTIVES

We have witnessed remarkable progress in the understanding of the microbiome and its enormous health impacts in recent years. Ever-increasing realization of health significance of eubiotic and dysbiotic microbiomes, and the factors which influence their transition from one state to another have prompted synthetic biologists to explore ways to engineer microbiome both at personalized and generalized levels. We expect to see translation of this novel approaches in the near future. HMs impact our microbiome and also our metabolome. With regard to the importance of bidirectional interaction between HMs and microbiome, we have the necessary experimental tools required for in-depth analysis. The gut microbiome cumulatively possesses extremely complex and diverse metabolic power to sequester and detoxify xenobiotics including HMs. The use of probiotics for "gut remediation" holds great promise in cases of both acute and chronic HM toxicity. Most countries have legislative policies for the presence and exposure of HMs as well as strategies for remediation and treatment. Soil and water monitoring is often done to prevent excessive consumption, but many of these initiatives and innovations are not readily available in developing countries (Ahsan et al., 2000; Levin et al., 2008; Monacheseet al., 2012). Ultimately people are globally vulnerable, so new tactics need to be sought to mitigate the detrimental effects of HM accumulation.

REFERENCES

Ahsan, H., Perrin, M., Rahman, A. et al. (2000). Associations between drinking water and urinary arsenic levels and skin lesions in Bangladesh. *Journal of Occupational and Environmental Medicine*. 42: 1195–1201.

Arumugam, M., Raes, J.E., Pelletier, D. et al. (2011). Enterotypes of the human gut microbiome. *Nature* 473: 174–180.

Assefa, S. and Köhler, G. (2020) Intestinal microbiome and metal toxicity. *Current Opinion in Toxicology.* 19: 21–27.ATSDR (2007). *Agency for Toxic Substances and Disease Registry. The Toxicological Profile for Lead.* Atlanta, GA: U.S. Department of Health and Human Services. Available from: http://www. atsdr.cdc.gov/ToxProfiles/tp13.pdf.

ATSDR (2015). Agency for toxic substances and disease registry. Secondary Agency for Toxic Substances and Disease Registry. http://www.atsdr.cdc.gov/.

Balik, E.M., Tatsioni, A., Lichtenstein, H., Lau, J. and Pittas, A. (2007). Effect of chromium supplementation on glucose metabolism and lipids. *Diabetes Care.* 30: 2154–2163.

Belizário, J.E., Faintuch, J. and Garay-Malpartida, M. (2018). New frontiers for treatment of metabolic diseases. *Mediators of Inflammation.* 2037838. doi: 10.1155/2018/2037838. eCollection.

Bifari, F., Ruocco, C., Decimo, I. et al. (2017). Amino acid supplements and metabolic health: A potential interplay between intestinal microbiota and systems control. *Genes and Nutrition.* 12: 27 doi: 10.1186/ s12263-017-0582-2.

Bisanz, J.E., Enos, M.K., Mwanga, J.R. et al. (2014). Randomized open-label pilot study of the influence of probiotics and the gut microbiome on toxic metal levels in Tanzanian pregnant women and school children. *mBio,* 5: 1–7. doi: 10.1128/mBio.01580-14.

Brendan A.D., Monachese, M., Trinder, M. et al. (2019). Immobilization of cadmium and lead by Lactobacillus rhamnosus GR-1 mitigates apical-to-basolateral heavy metal translocation in a Caco-2 model of the intestinal epithelium. *Gut Microbes.* 10: 321–333.

Breton, J., Massart, S., Vandamme, P. et al. (2013). Ecotoxicology inside the gut: Impact of heavy metals on the mouse microbiome. *BMC Pharmacology & Toxicology,* 14: 62. doi: 10.1186/2050-6511-14-62.

Bridges, K.N., Zhang, Y., Curran, T.E. et al. (2018). Alterations to the intestinal microbiome and metabolome of *Pimephales promelas* and *Mus musculus* following exposure to dietary methylmercury. *Environmental Science and Technology.* 52: 8774–8784.

Caesar, R., Tremaroli, V. Kovatcheva-Datchary, P. (2015). Crosstalk between gut microbiota and dietary lipids aggravates WAT inflammation through TLR signaling. *Cell Metabolism.* 22: 658–668.

Carabotti, M., Scirocco, A., Maselli, M.A. and Severi, C. (2015). The gut-brain axis: Interactions between enteric microbiota, central and enteric nervous systems. *Annals of Gastroenterology.* 28: 203–209.

Claus, S.P., Guillou, H., and Ellero-Simatos, S. (2016). The gut microbiota: A major player in the toxicity of environmental pollutants? *NPJ Biofilms and Microbiomes.* 2: 1–12. doi: 10.1038/npjbiofilms.2016.3.

Coryell, M., Roggenbeck, B.A., and Walk, S.T. (2019). The human gut microbiome's influence on arsenic toxicity. *Current Pharmacology Reports.* 5: 491–504.

Cummings, J. H. and G. T. Macfarlane. 1991. The control and consequences of bacterial fermentation in the human colon. *Journal of Applied Bacteriology.* 70: 443–459.

Derrien, M. (2004). *Akkermansia muciniphila,* a human intestinal mucin-degrading bacterium. *International Journal of Systematic and Evolutionary Microbiology.* 54: 1469–1476.

de Vos, W.M. (2017). Microbe Profile: *Akkermansia muciniphila*: A conserved intestinal symbiont that acts as the gatekeeper of our mucosa. *Microbiology.* 1635: 646–648.

Dong, X., Shulzhenko, N., Lemaitre, J. et al. (2017). Arsenic exposure and intestinal microbiota in children from Sirajdikhan, Bangladesh. *PLoS ONE,* 12 (12), 1–20. doi: 10.1371/journal.pone.0188487.

Earley, H., Lennon, G., Balfe, Á. et al. (2019). The abundance of Akkermansia muciniphila and its relationship with sulphated colonic mucins in health and ulcerative colitis. *Science Report* 9: 15683. doi: 10.1038./ s41598-019-51878-3.

EC (2006). Commission Regulation (EC) No 1881/2006 of 19 December 2006 setting maximum levels for certain contaminants in foodstuffs. p. 5–24. ELI: http://data.europa.eu/eli/reg/2006/1881/oj

EPA (1998). *U.S. Environmental Protection Agency. Toxicological review of trivalent chromium. National Center for Environmental Assessment.* Washington, DC: Office of Research and Development.

Everard, A., Belzer, C., Geurts, L. et al. (2013). Cross-talk between *Akkermansia muciniphila* and intestinal epithelium controls diet-induced obesity. *Proceeding of the National Academy of Sciences, USA.* 110: 9066–9071.

Fazeli, M., Hassanzadeh, P. and Alaei, S. (2011). Cadmium chloride exhibits a profound toxic effect on bacterial microflora of the mice gastrointestinal tract. *Human and Experimental. Toxicology.* 30: 152–159.

Feng, P., Ye, Z., Kakade, A., et al (2019). A review on gut remediation of selected environmental contaminants: Possible roles of probiotics and gut microbiota. *Nutrients,* 11: doi: 10.3390/nu11010022.

Gao, B., Chi, L. Mahbub, R. et al. (2017) Multi-omics reveals that lead exposure disturbs gut microbiome development, key metabolites, and metabolic pathways. *Chemical Research in Toxicology.* 30: 996–1005.

Garmaeva, S., Sinha, T., Kurilshikov, A. et al. (2019). Studying the gut virome in the metagenomic era: Challenges and perspectives. *BMC Biology* 17. doi:10.1186/s12915-019-0704-y.

Gokulan, K., Arnold, M.G., Jensen, J. et al. (2018). Exposure to arsenite in CD-1 mice during juvenile and adult stages: Effects on intestinal microbiota and gut-associated immune status. *mBio,* 9: 1–17. doi: 10.1128/mBio.01418-18.

Gomez, H.F., Borgialli, D.A., Sharman M (2019). Analysis of blood lead levels of young children in Flint, Michigan before and during the 18-month switch to Flint River water. *Clinical Toxicology.* 57: 790–797.

Grant, L.D. (2008). Lead and Compounds. In Environmental Toxicants, M. Lippmann (Ed.). doi:10.1002/9780470442890.ch20

Levin, R., Brown, M.J., Kashtock, M.E. et al. (2008). Lead exposures in U.S. children, 2008: Implications for prevention. *Environmental Health Perspectives,* 116: 1285–1293.

Güemes, A.G., Youle, M., Cantú, V.A. et al. (2016). Viruses as winners in the game of life. *Annual Reviews in Virology.* 3: 197–214. doi: 10.1146/annurev-virology-100114-054952.

Guo, X., Liu, S., Wang, Z. et al. (2014). Metagenomic profiles and antibiotic resistance genes in gut microbiota of mice exposed to arsenic and iron. *Chemosphere.* 112: 1–8.

Hasan, N. and Yang, H. (2019). Factors affecting the composition of the gut microbiota and its modulation. *Peer J.* 1–31. doi: 10.7717/peerj.7502.

Hoen, A.G., Madan, J. C., Li, Z. et al. (2018). Sex-specific associations of infants' gut microbiome with arsenic exposure in a US population. *Scientific Reports.* 8: 1–10. doi: 10.1038/s41598-018-30581-9.

Hooks, K. B. and O'Malley, M. A. (2017). Dysbiosis and its discontents. *mBio* 8: e01492.

Hossain, A., Akhtar, S. and Kabir, Y. (2015). Diet, microbiome and human health. In: *Health benefits of fermented foods and beverages.*edited by Jyoti Prakash Tamang, Boca Raton, FL: Taylor and Francis Publications.

Hossain, A., Menezes, G., Al-Mogbel, M. and Ashankyty, I. (2016). Modulation of environmental toxicants and therapeutic agents by gut microbiome. In: *Food Toxicology.* edited by Debasis Bagchi, Boca Raton, FL: Taylor and Francis Publications.

Hussain, M., Ijaz, M. U., Ahmad, M. I. et al. (2019). Meat proteins in a high-fat diet have a substantial impact on intestinal barriers through mucus layer and tight junction protein suppression in C57BL/6J mice. *Food and Nutrition.* 10: 6903–6914.

Iebba, V., Totino, V., Gagliardi, A. et al. (2016). Eubiosis and dysbiosis: The two sides of the microbiota. *New Microbiologica.* 39: 1–12.

Jarup, L. and Akesson, A. (2009). Current status of cadmium as an environmental health problem. *Toxicology and Applied Pharmacology.* 238: 201–208.

Johri, N., Jacquillet, G. and Unwin, R. (2010). Heavy metal poisoning: The effects of cadmium on the kidney. *Biometals.* 23: 783–792.

Kaji, M. and Nishi, Y. (2006). Lead and growth. *Clinical Pediatric Endocrinology* 15: 123–128.

Karusheva, Y., Koessler, T., Strassburger, K. et al. (2019). Short-term dietary reduction of branched-chain amino acids reduces meal-induced insulin secretion and modifies microbiome composition in type 2 diabetes: A randomized controlled crossover trial, *The American Journal of Clinical Nutrition,* 110: 1098–1107.

Kim, Y.-K. and Shin, C. (2018). The microbiota-gut-brain axis in neuropsychiatric disorders: Pathophysiological mechanisms and novel treatments. *Current Neuropharmacology.* 15: 559–573.

Lanphear, B. P., Hornung, R. and Khoury, J. (2005) Low-level environmental lead exposure and children's intellectual function: An international pooled analysis. *Environmental Health Perspectives* 113: 894–899.

Lapanje, A., Zrimec, A., Drobne, D., and Rupnik, M. (2010). Long-term Hg pollution-induced structural shifts of bacterial community in the terrestrial isopod (*Porcellio scaber*) gut. *Environmental Pollution.* 158: 3186–3193.

Lemire, J.A., Harrison, J.J., and Turner, R.J. (2013). Antimicrobial activity of metals: Mechanisms, molecular targets and applications. *Nature Reviews Microbiology,* 11: 371–384.

Lester G. (2020). "Lead and Compounds." Environmental Toxicants: Human exposures and their health effects, by Lippmann, M. and Leikauf, G. D. Wiley, 2020, pp. 627–663.

Ley, R. E., Turnbaugh, P. J., Klein, S. and Gordon, J. I. (2006). Microbial ecology: Human gut microbes associated with obesity. *Nature*. 444: 1022–1033.

Li, H., Lin, X., Zhao, J. et al. (2019). Intestinal methylation and demethylation of mercury. *Bulletin of Environmental Contamination and Toxicology*. 102: 597–604.

Li, J., Zhao, F., Wang, Y. et al. (2017). Gut microbiota dysbiosis contributes to the development of hypertension. *Microbiome*. 5: 1–19.

Liu, Q., Zhang, R., Wang, X. et al. (2019). Effects of sub-chronic, low-dose cadmium exposure on kidney damage and potential mechanisms. *Annals in Translational Medicine*. 7: 177. doi:10.21037/atm.2019.03.66.

Liu, Y., Li, Y., Liu, K. and Shen, J. (2014). Exposing to cadmium stress cause profound toxic effect on microbiota of the mice intestinal tract. *PLoS One*. 9(2):e85323. doi:10.1371/journal.pone.0085323.

Lloyd, N.A., Janssen, S.E., Reinfelder, J.R., and Barkay, T. (2016). Co-selection of mercury and multiple antibiotic resistances in bacteria exposed to mercury in the *Fundulus heteroclitus* gut microbiome. *Current Microbiology*. 73: 834–842.

Lu, K., Abo, R.P., Schlieper, K.A. et al. (2014). Arsenic exposure perturbs the gut microbiome and its metabolic profile in mice: An integrated metagenomics and metabolomics analysis. *Environmental Health Perspectives*. 122: 284–291.

Lu, K., Mahbub, R., and Fox, J.G. (2015). Xenobiotics: Interaction with the intestinal microflora. *Institute for Laboratory Animal Research Journal*. 56: 218–227.

Lynch, C. L. and Adams, S. H. (2014). Branched-chain amino acids in metabolic signalling and insulin resistance. *Nature Reviews Endocrinology*. 10: 723–736.

McConnell, J.R. and Edwards, R. (2008). Coal burning leaves toxic heavy metal legacy in the Arctic. *Proceedings of the National Academy of Sciences of the United States of America*, 105: 12140–12144.

Messner, B. and Bernhard, D. (2010). Cadmium and cardiovascular diseases: Cell biology, pathophysiology, and epidemiological relevance. *Biometals*. 23: 811–822.

Monachese, M., Burton, J.P., and Reid, G. (2012). Bioremediation and tolerance of humans to heavy metals through microbial processes: A potential role for probiotics? *Applied and Environmental Microbiology*. 78: 6397–6404.

Mukhopadhya, I., Segal, J.P., Carding, S.R. (2019). The gut virome: the 'missing link' between gut bacteria and host immunity? *Therapeutic Advances in Gastroenterology*. 12:1756284819836620. doi: 10.1177/1756284819836620.Naicker, N., Richter, L., Mathee, A., Becker, P., Norris, S.A. (2012). Environmental lead exposure and socio-behavioural adjustment in the early teens: the birth to twenty cohort. *Science of the Total Environment* 414: 120–125. doi:10.1016/j.scitotenv.2011.11.013

Naito, Y., Uchiyama, K. and Takagi, T. (2018). A next-generation beneficial microbe: *Akkermansia muciniphila*. *Journal of Clinical Biochemistry and Nutrition*. 63: 33–35.

NTP (National Toxicology Program). (2012). NTP Monograph: Health Effects of Low-Level Lead. https://ntp.niehs.nih.gov/go/36443

Panel, E. and Chain, F. (2012). Scientific opinion on the risk for public health related to the presence of mercury and methylmercury in food. *European Food Safety Association Journal*, 10(12): 2985.

Raghavendra, P. and Pullaiah, T. (2018). Pathogen identification using novel sequencing methods. *Advances in Cell and Molecular Diagnostics*. Academic Press. pp. 161–202. doi: 10.1016/B978-0-12-813679-9.00007-5.

Rhee, S.H., Pothoulakis, C. and Mayer, E.A. (2009). Principles and clinical implications of the brain-gut-enteric microbiota axis. *Nature Reviews Gastroenterology and Hepatology*. 6: 306–314.

Rice, K.M., Walker, E.M., Wu, M., Gillette, C., and Blough, E.R. (2014). Environmental mercury and its toxic effects. *Journal of Preventive Medicine and Public Health*. 47: 74–83.

Richardson, J.B., Dancy, B.C.R., Horton, C.L. et al. (2018). Exposure to toxic metals triggers unique responses from the rat gut microbiota. *Scientific Reports*. 8: 1–12. doi: 10.1038/s41598-018-24931-w.

Roggenbeck, B.A., Leslie, E. M., Walk, S. T. and Schmidt, E. E. (2019). Redox metabolism of ingested arsenic: Integrated activities of microbiome and host on toxicological outcomes. *Current Opinion in Toxicology*. 13: 90–98.

Rohwer, F., Prangishvili, D. and Lindell, D. (2009). Roles of viruses in the environment. *Environmental Microbiology*. 11: 2771–2774.

Rosenfeld, C.S. (2017). Gut dysbiosis in animals due to environmental chemical exposures. *Frontiers in Cellular and Infection Microbiology*, 7: doi: 10.3389/fcimb.2017.00396.

Ruan, Y., Wu, C., Guo, X. (2019). High doses of copper and mercury changed cecal microbiota in female mice. *Biological Trace Element Research*, 189: 134–144.

Schippa, S. and Conte, M. P. (2014). Dysbiotic events in gut microbiota: Impact on human health. *Nutrients*. 6: 5786–5805.

Sharma, M. and Devi, M. (2014). Probiotics: A comprehensive approach toward health foods. *Critical Reviews in Food Sciences and Nutrition*. 54: 537–552.

Skolarczyk, J., Pekar, J., Łabądź, D., Skórzyńska-Dziduszko, K. (2018). Role of heavy metals in the development of obesity: A review of research. *Journal of Elementology*. 23: 1271–1280.

Spor, A., Koren, O., and Ley, R. (2011). Unravelling the effects of the environment and host genotype on the gut microbiome. *Nature Reviews Microbiology*. 9: 279–290.

Starr, D.A. (2011). Oligofructose protects against arsenic-induced liver injury in a model of environment/obesity interaction. *Physiology and Behavior*. 176: 139–148.

Summers, A.O., Wireman, J., Vimy, M.J. et al. (1993). Mercury released from dental 'silver' fillings provokes an increase in mercury- and antibiotic-resistant bacteria in oral and intestinal floras of primates. *Antimicrobial Agents and Chemotherapy*. 37: 825–834.

Sun, H., Brocato, J. and Costa, M. (2015). Oral chromium exposure and toxicity. *Current Environmental Health Reports*. 2: 295–303.

Suttle, C. A. (2007). Marine viruses: Major players in the global ecosystem. *Nature Review Microbiology*. 5: 801–812.

Swaroop, A., Bagchi, M., Preuss, H.G., Zafra-Stone, S., Ahmad, T., Bagchi, D. (2019) Chapter 8 - Benefits of chromium(III) complexes in animal and human health. In: *The Nutritional Biochemistry of Chromium (III)* (Second Edition), edited by John B. Vincent, Elsevier, 251–278, doi:10.1016/B978-0-444-64121-2.00008-8.Telisman, S., Jurasović, J., Pizent. A. et al. (1997). Cadmium in the blood and seminal fluid of nonoccupationally exposed adult male subjects with regard to smoking habits. *International Archives of Occupational and Environmental Health*. 70: 243–248.

Thevenon, F., Graham, N.D., Chiaradia, M. et al. (2011). Local to regional scale industrial heavy metal pollution recorded in sediments of large freshwater lakes in central Europe (lakes Geneva and Lucerne) over the last centuries. *Science of the Total Environment*. 412: 239–247.

Thursby, E. and Juge, N. (2017). Introduction to the human gut microbiota. *Biochemical Journal*. 474: 1823–1836.

Tian, F., Zhai, Q., Zhao, J. et al. (2012). *Lactobacillus plantarum* CCFM8661 alleviates lead toxicity in mice. *Biological Trace Element Research*. 150: 264–271.

Tilg, H. and Kaser, A. (2011) Gut microbiome, obesity, and metabolic dysfunction. *Journal of Clinical Investigation*. 121: 2126–2132.

Upreti, R. K., Shrivastava, R. and Chaturvedi, U. C. (2008). Gut microflora and toxic metals: Chromium as a model. *Indian Journal of Medical Research*. 119: 49–55.

USEPA. (2002). Implementation guidance for the arsenic rule drinking water regulations for arsenic and clarifications to compliance and new source contaminants monitoring. *United States Environmental Protection Agency (USEPA)*, 83. https://www.epa.gov/dwreginfo/state-implementation-guidance-arsenic-rule.

WHO (2010). Exposure to lead: A major public health concern. Geneva, Switzerland. 2010. Available from: http://www.who.int/ipcs/features/lead.pdf.

WHO (2015). Lead exposure in African children: Contemporary sources and concerns. World Health Organization. Regional Office for Africa. Available from: who.int/iris/handle/10665/200168.

WHO (2017). Guidelines for drinking-water quality: Fourth edition incorporating the first addendum. Geneva: World Health Organization. Licence: CC BY-NC-SA 3.0 IGO. Cataloguing-in-Publication.

Wu, G., Xiao, X., Feng, P. et al. (2017). Gut remediation: A potential approach to reducing chromium accumulation using *Lactobacillus plantarum* TW1-1. *Scientific Reports*, 7: 1–12. doi: 10.1038/s41598-017-15216-9.

Wylie, K.M., Weinstock, G.M. and Storch, G.A. (2012). Emerging view of the human virome. *Translational Research*. 160: 283–290.

Zhai, Q., Wang, J., Cen, S. et al. (2019). Modulation of the gut microbiota by a galactooligosaccharide protects against heavy metal lead accumulation in mice. *Food and Function*. 10: 3768–3781.

Zhang, S., Jin, Y., Zeng, Z., Liu, Z., and Fu, Z., 2015. Subchronic exposure of mice to cadmium perturbs their hepatic energy metabolism and gut microbiome. *Chemical Research in Toxicology*. 8: 2000–2009.

PART IV

Radionuclides Toxicity

CHAPTER 7

A Review on Radionuclide Toxicity

Fco. Javier Guillén Gerada and **José Ángel Corbacho Merino**
University of Extremadura

CONTENTS

7.1 INTRODUCTION

Usually, some metals are elements that are beneficial and essential for vital processes of living beings, although in some cases, they are only found in trace amounts in body tissues. Thus, as a way of example, cobalt, copper, iron, manganese, molybdenum, vanadium, strontium, and zinc can be cited. However, other metals are very toxic, even in small quantities. They are usually called heavy metals, i.e. mercury, chromium, cadmium, arsenic, lead, etc. From the chemical point of view, radioactive isotopes of metals are totally indistinguishable from the corresponding stable isotopes of the same element. Regardless of their benefits or toxicity due to their accumulation in the living beings' tissues, the radiotoxicity due to its radioactive decay processes must be considered. In fact, in some situations, radiotoxicity could be a hazard, even when only trace amounts of radioactive metal are present in the inner tissues. As it will be described in the following sections, the radiotoxicity relies both on activity and the types of particles emitted in those decay. In this chapter, the main points of radiological protection for humans and wildlife biota are introduced.

7.2 BASIC CONCEPTS ABOUT RADIONUCLIDES

A radionuclide is an isotope of an element, which is unstable. It decays with time, losing its excess energy, into another isotope of the same element or transforms into another element. The decay probability of a radionuclide, λ, is a physical property unique of that radionuclide. This probability is expressed in terms of half-live, $T_{1/2}$. The half-life of a radionuclide is the amount of time to be elapsed so that the number of radionuclides is halved. Different radionuclides of the same element have different half-lives. The activity of a radionuclide, A, is its decay rate, and it is expressed by the following equation:

$$A = \lambda N \tag{7.1}$$

$$\lambda = \frac{\ln 2}{T_{\frac{1}{2}}} \tag{7.2}$$

where A is the activity, λ the decay probability and $T_{\frac{1}{2}}$ the half-life, and N is the number of atoms of the considered radionuclide. The SI unit is Becquerel (Bq), and expresses 1 decay per second. There are other units, such as dpm (disintegrations per minute), and Curie (Ci), which is the activity of 1 g of ^{226}Ra (1 Ci $= 3.7 \times 10^{10}$ Bq).

Radionuclides decay by means of emission of:

- **α-Particles**: are ^{2}He nuclei (2 protons + 2 neutrons) with kinetic energies in the range of 3.95–8.78 MeV[1]. Their interaction with matter is so strong that ever a sheet of paper can stop them, depositing all their energy in that thickness.
- **β-Particles**: it englobes several decay modes depending on the considered radionuclide (β^-, β^+, electronic capture or electron conversion), resulting in the emission of electrons or positrons with energies in the range of 0.01859–3.54 MeV. They can penetrate deeper in matter, and a few cm of aluminum are required to stop them.
- **γ-Rays**: are high-energy photons with energies in the range of a few keV up to about 3 MeV. They require several cm of lead to be fully absorbed.
- **Neutrons**: one or more neutrons are ejected from a radionuclide by spontaneous, photoneutron, beta-delayed neutron emissions or by spontaneous or induced fission. The free neutron kinetic energy is also called temperature and can be observed in the range from $2.5{\cdot}10^{-5}$ to >20 MeV.

7.3 DOSIMETRY QUANTITIES

When radiation interacts with matter, it losses partially or totally its energy and generates ion pairs depending on the composition of the material and the deposited energy. In this process, there are changes in atoms and molecules and chemical bonds may rupture. If these molecules are in a living cell, it may be damaged directly or indirectly, i.e. by the production of free radicals.

The amount of energy deposited by the radiation deposited in matter is quantified by the absorbed dose, D, is defined as:

$$D = \frac{d\varepsilon}{dm} \tag{7.3}$$

where $d\varepsilon$ is the amount of energy deposited in a volume element, and dm is the mass of matter in that volume element. The SI unit of absorbed dose is gray (Gy, 1 Gy $=$ 1 J/kg). There is also an old unit, the roentgen absorbed dose (rad, 1 Gy $=$ 100 rad).

The equivalent dose, H_T, to a tissue is the product of the absorbed dose delivered by a radiation R to a tissue or organ T, D_T, and the radiation weighting factor for the radiation R, w_R.

$$H_T = \sum_R w_R {\cdot} D_T \tag{7.4}$$

---

[1] An electronvolt (eV) is a usual unit of energy for ionizing radiation. It is equal to the kinetic energy gained by a single free electron accelerated by a difference in potential of 1 eV. 1eV $= 1.609 \times 10^{-19}$ J.

Table 7.1 Radiation Weighting Factors, w_R (ICRP, 1991)

Type and Energy Range	w_R
Photons, all energies	1
Electrons and muons, all energies	1
Neutrons (<1 keV)	5
Neutrons (10–100 keV)	10
Neutrons (>100 keV–2 MeV)	20
Neutrons (>2–20 MeV)	10
Neutrons (>20 MeV)	5
Protons, other than recoil protons, energy >2 MeV)	5
Alpha particles, fission fragments, heavy nuclei	20

where the sum is extended to all radiation types in the radiation field. The radiation weighting factor varies in the range 1–20, depending on the radiation type but independent of tissue or organ. Table 7.1 shows the values recommended by the International Commission on Radiological Protection (ICRP, 1991). The SI unit of equivalent dose is the Sievert (Sv, 1 Sv = 1 J/kg).

In order to consider the radiation sensitivities of different tissues, the effective dose is defined as the sum of the weighted equivalent doses in all tissues and organs of the body,

$$E = \sum_T w_T \cdot H_T = \sum_T w_T \sum_R w_R \cdot D_{T,R} \tag{7.5}$$

where H_T is the equivalent dose in tissue or organ T, and w_T is the weighting factor for tissue T. The tissue weighting factor, w_T, is independent of the type and energy of the radiation incident on the body. Table 7.2 shows the tissue weighting factors, which were developed from a reference population of equal numbers of both sexes and a wide range of ages. The initial values for the tissue weighting factors (ICRP, 1991) were updated in the 2007 ICRP recommendations (ICRP, 2007).

The dose rate is the variation of dose with time

$$\dot{D} = \frac{dD}{dt} \vee \dot{H} = \frac{dH}{dt} \tag{7.6}$$

where D is the dose and t time. The SI units are Gy/s or Sv/s, but nGy/h and nSv/h are frequently used.

Table 7.2 Tissue Weighting Factors, w_T (ICRP, 2007)

Tissue or Organ	w_T	Tissue or Organ	w_T
Gonads	0.08	Esophagus	0.04
Bone marrow (red)	0.12	Thyroid	0.04
Colon	0.12	Skin	0.01
Lung	0.12	Bone surface	0.01
Stomach	0.12	Brain	0.01
Bladder	0.04	Salivary glands	0.01
Breast	0.12	Remainder[a]	0.12
Liver	0.04		

[a] The remainder is composed of the following tissues and organs: adrenals, extrathoracic (ET) region, gall bladder, heart, kidneys, lymphatic nodes, muscle, oral mucosa, pancreas, prostate (for men), small intestine, spleen, thymus, and uterus/cervix (for women).

In case of intake of radionuclides, through ingestion, inhalation, or skin absorption, the committed dose is used

$$E(\tau) = \int_{t_0}^{t_0+\tau} \dot{E}(t)\,dt \tag{7.7}$$

where t_0 is the time of intake, $\dot{E}(t)$ the effective dose rate, and τ the time elapsed since the intake of radionuclides. Usually, it is considered 50 years for adults and 70 years for children. The SI for committed dose is Sv.

7.4 EXTERNAL AND INTERNAL EXPOSURES

The assessment of dose from exposure to radiation from external sources is usually done by the use of personal dosimeters or from the measurement or evaluation of the ambient dose equivalent, $H^*(10)$. The ambient dose equivalent at a point in a radiation field is the dose equivalent to that would be produced by the corresponding expanded and aligned field in the ICRU sphere at a depth of 10 mm on the radius opposing the direction of the aligned field (ICRP, 2007). If the radiation field is known (point source, cloud immersion, surface contamination, etc.), $H^*(10)$ can be estimated using conversion coefficients (ICRP, 2007). Regarding decay types, the external dose arises primarily from γ-emitting radionuclides, and secondarily from exposure to neutrons and β-particles to the skin. The influence of α-emitting radionuclides to the external exposure can be considered negligible, save for some very special cases, as a geometry extremely close to the eye.

The internal exposure comprises the intake of radionuclides by inhalation and ingestion. The committed effective dose from internal exposure is assessed by the following equation

$$E = \sum_j e_{j,\,\text{inh}}\left(\text{Sv/Bq}\right) I_{j,\,\text{inh}}\left(\text{Bq}\right) + \sum_j e_{j,\,\text{ing}}\left(\text{Sv/Bq}\right) I_{j,\,\text{ing}}\left(\text{Bq}\right) \tag{7.8}$$

where $e_{j,\text{inh}}$ is the dose effective coefficient for the inhalation intake of radionuclide j, $I_{j,\text{inh}}$ is the activity intake of radionuclide j by inhalation, $e_{j,\text{ing}}$ is the dose effective coefficient for the ingestion intake of radionuclide j, and $I_{j,\text{inh}}$ is the activity intake of radionuclide j by ingestion. The inhalation and ingestion dose coefficients for members of the public are age dependent. There are five age groups for children and one for adults. The inhalation dose coefficients are calculated assuming that clearance from the respiratory tract and absorption into blood is independent of age and sex. For particulate aerosols, an activity mean aerodynamic diameter (AMAD) of 1 µm was used. Daily air intake, averaged over a 24-h period, was taken to be 2.9, 5.2, 8.7, 15, 20, and 22 m^3 for the 3-month-old infant, 1-, 5-, 10-, and 15-year-old children, and adult, respectively (ICRP, 2012). Ingestion dose coefficients are calculated using higher gastrointestinal absorption factor (f_1) values for 3-month-old infants than older individuals. For most elements, adult values are applied to 1-, 5-, 10-, and 15-year-old children. However, intermediate values are used for children in the cases of calcium, iron, cobalt, strontium, barium, lead, and radium (ICRP, 2012). The incorporation of radionuclides through other events, such as wounds, is not considered in this approach (ICRP, 2007). Tables 7.3 and 7.4 show some dose coefficients for inhalation and ingestion, respectively (ICRP, 2012). It can also be observed that the dose coefficients for α-emitting radionuclides are higher than those γ-emitters, due to the short range in which α-particles deposit their energy. Therefore, α-emitting radionuclides, such as ^{238}U, ^{226}Ra, ^{210}Po, ^{232}Th, among others, are usually the main contributors to internal exposure.

Table 7.3 Inhalation Dose Coefficients for Members of the Public, e_{inh}, Expressed in *Sv/Bq* for Different Group Ages

Nuclide	$T_{1/2}$	Type	e_{inh} (Sv/Bq) (AMAD = 1 μm)					
			<1 year	1 year	5 years	10 years	15 years	Adult
^{26}Al	7.16E5 years	F	8.1E-08	6.2E-08	3.2E-08	2.0E-08	1.3E-08	1.1E-08
		M	8.8E-08	7.4E-08	4.4E-08	2.9E-08	2.2E-08	2.0E-08
^{40}K	1.28E9 years	F	2.4E-08	1.7E-08	7.5E-09	4.5E-09	2.5E-09	2.1E-09
^{54}Mn	312.5 days	F	5.2E-09	4.1E-09	2.2E-09	1.5E-09	9.9E-10	8.5E-10
		M	7.5E-09	6.2E-09	3.8E-09	2.4E-09	1.9E-09	1.5E-09
^{59}Fe	44.529 days	F	2.1E-08	1.3E-08	7.1E-09	4.2E-09	2.6E-09	2.2E-09
		M	1.8E-08	1.3E-08	7.9E-09	5.5E-09	4.6E-09	3.7E-09
		S	1.7E-08	1.3E-08	8.1E-09	5.8E-09	5.1E-09	4.0E-09
^{58}Co	70.80 days	F	4.0E-09	3.0E-09	1.6E-09	1.0E-09	6.4E-10	5.3E-10
		M	7.3E-09	6.5E-09	3.5E-09	2.4E-09	2.0E-09	1.6E-09
		S	9.0E-09	7.5E-09	4.5E-09	3.1E-09	2.6E-09	2.1E-09
^{60}Co	5.271 years	F	3.0E-08	2.3E-08	1.4E-08	8.9E-09	6.1E-09	5.2E-09
		M	4.2E-08	3.4E-08	2.1E-08	1.5E-08	1.2E-08	1.0E-08
		S	9.2E-08	8.6E-08	5.9E-08	4.0E-08	3.4E-08	3.1E-08
^{63}Ni	96 years	F	2.3E-09	2.0E-09	1.1E-09	6.7E-10	4.6E-10	4.4E-10
		M	2.5E-09	1.9E-09	1.1E-09	7.0E-10	5.3E-10	4.8E-10
		S	4.8E-09	4.3E-09	2.7E-09	1.7E-09	1.3E-09	1.3E-09
^{90}Sr	29.12 years	F	1.3E-07	5.2E-08	3.1E-08	4.1E-08	5.3E-08	2.4E-08
		M	1.5E-07	1.1E-07	6.5E-08	5.1E-08	5.0E-08	3.6E-08
		S	4.2E-07	4.0E-07	2.7E-07	1.8E-07	1.6E-07	1.6E-07
^{110m}Ag	249.9 days	F	3.5E-08	2.8E-08	1.5E-08	9.7E-09	6.3E-09	5.5E-09
		M	3.5E-08	2.8E-08	1.7E-08	1.2E-08	9.2E-09	7.6E-09
		S	4.6E-08	4.1E-08	2.6E-08	1.8E-08	1.5E-08	1.2E-08
^{131}I	8.04 days	F	7.2E-08	7.2E-08	3.7E-08	1.9E-08	1.1E-08	7.4E-09
		M	2.2E-08	1.5E-08	8.2E-09	4.7E-09	3.4E-09	2.4E-09
		S	8.8E-09	6.2E-09	3.5E-09	2.4E-09	2.0E-09	1.6E-09
^{137}Cs	30.0 years	F	8.8E-09	5.4E-09	3.6E-09	3.7E-09	4.4E-09	4.6E-09
		M	3.6E-08	2.9E-08	1.8E-08	1.3E-08	1.1E-08	9.7E-09
		S	1.1E-07	1.0E-07	7.0E-08	4.8E-08	4.2E-08	3.9E-08
^{210}Pb	22.3 years	F	4.7E-06	2.9E-06	1.5E-06	1.4E-06	1.3E-06	9.0E-07
		M	5.0E-06	3.7E-06	2.2E-06	1.5E-06	1.3E-06	1.1E-06
		S	1.8E-05	1.8E-05	1.1E-05	7.2E-06	5.9E-06	5.6E-06
^{226}Ra	1600 years	F	2.6E-06	9.4E-07	5.5E-07	7.2E-07	1.3E-06	3.6E-07
		M	1.5E-05	1.1E-05	7.0E-06	4.9E-06	4.5E-06	3.5E-06
		S	3.4E-05	2.9E-05	1.9E-05	1.2E-05	1.0E-05	9.5E-06
^{232}Th	1.405E10 years	F	2.3E-04	2.2E-04	1.6E-04	1.3E-04	1.2E-04	1.1E-04
		M	8.3E-05	8.1E-05	6.3E-05	5.0E-05	4.7E-05	4.5E-05
		S	5.4E-05	5.0E-05	3.7E-05	2.6E-05	2.5E-05	2.5E-05
^{238}U	4.468E9 years	F	1.9E-06	1.3E-06	8.2E-07	7.3E-07	7.4E-07	5.0E-07
		M	1.2E-05	9.4E-06	5.9E-06	4.0E-06	3.4E-06	2.9E-06
		S	2.9E-05	2.5E-05	1.6E-05	1.0E-05	8.7E-06	8.0E-06
^{239}Pu	24065 years	F	2.1E-04	2.0E-04	1.5E-04	1.2E-04	1.1E-04	1.2E-04
		M	8.0E-05	7.7E-05	6.0E-05	4.8E-05	4.7E-05	5.0E-05
		S	4.3E-05	3.9E-05	2.7E-05	1.9E-05	1.7E-05	1.6E-05
^{241}Am	432.2 years	F	1.8E-04	1.8E-04	1.2E-04	1.0E-04	9.2E-05	9.6E-05
		M	7.3E-05	6.9E-05	5.1E-05	4.0E-05	4.0E-05	4.2E-05
		S	4.6E-05	4.0E-05	2.7E-05	1.9E-05	1.7E-05	1.6E-05

F, M, and S types indicate fast, medium, and slow absorption (ICRP, 2012).

Table 7.4   Ingestion Dose Coefficients for Members of the Public, e_{inh}, Expressed in Sv/Bq for Different Group Ages (ICRP, 2012)

Nuclide	$T_{1/2}$	e_{ing} (Sv/Bq)					
		<1 year	1 year	5 years	10 years	15 years	Adult
^{26}Al	7.16E5 years	3.4E-08	2.1E-08	1.1E-08	7.1E-09	4.3E-09	3.5E-09
^{40}K	1.28E9 years	6.2E-08	4.2E-08	2.1E-08	1.3E-08	7.6E-09	6.2E-09
^{54}Mn	312.5 days	5.4E-09	3.1E-09	1.9E-09	1.3E-09	8.7E-10	7.1E-10
^{59}Fe	44.529 days	3.9E-08	1.3E-08	7.5E-09	4.7E-09	3.1E-09	1.8E-09
^{58}Co	70.80 days	7.3E-09	4.4E-09	2.6E-09	1.7E-09	1.1E-09	7.4E-10
^{60}Co	5.271 years	5.4E-08	2.7E-08	1.7E-08	1.1E-08	7.9E-09	3.4E-09
^{63}Ni	96 years	1.6E-09	8.4E-10	4.6E-10	2.8E-10	1.8E-10	1.5E-10
^{90}Sr	29.12 years	2.3E-07	7.3E-08	4.7E-08	6.0E-08	8.0E-08	2.8E-08
^{110m}Ag	249.9 days	2.4E-08	1.4E-08	7.8E-09	5.2E-09	3.4E-09	2.8E-09
^{131}I	8.04 days	1.8E-07	1.8E-07	1.0E-07	5.2E-08	3.4E-08	2.2E-08
^{137}Cs	30.0 years	2.1E-08	1.2E-08	9.6E-09	1.0E-08	1.3E-08	1.3E-08
^{210}Pb	22.3 years	8.4E-06	3.6E-06	2.2E-06	1.9E-06	1.9E-06	6.9E-07
^{226}Ra	1600 years	4.7E-06	9.6E-07	6.2E-07	8.0E-07	1.5E-06	2.8E-07
^{232}Th	1.405E10 years	4.6E-06	4.5E-07	3.5E-07	2.9E-07	2.5E-07	2.3E-07
^{238}U	4.468E9 years	3.4E-07	3.4E-07	1.2E-07	8.0E-08	6.8E-08	6.7E-08
^{239}Pu	24065 years	4.2E-06	4.2E-06	4.2E-07	3.3E-07	2.7E-07	2.4E-07
^{241}Am	432.2 years	3.7E-06	3.7E-07	2.7E-07	2.2E-07	2.0E-07	2.0E-07

7.5 RADIOLOGICAL PROTECTION FOR HUMANS

Everybody is exposed to ionizing radiation from natural and anthropogenic sources. The assessment of these exposures is complex, and some individuals may be subject to several categories, i.e. a worker is exposed to radiation as part of his/her work, but also is a member of the public and may be a patient subject to a treatment. Therefore, the Commission recommended to classify the exposures into three categories (ICRP, 1991):

- **Occupational Exposure**: all radiation exposure of workers as result of their work.
- **Medical Exposure of Patients**: occurs in diagnostic, interventional, and therapeutic procedures. The exposure is intentional and for the direct benefit of the patient.
- **Public Exposure**: encompasses all exposures of the public other than occupational and medical exposure of patients.

Within each category of exposure, individuals can be exposed to several sources. The Commission also recommended the following types of exposure (ICRP, 2007):

- **Planned Exposure Situations**: involving the deliberate introduction and operation of sources.
- **Emergency Exposure Situations**: may occur during the operation of a planned situation, or from a malicious act, requiring urgent attention.
- **Existing Exposure Situations**: which are exposure situations that already exist when a decision on control has to be taken, such as those caused by natural background radiation.

The three key principles of radiological protection are defined as follows (ICRP 1991, 2007):

- **The Principle of Justification**: any decision that alters the radiation exposure situation should do more good than harm.

- **The Principle of Optimization of Protection**: the likelihood of incurring exposure, the number of people exposed, and the magnitude of their individual doses should all be kept as low as reasonably achievable (ALARA), taking into account economic and societal factors.
- **The Principle of Application of Dose Limits**: the total dose to any individual from regulated sources in planned exposure situations other than medical exposure of patients should not exceed the appropriate limits specified by the Commission.

The principles of justification and optimization apply in all three exposure situations, whereas the principle of application of dose limits applies only for doses resulting from planned exposure situations. The concepts of dose constraint, dose limit, reference level, and diagnostic reference level derive from the use of these principles (ICRP, 2007).

- Dose constraint is a prospective and source-related restriction on the individual dose from a source, which provides a basic level of protection for the most highly exposed individuals from a source, and serves as an upper bound on the dose in optimization of protection for that source.
- Dose limit is the value of the effective dose or the equivalent dose to individuals from planned exposure situations that shall not be exceeded.
- **Reference Level**: in emergency or existing controllable exposure situations, this represents the level of dose or risk, above which it is judged to be inappropriate to plan to allow exposures to occur, and below which optimization of protection should be implemented.
- Diagnostic reference levels are used in medical diagnosis (i.e. planned exposure situations) to indicate whether, in routine conditions, the levels of patient dose or administered activity from a specified imaging procedure are unusually high or low for that procedure.

The chosen values for dose constraint or reference level depend upon the circumstances of the considered exposure. Table 7.5 shows the different types of dose restrictions used in the Commission's system of protection in relation to type and category of exposure. Dose limits apply only in planned exposure situations, but not to medical exposure of patients. Table 7.6 shows the recommended dose limits in planned exposure situations.

The annual worldwide average effective dose to individuals from ionizing radiation is 3.0 mSv (UNSCEAR, 2008). The main contributors (see Figure 7.1) are naturally occurring sources, such as radon inhalation, external exposure (mainly from γ-emitting radionuclides in soil), ingestion of water and foodstuffs, and cosmic radiation. About 19.8% is due to medical exposure; and only 0.5% from anthropogenic sources, comprising fallout from atmospheric nuclear testings (occurring in 1950–1960s, now banned), occupational exposure sources, Chernobyl accident, and nuclear fuel cycle. The annual average effective dose is above the recommended dose limit for members. There is no contradiction, as this dose limit, 1 mSv/year, is only applicable for planned exposures. The influence of normal natural sources of radiation is not considered as part of a planned exposure. Indoor radon concentration (either in dwellings or workplaces) and naturally occurring radioactive materials (NORM) are examples of existing exposure situations, and reference levels are recommended to be set in the 1–20 mSv band (ICRP, 2007).

Table 7.5 Use of Dose Constraints and Reference Levels in the Commission's System of Protection (ICRP, 2007)

Type of Exposure	Category of Exposure		
	Occupational	Public	Medical
Planned	Dose limit (Dose constraint)	Dose limit (Dose constraint)	Diagnostic reference level (Dose constraint)
Emergency	Reference level	Reference level	n.a.
Existing	n.a.	Reference level	n.a.

n.a.: not applicable.

Table 7.6   Recommended Dose Limits in Planned Exposure (ICRP, 2007)

| | Category of Exposure | |
Type of Limit	Occupational	Public
Effective dose	20 mSv/year[a]	1 mSv/year[b]
Annual Equivalent Dose in		
Lens of the eye	150 mSv	15 mSv
Skin[c]	500 mSv	50 mSv
Hands and feet	500 mSv	---

[a] Averaged over a defined period of 5 years, and with further provision that the effective dose should not exceed 50 mSv in a single year.
[b] In special circumstances, a higher value can be allowed in a single year provided that the average over 5-year period is below 1 mSv/year.
[c] Averaged over 1 cm² area of skin regardless of the area exposed.

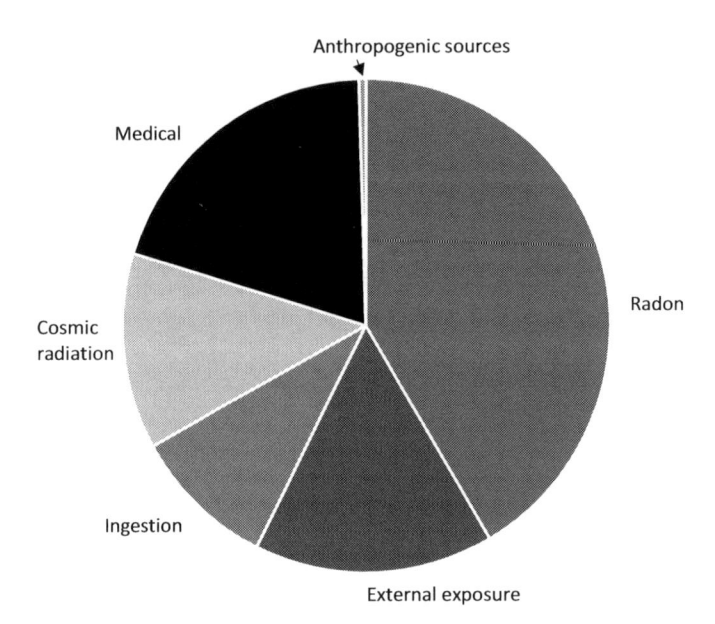

Figure 7.1   Percentages of source contributors to the annual average worldwide effective dose from ionizing radiation. Data from UNSCEAR (2008).

7.6 RADIOLOGICAL PROTECTION OF THE ENVIRONMENT

The radiological protection of the environment evolved from an anthropogenic point of view:

The Commission therefore believes that if man is adequately protected then other living things are also likely to be sufficiently protected

ICRP (1977)

to recommendations for the environment to be assessed in its own right (ICRP, 2008).

In order to establish a scheme for radiological protection of the environment similar to that used for humans with the concept of Reference Man, the concept of Reference Animal and Plant (RAP) was introduced. A RAP is defined as a hypothetical entity, with the assumed basic characteristics of a specific type of animal or plant, as described to the generality of the taxonomic level of Family, with defined anatomical, physiological, and life-history properties, that can be used for the purposes of relating exposure to dose, and dose to effects, for that type of living organism (ICRP, 2007). The aim of using RAPs is to prevent or reduce the frequency of deleterious ionizing radiation effects in the environment to a level where they would have a negligible impact on the maintenance of biological diversity, the conservation of species, or the health and status of natural habitats, communities, and ecosystems (ICRP, 2007, 2008). Different RAPs have been selected according to three main environments: terrestrial, freshwater, and marine ecosystems (see Table 7.7). The radiological protection scheme using RAPs is analogous to the scheme for man, with the use of Reference Man. But there is a major difference, whereas the latter is focused on the protection of the individual, the non-human biota protection is focused on the population (lifetime reproduction, gender ratios, fecundity, and total size), not the individual (IAEA, 2008).

The exposure to ionizing radiation is assessed in terms of dose rate, by models taking into account typical assumptions for energies, contaminated media, and organism sizes. Dose conversion coefficients (DCCs) were evaluated for external and internal exposure, modeling the whole organism in a very simple shape, a solid ellipsoid (ICRP, 2008, 2017). The DCC provides radionuclide specific conversion factors relating the radionuclide content in the surrounding medium (external exposure) or in the RAPs (internal exposure), expressed in Bq/kg, to absorbed dose rate, expressed in µGy/day or µGy/h.

The types of exposure for non-human biota are the same as stated for humans: planned, emergency, and existing exposure situations (ICRP 2007, 2014). Of the three fundamental principles underlying the system of radiological protection, the Commission does not recommend any general form of dose limitation for biota, because the objectives are different, and the highly variable nature of exposure situations makes it difficult to establish limits scientifically defensible (ICRP, 2014). In its place, Derived Consideration Reference Levels (DCRLs) are used for protection in the environment. The DCRLs were defined in terms of band of dose rates spanning over an order of magnitude relevant to each RAP (ICRP, 2014). Table 7.8 shows the DCLRs for environmental protection for each RAP.

Table 7.7 Reference Animal and Plants (RAPs) Considered in Terrestrial, Freshwater, and Marine Ecosystem

Terrestrial Ecosystem	Freshwater Ecosystem	Marine Ecosystem
Deer (*Cervidae*)	Trout (*Salmonidae*)	Crab (*Cancridae*)
Rat (*Muridae*)	Duck (*Anatidae*)	Flatfish (*Pleuronectidae*)
Bee (*Apidae*)	Frog (*Ranidae*)	Duck (*Anatidae*)
Earthworm (*Lumbricidae*)	Wild grass (*Poaceae*)	Seaweed (*Fucaceae*)
Duck (*Anatidae*)		
Frog (*Ranidae*)		
Wild grass (*Poaceae*)		
Pine tree (*Pinaceae*)		

The corresponding family is in brackets (ICRP, 2008).

Table 7.8 Derived Consideration Reference Levels, Expressed in nGy/day, for Environmental Protection for RAPs (ICRP, 2014)

RAP	Derived Consideration Reference Levels (DCRLs) (nGy/day)
Deer, Rat, Duck, Pine Tree	0.1–1
Grass, Frog, Trout, Flatfish, Seaweed	1–10
Bee, Crab, Earthworm	10–100

ACKNOWLEDGMENTS

We are grateful to the Autonomous Government of Extremadura (Junta de Extremadura) for financial support granted to the LARUEX research group (FQM001).

REFERENCES

ICRP, 1977. Recommendations of the International Commission on Radiological Protection. ICRP Publication 26. *Annals of the ICRP* 1(3).

ICRP, 1991. The 1990 Recommendations of the International Commission on Radiological Protection. ICRP Publication 60. *Annals of the ICRP* 21 (1–3).

ICRP, 2007. The 2007 Recommendations of the International Commission on Radiological Protection. ICRP Publication 103. *Annals of the ICRP* 37 (2–4).

ICRP, 2008. Environmental Protection - the Concept and Use of Reference Animals and Plants. ICRP Publication 108. *Annals of the ICRP* 38 (4–6).

ICRP, 2012. Compendium of Dose Coefficients based on ICRP Publication 60. ICRP Publication 119. *Annals of the ICRP* 41(Suppl.).

ICRP, 2014. Protection of the Environment under Different Exposure Situations. ICRP Publication 124. *Annals of the ICRP* 43(1).

ICRP, 2017. Dose coefficients for nonhuman biota environmentally exposed to radiation. ICRP Publication 136. *Annals of the ICRP* 46(2).

UNSCEAR, 2008. Sources and effects of ionizing radiation. United Nations Scientific Committee on the Effects of Atomic Radiation. UNSCEAR 2008. *Report to the General Assembly with Scientific Annexes.*

PART V

Benefits of Metals

CHAPTER 8

Contribution of Bioavailable Silicon in Human Health

Keith Robert Martin
University of Memphis

CONTENTS

8.1 INTRODUCTION

Silicon is the second most abundant chemical element in the earth's crust exceeded only by oxygen. Environmentally, silica is part of a cycle that is considerably intertwined with other major biogeochemical cycles, such as the carbon and nitrogen cycles, and is closely related to marine primary production, the efficiency of carbon export to the deep sea, and control of carbon dioxide in the atmosphere.[1-3] Silicon occurs most frequently as oxygen-bound silica and silicates, a family of anions consisting of silicon and oxygen, and accounts for approximately 27% of elemental mass with oxygen accounting for 45% of this total.[4-6] Regarding prevalence, silicon as silica is widespread and is part of most of the minerals found on earth as well as geological formations such as rocks, sands, and clays. Silicon also occurs in various, diverse chemical forms such as quartz, emerald, feldspar, serpentine, mica, talc, clay, asbestos, and glass all of which have different uses.[7] Collectively, quartz and aluminosilicates are the most prevalent naturally occurring silicates.[8] Silicon is used widely and routinely in industrial applications as a component of fabricated steel, a component of abrasives (silicon carbide), a basic unit for construction of transistors (along with boron, gallium, arsenic, etc.), solar cells, rectifiers, and other electronic solid-state devices.[9] Sand-derived silica is also used for synthesis of glass and in the production of computer chips, as well as filler for rubber and paint ceramics, and as a component of lubricants, concrete, and brick. Silicon is also used frequently for medical devices.[7,10] Although silicon is used in many technical applications, overall exposure including inhalation and/or ingestion in humans is limited and largely as chemical forms that are not readily absorbed nor bioavailable.

Silicon, as silica, occurs in the food supply ingested by humans since it is used widely in the food and beverage industry as a food additive, i.e., anti-caking agent in foods, clarifying agent in beverages, viscosity modulator, as an anti-foaming agent, dough modifier, and as an excipient in drugs and vitamins.[7] Thus, silicon as silica is a dietary component with potential demonstration of and/or conversion to biologically active forms, viz., orthosilicic acid (OSA), although largely assumed to be inert when provided in forms typically used in technical applications. Humans are exposed to diet-derived forms of silicon and silica suggesting the potential capacity for absorption and ultimate bioavailability, thus as a molecular component could exert beneficial, biological effects in humans. Currently, there is considerable debate as to the biological essentiality of dietary silicon although it is not recognized as a nutrient. Emerging research suggests benefit from consumption of water-soluble forms of silicon, viz., silicic acid, and, as a result, there is a growing interest in the potential beneficial effects of silica on human health, e.g., bone health.

Numerous studies have reported the effects of environmental exposure to silicon and its many chemical forms and the resultant detrimental effects on humans. Primarily inhaled crystalline silica and silica-derived asbestos are well-recognized carcinogens.[11] In fact, silicon has long been recognized as a pulmonary carcinogen with resultant silicosis or asbestosis developing upon prolonged and/or heavy exposure to airborne material.[12] As a disease of the lungs, silicosis is caused by chronic inhalation of mineral dust and is characterized by progressive fibrosis (excessive fibrous connective tissue) and chronic shortness of breath.[13] Asbestosis is similar in etiology and pathology but distinct as an exposure. There are clearly inherent dangers associated with inhalation of crystalline silica; however, there are multiple forms of silica in nature that are considered safe, absorbable, and potentially beneficial to human health.[14] Regarding non-toxic forms, questions remain as to the relative water solubility of different compounds, overall amounts of each ingested, efficiency of absorption and overall bioavailability (delivery to target tissues). Low-molecular-weight silica can dissolve in water as silicic acid rendering it bioavailable and potentially beneficial in humans. Collectively, the lack of understanding of the relative dependence of physicochemical structure of silica and silicates on water solubility for absorption and safety has limited research progress regarding silicic acid. Thus, a better understanding of the chemistry of silica, specifically of aqueous OSA, is critical to initiating, executing, and corroborating research on potential health benefits.

8.2 DISTRIBUTION AND PREVALENCE OF SILICON

Silica is an oxide of silicon and chemically represented as silicon dioxide (SiO_2). Silicon itself is a tetravalent metalloid occurring in group 14 of the periodic table and occurring naturally as a hard, brittle crystalline solid with a blue-gray metallic luster. Its chemical properties lie between that of a metal and non-metal element and its presence is second only to oxygen in its abundance on earth constituting almost a third of the earth's crust.[15] Pure silicon exerts an extreme propensity to react with oxygen and water, thus typically does not exist in a natural elemental state. For example, silica, SiO_2, and other oxides, are ubiquitously found in polymerized combinations with metals, as silicate minerals, and embedded in geologic rock formations. With silicon's widespread distribution and chemical reactivity with other elements, its various, diverse forms exert different physicochemical properties with some forms causing toxicity and others that are innocuous and perhaps essential for health.

As a component of geographical formations, silica is not readily released from these substrates unless by considerable environmental conditions, e.g., weathering, of these structures. Upon release, the forms and molecular sizes of polymers and aggregates are dependent on the pH of their micro-environment and concentrations in aqueous matrices. For example, at low concentrations silicon exists in a monomeric form as OSA with markedly enhanced water solubility greater than that of higher, molecular weight forms. As concentrations increase, polymerization will occur to form oligomers and eventually colloids (homogeneous noncrystalline substances), aggregates, and solid amorphous precipitates with concentration-dependent water solubility.[16] The increased molecular weights and structural complexity restrict interaction with water thus inherent water solubility and, as a result, limit potential absorption and bioavailability by humans and animals.

8.2.1 Dietary Sources

8.2.1.1 Dietary Silica Intake

Silica is prevalent in the typical human diet and generally considered safe, even if indigestible and non-absorbable. The lack of clear understanding of the myriad chemical forms of silica and significant, widely communicated likelihood of increased risk of cancer has unduly overshadowed the study of the potential protective effects of silica on human health. It is the intent of this chapter to provide insight into the chemical properties of silica that may render it bioavailable and beneficial to human health. Although there are dietary sources of silicon which are thought to exert beneficial effects in humans, there is no RDA for silicon and, in fact many do not recognize silicon as a micronutrient essential for life.[17] However, if one considers the risk assessment of amorphous silicon dioxide as a common silicon source, although non-absorbed, the safe tolerable upper intake level (TUL), a component of the Dietary Reference Intakes (DRIs), is estimated to be 700 mg/day for adults, which is equivalent to 12 mg silicon/kg body weight/day for a 60 kg adult.[18] Only minimal amounts of silicon become water soluble and ultimately absorbed, thus the systemic plasma concentration does not increase significantly. The mean dietary silicon intake reported for a Finnish population was 29 mg silicon/day, and for a typical British diet 20–50 mg/day corresponding to

0.3–0.8 mg silicon/kg body weight/day.[19–22] The estimated dietary intake in the United States is 24–33 mg silicon/day with males generally consuming more.[23]

8.2.1.2 Silicon Intake from Plants

Silica exists in the food chain with concentrations tending to be much higher in plant-based foods, i.e., phytolithic, than animal foods. In plants, there is considerable variation in silica concentrations due to varying capacities of plants to deposit biogenic silica (biosilicifiers or bioaccumulators). A proposed, potential mechanism is simply that of a passive process without the need for new models of silicon biochemistry with biological/ecological advantages that are clearly observed under certain conditions.[15,24] Indeed, a paradigm shift has been proffered involving a comparative, comprehensive approach to assess the mode of action of silicon between plant types (accumulators and non-accumulators) and between biotic and abiotic stressors, e.g., pathogens, herbivores, drought, salt, with a focus on metabolic processes and pathways and the use of cutting-edge technologies such as genomic, transcriptomic, proteomic, and metabolomic to understand the role of silicon in plants.[25,26] In contrast, it has been argued that much of the empirical evidence, viz., functional genomics, conflicts with mechanistic assertions regarding silicon's role. In short, data do not support reports that silicon modulates a wide range of molecular, genetic, biochemical, and physiological processes.[27] In fact, Coskun et al. have proposed the term "apoplastic obstruction hypothesis," which argues the fundamental role of silicon is as an extracellular prophylactic agent against biotic and abiotic stresses, not as an active cellular modulator, with significant downstream effects on plant form and function.[27] Although discussion continues regarding silicon's role in plants, the use of silicon in fertilizers will continue since it is an ecologically compatible and environmentally friendly technique that stimulates plant growth, mitigates and/or prevents various biotic and abiotic stresses in plants, and augments plant resistance to multiple stresses. Moreover, silicon is not harmful or damaging to plants.[28,29]

Silicon fertilization by natural silicates has the potential to mitigate environmental stresses and soil nutrient depletion and as a consequence is an alternative to the extensive use of phytosanitary (free from regulated pests) and NPK (nitrogen, phosphorous, potassium) fertilizers for maintaining sustainable agriculture. Seven among the ten most important crops are considered to be silicon accumulators, with concentration of silicon above 1% dry weight. There is a general agreement on an uptake of dissolved silica as silicic acid and precipitation as amorphous silica particles, viz., phytoliths, but the mechanism, either active or passive, is still a matter of debate. The benefits of silicon are well demonstrated when plants are exposed to abiotic and biotic stresses. Moreover, phytoliths that are located mainly in shoots of monocots return to the soil through litterfall if the plants are not harvested and contribute to the biogeochemical cycle of silicon.[30]

Silicon is extremely abundant in soil and occurs ubiquitously in all organisms including plants and humans. Plant silicon exerts a pivotal role in growth and development and is dependent on bioaccumulation by plants, which protects them against various forms of biotic and abiotic stresses. Consumption of silicon-rich plants by humans imparts health benefits such as strengthening bones and improving immune response, as well as supporting neuronal and connective tissue health.[31] Silicon is a non-essential plant nutrient although numerous reports have shown its beneficial effects in a variety of species and environmental circumstances. New insights have revealed that many plant species supplied with silicon possess phenylpropanoid and terpenoid pathways that are potentiated and exert faster and more robust transcription of defense genes and enhanced activities of defense enzymes as well as improved photosynthesis and antioxidant systems.[29,32]

Development and production of new silicon-rich plants to improve nutrition and eliminate micronutrient inadequacies are current areas of research interest. To that end, silicon biofortification may be an innovative and effective agricultural approach for improving health, e.g., bone mineralization.[33] In recent studies, inclusion of silicon in a nutrient solution for crops (50–100 mg/L) caused silicon accumulation in six edible leafy vegetables.[34] In another study, soilless cultivation also significantly increased silicon concentrations in green beans by threefold.[35] Regarding bioavailability, biofortified vegetables rendered more bioaccessible silicon capable of release from the food matrix compared to unbiofortified vegetables. In subsequent studies, the bioavailable fraction of biofortified purslane and Swiss chard, both improved the expression of osteoblast markers supporting increased bone mineralization, compared with other vegetables. Addition of leafy green digestate in vitro demonstrated increased total alkaline phosphatase activity and increased type 1 collagen, biochemical markers of bone formation.[36] Collectively, biofortification may be an effective means of increasing silicon concentrations in foods with enhanced bioavailability and ultimately beneficial effects on health in humans, e.g., bone formation.

8.2.1.3 Beverages

Beverages are the major contributor to dietary silica, or silicon, and resultant intake and include water, coffee, and beer (due to barley, hops, etc.).[37,38] Beer is the major source of bioavailable silicon for males with estimated concentrations of 9–39 mg/L.[21,23,39] Silica is also prevalent in municipal water supplies and is particularly high in bottled spring and artesian waters depending on the respective geological source. Beverages alone contribute 55% of total dietary intake of silicon as water-soluble silica. Dietary grains and grain products including cereals, oats, barley, white wheat flour, and polished rice contribute 14% and vegetables contribute 8%.[22] In the Western diet, major sources of silicon are cereal (30%) followed by fruits, beverages, and vegetables, which collectively account for 75% of total silicon intake.[40] Processing and refinement of grains remove silicon during the processing but silica-derived food additives, although considered inert, can replace the stripped silicon and increase the content although the absorptivity of silicon from additives is questionable. Estimation of dietary intake from all sources of silicon is approximately 20–50 mg silicon/day for Western populations but up to ~200 mg/day for populations consuming a more plant-based diet as in India, China, and Belgium.[23,37,41,42] The presence of large amounts of silica in geological formations contributes greatly to the silica content of water. For example, in the United Kingdom, silicon concentrations are <2.5 mg/L in north and west Britain but up to 14 mg/L in south and east Britain.[43–45] Silica occurs in fresh water with concentrations of 1–100 mg/L depending on the geographical location, e.g., soil content. Typical municipal water supplies can provide 4–11 mg/L of aqueous silica as noted in a study of the large cities of France and levels of ~18–20 mg/L in the United States. Bottled waters also contain modest concentrations of silica ranging from 8 to 36 mg/L as noted for the French brands Badoit, Vichy Celestian, and Volvic.[46] Interestingly, bottled water from Malaysia contains 30–40 mg/L silica and from the Fiji Islands contains 85 mg/L silica, more than four times the levels found in fresh water and municipal supplies and over twice that of other bottled waters. This difference occurs presumably due to the leaching of water-soluble silica from silicon-rich volcanic rock. Thus, aqueous sources provide a wide range of concentrations of water-soluble, bioavailable silica.

8.2.1.4 Dietary Supplements

Dietary supplements are an alternative silicon source containing OSA or other forms that are presumably modified to a form that is water-soluble, absorbed and bioavailable although this does not universally apply.[47,48] In fact, the estimated overall bioavailability of silicon from supplements ranges from <1% to >50%, a remarkably wide range, depending on the formulation and concentration.[8] As a result, many manufacturers of dietary supplements are diligently developing silicon supplements in chemical forms with greater water solubility. In addition to OSA and its stabilized formulations including choline chloride-stabilized OSA and sodium or potassium silicates, the most important sources of bioavailable OSA are colloidal silicic acid (hydrated silica gel), silica gel (amorphous silicon dioxide), and zeolites (microporous, aluminosilicate minerals). Ironically, these compounds are remarkably water insoluble, but they can release small amounts of biologically meaningful OSA in aqueous environments.[4]

Dietary supplements are an important source of silica with potential health benefits. For example, an inositol-stabilized arginine silicate dietary supplement (ASI; Nitrosigine) has been tested clinically in healthy adult males and emerging data support its capacity to increase plasma levels of arginine, silicon, and nitric oxide. The authors conclude that provision of ASI (1,500 mg/day) for 14 days significantly improved cognition and tests requiring mental flexibility, processing speed, and executive functioning.[49] Marcowycz et al. studied the bioavailability of silicon from OSA-stabilized by vanillin (OSA-VC), a phenolic aldehyde in vanilla bean extract. In this study, 14 healthy subjects consumed either OSA-VC or a placebo on two separate occasions with blood and urine sampling at 6 h post-dose. Plasma silicon was significantly higher after OSA-VC ingestion compared to placebo ingestion and higher urinary silicon excretion was also reported compared to placebo with ~21% of ingested silicon excreted.[50] Scholey et al. assessed uptake and potential efficacy of a novel, pH neutral form of a silicon supplement in vitro using broiler chickens as a model species. In vitro bioavailability of this supplement was significantly higher than other commercial supplements tested. Tibia and foot ash residues (% dry mass) were higher with inclusion of the dietary supplement with significant increases in tibia breaking strength. The authors demonstrate bioavailability and efficacy and posit that this is due to the novel non-condensed, monomeric form.[51] Incorporating silicon into Spirulina, an edible cyanobacteria, could be a way to produce a bioavailable food supplement. To that end, Vidé et al. assessed the potential toxic aspects of dietary exposure to silicon-rich Spirulina as a dietary supplement in rats.[52] Consumption of Spirulina (up to 285 mg/kg body weight/day) for 90 days was safe and exerted no adverse effects on behavior, biochemical parameters, markers of oxidative stress, inflammatory markers, or pathological or histological findings. Others have evaluated the digestive absorption of silicon after a single bolus dose in the form of methylsilanetriol salicylate.[53]

Monomethylsilanetriol (MMST), an organic silicon, is a novel food ingredient used in food supplements, and the bioavailability and safety of OSA from this source have been assessed by the EFSA Panel on Food Additives and Nutrient Sources.[54] According to this report, MMST was used as aqueous solution (4.1 mM) corresponding to 115 mg silicon/L with anticipated silicon intake (60–90 mL/day) corresponding to ~7–10 mg silicon/day. The no observed adverse effect level (NOAEL) was set at 20.5 mM of MMST, corresponding to 232 mg/kg body weight/day, the highest dose tested and technically achievable. EFSA therefore concluded that MMST as a source of silicon to be used in food supplements was safe although silicic acid could not be directly measured. Based on the indirect evidence provided, EFSA concluded that OSA was released from MMST and was bioavailable. Despite this report, there continues to be controversy regarding the safety of MMST.[55–57]

8.2.2 Non-Dietary Sources

With the documented prevalence and widespread use of silicon and silica, there are numerous, diverse sources of and exposures to non-dietary silica/silicon. These occur primarily from exposure to dust, pharmaceuticals, cosmetics, medical implants, and medical devices. Often the forms of silicon occur as "silicones," synthetic organosilicon compounds that, for the most part, are sparse in the human diet and contribute little silicon overall. Moreover, the forms that do result in exposure are not readily absorbed or biologically useful. For example, some pharmaceuticals can increase exposure of silicon to >1 g/day, but the molecular species are largely inert and not absorbed to any significant extent. This seems to be the case with other non-dietary sources such as toiletries, e.g., toothpaste, lipstick, etc., and detergents, tissue implants, etc.[8] The safety of nanostructured synthetic amorphous silica (SAS) as a food additive (E551) has also been addressed.[58] E551 is commonly used as an anti-caking agent in food products and is included at 2.7–14.5 µg silicon/g food product.[59]

The forms of SAS used as E551 include fumed silica and hydrated silica (precipitated silica, silica gel, and hydrous silica). Others have concluded, based on the current available database, that there was no indication for toxicity of E551 at the reported uses and use levels.[60]

8.2.2.1 Zeolites

One of the most important sources of bioavailable OSA is zeolites. Zeolites are microporous, aluminosilicate minerals commonly used as commercial adsorbents and catalysts.[61] Zeolites occur naturally but are also produced industrially on a large scale where they are used extensively in various technological applications, e.g., as molecular sieves, for separating and sorting various molecules, for water and air purification, including removal of radioactive contaminants, for harvesting waste heat and solar heat energy, for adsorption refrigeration, as detergents, etc.[62] Current estimates indicate that 245 unique zeolite frameworks exist and have been identified, and over 40 naturally occurring zeolite frameworks are known. In addition to numerous technological applications, zeolites have many applications in biotechnology and medicine. As components of scaffolds for bone tissue engineering, zeolites can deliver oxygen to cells, stimulate cellular differentiation of osteoblasts, and inhibit bone resorption.[62,63] In addition, zeolites can function as oxygen reservoirs and improve cellular characteristics associated with vascular and skin tissue engineering and wound healing. For example, silica-based films on silica substrates foster adhesion, growth, viability, and osteogenic differentiation of human osteoblast-like cells.[62] Others have shown that zeolites increase cellular proliferation, differentiation, and transforming growth factor beta production in normal adult human osteoblast-like cells in vitro.[64]

8.2.2.2 Silica Nanoparticles and Microparticles

Nanomaterials are increasingly becoming prevalent in various consumer products due to technical commercialization.[59,65] Recently, concerns have been raised about potential adverse effects of nanoparticles commonly used in food additives, since silica nanoparticles have been detected in food containing E551.[65,66] As a result, current research interests are geared toward the reported knowledge gaps involving risk assessment of nanosilica in food particularly regarding the dissolution and toxicity of different forms.[67] Silica nanoparticles are ultrafine, submicroscopic units with dimensions measured in nanometers (10^{-9} m) and occur in the human diet. In one study, detection of silicon dioxide in dietary supplements specifically targeted for women was determined and out of 12 commercial products randomly purchased from retailers with SiO_2 in the ingredient list but without claims regarding particle size, 11 of the 12 products contained isolated nanoscale materials many with a high degree of aggregation.[68] Given the increasing prevalence in the diet, safety concerns have arisen and a critical review of the safety assessment of nanostructured silica additives in food has been conducted.[69] In preliminary studies, Liang et al. have demonstrated that neither silica nanoparticles nor silica microparticles induced toxicological effects after subchronic oral exposure in rats.[70]

8.3 CHEMICAL SPECIATION OF SILICON

8.3.1 Silicon Chemistry

Silicon is the second most abundant element on earth exceeded only by oxygen with properties that are a mixture of both metals and nonmetals resulting in classification as an elemental metalloid. As stated previously, silicon is rarely found in its elemental form but rather complexed with oxygen and/or other elements forming silica and silicates. Silicon dioxide, SiO_2, is the oxide of silicon most commonly found in nature as sand or quartz. Generally, a silicate is any compound containing silicon as an anion, SiO_4^{4-}, with most occurring as hexafluorosilicate anion ($[SiF_6]^{2-}$). Chemically, silicate anions can form compounds with numerous, diverse cations, thus this chemical class of compounds is large with formation of aluminosilicates being the most prevalent in nature.

Silica, SiO_2, is a silicic acid anhydride of monomeric OSA (H_2SiO_4) which is water soluble and stable in aqueous solutions when relatively dilute. Several other low molecular weight, but hydrated forms, of silicic acid exist in aqueous solutions and include metasilicic acid (H_2SiO_3), lower molecular weight oligomers such as disilicic acid ($H_2Si_2O_5$) and trisilicic acid ($H_2Si_3O_7$), as well as their hydrated forms pentahydro- and pyrosilicic acids.[4] Depending on the environmental conditions and exposure time, formation of numerous potential polymerized silicic acids is possible through chemical condensation and crosslinking resulting in colloids and gels. It is the lower molecular weight forms, especially the OSA that is of the greatest research interest in exerting beneficial effects since this form preferentially provides silicon to cells. Interestingly, most aqueous silica, i.e., oceanic, occurs as OSA (H_4SiO_4) making it an important environmental exposure in the context of biological systems due both to its water solubility and bioavailability.[2]

OSA is water soluble at relatively low concentrations but polymerizes readily at higher concentrations to form colloids (homogeneous noncrystalline substance) and gels, (semirigid slab or cylinder of an organic polymer) which are less bioavailable. More concentrated solutions of OSA can be stabilized to avoid polymerization and, in fact, choline-stabilized OSA, a liquid formulation with considerably greater bioavailability than non-stabilized forms, has been developed and approved for human consumption. It is considered non-toxic at high doses with a lethal dose exceeding 5,000 mg/kg body weight in humans and 6,640 mg/kg body weight in animals.[71] For a 70 kg human, this translates to a safe level of consumption of 350 g. This stabilized form currently represents the most bioavailable form/source of supplemental silicon although there are other emerging formulations currently being generated.

8.3.2 Absorption, Accumulation, and Elimination

An important research pursuit is the elucidation of short-term biodistribution and clearance of administered silica nanoparticles. To that end, a single intravenous injection (20 mg/kg body weight) of SAS was administered to rats with observation at 6 and 24 h post-dose.[72] Silicon accumulated primarily in the liver and spleen and within 24 h, silicon was excreted largely through urine. In another study, healthy volunteers in two matched groups (three males and three females/group) ingested 500 mL water alone or containing OSA (28.9 mg silicon). After 48 h, serum silicon analysis confirmed the absorption of silicon. Mean total urinary and fecal silicon excretions over the 24 h post-dose period accounted for $57 \pm 9.5\%$ and $39 \pm 9.4\%$ of the ingested dose, respectively. Collectively, $96.3 \pm 5.8\%$ of the ingested dose was recovered in fecal plus urinary excretions over the 24 h post-dose period.[73] In a human uptake experiment, urine collections were made from 2 to 12 h over 2 days following ingestion by a single human subject of a neutral silicic acid solution.[74] Silicon uptake was essentially complete within 2 h of ingestion and elimination occurred by two processes with half-lives of 2.7 and 11.3 h, representing around 90% and 10%, respectively, of the total output. Elimination of absorbed silicon was complete after 48 h and was equivalent to 36% of the ingested dose.[75]

Others have also determined overall absorption of dietary silicon and the effects of pregnancy, age, or sex. For example, serum silicon concentrations were measured in 14 pregnant women (15–24 weeks of gestation) and compared with those of 17 non-pregnant, non-lactating female controls. Silicon concentrations in pregnant women were not significantly different from those of the female controls and showed little change over time with the authors concluding that a reduction in fasting serum silicon levels during pregnancy could not be confirmed and higher serum silicon levels in newborns compared to their mothers could not be ruled out.[76] In a separate study, urine samples were collected from 26 healthy adults at baseline and over a 6 h period following ingestion of 17.4 mg OSA. Fasting baseline serum and urinary silicon concentrations were marginally higher in older adults (51–66 years old) compared with young adults (20–47 years old), but there was no difference in the absorption of silicon into serum or the rate of elimination of silicon from serum with age or sex. Overall, host age and sex did not appear to influence silicon absorption or excretion in human adults.[77] The relationship between dietary intake and urinary excretion of silicon in free-living Korean adult men and women has also been tested.[78] In a study of 80 healthy Korean adults (40 males and 40 females), mean dietary silicon intake was 22.8 mg/day for men and 19.3 mg/day for women with no differences between gender and age. Silicon intake was highest from vegetables both in men and women, followed by grains, beverages, and fruits in men and by grains, fruits, and milks in women. Urinary silicon excretion was significantly higher in men (9.8 mg/day) compared to women (9.3 mg/day), and significantly increased with age only in women. The urinary excretion of silicon in healthy humans or those with uremia placed on a low silicon diet, a normal diet, and after ingestion of silicate antacid was determined. After 24 h, urinary silicon was primarily via dietary intake. In healthy adults, serum silicon was maintained within a narrow range, but a significant hypersilicemia occurred in individuals with uremia.[43,79]

8.4 SILICON AND ITS POTENTIAL HEALTH BENEFITS

Silicon is the third most abundant trace element in the human body.[23,80] Silicon concentrations are 1–10 ppm (mg/kg) in hair, nails, the epidermis and epicuticle of hair.[81–83] Considering the natural abundance of bioavailable chemical forms of silicon, dietary exposure in humans suggests likely benefit although whether silicon is an essential micronutrient continues to be debated. It has, however, been reported in the peer-reviewed literature that silicon is actively involved, and perhaps integral, in bone mineralization and prevention of osteoporosis, collagen synthesis, and prevention of the aging of skin, overall condition of hair and nails, reduced risk of atherosclerosis and Alzheimer's disease, as well as exerting other beneficial biological effects.[84–88] Interestingly, serum levels are similar to other trace elements and appear to be dependent on life stage, age, and sex with levels of 11–31 μg/dL depending on the population assessed and means of analysis.[43,89] Compelling evidence derived from human, animal, and in vitro studies has accumulated in support of silicon's role as an essential trace element.[90] Moreover, the discovery of a novel mammalian silicon transporter, a sodium–phosphate cotransporter, has been recently reported suggesting an important role in vertebrates and the need to further characterize silicon transport across cell membranes.[91–93] Furthermore, human aquaglyceroporins can also function as silicon transporters suggesting that body distribution is regulated.[92] Emerging evidence demonstrates that removal of dietary silicon has adverse effects on health, viz., bone health in particular, and dietary inclusion and/or supplementation imparts additional beneficial effects. It has been argued that bioactive trace elements, or ultra-trace elements, not recognized as essential, but with beneficial health effects, should have dietary intake recommendations, e.g., adequate intake (AI), such as 25 mg/day as suggested as adequate by others.[94,95]

8.4.1 Bone Health and Skeletal Development

The initial observations of the role of silicon in bone health were reported over four decades ago. Carlisle demonstrated that silicon was required for normal skull formation in chicks and was an "essential" element for this process.[96–98] Additionally, the biochemical and morphological changes due to silicon deficiency were associated with long bone abnormalities.[99] It was also shown that silicon is required in vivo for articular cartilage and connective tissue formation in the chick.[100] Later observations suggested silicon was an essential trace nutrient in overall animal nutrition.[101,102]

The results of studies have suggested that dietary silicon is "essential" in humans with beneficial effects particularly regarding bone health.

8.4.1.1 Osteoporosis

Osteoporosis is a leading cause of morbidity and mortality in the elderly and considerably affects overall quality of life, as well as life expectancy. Compelling evidence demonstrates that dietary silicon can strengthen bones and, as a result, reduce the risk of osteoporosis.[103,104] As a result, there is considerable interest in elucidation and use of dietary approaches using specific nutrients, non-nutritive, bioavailable dietary components, and/or bioactive compounds of natural origin singly or in combination as a strategy of mitigating osteoporosis and/or osteomalacia, as well as for maintenance of bone health. Calcium and vitamin D have largely been the primary focus of nutritional prevention of osteoporosis, however, supplementation with other vitamins and/or nutrients including B, C, and K has been an area of increased research as well as the use of silicon for proper maintenance of bone health.[105–108] Indeed, others have shown that silicon is a requirement for bone formation independent of vitamin D_1.[109] Spector et al. report that combined therapy of choline-stabilized OSA, as an adjunct to calcium and vitamin D_3 supplementation, stimulated markers of bone formation in osteopenic females more than provision of calcium and vitamin D_3 alone.[110] Others have shown that silicon has a physiological role in bone formation and that arginine intake can affect that role.[111,112] It has also been shown that silicon supplementation (as sodium metasilicate up to 0.075%) for 4 weeks to male mice did not affect weight gain or bone mineral density, but decreased magnesium retention suggesting perhaps an increased need for magnesium.[64,113]

Clinically, osteoporosis is defined as a progressive, debilitating skeletal disorder characterized by low bone mass and deterioration of the microarchitecture.[110,114] Numerous human studies support a role for dietary silicon in bone health including reduction of risk for osteoporosis. In a retrospective, clinical study by Eisinger and Clairet, dietary silicon administration induced significant increases in bone mass and bone mineral density of the femur in human females.[84] Moukarzel have also shown a direct, significant relationship between silicon intake and bone mineral density.[115] In osteoporotic participants, supplementation with silicon increased trabecular bone volume and femoral bone mineral density.[84,116] Spector et al. further showed an increase in bone formation markers and significant increases in femoral bone mineral density.[110] Maehira et al. have shown in mice fed five different calcium sources with differing silicon concentrations that soluble silicate and coral sand, with the highest silicon content, significantly improved bone biochemical and mechanical properties through induced gene expression encouraging osteoblastogenesis and suppressing osteoclastogenesis in agreement with others.[117–119] Jugdaohsingh et al. have shown a positive association between serum silicon levels and bone mineral density in female rats after supplementation with MMST.[120]

As mentioned before, stabilized preparations of silicic acid have been developed, e.g., choline-stabilized OSA, permitting water-soluble preparations with higher concentrations and also markedly enhanced bioavailability. In a randomized controlled animal study, long-term treatment with choline-stabilized OSA prevented partial femoral bone loss and exerted a positive, beneficial effect on bone turnover.[47] In this study, ovariectomized aged rodents were used suggesting a potential interrelationship between estrogen and bone health and silicon metabolism. Bae et al. showed that short-term administration of water-soluble silicon improved mineral density of the femur and tibia in ovariectomized rats.[121] In an in vitro study, the effect of OSA (0, 0.5, 1, 5, and $10\,\mu M$) on gene expression in human osteoblast cells isolated from trabecular bone was measured. Results showed that collagen type I mRNA expression was increased by the addition of OSA, alkaline phosphatase message was suppressed, and osteocalcin levels were decreased.[122] Macdonald et al. have reported that dietary silicon interacts with estrogen to influence bone health as noted in the Aberdeen Prospective Osteoporosis Screening study.[123] It has also been shown that silicon supplementation increased hip bone mineral density in men and pre-menopausal, but not post-menopausal, women although a subsequent study showed increases in the spine and femur of both pre- and post-menopausal women currently prescribed hormone replacement therapy.[123,124]

8.4.1.2 Silicon at the Cellular and Molecular Level

Soluble silica derived from the diet and bioactive glass (BG, 45S5) has been shown to inhibit osteoclast formation and bone resorption in vitro using bone marrow cultures and RAW264.7 macrophage cells. The authors show that silicon caused significant inhibition of osteoclast phenotypic gene expressions, osteoclast formation, and bone resorption in vitro supporting a dual role in bone metabolism via stimulating osteoblasts and inhibiting osteoclasts.[125] Reffitt et al. demonstrated as well that OSA stimulated collagen type 1 synthesis and osteoblastic differentiation in human osteoblast-like cells in vitro.[126] Schütze et al. have shown that zeolite A, a microporous, aluminosilicate mineral, and source of silicic acid, inhibits osteoclast-mediated bone resorption in vitro.[127] It is noteworthy that one study has shown that adsorption of amorphous silica nanoparticles onto hydroxyapatite surfaces can differentially alter surface properties and ultimately adhesion of osteoclasts.[128] In another study, the effect of silicon supplementation on bone status and gene expression related to bone metabolism and inflammatory mediators were tested in young estrogen-deficient rats under calcium-replete condition (0.5% diet). The authors showed 15-week supplementation of both doses of silicon (0.025% and 0.075% diet) did not restore the ovariectomy (OVX)-induced decrease of bone mineral density of vertebrae, femur, and tibia. Moreover, several bone biochemical markers were not significantly changed by silicon supplementation. However, the decrease in the ratio of receptor activator of nuclear factor kappa-B ligand (RANKL)/osteoprotegerin protein (OPG) back to normal suggested possible bone health benefits from silicon supplementation.[129] In a different study, the authors determined the short-term effects of silanol, a soluble organic silicon, on trabecular bone in mature ovariectomized rats. Three-month-old rats were either sham-treated or ovariectomized (OVX) and provided 10 µg/kg/day of 170 estradiol (E2) alone, or with 0.1 mg silicon/kg/day or 1.0 mg silicon/kg/day of silanol for 1 month. The results indicate that a short-term preventive treatment with the organic silicon silanol partially prevented the trabecular bone loss in mature OVX rats by reducing bone resorption and increasing bone formation, possibly through stimulatory effects on the formation and/or the stability of the organic bone matrix.[130] A subsequent study by Macdonald et al. found that dietary silicon interacts with estrogen to beneficially affect bone health.[123] Silicon has previously been shown to significantly enhance the rate of bone mineralization and calcification much like vitamin D.[96] There are potentially conflicting reports since Jugdaohsingh et al. found that silicon supplementation in drinking water did not significantly alter silicon concentrations in the bones of rodents suggesting an additional nutritional cofactor might be absent such as vitamin K in rodents fed a low silicon diet.[131]

Dietary silicon can interact synergistically with non-nutritive dietary components. For example, studies have demonstrated that genistein (GEN), a polyphenolic isoflavone, and silicon synergistically protected against ovariectomy-induced bone loss through upregulation of the OPG/RANKL ratio.[132,133] The effects of concomitant supplementation of genistein and silicon on bone mineral density and bone metabolism-related markers were determined in OVX female rats treated daily for 10 weeks either with dietary (5 mg genistein/g body weight) and soluble silicon in demineralized water (silicon 20 mg/kg body weight/day). The lumbar spine and femur bone mineral densities were significantly decreased after OVX surgery; however, these decreases were inhibited by the genistein and/or silicon, and the bone mineral density of the lumbar spine and femur was the highest in the OVX-GEN-Si-treated group. The authors also noted restored bone volume and trabecular thickness of femoral trabecular bone in the OVX group. At the same time, the treatment with genistein and/or silicon decreased serum alkaline phosphatase and osteocalcin, which were increased by ovariectomy; serum alkaline phosphatase and osteocalcin in the OVX-GEN-Si group were lower than those in the OVX-GEN and OVX-Si groups. The results above indicate that genistein and silicon have synergistic effects on bone formation in ovariectomized rats.

Biosilica, synthesized by the enzyme silicatein, induces hydroxyapatite formation in osteoblast-like SaOS-2 cells. During growth of SaOS-2 cells on biosiliceous matrices, hydroxyapatite formation is induced, while syntheses of cartilaginous proteoglycans and sulfated glycosaminoglycans are downregulated. Thus, it appears that biosilica-stimulated osteoprotegerin synthesis in osteoblast-like cells counteracts those pathways that control RANKL expression and function, e.g., maturation of pre-osteoclasts and activation of osteoclasts rendering considerable biomedical potential of bio-silica for treatment and prophylaxis of osteoporotic disorders.[134] Zhou et al. show that OSA stimulates osteoblast differentiation in vitro by upregulating miR-146a to antagonize NF-kB activation.[135] Dong et al. demonstrate that biological silicon stimulates collagen type 1 and osteocalcin synthesis in human osteoblast-like cells through BMP2/Smad/RUNX2 signaling pathway.[136] Collectively, there are several potential putative mechanisms for the molecular effects of silicon.

8.4.2 Vascular Disease and Atherosclerosis

There are higher incidences of sudden death, cerebrovascular diseases, arterial hypertension, and coronary heart disease in soft water areas (low concentrations of ions, e.g., calcium and magnesium) of the United States suggesting, in part, that the absence of components presence in hard water, i.e., minerals, may be contributors to vascular diseases. As a result, a major research effort has been devoted to identifying potential protective factors in hard water including calcium, magnesium, manganese, and silicon, as examples, all of which are considered potentially beneficial.[137]

Silicon is recognized by epidemiologic and biochemical studies as a protective trace element in atherosclerosis. The highest concentrations of silica in the human occur in connective and elastic tissues and especially the normal human aorta where it functions as a crosslinking agent stabilizing collagen and presumably strengthening the vasculature.[86] Atherosclerosis significantly decreases silicon levels in arterial walls. Moreover, silicon levels decrease just prior to plaque development, which may indicate that silicon deficiencies cause inherent weaknesses in blood vessel walls. In a study by Trincǎ et al., the antiatheromatous effect of sodium silicate was tested in rabbits fed a standard control diet, an atherogenic diet, and a sodium silicate-supplemented atherogenic diet. Levels of total lipids, cholesterol, triglycerides, free fatty acids, and phospholipids indicated that the sodium silicate supplemented to the atherogenic diet maintained lipid levels of free fatty acids and triglycerides in rabbits.[138] In a subsequent study, silicon administered intravenously or per os in rabbits inhibited experimental atheromas normally induced by an atheromatous diet, decreasing the number of atheromatous plaques and lipid deposits. It was proposed that the preservation of elastic fiber architecture, as well as of ground substance and the lack of free fatty acid accumulation in the aortic intima decreased plaque formation.[139] In contrast, it has been shown in female ApoE knockout mice, prone to atherosclerosis, that silicon deficiency does not exacerbate diet-induced fatty lesions suggesting that dietary silicon has no effect on atherosclerosis development and vascular health contrary to the reported findings in the cholesterol-fed rabbit model.[140] In a study by Maehira et al. using soluble silica and coral sand, as a natural silicon-containing material, the effect on hypertension, a contributing factor to atherosclerosis, was evaluated in spontaneously hypertensive rats. In rats fed 50 mg/kg dietary silicon for 8 weeks, systolic blood pressure was significantly lowered by 18 mmHg. In spontaneously hypertensive rats fed water-soluble silica or silica-rich coral sand for 8 weeks, systolic blood pressure (BP) was significantly lowered by 18 mm Hg for the silicon group and 16 mm Hg for the coral sand group compared with the control. Soluble silica and coral sand treatments suppressed the aortic gene expressions of angiotensinogen and growth factors related to vascular remodeling. Silicon stimulated the expression of peroxisome proliferator-activated receptor-γ, which has anti-inflammatory and antihypertensive effects on vascular cells.[141] In another study, the functional changes of the endothelial cells were determined in the aortic rings of rats subjected to 50 mg silicon/kg body weight in their drinking water for 8 days. Dietary silicon beneficially modified the characteristics of endothelial relaxants and attenuated smooth muscle cell responsiveness to nitric oxide.[142]

Silicon has also been suggested to exert a protective role in atherosclerosis through its effects on blood vessel-associated glycosaminoglycan (GAG) and collagen integrity and function as a cross-linker.[40] GAGs are long unbranched (linear) polysaccharides consisting of repeating disaccharide units including hyaluronan, chondroitin, dermatan, heparan, and keratan. Silicon is also a constituent of the enzyme prolylhydrolase, which synthesizes collagen and GAGs. Dietary silicon may also facilitate the formation of GAGs and collagen and/or serve a structural role as a component of GAGs where it crosslinks, and strengthens, polysaccharide chains. Nakashima et al. have noted that the GAGs content of the aorta was inversely correlated with the severity of atherosclerosis. Interestingly, Nakashima et al. showed that the silicon content in fatty streaks and/or atheroma was significantly higher than in normal human aortic intimal regions suggesting that the increase of silicon in the aortic intima is related to the occurrence and/or progression of atherosclerosis.[143]

In other studies, the effect of silicon, as silicon-enriched restructured pork, was tested on lipemia, lipoprotein profile, and oxidation markers of aged rats fed high-fat, high-energy, cholesterol-enriched atherogenic diets for 8 weeks, as well as the effects of silicon on liver antioxidant defense in aged rats fed cholesterol-enriched high saturated/high cholesterol diets as a model of non-alcoholic steatohepatitis.[144] CHOL-Si diets enhanced hepatic antioxidant status, reduced hepatosomatic index and increased superoxide dismutase (SOD) antioxidant activity. The authors suggest that silicon incorporated into restructured pork matrix counterbalanced the deleterious effect of consuming an atherogenic high-saturated/high-cholesterol diet, by improving hepatic antioxidant defenses. Further studies have demonstrated that silicon-enriched restructured pork significantly reduced VLDL-C levels, VLDL oxidation, and increased LDL receptor gene expression in aged rats fed the atherogenic diet.[145]

8.4.3 Neurodegenerative Disease

Metals, such as aluminum, that can cross the blood-brain barrier and generate directly or indirectly reactive oxygen (ROS) and/or reactive nitrogen species (RNS) can cause oxidative stress and induce significant damage to the neuronal structure of the brain. Aluminum is abundant in the environment but is not a micronutrient. However, ingestion and/or exposures can cause deposition and accumulation in the body, e.g., brain, where it can cause considerable damage.[146,147] Aluminum, a nonredox-active metal, is a well-known toxicant and its salts can accelerate oxidative damage of neurons.[148] Oxidative stress is one of the critical features in the pathogenesis of Alzheimer's disease and has been demonstrated in brain tissue from Alzheimer's patients as well as other neurodegenerative diseases. Aluminum is a contributing factor to oxidative stress, as it generates ROS shown to cause oxidative damage to neurons through interaction with iron, a redox active metal, and promotion of free radical-generating Fenton reactions, which can increase hallmark aggregation and accumulation of beta-amyloid peptides (of 36–43 amino acids) intimately involved in Alzheimer's disease. Collectively, studies clearly indicate that aluminum promotes oxidative stress capable of damaging neuronal cell.[149]

The molecular pathogenesis of Alzheimer's disease includes many risk factors including extracellular deposition of β-amyloid, accumulation of intracellular neurofibrillary tangles, oxidative neuronal damage, and activation of inflammatory cascades.[150] Although the subject of continuing scientific debate, aluminum has been detected in the neurons of the brain within neurofibrillary tangles in both Alzheimer's and Parkinson's disease patients with dementia and is proposed to play crucial roles as a crosslinker in β-amyloid oligomerization.[151–154] Given silica's propensity to bind aluminum, different forms were compared and it was determined that a soluble silica polymer, oligomeric silica, has a much higher affinity for aluminum than does monomeric silica. In fact, oligomeric silica reduced the availability of aluminum by 67% compared with the control, whereas monomeric silica had no effect. Monomeric silica was readily taken up from the gastrointestinal tract and then excreted in urine (53%), whereas oligomeric silica was not absorbed or excreted.[155] Although the neurotoxicity of aluminum is well documented, the association with neurodegenerative disorders is the subject of debate as is the potential benefit of consuming silica.[149,156] Some epidemiological studies, but not all, suggested that silica could be protective against aluminum damage because it complexes aluminum, reduces absorption, and/or enhances its excretion.[157–159] As stated previously, silicon readily complexes with aluminum in nature with aluminosilicates that are the most prevalent naturally occurring silicates. Aluminum silicates are water insoluble and although the processes involved in aluminum bioavailability are unclear regarding its transport into the central nervous system, numerous reports show that silicic acid can, in fact, reduce aluminum absorption and ultimate brain deposition.[21,160]

Total allowable concentrations of monomeric inorganic aluminum and hydrated aluminum silicates in drinking water have been determined.[148] Hydrated aluminum silicates seldom cause medical problems unless the daily doses consumed are substantially greater than those used clinically or as dietary supplements. A NOAEL of 13 mg/kg/day as total aluminum can be identified based on histologic osteomalacia seen in adult hemodialysis patients given aluminum hydroxide for up to 7 years as a phosphate binder. A chronic NOAEL for montmorillonite as representative of the hydrated aluminum silicates was identified from the highest dietary concentration (20,000 ppm) fed in a 28-week bioassay with male and female Sprague-Dawley rats. Since young rats consume standard laboratory chow at ~23 g/day, this concentration corresponds to 56 mg aluminum/kg/day. Application of 3× interspecies uncertainty factor and a 3× factor to account for study duration results in a chronic oral reference dose of 6 mg aluminum/kg/day.[148]

There have been several studies exploring the interaction of silicon and aluminum in cognitive impairment. In an epidemiological study, Rondeau et al. examined associations between exposure to aluminum or silica from drinking water and risk of cognitive decline, dementia, and Alzheimer's disease among 1,925 elderly subjects followed for 15 years. The authors concluded that cognitive decline with time was greater in subjects with a higher daily intake or geographic exposure to aluminum from drinking water. An increase of 10 mg/day in silica intake was significantly associated with a reduced risk of dementia.[161] In a population-based survey of 3,777 French subjects age 65 years and older, cognitive impairment was studied to determine any impact of silicon and aluminum concentration on cognition in the elderly. The association between cognitive impairment and aluminum depended on the pH and the concentration of silica. That is, high levels of aluminum appeared to exert deleterious effects when the silica concentration was low, but there was a protective effect when the pH and the silica level were high. The threshold for an aluminum effect was low (3.5 μg/L).[162,163] Thus, it appears that the relative concentration of both aluminum and silica in drinking water are important in determining benefit or detriment regarding the risk and/ or exacerbation of Alzheimer's disease.[46] For example, soft water contains less silica acid and more aluminum while the converse is true for hard water.[44] In a study by Exley et al., introduction of hard water rich in silica significantly reduced overall aluminum levels in the body presumably through reduced absorption of aluminum as supported by reduced urinary concentrations.[164] A subsequent study showed that drinking up to 1 L of a silicon-rich mineral water daily for 12 weeks fostered urinary removal of aluminum in both control and Alzheimer patient groups without increasing urinary excretion of iron and copper.[165] Moreover, there were clinically relevant increases in cognitive performance in 20% of participants.

González-Muñoz et al. have shown that beer consumption, a rich source of bioavailable silicic acid, can reduce cerebral oxidation caused by aluminum toxicity purportedly by modulating gene expression of pro-inflammatory cytokines and antioxidative enzymes.[88,166] Previous studies suggest that regular beer intake reverses the pro-oxidant and inflammatory statuses induced by aluminum nitrate intoxication. Non-alcoholic beer, silicon, and hops blocked the negative effects on in vivo antioxidant and inflammatory status induced by aluminum nitrate and improved swimming and rearing behaviorals.[167] Emerging evidence suggests that by affecting mineral balance, aluminum may enhance some events associated with neurodegenerative diseases particularly in the aged.[147,154] Administration of $Al(NO_3)_3$ to 6-week-old male rats induced metal imbalance, inflammation, and antioxidant status impairment in the brain. Those effects were blocked to a significant extent by silicic acid and beer administration.[168] Using similar conditions of parenteral nutrition, it was determined whether the reaction of silicon (SiO_2) with Al^{3+} to form hydroxyaluminosilicates (HAS) reduced the bioavailability and toxicity through intraperitoneal administrations of 0.5 mg aluminum/kg/day and/or 2 mg silicon/kg/day in Wistar rats. SiO_2 displayed a protective effect in the hippocampus and cerebellum against cellular damage caused by Al^{3+}-induced lipid peroxidation, thus may be an important protector.[169]

Aluminum exposure has also been linked to multiple sclerosis (MS) with affected individuals excreting elevated amounts of urinary aluminum. Progressive MS is a chronic autoimmune condition without a known, specific cause and few therapeutic strategies. Silicon-rich mineral waters foster aluminum removal thus may be efficacious in individuals with MS. In a study of 15 individuals diagnosed with secondary progressive MS, individuals consumed <1.5 L of a silicon-rich mineral water every day for 12 weeks. In those consuming silicon-rich mineral water, urinary aluminum excretion increased significantly compared to a 12-week baseline period suggesting efficacy for those with secondary progressive MS.[170]

8.4.4 Diabetes

Type 2 diabetes is a disorder of glycemia occurring largely via insulin resistance. Many micronutrients can regulate metabolism and gene expression associated with glycemia thus potentially influence the development of diabetes.[171] In a report by Oschilewski et al., administration of silica to Biobreeding (BB) rats, prone to spontaneous diabetic syndrome, completely prevented the development of diabetes.[172] Rats were treated with 100 mg silica/kg body weight via intraperitoneal and intravenous routes and observed for weight changes, glycosuria, and ketonuria. The authors showed marked inhibition of the development of diabetes (1 of 31 in treated group versus 9 of 31 for control group) and attribute the protection of silica to reduced infiltration of the endocrine (hormone [insulin]-producing) pancreatic islets by macrophages. Kahn et al. showed in a previous study exploring bone health that dietary silicon suppressed bone marrow-derived peroxisome-proliferator receptor-gamma, which regulates bone metabolism, but also regulates glucose metabolism where it is a ligand-activated transcription factor and a molecular target of a class of insulin-sensitizing drugs referred to as thiazolidinediones.[14,173] In the subsequent study, the anti-diabetic effects of silicon were investigated in obese diabetic KKAy mice prone to hyperleptinemia, hyperinsulinemia, and hyperlipidemia (50 ppm silicon for 8 weeks). Interestingly, silicon and coral sand, a rich source of silicon, displayed anti-diabetic effects through blood glucose reductions and increases in insulin responsiveness, as well as improvement in the responses to the adipokines leptin and adiponectin.[174] The authors report this as a novel function of anti-osteoporotic silicon and suggest use of silicon as a potential anti-diabetic agent capable of reducing plasma glucose and reducing the risk of diabetic glomerulonephropathy. Given the promising results from current studies, it is evident that more research is needed regarding the potential novel therapeutic applications of silicon, as silica, for prevention and management of diabetes.

8.4.5 Wound Healing

Silica is widely and routinely used in medical and surgical applications including engineering for regeneration of tissues, e.g., wound repair, and organs. This typically is in the form of collagen scaffolds, which are used as sponges, thin sheets, or gels. Collagen as a long fibrous structural protein possesses the appropriate properties for tissue regeneration including pore structure, permeability, hydrophilicity, and stability in vivo. Moreover, scaffolds permit deposition and growth of cells, e.g., osteoblasts and fibroblasts, promoting normal tissue growth and restoration.[175] There are studies that suggest that dietary silicon as well can exert beneficial effects on wound repair when used in surgical applications.

The successful healing of wounds requires local synthesis of significant amounts of collagen with its high hydroxyproline content using amino acid precursors such as proline and ornithine.[176] In animal studies, silica-deficient diets result in poor formation of connective tissues including collagen and, as a result, ultimate structural damage. Silica maintains the health of connective tissues due, in part, to its interaction with the formation of GAG where silicon is routinely found and presumed to have an integral role. As a result, a deficiency in silica could result in reduced skin elasticity and wound healing due to its role in collagen and GAG formation. Seaborn and Nielsen have reported that silicon deprivation decreases collagen formation in wounds and bone, and decreases ornithine transaminase enzyme activity in liver with reduced enzymatic conversion of ornithine to proline.[111,177] In a rodent study, silicon deprivation affected collagen formation at several different stages of bone development, the activities of collagen-forming enzymes, and consequent collagen deposition in other tissues. It has also been proposed that the decrease in silicon concentration of connective tissues in rats is a marker of connective tissue turnover.[120] This has major implications suggesting that silicon is important in wound healing and supports that dietary silicon, as silicic acid, can exert therapeutic effects for this use.

There is compelling evidence that some silicon-containing materials such as nanoparticles promote would healing via delivery of bioactive OSA either directly, or following metabolism, achieving concentrations up to 2 mM in aqueous microenvironments. In a study of the effect of silica nanoparticle suspensions with differing degrees of surface charge and dissolution rate on human dermal fibroblasts, investigators found that silica was non-toxic and stimulated fibroblast proliferation and migration.[178] Moreover, it was determined that amine-functionalized nanoparticles promoted wound closure more rapidly than soluble OSA alone. An arginine silicate inositol (ASI) complex exerted beneficial effects on wound healing in a study using a rat model. Full-thickness excision wounds were introduced followed by coverage with 4% or 10% ASI ointments twice a day for 5, 10, or 15 days. Granulation tissue (new vascular tissue indicative of healing) appeared significantly faster in the ASI-treated groups compared to controls. The average unhealed wound area was significantly smaller, and the average percentage of total wound healing was significantly higher in ASI-treated wounds compared to control wounds. Inducible nitric oxide synthase, endothelial nitric oxide synthase, collagen, matrix metalloproteinase-2, matrix metalloproteinase-9, vascular endothelial growth factor, fibroblast growth factor, epidermal growth factor, nuclear factor kappa-light-chain-enhancer of activated B cells, and various cytokines (TNF-α and IL-1β) measured in this study displayed significant reductions in expression levels in ASI-treated wounds.[179]

The chemical form of silicon determines its absorption and bioavailability with particulate and polymerized forms exhibiting minimal oral bioavailability and monomers (maltodextrin-stabilized orthosilicic acid, M-OSA) and organic compounds (MMST) displaying high absorption. Silicon is important for optimal collagen synthesis and activation of hydroxylating enzymes improving skin strength and elasticity.[180,181] Regarding hair, higher silicon content in hair has been suggested to reduce hair loss and confer increased brightness. As a result, there is therapeutic potential for silica as the dietary supplements MMST or choline-stabilized silicic acid (ch-OSA) in skin and hair health. In another study, the dermatological effects of oral ingestion of silicon, either as solid (M-OSA–SiliciuMax® Powder) or liquid (MMST, SiliciuMax® Liquid) were determined on the skin, hair, and nails of healthy volunteers. Patients were randomized to receive elemental silicon (5 mg), either M-OSA or MMST or placebo twice a day for 5 months. Use of M-OSA and MMST significantly improved facial wrinkles and ultraviolet spots and aluminum concentrations in hair significantly decreased with the treatments.[182] Oral intake of choline-stabilized OSA (10 mg silicon/day) by 48 women with fine hair in the form of ch-OSA beadlets ($n = 24$) for 9 months exerted a positive effect on tensile strength including elasticity and break load and resulted in thicker hair.[183,184]

8.5 TOXICOLOGY OF SILICON AND SILICA

8.5.1 Chemical Forms Contributing to Toxicity

As previously discussed, elemental silicon exists primarily as an oxide largely in the form of silicon dioxide. Silica, SiO_2, is a silicic acid anhydride of monomeric OSA (H_2SiO_4), which is water-soluble and stable in aqueous solutions when relatively dilute but can polymerize and complex with numerous minerals to form silicates with aluminum silicate being the most prevalent. Several other low molecular weight, but hydrated forms, of silicic acid exist in aqueous solutions and are non-toxic. Forms of silicon that are toxic include long fibrous crystalline forms such as asbestos, which is a group of crystalline 1:1 layer hydrated silicate fibers that are classified into six types based on different physicochemical features. These include chrysotile [$Mg_6Si_4O_{10}(OH)_8$], the most common and economically important asbestos in the Northern Hemisphere, and the amphiboles: crocidolite [$Na_2(Fe^{3+})_2(Fe^{2+})_3Si_8O_{22}(OH)_2$], amosite [$(Fe, Mg)_7Si_8O_{22}(OH)_2$], anthophyllite [$(Mg, Fe)_7Si_8O_{22}(OH)_2$], tremolite [$Ca_2Mg_5Si_8O_{22}(OH)_2$], and actinolite [$Ca_2(Mg, Fe)_5-Si_8O_{22}(OH)_2$].[184,185]

Silica occurs in both noncrystalline (amorphous) and crystalline forms where crystalline silica is a basic component of soil, sand, granite, and many other minerals. Crystalline forms technically are physical states in which the silicon dioxide molecules are arranged in a repetitive pattern with unique spacing, lattice structure, and angular relationship of the atoms. Crystalline silica forms, viz., polymorphs, include quartz, cristobalite, tridymite, keatite, coesite, stishovite, and moganite. Silicosis largely occurs due to inhalation of one of the forms of crystalline silica, most commonly quartz. All three forms may become respirable size particles when workers chip, cut, drill, or grind objects that contain crystalline silica.

8.5.2 Routes of Exposure and Safety

The most noted toxicity associated with silica and asbestos are silicosis and asbestosis, respectively, as discussed previously. The key route of exposure leading to toxicity is respiratory with progressive, debilitating damage from lengthy and/or heavy inhalation of silica dust.[186] In fact, the International Agency for Research on Cancer (IARC) classifies silica as a "known human carcinogen" based on inhalation as a route of exposure.[186] Regarding dietary exposure, there is no evidence of carcinogenesis when silica was fed to rodents for ~2 years (effectively the whole life span) supporting the route of exposure is more critical than chemical. However, there are reports that magnesium trisilicate (6.5 mg elemental silicon) when used as an antacid in large amounts for years may be associated with the development of urolithiasis due to formation, in vivo, of silicon-containing stones although fewer than 30 cases have been reported in the last 80 years.[187,188] There are other reports of toxicity from oral ingestion of crystalline and amorphous silicates. For example, nephropathy can result from finely ground silicates and nephritis from long-term use of high dose, silica-containing medications as well as kidney damage and kidney stones.[189] There are some reports of increased risk of cancer (esophagus and skin) from silica-rich millet and seeds.[21,190,191] It is noteworthy that urinary excretion of silicon increases with increasing dietary intake but ultimately reaches a maximum that is set by the rate and amount of absorption.[192] This overall limitation of absorption and efficient excretion work together to minimize the potential for toxicity when silica is consumed in the diet.

8.5.2.1 Inhalation and Asbestosis

Human exposure to certain forms of crystalline silica can result in adverse effects on human health. Since 1997, IARC has classified crystalline silica as a Group 1 carcinogen, which was confirmed. The role of inflammation driven by quartz surface in genotoxic and carcinogenic effects after inhalation is confirmed, and findings support a practical threshold. Classic in vitro genotoxicity studies demonstrate a NOAEL in cell cultures of 60–70 µg/cm^2; however, transformation frequency in SHE cells suggests a lower threshold around 5 µg/cm^2. These levels correlate to an in vivo dose of 2–4 mg beyond in vivo doses (>200 µg) that cause persistent inflammation and tissue remodeling in the rat lung.[11]

When asbestos fibers are inhaled, most fibers are expelled, but some can become lodged in the lungs and remain there throughout life increasing the risk of asbestosis. Asbestosis is a chronic inflammatory and fibrotic disease affecting the parenchymal tissue of the lungs, referred to as interstitial fibrosis, caused by the inhalation and deposition of fibrous asbestos.[193] Manifestation of the disease occurs typically after high intensity and/or long-term exposure to asbestos, or silica, as a specific group of airborne crystalline silicate fibers. Asbestos fibers are invisible without magnification because their size is approximately 3–20 µm wide but as small as 0.01 µm.[184] For reference, human hair has a width of ~20–180 µm. Given the omnipotence of asbestosis in technical applications, it is considered an occupational lung disease.

8.5.2.2 Inhalation and Silicosis

Silicosis is also a form of irreversible occupational lung disease, technically a type of pneumoconiosis that is caused by inhalation of small particles of crystalline silica dust.[194,195] Inhaling finely divided crystalline silica dust even in small quantities (OSHA allows 0.1 mg/m^3) over time can lead to silicosis, bronchitis, or cancer, as the dust becomes lodged in the lungs and continuously irritates them, reducing lung capacities. It is marked by inflammation, pulmonary edema, scarring of the lungs, and formation of nodular lesions in the upper lobes of the lungs with resultant difficulty in breathing. There are several different clinical and pathologic varieties of silicosis, including simple (nodular) silicosis, acute silicosis (silicoproteinosis), complicated pneumoconiosis (progressive massive fibrosis), and true diffuse interstitial fibrosis.

8.5.2.3 Ingestion and Effect of Silica on the Intestinal Tract

Liang et al. have investigated the subchronic oral toxicity of silica nanoparticles and silica microparticles in rats to compare the difference in toxicity between two particle sizes. The authors concluded that there were no toxicologically significant changes in mortality, clinical signs, body weight, food consumption, necropsy findings, and organ weights. The tissue distribution of silicon was comparable across all groups and neither silica nanoparticles nor silica microparticles induced toxicological effects after subchronic oral exposure.[70] In a systematic review of scientific papers (2010–2016) regarding silica nanoparticle toxicity, the conclusion was that toxicity in vitro was size, dose, and cell type dependent.[196] In vivo, route of administration and physicochemical properties influenced the toxicokinetics and adverse effects were largely noted in acutely, but not chronically, exposed animals.[197] In a study of oral consumption of either magnesium trisilicate, crushed quartz, or crushed Arran granite, male guinea pigs fed magnesium trisilicate for 4 months displayed focal tubulo-interstitial nephritis affecting the distal nephron.[189] Quartz consumption showed similar but less intense lesions.

Consumer intake of silica from food has been estimated at 9.4 mg/kg body weight/day, of which 1.8 mg/kg body weight/day was estimated to be in the nano-size range (i.e., 5–200 nm) up to 43% of the total silica content. The behavior and biological effects in the gastrointestinal tract associated with oral administration remains unknown. In chronic toxicity study, rats were exposed to 100, 1000, or 2500 mg/kg body weight/day per os of SAS, or to 100, 500, or 1000 mg/kg body weight/day of NM-202 (a representative nanostructured silica) for 28 days, or to the highest dose of SAS or NM-202 for 84 days. After in vitro digestion, the intestinal content of the mid/high-dose groups had stronger gel-like properties than the low-dose groups, implying low gelation and high bioaccessibility of silica in the human intestine at relevant consumer exposure levels. After 84 days of SAS exposure, but not NM-202, silica accumulated in the spleen. Histopathological analysis revealed increased liver fibrosis after 84 days of exposure with a significant increase in hepatic expression of fibrosis-related genes.

Caco-2 cells are a well-accepted in vitro model of the human intestinal epithelium to investigate the transport across the intestinal barrier in both the absorption and excretion directions. In this model, amorphous silica particles with diameters of 50, 100, and 200 nm were incubated in simulated fasted-state and fed-state gastric and intestinal fluids. In the fed-state, silica particles agglomerated with the increased particle size inhibiting the particles' absorption into Caco-2 cells and particles' transport through the cells. There was no cytotoxicity when the average particle size was >100 nm, independent of the fluid and the concentration.[198] In a separate study, silicon supplied as OSA stabilized by vanillin complex (OSA-VC) was incubated with Caco-2 cells. OSA-VC transport was concentration-dependent and increased with duration of incubation. Absorption and excretion rates were similar indicating paracellular diffusion. Cellular accumulation of Si, polarized from the apical side of cells, was furthermore detected suggesting silicon ingested as a food supplement as OSA-VC, crosses the intestinal mucosa by passive diffusion, viz., paracellular pathway through intercellular tight junctions and accumulates by facilitated diffusion.[199] The cytotoxicity and genotoxicity of SiO_2 nanoparticles were also determined in HT29 human intestine cells. The results showed that SiO_2-25 nm and SiO_2-100 nm induced cytotoxic and genotoxic effects after a 24 h exposure. However, regarding cell viability and genotoxicity, nanoparticles (SiO_2-100 nm) displayed an inversely proportional, dose-dependent effect. That is, the higher the dose of SiO_2-100 nm used then the lower the cytotoxic and genotoxic effects observed.[200] In an additional study, the uptake of colloidal SiO_2 nanoparticles (15 and 55 nm) in Caco-2 cells was determined. The results indicated size- and concentration-dependent effects on cell death and chromosome damage following exposure to nanoparticles, concomitantly with generation of ROS with smaller particles being the most potent. The smaller particles also increased pro-inflammatory IL-8 secretion at the highest tested dose (32 μg/mL). All nanoparticles localized within the cytoplasm and not the nucleus suggesting observed genotoxic effects were likely mediated through oxidative stress rather than a direct DNA interaction.[201,202]

8.5.3 Mechanisms of Toxicity

The molecular mechanism of silica and asbestos-induced carcinogenesis is complex and unclear. Inhalation is the primary route of exposure leading to toxicity and depends on the shape, size, and absorptivity of the silica fibers as well as lung clearance and genetics.[203,204] Several mechanisms have been proposed including the adsorption, chromosome tangling, and oxidative stress hypotheses. The adsorption theory posits that the surface of asbestos has a high natural affinity for proteins and other biomolecules and presumably disrupts cell function. The chromosome tangling hypothesis argues that asbestos can interact with chromosomes and "tangle" them during cellular division causing clastogenic (chromosomal) damage. Probably the most compelling mechanism at this time is the oxidative stress theory, which purports that iron associated with asbestos fibers, once internalized, can contribute to Fenton chemistry with generation of reactive, damaging free radicals and reactive oxygen species (ROS). Moreover, deposition of asbestos and silica particles in the lungs can initiate chronic inflammation via involvement of phagocytic, ROS-producing macrophages. Although discussed separately, oxidative stress and inflammation are intimately linked and often occur concurrently, with both present in lung disease.

8.5.3.1 Oxidative Stress

Cumulative supporting evidence suggests a role for ROS and RNS in the pathogenesis of asbestos- and silica-induced diseases.[205,206] Oxidative damage to the lungs can occur directly through highly reactive hydroxyl radical formation via the Fenton and Haber–Weiss reactions with fiber surface iron, and indirectly through inflammation.[207,208] This route involves the recruitment and activation of ROS-producing inflammatory cells, such as macrophages. Other cell types also participate in the process including mesothelial cells and lung fibroblasts, which also produce ROS species in response to silica and/or asbestos. Results also suggest that E551 induced a dose-dependent cytotoxicity and changed ROS levels and altered gene expression and cell cycle in human lung fibroblasts.

Numerous in vitro studies have shown the involvement of oxidative stress in damage caused by silica. For example, Liu et al. tested the effects of silica nanoparticles on endothelial cells by measuring ROS generation, apoptosis and necrosis, proinflammatory and prothrombic properties and the levels of the apoptotic signaling proteins, and the transcription factors after exposure to silica nanoparticles (25–200 μg/mL) for 24 h.[204] Silica nanoparticles markedly induced ROS production, mitochondrial depolarization, and apoptosis in endothelial cells. Others have shown similar results with primary endothelial cells exposed to silica nanoparticles with activated and dysfunction of endothelial cells as shown by release of Von Willebrand factor and necrotic cell death.[209] In a study of mesothelial cells, exposure to crocidolite asbestos was shown to induce oxidative stress, cause DNA damage, and induce apoptosis demonstrating that phagocytosis was important for asbestos-induced injury to mesothelial cells.[210] Inflammatory and oxidative stress parameters as potential early biomarkers for silicosis.

Several human studies have been conducted to determine if oxidative stress results from asbestos exposure. Pelclova et al. measured 8-isoprostane, an oxidative stress marker, in 92 former asbestos workers with an average exposure of 24 years.[211] The results indicated higher levels of 8-isoprostane in exposed subjects compared to control subjects (69.5 versus 47.0 pg/mL) supporting asbestos-induced oxidative stress. An increase in the exhaled breath condensate concentrations of 8-isoprostane has been observed in patients with idiopathic pulmonary fibrosis and in a limited study with asbestos-exposed subjects. Measurement of exhaled breath condensate for markers of oxidative stress is one of the most promising methods available for determining pulmonary damage from environmental exposures.[212] In a study involving 83 patients (45 with asbestosis and hyalinosis and 37 with silicosis) 8-isoprostane and hydroxynonenal in exhaled breath condensate were measured, as well as hydroxynonenal, an oxidative degradation product, in urine. The results indicated that most markers correlated positively and significantly with lung function impairment.[207] In a study of 92 individuals that had either been exposed to crystalline silica or diagnosed with silicosis, and a control group of unexposed workers, samples were evaluated using numerous inflammatory and oxidative stress tests as potential early biomarkers for silicosis.[213] Significant associations were observed among inflammatory and oxidative stress biomarkers in agreement with others.[214] Moreover, L-selectin surface protein, expressed on lymphocytes, was significantly decreased in those exposed to crystalline silica indicating the importance of this immune system component as a potential marker of crystalline silica-induced toxicity.[215] These markers as well as others have been effectively developed to detect and confirm oxidative stress in patients with asbestosis and silicosis.[216,217]

Reactive substances may occur on the surfaces of silica source material. Reactive radicals at the surface of quartz or other silicon dioxide polymorphs have been studied regarding a potential role in pathogenicity. All examined dusts displayed the characteristic array of radicals of silica ground in air: silicon, silicon monoxide (SiO), SiO_2 (peroxyradical), and O^{2-} (superoxide ion), but some dusts reveal additional silyl radical forms. Comparison of standard quartz dusts with a natural quartz, grinding higher purity results in a higher radical population. As a result, a possible mechanism for fibrogenicity is proposed whereby, within the activated macrophage, a catalytic reaction occurs between surface functionalities and macrophage oxygen metabolites produced by macrophages, which ultimately triggers abnormal production of fibroblast-stimulating factors, elaborating silicosis.[218]

8.5.3.2 Inflammation

In 1987, IARC classified crystalline silica as a probable carcinogen, and in 1997, reclassified it as a Group 1 carcinogen, i.e., there was sufficient evidence for carcinogenicity in experimental animals and sufficient evidence for carcinogenicity in humans.[11] Moreover, IARC confirmed the carcinogenicity of "silica dust, crystalline in the form of quartz or cristobalite" as a Group 1 agent, with the lung as the sole tumor site. Of special relevance to the present review is that the cited "established mechanism events" for crystalline silica are restricted to the words "impaired particle clearance leading to macrophage activation and persistent inflammation."

8.5.3.2.1 Silica and Pulmonary Function

There is growing evidence that amorphous silica can cause an inflammatory response in the lung. These crystalline silicates are phagocytosed by macrophages that then release cytokines that attract and stimulate other immune cells including fibroblasts, which are responsible for the excessive production of collagen (fibrotic tissue) that is characteristic of silicosis.[219] In a study by McCarthy et al., exposure of human lung submucosal cells to SiO_2 nanoparticles (10–500 nm) for up to 24 h increased cytotoxicity and cell death, induced proinflammatory and immune system modulator gene expression, and release of IL-6 and IL-8, as well as upregulation of pro-apoptotic genes indicating oxidative stress-associated injury.[220] Bauer et al. also showed that silica nanoparticles caused dysfunction and cytotoxicity through exocytosis of von Willebrand factor and necrotic cell death in primary human endothelial cells.[209] In the study by Liu et al., incubation of endothelial cells with 200 μg/mL silica caused cell death and the release of numerous, diverse proinflammatory mediators (TNF, IL-6, IL-8, and MCP-1).[221] Silica nanoparticles also activated proinflammatory gene expression, e.g., NF-kB, and suppressed anti-inflammatory gene expression, e.g., Bcl-2. The study collectively showed that silica nanoparticles damaged endothelial cells through oxidative stress via changes in gene expression associated with inflammation.

8.5.3.2.2 Silica and the Immune System

To test the biodistribution and excretion of a well-characterized nanostructured silica representative for food applications, viz., NM-200, a single intravenous injection of NM-200 to female Sprague-Dawley rats (20 mg/kg body weight) was given with evaluation after 6 and 24 h. Silicon accumulated in the liver and spleen with concentrations decreasing in both over 24 h. Silicon induced a prominent increase of hepatic macrophages, a potentially proinflammatory cell, between both evaluation times.[72] Vis et al. have also shown that non-functionalized ultrasmall silica nanoparticles directly and size-selectively activate T cells, a proinflammatory immune cell.[222] Dendritic cells (DCs), embedded in the intestinal mucosa, are specialized first-line sensors of foreign materials invading the organism where they may initiate immune reactions against pathogens.[69] An additional function is to mediate tolerance to self-antigens and, in the intestinal milieu, to nutrients and commensal microorganisms. Co-incubation of primed DCs with food-grade silica particles and their internalization by endocytic uptake did not cause cytotoxicity or release of proinflammatory cytokines. However, silica activated immature DCs as noted via the expression of maturation markers on the cell surface capable of pathogen pattern recognition and cell signaling initiation. The authors conclude that the massive use of silica particles as a food additive should be reconsidered particularly given the observations that nanoparticles appear to activate the immune system with ensuing inflammation.[223]

8.6 POTENTIAL MEDICINAL USES OF SILICON AND SILICATES

The potential medicinal uses of silicon in the form of silica have only recently been recognized particularly with respect to bone health and prevention of neurodegenerative diseases. Data are preliminary yet supportive of potential roles in reducing the occurrence of type 2 diabetes and preservation and production of collagen, e.g., wound repair. Silicon is environmentally prevalent representing the second most abundant element yet the biological availability of silica is limited and distributed unevenly based largely on geographic location and source. As discussed previously, it is the OSA that is water soluble and bioavailable yet overall intake and absorption could be improved. Thus, OSA will likely be a prominent therapeutic medicinal agent and, in fact, many potential therapeutic applications have already been presented. For example, silicon appears to play a significant role in maintaining bone health through increased bone formation and increased bone mineral density and maintenance of connective tissues. Silicon, as dietary silica, also inhibits of absorption of toxic aluminum, which may contribute to the development of Alzheimer's disease. This occurs at a time when there is increased prevalence of osteoporosis and Alzheimer's disease as populations worldwide become older. Other potential uses include enhancement of immune function, preservation, and health of skin, hair, and nails, and use as potential anti-diabetic and anti-cancer agents.

Choline-stabilized OSA is a newly developed, concentrated solution of OSA in a choline and glycerol matrix and is promoted as biologically active and the most bioavailable form of silicon. Moreover, choline-stabilized OSA has been approved for human consumption and is considered relatively non-toxic with a tolerable upper limit exceeding 5 g/kg body weight.[17,224] There are many other supplements available including extracts of horsetail, which contains 12 mg silicon per tablet of which 85% is suggested to be bioavailable. There are many more formulations, e.g., water-soluble choline-stabilized OSA, available or currently being developed for potential marketing. NHANES III indicates median intake from supplements to be 2 mg/day, but with preparations such as the aforementioned could markedly increase.

A particularly interesting area of research and development has been the emergence and/or use of OSA-releasing compounds. Specifically, certain types of zeolites, a class of aluminosilicates with well-described ion (cation)-exchange properties have been shown to release OSA. These are already widely employed in chemical and food industries, agriculture, and environmental technologies but could find much greater use as medicinal and/or nutritional agents. In fact, the biomedical applications of zeolites include, in part, modulation of enzyme kinetics, use in hemodialysis, prevention of diabetes, increased bone formation, function as an antidiarrheal and antibacterial agent and as vaccine and tumor adjuvants.[225] The numerous biological activities of some types of zeolites documented so far is thought to be due, in large part, to the OSA-releasing property.

8.7 SUMMARY AND FUTURE DIRECTION

In conclusion, silicon, as silica and silicates, represents a very large family of molecules with potential health benefits but also with potential toxic effects depending on the form, water solubility, route of exposure, and amount consumed. For example, inhaled particulate fibrous crystalline silica can be toxic and depends heavily on route of exposure and chemical form. Silica can also dissolve in water to form non-toxic bioavailable silicic acids and specifically OSA. This form of absorbable silica found in foods and water sources is readily absorbed, reaches key tissue and organ target sites of action, and is efficiently excreted. The lack of apparent toxicity of water-soluble forms that are consumed, as opposed to inhaled, and the ongoing debate regarding essentiality as a micronutrient have obscured the relative importance of chemical speciation and potential contributions of silica.

Even though water soluble to some degree, there are limitations to absorption dictated largely by chemical instability, e.g., propensity to polymerize, and maximum allowable concentrations of water-soluble OSAs. However, there has been development of acid forms with markedly increased stability and, as a result, significant increased concentrations and bioavailability of silicon. Choline chloride-stabilized OSA is a pharmaceutical formulation that is particularly promising but other forms exist including sodium or potassium silicates, and OSA-releasing forms such as zeolites.

Further research on silicon is critically needed particularly focusing on the physiological roles of silicon and how this relates to human health, as well as the dependence on chemical speciation. Specifically, ample data exist to support a possible role of silicon in wound repair, atherosclerosis and hypertension, diabetes, several bone and connective tissue disorders, neurodegenerative diseases, e.g., Alzheimer's and Parkinson's diseases, and other conditions that occur particularly in the aging population. It is also important to further elucidate biochemical mechanisms of action of silicon-containing molecules, as silicic acids, and to extend testing more into whole body systems. Specifically, larger studies with humans are needed to explore the medicinal and nutritional potential of silicon.

ABBREVIATIONS

ASI	arginine silicate dietary supplement
DC	dendritic cells
DRI	dietary reference intake
EFSA	European Food Safety Authority
IARC	International Agency for Research on Cancer
MMST	monomethylsilanetriol
MP	microparticles
NOAEL	no observed adverse effects level
NP	nanoparticles
OSA	orthosilicic acid
OSA-VC	OSA-vanillin
RANKL	receptor activator of nuclear factor beta
OPG	osteoprotegerin
OVX	ovariectomized
RDA	recommended dietary allowance
ROS	reactive oxygen species
RNS	reactive nitrogen species
SAS	synthetic amorphous silica
TUL	tolerable upper limit

REFERENCES

1. P. J. Tréguer and C. L. De La Rocha, The world ocean silica cycle, *Annual Review of Marine Science,* 2013, **5**, 477–501.
2. P. Treguer, D. M. Nelson, A. J. Van Bennekom, D. J. DeMaster, A. Leynaert and B. Queguiner, The silica balance in the world ocean: a reestimate, *Science,* 1995, **268**, 375–379.
3. P. Pondaven and P. Tréguer, Global change silica control of carbon dioxide, *Nature,* 2000, **406**, 358–359.
4. L. M. Jurkić, I. Cepanec, S. K. Pavelić and K. Pavelić, Biological and therapeutic effects of orthosilicic acid and some ortho-silicic acid-releasing compounds: new perspectives for therapy, *Nutrition & Metabolism,* 2013, **10**, 2.
5. C. Exley, Silicon in life. A bioinorganic solution to bioorganic essentiality, *Journal of Inorganic Biochemistry,* 1998, **69**, 139–144.
6. S. Sjöberg, Silica in aqueous environments, *Journal of Non-Crystalline Solids,* 1996, **196**, 51–57.

7. K. R. Martin, The chemistry of silica and its potential health benefits, *Journal of Nutrition, Health and Aging*, 2007, **11**, 94–97.
8. R. Jugdaohsingh, Silicon and bone health, *Journal of Nutrition, Health and Aging*, 2007, **11**, 99–110.
9. A. R. Elmore, Cosmetic ingredient rev exp panel, final report on the safety assessment of aluminum silicate, calcium silicate, magnesium aluminum silicate, magnesium silicate, magnesium trisilicate, sodium magnesium silicate, zirconium silicate, attapulgite, bentonite, fuller's earth, hectorite, kaolin, lithium magnesium silicate, lithium magnesium sodium silicate, montmorillonite, pyrophyllite, and zeolite, *International Journal of Toxicology*, 2003, **22**, 37–102.
10. R. Villota and J. G. Hawkes, Food applications and the toxicological and nutritional implications of amorphous silicon dioxide, *Critical Reviews in Food Science and Nutrition*, 1986, **23**, 289–321.
11. P. J. A. Borm, P. Fowler and D. Kirkland, An updated review of the genotoxicity of respirable crystalline silica, *Particle and Fibre Toxicology*, 2018, **15**, 23–17.
12. R. Merget, H. Bauer, S. Kupper, H. Philippou, R. Bauer, R. Breitstadt, et al., Health hazards due to the inhalation of amorphous silica, *Archives of Toxicology*, 2002, **75**, 625–634.
13. U. Saffiotti, L. N. Daniel, Y. Mao, X. Shi, A. O. Williams and M. E. Kaighn, Mechanisms of carcinogenesis by crystalline silica in relation to oxygen radicals, *Environmental Health Perspectives*, 1994, **102 Suppl 10**, 159–163.
14. K. R. Martin, Silicon: the health benefits of a metalloid, *Metal Ions in the Life Sciences*, 2013, **13**, 451–473.
15. C. Exley, G. Guerriero and X. Lopez, Silicic acid: the omniscient molecule, *Science of the Total Environment*, 2019, **665**, 432–437.
16. C. C. Perry, An overview of silica in biology: its chemistry and recent technological advances, *Progress in Molecular and Subcellular Biology*, 2009, **47**, 295.
17. Scientific Panel, Opinion of the scientific panel on dietetic products, nutrition and allergies on a request from the commission related to the tolerable upper intake level of silicon, *The EFSA Journal*, 2004, **60**, 1–11.
18. Institute of Medicine, *Dietary Reference Intakes for Vitamin A, Vitamin K, Arsenic, Boron, Chromium, Copper, Iodine, Iron, Manganese, Molybdenum, Nickel, Silicon, Vanadium, and Zinc*. Washington, DC: The National Academies Press, 2001.
19. S. Sripanyakorn, R. Jugdaohsingh, W. Dissayabutr, S. H. C. Anderson, R. P. H. Thompson and J. J. Powell, The comparative absorption of silicon from different foods and food supplements, *British Journal of Nutrition*, 2009, **102**, 825–834.
20. H. J. M. Bowen and A. Peggs, Determination of the silicon content of food, *Journal of the Science of Food and Agriculture*, 1984, **35**, 1225–1229.
21. J. P. Bellia, J. D. Birchall and N. B. Roberts, Beer: a dietary source of silicon, *The Lancet*, 1994, **343**, 235.
22. J. A. Pennington, Silicon in foods and diets, *Food Additives and Contaminants*, 1991, **8**, 97–118.
23. R. Jugdaohsingh, S. H. C. Anderson, K. L. Tucker, H. Elliott, D. P. Kiel, R. P. H. Thompson, et al., Dietary silicon intake and absorption, *The American Journal of Clinical Nutrition*, 2002, **75**, 887–893.
24. C. Exley, A possible mechanism of biological silicification in plants, *Frontiers in Plant Science*, 2015, **6**, 853.
25. A. Frew, L. A. Weston, O. L. Reynolds and G. M. Gurr, The role of silicon in plant biology: a paradigm shift in research approach, *Annals of Botany*, 2018, **121**, 1265–1273.
26. C. C. Perry and T. Keeling-Tucker, Biosilicification: the role of the organic matrix in structure control, *Journal of Biological Inorganic Chemistry*, 2000, **5**, 537–550.
27. D. Coskun, R. Deshmukh, H. Sonah, J. G. Menzies, O. Reynolds, J. F. Ma, et al., The controversies of silicon's role in plant biology, *New Phytologist*, 2019, **221**, 67–85.
28. H. Etesami and B. R. Jeong, Silicon (Si): review and future prospects on the action mechanisms in alleviating biotic and abiotic stresses in plants, *Ecotoxicology and Environmental Safety*, 2018, **147**, 881–896.
29. D. Debona, F. A. Rodrigues and L. E. Datnoff, Silicon's role in abiotic and biotic plant stresses, *Annual Review of Phytopathology*, 2017, **55**, 85–107.
30. F. Guntzer, C. Keller and J. Meunier, Benefits of plant silicon for crops: a review, *Agronomy for Sustainble Development*, 2012, **32**, 201–213.

31. M. A. Farooq and K. J. Dietz, Silicon as versatile player in plant and human biology: overlooked and poorly understood, *Frontiers in Plant Science,* 2015, **6**, 994.
32. A. Manivannan and Y. Ahn, Silicon regulates potential genes involved in major physiological processes in plants to combat stress, *Frontiers in Plant Science,* 2017, **8**, 1346.
33. K. R. Martin, Dietary silicon: is biofortification essential? *Journal of Nutrition and Food Science Forecast,* 2018, **1**, 1006.
34. M. D'Imperio, M. Renna, A. Cardinali, D. Buttaro, P. Santamaria and F. Serio, Silicon biofortification of leafy vegetables and its bioaccessibility in the edible parts, *Journal of the Science of Food and Agriculture,* 2016, **96**, 751–756.
35. F. F. Montesano, M. D'Imperio, A. Parente, A. Cardinali, M. Renna and F. Serio, Green bean biofortification for Si through soilless cultivation: plant response and Si bioaccessibility in pods, *Scientific Reports,* 2016, **6**, 31662.
36. M. D'Imperio, G. Brunetti, I. Gigante, F. Serio, P. Santamaria, A. Cardinali, et al., Integrated in vitro approaches to assess the bioaccessibility and bioavailability of silicon-biofortified leafy vegetables and preliminary effects on bone, *In Vitro Celular and Developmental Biology,* 2017, **53**, 217–224.
37. S. A. McNaughton, C. Bolton-Smith, G. D. Mishra, R. Jugdaohsingh and J. J. Powell, Dietary silicon intake in post-menopausal women, *British Journal of Nutrition,* 2005, **94**, 813–817.
38. L. Burns, M. Ashwell, J. Berry, C. Bolton-Smith, A. Cassidy, M. Dunnigan, et al., UK Food Standards Agency Optimal Nutrition Status Workshop: environmental factors that affect bone health throughout life, *British Journal of Nutrition,* 2003, **89**, 835–840.
39. S. Sripanyakorn, R. Jugdaohsingh, H. Elliott, C. Walker, P. Mehta, S. Shoukru, et al., The silicon content of beer and its bioavailability in healthy volunteers, *British Journal of Nutrition,* 2004, **91**, 403–409.
40. Mancinella, Silicon, a trace element essential for living organisms. Recent knowledge on its preventive role in atherosclerotic process, aging and neoplasms, *La Clinica terapeutica,* 1991, **137**, 343–350.
41. A. Anasuya, S. Bapurao and P. K. Paranjape, Fluoride and silicon intake in normal and endemic fluorotic areas, *Journal of Trace Elements in Medicine and Biology,* 1996, **10**, 149–155.
42. H. Robberecht, R. Van Cauwenbergh, V. Van Vlaslaer and N. Hermans, Dietary silicon intake in Belgium: sources, availability from foods, and human serum levels, *Science of the Total Environment,* 2009, **407**, 4777–4782.
43. J. W. Dobbie and M. J. B. Smith, The silicon content of body fluids, *Scottish Medical Journal,* 1982, **27**, 17–19.
44. G. A. Taylor, A. J. Newens, J. A. Edwardson, D. Kay and D. P. Forster, Alzheimer's disease and the relationship between silicon and aluminium in water supplies in northern England, *Journal of Epidemiology and Community Health,* 1995, **49**, 323–324.
45. J. J. Powell, S. A. McNaughton, R. Jugdaohsingh, S. H. C. Anderson, J. Dear, F. Khot, et al., A provisional database for the silicon content of foods in the United Kingdom, *British Journal of Nutrition,* 2005, **94**, 804–812.
46. S. Gillette-Guyonnet, S. Andrieu, F. Nourhashemi, V. de La Guéronnière, H. Grandjean and B. Vellas, Cognitive impairment and composition of drinking water in women: findings of the EPIDOS study, *The American Journal of Clinical Nutrition,* 2005, **81**, 897–902.
47. M. Calomme, P. Geusens, N. Demeester, G. Behets, P. D'Haese, J. Sindambiwe, et al., Partial prevention of long-term femoral bone loss in aged ovariectomized rats supplemented with choline-stabilized orthosilicic acid, *Calcified Tissue International,* 2006, **78**, 227–232.
48. A. Barel, M. Calomme, A. Timchenko, K. Paepe, N. Demeester, V. Rogiers, et al., Effect of oral intake of choline-stabilized orthosilicic acid on skin, nails and hair in women with photodamaged skin, *Archives of Dermatological Research,* 2005, **297**, 147–153.
49. D. S. Kalman, S. Feldman, A. Samson and D. R. Krieger, A clinical evaluation to determine the safety, pharmacokinetics, and pharmacodynamics of an inositol-stabilized arginine silicate dietary supplement in healthy adult males, *Clinical Pharmacology: Advances and Applications,* 2015, **7**, 103–109.
50. A. Marcowycz, B. Housez, C. Maudet, M. Cazaubiel, G. Rinaldi and K. Croizet, Digestive absorption of silicon, supplemented as orthosilicic acid–vanillin complex, *Molecular Nutrition & Food Research,* 2015, **59**, 1584–1589.
51. D. V. Scholey, D. J. Belton, E. J. Burton and C. C. Perry, Bioavailability of a novel form of silicon supplement, *Scientific Reports,* 2018, **8**, 1–8.

52. J. Vidé, A. Virsolvy, C. Romain, J. Ramos, N. Jouy, S. Richard, et al., Dietary silicon-enriched spirulina improves early atherosclerosis markers in hamsters on a high-fat diet, *Nutrition,* 2015, **31**, 1148–1154.

53. P. Allain, A. Cailleux, Y. Mauras and J. C. Renier, Digestive absorption of silicon after a single administration in man in the form of methylsilanetriol salicylate, *Therapie,* 1983, **38**, 171.

54. F. Aguilar, A. Oskarsson and M. Younes, Safety of organic silicon (monomethylsilanetriol, MMST) as a novel food ingredient for use as a source of silicon in food supplements and bioavailability of orthosilicic acid from the source, *EFSA Journal,* 2016, **14**, n/a.

55. R. Jugdaohsingh, S. H. Anderson, S. D. Kinrade and J. J. Powell, Response to Prof D. Vanden Berghe letter: there are not enough data to conclude that monomethylsilanetriol is safe, *Nutrition & Metabolism,* 2013, **10**, 65.

56. R. Jugdaohsingh, M. Hui, S. H. Anderson, S. D. Kinrade and J. J. Powell, The silicon supplement 'Monomethylsilanetriol' is safe and increases the body pool of silicon in healthy pre-menopausal women, *Nutrition & Metabolism,* 2013, **10**, 37.

57. D. A. Vanden Berghe, There are not enough data to conclude that monomethylsilanetriol is safe, *Nutrition & Metabolism,* 2013, **10**, 66.

58. C. Fruijtier-Pölloth, The safety of nanostructured synthetic amorphous silica (SAS) as a food additive (E 551), *Archives of Toxicolicology,* 2016, **90**, 2885–2916.

59. J. Athinarayanan, V. Periasamy, M. Alsaif, A. Al-Warthan and A. Alshatwi, Presence of nanosilica (E551) in commercial food products: TNF-mediated oxidative stress and altered cell cycle progression in human lung fibroblast cells, *Cellular Biology and Toxicology,* 2014, **30**, 89–100.

60. M. Younes, P. Aggett, F. Aguilar, R. Crebelli, B. Dusemund, M. Filipič, et al., Re-evaluation of silicon dioxide (E 551) as a food additive, *EFSA Journal,* 2018, **16**, n/a.

61. C. Laurino and B. Palmieri, Zeolite: "The Magic Stone"; main nutritional, environmental, experimental and clinical fields of application, *Nutricion Hospitalaria,* 2015, **32**, 573–581.

62. L. Bacakova, M. Vandrovcova, I. Kopova and I. Jirka, Applications of zeolites in biotechnology and medicine - a review, *Biomaterials Science,* 2018, **6**, 974–989.

63. E. A. Cefali, J. C. Nolan, W. R. McConnell and D. L. Walters, Pharmacokinetic study of zeolite A, sodium aluminosilicate, magnesium silicate, and aluminum hydroxide in dogs, *Pharmaceutical Research,* 1995, **12**, 270–274.

64. P. E. Keeting, M. J. Oursler, K. E. Wiegand, S. K. Bonde, T. C. Spelsberg and B. L. Riggs, Zeolite a increases proliferation, differentiation, and transforming growth factor β production in normal adult human osteoblast-like cells in vitro, *Journal of Bone and Mineral Research,* 1992, **7**, 1281–1290.

65. S. Dekkers, P. Krystek, R. J. B. Peters, D. P. K. Lankveld, B. G. H. Bokkers, P. H. van Hoeven-Arentzen, et al., Presence and risks of nanosilica in food products, *Nanotoxicology,* 2011, **5**, 393–405.

66. L. Xu, Y. Liu, R. Bai and C. Chen, Applications and toxicological issues surrounding nanotechnology in the food industry, *Pure and Applied Chemistry,* 2010, **82**, 349–372.

67. S. Dekkers, H. Bouwmeester, P. M. J. Bos, R. J. B. Peters, A. G. Rietveld and A. G. Oomen, Knowledge gaps in risk assessment of nanosilica in food: evaluation of the dissolution and toxicity of different forms of silica, *Nanotoxicology,* 2013, **7**, 367–377.

68. J. H. Lim, P. Sisco, T. K. Mudalige, G. Sánchez-Pomales, P. C. Howard and S. W. Linder, Detection and characterization of SiO_2 and TiO_2 nanostructures in dietary supplements, *Journal of Agricultural and Food Chemistry,* 2015, **63**, 3144–3152.

69. H. C. Winkler, M. Suter and H. Naegeli, Critical review of the safety assessment of nano-structured silica additives in food, *Journal of Nanobiotechnology,* 2016, **14**, 6.

70. C. L. Liang, Q. Xiang, W. M. Cui, J. Fang, N. N. Sun, X. P. Zhang, et al., Subchronic oral toxicity of silica nanoparticles and silica microparticles in rats, *Biomedical and Environmental Sciences,* 2018, **31**, 197–207.

71. G. M. Berlyne, A. J. Adler, N. Ferran, S. Bennett and J. Holt, Silicon metabolism. I. Some aspects of renal silicon handling in normal man, *Nephron,* 1986, **43**, 5.

72. N. Waegeneers, A. Brasseur, E. Van Doren, S. Van der Heyden, P. Serreyn, L. Pussemier, et al., Short-term biodistribution and clearance of intravenously administered silica nanoparticles, *Toxicology Reports,* 2018, **5**, 632–638.

73. S. Pruksa, A. Siripinyanond, J. J. Powell and R. Jugdaohsingh, Silicon balance in human volunteers; a pilot study to establish the variance in silicon excretion versus intake, *Nutrition & Metabolism,* 2014, **11**, 4.

74. N. B. Roberts and P. Williams, Silicon measurement in serum and urine by direct current plasma emission spectrometry, *Clinical Chemistry,* 1990, **36**, 1460–1465.
75. J. F. Popplewell, S. J. King, J. P. Day, P. Ackrill, L. K. Fifield, R. G. Cresswell, et al., Kinetics of uptake and elimination of silicic acid by a human subject: a novel application of 32Si and accelerator mass spectrometry, *Journal of Inorganic Biochemistry,* 1998, **69**, 177–180.
76. R. Jugdaohsingh, S. Anderson, L. Lakasing, S. Sripanyakorn, S. Ratcliffe and J. J. Powell, Serum silicon concentrations in pregnant women and newborn babies, *The British Journal of Nutrition,* 2013, **110**, 2004–2010.
77. R. Jugdaohsingh, S. Sripanyakorn and J. J. Powell, Silicon absorption and excretion is independent of age and sex in adults, *The British Journal of Nutrition,* 2013, **110**, 1024–1030.
78. Y. Kim, M. Kim and M. Choi, Relationship between dietary intake and urinary excretion of silicon in free-living Korean adult men and women, *Biological Trace Element Research,* 2019, **191**, 286–293.
79. J. W. Dobbie and M. J. B. Smith, Urinary and serum silicon in normal and uraemic individuals. In: D. Evered and M. O'Connor, editors, *Ciba Foundation Symposium 121 - Silicon Biochemistry.* Chichester, UK: John Wiley & Sons, Ltd, 2007: 194–213.
80. D. M. Reffitt, R. Jugdaohsingh, R. P. H. Thompson and J. J. Powell, Silicic acid: its gastrointestinal uptake and urinary excretion in man and effects on aluminium excretion, *Journal of Inorganic Biochemistry,* 1999, **76**, 141–147.
81. B. L. Smith, Analysis of hair element levels by age, sex, race, and hair color. In: M. Anke, D. Meissner and C. F. Mills, editors, *Trace Elements in Man and Animals,* New York: Kluwer, 1993: 1091–1093.
82. J. H. Austin, Silicon levels in human tissues, *Nobel Symposium,* 1997, 255–268.
83. S. Fregert, Studies on silicon in tissues with special reference to skin, *Journal of Investigative Dermatology,* 1958, **31**, 95–96.
84. J. Eisinger and D. Clairet, Effects of silicon, fluoride, etidronate and magnesium on bone mineral density: a retrospective study, *Magnesium Research,* 1993, **6**, 247.
85. A. Lassus, Colloidal silicic acid for oral and topical treatment of aged skin, fragile hair and brittle nails in females, *Journal of International Medical Research,* 1993, **21**, 209–215.
86. K. Schwarz, B. Ricci, S. Punsar and M. Karvonen, Inverse relation of silicon in drinking water and atherosclerosis in Finland, *The Lancet,* 1977, **309**, 538–539.
87. K. Schwarz and D. B. Milne, Growth-promoting effects of silicon in rats, *Nature,* 1972, **239**, 333–334.
88. M. J. González-Muñoz, I. Meseguer, M. I. Sanchez-Reus, A. Schultz, R. Olivero, J. Benedí, et al., Beer consumption reduces cerebral oxidation caused by aluminum toxicity by normalizing gene expression of tumor necrotic factor alpha and several antioxidant enzymes, *Food and Chemical Toxicology,* 2008, **46**, 1111–1118.
89. E. Bissé, T. Epting, A. Beil, G. Lindinger, H. Lang and H. Wieland, Reference values for serum silicon in adults, *Analytical Biochemistry,* 2005, **337**, 130–135.
90. F. H. Nielsen, Update on the possible nutritional importance of silicon, *Journal of Trace Elements in Medicine and Biology,* 2014, **28**, 379–382.
91. S. Ratcliffe, R. Jugdaohsingh, J. Vivancos, A. Marron, R. Deshmukh, J. F. Ma, et al., Identification of a mammalian silicon transporter, *American Journal of Physiology: Cell Physiology,* 2017, **312**, C561.
92. A. P. Garneau, G. A. Carpentier, A. A. Marcoux, R. Frenette-Cotton, C. F. Simard, W. Rémus-Borel, et al., Aquaporins mediate silicon transport in humans, *PLoS One,* 2015, **10**, e0136149.
93. A. P. Garneau, A. Marcoux, R. Frenette-Cotton, R. Bélanger and P. Isenring, A new gold standard approach to characterize the transport of Si across cell membranes in animals, *Journal of Cellular Physiology,* 2018, **233**, 6369–6376.
94. F. H. Nielsen, How should dietary guidance be given for mineral elements with beneficial actions or suspected of being essential? *Journal of Nutrition,* 1996, **126**, 2385S.
95. F. H. Nielsen, Should bioactive trace elements not recognized as essential, but with beneficial health effects, have intake recommendations, *Journal of Trace Elements in Medicine and Biology,* 2014, **28**, 406–408.
96. E. M. Carlisle, Silicon: a possible factor in bone calcification, *Science,* 1970, **167**, 279–280.
97. E. M. Carlisle, Silicon: an essential element for the chick, *Science,* 1972, **178**, 619–621.
98. E. M. Carlisle, The nutritional essentiality of silicon, *Nutrition Reviews,* 1982, **40**, 193–198.
99. E. M. Carlisle, Biochemical and morphological changes associated with long bone abnormalities in silicon deficiency, *The Journal of Nutrition,* 1980, **110**, 1046–1056.

100. E. M. Carlisle, In vivo requirement for silicon in articular cartilage and connective tissue formation in the chick, *The Journal of Nutrition,* 1976, **106**, 478–484.

101. E. M. Carlisle, Silicon as a trace nutrient, *Science of the Total Environment,* 1988, **73**, 95–106.

102. E. M. Carlisle, Silicon as an essential trace element in animal nutrition. In: D. Evered and M. O'Connor, editors, *Ciba Foundation Symposium 121 - Silicon Biochemistry.* Chichester, UK: John Wiley & Sons, Ltd, 2007: 123–139.

103. S. P. Robins and S. A. New, Markers of bone turnover in relation to bone health, *Proceedings of the Nutrition Society,* 1997, **56**, 903–914.

104. L. F. Rodella, V. Bonazza, M. Labanca, C. Lonati and R. Rezzani, A review of the effects of dietary silicon intake on bone homeostasis and regeneration, *Journal of Nutrition, Health, and Aging,* 2014, **18**, 820–826.

105. M. Rondanelli, A. Opizzi, S. Perna, and M. A. Faliva, Update on nutrients involved in maintaining healthy bone, *Endocrinología y Nutrición,* 2012, **60**, 197–210.

106. J. Nieves, Skeletal effects of nutrients and nutraceuticals, beyond calcium and vitamin D, *Osteoporos International,* 2013, **24**, 771–786.

107. C. T. Price, J. R. Langford and F. A. Liporace, Essential nutrients for bone health and a review of their availability in the average North American diet, *The Open Orthopaedics Journal,* 2012, **6**, 143–149.

108. C. T. Price, K. J. Koval and J. R. Langford, Silicon: a review of its potential role in the prevention and treatment of postmenopausal osteoporosis, *International Journal of Endocrinology,* 2013, **2013**, 316783.

109. E. M. Carlisle, Silicon: a requirement in bone formation independent of vitamin D1, *Calcified Tissue International,* 1981, **33**, 27–34.

110. T. D. Spector, M. R. Calomme, S. H. Anderson, G. Clement, L. Bevan, N. Demeester, et al., Choline-stabilized orthosilicic acid supplementation as an adjunct to calcium/vitamin D3 stimulates markers of bone formation in osteopenic females: a randomized, placebo-controlled trial, *BMC Musculoskeletal Disorders,* 2008, **9**, 85.

111. C. Seaborn and F. Nielsen, Dietary silicon and arginine affect mineral element composition of rat femur and vertebra, *Biological Trace Element Research,* 2002, **89**, 237–246.

112. C. D. Seabron and F. H. Nielsen, Dietary silicon and arginine affect acid and alkaline phosphatase and calcium 45 uptake in bone of rats, *Journal Trace Elements Experimental Medicine,* 1994, **7**, 1–11.

113. M. Kim, E. Kim, J. Jung and M. Choi, Effect of water-soluble silicon supplementation on bone status and balance of calcium and magnesium in male mice, *Biological Trace Element Research,* 2014, **158**, 238–242.

114. R. Marcus, D. W. Dempster and M. L. Bouxsein, The nature of osteoporosis. In: R. Marcus, D. Dempster, J. Cauley, D. Feldman, editors, *Osteoporosis.* 2013: 21–30.

115. A. A. Moukarzel, Silicon deficiency may be involved in bone disease of parenteral nutrition, *Journal of the American College of Nutrition,* 2012, **50**, 681–687.

116. A. Schiano, F. Eisinger, P. Detolle, A. M. Laponche, B. Brisou and J. Eisinger, Silicon, bone tissue and immunity, *Revue du rhumatisme et des maladies osteo-articulaires,* 1979, **46**, 483.

117. F. Maehira, I. Miyagi and Y. Eguchi, Effects of calcium sources and soluble silicate on bone metabolism and the related gene expression in mice, *Nutrition,* 2009, **25**, 581–589.

118. M. Kim, Y. Bae, M. Choi and Y. Chung, Silicon supplementation improves the bone mineral density of calcium-deficient ovariectomized rats by reducing bone resorption, *Biological Trace Element Research,* 2009, **128**, 239–247.

119. E. Kim, S. Bu, M. Sung and M. Choi, Effects of silicon on osteoblast activity and bone mineralization of MC3T3-E1 cells, *Bioogical Trace Element Research,* 2013, **152**, 105–112.

120. R. Jugdaohsingh, A. I. E. Watson, P. Bhattacharya, G. H. van Lenthe and J. J. Powell, Positive association between serum silicon levels and bone mineral density in female rats following oral silicon supplementation with monomethylsilanetriol, *Osteoporosis International: A Journal Established as Result of Cooperation between the European Foundation for Osteoporosis and the National Osteoporosis Foundation of the USA,* 2015, **26**, 1405–1415.

121. Y. Bae, J. Kim, M. Choi, Y. Chung and M. Kim, Short-term administration of water-soluble silicon improves mineral density of the femur and tibia in ovariectomized rats, *Biological Trace Element Research,* 2008, **124**, 157–163.

122. M. Q. Arumugam, D. C. Ireland, R. A. Brooks, N. Rushton and W. Bonfield, The effect of orthosilicic acid on collagen type I, alkaline phosphatase and osteocalcin mRNA expression in human bone-derived osteoblasts in vitro, *Key Engineering Materials,* 2006, **309–311**, 121–124.

123. H. M. Macdonald, A. C. Hardcastle, R. Jugdaohsingh, W. D. Fraser, D. M. Reid, J. J. Powell, Dietary silicon interacts with oestrogen to influence bone health: evidence from the Aberdeen Prospective Osteoporosis Screening Study, *Bone,* 2011, **50**, 681–687.

124. R. Jugdaohsingh, K. L. Tucker, N. Qiao, L. A. Cupples, D. P. Kiel and J. J. Powell, Dietary silicon intake is positively associated with bone mineral density in men and premenopausal women of the Framingham Offspring Cohort, *Journal of Bone and Mineral Research,* 2004, **19**, 297–307.

125. Ž. Mladenović, A. Johansson, B. Willman, K. Shahabi, E. Björn and M. Ransjö, Soluble silica inhibits osteoclast formation and bone resorption in vitro, *Acta Biomaterialia,* 2014, **10**, 406–418.

126. D. M. Reffitt, N. Ogston, R. Jugdaohsingh, H. F. J. Cheung, B. A. J. Evans, R. P. H. Thompson, et al., Orthosilicic acid stimulates collagen type 1 synthesis and osteoblastic differentiation in human osteoblast-like cells in vitro, *Bone,* 2003, **32**, 127–135.

127. N. Schütze, M. J. Oursler, J. Nolan, B. L. Riggs and T. C. Spelsberg, Zeolite a inhibits osteoclast-mediated bone resorption in vitro, *Journal of Cellular Biochemistry,* 1995, **58**, 39–46.

128. P. Kalia, R. A. Brooks, S. D. Kinrade, D. J. Morgan, A. P. Brown, N. Rushton, et al., Adsorption of amorphous silica nanoparticles onto hydroxyapatite surfaces differentially alters surfaces properties and adhesion of human osteoblast cells, *PLoS One,* 2016, **11**, e0144780.

129. S. Y. Bu, M. Kim and M. Choi, Effect of silicon supplementation on bone status in ovariectomized rats under calcium-replete condition, *Biological Trace Element Research,* 2016, **171**, 138–144.

130. M. Hott, C. de Pollak, D. Modrowski and P. J. Marie, Short-term effects of organic silicon on trabecular bone in mature ovariectomized rats, *Calcified Tissue International,* 1993, **53**, 174–179.

131. R. Jugdaohsingh, M. R. Calomme, K. Robinson, F. Nielsen, S. H. C. Anderson, P. D'Haese, et al., Increased longitudinal growth in rats on a silicon-depleted diet, *Bone,* 2008, **43**, 596–606.

132. C. Chen, H. Zheng and S. Qi, Genistein and silicon synergistically protects against ovariectomy-induced bone loss through upregulating OPG/RANKL ratio, *Biological Trace Element Research,* 2018, 188, 1–10.

133. S. Qi and H. Zheng, Combined effects of phytoestrogen genistein and silicon on ovariectomy-induced bone loss in rat, *Biological Trace Element Research,* 2017, **177**, 281–287.

134. M. Wiens, X. Wang, H. C. Schröder, S. Ute, U. Ute, M. Hiroshi, et al., The role of biosilica in the osteoprotegerin/RANKL ratio in human osteoblast-like cells, *Biomaterials,* 2010, **31**, 7716–7725.

135. X. Zhou, F. M. Moussa, S. Mankoci, P. Ustriyana, N. Zhang, S. Abdelmagid, et al., Orthosilicic acid, Si(OH)$_4$, stimulates osteoblast differentiation in vitro by upregulating miR-146a to antagonize NF-κB activation, *Acta Biomaterialia,* 2016, **39**, 192–202.

136. M. Dong, G. Jiao, H. Liu, W. Wu, S. Li, Q. Wang, et al., Biological silicon stimulates collagen type 1 and osteocalcin synthesis in human osteoblast-like cells through the BMP-2/Smad/RUNX2 signaling pathway, *Biological Trace Element Research,* 2016, **173**, 306–315.

137. S. Tubek, Role of trace elements in primary arterial hypertension, *Biological Trace Element Research,* 2007, **115**, 301.

138. L. Trincă, O. Popescu and I. Palamaru, Serum lipid picture of rabbits fed on silicate-supplemented atherogenic diet, *Revista medico-chirurgicala a Societatii de Medici si Naturalisti din Iasi,* 1999, **103**, 99.

139. J. Loeper, J. Goy-Loeper, L. Rozensztajn and M. Fragny, The antiatheromatous action of silicon, *Atherosclerosis,* 1979, **33**, 397–408.

140. R. Jugdaohsingh, K. Kessler, B. Messner, M. Stoiber, L. D. Pedro, H. Schima, et al., Dietary silicon deficiency does not exacerbate diet-induced fatty lesions in female ApoE knockout mice, *The Journal of Nutrition,* 2015, **145**, 1498–1506.

141. F. Maehira, K. Motomura, N. Ishimine, I. Miyagi, Y. Eguchi and S. Teruya, Soluble silica and coral sand suppress high blood pressure and improve the related aortic gene expressions in spontaneously hypertensive rats, *Nutrition Research,* 2011, **31**, 147–156.

142. G. Oner, S. Cirrik, M. Bulbul and S. Yuksel, Dietary silica modifies the characteristics of endothelial dilation in rat aorta, *Endothelium,* 2006, **13**, 17–23.

143. Y. Nakashima, A. Kuroiwa and M. Nakamura, Silicon contents in normal, fatty streaks and atheroma of human aortic intima: its relationship with glycosaminoglycans, *British Journal of Experimental Pathology,* 1985, **66**, 123.

144. J. A. Santos-López, A. Garcimartín, P. Merino, M. E. López-Oliva, S. Bastida, J. Benedí, et al., Effects of silicon vs. hydroxytyrosol-enriched restructured pork on liver oxidation status of aged rats fed high-saturated/high-cholesterol diets, *PloS One,* 2016, **11**, e0147469.

145. A. Garcimartín, J. A. Santos-López, S. Bastida, J. Benedí and F. J. Sánchez-Muniz, Silicon-enriched restructured pork affects the lipoprotein profile, VLDL oxidation, and LDL receptor gene expression in aged rats fed an atherogenic diet, *The Journal of Nutrition,* 2015, **145**, 2039–2045.

146. A. C. Miu, The silicon link between aluminium and Alzheimer's disease, *Journal of Alzheimers Disease,* 2006, **10**, 39–42.

147. J. D. Birchall, The interrelationship between silicon and aluminium in the biological effects of aluminium. In: D. J. Chadwick and J. Whelan, editors, *Ciba Foundation Symposium 169 - Aluminium in Biology and Medicine.* Chichester, UK: John Wiley & Sons, Ltd, 2007: 50–68.

148. C. C. Willhite, G. L. Ball and C. J. McLellan, Total allowable concentrations of monomeric inorganic aluminum and hydrated aluminum silicates in drinking water, *Critical Reviews in Toxicology,* 2012, **42**, 358–442.

149. V. Gupta, S. Anitha, M. Hegde, L. Zecca, R. Garruto, R. Ravid, et al., Aluminium in Alzheimer's disease: are we still at a crossroad? *CMLS, Cellular and Molecular Life Sciences,* 2005, **62**, 143–158.

150. K. Chopra, S. Misra and A. Kuhad, Neurobiological aspects of Alzheimer's disease, *Expert Opinion on Therapeutic Targets,* 2011, **15**, 535–555.

151. M. Kawahara and M. Kato-Negishi, Link between aluminum and the pathogenesis of Alzheimer's disease: the integration of the aluminum and amyloid cascade hypotheses, *International Journal of Alzheimer's Disease,* 2011, **2011**, 276393-17.

152. D. P. Perl and A. R. Brody, Alzheimer's disease: X-ray spectrometric evidence of aluminum accumulation in neurofibrillary tangle-bearing neurons, *Science,* 1980, **208**, 297–299.

153. D. P. Perl, Relationship of aluminum to Alzheimer's disease, *Environmental Health Perspectives,* 1985, **63**, 149–153.

154. E. M. Carlisle and M. J. Curran, Effect of dietary silicon and aluminum on silicon and aluminum levels in rat brain, *Alzheimer Disease and Associated Disorders,* 1987, **1**, 83–89.

155. R. Jugdaohsingh, D. M. Reffitt, C. Oldham, J. P. Day, L. K. Fifield, R. P. Thompson, et al., Oligomeric but not monomeric silica prevents aluminum absorption in humans, *The American Journal of Clinical Nutrition,* 2000, **71**, 944–949.

156. V. Rondeau, A review of epidemiologic studies on aluminum and silica in relation to Alzheimer's disease and associated disorders, *Reviews on Environmental Health,* 2002, **17**, 107–121.

157. V. Frisardi, V. Solfrizzi, C. Capurso, P. G. Kehoe, B. P. Imbimbo, A. Santamato, et al., Aluminum in the diet and Alzheimer's disease: from current epidemiology to possible disease-modifying treatment, *Journal of Alzheimer's Disease,* 2010, **20**, 17–30.

158. J. L. Domingo, M. Gomez and M. T. Colomina, Oral silicon supplementation: an effective therapy for preventing oral aluminum absorption and retention in mammals, *Nutrition Reviews,* 2011, **69**, 41–51.

159. J. A. Edwardson, P. B. Moore, I. N. Ferrier, J. S. Lilley, J. Barker, J. Templar, et al., Effect of silicon on gastrointestinal absorption of aluminium, *The Lancet,* 1993, **342**, 211–212.

160. E. M. Carlisle, A silicon aluminum relationship in aged brain, *Neurobiology of Aging,* 1986, **7**, 545–546.

161. V. Rondeau, H. Jacqmin-Gadda, D. Commenges, C. Helmer and J. Dartigues, Aluminum and silica in drinking water and the risk of Alzheimer's disease or cognitive decline: findings from 15-year follow-up of the PAQUID cohort, *American Journal of Epidemiology,* 2009, **169**, 489–496.

162. H. Jacqmin-Gadda, D. Commenges, L. Letenneur and J. F. Dartigues, Silica and aluminum in drinking water and cognitive impairment in the elderly, *Epidemiology,* 1996, **7**, 281–285.

163. V. Rondeau, H. Jacqmin-Gadda, D. Commenges and J.-F. Dartigues, RE: aluminum in drinkgin water and cognitive decline in elderly subjects: the Paquid Cohort, *American Journal of Epidemiology,* 2001, **154**, 288–290.

164. C. Exley, C. Schneider and F. J. Doucet, The reaction of aluminium with silicic acid in acidic solution: an important mechanism in controlling the biological availability of aluminium? *Coordination Chemistry Reviews,* 2002, **228**, 127–135.

165. S. Davenward, P. Bentham, J. Wright, P. Crome, D. Job, A. Polwart, et al., Silicon-rich mineral water as a non-invasive test of the 'Aluminum Hypothesis' in Alzheimer's disease, *Journal of Alzheimer's Disease,* 2013, **33**, 423–430.

166. M. J. González-Muñoz, A Peña and I Meseguer, Role of beer as a possible protective factor in preventing Alzheimer's disease, *Food and Chemical Toxicology: An International Journal Published for the British Industrial Biological Research Association,* 2008, **46**, 49–56.

167. P. Merino, J. A. Santos-López, C. J. Mateos, I. Meseguer, A. Garcimartín, S. Bastida, et al., Can nonalcoholic beer, silicon and hops reduce the brain damage and behavioral changes induced by aluminum nitrate in young male Wistar rats? *Food and Chemical Toxicology,* 2018, **118**, 784–794.

168. M. J. González-Muñoz, A. Garcimartán, I. Meseguer, C. J. Mateos-Vega, J. M. Orellana, A. Peña-Fernández, et al., Silicic acid and beer consumption reverses the metal imbalance and the prooxidant status induced by aluminum nitrate in mouse brain, *Journal of Alzheimer's Disease,* 2017, **56**, 917–927.

169. S. Noremberg, D. Bohrer, M. Schetinger, A. Bairros, J. Gutierres, J. Gonçalves, et al., Silicon reverses lipid peroxidation but not acetylcholinesterase activity induced by long-term exposure to low aluminum levels in rat brain regions, *Biological Trace Element Research,* 2016, **169**, 77–85.

170. K. Jones, C. Linhart, C. Hawkins and C. Exley, Urinary excretion of aluminium and silicon in secondary progressive multiple sclerosis, *EBioMedicine,* 2017, **26**, 60–67.

171. B. S. O'Connell, Select vitamins and minerals in the management of diabetes, *Diabetes Spectrum,* 2001, **14**, 133–148.

172. U. Oschilewski, U. Kiesel and H. Kolb, Administration of silica prevents diabetes in BB-rats, *Diabetes,* 1985, **34**, 197–199.

173. S. E. Kahn and B. Zinman, Point: recent long-term clinical studies support an enhanced role for thiazolidinediones in the management of type 2 diabetes, *Diabetes Care,* 2007, **30**, 1672–1676.

174. F. Maehira, N. Ishimine, I. Miyagi, Y. Eguchi, K. Shimada, D. Kawaguchi, et al., Anti-diabetic effects including diabetic nephropathy of anti-osteoporotic trace minerals on diabetic mice, *Nutrition,* 2011, **27**, 488–495.

175. S. M. Oliveira, R. A. Ringshia, R. Z. Legeros, E. Clark, M. J. Yost, L. Terracio, et al., An improved collagen scaffold for skeletal regeneration, *Journal of Biomedical Materials Research. Part A,* 2010, **94**, n/a.

176. J. E. Albina, J. A. Abate and B. Mastrofrancesco, Role of ornithine as a proline precursor in healing wounds, *Journal of Surgical Research,* 1993, **55**, 97–102.

177. C. Seaborn and F. Nielsen, Silicon deprivation decreases collagen formation in wounds and bone, and ornithine transaminase enzyme activity in liver, *Biological Trace Element Research,* 2002, **89**, 247–258.

178. S. Quignard, T. Coradin, J. J. Powell and R. Jugdaohsingh, Silica nanoparticles as sources of silicic acid favoring wound healing in vitro, *Colloids and Surfaces B: Biointerfaces,* 2017, **155**, 530–537.

179. A. Durmus, M. Tuzcu, O. Ozdemir, C. Orhan, N. Sahin, I. Ozercan, et al., Arginine silicate inositol complex accelerates cutaneous wound healing, *Biological Trace Element Research,* 2017, **177**, 122–131.

180. L. A. Araujo, F. Addor and P. M. Campos, Use of silicon for skin and hair care: an approach of chemical forms available and efficacy, *Annals of Dermatolology,* 2016, **91**, 331–335.

181. A. Lassus, Colloidal silicic acid for the treatment of psoriatic skin lesions, arthropathy and onychopathy. A Pilot Study, *Journal of International Medical Research,* 1997, **25**, 206–209.

182. A. Ferreira, É. Freire, H. Polonini, P. da Silva, M. Brandão and N. Raposo, Anti-aging effects of monomethylsilanetriol and maltodextrin-stabilized orthosilicic acid on nails, skin and hair, *Cosmetics,* 2018, **5**, 41.

183. R. Wickett, E. Kossmann, A. Barel, N. Demeester, P. Clarys, D. Vanden Berghe, et al., Effect of oral intake of choline-stabilized orthosilicic acid on hair tensile strength and morphology in women with fine hair, *Archives of Dermatological Research,* 2007, **299**, 499–505.

184. T. A. Sporn, Mineralogy of asbestos, *Recent Results in Cancer Research. Fortschritte der Krebsforschung. Progres dans les recherches sur le Cancer,* 2011, **189**, 1.

185. D. M. Bernstein, The health risk of chrysotile asbestos, *Current Opinion in Pulmonary Medicine,* 2015, **20**, 366–370.

186. I. T. S. Yu, L. A. Tse, T. W. Wong, C. C. Leung, C. M. Tam and A. C. Chan, Further evidence for a link between silica dust and esophageal cancer, *International Journal of Cancer,* 2005, **114**, 479–483.

187. F. S. Haddad and A. Kouyoumdjian, Silica stones in humans, *Urologia Internationalis,* 1986, **41**, 70–76.

188. C. A. Osborne, F. Jacob, J. P. Lulich, M. J. Hansen, C. Lekcharoensul, L. K. Ulrich, et al., Canine silica urolithiasis. Risk factors, detection, treatment, and prevention, *Veterinary Clinics of North America, Small Animal Practice,* 1999, **29**, 213–230.

189. J. W. Dobbie and M. J. B. Smith, Silicate nephrotoxicity in the experimental animal: the missing factor in analgesic nephropathy, *Scottish Medical Journal,* 1982, **27**, 10–16.
190. C. O'Neill, P. Jordan, T. Bhatt and R. Newman, Silica and oesophageal cancer. In: D. Evered and M. O'Connor, editors, *Ciba Foundation Symposium 121 - Silicon Biochemistry.* Chichester, UK: John Wiley & Sons, Ltd, 2007: 214–230.
191. R. Newman, Association of biogenic silica with disease, *Nutrition and Cancer,* 1986, **8**, 217–221.
192. F. H. Nielsen, Micronutrients in parenteral nutrition: boron, silicon, and fluoride, *Gastroenterology,* 2009, **137**, 55.
193. Y. Haibing, Y. Lei, Z. Junyue and C. Jingqiong, Natural course of silicosis in dust-exposed workers, *Journal of Huazhong University of Science and Technology,* 2006, **26**, 257–260.
194. U. Saffiotti, Silicosis and lung cancer: a fifty-year perspective, *Acta Bio-Medica: Atenei Parmensis,* 2005, **76 Suppl 2**, 30–37.
195. C. C. Leung, I. T. Yu and W. Chen, Silicosis, *Lancet,* 2012, **379**, 2008–2018.
196. S. Murugadoss, D. Lison, L. Godderis, S. Van Den Brule, J. Mast, F. Brassinne, et al., Toxicology of silica nanoparticles: an update, *Archives of Toxicology,* 2017, **91**, 2967–3010.
197. M. van der Zande, R. J. Vandebriel, M. J. Groot, E. Kramer, Z. E. H. Rivera, K. Rasmussen, et al., Sub-chronic toxicity study in rats orally exposed to nanostructured silica, *Particle and Fibre Toxicology,* 2014, **11**, 8.
198. K. Sakai-Kato, M. Hidaka, K. Un, T. Kawanishi and H. Okuda, Physicochemical properties and in vitro intestinal permeability properties and intestinal cell toxicity of silica particles, performed in simulated gastrointestinal fluids, *BBA - General Subjects,* 2014, **1840**, 1171–1180.
199. T. Sergent, K. Croizet and Y. Schneider, In vitro investigation of intestinal transport mechanism of silicon, supplied as orthosilicic acid-vanillin complex, *Molecular Nutrition & Food Research,* 2017, **61**, 1–6.
200. J. Sergent, V. Paget and S. Chevillard, Toxicity and genotoxicity of nano-SiO$_2$ on human epithelial intestinal HT-29 cell line, *The Annals of Occupational Hygiene,* 2012, **56**, 622–630.
201. A. Tarantini, S. Huet, G. Jarry, R. Lanceleur, M. Poul, A. Tavares, et al., Genotoxicity of synthetic amorphous silica nanoparticles in rats following short-term exposure. Part 1: Oral route, *Environmental and Molecular Mutagenesis,* 2015, **56**, 218–227.
202. A. Tarantini, R. Lanceleur, A. Mourot, M. -. Lavault, G. Casterou, G. Jarry, et al., Toxicity, genotoxicity and proinflammatory effects of amorphous nanosilica in the human intestinal Caco-2 cell line, *Toxicology in Vitro,* 2015, **29**, 398–407.
203. S. Toyokuni, L. Jiang, Q. Hu, H. Nagai, Y. Okazaki, S. Akatsuka, et al., Mechanisms of asbestos-induced carcinogenesis, *Nihon eiseigaku zasshi. Japanese Journal of Hygiene,* 2011, **66**, 562–567.
204. G. Liu, P. Cheresh and D. W. Kamp, Molecular basis of asbestos-induced lung disease, *Annual Review of Pathology,* 2013, **8**, 161–187.
205. B. T. Mossman and A Churg, Mechanisms in the pathogenesis of asbestosis and silicosis, *American Journal of Respiratory and Critical Care Medicine,* 1998, **157**, 1666–1680.
206. A. Shukla, M. Gulumian, T. K. Hei, D. Kamp, Q. Rahman and B. T. Mossman, Multiple roles of oxidants in the pathogenesis of asbestos-induced diseases, *Free Radical Biology and Medicine,* 2003, **34**, 1117–1129.
207. D. Pelclova, Z. Fenclova, K. Syslova, K. Vlckova, J. Lebedova, O. Pecha, et al., Oxidative stressmarkers in exhaled breath condensate in lung fibroses are not significantly affected by systemic diseases, *Industrial Health,* 2011, **49**, 746–754.
208. D. W. Kamp and S. A. Weitzman, The molecular basis of asbestos induced lung injury, *Thorax,* 1999, **54**, 638–652.
209. A. T. Bauer, E. A. Strozyk, C. Gorzelanny, C. Westerhausen, A. Desch, M. F. Schneider, et al., Cytotoxicity of silica nanoparticles through exocytosis of von Willebrand factor and necrotic cell death in primary human endothelial cells, *Biomaterials,* 2011, **32**, 8385–8393.
210. W. Liu, J. D. Ernst and V. Courtney Broaddus, Phagocytosis of crocidolite asbestos induces oxidative stress, DNA damage, and apoptosis in mesothelial cells, *American Journal of Respiratory Cell and Molecular Biology,* 2000, **23**, 371–378.
211. D. Pelclova, Z. Fenclova, P. Kacer, M. Kuzma, T. Navratil and J. Lebedova, Increased 8-isoprostane, a marker of oxidative stress in exhaled breath condensate in subjects with asbestos exposure, *Industrial Health,* 2008, **46**, 484–489.

212. M. Corradi, P. Gergelova and A. Mutti, Use of exhaled breath condensate to investigate occupational lung diseases, *Current Opinion in Allergy and Clinical Immunology,* 2010, **10**, 93–98.

213. J. Nardi, S. Nascimento, G. Göethel, B. Gauer, E. Sauer, N. Fão, et al., Inflammatory and oxidative stress parameters as potential early biomarkers for silicosis, *Clinica Chimica Acta,* 2018, **484**, 305–313.

214. R. M. Park and W. Chen, Silicosis exposure–response in a cohort of tin miners comparing alternate exposure metrics, *American Journal of Industrial Medicine,* 2013, **56**, 267–275.

215. C. Peruzzi, S. Nascimento, B. Gauer, J. Nardi, E. Sauer, G. Göethel, et al., Inflammatory and oxidative stress biomarkers at protein and molecular levels in workers occupationally exposed to crystalline silica, *Enviromental Science and Pollution Research,* 2019, **26**, 1394–1405.

216. K. Syslová, P. Kačer, M. Kuzma, V. Najmanová, Z. Fenclová, Š. Vlčková, et al., Rapid and easy method for monitoring oxidative stress markers in body fluids of patients with asbestos or silica-induced lung diseases, *Journal of Chromatography B,* 2009, **877**, 2477–2486.

217. K. Syslová, P. Kačer, M. Kuzma, A. Pankrácová, Z. Fenclová, Š. Vlčková, et al., LC-ESI-MS/MS method for oxidative stress multimarker screening in the exhaled breath condensate of asbestosis/silicosis patients, *Journal of Breath Research,* 2010, **4**, 017104.

218. B. Fubini, E. Giamello, M. Volante and V. Bolis, Chemical functionalities at the silica surface determining its reactivity when inhaled. Formation and reactivity of surface radicals, *Toxicology and Industrial Health,* 1990, **6**, 571–598.

219. R. K. Iler, *The Chemistry of Silica: Solubility, Polymerization, Colloid and Surface Properties and Biochemistry of Silica.* New York: John Wiley & Sons, 1979.

220. J. McCarthy, I. Inkielewicz-Stępniak, J. J. Corbalan and M. W. Radomski, Mechanisms of toxicity of amorphous silica nanoparticles on human lung submucosal cells in vitro: protective effects of fisetin, *Chemical Research in Toxicology,* 2012, **25**, 2227–2235.

221. X. Liu, Y. Xue, T. Ding and J. Sun, Enhancement of proinflammatory and procoagulant responses to silica particles by monocyte-endothelial cell interactions, *Particle and Fibre Toxicology,* 2012, **9**, 36.

222. B. Vis, R. E. Hewitt, N. Faria, C. Bastos, H. Chappell, L. Pele, et al., Non-functionalized ultrasmall silica nanoparticles directly and size-selectively activate T cells, *ACS Nano,* 2018, **12**, 10843–10854.

223. H. C. Winkler, J. Kornprobst, P. Wick, L. M. Von Moos, I. Trantakis, E. M. Schraner, et al., MyD88-dependent pro-interleukin-1 beta induction in dendritic cells exposed to food-grade synthetic amorphous silica, *Particle and Fibre Toxicology,* 2017, **14**, 1–23.

224. K. Van Dyck, R. Van Cauwenbergh, H. Robberecht and H. Deelstra, Bioavailability of silicon from food and food supplements, *Fresenius Journal of Analytical Chemistry,* 1999, **363**, 541–544.

225. K. Pavelić, M. Hadžija, L. Bedrica, J. Pavelić, I. Đikić, M. Katić, et al., Natural zeolite clinoptilolite: new adjuvant in anticancer therapy, *Jouornal of Molecular Medicine,* 2001, **78**, 708–720.

CHAPTER 9

Benefits of Trivalent Chromium in Human Nutrition?

John B. Vincent
The University of Alabama

CONTENTS

9.1 INTRODUCTION

The answer to the title question is straightforward currently—no conclusively demonstrated benefits accompany supplementation of the diets of humans. This is true regardless of the health status of the individual. The European Food Safety Authority's (EFSA) Panel on Dietetic Products, Nutrition and Allergies in 2014 determined "there is no convincing evidence for a role of chromium in human metabolism and physiology," "there is no evidence that the general population is chromium deficient or has Cr(III)-responsive metabolic effects," and "there is no proof that chromium is an essential element" (1). However, such an understanding was not always the case. This chapter will examine how and why the current understanding was achieved and what future potential may exist for chromium-based neutraceuticals or pharmaceuticals.

The stable form of chromium in an aqueous environment exposed to dioxygen is the +3 oxidation state. The trivalent or chromic ion is abbreviated as Cr^{3+}; however, this notation is specifically for the ion itself. In contrast, when part of a coordination complex, the ion is written as Cr(III). In this chapter, the chromic ion can always be assumed to be part of a coordination complex and will be written as Cr(III). Even chromic chloride as added to human and animal diets is actually *trans*-$[Cr(H_2O)_4Cl_2]Cl \cdot 2H_2O$, not truly an inorganic salt but a coordination complex. While the term inorganic salts will be used to represent species of the general formula $CrX_3 \cdot nH_2O$ where X is an anion, these are all coordination complexes.

9.2 GLUCOSE TOLERANCE FACTOR AND ESSENTIAL CHROMIUM

The story of how chromium was proposed to be essential and the problems with those proposals have been reviewed many times (2) and will only be briefly summarized. In 1955, Mertz and Schwarz reported feeding rats a *Torula* yeast-based diet that resulted in the rats apparently developing impaired glucose tolerance in response to an intravenous glucose load (3). The authors believed they had identified a new dietary requirement absent from the *Torula* yeast-based diet and responsible for the glucose intolerance, for which they coined the term glucose tolerance factor or GTF (4). They (5) followed their report in 1959 by identifying the active ingredient of "GTF" as Cr(III). Addition of several inorganic compounds containing several elements (200–500 µg/kg body mass) could not restore glucose tolerance, while some inorganic Cr(III) complexes (200 µg/kg body mass) restored glucose tolerance; Brewer's yeast and acid-hydrolyzed porcine kidney powder were identified as natural sources of "GTF" (5).

However, these studies contain several flaws. The chromium content of the diet was not determined. Additionally, the rats were maintained in wire-mesh cages, possibly with stainless-steel components, allowing the rats to obtain chromium by chewing on these components. Consequently, the actual chromium intake of the rats in these studies is impossible to gauge, putting into great question the suggestion that the rats were chromium deficient. The use of the large amounts of metal ions is also of concern. As will be discussed, large doses of chromium may have pharmacological effects, not related to any nutritional requirements. Questions about data handling and the significance of the effect observed from the chromium treatment have also been raised (2, 6).

These initial studies led to efforts to isolate "GTF." Mertz and coworkers (7) reported the details of the isolation of Brewer's yeast "GTF" in 1977. This material from Brewer's yeast became equated with the term GTF after this time. The isolation procedure needs to be examined here. Brewer's yeast was extracted with boiling 50% ethanol. The ethanol was removed under vacuum, and the aqueous residual was applied to activated charcoal. Material active in bioassays (the ability of the species to potentiate the action of insulin to stimulate *in vitro* the metabolism of rat epididymal fat tissue) was eluted from the charcoal with a 1:1 mixture of concentrated ammonia and diethyl ether. After removal of the ammonia and ether under vacuum, the resulting solution was hydrolyzed by *refluxing for 18h in 5M HCl*. Finally, the HCl was removed under vacuum, the solution was extracted with ether, and the pH of the solution was adjusted to three. The cationic orange-red material was further purified by ion-exchange chromatography (7). Unfortunately, these incredibly harsh conditions would have destroyed any proteins, peptides, complex sugars, or nucleic acids that initially could have been associated with the chromium. Thus, the possibility that the form of chromium recovered after the treatment resembles the form in the yeast is remote at best.

The chemical characterization of GTF was equally problematic. The isolated "GTF" possessed a distinct feature at 262 nm in its ultraviolet spectrum, while mass spectral studies (no data present) indicated the presence of a pyridine moiety (7). This led to identification of nicotinic acid as a component of "GTF"; nicotinic acid was sublimed from the material (no experimental information given) and identified by extraction with organic solvents (no data presented). Consequently, no data were actually presented that nicotinic acid is associated with "GTF." Amino acid analyses indicated the presence of glycine, glutamic acid, and cysteine, as well as other amino acids, although neither the absolute or relative amounts were reported. Thus, the ratio of chromium to any of the amino acids or to nicotinic acid (i.e., any of the organic components) cannot be determined. Yet, the results were interpreted to indicate that "GTF" was a complex of chromium, nicotinate, glycine, cysteine, and glutamate (7).

In subsequent paper chromatography experiments (7), the material gave several bands, only one of which was active in the bioassays. The chromium in the active band represented only 6% of the total chromium. Thus, the chemical characterization was performed on an impure material, of which only a tiny fraction was active. Hence, the chemical data in no way is likely to reflect the makeup of the active species. Subsequent attempts to isolate "GTF" have found that the biologically active species from Brewer's yeast can be separated from the chromium (reviewed in Ref. (2)), i.e., "GTF" from Brewer's yeast does not contain chromium. The use of the term glucose tolerance factor (or GTF) should be discontinued. Curiously, the proposed presence of nicotinic acid, 3-carboxypyridine, inspired examination of the use of complexes made from chromium and picolinic acid (2-carboxypyridine), most notably Cr(III) picolinate (normally used as the monohydrate [Cr(picolinate)$_3$]·H$_2$O), as nutritional supplements and therapeutic agents.

9.3 CHROMIUM PICOLINATE AND BODY MASS GAIN

The history leading to the first publication of beneficial effects from chromium picolinate on body mass loss, muscle mass increase, and improvements in symptoms of type 2 diabetes in 1989 has been excellently summarized by Forrest Nielsen of the United States Department of Agriculture (USDA) (Grand Forks Human Nutrition Research Center) (8). While he was examining the use of zinc picolinate to treat children with acrodermatitis enteropathica, a genetic disorder that results in the inability to absorb zinc from cow's milk, Gary Evens, employed by the Grand Forks Human Nutrition Research center, found that zinc and other metal picolinates were absorbed better than the corresponding inorganic salts. Evans patented the supplementation of the diet with synthetic metal complexes of picolinate (9). Nutrition 21, a neutraceutical company that was to become the major supplier of chromium picolinate, licensed the patent from the USDA in 1986 and supported Evans' research leading to the 1989 publications.

> In other words, contrary to what many advertisements touting chromium picolinate led many to believe, the USDA patent is not specific to chromium, nor does it mention that chromium picolinate has any beneficial effects claimed for this form of chromium supplement.
>
> **Nielsen (8)**

Curiously, for Cr(III) picolinate, the synthesis described in the patent is basically the same as that of Ley and Fricken published in 1917 (10).

In 1989, two studies conducted by Evans with young men participating in weight training reported significant losses in body fat and increases in lean muscle mass with Cr(III) picolinate supplementation (11). In one study, ten male subjects between 18 and 32 years of age were included. Half of the student subjects received a supplement containing 200 µg chromium as chromium picolinate for 40 days; the other half received a placebo. The subjects engaged in 40-min exercise periods twice a week. Body composition was estimated by measuring the thickness of skin folds and bicep and calf circumferences. Subjects in the supplement group on average gained 2.2 kg body mass, had no significant change in percentage body fat, and had an increase of 1.6 kg in lean body mass. In contrast subjects on the placebo on average gained 1.25 kg body mass had an increase in body mass of 1.1% and increased their lean body mass 0.04 kg. The increase in lean body mass for the subjects receiving chromium was said to be statistically greater than that for the control (or placebo) group ($P=0.019$). No method to determine compliance of subjects was indicated. The error in the data (neither standard error nor standard deviation) was not presented, not allowing for the statistical data to be used in meta-analyses.

In the second study (11), 31 (of an initial 40) college football players completed a 42-day program. Half of the players were given 200 µg chromium as chromium picolinate while the other half received a placebo. The subjects exercised 1 h/day for 4 days/week. Body composition was estimated by measuring thigh, abdomen, and chest skin folds and thigh, bicep, and calf circumference. After 14 days, subjects receiving chromium on average lost 2.7% of their body fat and had an increase of lean body mass of 1.8 kg, while no changes were observed in the control group. After 6 weeks, the chromium group on average lost 1.2 kg, lost 3.6% (or 3.4 kg) of their body fat, and had an increase in lean body mass of 2.6 kg. In contrast, the control group had a loss of 1 kg of body fat and a 1.8 kg increase in lean body mass. Both the loss of body fat and the increase in lean body mass were said to be significantly greater for the chromium group ($P=0.001$ and $P=0.031$, respectively). No method to determine compliance of subjects was indicated. The error in the data was again not presented.

In 1993, Evans and Pouchnik (12) reported a study of the effects of chromium picolinate on body composition. Twelve male and twelve female participants were involved in a weekly aerobics class and were between 25 and 36 years of age. Male received 400 µg Cr/day as chromium picolinate or 400 µg Cr/day as chromium nicotinate. Females received half as much of either chromium source. Lean body mass was measured by resistivity. Data were presented with standard errors. For males receiving chromium picolinate, the lean body mass increased 2.1 kg, and the final body mass was statistically equivalent to the initial values and to the values for the group receiving chromium nicotinate. For females, the lean body mass increased 1.8 kg, and the final body mass was statistically equivalent to the initial value and to those of the group receiving chromium nicotinate. Yet, Evans claimed that despite this, the change in lean body mass for both males and females on chromium picolinate was significant ($P<0.01$). The statistical analysis, which indicated that while final and initial values were equivalent but that the difference was significant, failed to incorporate the error from both the initial and final values in the calculation of the error of the difference.

The results of the first human studies by Evans and coworkers (11, 12) were quickly questioned. For example, Lefavi (13) pointed out that

> It is likely that reviewers well-read in exercise physiology would find the notion of a 4.6-lb *lean* body mass (LBM) increase in males and a 4.0-lb LBM in females resulting from 12 weeks of a weekly aerobics class preposterous. A LBM increase that dramatic is not typically seen in subjects who are weight training three times per week for 12 weeks, no matter what they are taking…Investigators familiar with this type of research would suggest either (a) that this was one great aerobics class, or (b) people in Bemidji, MN respond in a highly unusual manner to aerobic exercise and/or are extremely chromium deficient, or (c) Dr. Evan's group is consistently having difficulty accurately measuring LBM.

The first of the two studies appears to be the basis for claims in a US patent granted to Nutrition 21 (Herb Boynton, founder of Nutrition 21, and Gary W. Evans, inventors) in 1992 regarding the use of chromium picolinate as a method to increase lean body mass and reducing percentage body fat (14). The data for the placebo group are identical between the paper and patent, but only nine subjects are included in the data for the chromium group in the patent.

The Federal Trade Commission (FTC) of the United States in 1997 ordered entities associated with the nutritional supplement chromium picolinate (specifically Herbert H. Boynton, Nutrition 21), and Selene Systems (a general partner of Nutrition 21) to stop making several representations because of the lack of "competent and reliable scientific evidence" (15). These representations include that chromium picolinate reduces body fat; chromium picolinate causes weight loss; chromium picolinate causes weight loss without dieting or exercise; chromium picolinate increases lean body mass or builds muscle; chromium picolinate increases human metabolism; chromium picolinate lowers elevated blood sugar levels; chromium picolinate is effective in the treatment or prevention of diabetes; or 90% of US adults do not consume diets with sufficient chromium to support normal insulin function, resulting in increased risk of obesity, heart disease, elevated blood sugar, high blood pressure, diabetes, or some other adverse effect on health.

The state of the field of chromium nutritional research at this time was summarized in an invited review in *Nutrition Reviews* by Hellerstein (16), whose research is outside the field of chromium:

> To an outside reviewing the literature on chromium and diabetes/obesity, the field is most striking for two features: its nearly complete lack of biomedical or clinical understanding and its high degree of

polarization...As in all field with more heat than light, the reason has been the incomplete ability to measure and test key factors...The high degree of politicization and polarization in this field is characterized by unproven claims and counterclaims and suspicion among investigators. Concerns about possible bias and potential conflict of interest have naturally emerged. Reports of benefits of supplementation (e.g., that lean tissue is increased and fat decreased by chromium in athletes in training) that were not confirmed by several subsequent studies have further these concerns.

Little has changed since 1997. For example, Vincent (17) thoroughly reviewed studies on body mass gain in 2003 (and updated in 2013 (2)) and found supplementation had not demonstrated effects on the body composition of healthy individuals, even when taken in combination with an exercise training program. Also in 2003, Pittler et al. (18) performed a meta-analysis using ten trials of normal and overweight subjects receiving chromium picolinate. They found that chromium picolinate had a statistically significant effect on body mass loss ($1.1 \pm 0.7\,kg$), although the clinical meaningfulness was questionable. No so significant statistical heterogeneity was found; however, the funnel plot was asymmetrical, suggesting the possibility of publication bias. The two largest clinical trials included in the meta-analysis reported the largest body mass gain and were performed by the same research group. These two studies along with five others of the ten trials utilized in the meta-analyses were funded by the same supplier of chromium picolinate, Nutrition 21. In fact, sensitivity analyses suggested that the results were not robust; after excluding one trial that accounted for 58% of the overall effect, the body mass change was no longer statistically significant. No effect was found for lean body mass. Body fat was found to be statistically reduced; however, excluding one study that accounted for 61% of the overall effect removed the statistical significance (18).

This trial that potentially skewed the results as it was the largest study was by Kaats et al. (19) and is worth a more critical examination. Three groups were examined: a control ($n=55$), a group receiving 200 µg chromium as chromium picolinate daily ($n=33$), and a group receiving 400 µg chromium as chromium picolinate daily ($n=66$). The data reported are the initial body mass and associated standard deviation for each experimental group and the average changes and associated standard deviations over the 72 day trial; the final data after 72 days are not reported. In addition, data from both groups receiving chromium picolinate were combined; however, the averages reported are calculated incorrectly as the average body mass initially for the combined group is lower than that of either individual group. The average changes were used in the statistical treatment of the data using one-way ANOVA, and pairs of results were compared using Student's t-tests. (Curiously, the body masses are reported to two decimal places while the changes in body mass are reported to three.)

The current author has attempted to analyze the data. One-way ANOVA's using the mean body mass changes and standard deviations including all three groups or all three groups plus the combined chromium picolinate group find the differences in the mean values among the treatment groups are not statistically significant ($P=0.062$ and $P=0.085$, respectively). However, even this comparison is questionable as the standard deviations reported for the average body mass changes appear far too small, failing to include the error from both the initial and final mass measurements. Thus, the P values are probably actually significantly higher. To attempt to check this, the final average body mass for each group was calculated using the initial values and the changes. As the final standard deviations were not reported, the error was assumed to be the same as those for the initial measurements. Using those numbers, a Student t-test was then performed individually on each group comparing initial body mass to final body mass. The initial and final body masses were not statistically different for any group (control, $P=0.951$; 200 µg chromium, $P=0.791$; 400 µg chromium, $P=0.598$). If no group experiences a significant change in body mass, then what is the use of comparing the magnitude of these insignificant changes? This study simply did not observe an effect from chromium picolinate on body mass.

The second largest trial was also performed by Kaats et al. (20). It does not report the final masses and associated standard deviations. An examination of the results of this paper also reveals that no significant change in body mass occurred for the placebo group or group receiving 400 mg chromium daily as chromium picolinate over the 90-day trail ($P=0.84$ or 1.000 with the assumption that the error in the final mean is similar to that of the initial mean). Similarly, a comparison of the average changes between the groups over the 90 days reveals no statistically significant difference ($P=0.072$), although this uses standard deviations that are probably too small again.

In 2013, two meta-analyses looked at the effect of chromium on body mass of overweight or obese subjects. Onakpoya et al. (21) analyzed 11 studies finding a statistically significant, but clinically insignificant, body mass loss of 0.50 ± 0.47 kg from chromium supplementation. The analysis was accompanied by significant heterogeneity. No significant relationship existed between dose and body mass change although doses ranged from 150 μg chromium to 1,000 μg chromium/day. No effects were found for body mass index, percentage body fat, waist circumference, or waist to hip ratio. Tian et al. (22) analyzed nine studies also found a statistically significant, but clinically insignificant, body mass loss of 1.1 ± 0.7 kg from chromium picolinate supplementation. No dose effect existed (200–1,000 mg chromium/day). The overall quality of evidence was considered low. No effects were found for body mass index, percentage body fat, or waist circumference. Both of these papers include the data of Kaats et al. (19, 20) with probably incorrect standard deviations; the results of the meta-analyses if the appropriate standard deviations for the two large studies were used would be most interesting.

Thus, chromium supplementation in doses up to 1,000 μg chromium daily does not appear to have any clinically significant effect on body mass and related parameters for healthy, overweight, or obese subjects, if it has any effect at all. The American population and populations of other developed countries have wasted millions of dollars on chromium picolinate and other chromium supplements as body mass loss agents for muscle development agents.

9.4 DIABETES AND RELATED CONDITIONS

9.4.1 Type 2 Diabetes

In a similar fashion to studies on body composition, clinical trials of chromium supplementation took off after the 1989 paper of Evans (11), which also reported the results of a clinical trial using diabetic subjects. Eleven subjects with type 2 diabetes (six men and five women) participated in the study in which they received 200 μg chromium as chromium picolinate daily for two 42-day periods with a 14-day period between the treatments. Student's t-tests were used to test for statistically significant differences. Chromium supplementation was reported to lower serum glucose, glycated hemoglobin, total cholesterol, and LDL cholesterol. No approval from an institutional review board, declaration of conflict of interest, or source of funding was provided, although approval by a review board was noted for the two trials looking at body composition described in the same paper. The results of this trial became part of a patent application to Nutrition 21 (Herb Boynton and Gary Evans, inventors) (23). However, the current author performed Student's t-tests comparing the initial and final means for every group, whether starting with the placebo or chromium, and variable; no change in any variable for any group was found to be statistically significant. Then, the changes from the initial to final value for all groups for each of the variables were analyzed by one-way ANOVA; the standard error for the changes was calculated using the reported standard error associated with each initial and final value. None of the changes were found to be statistically significant.

Agencies of the US government have commissioned two meta-analyses that have generated inconclusive results on whether chromium affects symptoms of type 2 diabetes. In the first, published in 2002 with funding from the Office of Dietary Supplements of the National Institutes of

Health (NIH), only quality four studies were identified for analysis, leading to the results being inconclusive as the combined results except for whose of one study found no effects from chromium supplementation (24). In the second, published in 2007 with funding from the Department of Health and Human Services, 18 studies were identified. The authors concluded that chromium supplementation "may have a modest effect" on glucose metabolism in type 2 diabetics but that "large heterogeneity and the overall poor quality limit the strength of our conclusions" (25). Unfortunately, the positive effects of the greatest magnitude used in the analysis came from the 12 studies ranked lowest in quality, including one large study reporting positive effects used by the first meta-analysis. Additionally, a trend was observed that commercial industry-sponsored studies were more likely to observe beneficial effects.

Clinical trials utilizing type 2 diabetic patients have been the subject of seven meta-analyses in the last decade (26–32). These analyses have come to conflicting conclusions. For example, while all examined fasting blood glucose, four found chromium supplementation resulted in lower glucose levels, and three found no statistically significant effect. The difference results largely from the criteria used to select which publications to include. The meta-analysis providing the most comprehensive literature search, finding trials missed by the other studies, and using the most rigorous inclusion standards by, for example, requiring sufficient data to calculate pre-intervention standard errors or deviations and pre- and post-intervention means found no effect from chromium supplementation (31). Even the meta-analyses finding modest beneficial effects fail to find any dependence of the effects on dose or length of time of chromium supplementation (32).

The nutraceutical company Nutrition 21 petitioned the US Food and Drug Administration (FDA) for eight qualified health claims in December 2003. They were as follows: chromium picolinate may reduce the risk of insulin resistance; chromium picolinate may reduce the risk of cardiovascular disease when caused by insulin resistance; chromium picolinate may reduce abnormally elevated blood sugar levels; chromium picolinate may reduce the risk of cardiovascular disease when caused by abnormally elevated blood sugar levels; chromium picolinate may reduce the risk of type 2 diabetes; chromium picolinate may reduce the risk of cardiovascular disease when caused by type 2 diabetes; chromium picolinate may reduce the risk of retinopathy when caused by abnormally high blood sugar level; and chromium picolinate may reduce the risk of kidney disease when caused by abnormally high blood sugar levels. After extensive review, the FDA issued a letter of enforcement discretion allowing only one qualified health claim for the labeling of dietary supplements. The claim states: "One small study suggests that chromium picolinate may reduce the risk of type 2 diabetes. FDA concludes that the existence of such a relationship between chromium picolinate and either insulin resistance or type 2 diabetes is highly uncertain" (33).

In 2014 and 2015, the American Diabetes Association's (ADA) position was "There is insufficient evidence to support the routine use of micronutrients such as chromium…to improve glycemic control in people with diabetes" (34, 35). Since 2015, the ADA's basic position has not been changed although chromium is no longer explicitly mentioned (36).

Thus, chromium supplementation in doses up to 1,000 μg chromium/day does not appear to have any clinically significant effect on symptoms of type 2 diabetes, if it has any effect at all. The American population and populations of other developed countries have wasted millions of dollars on chromium picolinate and other chromium supplements as adjuvant therapies for type 2 diabetes.

9.4.2 Polycystic Ovarian Syndrome (PCOS)

Other conditions related to insulin resistance have drawn attention for investigation of potential benefits from chromium supplementation. The use of chromium supplementation to improve symptoms of PCOS has been examined a few times in the last 20 years, although less than ten papers have appeared. Recently, three meta-analyses have been published, all using different papers and coming to different conclusions (37–39). The small number of studies with small sample sizes and mixed results do not allow for any definitive determination on whether chromium could potentially have beneficial (or deleterious) effects on patients with PCOS. For a more detailed review, see Ref. (40).

9.4.3 Depression and Related Conditions

A handful of papers in the last two decades have been focused on the potential effects of chromium supplementation on depression and related conditions where these conditions have an association with insulin resistance. However, rigorously controlled clinical trials are lacking in this area so that the results must be viewed with great caution (40).

9.5 PHARMACOLOGICAL EFFECTS

Recently, studies with rats on low chromium diets have proposed that the effects of chromium supplementation on insulin sensitivity and potentially other variables are not nutritionally relevant, but pharmacologically relevant (requiring supra-nutritional doses of chromium(III) (*vide infra*)). To understand this, one must look at exactly what a nutritionally relevant dose of chromium(III) would be.

The National Research Council (US) in 2001 established a new guideline for chromium intake. The daily adequate intake (AI, new term replacing ESADDI, estimated safe and adequate daily dietary intake) of chromium was 35 µg/day for an adult male and 25 µg/day for an adult female (41), as the old ESADDI from 1980 was clearly set too high at 50–200 µg chromium. This number sets an important guideline for assessing human studies of chromium supplementation. However, this number should be carefully understood. A Recommended Daily Allowance (RDA) is set so that >98% of Americans would get the required amount of the nutrient at this daily dose and not show signs of deficiency. The AI is more conservative, meaning that more than greater than 98% of Americans would be sufficient at this intake. Thirty µg Cr/day is the amount of Cr found in both a nutritionist designed or self-selected American diet (42, 43). Also important is realizing that the majority of Cr in the diet of people of developed nations probably comes from processing of food in stainless steel equipment, so that humans probably evolved on considerably less Cr in their diet than ~30 µg/day (44). In other words, chromium deficiency is not a problem in at least developed countries (if in any country at all). With research since 2001 failing to establish any definitive instances or symptoms of chromium deficiency, the subsequent action of the EFSA in 2014 to not set an AI for chromium is readily understandable. Given the average body mass of an American adult is about 80 kg, the AI corresponds to just under 0.40 µg chromium/kg body mass.

Several rat models of type 2 diabetes have been examined for effects from chromium administration (45). The results are dependent on the model utilized. Models of hepatic insulin resistance are unaffected by chromium supplementation. However, in contrast to human clinical trials, beneficial effects on insulin sensitivity were generally observed for models of peripheral insulin resistance, suggesting chromium(III)'s effects derive primarily from skeletal muscle (45). For the supra-nutritional doses utilized, effects from chromium supplementation did not appear to be dependent on dosage, form of chromium, or length of time of administration. As seems to happen with all chromium studies, heterogeneity of results is observed, making further interpretation difficult. A comprehensive meta-analysis of these studies has been suggested, particularly to examine potential industrial-based funding bias in the results (45). Rats in these studies received 4–100,000 µg chromium/kg body mass. Most doses far exceeded the <0.4 µg/kg corresponding to the AI in humans (even when the data are corrected for different metabolic rates or surface area between rats and humans), potentially suggesting that higher doses used in the rat studies may explain the effects in rats and lack of effects in humans from chromium(III) supplementation.

A study providing rats in metal-free cages a diet with as little chromium as reasonably possible has observed supra-nutritional doses of chromium(III) are clearly required for beneficial effects on insulin sensitivity (46). Rats were provided the AIN-93G purified diet with no added chromium in the mineral mix (<20 µg chromium/kg diet, the lowest quantity of chromium used in a rat diet examined to date) and were kept in cages with no access to metal for 6 months. No differences in body mass, insulin sensitivity, or response to a glucose challenge were observed compared to rats on the complete AIN-93G diet (with 1,000 µg chromium/kg). Adding additional Cr(III) to the diet (200 or 1,000 µg/kg), clearly supra-nutritional or pharmacological doses, also had no effects on body mass but resulted in increased insulin sensitivity. Insulin sensitivity was shown to increase as a function of added chromium. This study suggests that rats on the AIN-93G diet are already obtaining enough chromium to begin to have beneficial effects from supplementation. Thus, basal chromium levels in rodent diets must be considered when designing experiments, and effects from supplemental chromium could be difficult to detect on top of beneficial effects from high chromium content in the basal diets. In fact, studies have appeared that appear to have misinterpreted the beneficial effect from the supra-nutritional dose of chromium in the AIN-93G diet (see, for example, Refs. (47) and (48)). The chromium content of the AIN-93 diets needs to be reexamined, and chromium probably should be removed from the mineral mix.

Studies of subjects on total parenteral nutrition (TPN) have often been cited as evidence for chromium(III) being essential. Subjects that developed glucose intolerance and insulin resistance and appeared to benefit from chromium supplementation were on TPN solutions providing 5–16 µg chromium/day (for cases where the chromium concentration was reported). Because TPN is an intravenous diet, all the chromium in the TPN is introduced into the bloodstream, while only ~0.5% of chromium in a regular human diet is absorbed into the bloodstream (2, 42, 49). Therefore, the 30 µg of chromium in a typical daily American diet results in only ~0.15 µg chromium passing into the bloodstream. Consequently, the TPN solutions provided an order of magnitude to two orders of magnitude more chromium than the typical (chromium sufficient) American diet, even before treatment with additional chromium. Subjects on the TPN diets that developed insulin resistance were then treated with TPN supplemented with an additional 12, 40, or 15–250 µg chromium/day, an ~100- to 1,000-fold increase compared to chromium provided to the bloodstream from the typical diet, to alleviate their conditions. These undoubtedly represent pharmacological, rather than nutritional, doses. These studies, hence, do not provide evidence for chromium being an essential element, but they do suggest that large doses of chromium may have pharmacological effects in humans with altered glucose and carbohydrate metabolism, not just in rodents. Of course, even this conclusion must be tempered as an approximately equal number of subjects have been reported that did not respond to supplemental chromium in TPN as those that did. Additionally, the condition of the subjects before they entered the studies varied considerably, as did their time on TPN, the amount of chromium received, and several other variables. For case reports of patients on TPN, the EFSA determined that "it is unclear on the basis of these case reports whether deficiency of chromium could be considered the only cause of glucose intolerance in these patients, whether deficiency of chromium has occurred in these patients and whether chromium deficiency occurs in healthy populations" (1).

Recent case studies examining intravenous infusions of chromium (generally 3 µg/h) as a treatment for glucose intolerance have found chromium reduced insulin requirements for subjects with hyperglycemia. These studies are suggestive of a pharmacological effect of chromium as supra-nutritional quantities of chromium were utilized, but only a small number of subjects have been examined to date (50).

These case study results in combination with the rodent studies showing that supra-nutritional doses of chromium are required before any effects are observed on insulin sensitivity may suggest that beneficial effects from chromium supplementation in humans are possible at higher doses that have been used to date in clinical trials. However, clinical trials using probably ~10 mg chromium/day (or perhaps greater) will absolutely be required before any definitive conclusions regarding potential beneficial effects of pharmacological doses of chromium in humans can be reached.

9.6 TOXICITY

As no adverse effects have been convincingly associated with excess intake of chromium from food or supplements, no Upper Tolerable Limit was established for chromium as the trivalent ion in 2001 by the Institutes of Medicine of the National Research Council (US) (41). Other agencies have taken similar positions. In 2005, the US FDA found that the use of chromium supplements was safe up to 1 mg chromium/day, the highest amount used in clinical trials (32). In 2003, the Scientific Committee on Food (European Union, now replaced by EFSA) found insignificant evidence to set a Tolerable Upper Intake Level. From the limited number of studies, no evidence of adverse effects was found to be associated with supplemental chromium intake up to a dose of 1 mg/day; this recommendation was specifically for sources of chromium(III) other than chromium picolinate (51). In 2014, the EFSA Panel on Contaminants in the Food Chain (CONTAM Panel) derived a Tolerable Daily Intake (TDI) of 300 µg chromium/kg body mass daily from the lowest No Observed Adverse Effect Level (NOAEL) identified in a chromic oral toxicity study in rats (52). In 2003, the Food Standard Agency's Expert Group on Vitamins and Minerals (United Kingdom) determined that doses up to 10 mg of chromium daily should be safe for humans (53). This does not mean that no toxic effects might be associated with high intakes of trivalent chromium. No significant health concerns have been found for current commercial chromium supplements at current doses, although questions about the use of chromium picolinate have been raised.

Concerns about the potential toxicity of chromium picolinate were first raised in studies by Stearns and coworkers (54) who demonstrated the compound as a solid suspension in acetone or the mother liquor from the synthesis of chromium picolinate generated chromosomal aberrations in Chinese hamster ovary (CHO) AA8 cells. Particulate doses of chromium picolinate of 8.0 and 40 µg/cm^2, but not 4.0 µg/cm^2, were found to lead to more total aberrations than controls treated with only acetone. The number of aberrations was dose responsive. Subsequent studies by this group demonstrated that chromium picolinate was mutagenic at the hypoxanthine (guanine) phosphoribosyltransferase locus in CHO AA8 cells (55) and generates mitochondrial damage and apoptosis (56). The clastogenicity of chromium picolinate has also been studied in CHO K1 cells, and no chromosome damage was found with doses up to 770 µg/mL, which is equivalent to 123 µg/cm^2 (57). Related studies in CHO K1 cells found chromium picolinate was not mutagenic at the hypoxanthine (guanine) phosphoribosyltransferase locus (58). These studies, funded by Nutrition 21, contrast starkly with the results of Stearns and coworkers. However, the hydroxyl radical scavenger dimethylsulfoxide, DMSO, was used as the solvent for chromium picolinate in the latter studies, indicating the possibility that free radicals produced by chromium picolinate were being trapped by DMSO. Coryell and Stearns (59) have actually demonstrated the quenching of chromium picolinate's mutagenicity by DMSO. They also found that substitutions comprised 33% of chromium picolinate-derived DNA mutations, with transversions being predominate; 62% were deletions with one-exon deletions predominating. Insertions of 1–4 base pairs comprised 5%. Hence, chromium picolinate appears to be mutagenic in mammalian cells. As noted above in 2003, the UK's Expert Group on Vitamins and minerals (53) found that doses of chromium(III) up to 10 mg/day were expected to be without adverse health effects but chromium picolinate was excluded from this guidance as it had been shown to cause DNA damage in mammalian cells *in vitro*. This group put out a subsequent statement on chromium and chromium picolinate that removed the exclusion of chromium picolinate; the change was based largely on the flawed *in vitro* studies using chromium picolinate dissolved I DMSO (60). Commercial bias in results of cell culture studies with chromium supplements has been noted in a review of the subject (61).

A recent study has attempted to independently access the effects of chromium picolinate on cultured mammalian cells (62). Clastogenicity was observed after treatments with 4.0 and 80 µg/cm^2 of chromium picolinate, and the response was concentration dependent from 4.0 to 80 µg/cm^2. The current results were consistent with the work by Stearns et al. (54), who reported the exposure to chromium picolinate at 40 µg/cm^2 produced chromosomal aberrations 16-fold above control levels.

These chromosome aberrations are only significant if the body is unable to repair this damage. Hepburn and coworkers using *Drosophila* as a model organism have found that chromium picolinate, but not other forms of chromium(III), at 260 µg chromium/kg food generated developmental delays and decreases in success rates of hatching and eclosion (63). Chromium picolinate was found to generate approaching one mutation per chromosome per individual and 12% sterility. Subsequently, the ability of chromium picolinate to generate chromosomal aberrations in polytene chromosomes of salivary glands of *Drosophila* larvae was examined. In the chromium picolinate-treated group, 53% of the identified chromosomal arms were found to contain one or more aberrations, while no aberrations were observed for the identified chromosomal arms of the control group (64). Thus, cell culture studies and *in vivo* studies using mammals where the compound was administered intravenously or using fruit flies suggest chromium picolinate is toxic and clastogenic, mutagenic, and possibly carcinogenic (65).

However, a study commissioned by the National Toxicology Program (NIH) has found that providing chromium picolinate up to 5% of the diet of male and female rats and mice for 2 years had no conclusive deleterious health effects (66). This is probably the result of chromium picolinate dissociating in the gastrointestinal tract when provided orally. In the cell culture study and *in vivo* studies, the chromium picolinate probably reached cells intact (65). Hence, the supplement should not be used intravenously, for example, in TPN. The unique potential for deleterious effects from chromium picolinate apparently arises from its redox potential, which could possibly allow the chromic center to undergo redox chemistry *in vivo*.

In the last few years, concerns have arisen over the use of supplemental chromium in TPN solutions as subjects on TPN solutions that provide 10 or more µg chromium/day accumulate chromium. The American Society for Parenteral and Enteral Nutrition (ASPEN) has made examining chromium requirements in parenteral nutrition their number one urgent priority (67). While no reported cases of chromium toxicity in patients on long-term TPN are known, ASPEN has recommended that TPN without added chromium should be made available.

9.7 CONCLUSION

Chromium can no longer be considered an essential trace element. No benefits for humans have convincingly been observed from chromium supplementation so that chromium nutritional supplements cannot be recommended. Chromium is probably safe as a component of current multivitamins. Chromium(III) might have beneficial effects at larger doses than have been examined to date in clinical trials; however, quality randomized clinical trials are required to access this.

REFERENCES

1. EFSA Panel on Dietetic Products, Nutrition and Allergies. 2014. Scientific opinion on dietary reference values for chromium. *EFSA J.* 12, no. 10:3845. https://www.efsa.europa.eu/en/efsajournal/pub/3845.
2. Vincent, J. B. 2013. *The Bioinorganic Chemistry of Chromium.* Chichester, UK: John Wiley & Sons, Ltd.
3. Mertz, W., and K. Schwarz. 1955. Impaired intravenous glucose tolerance as an early sign of dietary necrotic liver degeneration. *Arch. Biochem. Biophys.* 58:504–506.
4. Schwarz, K., and W. Mertz. 1957. A glucose tolerance factor and its differentiation from factor 3. *Arch. Biochem. Biophys.* 72:515–518.
5. Schwarz, K., and W. Mertz. 1959. Chromium(III) and the glucose tolerance factor. *Arch. Biochem. Biophys.* 85:292–295.

6. Woolliscroft, J., and J. Barbosa. 1977. Analysis of chromium induced carbohydrate intolerance in the rat. *J. Nutr.* 107:1702–1706.

7. Toepfer, E. W., Mertz, W., Polansky, M. M., Roginski, E. E., and W. R. Wolf. 1977. Preparation of chromium-containing material of glucose tolerance factor activity from brewer's yeast extracts and by synthesis. *J. Agric. Food Chem.* 25:162–166.

8. Nielsen, F. H. 1996. Controversial chromium: does the superstar mineral of the Montebanks receive appropriate attention from clinicians and nutritionists? *Nutr. Today* 31:226–233.

9. Evans, G. W. 1982. Dietary supplementation with essential metal picolinates. US patent No. 4,315,927. Feb. 16.

10. Ley, H., and K. Ficken. 1917. Isomerie und lichtabsorption bei inneren komplexsalzen. (Uber innere komplexsalze. XV.) *Berichte* 50:1123.

11. Evans, G. W. 1989. The effect of chromium picolinate on insulin controlled parameters in humans. *Int. J. Biosci. Med. Res.* 11:163–180.

12. Evans, G. W., and D. J. Pouchnik 1993. Composition and biological activity of chromium-pyridine carboxylate complexes. *J. Inorg. Biochem.* 49:177–187.

13. Lefavi, R. G. 1993. Response. Chromium picolinate is an efficacious and safe supplement. *Int. J. Sport Nutr.* 3:120–122.

14. Boynton, H., and G. W. Evans. 1992. Chromic picolinate treatment. US patent No. 5,087,624. Feb. 11.

15. Federal Trade Commission. 1997. Docket No. C-3758 Decision and Order, 1997. https://www.ftc.gov/sites/default/files/documents/cases/1997/07/nutrit2.htm (accessed 14 June 2018).

16. Hellerstein, M. K. 1998. Is chromium supplementation effective in managing type II diabetes? *Nutr. Rev.* 56:302–306.

17. Vincent, J. B. 2004. The potential value and potential toxicity of chromium picolinate as a nutritional supplement, weight loss agent, and muscle development agent. *Sports Med.* 33:213–230.

18. Pittler, M. H., Stevinson, S. C., and E. Ernst. 2003. Chromium picolinate for reducing body weight: meta-analysis of randomized trials. *Int. J. Obes. Relat. Metab. Disord.* 27:522–529.

19. Kaats, G. R., Blum, K., Fisher, J. A., and J. A. Adelman. 1996. Effects of chromium picolinate supplementation on body composition: a randomized, double-maked, placebo-controlled study. *Curr. Ther. Res.* 57:747–756.

20. Kaats, G. R., Blum, K., Pullin, D., Keith, S. C., and R. Wood. 2018. A randomized, double-masked, placebo-controlled study of the effects of chromium picolinate supplementation on body composition: a replication and extension of a previous study. *Curr. Ther. Res.* 59:379–388.

21. Onakpoya, I., Posadzki, P., and E. Ernst. 2013. Chromium supplementation in overweight and obesity: a systematic review and meta-analysis of randomized clinical trials. *Obes. Rev.* 14:496–507.

22. Tian, H., Guo, X., Wang, X., He, Z., Sun, R., Ge, S., and Z. Zhang. 2013. Chromium picolinate supplementation for overweight or obese adults. *Cochrane Database Syst. Rev.* 2013, no. 11:1–45. https://www.cochranelibrary.com/cdsr/doi/10.1002/14651858.CD010063.pub2/epdf/full.

23. Boynton, H., and G. W. Evans. 1992. Chromic picolinate treatment. US patent No. 5,087,623. Feb. 11.

24. Althius, M. D., Jordan, N. E., Ludington, E. A., and J. T. Wittes. 2002. Glucose and insulin responses to dietary chromium supplements: a meta-analysis. *Am. J. Clin. Nutr.* 76:148–155.

25. Balk, E. M., Tatsioini, A., Lichtenstein, A. H., Lau, J., and A. G. Pittas. 2007. Effect of chromium supplementation on glucose metabolism and lipids: a systemic review of randomized controlled trials. *Diabetes Care* 30:2154–2163.

26. Patal, P. C., Cardino, M. T., and C. A. Jimeo. 2010. A meta-analysis on the effect of chromium picolinate on glucose and lipid profiles among patients with type 2 diabetes mellitus. *Philipp. J. Int. Med.* 48:32–37.

27. Abdollahi, M., Farshchi, A., Nikfar, S., and M. Seyedifar. 2013. Effect of chromium on glucose and lipid profiles in patients with type 2 diabetes: a meta-analysis review of randomized trials. *J. Pharm. Pharm. Sci.* 16:99–114.

28. Suksomboon, N., Poolsup, N., and A. Yuwanakorn. 2014. Systematic review and meta-analysis of the efficacy and safety of chromium supplementation of diabetes. *J. Clin. Pharm. Ther.* 39:292–306.

29. Yin, R. V., and O. J. Phung. 2015. Effect of chromium supplementation on glycated hemoglobin and fasting plasma glucose in patients with diabetes mellitus. *Nutr. J.* 14:14.

30. San Mauro-Martin, I., Ruiz-Leon, A. M., camina-Martin, M. A., Garicano-Vilar, E., Collado-Yurrita, L., Mateo-Silleras, Bd., and Mde. P. Redondo Del Rio. 2016. Chromium supplementation in patients with type 2 diabetes and high risk of type 2 diabetes: a meta-analysis of randomized controlled trials. *Nutr. Hosp.* 33:156–161.

31. Bailey, C. H. 2014. Improved meta-analytic methods show no effect of chromium supplements on fasting glucose. *Biol. Trace Elem. Res.* 157:1–8.

32. Huang, C. H., Chen, G., Dong, Y., Zhu, Y., and H. Chen. 2018. Chromium supplementation for adjuvant treatment of type 2 diabetes mellitus: results from a pooled analysis. *Mol. Nutr. Food Res.* 62, no. 1:1700438. https://onlinelibrary.wiley.com/doi/epdf/10.1002/mnfr.201700438.

33. Food and Drug Administration. 2005. Qualified health claims: letter of enforcement discretion – chromium picolinate and insulin resistance (Docket No. 2004Q0144). http://wayback.archive-it.org/7993/20171114183739/https://www.fda.gov/Food/IngredientsPackagingLabeling/LabelingNutrition/ucm073017.htm (accessed 14 June 2018).

34. American Diabetes Association. 2014. Standard in medical care in diabetes - 2014. *Diabetes Care* 37, Suppl. 1:S14–S80.

35. American Diabetes Association. 2015. 4. Foundations of care: education, nutrition, physical activity, smoking cessation, psychosocial care, and immunization. *Diabetes Care* 38, Suppl. 1:S20–S30.

36. American Diabetes Association. 2018. 4. Lifestyle management: standards of medical care in diabetes – 2018. *Diabetes Care* 41, Suppl. 1:S38–S50.

37. Fazelian, S., Rouhani, M. H. Ban, S. S., and R. Amani. 2017. Chromium supplementation and polycystic ovary syndrome: a systematic review and meta-analysis. *J. Trace Elem. Med. Biol.* 42:92–96.

38. Tang, X.-L., Sun, Z., and L. Gong. 2018. Chromium supplementation in women with polycystic ovary syndrome: systematic review and meta-analysis. *J. Obstet. Gynaecol. Res.* 44:134–143.

39. Heshmati, J., Omani-Samani, R., Vesali, S., Maroufizadeh, S., Rezaeinejad, M., Razavi, M., and M. Sepidarkish. 2018. The effects of supplementation with chromium on insulin resistance indices in women with polycystic ovarian syndrome: a systematic review and meta-analysis of randomized clinical trials. *Horm. Metab. Res.* 50:193–200.

40. Costello, R. B., Dwyer, J. T., and J. M. Merkel. 2019. Chromium supplementation in health and disease. In *Nutritional Biochemistry of Chromium(III)*, 2nd ed., ed. J. B. Vincent, 219–250. Amsterdam: Elsevier.

41. Institute of Medicine. 2001. *Dietary Reference Intakes for Vitamin A, Arsenic, Boron, Chromium, Copper, Iodine, Iron, Manganese, Molybdenum, Nickel, Silicon, Vanadium, and Zinc*. Washington, DC: National Academies Press.

42. Anderson, R. A., and A. S. Kozlovsky. 1985. Chromium, intake, absorption and excretion of subjects consuming self-selected diets. *Am. J. Clin. Nutr.* 41:1177–1183.

43. Anderson, R. A., Bryden, N. A., and M. M. Polansky. 1992. Dietary Cr intake. Freely chosen diets, institutional diets, and individual foods. *Biol. Trace Elem. Res.* 32:117–121.

44. Vincent, J. B. 2016. Chromium: properties and determination. In *Encyclopedia of Food and Health*, ed. B. Caballero, P. Finglas, and F. Toldra, Vol. 2, 114–118. Oxford: Academic Press.

45. Vincent, J. B. 2014. Is chromium pharmacologically relevant? *J. Trace Elem. Biol. Med.* 28:397–405.

46. Di Bona, K. R., Love, S., Rhodes, N. R., McAdory, D., Sinha, S. H., Kern, N., Kent, J., Strickland, J., Wilson, J., Beaird, J., Ramage, J., Rasco. J. F., and J. B. Vincent. 2011. Chromium is not an essential element for mammals: effects of a "low-chromium" diet. *J. Biol. Inorg. Chem.* 16:381–390.

47. Padmavathi, I. J., Rao, K. R., Ganeshan, M., Kumar, K. A., Rao, Ch. N., Harishankar, N., Ismail, A., and M. Raghunath. 2010. Chromic maternal dietary chromium restriction modulates visceral adiposity: probable underlying mechanisms. *Diabetes* 59:98–104.

48. Vincent, J. B., and J. F. Rasco. 2010. Comment on Padmavathi et al. (2010) Chromic maternal dietary chromium restriction modulates visceral adiposity: probable underlying mechanisms. *Diabetes* 59:e2.

49. Anderson, R. A., Polansky, M. M., Bryden, N. A., Patterson, K. Y., Veillon, C., and W. H. Glinsmann, 1983. Effects of chromium supplementation on urinary Cr excretion of human subjects and correlation with selected clinical parameters. *J. Nutr.* 113:276–281.

50. Vincent, J. B. 2017. New evidence against chromium as an essential trace element. *J. Nutr.* 12:2212–2219.

51. Scientific Committee on Food. 2003. *Opinion on the Scientific Committee on Food on the Tolerable Upper Intake Level of Trivalent Chromium*. SCF/CS/NUT/UPPLEV/67 Final. https://ec.europa.eu/food/sites/food/files/safety/docs/sci-com_scf_out197_en.pdf (accessed 14 June 2018).

52. EFSA Panel on Contaminants in the Food Chain. 2014. Scientific opinion on the risks to public health related to the presence of chromium in food and drinking water. *EFSA J.* 12, no. 3:3595. https://efsa.onlinelibrary.wiley.com/doi/epdf/10.2903/j.efsa.2014.3595.

53. Expert Group on Vitamins and Minerals. 2003. *Safe Upper Levels for Vitamins and Minerals*. London: Food Standards Agency.

54. Stearns, D. M., Wise Sr., J. P., Patierno, S. R., and K. E. Wetterhahn. 1995. Chromium(III) picolinate produces chromosome damage in Chinese hamster ovary cells. *FASEB J.* 9:1643–1648.

55. Stearns, D. M., Silveeira, S. M., Wolk, K. K., and A. M. Luke. 2002. Chromium(III) tris(picolinate) is mutagenic at the hypoxanthine (guanine) phosphoribosyltransferase locis in Chinese hamster ovary cells. *Mutat. Res.* 513:135–142.

56. Manygoats, K. R, Yazzie, M., and D. M. Stearns. 2002. Ultrastructural damage in chromium picolinate-treated cells: a TEM study. *J. Biol. Inorg. Chem.* 7:791–798.

57. Gudi, R., Slesinski, R. S., Clarke, J. J., and R. H. san. 2005. Chromium picolinate does not produce chromosome damage in CHO cells. *Mutat. Res.* 589:140–146.

58. Slesinski, R. S., Clarke, J. J., San, R. H. C., and R. Gudi. 2005. Lack of mutagenicity of chromium picolinate in the hypoxanthine phosphoribosylransferease gene assay in Chinese hamster ovary cells. *Mutat. Res.* 585:86–95.

59. Coryell, V. H., and D. M. Stearns. 2006. Molecular analysis of hprt mutations induced by chromium pciolinate in CHO AA8 cells. *Mutat. Res.* 610:114–123.

60. Expert Group on Vitamins and Minerals. 2004. Statement on the mutagenicity of trivalent chromium and chromium picolinate. https://cot.food.gov.uk/sites/default/files/cot/comsection.pdf (accessed 14 June 2018).

61. Stallings, D., and J. B. Vincent. 2006. Chromium: a case study in how not to perform nutraceutical research. *Curr. Top. Nutraceutical Res.* 4:89–112.

62. Jiang, L., Vincent, J. B., and M. M. Bailey. 2018. $[Cr_3O(O_2CCH_2CH_3)_6(H_2O)_3]^+$ (Cr_3) toxicity potential in bacterial and mammalian cells. *Biol. Trace Elem. Res.* 183:342–350.

63. Hepburn, D. D. D., Xiao, J., Bindom, S., Vincent, J. B., and J. O'Donnell. 2003. Nutritional supplement chromium picolinate causes sterility and lethal mutations in *Drosphila melanogaster*. *Proc. Natl. Acad. Sci. U. S. A.* 100:3766–3771.

64. Stallings, D. M., Hepburn, D. D., Hannah, M., Vincent, J. B., and J. O'Donnell. 2006. Nutritional supplement chromium picolinate generates chromosomal aberrations and impedes progeny development in *Drosophila melanogaster*. *Mutat. Res.* 610:101–113.

65. Vincent, J. B. 2010. Chromium: celebrating 50 years as an essential element? *Dalton Trans.* 39:3787–3794.

66. Stout, M. D., Nyska, A., Collins, B. J., Witt, K. L., Kissling, G. E., Malarkey, D. E., and M. J. Hooth. 2009. Chronic toxicity and carcinogenicity studies of chromium picolinate monohydrate administered in feed to F344/N rats and B6C3F1 mice for 2 years. *Food Chem. Toxicol.* 47:729–733.

67. Vanek, V. W., Borum, P., Buchman, A., Fessler, T. A., Howard, L., Jeejeebhoy, K., Kochevar, M., Shenkin, A., and C. J. Valentine. 2012. A.S.P.E.N. position paper: recommendations for changes in commercially available parenteral multivitamin and multi-trace element products. *Nutr. Clin. Pract.* 27:440–491.

CHAPTER 10

Trivalent Chromium
A Vital Insulin Sensitizer

Harry G. Preuss
Georgetown University Medical Center

Debasis Bagchi
University of Houston

CONTENTS

10.1 GENERAL BACKGROUND OF TRIVALENT CHROMIUM

Trivalent chromium is an essential nutrient required for sugar and fat metabolism that functions mainly through optimizing insulin sensitivity [1–3]. The precise mechanism(s) behind its effect on insulin function is uncertain, however, less than optimal intakes of chromium by individuals ingesting average diets have been associated with diminished insulin binding and receptor number [2]. Despite a shortcoming of detailed knowledge, the propensity to find harmful intensified insulin resistance (IR) during aging can potentially be prevented, delayed, and/or ameliorated by trivalent chromium replacement which, in turn, should prevent the onset of a number of common chronic disorders linked to IR [4–9]. Aside from diabetes type 2, chromium can be useful to direct overall metabolism toward fat loss with a proclivity to spare lean body mass [10–12]. Another potential benefit could be a longer, more healthful, and productive lifespan [13].

Attempts to discuss proper daily dosing for trivalent chromium lead one into murky waters. The Tolerable Upper Intake Level (UL) is the *maximum* daily intake of a nutrient that is unlikely to cause adverse health effects. Since few serious adverse effects have been linked to high intakes of chromium, the Institute of Medicine has not definitively established a UL for this mineral. But then again, it is generally accepted that individuals in the Western world consume less than the upper limit of the currently estimated safe and adequate intake (ESADDI) of 50–200 mcg/day [14]. An adequate daily intake (AI) of chromium for most adults has been established and would range between 20 and 35 mcg for males compared to 20–25 mcg for females [15]. The average AI's are usually established when there is insufficient research to ascertain a recommended daily intake (RDI and is generally set at a value that healthy people typically consume. Despite the fairly solid evidence for the overall safety of trivalent chromium, still fears concerning toxicity from chronic use persist and have psychologically limited its utility. Another unfortunate outcome of this fear is that it has contributed to less general knowledge in the chromium field being gained over the last couple of decades.

Can enough daily chromium be ingested using the average Western diets? The answer is a solid "No"! Even the best of meal planning by professionals cannot provide the minimum ESADDI of 50 mcg/day. Anderson estimated that it would take roughly 3,000–10,000 kcal to do such [1,3]. Even then, common stresses such as glucose loading, consuming diets high in simple sugars, lactation, infection, acute and chronic exercising, and physical trauma would necessitate even higher caloric intakes to allow the needed additional chromium [1].

10.2 ROLE OF INSULIN RESISTANCE IN MANY MEDICAL PERTURBATIONS

Approximately three decades ago, DeFronzo and Ferrannini wrote a classic article entitled *Insulin resistance. A multifaceted syndrome responsible for NIDDM, obesity, hypertension, dyslipidemia, and atherosclerotic cardiovascular disease* [7]. Although the title given to the defined pattern of maladies in their paper, "insulin resistance syndrome," is somewhat appropriate; still, the collection of medical conditions described above is more commonly referred to as the "metabolic syndrome" (MS) [16]. Even so, it is generally acknowledged by most medical professionals that IR is a major driving force behind MS [4–9]. Up to now, thousands of papers have been written concerning the role of IR in this commonly occurring syndrome that is often associated with the aging process. In fact, many elements comprising MS are frequently associated with the term "epidemic"—as examples, obesity, diabetes, many cardiovascular disorders, hypertension, and non-alcoholic fatty liver disease (NAFLD). Suffice it to say, the role that this syndrome plays in a long-term healthy existence is becoming more noticeable by most observers with the passing of time [17,18].

10.3 PRESENCE OF NATURAL DIETARY
SUPPLEMENTAL INSULIN SENSITIZERS

Interestingly, in discussing prevention, amelioration, and even cure for the collection of maladies making up MS, a comment was offered by DeFronzo and Ferrannini regarding the need to develop a special class of drugs that could directly and favorably overcome IR—so-called "insulin sensitizers" [7]. Despite the obvious value that such a class of pharmaceuticals would provide, attempts to develop these agents up to now have failed most likely due to the ensuing adversities that commonly occur subsequent to their use. Drug toxicity has been especially noted both in the past and present with agents designed to treat obesity and diabetes [19–21]. Nevertheless, there is hope, because overlooked in the above-mentioned article by DeFronzo and Ferrannini is the fact that certain effective natural dietary insulin sensitizers were and still are available that can provide effectiveness and the safety that current toxic drugs cannot. These natural supplements include bitter melon, fraction SX of maitake mushroom, vitamin D3, and the subject of this chapter trivalent chromium.

10.4 FOCUSING ON MORE FACTS CONCERNING TRIVALENT CHROMIUM

Of all the possible, natural insulin sensitizers, the focus of this report will be on trivalent chromium that is recognized universally for its prominent role in glucose-insulin metabolism [1–3]. As a dietary supplement, trivalent chromium has been available for a number of years, but after an initial period of popularity, its prominence lessened considerably—too much popularity at one time based largely on some ludicrous claims. Ads early on overstated the possibility that if an individual started using it, practically in no time the body would develop superhuman qualities. Think of it, optimal health overnight! This possibility was quite appealing, because unfortunately, countless humans cannot delay gratification. The difficulty here is that success with chromium works gradually over time—years to months rather than overnight. One should not expect great results in hours, days, or even weeks from trivalent chromium usage. A more appropriate question would be to ask how the consumers feel after taking chromium for 5 years rather than 1 week? With all the above-mentioned hype, can you imagine average adults maintaining chromium supplementation that long without immediate, overly impressive happenings? However, that unusual persistent individual who can comply might be able to say "little or no difference over the five years." Realistically, would he/she realize the importance of that revelation toward a healthy lifespan? Unfortunately, in regard to our information base, consistent taking of trivalent chromium over long periods is not a common happening. Therefore, reliable exact information is not readily available dealing with this particular matter.

Trivalent chromium is perhaps the best-recognized natural supplement that enhances insulin sensitivity [2,23,24]. Over 60 years ago, Schwartz and Mertz established that an extract derived from pork kidney improved glucose tolerance in rats [25]. Soon after, the same team reported that chromium was a key component of the so-called "glucose tolerance factor" (GTF) [26]. What is known about GTF? The association of GTF to human metabolism was initially uncovered in individuals receiving total parenteral nutrition (TPN), because this procedure was carried out in the beginning when parenteral fluid was lacking in chromium [27–29]. Early accounts revealed that numerous patients ultimately showed characteristic signs and symptoms of chromium deficiency that in many cases resembled type 2 diabetes—a severe form of IR. [1]. Noted in the afflicted were elevated circulating glucose concentrations, a requirement for added insulin replacement, high circulating lipids, and even peripheral neuropathy and encephalopathy that remitted to a reasonable extent after trivalent chromium replacement [27–29]. As might be expected, a cascade of knowledge concerning chromium metabolism began to appear shortly after these observations [1,2].

10.5 PERSISTENT INSULIN RESISTANCE, A HINDRANCE TO OPTIMAL HEALTH OVER MANY YEARS

At this point, it seems somewhat necessary to describe the background behind IR in more detail, because history reveals in actuality the only real well-developed role of chromium is to enhance insulin sensitivity—using other words, to ameliorate IR. Despite limitations in applicability, this single recognized role is extremely important. IR has received much recently well-deserved respect for its role in optimal health by both professionals and the lay public. Many differing chronic metabolic disorders are becoming more widespread and severely damaging in the aging populations, and much of this can be attributed to the progression of the health perturbation referred to as IR [5,6,30–32].

A century ago after the isolation and purification of insulin by Banting and colleagues, it was largely insinuated that diabetes, an abnormal escalation in circulating glucose, was wholly due to a deficiency of circulating insulin [33]. So, the general conviction was that diabetes was quite curable because lack of insulin could now be handled. However, another form of diabetes was soon discovered when it was noted that an unexpected increase of insulin in the bloodstream was frequently present despite abnormally high circulating glucose concentrations [4]. Ironically, this picture eventually became more common than diabetes originally described with diminished insulin circulation—type 1. Thus, this more common form attributed greatly to the above-mentioned IR was naturally referred to as diabetes mellitus type 2. Worth repetition, IR is generally accepted to be a critical driving force behind the linked constituents of MS including NAFLD [5,6,34–36].

In "endocrine terms," what precisely is IR? Similar to the thyroid and adrenals as endocrine systems, the glucose-insulin system has many compensatory capabilities. Specifically, it also has distinctive responses to any diminution in hormonal activity. In the case of IR, a general tendency exists under certain conditions for peripheral target tissues such as liver, fat, and muscle to experience a lessened, apparently sub-optimal response to insulin over time. To compensate for the lower activity, more insulin is produced and released from the pancreas as its compensatory response. While this benefit has some merit by returning circulating glucose more toward optimum, the final outcome is to impose higher circulating glucose and insulin levels than prior that can be potentially harmful presently and in the future [17,18]. Such occurrences have been given a unique nomenclature, explicitly "trade off" because they can create instant benefits but future harms [37]. As the organ resistance to insulin continues to chronically become more severe, the prolongation of the interplay between glucose and insulin can continue to unfold. Hence, both levels of circulating glucose and insulin may elevate more and more and be associated with increasing prevalence and severity of the risk factors leading to the plethora of disorders associated with diabetes type 2, MS, and NAFLD [7,17]. In fact, there is some possibility that not maintaining an optimal glucose-insulin system could lead to a shorter, unhealthier lifespan [17,18]. So, it should be obvious that the desire of DeFronzo and Ferrannini to develop a safe, effective insulin sensitizer would be most beneficial [7].

10.6 FOCUSING ON BIOAVAILABILITY OF TRIVALENT CHROMIUM

Whatever the basic cause behind it, knowledge of the benefits and risks of trivalent chromium has lagged. Although the hexavalent form of chromium is undoubtedly toxic, sparse evidence to date suggests the same prospective for the trivalent form [23]. Trivalent chromium, such as the nicotinic, picolinic, and histidinate formulae, appears to be quite nontoxic, yet still effective [38,39]. An early study that pointed out possible mutagenic effects of chromium picolinate today seems to lack significance [40] inasmuch as further studies with both animals and humans have failed to corroborate noxious effects [41,42]. Regrettably, little evidence is available among humans concerning adverse effects from long-term (lifetime) use of trivalent chromium compounds.

A major inconvenience arises because of the typically poor bioavailability of trivalent chromium [2,22,23,24]. It appears that not all forms of trivalent chromium possess equal bioavailability according to a multitude of laboratory and clinical studies [2,24,43,44]. When interest in chromium began, the most common form of the supplement was chromium chloride. However, Evans revealed the reality regarding the poor bioavailability of chromium chloride (inorganic ligand) when compared to chromium picolinate (organic ligand) as early as 1989 [24]. The normal absorption for chromium chloride is somewhere near 0.4%–1.0% of the oral dose [2]. This fact necessitated development of more bioavailable, organic forms of chromium—nicotinate, picolinate, and histidinate as ligands [2,24,43,44]. In time, Seaborn and Stoecker performed studies examining the effects of CHO, acids, and agents affecting prostaglandin metabolism to alter absorption [45–47]. Of utmost importance, the provider must be sure to use the best bioavailable chromium agents to obtain the best results [43].

10.7 OVERVIEW OF USE OF CHROMIUM IN A VARIETY OF HEALTH DISORDERS

In keeping with the concept of MS, the ability of chromium supplementation to favorably influence IR and central fat accumulation implies that this trace element should beneficially impact other health parameters linked to the syndrome [7,9,16]. However, caution is necessary as a great deal of the support for the above information emanates heavily from animal studies, particularly regarding blood pressure. In the remaining portion of this overview, we will focus on major areas where a modicum of work has been carried out, namely (a) IR and type 2 diabetes, (b) overweight and obesity, (c) blood pressure regulation, and (d) aging. Because of the dearth of information derived from human clinical trials dealing with chromium, this chapter will concentrate more on the reports with positive findings to provide some idea of what "could be" rather than what "is."

10.7.1 Insulin Resistance and Diabetes Type 2

Our initial selection concerning discussion of therapeutic uses for trivalent chromium logically deals with diabetes, particularly type 2, the form linked closely to IR [5–7,48]. Regrettably, many early studies examining efficacy of chromium were compromised for numerous reasons—some of which usually go along with any clinical trial [11,49]. First, there was and still is widespread discouragement among investigators to use levels of chromium exceeding 200 mcg/day in situations where it is more than likely that larger doses are needed. Second, the type of chromium taken per os influences the results, and several early agents possessed extremely poor bioavailability. Third, as implied, poor compliance in any clinical studies obviously affects the results negatively. What is behind the poor compliance? Chromium works chronically rather than acutely. Many subjects overlooking this fact tend to get discouraged when early results do not produce immediate success. These volunteers then believe that they are receiving placebo whether true or not and compliance goes awry. The list of difficulties against successful achievement goes on and on. Accordingly, these variables make it near impossible to analyze different investigative results with total confidence.

An example of all difficulties in chromium studies can be seen in a recent chapter examining chromium therapy for type 2 diabetes by Costello et al. [50]. In their meta-analysis, the investigators evaluated the results of chromium supplementation on fasting blood glucose and HbA1C concentrations. Not all, but slightly more than half of the accepted trials found a statistically significant lowering of fasting glucose by chromium (4 of 6), HbA1C (3 of 5) or both (3 of 5). Nevertheless, after discussing the limitations of their review in detail, the following statement appears, "it was not appropriate to perform a new meta-analysis of the effects of chromium supplementation on patients with diabetes type 2, because of the significant heterogeneity between studies" [50].

Nevertheless, using higher dosing in some clinical studies has supported claims of benefits from oral organic trivalent chromium compounds. Richard Anderson and colleagues performed the most persuasive of these studies in China [51]. About 180 men and women with diabetes type 2 were randomly divided into three groups: placebo, the most accepted replacement dose of chromium 200 mcg daily, and the uncommon dose at that time of chromium 1,000 mcg daily. Subjects continued their regular medications, nutritional support, and additional lifestyle habits. After 4 months, the fortunate individuals in the higher chromium dose group attained the earliest and greatest benefits, although the group receiving the lower dose also eventually benefited. Fasting blood glucose, circulating insulin levels, HbA1C, and plasma total cholesterol levels improved significantly in the active arms. Most important to note, the beneficial effects in individuals with diabetes occurred best at levels of chromium exceeding the upper limit of the ESADDI.

Martin et al. assessed subjects with diabetes, type 2 who had previously received glipizide and placebo during a 3-month run-in period [52]. Over another 6 months, 12 continued with the sulfonylurea and placebo regimen, whereas 17 received 1,000 mcg of chromium picolinate along with the glipizide. Compared to placebo, volunteers in the chromium group showed meaningful improvements in insulin sensitivity and glucose control. Interestingly, this group also developed less body weight gain and visceral fat accumulation—important components of MS.

Thus, many feel that chromium supplementation at proper dosing to some extent provides an exceptional tool to overcome and/or at least ameliorate diabetes replacing more toxic drugs developed for this same purpose [1–3]. Linday remarked in 1997 about the Diabetes Prevention Program developed to determine whether and how diabetes type 2 could be prevented or delayed in persons with impaired glucose tolerance [53]. Having perceived that popular medications such as metformin and troglitizone had already been assessed in previous clinical studies, the author questioned why trivalent chromium with its ability to ameliorate IR as well its acceptable adverse action profile and reasonable costs had not been an arm of the study. In the same time frame, Preuss and Anderson came to similar conclusions regarding the unfortunate choice of excluding chromium as an arm of the study [3].

10.7.2 Overweight and Obesity

It is obvious from multiple reliable sources that a significant increase in body weight and fat mass persist among too many individuals universally; and in fact, the proportion of such is by and large increasing [54,55]. Since IR has been linked to increased fat accumulation [7], it is not a stretch to postulate that chromium which can ameliorate IR may favorably influence overweight and obesity. Studies along these lines have been met with mixed results. Kaats et al. examined the effects of chromium picolinate consumption on body composition. Using both underwater testing [56] and dual energy X-ray absorptiometry methodology [57]. In the second study using dual energy X ray absorptometry (DEXA) measurements, Kaats and colleagues showed a statistically significant beneficial change in body composition in those individuals receiving chromium supplementation 400 mcg daily [57]. Sixty-two subjects were randomized to receive chromium and 60 received placebo over a 90-day period. The active group lost significantly more weight ($-7.79\,\mathrm{kg} \pm 1.23$ (SEM) vs. $1.81\,\mathrm{kg} \pm 0.39$ (SEM), respectively, $p = <0.001$) and fat mass ($7.71\,\mathrm{kg} \pm 1.21$ (SEM) vs. $1.53\,\mathrm{kg} \pm 0.36$ (SEM), $p < 0.001$, respectively). It was concluded that this study with DEXA [57] replicated the earlier findings with underwater testing [56] and that chromium picolinate can lead to a significant improvement in body composition.

A study by Crawford et al. was designed to determine whether 600 mcg niacin-bound chromium ingested daily over 2 months by African-American women undergoing a modest dietary and exercise regimen influences weight loss and body composition [12]. Twenty African-American females were examined in a crossover-designed study. Niacin-bound chromium given to modestly dieting-exercising African-American women caused a significant loss of fat compared to placebo (-2.0 lbs vs. -0.2 lbs) and sparing of muscle compared to placebo (-0.7 lbs vs. -2.4 lbs). Although weight loss was essentially comparable in both the active and placebo groups (-2.7 lbs vs. -2.6 lbs), loss of fat was significantly greater in the former, while the loss of fat-free mass was decidedly less. Blood chemistries revealed no significant adverse effects from the ingestion of 600 mcg of niacin-bound chromium daily over 2 months. The investigators were more impressed with the ability of the chromium supplementation to spare muscle mass while losing fat more so than its ability to cause marked scale weight loss.

Not all studies for weight and/or fat loss have been positive [58]. Forty-four women, 27–51 years of age, were randomly assigned to two groups. Subjects received either 400 mcg/day of chromium as a picolinate supplement or a placebo in double-blind fashion. In addition, the subjects took part in a supervised weight training and walking program 2 days/week for 12 weeks. Body composition was evaluated at baseline and 12 weeks. This regimen did not significantly influence body composition, plasma glucose/insulin, and lipid concentrations. The lack of effect on the glucose-insulin system and presumably IR might explain the lack of effect on body composition.

10.7.3 Blood Pressure Regulation

Similar to the other fields, significant human clinical investigations dealing with chromium and blood pressure regulation are lacking. This is quite interesting since there are a number of cogent opinions claiming that IR plays a significant role in essential hypertension [59,60]. Nevertheless, much information has been gleaned *via* animal studies. The author of this chapter first became interested in the subject after showing that sucrose feeding in a variety of different rat strains increased blood pressure readings significantly [61]. To determine if IR played an important role in these happenings, the influence of chromium on the sugar-induced response was investigated, since the only known major effect of chromium is to increase insulin sensitivity. Sure enough, chromium could completely overcome sugar-induced blood pressure increases suggesting that IR played a significant role in the response [62]. Many studies that followed corroborated the fact that trivalent chromium could overcome the sugar-induced portion and ONLY the sugar-induced portion of the blood pressure elevations in rats, even in a genetically hypertensive strain (SHR) [62–66].

Studies with rats were undertaken to examine the interplay between sucrose and chromium consumption in spontaneously hypertensive rats (SHRs). The design for the following study is summarized in Table 10.1, and the overall average systolic blood pressure measurements are illustrated in Figure 10.1. Data were gathered from a number of studies when the rats were in a steady state [62–66]. SHRs were divided into four groups to determine the effects of various combinations of sucrose (Sucr) (either 18% or 52% w/w) and chromium (Chr) (either 5 or 25 ppm) on sucrose-induced blood pressure elevations (see Table 10.1). This led to four groupings based on the combinations at final testing (Third Arm): Group I—Sucr 18%/Chr 5 ppm; Group II—Sucr 18%/Chr 25 ppm; Group III—Sucr 52%/Chr 5 ppm; and Group IV—Sucr 52% /Chr 25 ppm. In other words, related to the two concentrations of Sucr and Chr individually, groups were: I=Lo/Lo, II=Lo/Hi, III=Hi/Lo and IV=Hi/Hi.

To compare the effects of the four combinations, each of the four groups was examined separately by using three arms (Table 10.1). The first arm common to all groups served as control, i.e., all four groups consumed only standard rat chow. The groups in the second arm were challenged with one of the two concentrations of sucrose (Sucr) added to the chow (18% w/w—I and II or 52% w/w—III and IV). The 18% is near the average intake of calories from refined carbohydrates in the American diet and the 52% might approach that of an 18-year-old male teenager [61]. The final third arm represented the addition of chromium (Chr) either 5 ppm (I and III) or 25 ppm (II and IV) to the chow containing the sucrose percentage indicate in the second arm. (This made for the four groupings listed in the paragraph above.)

In Figure 10.1, the first clear bar in each grouping depicts the average systolic blood pressure of the SHR receiving the standard diet alone. No significant statistical differences were found among

Table 10.1 Designation of Groups Based on Final Challenge and Description of Arms for Each

	Sucrose % W/W	Chromium ppm	Mix (Sucr/Chrom)
Group I	18%	5 ppm	Low/Low
Group II	18%	25 ppm	Low/High
Group III	52%	5 ppm	High/Low
Group IV	52%	25 ppm	High/High

	First Arm	Second Arm (w/w)	Third Arm
Group I	Basic chow	Sucrose 18%	Sucrose 18% chromium 5 ppm
Group II	Basic chow	Sucrose 18%	Sucrose 18% chromium 25 ppm
Group III	Basic chow	Sucrose 52%	Sucrose 52% chromium 5 ppm
Group IV	Basic chow	Sucrose 52%	Sucrose 52% chromium 25 ppm

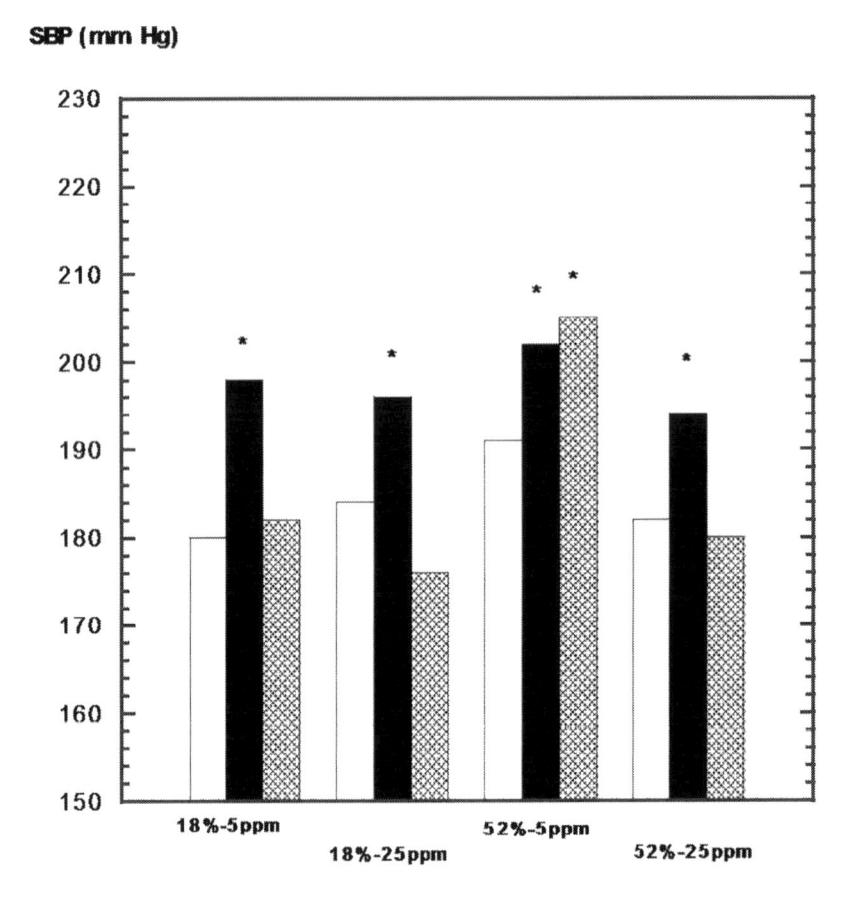

SBP (mm Hg)

18%-5ppm 18%-25ppm 52%-5ppm 52%-25ppm

Figure 10.1 Effects of different concentrations of sucrose alone (black bars) and combined with chromium (gray bars) on average systolic blood pressure of SHRs. The information below the four groups indicates the combination received in the last subset. * indicates statistical significance compared to baseline. See text for more details.

the four control groups receiving diet alone. The second black bar in each group represents the average steady-state blood pressure weeks after the addition of sucrose at the designated percentage weight of sucrose in the feed. This average blood pressure rose with the sucrose challenge whether it be 18% w/w in groups I and II or 52% w/w in groups III and IV. In each of the four groups, while the sucrose addition to the feed raised the systolic blood pressure significantly over controls, there was not that much more with the higher dosing of sucrose (18% vs. 52%). We have previously found a ceiling for the sucrose effect on blood pressure in earlier studies [62–66].

The third arm of the four groups (Table 10.1) was developed to test our main objective—the ability of different chromium challenges (Lo-5 ppm and Hi-25 ppm) to overcome the sugar-induced (Lo 18% sucrose and Hi 52% sucrose) elevations in systolic blood pressure. Both the 5 and 25 ppm chromium addition to the first two groups consuming 18% sucrose lowered the systolic blood pressure virtually back to baseline. However, at 52% sucrose in groups III and IV, only the higher dose of chromium (25 ppm) in group IV did so. Thus, positive effects from chromium are dependent on the breadth of the sugar challenge, i.e., the extent of IR.

In working with SHRs, only the sugar-induced elevation of blood pressure is influenced by chromium, the rest of the genetic effects influencing pressure are not as eluded to earlier. Thus, in the case of SHR, the basic inherited hypertension is not likely to be dependent on IR to any extent, as is believed in the case of human hypertension [59,60]. Important to note: even though chromium challenge does not essentially affect systolic blood pressure of rats consuming normal chow devoid of excess sucrose, nevertheless, it is possible based on previous findings in humans suggesting a role for chromium that this trace metal might reduce blood pressure significantly in them [59,60].

Conclusions reached from sucrose-induced rat studies are systolic blood pressure (SBP) elevations induced by sucrose ingestion can be suppressed by concomitant ingestion of chromium. With higher sucrose intake, higher intake of chromium is necessary to overcome elevated systolic blood pressure. Not revealed in Figure 10.1 are the following facts. Chromium can overcome sucrose-induced elevations in SBP of young and old rats of either gender. The effects of chromium loading on SBP can persist over time, even after chromium loading is halted. We postulate that the mechanism allowing chromium to overcome elevated SBP is due to enhancement of insulin sensitivity. We conclude that chromium and other agents are known to sensitize insulin response can markedly lower SBP in normotensive rats for at least 1-year period when taken consistently [62–66].

10.7.4 Aging

Solid evidence that IR in humans influences lifespan negatively would favor the likelihood that chromium can forestall the process at least to some extent [17,18]. Because diabetes mellitus has been regularly linked to premature aging [67–70], this fact strengthens the possibility that a perturbation in the glucose-insulin system, namely IR even in milder form over an extended period could hasten the aging phenomenon [71,72]. So, we began examining IR in non-diabetics (FBG range < 125 mg/dL). In this group of individuals, there is a positive correlation between fasting blood glucose (FBG) levels when circulating insulin levels and HbA1C are the independent variables [8]. Because of practicality and some question concerning the precision of insulin measurements [59], FBG became the preferred surrogate for IR. The fact that initial findings showed positive linkages with elements making up MS even in non-diabetics strengthens the use of this surrogate. Significant positive links with MS included *Body Composition Alterations* (body weight, body fat mass), *Cardiac Dynamics* (systolic and diastolic BP), *Blood Chemistries* (triglycerides), *Non-alcoholic Fatty Liver Disease—NAFLD* (ALT), and *Inflammatory Markers* (hsCRP and WBC and neutrophil counts). An appropriate significant negative link for MS included HDL cholesterol.

The first linkages with aging showed a highly significant linear regression line between age and FBG levels ($R=0.69$, $p < 0.00$) (Figure 10.2). However, when we examined the same data employing a weighted line, a distinctly different pattern was seen. The positive line of regression occurred up to approximately age 65 years but descended significantly negative thereafter until the 80s ages were reached. In other reports, this change in direction was attributed to "survivor bias" rather than a sudden improvement in metabolism [17,18]. Previously, we have referred to this phenomenon as the "aging paradox" [17,18].

Survival bias proposes that those individuals with the near optimal IR status over the years reflected in relatively low FBG measurements are the healthy survivors in later years. Generally those with the higher FBG are diseased or inactive. Suffice it to say, in line with improved IR (FBG levels) after age 65 years depicted in Figure 10.2, triglyceride levels and body fat mass decrease while HDL cholesterol increases [17,18].

An animal study supports the above human correlation between aging and FBG levels used as a surrogate for IR (Figure 10.3). In the classic study by Masoro et al. [73], two sets of 21 rats each were followed over their lifespan. One set ate their normal daily intake of rat chow (*ad libitum*) while the second group was limited to caloric restriction (60% bulk of same diet). The average daily glucose levels revealed the *ad libitum* group to have an average daily glucose level that ran roughly 20 mg/dL higher than the caloric restricted rats. The bold line in between the two others represents the combined estimates for average FBG of the two sets of rats. All rats survived up to the 16,17th month and a steady increase in average FBGs could be seen. Eight rats died by the 22,23rd month,

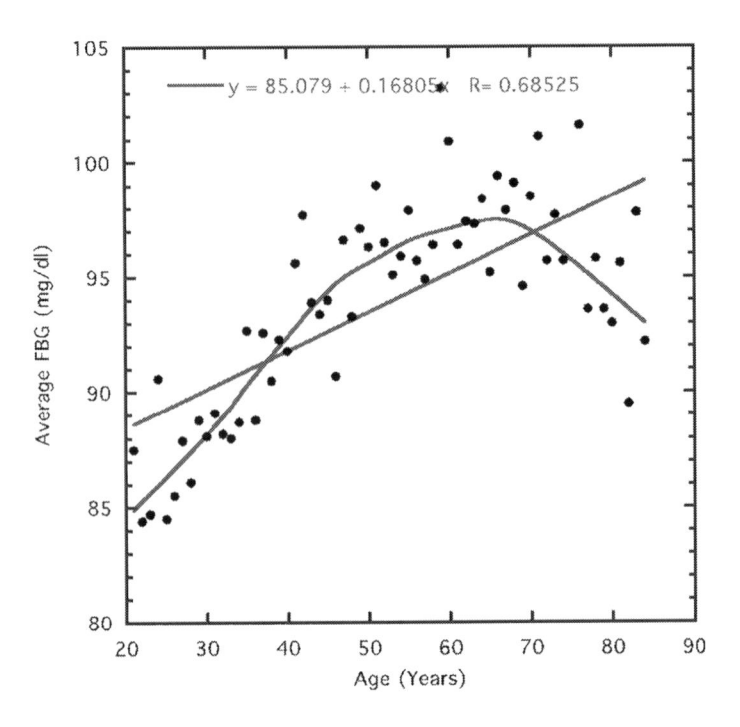

Figure 10.2   Correlation of chronological age with average yearly FBG. Both linear regression and weighted lines are shown.

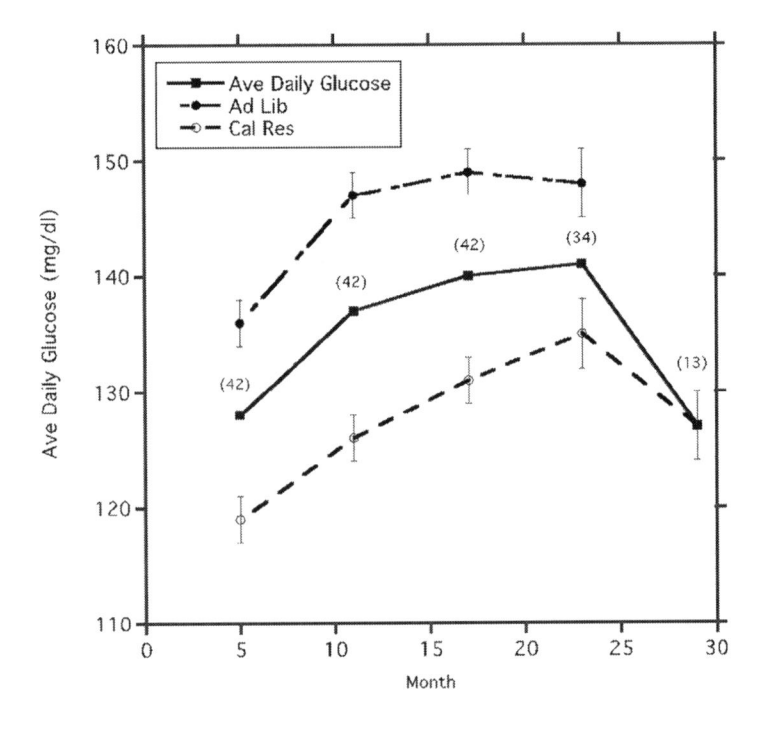

Figure 10.3   Example of survivor bias. Average mean 24-h glucose levels over the combined lifespan of *ad libitum* fed and caloric-restricted male F344 rats depicted in bold. Figures in parentheses indicate number of rats surviving at that point. (Data derived from Masoro EJ, McCarter RJM, Katz MS, McMahan CA: Dietary restriction alters characteristics of glucose fuel use. *J Gerontol* 47:B202–B208, 1992.)

and the combined average FBG had leveled off from the previous reading. After week 22, all the *ad libitum* rats had died leaving only the healthier caloric restrict ones with the lower FBG reading to survive. The bold line resembles survivor bias: the healthier rats with the lower daily glucose readings lived longer. Notice that the lower line representing only readings from caloric restricted rats seems to show its own survivor bias.

As a final point, a study performed in rats reported an increased lifespan in those consuming chromium when compared to controls that did not receive this trace metal [13]. Other studies using different insulin-sensitizing agents to ameliorate IR have also reported a proclivity to extend lifespan, strengthening deductions from the "chromium study" [74–76].

REFERENCES

1. Anderson RA: Chromium as an essential nutrient for humans. *Reg Toxicol Pharmacol* 26:S35–S41, 1997.
2. Anderson RA: Nutritional factors influencing the glucose/insulin system: chromium. *J Am Coll Nutr* 16:404–410, 1997.
3. Preuss HG, Anderson RA: Chromium update: examining recent literature 1997–1998. *Curr Opin Nutr Metab Care* 1:509–512, 1998.
4. Himsworth H: Diabetes mellitus: a differentiation into insulin-sensitive and insulin-insensitive types. *Lancet* 1:127–130, 1936.
5. Ginsberg H, Kimmerling G, Olefsky JM, Reaven GM: Further evidence that insulin resistance exists in patients with chemical diabetes. *Diabetes* 23:674–678, 1974.
6. Olefsky JM: Diabetes mellitus. In: Wyngaarden JB, Smith LH Jr, Bennett JC (eds). Cecil Textbook of Medicine, 19th edn. WB Saunders Co., Philadelphia, PA, pp 1291–1310, 1992.
7. DeFronzo RA, Ferrannini E: Insulin resistance: a multifaceted syndrome responsible for NIDDM, obesity, hypertension, dyslipidemia, and atherosclerotic cardiovascular disease. *Diabetes Care* 14:173–194, 1991.
8. Preuss HG, Mrvichin N, Bagchi D, Preuss J, Perricone N, Kaats GR: Fasting circulating glucose levels in the non-diabetic range correlate appropriately with many components of the metabolic syndrome. *Orig Intern* 23:78–89, 2016.
9. Reaven GM: The individual components of the metabolic syndrome: is there a raison d'etre? *J Am Coll Nutr* 26:191–195, 2007.
10. Bagchi M, Preuss HG, Zafra-Stone S, Bagchi D: Chromium III in promating weight loss and lean body mass. In: D Bagchi, HG Preuss (eds). Obesity. Epidemiology, Pathophysiology, and Prevention, second edition, CRC Press, Boca Raton, FL, pp 501–510, 2012.
11. Kaats GR, Preuss HG: Challenges to the conduct and interpretation of weight loss research. In: D Bagchi, HG Preuss (eds). Obesity. Epidemiology, Pathophysiology, and Prevention, second edition CRC Press, Boca Raton, FL, pp 833–852, 2012.
12. Crawford V, Scheckenbach R, Preuss HG: Effects of niacin-bound chromium supplementation on body composition of overweight African-American women. *Diabetes Obes Metab* 1:331–337, 1999.
13. Preuss HG, Echard B, Clouatre D, Bagchi D, Perricone NV: Niacin-bound chromium (NBC) increases life span in Zucker rats. *J Inorg Chem* 105:1344–1349, 2011.
14. National Research Council, Food and Nutrition Board. Recommended Dietary Allowances, 10th edn. National Academy Press, Washington, DC, 1989.
15. ods.od.nih.gov: National Institutes of Health, Office of Dietary Supplements, Chromium Dietary Supplement Fact Sheet (Updated: July 9, 2019).
16. Chan JC, Tong PC, Critchley JA: The insulin resistance syndrome: mechanism of clustering of cardiovascular risks. *Semin Vasc Med* 2:45–57, 2002.
17. Preuss HG, Mrvichin N, Kaats GR, Bagchi D: Reflecting on concepts relevant to contemplating the relationship between glucose/insulin perturbations and aging. *J Am Coll Nutr.* 38:463–469, 2019.
18. Preuss HG, Kaats GR, Mrvichin N, Bagchi D, Preuss JM: Cross-sectional examination of circulating ALT and AST levels in relatively healthy volunteers over their lifespan: implications. *J Am Coll Nutr* 10:1–9, 2019. doi: 10.1080/07315724.2019.1580169. [Epub ahead of print].

19. Kang JG, Park CY: Anti-obesity drugs: a review about their effects and safety. *Diabetes Metab J* 36:13–25, 2012.

20. Onakpoya IJ, Heneghan CJ, Aronson JK: Postmarketing withdrawal of human medicinal products. *BMC Med* 14:191, 2016. doi: 10.1186/s12916-016-0735-y.

21. Spiller HA, Sawyer TS: Toxicology of oral antidiabetic medications. *Am J Health Syst Pharm* 63:929–938, 2006.

22. Preuss HG, Gottlieb W: The Natural Fat Loss Pharmacy. Broadway Books, New York, NY, pp. 1–260, 2007.

23. Mertz W: Chromium. History and nutritional importance. *Biol Trace Element Res* 32:3–8, 1992.

24. Evans GW: The effect of chromium picolinate on insulin controlled parameters in humans. *Int J Biosocial Med Res* 11:163–180, 1989.

25. Schwartz K, Mertz W: A glucose tolerance factor and its differentiation from factor 3. *Arch Biochem Biophy* 72:515–518, 1957.

26. Schwartz K, Mertz W: Chromium (III) and the glucose tolerance factor. *Arch Biochem Biophys* 85:292–295, 1959.

27. Jeejeebhoy KN, Chu RC, Marliss EB, et al.: Chromium deficiency, glucose intolerance, and neuropathy reversed by chromium supplementation in a patient receiving long-term total parenteral nutrition. *Am J Clin Nutr* 30:531–538, 1977.

28. Freund H, Atamian S, Fischer JE: Chromium deficiency during total parenteral nutrition. *JAMA* 241:496–498, 1979.

29. Brown RO, Forloines-Lynn S, Cross RE, Heizer WD: Chromium deficiency after long-term parenteral nutrition. *Dig Dis Sci* 31:661–664, 1986.

30. Preuss HG, Clouatre D: Potential of diet and dietary supplementation to ameliorate the chronic clinical perturbations of the metabolic syndrome. In: S Sinatra and M Houston (eds). Nutritional and Integrative Strategies in Cardiovascular Medicine. CRC Press, Boca Raton, FL, pp 148–178, 2015.

31. Preuss HG, Preuss JM: The global diabetes epidemic: focus on the role of dietary sugars and refined carbohydrates in strategizing prevention. In: MM Rothkopf, MJ Nusbaum, LP Haverstick (eds). Metabolic Medicine and Surgery. CRC Press, Boca Raton, FL, pp 183–206, 2014.

32. Reaven GM: Role of insulin resistance in human disease (Banting Lecture 1988). *Diabetes* 37:1595–1607, 1988.

33. Banting FG, Best CH, Collip JB, Campbell WR, Fletcher AA: Pancreatic extracts in the treatment of diabetes mellitus. *Can Med Assoc J* 12:141–146, 1922.

34. Petersen KF, Dufour S, Feng J, et al.: Increased prevalence of insulin resistance and nonalcoholic fatty liver disease in Asian-Indian men. *Proc Natl Acad Sci* 103:18273–18274, 2006.

35. Preuss HG, Kaats GR, Mrvichin N, et al.: Examining the relationship between nonalcoholic fatty liver disease and the metabolic syndrome in nondiabetic subjects. *J Am Coll Nutr* 13:1–9, 2018. doi: 10.1080/07315724.2018.1443292. [Epub ahead of print]. PMID:29652564.

36. Hamaguchi M, Takeda N, Kojima T, et al.: Identification of individuals with non-alcoholic fatty liver disease by the diagnostic criteria for the metabolic syndrome. *World J Gastroenterol* 18:1508–1519, 2012.

37. Bricker NS: On the pathogenesis of the uremic state. An exposition of the "trade-off hypothesis". *New Eng J Med* 286:1093–1099, 1972.

38. Anderson RA, Bryden NA, Polansky MM: Lack of toxicity of chromium chloride and chromium picolinate in rats. *J Am Coll Nutr* 18:273–279, 1997.

39. Lamson DW, Plaza SM: The safety and efficacy of high-dose chromium. *Altern Med Rev.* 7:218–235, 2002.

40. Stearns DM, Wise JP Sr, Patierno SR, Wetterhahn KE: Chromium (III) picolinate produces chromosome damage in Chinese hamster ovary cells. *FASEB J* 9:1643–1648, 1995.

41. Hininger I, Benaraba R, Osman M, et al.: Safety of trivalent chromium complexes: no evidence for DNA damage in human HaCaT keratinocytes. *Free Radic Biol Med* 42:1759–1765, 2007.

42. Talpur N, Echard B, Yasmin D, et al.: Effects of niacin-bound chromium, maitake mushroom fraction SX and a novel (-)-hydroxycitric acid extract on the metabolic syndrome in aged diabetic Zucker fatty rats. *Molec Cell Biochem* 252:369–377, 2003.

43. Preuss HG, Echard B, Perricone NV, et al.: Comparing metabolic effects of six different commercial trivalent chromium compounds. *J Inorg Biochem* 102:1986–1990, 2008.

44. Anderson RA, Polansky MM, Bryden NA: Stabilitiy and absorption of chromium histidinate complexes by humans. *Biol Trace Elem Res* 101:211–218, 2004.

45. Seaborn CD, Stoecker BJ: Effects of starch, sucrose, fructose on chromium absorption and tissue concentrations in obese and lean mice. *J Nutr* 119:1444–1451, 1989.

46. Seaborn CD, Stoecker BJ: Effects of antacid or ascorbic acid on tissue accumulation of [51]chromium. *Nutr Res* 10:1401–1407, 1989.

47. Seaborn CD, Stoecker BJ: Effects of ascorbic acid depletion and chromium status on retention and urinary excretion of [51]chromium. *Nutr Res* 12:1229–1234, 1992.

48. Kuritzky L, Samraj GPN, Quillen DM: Improving management of type 2 diabetes mellitus: 6. Chromium. *Hosp Pract* 35:113–116, 2000.

49. Kaats GR, Michalek J, Ingram C, Preuss HG: Konjac glucomannon dietary supplementation causes fat loss in compliant overweight adults. *J Am Coll* 22:1–7, 2015.

50. Costello RB, Dwyer JT, Bailey RL: Chromium supplements for glycemic control in type 2 diabetes. *Nutr Rev* 74:455–468, 2016.

51. Anderson RA, Cheng N, Bryden NA, et al.: Elevated intakes of supplemental chromium improve glucose and insulin variables in individuals with type 2 diabetes. *Diabetes* 46:1786–1791, 1997.

52. Martin J, Wang ZQ, Zhang XH, et al.: Chromium picolinate supplementation attenuates body weight gain and increases insulin sensitivity in subjects with type 2 diabetes.

53. Linday LA: Trivalent chromium and the diabetes prevention program. *Med Hypotheses* 49:47–49, 1997.

54. Mokdad AH, Serdula MK, Dietz WH, Bowman BA, Marks JS, Koplan JP: The spread of the obesity epidemic in the United States, 1991–1998. *KAMA* 282:1519–1522, 1999.

55. Wang Y, Beydoun MA: The obesity epidemic in the United States – gender, age, socioeconomics, racial/ethnic and geographic characteristics: a systematic review and meta-regression analysis. *Epidemiol Rev* 29:6–28, 2007.

56. Kaats GR, Blum K., Fisher JA, Adelman JA: Effects of chromium picolinate supplementation on body composition: a randomized doublemasked, placebo-controlled study. *Curr Ther Res* 57:747–756, 1996.

57. Kaats GR, Blum K, Pullin D, Keith SC, Wood R: A randomized, double-masked, placebo-controlled study of the effects of chromium picolinate supplementation on body composition: a replication and extension of a previous study. *Curr Ther Res* 59:379–388, 1998.

58. Volpe SL, Huang HW, Larpadisorn K, Lesser II: Effect of chromium supplementation and exercise on body composition, resting metabolic rate and selected biochemical parameters in moderately obese women following and exercise program. *J Am Coll Nutr* 20:293–306, 2001.

59. Ferrannini E, Haffner SM, Stern MP: Essential hypertension: an insulin resistant state. *J Cardiovac Pharmacol* 15(Suppl 3):S18–S25, 1990.

60. Sowers JR: Is hypertension an insulin-resistant state? Metabolic changes associated with hypertension and anti hypertensive therapy. *Am Heart J* 122:932–935, 1991.

61. Preuss MB, Preuss HG: Effects of sucrose on the blood pressure of various strains of Wistar rats. *Lab Invest* 43:101–107, 1980.

62. Preuss HG, Gondal JA, Bustos E, et al.: Effect of chromium and guar on sugar-induced hypertension in rats. *Clin Neph* 44:170–177, 1995.

63. Preuss HG, Anderson RA, Gondal J: Comparative effects of chromium, vanadium, and *Gymnema Sylvestre* on sugar-induced blood pressure elevations in SHR. *J Am Coll Nutr* 17:116–123, 1998.

64. Preuss HG, Montamarry S, Echard B, Scheckenbach R, Bagchi D: Long-term effects of chromium, grape seed extract, and zinc on various metabolic parameters of rats. *Mol Cell Biochem* 223:95–102, 2001.

65. Preuss HG, Talpur N, Venkataramiah N, Manohar V, Anderson R: Chromium and hypertension. *J Trace Elem Exp Med* 12:125–130, 1999.

66. Perricone NV, Bagchi D, Echard B, Preuss HG: Long-term metabolic effects of different doses of niacin-bound chromium on Sprague-Dawley rats. *Mol Cell Biochem* 338:91–103, 2009.

67. Loukine L, Waters C, Choi BCK, Ellison J: Impact of diabetes mellitus on life expectancy in Canada. Population Health Metrics 2012, 10:7. http://www.pophealthmetrics.com/content/10/1/7

68. Franco OH, Steyerberg EW, Hu FB, Mackenbach J, Nusselder W: Associations of diabetes mellitus with total life expectancy and life expectancy with and without cardiovascular disease. *Arch Intern Med* 167:1145–1151, 2007.

69. Gu K, Cowie CC, Harris MI: Mortality in adults with and without diabetes in a national cohort of the U.S. population, 1971–1993. *Diabetes Care* 21:1138–1145, 1998.

70. Paolisso G, Barbieri M, Rizzo MR, et al.: Low insulin resistance and preserved beta-cell function contribute to human longevity but are not associated with TH-INS genes. *Exp Gerontol* 37:149–156, 2001.

71. Preuss HG, Bagchi D, Clouatre D: Insulin resistance: a factor of aging. In: Ghen MJ, Corso N, Joiner-Bey H, Klatz R, Dratz A (eds). The Advanced Guide to Longevity Medicine. Ghen, Landrum, SC, pp. 239–250, 2001.

72. Preuss HG: Effects of glucose/insulin perturbations on aging and chronic disorders of aging: the evidence. *J Am Coll Nutr* 16:397–403, 1997.

73. Masoro EJ, McCarter RJM, Katz MS, McMahan CA: Dietary restriction alters characteristics of glucose fuel use. *J Gerontol* 47:B202–B208, 1992.

74. Dilman VM, Anisimov VN: Effect of treatment with phenformin, diphenylhydantoin or L-dopa on life span and tumour incidence in C_3H/Sn mice. *Gerontology* 26:214–246, 1980.

75. Anisimov VN, Semenchenko AV, Yashin AI: Insulin and longevity: antidiabetic biguanides as geroprotectors. *Biogerontology* 4:297–307, 2003.

76. Martin-Montalvo A, Mercken EM, Mitchell SJ, et al.: Metformin improves healthspan and lifespan in mice. *Nat Commun* 4:2192, 2013. doi: 10.1038/ncomms3192.

CHAPTER 11

Differential Impact of Organic Vanadium Compounds on Human Health
Deficiency States, Toxic Effects, and Potential Therapeutic Use

Satinath Mukhopadhayay and Bidisha Mukherjee
Institute of Post Graduate Medical Education and Research

CONTENTS

11.1 INTRODUCTION

Vanadium is a transition element occupying group Vb in the periodic table. Its oxidation state varies from $-III$ to $+V$, the most stable oxidation state being the tetravalent oxovanadium ion (1, 2).

At a pH below 3.5, vanadium exists as vanadyl (VO^{+2}) compounds. In basic solutions, it forms orthovanadate (VO_4^{-3}), chemically similar to orthophosphate (PO_4^{-3}) (3). In neutral solutions, it usually occurs as $H_2VO_4^-$, which, at a higher concentration, forms the tetramer ($V_4O_{12}^{4-}$); further polymerization leads to the formation of isopolyanion, $V_{10}O_{28}^{6-}$ (4). Presence of vanadium is quite frequent in the environment, as a result it is absorbed by plants and moves along the food chain into the bodies of animals and humans. While soil, water, and air contain a high concentration of vanadium, it has low dietary (10 µg/kg) and tissue contents [5–20 nM] (5).

Small amounts of dietary vanadium are enough for growth and development of the mammalian species; deficiency is associated with impaired growth, defective bone mineralization and deranged thyroid hormone, lipid and carbohydrate metabolism (1, 6, 7). The insulinomimetic effects of vanadates are manifested by their ability to improve hyperglycemia, dyslipidemia, and insulin sensitivity. Vanadium is considered as a toxic element in both its cationic and anionic forms but the latter is responsible for more of its adverse effects, presumably due to its faster absorption (8). There appears to be a clear need therefore to identify and/or develop new organovanadium compounds with higher efficacy and fewer toxicity for potential therapeutic application as insulin sensitizers. Our group has recently identified a small, orally bioavailable, nontoxic insulinomimetic small molecule [3,5-dimethylpyrazole-peroxy-vanadate (dmp)], which improves insulin sensitivity not only by activating the insulin receptor signaling pathway but also by upregulating the expression of PPARγ and its target genes. This monograph summarizes the physiological effects, potential therapeutic applications, and toxicity issues related to stable organovanadium derivatives in humans.

11.2 VANADIUM METABOLISM AND INTRACELLULAR TRANSPORT

While water-soluble vanadium salts enter the body via the alimentary canal, vanadium pentoxide [V_2O_5] is absorbed mainly via the respiratory system (9, 10). Around 5%–10% of dietary vanadium eventually enters the human body (8). In the stomach, most vanadium compounds are transformed into the cationic (vanadyl) form and then absorbed through the duodenal mucous membrane into the blood, where 90% of them are bound to transferrin or albumin, which maintain their stability in the circulation. Free vanadyl ions are spontaneously oxidized into the anionic vanadates in the gastrointestinal tract (GIT) from where they enter the portal circulation in a bound form with transferrin. While both vanadyl and vanadate ions are widely distributed in the body, half of the "bound" vanadium pool accumulates in the liver, spleen, kidneys, and bones (11). The intracellular fate of vanadium is decided by a series of redox steps involving reduced glutathione, cysteine, catechol, NADH, NADPH, ascorbic acid, and glutathione reductase that eventually determine the form in which vanadium will be found inside the cell (12).

11.3 VANADIUM ABSORPTION, DISTRIBUTION, AND ELIMINATION

The absorption of vanadium compounds depends on their solubility and the route of entry. Although the rates of absorption are not well defined, soluble vanadium compounds including V_2O_5 are well absorbed following instillation into the lungs. In contrast, vanadium salts (e.g., ammonium vanadyl tartrate) are poorly absorbed from the GIT (13,14). Dietary vanadium exists as either vanadyl (+4) or vanadate (+5), the absorption of the latter from the GIT being about three times more than the former (15).

The distribution of vanadium from the blood is rapid ($T_{1/2}$ about 1 h) and initially the highest concentrations of vanadium appear in the kidney, liver, and lungs. Approximately 90% of the vanadium circulates in the vanadyl form bound to transferrin and albumin (16). The concentration of vanadium in the brain is approximately 5% of that in the blood (17). The total amount of vanadium in the human body is low with the total body burden averaging about 100 µg V (18). Based on animal studies, the bones and teeth retain the highest concentrations of vanadium, and these tissues provide a major reservoir for retained vanadium.

The principal route of elimination of absorbed vanadium is through the kidneys with only a minor amount (<10%) being excreted through feces. About 40%–60% of an absorbed dose of vanadium is excreted by the kidneys within 1–3 days (19). The biological half-life of vanadium in the urine is 20–40 h (20). The elimination of vanadium following inhalation of vanadium oxides is biphasic with an initial rapid (10–20 h) and a longer terminal phase (40–50 days) (21).

11.4 VANADIUM TOXICITY

Stability and toxicity have remained major hurdles in the therapeutic development of organic vanadium compounds as insulin sensitizers. Industries that use fossil fuels as raw materials cause the most widespread discharge of vanadium into the environment (22–24). Studies looking at the impact of workplace exposure to and emission sources of vanadium pentoxide are limited. Fuel-oil ash is a significant source of exposure to boiler makers (25). During coal combustion, flue gases may contain dangerous levels of vanadium pentoxide and in a poorly ventilated environment, may result in significant exposure to vanadium (26, 27).

The biological monitoring of vanadium is widely used. Urine testing, in preference for plasma testing, assumes that vanadium is excreted with a half-life of 15–40 h (28). Pre- and post-shift urine vanadium levels measured at the beginning and at the end of a working week give a measure of daily absorption and accumulated dose (29). Normal serum concentrations in the unexposed population are, 1–2 mg V/L. In the general population, mean concentrations in the urine average about 0.1–0.2 mg V/L and are usually, 1 mg V/g of creatinine. After adjusting for creatinine concentrations, urine vanadium concentrations are probably a more reliable measure of exposure for biological monitoring than blood vanadium levels (30). In a study of 20 workers exposed to low (0.36–32 mg V/m^3) air concentrations of vanadium during the overhaul of an oil-fired boiler, the urine vanadium concentrations ranged from 0.19–4.3 mg V/g creatinine (31).

The common symptoms of vanadium poisoning are cough with expectoration, ear/nose/throat irritation, headache, palpitation, etc. Greenish discoloration of the tongue is a reliable sign of vanadium toxicity in humans. Organic vanadium compounds are much less toxic than the inorganic compounds. Diabetic rats receiving organic compounds did not show any gastrointestinal side-effects and did not develop diarrhea (32). In addition to gastrointestinal discomfort, other toxic effects of vanadium salts include hepatotoxicity, nephrotoxicity, teratogenicity, and reproductive dysfunction (33, 34). While vanadium compounds are potentially mitogenic and are reported to exert tumorigenic/carcinogenic activity, more recent studies suggest anti-mitogenic and anti-tumor activity of certain vanadium compounds (35–39).

Ascorbic acid, used as an antidote for vanadium poisoning, has the ability to reduce toxic, poisonous pentavalent vanadate compounds to the less toxic quadrivalent vanadyl compounds (40,41). Ascorbic acid is believed to be more effective in this context than dimercaprol (also used to remove a heavy metal such as lead or mercury), the other agent used to mitigate the effects of vanadium toxicity.

11.5 VANADIUM AS AN INSULIN SENSITIZER

Incidence of diabetes is threateningly rising over the globe. It is strongly associated with cardiovascular disease, retinopathy, and kidney failure. Diminution in insulin production or loss of insulin sensitivity in insulin target tissues causes diabetes mellitus. Type 1 diabetes occurs when pancreatic β-cells are destroyed due to autoimmune disorder affecting considerable depletion in insulin secretion (42). Hence, there remains only one treatment option for Type 1 diabetes, i.e., insulin injection. Type 2 diabetes, on the other hand, is affected because of decreased tissue sensitivity to insulin. Overabundance of lipids is primarily responsible for producing this defect and leads to insulin resistance (43).

Majority of the presently available drugs for Type 2 diabetes treatment target stimulation of insulin secretion from β-cells, i.e., sulfonylureas, meglitinide analogs, DPP-IV inhibitors, and GLP-1 receptor agonists, while metformin reduces hepatic glucose production and increases glucose utilization. Most relevant group of drugs that address the problem of insulin resistance in Type 2 diabetes are thiazolidinediones (TZDs) as they increase insulin sensitivity, however, their use has been restricted because of considerable adverse side effects (44). Demand for insulin therefore is substantially increasing for insulin replacement therapy. Insulin injection, may be more than one shot each day, is often inconvenient (45). Since 1899, vanadium had been used for diabetes treatment (46) and it was the only compound available for this purpose until the discovery of insulin (47, 48). However, toxicity of vanadium compounds is now a concern. Some potentially nontoxic peroxyvanadium compounds were prepared to circumvent these toxic effects (49, 50).

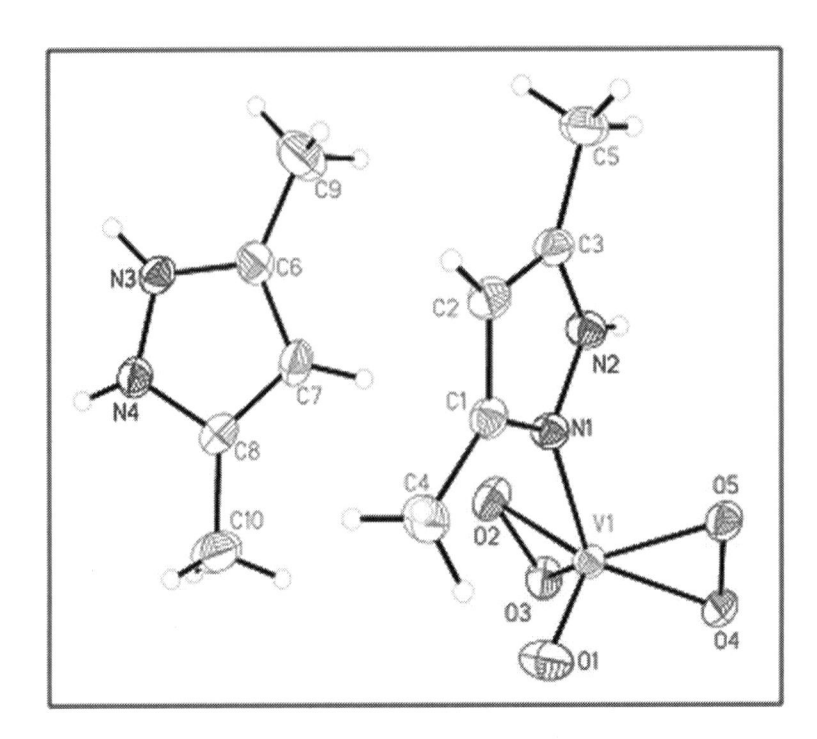

Figure 11.1 Structure of dmp: Oak Ridge Thermal Ellipsoid Plot with 35% probability ellipsoid of the anion of [VO(O2)2dmpz] ± and dmpzH+ cation, with selected bond distances (Å): V1–O1, 1.5919 (1); V1–O2, 1.861 (1); V1–O3, 1.853 (1); V1–O4, 1.894 (1);V1–O6, 1.1.588 (1); V1–N1, 2.104 (1); O1–O2, 1.480 (2); O4–O3, 1.461 (2); N1–N2, 1.360 (2); N1–C3, 1.335 (2); N2–C1, 1.341 (2). (Sandip Mukherjee, et al. *Plos ONE* January 10, 2017.)

The compound prepared by Crans et al. (51), while showing considerable anti-diabetic activity with minimal toxicity, could not be developed as a drug as it was not stable at room temperature and denatured without deep refrigeration. Hence, to retain the anti-diabetic activity, ensure stability at room temperature, and render it free from toxicity became the major challenges to come up with a vanadium compound as a suitable alternative to insulin. We recently synthesized a peroxyvanadate compound, DmpzH[VO(O2)2(dmpz)], referred to as dmp hereafter, which is soluble in water, stable at room temperature, and free from toxicity (52). This compound is a functionally insulinomimetic compound. It mimics the effect of native insulin with respect to insulin receptor activation, tyrosine kinase activity, and GLUT4 translocation. Inhibition of phosphotyrosine phosphatase 1b (PTP1b), one such phosphatase, is believed to be responsible for the postreceptor insulin-sensitizing effects of vanadium compounds. The same trend was reflected in glucose uptake and fatty acid uptake also in muscle cells, L6 myotubes, and adipocytes, respectively.

Excess expression of PPARγ has been associated with increase in insulin sensitivity (53), a function known to be performed by the PPARγ ligands. Our group has shown that dmp induces expression of PPARγ and its target genes through the suppression of Wnt3a thereby improving insulin sensitivity (53) (Figures 11.1 and 11.2).

11.6 OTHER BIOLOGICAL ACTIONS OF VANADIUM

Besides the action of vanadium as an insulin sensitizer, it's derivatives have also shown antineoplastic activity in some studies (54, 55). Derivatives of vanadium exert their anti-carcinogenic effect by modifying various xenobiotic enzymes (such as glutathione transferase, cytochrome P450, etc.), and also

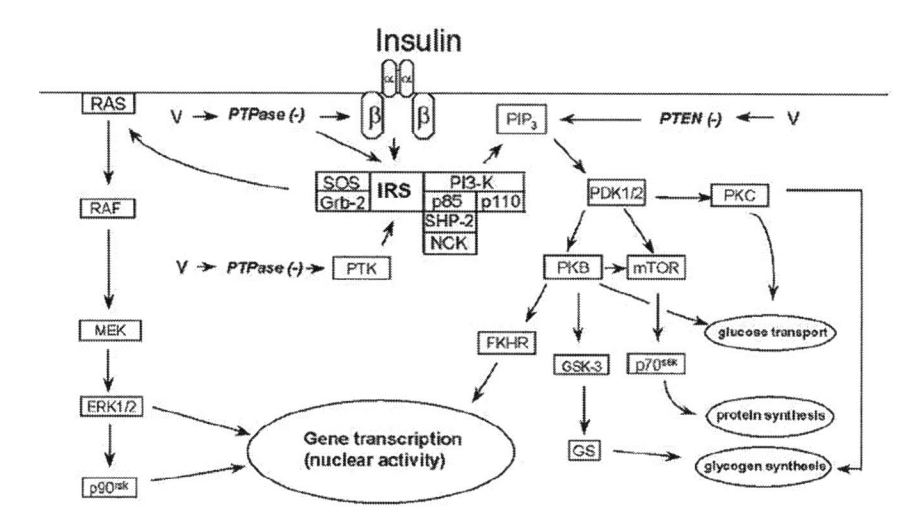

Figure 11.2 Insulinomimetic and anti-diabetic effects of vanadium compounds. (A. K. Srivastava and M. Z. Mehdi. *Diabetic Medicine*, 22, 2–13 (2004).) (56)

through inhibition of cellular tyrosine phosphatases and/or activation of tyrosine phosphorylases. It has also been reported that their cytotoxic effects are mainly due to regulation of DNA cleavage, fragmentation, and plasma membrane lipoperoxidation. Inorganic vanadium compounds, i.e., ammonium metavanadate, at its lower doses show significant reduction of tumor cell proliferation while at higher doses show reverse action in albino mice model (57). On the other hand, most organic vanadates and some inorganic vanadium compounds such as orthovanadate, vanadyl sulfate at low doses show significant anticancer activity in tumor-bearing mice (57). In view of their antiproliferative, cytostatic/cytotoxic, and antimetastatic activities, it has been suggested to use these compounds as adjuvants in cancer chemotherapy (55,58) (Figure 11.3).

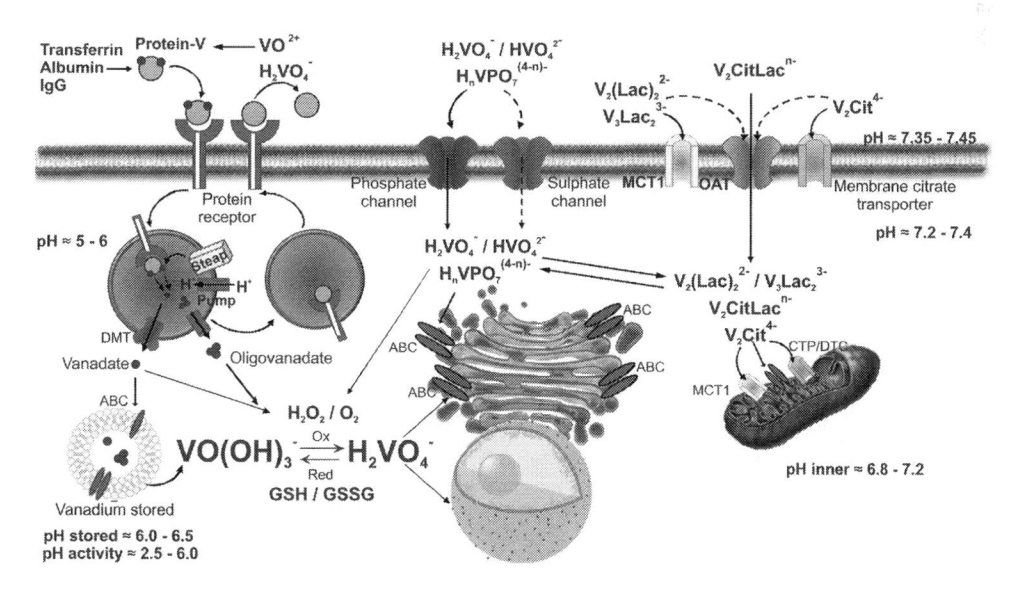

Figure 11.3 Vanadium species uptake and cellular compartmentalization. (Samuel Treviño et al. *Biol Trace Element Res*, 188, 68–98 (2019) (59).)

11.7 CONCLUSION

Vanadium compounds, organic as well as inorganic, have been shown to have differential effects on various pathophysiological states in the human body. While they have many salutary effects on health, potential for toxicity or adverse effects has restricted their use clinically. Vanadium has both organic and inorganic derivatives. While most soluble, inorganic vanadium derivatives are reported to be carcinogenic in nature, some of them, in low doses, have shown antineoplastic effects in animal models. Organic vanadates, on the other hand, not only have significant antineoplastic effects but they also have the potential to act as powerful insulin sensitizers. Since it's the cationic derivatives that have the beneficial effects in humans, the pH at which vanadates exist in the cationic form becomes an important determinant of the beneficial *vis-a-vis* harmful effects of vanadium compounds. The differential biological functions of the same vanadium compound are largely dose dependent; low doses inhibiting while high doses promoting tumor proliferation. Industrial workers exposed to vanadium are at risk of toxic effects involving multiple organ systems, i.e., respiratory, cardiovascular, digestive, renal, central nervous system, and skin. Vanadium is well absorbed at lungs and poorly absorbed at GIT. Its distribution is rapid from blood and with the highest tissue concentrations achieved in the kidney, liver, and lungs. Besides their toxicity, vanadium compound has potential therapeutic effect as insulin sensitizers and antineoplastic.

dmp, an insulinomimetic small molecule that is nontoxic, orally bioavailable and stable at room temperature for months, could satisfy all the requirements of a potential insulin sensitizer. dmp is a novel vanadium compound, distinctly varies from other vanadium compounds of the same group examined so far for insulin-like activities. Unlike insulin, which exerts both metabolic as well as mitogenic effects, dmp only showed metabolic activities without significant mitogenic activity as evident from unaltered epidermal growth factor receptor expression and activation. dmp binding to IR causes IR activation which, in turn, triggers downstream signaling cascade including translocation of GLUT4 to the skeletal muscle cell membrane. While organic vanadium compounds show powerful blood glucose lowering effects due to their potent insulin-sensitizing action, stability at room temperature and potential for toxicity remain significant stumbling blocks in their further development as useful anti-hyperglycemic medications. Future clinical trials using more potent and safer organo-vanadium complexes should be directed addressing the above concerns and also to explore the insulin-sensitizing/enhancing effects of organic vanadium compound such as dmp and others in persons with Type 2 diabetes.

REFERENCES

1. Moskalyk RR, Alfantazi AM. Processing of vanadium: a review. *Miner Eng* 2003; 16:793–805.
2. Verma S, Cam MC, McNeill JH. Nutritional factors that can favorably influence the glucose/insulin system: vanadium. *J Am Coll Nutr* 1998; 17:11–18.
3. Dafnis E, Sabatini S. Biochemistry and pathophysiology of vanadium. *Nephron* 1994; 67:133–143.
4. Phillips TD, Nechay BR, Heidelbaugh ND. Vanadium: chemistry and the kidney. *Fed Proc* 1983; 42:2969–2973.
5. Seńczuk W. Toksykologia, Państwowy Zak lad Wydawnictw Lekarskich, Warszawa, 1994 (in Polish).
6. French RJ, Jones PJH. Role of vanadium in nutrition: metabolism, essentiality and dietary considerations. *Life Sci* 1993; 52:339–346.
7. Rojas E, Herrera LA, Poirier LA, Ostrosky-Wegman P. Are metals dietary carcinogens? *Mutat Res* 1999; 443:157–181.
8. Goc A. Biological activity of vanadium compounds. *CEJB* 2006; 1(3):314–332.
9. Mukherjee B, Patra B, Mahapatra S, Banerjee P, Tiwari A, Chatterjee M. Vanadium - an element of biological significance. *Toxicol Lett* 2004; 150:135–143.
10. Thompson KH, Orvig C. Coordination chemistry of vanadium in metallophar-maceutical candidate compounds. *Coord Chem Rev* 2001; 219–221:1033–1053.

11. Yang XG, Wang K, Lu J, Crans DC. Membrane transport of vanadium compounds and interaction with the erythrocyte membrane. *Coord Chem Rev* 2003; 273:103–111.

12. Hayes GR, Lockwood DH. Role of insulin receptor phosphorylation in the insulinomimetic effects of hydrogen peroxide. *Proc Natl Acad Sci USA* 1987; 84:8115–8119.

13. Dimond EG, Caravaca J, Benchimol A. Vanadium: excretion, toxicity, lipid effect in man. *Am J Clin Nutr* 1963; 12:49–53.

14. Roshchin AV, Ordzhonikidze EK, Shalganova IV. Vanadium – toxicity, metabolism, carrier state. *J Hyg Epidemiol Microbiol Immunol* 1980; 24:377–383.

15. Nielsen FH. Vanadium in mammalian physiology and nutrition. *Met Ions Biol Syst* 1995; 31:543–573.

16. Chasteen ND, Lord EM, Thompson HJ, Grady JK. Vanadium complexes of transferrin and ferritin in the rat. *Biochim Biophys Acta* 1986; 884:84–92.

17. Al-Bayati MA, Culbertson MR, Schreider JP, Rosenblatt LS, Raabe OG. The lymphotoxic action of vanadate. *J Environ Pathol Toxicol Oncol* 1992; 11:83–91.

18. Nielsen FG. Nutritional requirements for boron, silicon, vanadium, nickel, and arsenic: current knowledge and speculation. *FASEB J* 1991; 5:2661–2667.

19. Talvitie NA, Wagner WD. Studies in vanadium toxicology II. Distribution and excretion of vanadium in animals. *Arch Ind Hyg* 1954; 9:414–422.

20. Lauwerys RR, Hoet P. Industrial Chemical Exposure. Guidelines for Biological Monitoring, 2nd edn. Boca Raton, FL: Lewis Publishers, 1993.

21. Rhoads K, Sanders CL. Lung clearance, translocation and acute toxicity of arsenic, beryllium, cadmium, cobalt, lead, selenium, vanadium, and ytterbium oxides following deposition in rat lung. *Environ Res* 1985; 36: 359–378.

22. Malabu UH, Dryden S, McCarthy HD, Kilpatrick A, Williams G. Effects of chronic vanadate administration in the STZ-induced diabetic rat. The antihyperglycemic action of vanadate is attributable entirely to its suppression of feeding. *Diabetes* 1994; 43:9–15.

23. Strout HV, Vicario PP, Biswas C, Saperstein R, Brady EJ, Pilch PF, et al. Vanadate treatment of streptozotocin diabetic rats restores expression of the insulin-responsive glucose transporter in skeletal muscle. *Endocrinology* 1990; 126:2728–2732.

24. Venkataraman BV, Sudha S. Vanadium toxicity. *Asian J Exp Sci* 2005; 19(2):117–134.

25. Liu Y, Woodin MA, Smith TJ, Herrick RF, Williams PL, Hauser R, et al. Exposure to fuel-oil ash and welding emissions during the overhaul of an oil-fired boiler. *J Occup Environ Hyg* 2005; 2:435–443.

26. Lee CW, Srivastava RK, Ghorishi SB, Hastings TW, Stevens FM. Investigation of selective catalytic reduction impact on mercury speciation under simulated NOx emission control conditions. *J Air Waste Manag Assoc* 2004; 54:1560–1566.

27. Hirtle B, Teschke K, van Netten C, Brauer M. Kiln emissions and potters' exposures. *Am Ind Hyg Assoc J* 1998; 59:706–714.

28. Sabbioni E, Moroni M. A study on vanadium in workers from oil-powered fire plants. Luxembourg, Commission of European Communities (EU Report -EN) 1983.

29. Costigan M, Cary R, Dobson S. Vanadium pentoxide and other inorganic vanadium compounds. Stuggart: Concise International Chemical Association Document 29, Wissenschaftliche Verlagsgesellschaft mbH. 2001: 54.

30. Stroop SD, Helinek G, Greene HL. More sensitive flameless atomic absorption analysis of vanadium in tissue and serum. *Clin Chem* 1982; 28:79–82.

31. Gylseth B, Leira H, Steines E, Thomassen Y. Vanadium in the blood and urine of workers in a ferroalloy plant. *Scand J Work Environ Health* 1979; 5:188–194.

32. Hauser R, Elreedy S, Ryan PB, Christiani DC. Urin vanadium concentrations in workers overhauling an oilfired boiler. *Am J Ind Med* 1998; 33:55–60.

33. Reul BA, Amin SS, Buchet JP, Ongemba LN, Crans DC, Brichard SM, et al. Effects of vanadium complexes with organic ligands on glucose metabolism: a comparison study in diabetic rats. *Br J Pharmacol* 1999; 126:467–477.

34. Domingo JL, Gomez M, Sanchez DJ, Llobet JM, Keen CL. Toxicology of vanadium compounds in diabetic rats: the action of chelating agents on vanadium accumulation. *Mol Cell Biochem* 1995; 153:233–240.

35. Domingo JL. Vanadium and tungsten derivatives as antidiabetic agents: a review of their toxic effects. *Biol Trace Elem Res* 2002; 88:97–112.

36. Wang H, Scott RE. Unique and selective mitogenic effects of vanadate on SV40-transformed cells. *Mol Cell Biochem* 1995; 153:59–67.

37. Faure R, Vincent M, Dufour M, Shaver A, Posner BI. Arrest at the G2/M transition of the cell cycle by protein-tyrosine phosphatase inhibition: studies on a neuronal and a glial cell line. *J Cell Biochem* 1995; 59:389–401.

38. Fantus IG, Tsiani E. Multifunctional actions of vanadium compounds on insulin signaling pathways: evidence for preferential enhancement of metabolic versus mitogenic effects. *Mol Cell Biochem* 1998; 182:109–119.

39. Djordjevic C. Antitumor activity of vanadium compounds. *Met Ions Biol Syst* 1995; 31:595–616.

40. Liasko R, Kabanos TA, Karkabounas S, Malamas M, Tasiopoulos AJ, Stefanou D, et al. Beneficial effects of a vanadium complex with cysteine, administered at low doses on benzo (alpha) pyreneinduced leiomyosarcomas in Wistar rats. *Anticancer Res* 1998; 18:3609–3613.

41. Hansen TV, Aaseth J, Alexander J. The effect of chelating agents on vanadium distribution in the rat body and on uptake by human erythrocytes. *Arch Toxicol* 1982; 50:195–202.

42. Jones MM, Basinger MA. Chelate antidotes for sodium vanadate and vanadyl sulfate intoxication in mice. *J Toxicol Environ Health* 1983; 12:749–756.

43. Bluestone JA, Herold K, Eisenbarth G. Genetics, pathogenesis and clinical interventions in type 1 diabetes. *Nature* 2010; 464:1293–1300. PMID: 20432533.

44. Boden G. Obesity and free fatty acids. *Endocrinol Metab Clin N Am* 2008; 37:635–646.

45. Nesto RW, Bell D, Bonow RO, Fanseca V, Grundy SM, Horton E, et al. Thiazolidinedione use, fluid retention, and congestive heart failure: a consensus statement from the American Heart Association and American Diabetes Association. *Circulation* 2003; 108:2941–2948.

46. Richardson T, Kerr D. Skin-related complications of insulin therapy: epidemiology and emerging management strategies. *Am J Clin Dermatol* 2003; 4:661–667. PMID: 14507228.

47. Lyonnet B, Martz ME, Martin E. L'emploi therpeutique des derives du vanadium. *La Presse Med* 1899; 1:191–192.

48. Levina A, Lay PA. Metal-based anti-diabetic drugs: advances and challenges. *Dalton Trans* 2011; 40:11675–11686. doi: 10.1039/c1dt10380f. PMID: 21750828.

49. Mordes JP, Poussier P, Rossini A, Blankenhorn EP, Greiner DL. Rat models in type 1 diabetes: genetics, environment and auto-immunity. In: Shafrir E, editor. Animal Models of Diabetes: Frontiers in Research. Oxford Academy, CRC press, Taylor and Francis group; 1992, pp. 1–40.

50. Posner BI, Faure R, Burgess JW, Bevan AP, Lachance D, Zhang-Sun G, et al. Peroxovanadium compounds. A new class of potent phosphotyrosine phosphatase inhibitors which are insulin mimetics. *J Biol Chem* 1994; 269:4596–4604. PMID: 8308031.

51. Crans DC, Keramidas AD, Hoover-Litty H, Anderson OP, Miller MM, Lemoine LM, et al. Synthesis, structure, and biological activity of a new insulinomimetic peroxovanadium compound: bisperoxovanadium imidazole monoanion. *J Am Chem Soc*. 1997; 119:5447–5448.

52. Mukherjee J, Ganguly S, Bhatterjee M. Synthesis, characterization and reactivity of vanadium(v) complexes containing coordinated peroxide and histidine: a model for the active site of enzyme bromoperoxidase. *Ind J Chem* 1996; 35A:471–474.

53. Mukherjee S, Chattopadhyay M, Bhattacharya S, Dasgupta S, Mukhopadhyay S, Bhattacharya S, et al. A small insulinomimetic molecule also improves insulin sensitivity in diabetic mice. *Plos One* 2017. doi: 10.1371/journal.pone.0169809.

54. Crans DC, Keramidas AD, Hoover-Litty H, Anderson OP, Miller MM, Lemoine LM, et al. Synthesis, structure, and biological activity of a new insulinomimetic peroxovanadium compound: bisperoxovanadium imidazole monoanion. *J Am Chem Soc* 1997; 119:5447–5448.

55. Chakraborty T, Ghosh S, Datt S, Chakraborty P, Chatterjee M. Vanadium suppress sister-chromatid exchange and DNA-protein cross-link formation and restores antioxidant status hepatocellular architecture during 2-acetylaminfluorene induced experimental rat hepatocarcinogenesis. *J Exp Ther Oncol* 2003; 3:346–362.

56. Srivastava AK, Mehedi MZ. Insulino-mimetic and anti diabetic effects of vanadium compounds. *Diabetic Medicine* 2005; 22:2–13.

57. Angelos M. Evangelou. Vanadium in cancer treatment. *Crit Rev Oncol Hematol* 2002; 42:249–265.

58. Djordjevic C, Wampler GL. Antitumor activity and toxicity of peroxo hetero ligand vanadates(V) in relation to biochemistry of vanadium. *J Inorg Biochem* 1985; 25:51–55.

59. Treviño S, Díazl A, Sánchez-Lara E, Sanchez-Gaytan BL, Perez-Aguilar JM, González-Vergara E. Vanadium in biological action: chemical, pharmacological aspects, and metabolic implications in diabetes mellitus. *Biol Trace Elem Res* 2019; 188:68–98.

CHAPTER 12

Vanadium—Speciation Chemistry Can Be Important When Assessing Health Effects on Living Systems

Debbie C. Crans
Colorado State University

Kahoana Postal
Colorado State University
Universidade Federal do Paraná

Judith A. MacGregor
Toxicology Consulting Services

CONTENTS

12.1 INTRODUCTION

Vanadium has been reported to have many biological effects, some beneficial and others detrimental (Treviño et al. 2019, Crans et al. 2018). Vanadium is a first-row transition metal that is present on the earth's surface as a trace metal. For the last couple of decades, vanadium compounds have continued to be developed for therapeutic applications, including applications for the treatment of several cancers and diabetes mellitus (Treviño et al. 2019, Crans et al. 2004, Evangelou 2002). Although the mode of action of many vanadium compounds is not well understood, the effects of many different vanadium salts and compounds have been reported (Evangelou 2002, Crans, LaRee et al. 2019).

While these beneficial applications of vanadium compounds are developed, regulatory bodies are interested in assessing their safety because of the concerns regarding the toxicity of vanadium compounds (Ghosh, Saha, and Saha 2015, Cooper 2007), such as V_2O_5 (Figure 12.1) which is used

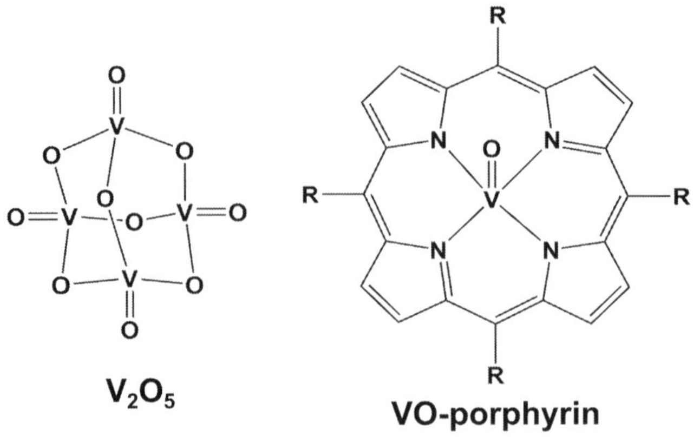

Figure 12.1 Structure representation of vanadium pentoxide (V_2O_5 also referred to as V^4O^{10}) and a generic VO-porphyrin, where R represents possible organic substitutes.

primarily in the production of the iron alloy ferrovanadium and as an industrial catalyst. Here, we present a mini-review describing some of these biological effects, in the context of important chemistry concepts relevant to this first-row transition metal.

Since vanadium is a metal, it forms a range of different metal ions depending on conditions. The existence of these different species is a phenomenon referred to as speciation, conferring this element fundamental properties to be significantly different from that of a non-metal element such as carbon (Crans and Smee 2003). Vanadium is naturally occurring in several minerals and in coal and crude oil as a V=O–porphyrin derivative (Zhao et al. 2014, Dechaine and Gray 2010) (Figure 12.1). Considering these fundamental differences, it is therefore not surprising that drugs containing vanadium are very different from the typical organic drugs. Although biological responses of different vanadium compounds are commonly attributed to "vanadium," this language is not properly reflecting the effects of the different vanadium compounds that have biological activities. Such imprecise language is problematic and is strongly discouraged particularly because it creates biases for the non-expert, which can be problematic regarding the development of beneficial uses of the vanadium compounds, as well as for the management and efforts to reduce undesirable environmental effects of different vanadium compounds. Because measurement of the vanadium species is rarely done, and elemental vanadium and total vanadium is typically analyzed instead, this creates difficulty in the assessment of health effects, which can vary with the specific vanadium compound.

In the following, we will briefly summarize concepts relating to understanding the beneficial and toxic effects of vanadium beginning with the speciation chemistry of this element. We will also describe the natural occurrence of vanadium, occupational studies, and industrial activities relating to vanadium as well as some safety assessments associated with vanadium compounds. We will also summarize some of the potential applications of vanadium compounds in medicine.

12.2 SPECIATION OF VANADIUM

Vanadium can form compounds in several oxidation states ranging from I– up to V+ (Wilkinson and Gillard 1987). The most common vanadium oxidation states are III, IV, and V and the speciation chemistry of these oxidation states will be summarized here. Generally, the system is described using a Pourbaix diagram (Kanamori and Tsuge 2012, Crans et al. 2004, Povar et al. 2019), which defines the conditions of the system in terms of redox potential and pH (Figure 12.2) distinguishing conditions in which vanadium(IV) and (V) oxides are formed in solution.

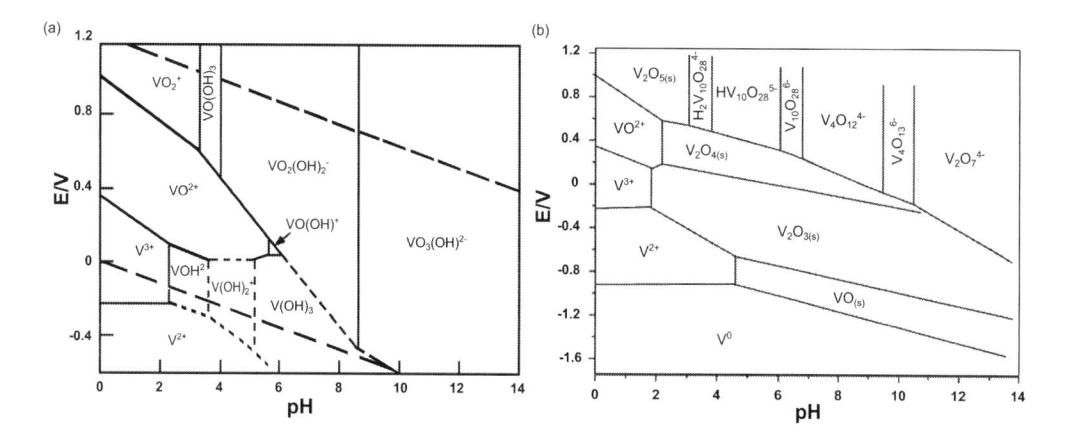

Figure 12.2 Representation of the Pourbaix diagram of vanadium of ionic strength of (a) 10^{-6} M and (b) 1 M at 25°C. (Adapted with permission from Crans et al. 2004. Copyright 2004 American Chemical Society.)

Vanadium compounds in the highest oxidation state generally are in the form of oxides as oxovanadium (VO^{2+}) or cisdioxovanadium (VO_2^+). However, the specific form depends on the concentration and the pH of the aqueous solution (Crans et al. 2004). Speciation diagrams are also used to describe the form of the vanadium present in aqueous solutions in specific ionic strength and concentration. Several different discrete vanadium(V) oxides form and the most common ones are generally referred to as oxovanadates which contain different nuclearities that can go from one to more than ten vanadium atoms in the structure (Aureliano and Crans 2009) as shown in Figure 12.3. Two speciation diagrams showing the distribution of oxovanadates are presented for 0.010 and 10 mM of an $H_2VO_4^-$ (V_1) solution, Figure 12.4. As seen in Figure 12.4, higher nuclearity oxovanadates are dominant at the 10 mM concentration, whereas at 0.010 mM, the presence of these higher nuclear species is negligible.

The solution chemistry of vanadium complexes can be hard to fully comprehend, due to the various oxidation states possible and the dependence of the media and conditions applied. For this reason, the conversion of vanadium compounds to other species can readily take place after the

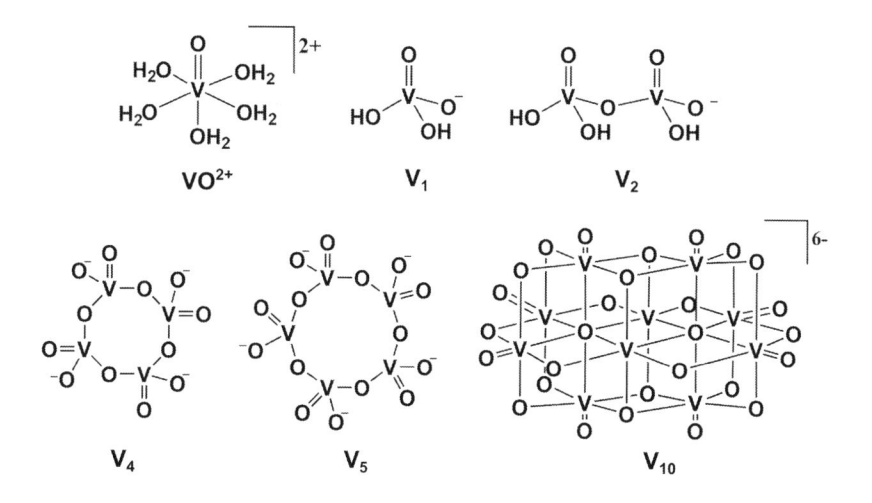

Figure 12.3 Structure representation of the most common oxovanadates in aqueous solution.

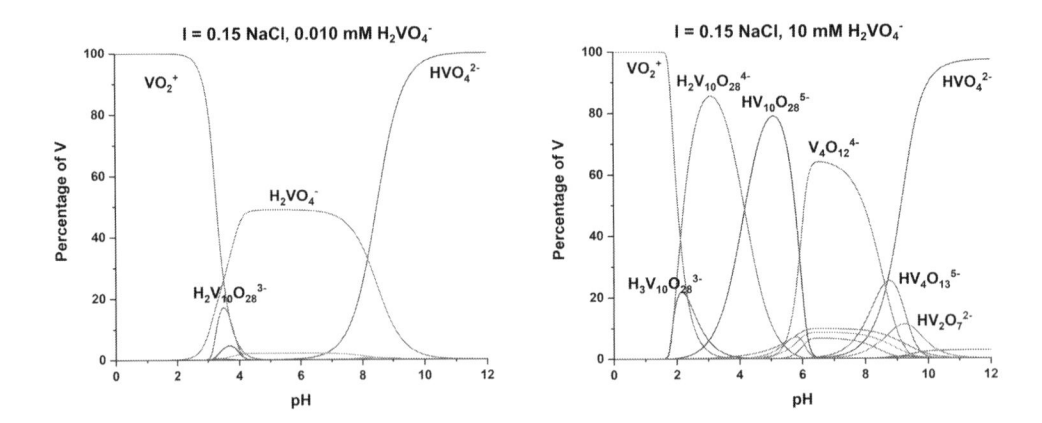

Figure 12.4 Solution speciation curves for 0.010- and 10-mM solutions of $H_2VO_4^-$ at an ionic strength of 0.15 NaCl.

administration of these compounds to any specific vanadium species (Crans et al. 2004, Rehder 2013). Although in general, the observed response is due to one specific form of the vanadium compound, it can be difficult to attribute the effects to such compound unless speciation studies have accompanied the biological studies. However, generally, such studies have not been carried out and the observed effects are attributed to the tested compound and not to the species formed after administration of the test material.

12.3 MINING AND INDUSTRIAL APPLICATIONS OF VANADIUM

Pure elemental vanadium does not exist naturally on the surface of the earth (Rehder 2015, McDonough and Sun 1995), instead, the vanadium is present as some of the many vanadium oxide-containing minerals that exist in the surface of the earth (Fortoul et al. 2014, Gustafsson 2019). Vanadium is produced as a by-product of mining processes, obtained from different vanadium oxides, including simple and more complex vanadates, as decavanadate (Cooper et al. 2019, Crans et al. 2017) and mixed-valence oxovanadates, present in minerals (Fortoul et al. 2014, Gustafsson 2019). Reports exist describing the problems miners and various occupational groups experience when exposed to high concentrations of vanadium (Cooper 2007, Assem and Levy 2012, ACGIH(R) 2009), Europe 2000). The vanadium is found in the form of different metal oxides, the mining and processing of these vanadium-containing minerals result in the vanadium products commercially available (Fortoul et al. 2014), which include ferrovanadium alloy, vanadium-metal, $NaVO_3$, NH_4VO_3, Na_3VO_4, and V_2O_5. A significant source of vanadium is in recycling of steel scrap and spent catalysts (Petranikova et al. 2020).

One of the primary uses of vanadium is in the steel alloys industry, as ferrovanadium alloy, where it is used to increase resistance to corrosion and strength of the final product (Schlesinger, Klein, and Vengosh 2017). Vanadium compounds have also been increasingly used to produce industrial energy storage, as renewable energy storage (Lourenssen et al. 2019). Other uses of vanadium include the V_2O_5 being involved in the production of maleic anhydride, widely used in the chemical industry, which is carried out using a process involving the catalyst P_2O_5 doped with V_2O_5 (Burnett, Keppel, and Robinson 1987). Considering the high volume of V_2O_5 needed in various industrial processes, an entire industry now exists. The burning of coal and particularly oil generates the major emissions of vanadium to the atmosphere, which accounts for the regulatory interest in vanadium (Schlesinger, Klein, and Vengosh 2017). Mining and production of vanadium can also release vanadium to the atmosphere, but to a lesser extent.

12.4 ENVIRONMENTAL PRESENCE OF VANADIUM

Vanadium is an abundant transition metal in the earth's crust and in certain mountainous areas the minerals in the rocks contain vanadium (Rehder 2015, Fortoul et al. 2014, Cooper et al. 2019, Crans et al. 2017). These minerals range from containing mainly vanadium to having vanadium as a minor component, with the most common being vanadinite ($[Pb_5(VO_4)_3Cl]$). In addition, these minerals contain different types of metal oxides. As a result, it is not surprising that there is a wide range of properties described to these minerals. In some minerals, the vanadium can be more easily leached to the water table, whereas in other cases, it is rather insoluble (Gustafsson 2019).

High concentrations of vanadium are present in some coal deposits because the VO-porphyrin is a very stable material only surpassed by the stability of the Ni-porphyrins (Dechaine and Gray 2010). Several coal deposits around the world (China, Russia, South Africa, and Venezuela) contain high vanadium content that upon combustion is likely to generate airborne particulates also with high vanadium levels (Nriagu 1998, Zhao et al. 2014). Such particulates can result in deposits in the lakes, rivers, and oceans with high vanadium contents, with an estimated mean residence time for vanadium in oceans of 100,000 years (Whitfield and Turner 1979). Moreover, vanadium can enter the ground-water by leaching from the vanadium-containing minerals present in the soil (Fortoul et al. 2014). For example, soils rich in vanadium such as those found in the Rocky Mountain region in the United States can transfer vanadium to the crops grown in the region, affecting the concentration of vanadium found in the generated products, like produce and even beer (May et al. 2019, V'Yuillaud et al. 2001).

This element is present in the oceans at a mean concentration of 1.8 μg/L levels (Costa Pessoa 2015, Schlesinger, Klein, and Vengosh 2017). In surface waters, vanadium is found as in the oxidation states V+ and IV+, mainly as the $H_2VO_4^-$ and HVO_4^{2-} species, respectively (Crans et al. 2004, Gustafsson 2019). Some seaweeds contain vanadium, and it has been found to be an essential co-factor for chloroperoxidases and bromoperoxidases which are proteins known to oxidize or functionalize organic materials (Rehder 2015, Fay et al. 2010, Rehder 2000, Butler and Walker 1993). Since most of these studies were done on surface waters, it is likely that as information on the composition of deep-sea waters in more reducing environments becomes available that vanadium will be found to exist in oxidation state IV+ (Crans et al. 2004).

Vanadium is also often reported as a minor component in other minerals, particularly in iron sources (Gustafsson 2019). It is, therefore, expected that there will be additional vanadium deposits discovered as our analytical techniques are being improved measuring lower levels of vanadium. In addition, more inhabitable areas of the earth and space are investigated, which are also likely to contain other sources of vanadium.

12.5 THERAPEUTIC AND BIOMEDICAL DEVELOPMENTS WITH VANADIUM COMPOUNDS

Numerous studies probing the fundamental biological activities of vanadium have been reported, indicating that it exerts many different activities (Costa Pessoa 2015). Vanadate ($H_2VO_4^-$) is a structural and electronic analog of phosphate ($H_2PO_4^-$) and can replace phosphate in some enzymes, acting as a substrate analog or as a transition state inhibitor (Crans et al. 2018, McLauchlan et al. 2015). Proteins such as phosphatases (McLauchlan et al. 2015) and serum albumin or transferrin (Pessoa and Tomaz 2010) are known to bind vanadium with high affinity. Vanadium bound to the blood proteins represents a key mechanism for distribution throughout the mammal and vanadium bound to these proteins can occur in oxidation states III, IV, and V (Pessoa and Tomaz 2010, Levina, Crans, and Lay 2017). Indeed, redox chemistry presumably in the form or Fenton chemistry has been reported with the formation of reactive oxygen species (ROS) (Willsky et al. 2011, Crans et al. 2013, Crans 2015).

Investigations into the potential for therapeutic applications of vanadium compounds extended from 1899 (Lyonnet, Martz, and Martin 1899) and are ongoing (Barrio and Etcheverry 2010, Treviño et al. 2019, Cam, Brownsey, and McNeill 2000, Thompson et al. 2009, Selman et al. 2018). The ability of vanadium salts to lower the elevated blood glucose levels on mice (Heyliger, Tahiliani, and McNeill 1985) was continuously explored for several decades (Costa Pessoa 2015, Sakurai 2002) with a range of different vanadium compounds (Rehder et al. 2002). The initial vanadium studies in humans were done with salts (Cusi et al. 2001, Halberstam et al. 1996, Goldfine et al. 1995, 2000), but in 1992, John McNeill and Chris Orvig team expanded into coordination complexes, as bis(2-ethyl-3-hydroxy-4-pyronato)oxovanadium(IV) abbreviated as BEOV, bis(3-Hydroxy-2-methyl-4H-pyranato)oxovanadium(IV) abbreviated as BMOV, and its derivatives (Thompson et al. 2009, Thompson and Orvig 2006, McNeill et al. 1992). The coordination complexes contain an organic ligand that acts to modify the properties of the vanadium and thus has also been referred to as organic vanadium derivatives (Crans 2015). For several years, many new reports of vanadium compounds were discussed in the literature until September 30, 2011, when the vanadium compound went off patent (Thompson et al. 2009, Thompson and Orvig 2006, McNeill et al. 1992). Since then a human study was reported documenting that observable vanadium levels in the blood protect the patient toward the development of diabetes (Wang et al. 2014). However, for future applications, human studies will need to be done using more subjects and longer observation times, since the FDA regulations to get compounds accepted as a drug have increased since the earlier studies (Smith, Pickering, and Lewith 2008, Crans, Henry, et al. 2019).

In addition, the number of studies reporting the anticancer properties of the vanadium compounds has increased (Evangelou 2002, Das et al. 2012, Pessoa, Etcheverry, and Gambino 2015). These studies demonstrate that both salt and organic vanadium compounds exert promising anticancer effects and the potential for treatment has been suggested in multiple animal studies (Evangelou 2002). However, no human studies for chemotherapies have been published. Recently, though, a particular potent application of vanadium compounds has been reported using oncolytic viruses (Bergeron et al. 2019, Selman et al. 2018), showing that the vanadium compounds are enhancing the anticancer effects of the oncolytic viruses up to 100-fold. Although little is known about this system, and few literature reports exist at this point, the work demonstrates that the immune system is involved in this effect and that there is selectivity (Bergeron et al. 2019, Selman et al. 2018). Considering that these effects are very potent and that salts and one organic vanadium complex have already been in clinical trials for diabetes (Thompson and Orvig 2006, Thompson et al. 2009), one can remain hopeful with future applications of vanadium compounds for cancer treatment.

12.6 TOXICITY STUDIES ASSESSING ADVERSE HEALTH EFFECTS

Safety studies for potential therapeutic vanadium compounds discussed above are relatively straight forward as the compound under investigation can be administered orally. However, for assessment of the adverse effects of vanadium compounds to humans from inadvertent exposures, the design of such studies for vanadium is fraught with challenges. Such safety assessments are done to derive safe exposure standards for humans from exposures to ambient air, drinking water and occupationally.

Hazard assessment studies are done using healthy animals, typically rodents. Furthermore, studies are divided based on the administration route, whether the compounds were administered in solution, through food or drinking water, or in the solid state as an airborne particulate matter (Parasuraman 2011). Depending on the vanadium compound evaluated, the studies may be difficult to interpret. V_2O_5, for example, when administered through food or drinking water, is dissolved, and at that point, due to the species described by the Pourbaix diagrams above, the administered compound is no longer V_2O_5. Therefore, there are many gaps in our knowledge of the toxicity of V_2O_5 and other vanadium compounds that change depending on how it is administered. It remains challenging to interpret the biological results, as evidenced by several *in vitro* studies with vanadium compounds in which speciation studies accompanied the biological work (Samart et al. 2018, Levina, Crans, and Lay 2017, Crans, Koehn, et al. 2019, Willsky, White, and McCabe 1984, Levina et al. 2015, Rehder et al. 2002). As vanadium generally support reactions with components in the medium and results depending on the form of the substance added, the concentration tested, and the composition and pH of the media used. Similarly, the bioprocessing of the vanadium compounds when administered to mammals will depend on the form of the vanadium substance added, the concentration tested, and the composition and pH (Willsky et al. 2011, 2013, Buglyó et al. 2005).

The exposure to vanadium is common by airborne-vanadium particulates formed when fossil fuels are burned, such as Venezuelan coal (Europe 2000). While it has been assumed that ambient exposure to airborne vanadium is in the form of V(V) and often stated to be V_2O_5, a recent study using chemical methodology allowing for determination of oxidation states of the vanadium has shown that in some cases the oxidation state IV+ is very stable and persists in the presence of oxygen (Shafer et al. 2012). Therefore, it is important not to assume that all airborne vanadium will be in oxidation state V+, and certainly not that it is the simple oxide V_2O_5, for proper assessment of the biological and toxicological effects of the compounds. However, more fundamental research is needed to understand the environmental chemistry of vanadium compounds so that appropriate safety studies can be designed.

Few studies exist which compare the relative toxicity of vanadium compounds. In a series of acute exposure studies of the inhalation of vanadium compounds with different oxidation states and solubilities, using compounds of similar size and exposure parameters, demonstrated differences in the effect due to oxidation state, solubility, and rodent species (Rajendran et al. 2016). The effects noted were on the lung, the target organ from inhalation exposures. This is consistent with the effects noted from both occupational exposures and controlled human exposure studies (ACGIH(R) 2009). Inhalation studies were also done in rats with a series of soluble and insoluble vanadium oxides, salts, and coordination compounds, $NaVO_3$ and V(IV) dipicolinate and V(III) bis(dipicolinate) and V_2O_5. The studies assessed the immunomodulation induce by the metal oxide, through alteration in local bacteria resistance, showing that the redox state and solubility are important physicochemical properties for pulmonary immunotoxicity. Where, soluble V(V) had the bigger effect than V(III), while V(IV) and insoluble V(V) showed no effect (Cohen et al. 2006, 2007). Further studies showed that the difference in the physicochemical properties of these compounds manifests as an effect via differential shifts in lung iron homeostasis (Cohen et al. 2010). Other than V_2O_5, data from inhalation safety assessment toxicity studies with longer exposure periods are lacking.

V_2O_5 has been the most widely studied vanadium compound in inhalation studies in rodents, following inhalation of crystalline respirable particulate V_2O_5 to rats and mice of both sexes for 6 h/day, 5 days a week for 16 days, 3 months and 2 years, respectively (NTP 2002). These studies confirm that the target organ is the respiratory system, the site of contact, based on observed lesions in the nose, larynx, and lung in both species. While pulmonary inflammation and hyperplasia were noted in both species at all levels tested in the lifetime studies (0.5–2 mg/m^3 in rats and 1–4 mg/m^3 in mice), differences were apparent with the development of lung tumors which were significantly increased in mice, whereas an oncogenic response was not confirmed in rats (NTP 2002, Starr et al. 2012). The reason for the difference between the two rodent species, mice and rats, is not clear as the mode of action of V_2O_5 in the two animals has not been determined. Results of several *in vivo* studies in the lungs of exposed mice do not support a genotoxic mode of action (Rondini, Walters, and Bauer 2010, Schuler et al. 2011, Banda et al. 2015, Manjanatha et al. 2015). An association of vanadium and lung cancer has furthermore not been demonstrated in humans (Boice, Mumma, and Blot 2007, Bai et al. 2019). Based on the NTP 2002 findings, The International Agency for Research on Cancer, (IARC), classified V_2O_5 as "2b Possibly Carcinogenic to Humans" (IARC 2006). This would appear reasonable as a species difference in tumor response is apparent. The ACGIH in 2009, when revising the threshold limit value for V_2O_5, classified the carcinogenicity as category A3, which is a confirmed animal carcinogen with unknown relevance for humans. Remarkably, in the extensive NTP rodent studies, there was no evidence of systemic organ toxicity or oncogenicity after inhalation of V_2O_5 at the levels toxic to the respiratory system however effects on the hematologic parameters noted by other investigators were apparent (NTP 2002, Ress et al. 2003). These results would indicate a difference in the form of vanadium initially inhaled as V_2O_5 and the form circulating following exposure.

Toxicity studies have also been done with other vanadium compounds by oral administration. A recent comparative study of a vanadium(V) compound, $NaVO_3$, and a vanadium(IV) compound, $VOSO_4$, administered continuously in the drinking water to rats and mice for 14 days (Roberts et al. 2016) and for 3 months (Roberts et al. 2019) has been conducted following detailed chemical analysis of the form of vanadium administered under the test conditions (Mutlu et al. 2017). The toxicity of both compounds was found to be dependent on relative GI absorption expressed as blood vanadium concentration and a comprehensive National Toxicology Program report is currently in preparation. These findings will be informative as they will compare the most common oxidation states of vanadium under physiological conditions, vanadium(V) and (IV); that is the $V^{IV}O^{2+}$, cationic form, and a $V^{V}O_3^-$, anionic forms of vanadium.

Toxicity studies with vanadium compounds have, generally, not included detailed speciation studies. In studies with miners, for example, the nature of the vanadium compounds that the workers were exposed to is not generally specified, once the mining location usually presents a mixture of vanadium oxide-containing minerals (Cooper 2007, Assem and Levy 2012, Cooper et al. 2019). Occupational exposure studies in vanadium workers have measured air or blood levels of vanadium in workers but lack measurement of the form of vanadium in workroom air or the presence of other substances present when processing vanadium (Ehrlich et al. 2008, Kiviluoto, Pyy, and Pakarinen 1979). Toxicity studies were done on healthy humans, as part of human trials, have demonstrated that as the dose of vanadate, vanadyl sulfate, or bis(maltolato)oxovanadium(IV) increases, some GI irritation and undesirable renal effects have been observed (Thompson et al. 2009, McNeill et al. 1992, Goldfine et al. 1995, 2000, Willsky et al. 2013). These studies also show that these symptoms happen at the levels needed for the antidiabetic effects to be observed (Ghosh, Saha, and Saha 2015).

12.7 REGULATION OF VANADIUM IN THE FOODS AND THE ENVIRONMENT

Trace concentrations of vanadium are found in various types of food and the daily intake of vanadium ranges from 0.010 to 0.020 mg. Vanadium also may be found in commercial nutritional supplements and multivitamins with amounts ranging from sub-micrograms to milligrams, depending on whether it is vanadyl sulfate, other complexes or a specific formulation (Doucette, Hassell, and Crans 2016, Crans et al. 2018). Consumption of some vanadium-containing supplements can result in a higher intake of vanadium when compared with food and water. The average vanadium concentrations in tap water are approximately 0.001 mg/L with vanadium levels ranging from about 0.20 to more than 100 µg/L with typical values being between 1 and 6 µg/L (Nordberg, Fowler, and Nordberg 2014). The intermediate minimal risk levels (MRLs) in drinking water have been set to an upper limit of 0.01 mg/kg/day by the Agency for Toxic Substances and Disease Control in the United States (ATSDR 2012) and based on oral administration of vanadyl sulfate to humans for 12 weeks (Fawcett et al. 1997). Although the recommended upper levels have been decreasing for the past several decades, these levels are generally not surpassed, except for regions in which there are deposits of minerals rich in vanadium, such as that found in the Rocky Mountain Plateau (Weeks, 1961). Areas where these levels are surpassed also include regions near power plants where vanadium-rich coal is processed, vanadium minerals are mined, and other sources of fly ash occur. A significant amount of vanadium also enters the environment from crustal weathering and volcanic emission, although the amount of vanadium entering the atmosphere is greater from anthropogenic sources (Gustafsson 2019). In a study by Nedrich et al., the toxicity of vanadium to environmental organisms was measured in river water sediments, showing less toxicity than expected due to the complexation and speciation of the vanadium, and sorption to metal oxides (Nedrich et al. 2018). In the two latter cases, soluble vanadium is likely to be removed from the river water by precipitation and forming sediments. Additional speciation studies of the forms of vanadium present, the composition of the water, and the settling of the vanadium precipitates in the aqueous river waters are needed to properly design toxicity studies to measure adverse effects from oral exposures to vanadium.

The Federal Office of Occupational Safety and Health Administration (abbreviated OSHA) is responsible for the enforcement of occupational safety and working conditions for working men and women in the United States. OSHA sets a legal ceiling limit of 0.5 mg/m^3 for respirable dust as V_2O_5, and 0.1 mg/m^3 for V_2O_5 fume in workroom air (OSHA). Since the OSHA information needs to be updated the American Conference of Governmental Industrial Hygienists, (ACGIH), has set Threshold Limit Values (TLVs) that are widely accepted and followed. This group of ACGIH has established a TLV of 0.05 mg V/m^3 measured as V_2O_5 for a time-weighted average exposure for V_2O_5 (ACGIH(R) 2009).

12.8 CONCLUSION AND FUTURE

Vanadium is a first-row transition metal that can present both beneficial and toxic health effects depending on the chemical forms, concentrations, and route of exposure. Several applications of vanadium in alloys and as an industrial catalyst are very valuable; furthermore, nutritional additives of vanadium salts or coordination compounds are available. Toxicity has been reported in occupational studies with workers mining vanadium, as well as workers involved in the production of vanadium alloys and catalysts. The toxicity of vanadium is strongly dependent on the specific compound, the concentration one is exposed to and the specific exposure conditions. However, the available studies are limited by the complexities of vanadium chemistry, which makes it difficult to determine the active form of the compounds administered. Accompanying speciation studies can aid in the interpretation of biological studies and the nature of the active species. However, such studies are difficult and rarely done. Regardless, vanadium compounds have been reported with biological properties and have been considered for therapeutic use, in the treatment of diabetes and some types of cancer (Barrio and Etcheverry 2010). Additional research is needed to better understand the effects of speciation on the health effects of vanadium compounds.

ACKNOWLEDGMENTS

We thank Heide Murakami for providing us with the speciation figures. KP and GGN thank the Coordenação de Aperfeiçoamento de Pessoal de Nível Superior (CAPES) and CAPES/PrInt program for the scholarship supporting KP's visit to Colorado State University.

REFERENCES

ACGIH(R). 2009. Vanadium Pentoxide: TLV® Chemical Substances 7th Edition Documentation. https://www.acgih.org/forms/store/ProductFormPublic/vanadium-pentoxide-tlv-r-chemical-substances-7th-edition-documentation.

Assem, Farida L., and Leonard S. Levy. 2012. "Inhalation Toxicity of Vanadium." In *Vanadium: Biochemical and Molecular Biological Approaches*, edited by Hitoshi Michibata, 209–224. Dordrecht: Springer Netherlands.

ATSDR. 2012. "Toxicological Profile for Vanadium." U.S. Department of Health and Human Services. https://www.atsdr.cdc.gov/ToxProfiles/TP.asp?id=276&tid=50.

Aureliano, Manuel, and Debbie C. Crans. 2009. "Decavanadate (V10O286-) and Oxovanadates: Oxometalates with Many Biological Activities." *Journal of Inorganic Biochemistry* 103 (4):536–546.

Bai, Yansen, Gege Wang, Wenshan Fu, Yanjun Lu, Wei Wei, Weilin Chen, Xiulong Wu, Hua Meng, Yue Feng, Yuhang Liu, Guyanan Li, Suhan Wang, Ke Wang, Juanxiu Dai, Hang Li, Mengying Li, Jiao Huang, Yangkai Li, Sheng Wei, Jing Yuan, Ping Yao, Xiaoping Miao, Meian He, Xiaomin Zhang, Handong Yang, Tangchun Wu, and Huan Guo. 2019. "Circulating Essential Metals and Lung Cancer: Risk Assessment and Potential Molecular Effects." *Environment International* 127:685–693.

Banda, Malathi, Karen L. McKim, Lynne T. Haber, Judith A. MacGregor, Bhaskar Gollapudi, and Barbara L. Parsons. 2015. "Quantification of Kras Mutant Fraction in the Lung DNA of Mice Exposed to Aerosolized Particulate Vanadium Pentoxide by Inhalation." *Mutation Research/Genetic Toxicology and Environmental Mutagenesis* 789–790:53–60.

Barrio, Daniel A., and Susana B. Etcheverry. 2010. "Potential Use of Vanadium Compounds in Therapeutics." *Current Medicinal Chemistry* 17 (31):3632–3642.

Bergeron, Anabel, Kateryna Kostenkova, Mohammed Selman, Heide A. Murakami, Elizabeth Owens, Naveen Haribabu, Rozanne Arulanandam, Jean-Simon Diallo, and Debbie C. Crans. 2019. "Enhancement of Oncolytic Virotherapy by Vanadium(V) Dipicolinates." *BioMetals* 32 (3):545–561.

Boice, John D., Michael T. Mumma, and William J. Blot. 2007. "Cancer and Noncancer Mortality in Populations Living Near Uranium and Vanadium Mining and Milling Operations in Montrose County, Colorado, 1950–2000." *Radiation Research* 167 (6):711–726.

Buglyó, Péter, Debbie C. Crans, Eszter M. Nagy, Ruby L. Lindo, Luqin Yang, Jason J. Smee, Wenzheng Jin, Lai-Har Chi, Michael E. Godzala, and Gail R. Willsky. 2005. "Aqueous Chemistry of the Vanadium^{III} (V^{III}) and the V^{III}–Dipicolinate Systems and a Comparison of the Effect of Three Oxidation States of Vanadium Compounds on Diabetic Hyperglycemia in Rats." *Inorganic Chemistry* 44 (15):5416–5427.

Burnett, J. C., R. A. Keppel, and W. D. Robinson. 1987. "Commercial Production of Maleic Anhydride by Catalytic Processes Using Fixed Bed Reactors." *Catalysis Today* 1 (5):537–586.

Butler, Alison, and J. V. Walker. 1993. "Marine Haloperoxidases." *Chemical Reviews* 93 (5):1937–1944.

Cam, Margaret C., Roger W. Brownsey, and John H. McNeill. 2000. "Mechanisms of Vanadium Action: Insulin-Mimetic or Insulin-Enhancing Agent?" *Canadian Journal of Physiology and Pharmacology* 78 (10):829–847.

Cohen, Mitchell D., Colette Prophete, Maureen Sisco, Lung-chi Chen, Judith T. Zelikoff, Jason J. Smee, Alvin A. Holder, and Debbie C. Crans. 2006. "Pulmonary Immunotoxic Potentials of Metals Are Governed by Select Physicochemical Properties: Chromium Agents." *Journal of Immunotoxicology* 3 (2):69–81.

Cohen, Mitchell D., Maureen Sisco, Colette Prophete, Lung-chi Chen, Judith T. Zelikoff, Andrew J. Ghio, Jacqueline D. Stonehuerner, Jason J. Smee, Alvin A. Holder, and Debbie C. Crans. 2007. "Pulmonary Immunotoxic Potentials of Metals Are Governed by Select Physicochemical Properties: Vanadium Agents." *Journal of Immunotoxicology* 4 (1):49–60.

Cohen, Mitchell D., Maureen Sisco, Colette Prophete, Kotaro Yoshida, Lung-chi Chen, Judith T. Zelikoff, Jason Smee, Alvin A. Holder, Jacqueline Stonehuerner, Debbie C. Crans, and Andrew J. Ghio. 2010. "Effects of Metal Compounds with Distinct Physicochemical Properties on Iron Homeostasis and Antibacterial Activity in the Lungs: Chromium and Vanadium." *Inhalation Toxicology* 22 (2):169–178.

Cooper, Mark A., Frank C. Hawthorne, Anthony R. Kampf, and John M. Hughes. 2019. "Determination of V4+:V5+ Ratios in the $[V_{10}O_{28}]_n$ – Decavanadate Polyanion." *The Canadian Mineralogist* 57 (2):235–244.

Cooper, Ross G. 2007. "Vanadium Pentoxide Inhalation." *Indian Journal of Occupational and Environmental Medicine* 11 (3):97–102.

Costa Pessoa, J. 2015. "Thirty Years through Vanadium Chemistry." *Journal of Inorganic Biochemistry* 147:4–24.

Crans, D. C. 2015. "Antidiabetic, Chemical, and Physical Properties of Organic Vanadates as Presumed Transition-State Inhibitors for Phosphatases." *Journal of Organic Chemistry* 80 (24):11899–11915.

Crans, D. C., L. Henry, G. Cardiff, and B. I. Posner. 2019. "Developing Vanadium as an Antidiabetic or Anticancer Drug: A Clinical and Historical Perspective." *Metal Ions in Life Sciences* 19:203–230.

Crans, Debbie C., Jordan T. Koehn, Stephanie M. Petry, Caleb M. Glover, Asanka Wijetunga, Ravinder Kaur, Aviva Levina, and Peter A. Lay. 2019. "Hydrophobicity may Enhance Membrane Affinity and Anti-Cancer Effects of Schiff Base Banadium(v) Catecholate Complexes." *Dalton Transactions* 48 (19):6383–6395.

Crans, D. C., H. LaRee, G. Cardiff, and B. I. Posner. 2019. "Developing Vanadium as an Antidiabetic or Anticancer Drug: A Clinical and Historical Perspective." *Essential Metals in Medicine: Therapeutic Use and Toxicity of Metal Ions in the Clinic* 19:203–230.

Crans, Debbie C., Benjamin J. Peters, Xiao Wu, and Craig C. McLauchlan. 2017. "Does Anion-Cation Organization in Na+-Containing X-ray Crystal Structures Relate to Solution Interactions in Inhomogeneous Nanoscale Environments: Sodium-Decavanadate in Solid State Materials, Minerals, and Microemulsions." *Coordination Chemistry Reviews* 344:115–130.

Crans, D. C., and J. J. Smee. 2003. "4.4- Vanadium." In *Comprehensive Coordination Chemistry II*, edited by Jon A. McCleverty and Thomas J. Meyer, 175–239. Oxford: Pergamon.

Crans, Debbie C., Jason J. Smee, Ernestas Gaidamauskas, and Luqin Yang. 2004. "The Chemistry and Biochemistry of Vanadium and the Biological Activities Exerted by Vanadium Compounds." *Chemical Reviews* 104 (2):849–902.

Crans, Debbie C., Kellie A. Woll, Kestutis Prusinskas, Michael D. Johnson, and Eugenijus Norkus. 2013. "Metal Speciation in Health and Medicine Represented by Iron and Vanadium." *Inorganic Chemistry* 52 (21):12262–12275.

Crans, Debbie, Lining Yang, Allison Haase, and Xiaogai Yang. 2018. "Health Benefits of Vanadium and Its Potential as an Anticancer Agent." *Metal Ions in Life Sciences* 18:251–279.

Cusi, K., S. Cukier, R. A. DeFronzo, M. Torres, F. M. Puchulu, and J. C. Pereira Redondo. 2001. "Vanadyl Sulfate Improves Hepatic and Muscle Insulin Sensitivity in Type 2 Diabetes." *The Journal of Clinical Endocrinology & Metabolism* 86 (3):1410–1417.

Das, Subhadeep, Mary Chatterjee, Muthumani Janarthan, Hari Ramachandran, and Malay Chatterjee. 2012. "Vanadium in Cancer Prevention." In *Vanadium: Biochemical and Molecular Biological Approaches*, edited by Hitoshi Michibata, 163–185. Dordrecht: Springer Netherlands.

Dechaine, Greg P., and Murray R. Gray. 2010. "Chemistry and Association of Vanadium Compounds in Heavy Oil and Bitumen, and Implications for Their Selective Removal." *Energy & Fuels* 24 (5):2795–2808.

Doucette, Kaitlin A., Kelly N. Hassell, and Debbie C. Crans. 2016. "Selective Speciation Improves Efficacy and Lowers Toxicity of Platinum Anticancer and Vanadium Antidiabetic Drugs." *Journal of Inorganic Biochemistry* 165:56–70.

Ehrlich, Veronika A., Armen K. Nersesyan, Christine Hoelzl, Franziska Ferk, Julia Bichler, Eva Valic, Andreas Schaffer, Rolf Schulte-Hermann, Michael Fenech, Karl-Heinz Wagner, and Siegfried Knasmüller. 2008. "Inhalative Exposure to Vanadium Pentoxide Causes DNA Damage in Workers: Results of a Multiple End Point Study." *Environmental Health Perspectives* 116 (12):1689–1693.

Europe, World Health Organization. Regional Office for. 2000. "Vanadium." In *Air Quality Guidelines for Europe*, edited by Frank Theakston, 170–172. Copenhagen: World Health Organization, Regional Office for Europe.

Evangelou, Angelos M. 2002. "Vanadium in cancer treatment." *Critical Reviews in Oncology/Hematology* 42 (3):249–265.

Fawcett, J. P., S. J. Farquhar, T. Thou, and B. I. Shand. 1997. "Oral Vanadyl Sulphate Does not Affect Blood Cells, Viscosity or Biochemistry in Humans." *Pharmacology & Toxicology* 80 (4):202–206.

Fay, Aaron W., Michael A. Blank, Chi Chung Lee, Yilin Hu, Keith O. Hodgson, Britt Hedman, and Markus W. Ribbe. 2010. "Characterization of Isolated Nitrogenase FeVco." *Journal of the American Chemical Society* 132 (36):12612–12618.

Fortoul, T. I., M. Rojas-Lemus, V. Rodriguez-Lara, A. Gonzalez-Villalva, M. Ustarroz-Cano, G. Cano-Gutierrez, S. E. Gonzalez-Rendon, L. F. Montaño, and M. Altamirano-Lozano. 2014. "Overview of Environmental and Occupational Vanadium Exposure and Associated Health Outcomes: An Article Based on a Presentation at the 8th International Symposium on Vanadium Chemistry, Biological Chemistry, and Toxicology, Washington DC, August 15–18, 2012." *Journal of Immunotoxicology* 11 (1):13–18.

Ghosh, Sumanta K., Rumpa Saha, and Bidyut Saha. 2015. "Toxicity of Inorganic Vanadium Compounds." *Research on Chemical Intermediates* 41 (7):4873–4897.

Goldfine, Allison B., Mary-Elizabeth Patti, Lubna Zuberi, Barry J. Goldstein, Raeann LeBlanc, Edwin J. Landaker, Zhen Y. Jiang, Gail R. Willsky, and C. Ronald Kahn. 2000. "Metabolic effects of vanadyl sulfate in humans with non-Insulin-dependent diabetes mellitus: In vivo and in vitro studies." *Metabolism* 49 (3):400–410.

Goldfine, A. B., D. C. Simonson, F. Folli, M. E. Patti, and C. R. Kahn. 1995. "Metabolic Effects of Sodium Metavanadate in Humans with Insulin-Dependent and Noninsulin-Dependent Diabetes Mellitus In Vivo and In Vitro Studies." *Journal of Clinical Endocrinology and Metabolism* 80 (11):3311–3320.

Gustafsson, Jon P. 2019. "Vanadium Geochemistry in the Biogeosphere – Speciation, Solid-Solution Interactions, and Ecotoxicity." *Applied Geochemistry* 102:1–25.

Halberstam, Meyer, Neil Cohen, Pavel Shlimovich, Luciano Rossetti, and Harry Shamoon. 1996. "Oral Vanadyl Sulfate Improves Insulin Sensitivity in NIDDM but Not in Obese Nondiabetic Subjects." *Diabetes* 45 (5):659–666.

Heyliger, C. E., A. G. Tahiliani, and J. H. McNeill. 1985. "Effect of Vanadate on Elevated Blood Glucose and Depressed Cardiac Performance of Diabetic Rats." *Science* 227 (4693):1474.

International Agency for Cancer Research (IARC). 2006. "Cobalt in Hard Metals and Cobalt Sulfate, Gallium Arsenide, Indium Phosphide and Vanadium Pentoxide." *IARC Monographs on the Evaluation of Carcinogenic Risks to Humans*, Vol. 86, 227–292. Lyon, France: IARC Press.

Kanamori, Kan, and Kiyoshi Tsuge. 2012. "Inorganic Chemistry of Vanadium." In *Vanadium: Biochemical and Molecular Biological Approaches*, edited by Hitoshi Michibata, 3–31. Dordrecht: Springer Netherlands.

Kiviluoto, Markku, Lauri Pyy, and Arto Pakarinen. 1979. "Serum and Urinary Vanadium of Vanadium-Exposed Workers." *Scandinavian Journal of Work, Environment & Health* 5(4):362–367.

Levina, Aviva, Debbie C. Crans, and Peter A. Lay. 2017. "Speciation of Metal Drugs, Supplements and Toxins in Media and Bodily Fluids Controls In Vitro Activities." *Coordination Chemistry Reviews* 352:473–498.

Levina, Aviva, Andrew I. McLeod, Sylvia J. Gasparini, Annie Nguyen, W. G. Manori De Silva, Jade B. Aitken, Hugh H. Harris, Chris Glover, Bernt Johannessen, and Peter A. Lay. 2015. "Reactivity and Speciation of Anti-Diabetic Vanadium Complexes in Whole Blood and Its Components: The Important Role of Red Blood Cells." *Inorganic Chemistry* 54 (16):7753–7766.

Lourenssen, Kyle, James Williams, Faraz Ahmadpour, Ryan Clemmer, and Syeda Tasnim. 2019. "Vanadium Redox Flow Batteries: A Comprehensive Review." *Journal of Energy Storage* 25:100844.

Lyonnet, B., S. Martz, and E. Martin. 1899. "L'emploi therapeutique des Derives du Vanadium." *La Presse médicale* 1:191–192.

Manjanatha, Mugimane G., Sharon D. Shelton, Lynne Haber, Bhaskar Gollapudi, Judith A. MacGregor, Narayanan Rajendran, and Martha M. Moore. 2015. "Evaluation of cII Mutations in Lung of Male Big Blue Mice Exposed by Inhalation to Vanadium Pentoxide for up to 8 Weeks." *Mutation Research/Genetic Toxicology and Environmental Mutagenesis* 789–790:46–52.

May, Bianca, Tim Dreifke, Claus-Dieter Patz, Christian L. Schütz, Ralf Schweiggert, and Helmut Dietrich. 2019. "Filter Aid Selection Allows Modulating the Vanadium Concentration in Beverages." *Food Chemistry* 300:125168.

McDonough, W. F., and S. s. Sun. 1995. "The Composition of the Earth." *Chemical Geology* 120 (3):223–253.

McLauchlan, Craig C., Benjamin J. Peters, Gail R. Willsky, and Debbie C. Crans. 2015. "Vanadium–Phosphatase Complexes: Phosphatase Inhibitors Favor the Trigonal Bipyramidal Transition State Geometries." *Coordination Chemistry Reviews* 301–302:163–199.

McNeill, John H., V. G. Yuen, H. R. Hoveyda, and Chris Orvig. 1992. "Bis(maltolato)oxovanadium(IV) Is a Potent Insulin Mimic." *Journal of Medicinal Chemistry* 35 (8):1489–1491.

Mutlu, Esra, Tim Cristy, Steven W. Graves, Michelle J. Hooth, and Suramya Waidyanatha. 2017. "Characterization of Aqueous Formulations of Tetra- and Pentavalent Forms of Vanadium in Support of Test Article Selection in Toxicology Studies." *Environmental Science and Pollution Research* 24 (1):405–416.

Nedrich, Sara M., Anthony Chappaz, Michelle L. Hudson, Steven S. Brown, and G. Allen Burton. 2018. "Biogeochemical Controls on the Speciation and Aquatic Toxicity of Vanadium and Other Metals in Sediments from a River Reservoir." *Science of the Total Environment* 612:313–320.

Nordberg, G. F., B. A. Fowler, and M. Nordberg. 2014. *Handbook on the Toxicology of Metals*. United States: Academic Press/Elsevier Science.

Nriagu, J. O. 1998. *Vanadium in the Environment, Chemistry and Biochemistry*. United States: John Wiley & Sons Inc.

NTP. 2002. "NTP Toxicology and Carcinogensis Studies of Vanadium Pentoxide (CAS No. 1314-62-1) in F344/N Rats and B6C3F1 Mice (Inhalation)." *National Toxicology Program Technical Report Series* 507:1–343.

OSHA. "OSHA Annotated Table Z-1." https://www.osha.gov/dsg/annotated-pels/tablez-1.html.

Parasuraman, S. 2011. "Toxicological Screening." *Journal of Pharmacology & Pharmacotherapeutics* 2 (2):74–79.

Pessoa, Joao C., Susana Etcheverry, and Dinorah Gambino. 2015. "Vanadium Compounds in Medicine." *Coordination Chemistry Reviews* 301–302:24–48.

Pessoa, J. C., and I. Tomaz. 2010. "Transport of Therapeutic Vanadium and Ruthenium Complexes by Blood Plasma Components." *Current Medicinal Chemistry* 17 (31):3701–3738.

Petranikova, M., Tkaczyk. A.H., Bartl, A., Amato, A., Lapkovskis, V., and C. Tunsu. 2020. "Vanadium Sustainability in the Context of Innovative Recycling and Sourcing Development." *Waste Manag.* 113:521–544.

Povar, I., O. Spinu, I. Zinicovscaia, B. Pintilie, and S. Ubaldini. 2019. "Revised Pourbaix Diagrams for the Vanadium - Water System." *Journal of Electrochemical Science and Engineering* 9 (2):75–84.

Rajendran, N., J. C. Seagrave, L. M. Plunkett, and J. A. MacGregor. 2016. "A Comparative Assessment of the Acute Inhalation Toxicity of Vanadium Compounds." *Inhalation Toxicology* 28 (13):618–628.

Rehder, Dieter. 2000. "Vanadium Nitrogenase." *Journal of Inorganic Biochemistry* 80 (1):133–136.

Rehder, Dieter. 2013. "Vanadium. Its Role for Humans." In *Interrelations between Essential Metal Ions and Human Diseases*, edited by Astrid Sigel, Helmut Sigel and Roland K. O. Sigel, 139–169. Dordrecht: Springer Netherlands.

Rehder, Dieter. 2015. "The Role of Vanadium in Biology." *Metallomics* 7 (5):730–742.

Rehder, D., J. C. Pessoa, Cfgc Geraldes, Mmca Castro, T. Kabanos, T. Kiss, B. Meier, G. Micera, L. Pettersson, M. Rangel, A. Salifoglou, I. Turel, and D. R. Wang. 2002. "In Vitro Study of the Insulin-Mimetic Behaviour of Vanadium(IV, V) Coordination Compounds." *Journal of Biological Inorganic Chemistry* 7 (4–5):384–396.

Ress, N. B., B. J. Chou, R. A. Renne, J. A. Dill, R. A. Miller, J. H. Roycroft, J. R. Hailey, J. K. Haseman, and J. R. Bucher. 2003. "Carcinogenicity of Inhaled Vanadium Pentoxide in F344/N Rats and B6C3F1 Mice." *Toxicological Sciences* 74 (2):287–296.

Roberts, Georgia K., K. Elsass, Dawn M. Fallacara, K. Levine, J. Harrington, Suramya Waidyanatha, Michelle J. Hooth, V. Godfrey-Robinson, B. Sparrow, and Matthew D. Stout. 2019. "3-Month Toxicity Studies of Tetravalent and Pentavalent Vanadium Compounds in Hsd:Sprague Dawley SD Rats and B6C3F1/N Mice via Drinking Water Exposure." *The Toxicologist, Supplement to Toxicological Sciences*, 168 (1):Abstract # 2374, Baltimore, Maryland, US.

Roberts, Georgia K., Matthew D. Stout, Brian Sayers, Dawn M. Fallacara, Milton R. Hejtmancik, Suramya Waidyanatha, and Michelle J. Hooth. 2016. "14-Day Toxicity Studies of Tetravalent and Pentavalent Vanadium Compounds in Harlan Sprague Dawley Rats and B6C3F1/N Mice via Drinking Water Exposure." *Toxicology Reports* 3:531–538.

Rondini, Elizabeth A., Dianne M. Walters, and Alison K. Bauer. 2010. "Vanadium Pentoxide Induces Pulmonary Inflammation and Tumor Promotion in a Strain-Dependent Manner." *Particle and Fibre Toxicology* 7 (1):9.

Sakurai, Hiromu. 2002. "A New Concept: The Use of Vanadium Complexes in the Treatment of Diabetes Mellitus." *The Chemical Record* 2 (4):237–248.

Samart, Nuttaporn, Zeyad Arhouma, Santosh Kumar, Heide A. Murakami, Dean C. Crick, and Debbie C. Crans. 2018. "Decavanadate Inhibits Mycobacterial Growth More Potently Than Other Oxovanadates." *Frontiers in Chemistry* 6:519.

Schlesinger, William H., Emily M. Klein, and Avner Vengosh. 2017. "Global Biogeochemical Cycle of Vanadium." *Proceedings of the National Academy of Sciences* 114 (52):E11092–E11100.

Schuler, Detlef, Hans-J. Chevalier, Mandy Merker, Katja Morgenthal, Jean-Luc Ravanat, Peter Sagelsdorff, Marc Walter, Klaus Weber, and Douglas McGregor. 2011. "First Steps Towards an Understanding of a Mode of Carcinogenic Action for Vanadium Pentoxide." *Journal of Toxicologic Pathology* 24 (3):149–162.

Selman, Mohammed, Christopher Rousso, Anabel Bergeron, Hwan Hee Son, Ramya Krishnan, Nader A. El-Sayes, Oliver Varette, Andrew Chen, Fabrice Le Boeuf, Fanny Tzelepis, John C. Bell, Debbie C. Crans, and Jean-Simon Diallo. 2018. "Multi-modal Potentiation of Oncolytic Virotherapy by Vanadium Compounds." *Molecular Therapy* 26 (1):56–69.

Shafer, Martin M., Brandy M. Toner, Joel T. Overdier, James J. Schauer, Sirine C. Fakra, Shaohua Hu, Jorn D. Herner, and Alberto Ayala. 2012. "Chemical Speciation of Vanadium in Particulate Matter Emitted from Diesel Vehicles and Urban Atmospheric Aerosols." *Environmental Science & Technology* 46 (1):189–195.

Smith, D. M., R. M. Pickering, and G. T. Lewith. 2008. "A Systematic Review of Vanadium Oral Supplements for Glycaemic Control in Type 2 Diabetes Mellitus." *QJM: An International Journal of Medicine* 101 (5):351–358.

Starr, Thomas B., Judith A. MacGregor, Kimberly D. Ehman, and Andrey I. Nikiforov. 2012. "Vanadium Pentoxide: Use of Relevant Historical Control Data Shows No Evidence for a Carcinogenic Response in F344/N Rats." *Regulatory Toxicology and Pharmacology* 64 (1):155–160.

Thompson, Katherine H., Jay Lichter, Carl LeBel, Michael C. Scaife, John H. McNeill, and Chris Orvig. 2009. "Vanadium Treatment of Type 2 Diabetes: A View to the Future." *Journal of Inorganic Biochemistry* 103 (4):554–558.

Thompson, Katherine H., and Chris Orvig. 2006. "Vanadium in Diabetes: 100 Years from Phase 0 to Phase I." *Journal of Inorganic Biochemistry* 100 (12):1925–1935.

Treviño, Samuel, Alfonso Díaz, Eduardo Sánchez-Lara, Brenda L. Sanchez-Gaytan, Jose M. Perez-Aguilar, and Enrique González-Vergara. 2019. "Vanadium in Biological Action: Chemical, Pharmacological Aspects, and Metabolic Implications in Diabetes Mellitus." *Biological Trace Element Research* 188 (1):68–98.

V'Yuillaud, R., E. Marchevskij, R. Olsina, and L. Martines. 2001. "Rapid and Simple Method for Determination of Vanadium in Beer by ICP-AES with Ultrasonic Nebulization." *Zhurnal Analiticheskoj Khimii* 56 (1):89–92.

Wang, Xia, Taoping Sun, Jun Liu, Zhilei Shan, Yilin Jin, Sijing Chen, Wei Bao, Frank B. Hu, and Liegang Liu. 2014. "Inverse Association of Plasma Vanadium Levels with Newly Diagnosed Type 2 Diabetes in a Chinese Population." *American Journal of Epidemiology* 180 (4):378–384.

Weeks, Alice D. 1961. "Mineralogy and Geochemistry of Vanadium in the Colorado Plateau." *Journal of the Less Common Metals* 3 (6):443–450.

Whitfield, M., and D. R. Turner. 1979. "Water–Rock Partition Coefficients and the Composition of Seawater and River Water." *Nature* 278 (5700):132–137.

Wilkinson, Geoffrey, and R. D. Gillard. 1987. *Comprehensive Coordination Chemistry: The Synthesis, Reactions, Properties and Applications of Coordination Compounds V3 Main Group and Early Transition Elements.* Oxford, UK: Pergamon Press.

Willsky, Gail R., Lai-Har Chi, Michael Godzala, 3rd, Paul J. Kostyniak, Jason J. Smee, Alejandro M. Trujillo, Josephine A. Alfano, Wenjin Ding, Zihua Hu, and Debbie C. Crans. 2011. "Anti-diabetic Effects of a Series of Vanadium Dipicolinate Complexes in Rats with Streptozotocin-Induced Diabetes." *Coordination Chemistry Reviews* 255 (19–20):2258–2269.

Willsky, Gail R., Katherine Halvorsen, Michael E. Godzala, Iii, Lai-Har Chi, Mathew J. Most, Peter Kaszynski, Debbie C. Crans, Allison B. Goldfine, and Paul J. Kostyniak. 2013. "Coordination Chemistry may Explain Pharmacokinetics and Clinical Response of Vanadyl Sulfate in Type 2 Diabetic Patients." *Metallomics* 5 (11):1491–1502.

Willsky, G. R., D. A. White, and B. C. McCabe. 1984. "Metabolism of Added Orthovanadate to Vanadyl and High-Molecular-Weight Vanadates by *Saccharomyces cerevisiae*." *Journal of Biological Chemistry* 259 (21):13273–13281.

Zhao, Xu, Quan Shi, Murray R. Gray, and Chunming Xu. 2014. "New Vanadium Compounds in Venezuela Heavy Crude Oil Detected by Positive-ion Electrospray Ionization Fourier Transform Ion Cyclotron Resonance Mass Spectrometry." *Scientific Reports* 4 (1):5373.

PART VI

Toxic Manifestations by Diverse Heavy Metals and Metalloids

CHAPTER 13

Clinical Toxicology of Copper
Source, Toxidrome, Mechanism of Toxicity, and Management

Sonal Sekhar Miraj and Mahadev Rao
Manipal College of Pharmaceutical Sciences

CONTENTS

13.1 INTRODUCTION

Copper (Cu) is a ubiquitous essential trace element, which presents in almost all body tissues. Cu is the third most abundant trace metal (after iron [Fe] and zinc [Zn]) in the body. However, the total quantity of Cu in the body is only 75,000–100,000 mcg (Willis et al. 2005). This noble metal is closely related to silver as well as gold, with many properties being shared among these metals (Ashish et al. 2013). The transitional metal is stable in its metallic states and produces cuprous (monovalent) and cupric (divalent) cationic forms. Cu in a metallic form is not poisonous but some of their salts are poisoned nature. Particularly, the most common salts of the Cu are the sulfate (Blue vitriol) and the subacetate (Verdigris). Modern days have a large number of applications for Cu. This ranges from coins to pigments. Therefore, demand for Cu metals remains high, particularly in industrialized countries. In day-to-day basis, many individuals contact with Cu in various situations. Cu can be found in numerous electronic items and in the wiring. It is also used to make cooking pots. The metal is also relatively corrosion resistant. Because of this, it is often mixed with other metals to form alloys such as bronze and brass. Cu is a constituent of intrauterine devices (IUDs) for contraception. Cu toxicity is increasingly becoming common nowadays. These attributes to widespread presence of Cu in our diet, hot water pipe, nutritional supplements, and contraceptive pills (Ashish et al. 2013). Many nations try to regulate their Cu industries for the prevention of wide-ranging pollution and health hazards related to it. Cu is accumulated primarily in the liver, however, low quantity can be seen in the brain, heart, kidney, and muscle (Osredkar and Sustar 2011). Although Cu is a vital dietary nutrient in a small amount, it can lead to toxicity at the levels of cells, tissue, and organ, if present in excess. Cu is a powerful inhibitor of enzymes, on the other hand, it is an essential catalyst for heme synthesis as well as Fe absorption (Barceloux 1999; Ashish et al. 2013). The poisoning effect of Cu will commence within 15–30 min.

13.1.1 History

Cu, atomic number 29 is a reddish-brown nonferrous mineral that has been used by different cultures for thousands of years (Ashish et al. 2013). The name of the metal has come from an Ancient Greek name "Kypros" for Cyprus, an Island with highly productive Cu mines (Ashish et al. 2013). The Latin name of Cu was "cyprium," which itself derived from "kypros," this name was gradually simplified to "cuprum," and this subsequently morphed into the English version, "copper." Archaeological evidence shows that Cu is one of the earliest metals used by humans to make utensils, ornaments, and weapons. Cold metalworking process of native Cu dates back to 9000 BC in Iran and Anatolia (Harper 1987). Hot-working became popular by 3500 BC, which produces relatively pure Cu from its oxidized form of ores with low content of arsenic. Ancient Assyrian, Egyptian, and Hindu alchemists have been used Cu compounds for several ailments. Earlier, copper sulfate was used as an emetic, astringent, and anthelmintic agent. Even now, Cu toxicity happens due to the usage of "spiritual green water" for religious reasons (Akintonwa 1989).

First records of the use of Cu in agriculture in 1761, when it was observed that bean soaked in a weak solution of copper sulfate protected the plant from seed boring fungi. Since then it has been widely used in various sectors including viticulture. This has led to a condition known as "vineyard sprayer's lung," a chronic Cu toxicity after the inhalation of copper sulfate mists. This was first reported in 1969 in Portuguese vineyard workers. Copper sulfate was one of the first chemicals used for fighting against plant diseases. This property was discovered by chance when Ireland potato crops were suffering from phytophthora, a potato disease. It was noticed that potatoes grown in the fields close to Cu-smelting industries were not damaged by the disease. Bordeaux mixture (composed of equal quantities of copper sulfate and calcium oxide in water) a fungicide for vines, fruit trees, and other plants were discovered merely accidentally. A French chemist, Joseph Louis Proust, who was best known for his discovery of the law of constant composition in 1794, addressed a request from winemakers to deter thieves from stealing ripen grapes from the vineyards. He formulated a thick mixture of copper sulfate, lime, and water, whose unappetizing appearance discouraged thieves from stealing the grapes. Long years after, a French botanist Pierre Marie Alexis Millardet while traveling in a vineyard where Proust's mixture was widely used, noticed no traces of grape decay. In October 1882, Millardet observed that this mixture also controlled the downy mildew, suggested its application as a fungicide. This blend of chemicals that was famous as the Bordeaux mixture was the earliest fungicide to obtain a large-scale application worldwide and the invention recognized as the start of a new era in the technology of agriculture (Encyclopaedia Britannica 2019).

13.1.2 Properties

Cu has a melting point of $1,083.4 \pm 0.2°C$, the boiling point of $2,567°C$, and the specific gravity of 8.96 (20°C). The industrial importance of Cu is associated with its useful physical properties. These properties include appearance, the ability of alloying, resistant to corrosion, malleability, and good electrical and thermal conductivity. The surface of pure Cu, which is freshly exposed, has color of a pinkish-orange. An oxidizing agent is required for dissolution of Cu in an acidic media. The resultant greenish layer of the carbonate (Cu_2CO_3) on the surface of the metal prevents further oxidation (Barceloux 1999). Cu dissolved in drinking water can impart a light blue or blue-green color with an unpleasant metallic and bitter taste (World Health Organization 2004).

In aqueous solutions, cuprous ions (I) dissociate quickly to cupric (II) and metallic Cu. Insoluble cuprous forms such as cuprous chloride (CuCl) are stable in aqueous media. Compounds of cupric forms are either bluish or greenish in color. These present often in water and have good solubility in aqueous media. However, Cu(III) form is a powerful oxidant, but, lack significant industrial as well as environmental importance (Agency for Toxic Substances and Disease Registry 1997). Cu consist of approximately 25% copper sulfate ($CuSO_4·5H_2O$) mass. The water acidification enhances leaching of Cu from its piping (Sharpe and DeWalle 1985).

13.1.3 Uses

13.1.3.1 Commercial

The major application of Cu is as a metal or its alloy (as bronze, brass, etc.), particularly in industry. Maximum Cu metal products are recycled. Cu has numerous commercial applications due to its versatility. Cu is used to produce electrical appliances such as electrical wires, pipes, valves, fittings; and coins, cooking utensils, and materials for buildings. Additionally, it also used in various machinery, weapons, and coatings. Cu is used for electroplating, azo dye preparation, engraving, lithography, petroleum refinery, mineral froth flotation, and pyrotechnics. Cu is a vital component of white gold as well as new alloys using for ornaments. Agricultural uses include fungicides, algaecides, insecticides, and wood preservatives, mainly accounted for copper sulfate. Cu compounds can be added to fertilizers and animal feeds as a nutritional supplement to promote the growth of plants and animals (Landner and Lindestrom 1999). Copper sulfate pentahydrate is sometimes added to surface water for the control of algae (NSF 2000).

13.1.3.2 Medical

Cu compounds are also used as food additives as nutrient as well as coloring agents (US FDA 1994). Cu-based alloys (with aluminum, cobalt, etc.) have used in the dental crown (Lucas and Lemons 1992). Cu is a key component of IUDs for contraception, since the liberation of Cu from these devices is vital for its action (Okereke et al. 1972). Typically, these devices have approximately 100,000–150,000 mcg of Cu, which is almost equal to the average total body content of Cu. Wearing of Cu bracelets is considered as a folk remedy for arthritis, however, scientific evidence is lacking to support this use (Walker et al. 1981; Whitehouse and Walker 1977). In African rituals, spiritual water (green water) containing approximately 1,300,000 mcg Cu/L is used as cathartics (Sontz and Schwieger 1995). Copper sulfate was earlier prescribed as an emetic agent; however, this use was discontinued owing to health-related adverse effects (Ellenhorn and Barceloux 1988).

Cu plays a vital role in our metabolic functions because it is essential for various key enzymes activities (Harris 2001). Cu is required for maintaining the strength of the skin, blood vessels, the epithelial and connective tissue within the body. It plays a role in the synthesis of hemoglobin, myelin, melanin, and the normal function of the thyroid gland (Groff et al. 1995). As an antioxidant, Cu scavenges free radicals and may reduce or protect certain damages they cause in cell walls, genetic materials, etc. (Bonham et al. 2002; Davis 2003).

13.2 SOURCES OF COPPER

13.2.1 Air

The concentration of Cu in the atmosphere ranges from 0.005 to 0.2 mcg Cu/m^3 (Agency for Toxic Substances and Disease Registry 1997). Emission of Cu into the environment accounts for only a very small fraction (0.4%) (Davies and Bennett 1985; Barceloux 1999). These emissions result from both natural sources such as windblown dust, volcanoes, forest fires, sea spray; and anthropogenic sources such as non-ferrous metal production, Cu smelters, Fe and steel production, municipal incinerators (Weant 1985). Usually, combustion processes generate Cu in the form of fine particles (<1 mm) of oxides and elemental Cu, whereas, Cu released from particulate matter (windblown dust) happens as large particles (<10 mm) of Cu compounds such as carbonate, oxide, and sulfate (Schroeder et al. 1987).

13.2.2 Soil

In soil, the natural concentration is around 50 parts per million (ppm) Cu. Cu in tailings from Cu mines and mills usually occurs in the form of insoluble sulfides as well as silicates. The majority of this Cu is incorporated into mineral lattices and thereby biologically inert (Barceloux 1999).

13.2.3 Water

Water is one of the main sources of Cu exposure in developed countries. Cu is present in different water sources such as in surface, ground, sea, and drinking water. However, it is primarily seen in complexes or as particulate matter. The majority of Cu in water results from natural run-off of Cu from soil (Agency for Toxic Substances and Disease Registry 1997). The concentration of Cu in aqueous solutions depends on factors such as pH, presence of competing cations, levels of compounds that complex organic and inorganic Cu substances, and concentrations of anions of insoluble cupric salts. Cu concentrations in surface waters range from 0.5 to 1,000 mcg/L (median value was 10 mcg/L) in various studies in the United States. The median concentration of Cu in natural water is 4–10 mcg Cu/L, with most of Cu tightly bound to organic matter and not readily exchangeable. The presence of insoluble organic and inorganic Cu compounds reduces the bioavailability of Cu in drinking water. Municipal treatment facilities remove around 80% of the total Cu in pretreated water and the Cu concentration in effluent usually is <100 mcg Cu/L of Cu (Stephenson and Lester 1987). Concentrations of Cu in drinking water are widely variable because of variations in features of water, such as pH, hardness, and availability of Cu in the distribution system. Drinking water provides 100–1,000 mcg of Cu per day in the majority of the time. Cu from pipes in the water distribution system may provide up to 1,000 mcg of Cu daily depending on the acidity and softness of the water, the extent of the Cu pipes and brass faucets. Drinking of either standing or partially flushed water from a distribution system which has Cu pipes or accessories can significantly increase total daily Cu exposure, particularly for infants fed formula reconstituted with using such tap water. Levels of Cu in running or fully flushed water tend to be low, whereas those of standing or partially flushed water are more variable and considerably higher. Concentrations of Cu in drinking water often increase during distribution, particularly in systems with an acid pH or high-carbonate waters with an alkaline pH (US EPA 1995).

The biological significance of the quantity of Cu rests on its bioavailability. A survey reported that Cu-contaminated water exhibited gastrointestinal (GI) disorders among residents of recently remodeled homes in an area where the water supplies were naturally corrosive (Knobeloch et al. 1998). However, there was no clear dose-response relationship between the Cu concentration in "first draw" tap water and the presence of GI symptoms. Water left overnight in these fixtures may contain up to 60,000 mcg Cu/L. The taste threshold ranges from 1,000 to 5,000 mcg Cu/L and a slight blue or green color appears in the water at levels >5,000 mcg Cu/L (California Department of Health Services 1991).

13.2.4 Food and Dietary Supplements

Like water, food is also a primary source of Cu exposure for human. Daily consumptions of Cu are necessary to maintain a steady state since the body has no specialized storage system for this element (Araya et al. 2006). Recommended daily intakes and tolerable intake levels for Cu is shown in Table 13.1. The foods rich in Cu include mushrooms, shellfish (oysters) and other seafood, beef and organ meats (particularly liver), nuts and sunflower seeds, peanut butter, soybeans, legumes, cereals, wheat, coconut, avocado, chocolate, cocoa, coffee, dark green leafy vegetables, and black pepper (Ashish et al. 2013; Araya et al. 2006; Groff et al. 1995). Concentrations of Cu in diet range 0.2–44 ppm Cu by wet weight (California Department of Health Services 1991).

Multivitamins that contain minerals usually have Cu. On the other hand, Cu is also available as an individual oral supplement. Ideally, it should be received from diet to children, instead of supplementation (Araya et al. 2006). However, infant formula is available and contains 0.6–2 mcg of Cu per kcal (Olivares and Uauy 1996). Vitamin or mineral formulations for children and adults mostly contain 2,000 mcg of Cu per preparation (tablet or capsule), mainly as copper oxide. The supplements may contain other different forms of Cu, including cupric sulfate, copper amino acid

Table 13.1   Recommended Daily and Tolerable Intake Levels for Copper

Age/Pregnancy	Recommended Daily Intake Levels (mcg)	Maximum Tolerable Intake Levels (mcg)
≤6 months	200	–
7–12 months	220	–
1–3 years	340	1,000
4–8 years	440	1,000
9–13 years	700	5,000
14–18 years	890	8,000
≥19 years	900	10,000
Pregnancy 14–18 years ≥19 years	1,000	8,000 10,000
Lactation 14–18 years ≥19 years	1,300	8,000 10,000

Source: Modified from National Institutes of Health. 2019. Copper: Fact Sheet for Health Professionals. Office of Dietary Supplements. https://ods.od.nih.gov/factsheets/Copper-HealthProfessional/#en3.

chelates, and copper gluconate. The amount of Cu in dietary supplements typically ranges from a few micrograms to 15,000 mcg (National Institutes of Health 2019). A survey in the United States shows that approximately 15% of the population takes a nutritional supplement having Cu (Institute of Medicine 2001). The average Cu requirements are 12.5 mcg/kg of body weight per day for adults and around 50 mcg/kg of body weight per day for infants. The Institute of Medicine (2001) recommends 10,000 mcg/day as a tolerable upper intake level for adults from foods and supplements. In the majority of dietary, Cu is found bound to macromolecules instead of free ion.

13.3 TOXIC DOSE

13.3.1 Acute (In Human)

The emetic dose of copper sulfate in adults is 250,000–500,000 mcg adult (Stein et al. 1976). Cu toxicity can occur even after the ingestion of 100,000 mcg of Cu. The consumption of beverages or foods having ≥25,000 mcg Cu/L has been related to acute gastroenteritis (Hopper and Adams 1958; Semple et al. 1960). A case reported an 18-month-old child developed hemolytic anemia and renal tubular damage after the estimated ingestion of 3 g cupric sulfate (≈120,000 mcg elemental Cu/kg) (Walsh et al. 1977). The lethal dose of Cu in an untreated adult is about 1,000,000–2,000,000 mcg of Cu. Based on data from accidental ingestion and suicide cases, the acute lethal dose for adults lies between 4,000 and 400,000 mcg of Cu(II) ion per kg of body weight (Chuttani et al. 1965; Agarwal et al. 1993). Copper sulfate is about 40% Cu by weight. Severe hemolysis and renal dysfunction occurred in an adult who ingested 17,500,000 mcg copper sulfate (Mittal 1972). Transient hepatic dysfunction as well as rhabdomyolysis developed in an adult, who underwent chelation therapy after estimated ingestion of 25,000,000 mcg copper sulfate (Jantsch et al. 1984).

13.3.2 Acute (In Animals)

In animals, the daily maximum tolerable dose is around 250,000 mcg Cu/kg (Haywood and Loughran 1985). The dose of Cu exceeding this causes severe hepatic centrilobular necrosis. Both species of animal and type of Cu compounds determine toxicity with Cu. Sheep, dogs, and cats are the most sensitive species compared to rodents, pigs, and poultry (Ishmael et al. 1971;

Andrews et al. 1990). On the other hand, compared with rats, mice exhibit more resistance toward cupric sulfate toxicity (Hebert et al. 1993). Moreover, dogs have higher susceptible to Cu overload than humans because of variations in the metabolism of Cu (Twedt et al. 1979).

13.3.3 Chronic (In Human)

The average daily intake of Cu in the United States is approximately 1,000 mcg (Turnlund 1988). At an average concentration of 130 mcg/L of Cu containing drinking water provides approximately 6%–13% of the average daily intake of Cu. The estimated absorption of 7,700 mcg of Cu did not produce evidence of Cu toxicity (Bentur et al. 1988). The rate of Cu loss from Cu-containing IUDs depends on the time in the body as well as the weight and surface area of IUD (Chantler et al. 1984). Oxidation of the Cu and the deposit of calcium limit the amount of Cu released from the IUDs. During the first month following insertion, mean doses of Cu released from IUDs with a surface area of 100–400 mm^2 ranged from 26 to 74 mcg of Cu daily (Timonen 1976.). Over the course of 10 years, the loss of Cu from a copper T averages 8 mcg of Cu per day for IUDs covered with calcareous deposits and 23 mcg of Cu per day for devices without visible coating (Thiery and Kosonen 1987). The chronic intake of an excessive quantity of Cu based on an early morning sample demonstrating 7,800 mcg Cu/L was associated with the development of acute GI symptoms, which resolves after stopping drinking the Cu-contaminated water (Spitalny et al. 1984).

13.4 TOXICOKINETICS

13.4.1 Absorption

Absorption of Cu occurs primarily in the upper GI tract after oral exposure in mammals and is regulated by a complex homeostatic process, which involves active (serosal) as well as passive (mucosal) transport (Linder and Hazehg-Azam 1996). The mucosal mechanism relies principally upon facilitated transport, whereas, the serosal mechanism is mediated by the saturable, energy-dependent pathway that acts rate-limiting for typical Cu intake (Linder and Hazehg-Azam 1996). Uptake of Cu from the intestine is susceptible to competitive inhibition by other transition metals (mainly Zn or Fe) and chemical form of Cu. The higher concentration of Zn prompts the formation of intestinal metallothionein, which blocks Cu absorption. Metallic Cu absorption from the GI tract is relatively small due to its insoluble nature. The presence of dietary proteins and amino acids, complexing or precipitating anions, fructose, ascorbic acid, phytate, fulvic acid, and fiber may also influence Cu uptake from the GI tract (Lönnerdal 1996; Institute of Medicine 2001). Preclinical data shows high dietary content of ascorbic acid decreases the fraction of Cu absorbed from the food (Smith and Bidlack 1980). Developmental age may influence Cu absorption. In the neonatal rat, intestinal absorption is high, however, by the time of weaning it decreases. The higher amount of Cu is transported to the liver and less remains in the intestine. The quantity of Cu stored in the body does not affect the absorption of Cu. The report shows the absorption of Cu from the ingestion of 275 US coins is sufficient to produce fatal Cu poisoning (Yelin et al. 1987). The GI tract absorbs approximately 60% of an ingested dose of copper acetate salt, with an inter-individual variation of 15%–97% (Weber et al. 1969). The absorption of Cu from the diet was approximately 65%–70% (≈1,000 mcg Cu per day) (Johnson et al. 1992). There is a paucity of data on the absorption of Cu from the lungs. Metal fume fever can happen after inhalation of volatilized Cu, which indicates lung absorbed Cu (Committee on Medical and Biological Effects of Environmental Pollutants 1977). However, the occurrence of Cu-induced metal fume fever is less due to the requirement of high temperature to volatilize Cu. Data on dermal absorption of Cu compounds are limited. Copper sulfate and azide can be absorbed via damaged skin after chronic exposure.

13.4.2 Distribution

In an adult male, the average total body content of Cu ranges from 50,000 to 120,000 mcg (Meret and Henkin 1971). Within the mucosal cell, the majority of the Cu is seen in the cytosol bound to transport proteins (chaperones) and storage protein (metallothionein). Intracellular Cu transport is tightly controlled, which involves a large number of Cu-binding proteins. This protects against free radical reactions initiated by Cu^+/Cu^{2+} oxidation-reduction reactions (Peña et al. 1999). A p-type ATPase active transport system mediates serosal transport from the mucosal cells. Cu in the portal blood is bound to albumin or transcuprein, and a small quantity may be chelated by histidine (Linder and Hazehg-Azam 1996; Camakaris et al. 1999). The liver releases Cu attached mainly to ceruloplasmin. This ∞_2-globulin binds approximately 95% of the serum Cu while the remainder of the Cu in the serum complexes with albumin (Cousins 1985). The Cu–albumin complex reflects the toxicologically active form of the serum Cu (Cartwright and Wintrobe 1964). Unbound ionic Cu probably does not exist in humans, except in the stomach. In acute poisoning, albumin, rather than ceruloplasmin, binds the excess Cu.

Under normal conditions, the highest concentrations of Cu present in the bile, liver, brain heart, kidneys, with moderate concentrations seen in the intestine, lung, and spleen (Evans 1973; Barceloux 1999). On the other hand, due to the highest mass of muscle and bone, the largest quantity of Cu (approximately 50%) present there, whereas the liver holds 8%–10% (Luza and Speisky 1996). In newborn infants, however, the liver contains 50%–60% of the body's Cu. The liver is the major site of deposition of Cu following large ingestion with the majority of the Cu bound to metallothionein. The Cu burden in normal adult liver ranges from 18,000 to 45,000 mcg Cu/g dry weight. The liver releases large quantities for Cu in the blood at concentrations exceeding 50,000 mcg Cu/g dry weight when liver necrosis occurs (Yelin et al. 1987). This release of a large quantity of Cu by the liver causes the rapid accumulation of the Cu in erythrocytes and the subsequent production of oxidative damage to the red blood cells.

13.4.3 Metabolism

As discussed earlier, Cu is absorbed in the gut and transported to the liver bound to plasma proteins. It enters the bloodstream via the plasma protein (ceruloplasmin), where its metabolism is controlled, and finally excreted in bile (Adelstein and Vallee 1961). Cu membrane transporter 1 (CMT1), a transporter protein present on the cells of the small intestine, carries Cu inside the cells, where some are bound to metallothioneins and part is carried by Cu transport protein (ATOX1) to the trans-Golgi network. Here, in response to rising concentrations of Cu, an enzyme called ATP7A to release Cu into the portal vein to the liver. Hepatocytes also have the CMT1 protein and metallothioneins and ATOX1 bind Cu inside the hepatocytes. On the other hand, ATP7B links Cu to ceruloplasmin and thereby releases it into the bloodstream, as well as get rid of excess Cu by secreting it into bile (Harris et al. 1998). Approximately 90% of the Cu in the blood is incorporated into ceruloplasmin, which is responsible for carrying Cu to tissues that need the mineral (Harris 2001; Groff et al. 1995). Since the elimination of Cu is so slow (10% in 72 h) an excessive dose of Cu is a lingering problem (Adelstein and Vallee 1961). Proper absorption and metabolism of Cu require an appropriate balance with minerals Zn and manganese. Zn has the ability to compete with Cu in the small bowel and can interfere with Cu absorption; therefore, individuals who received supplementation with an improperly high quantity of Zn and lower quantity of Cu may have an increasingly high risk of Cu deficiency (Harris 2001; Groff et al. 1995). In addition to the role of ceruloplasmin as a transport protein, it also acts as an enzyme; catalyzing the oxidation of minerals, most notably Fe (Adelstein and Vallee 1961). The oxidation of Fe by ceruloplasmin is required for Fe to be bound to its transport protein-transferrin. Therefore, Fe deficiency anemias may be a symptom of Cu deficiency (Araya et al. 2006; Groff et al. 1995). Mutations in Cu transport proteins,

ATP7A and ATP7B, can disable these transport systems leading to Menkes disease and Wilson's disease, respectively (Strausak et al. 2001; Schaefer and Gitlin 1999).

13.4.4 Elimination

Cu is excreted from the body in a different route such as bile, feces, sweat, hair, menses, and urine (Luza and Speisky 1996; Cox 1999). However, in humans, the main excretory pathway for absorbed Cu is bile, where Cu is bound to both low-molecular-weight as well as macromolecular substances. Biliary elimination seems to involve glutathione-dependent and glutathione-independent process (US NRC 2000). Biliary Cu is discharged to the intestine, where, after minimal absorption, it is eliminated in the feces. Preclinical studies showed little resorption of biliary Cu happens (Linder and Hazehg-Azam 1996; Johnson 1989). In normal humans, less than 3% of the daily Cu intake is excreted in the urine (Luza and Speisky 1996). The biological half-life of Cu in the blood after the ingestion of 290 mcg Cu ranged from 13 to 33 days (Johnson 1989; Johnson et al. 1992). Excretion of Cu in bile may be even more important than absorption in regulating the total body level of Cu (Turnlund et al. 1998).

13.5 MECHANISM OF TOXICITY

Cu works as a co-factor for the function of cellular enzymes, such as catalase, cytochrome oxidase, dopamine-beta-hydroxylase, and peroxidase. Cu toxicity usually affects in the order: erythrocytes, liver, and kidney. Excessive levels of Cu inhibit sulfhydryl groups on enzymes, such as glucose-6-phosphate (G6PD) and glutathione reductase, which protect cells from oxidative stress-induced damage by free radicals. Inhibition of G6PD causes hemolysis. Intravascular hemolysis happens within 12–24 h after the ingestion of copper sulfate. Cu ions can oxidize Fe to form methemoglobin, thereby oxygen-carrying capacity of blood declines. Clinically, this state is exhibited by cyanosis and chocolate brown blood. Acute Cu poisoning leads to erosion of the epithelial lining of the GI tract along with centrilobular necrosis of the liver and acute tubular necrosis (ATN) in the kidney. Cu produces direct damage to the proximal renal tubules, and ATN occurs following Cu poisoning without the appearance of hypotension or severe hemolysis (Dash 1989). Additionally, intravascular hemolysis plays a major role in the pathogenesis of renal failure. The heme pigment generated from hemolysis and direct toxicity of Cu released from lysis of red cells attribute to damage of tubular epithelium of both kidneys. Moreover, GI manifestations such as severe vomiting, diarrhea, lack of replacement of fluid and GI bleed, leading to hypotension, which could also contribute to renal failure. Renal complications can appear on the third or fourth day or onward following the poisoning. Copper sulfate, because of its corrosive nature, causes caustic burns of esophagus, superficial and deep ulcers in the stomach and the small bowel. Metallothionein is a cysteine-rich, low-molecular-weight protein that binds to Cu and provides some protection against Cu toxicity. The formation of metallothionein occurs early in acute Cu toxicity both in the liver as well as in the kidney (Kurisaki et al. 1988). Chronic Cu toxicity does not usually happen in humans due to transport systems, which regulate absorption as well as excretion (Turnlund et al. 2005). Cu toxicity is most likely to happen in people with the liver disorder or other disease conditions in which excretion of bile is compromised (Araya et al. 2006). Evidence shows that a link between chronic exposure to large concentrations of Cu and a decline in intelligence in young adolescence (Tamura and Turnlund 2004). Postpartum depression has also been linked with high levels of Cu, since Cu concentrations increase throughout pregnancy to almost double normal values, and it may take up to 3 months after delivery for Cu concentrations to normalize (Crayton and Walsh 2007).

No standard animal studies correlate Cu toxicity with reproductive function. However, available evidence shows that Cu can be a developmental toxicant at higher doses. On the other hand, low concentrations of supplemental Cu have a beneficial effect on development. Massive DNA damage was observed in hepatocytes from patients with Indian childhood cirrhosis and was postulated to result from excessive accumulation of Cu in the nucleus, leading to the production of free radicals that cause DNA strand breakage (Prasad et al. 1996). At high concentrations, Cu may be genotoxic or enhance the genotoxicity of other agents, possibly through the generation of reactive oxygen species/free radicals or through effects on DNA-related enzyme processes. Although a number of older studies have examined the carcinogenicity of various Cu compounds in animals, none of them are adequate by current methodological standards (US NRC 2000).

Certain genetic disorders can affect Cu utilization, which involves Menkes syndrome, a deficiency disorder, and Wilson disease, a toxicity disorder. Both of these have been identified as defects in p-type ATPases (US NRC 2000). Evidence suggests that an autosomal recessive gene can be a predisposing factor for Cu-related cases of infant or childhood cirrhosis (Müller et al. 1996; Tanner 1999). The etiologies of Indian childhood cirrhosis, endemic Tyrolean infantile cirrhosis, and idiopathic Cu toxicosis are complex and may involve a combination of genetic, developmental, and environmental factors (Müller et al. 1996). These disorders are characterized by liver enlargement, elevated Cu deposits in liver cells, pericellular fibrosis and necrosis and are generally fatal. Poor biliary excretion of Cu may play a role in the etiology of the disease.

Wilson's disease is a rare, progressive, autosomal recessive disorder characterized by impaired transport and excessive accumulation of Cu in the liver, brain, and other tissue (Popević et al. 2011). The condition is due to mutations in Wilson's disease protein of ATP7B gene, by impaired Cu incorporation to ceruloplasmin and biliary Cu excretion. It is characterized by hepatic cirrhosis, neurological manifestations, psychiatric manifestations, renal diseases, and Cu deposition in the cornea (Kayser-Fleischer ring) (Strausak et al. 2001; Schaefer and Gitlin 1999). In a study of Cu toxicosis in humans, lipid peroxidation and Cu content were significantly increased in hepatic mitochondria from patients with Wilson's disease. Modest elevation in lipid peroxidation was existed in microsomes of Wilson's disease patients. Mitochondrial Cu concentrations correlated strongly with the severity of mitochondrial lipid peroxidation. These data suggest that the hepatic mitochondria are an important target in hepatic Cu toxicity and that oxidative damage to the liver may be involved in the pathogenesis of Cu-induced injury. A significant decrease (37%) in the vitamin E/lipid ratio was also detectable in patients with Wilson's disease showing high free serum Cu (>10 mcg/dL). The data support a role for free radicals in the pathogenesis of active liver diseases (von Herbay et al. 1994). Another study revealed that impaired conversion of 25 (OH)D to 1.25 (OH)2D occurs in Cu intoxication and suggests that altered vitamin D metabolism is a potential factor in the development of bone and mineral abnormalities in Wilson's disease (Carpenter et al. 1988).

13.6 CLINICAL MANIFESTATIONS

13.6.1 Acute

In lower doses, Cu ions can produce symptoms typical of food poisonings such as headache, nausea, vomiting, and diarrhea. On the other hand, large doses of Cu causes GI bleeding, hematuria, intravascular hemolysis, methemoglobinemia, hepatocellular toxicity, acute renal failure, and oliguria (Agarwal et al. 1993) (Figure 13.1).

13.6.1.1 Gastrointestinal

Copper sulfate, a gastric irritant, causes a metallic taste, nausea, vomiting, crampy abdominal pain, epigastric burning, and hematemesis which occur immediately following ingestion. Vomiting generally appears within 15 min of intake. The vomitus normally is a greenish-blue color. Hemorrhagic gastroenteritis is related to mucosal erosions. Severe cases can cause melena or hematemesis, whereas diarrhea is less common (Chuttani et al. 1965). A large quantity of intravenous Cu via Cu tubing in a hemodialysis unit may result in acute necrotizing hemorrhagic pancreatitis (Klein et al. 1972).

13.6.1.2 Cardiovascular

In severe cases, cardiovascular collapse, hypotension, and tachycardia can happen initially within few hours after poisoning. This may be the reason for early fatal reactions or can happen late with other complications. GI symptoms are the factors usually contributing to hypovolemia. Resultant cardiac dysrhythmia and hypoxia due to severe methemoglobinemia, in turn, cause cardiovascular collapse. Additional factors are the direct effect of Cu on vascular and cardiac cells as well as sepsis because of transmucosal invasion.

13.6.1.3 Renal

Kidney-related complications are generally seen after 48 h. Anuria, oliguria, hematuria, and albuminuria may have observed, however, renal dysfunction is most often mild. About 20%–40% of victims may develop acute renal failure with acute copper sulfate poisoning. Damage to the proximal tubules of the kidney results either from direct effects of the copper sulfate or from the effects of hypotension and/or hemoglobinuria. In acute Cu poisoning, myoglobinuria is seldom present as a complication (Ashish et al. 2013).

13.6.1.4 Hematological

Intravascular hemolysis happens 12–24 h following ingestion. Severe methemoglobinemia occurs early in the victim's clinical course and is immediately followed by hemolysis. Hepatic injury or a direct effect of free Cu ions on the coagulation cascade can cause coagulopathy.

13.6.1.5 Hepatic

Since the liver is the primary site of deposition of Cu and acute Cu toxicity may produce jaundice as well as centrilobular necrosis of the liver. Jaundice develops following 24–48 h in severe poisonings, which may have related to either hemolytic or hepatocellular. This may be associated with tender hepatomegaly.

13.6.1.6 Central Nervous System

CNS depression varies from lethargy to coma. The seizure is often epiphenomenon associated with the involvement of other organs (Ashish et al. 2013).

13.6.1.7 Muscular

Highly elevated creatine phosphokinase (CPK)-related rhabdomyolysis has been detected. A case study reported myoglobinuria in the second day, whereas peak CPK level was seen on the sixth day.

13.6.1.8 Respiratory

Cu is considered as a respiratory irritant that can cause mucosal irritation of the nose, mouth, and eye (Askergren and Mellgren 1975). Nasal septal perforation can be induced by severe inhalation of Cu dust (Cohen 1974). Symptoms related to metal fume fever such as fever, chills, myalgia, headache, malaise, and dry throat may occur from the inhalation of Cu fumes from various sources.

However, the incidence of Cu-induced metal fume fever is very rare due to the requirement of high temperature to generate Cu fumes.

13.6.1.9 Dermal

Hair shafts, especially, damaged portions of the hair, stains a green hue after increased absorption of Cu. Apart from this, there will be loss of cuticle as well as the appearance of scattered micropits (Burnett 1989). A localized irritation and greenish discoloration of the skin area can occur after the corrosion of Cu glasses as well as rings by body sweat. Hypersensitivity reactions to Cu-based IUDs and metallic Cu are uncommon. However, very few cases of allergic contact dermatitis with Cu-based ornaments and urticaria with the use of a Cu-based IUD (Barkoff 1976).

13.6.1.10 Systemic

Lethargy, coma, and refractory hypotension, in addition to fever, chills, and pain may present in severe acute Cu intoxications (Schwartz and Schmidt 1986).

13.6.2 Chronic

Chronic exposure usually causes metallic taste, a green line on the gums at the base of the teeth, GI symptoms such as nausea, vomiting, colic, diarrhea, constipation, and general signs of progressive emaciation, anemia, malaise, and debility (Figure 13.1).

In Wilson's disease low levels of ceruloplasmin allow the accumulation of a large quantity of Cu initially in the liver and subsequently in the brain and other organs. The clinical presentations consist of altered mental status, abnormalities in motor function such as dysarthria, dysphagia, ataxia, difficulty in writing; hemolytic anemia, renal tubular dysfunction causing uricosuria, hypercalciuria; renal stones and fulminant hepatic failure. The liver disorder appears before extra-hepatic manifestations. Kayser–Fleischer rings, corneal deposits of Cu within descent's membrane, often present in symptomatic patients. In the absence of lifelong chelation therapy, progressive pathological changes happen, in the liver as well as in the brain (Scheinberg et al. 1987).

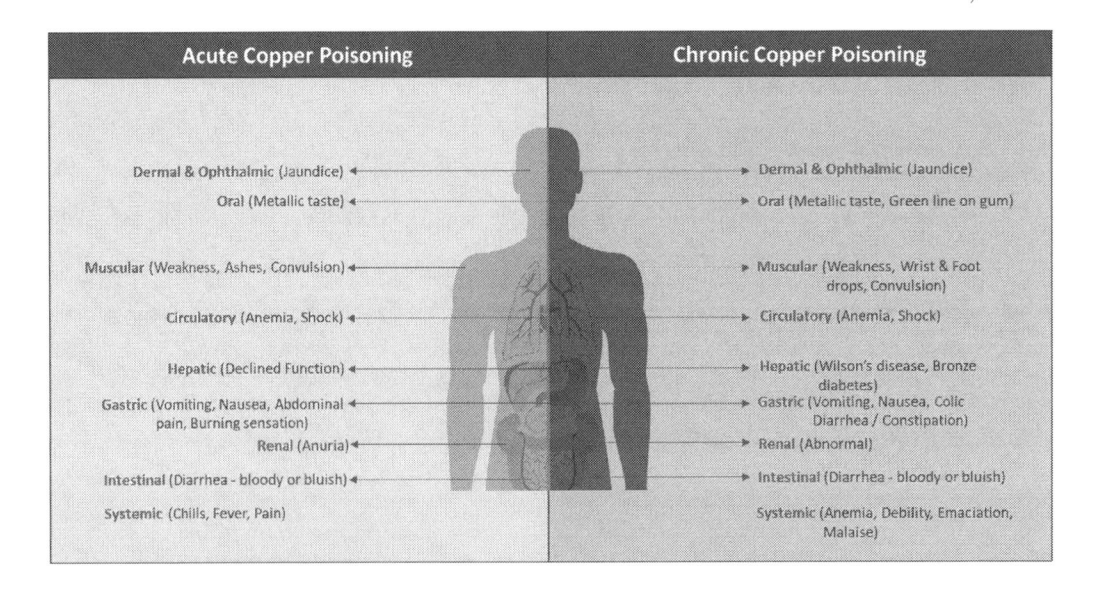

Figure 13.1 Clinical manifestations of acute and chronic copper poisoning.

Indian childhood cirrhosis causes elevated Cu levels without a decrease in the levels of ceruloplasmin unlike Wilson's disease. Histopathological changes in the Indian childhood cirrhosis comprise of prominent degenerative ballooning of the hepatocytes, diffuse Mallory bodies, marked neutrophilic infiltrate in acini, and severe micronodular fibrosis. The presence of coarse dark-brown orcein staining and intralobular pericellular fibrosis differentiates Indian childhood cirrhosis from Wilson's diseases. The absence of steatosis and the presence of cholestasis late in the disease course features in India childhood cirrhosis.

13.7 DIAGNOSIS

The concentration of Cu can be detected particularly in the blood and urine and limited data of hair. The most common methods used to determine Cu levels in the biological or environmental samples are atomic absorption spectrometry and inductively coupled plasma-atomic emission spectroscopy. This method is very sensitive to detect metals lower than 1 ppm. Rapid colorimetric assays can detect ranges near 100 mcg Cu/L (Abe et al. 1989).

13.7.1 Blood

A mean serum concentration of Cu in healthy individuals was found to be 860 mcg Cu/L (Gonzalez-Revalderia et al. 1990). In a series of events with Cu intoxication with copper sulfate, GI symptoms happened at whole blood concentrations >2,870 mcg Cu/L and hepatorenal dysfunction and/or shock occurred at the blood levels >7,980 mcg Cu/L (Chuttani et al. 1965). Renal failure increases Cu levels only slightly > the concentrations present in healthy adults (Zumkley et al. 1979). Approximately 95% of Wilson's disease patients have decreased concentrations (<20,000 mcg/L) of oxidase active ceruloplasmin (Scheinberg and Gitlin 1952). However, decreased concentrations of serum ceruloplasmin alone are not diagnostic of Wilson's disease, since nearly 20% of heterozygotes have reduced concentrations of ceruloplasmin. Similarly, other abnormal conditions also have low ceruloplasmin, which includes protein-losing enteropathies and nephropathies; and defective synthesis of ceruloplasmin peptide. Less than 6 months of infants usually have low serum ceruloplasmin concentration.

13.7.2 Urine

Urine Cu levels of the general population range up to 50 mcg Cu/L (Lauwerys and Hoet 1992). Cu concentration in urine of symptomatic Wilson's disease is often >100 mcg Cu/24-h urine collection (Martins da Costa et al. 1992).

13.7.3 Hair

Mean Cu levels are 17.7 ppm Cu in initial three centimeters of the proximal end of scalp hair and 11.9 ppm Cu in pubic hair (Wilhelm et al. 1990).

13.8 MANAGEMENT

13.8.1 Supportive Care

In pre-hospitalization, following acute ingestion with copper sulfate, immediate dilution with water or milk should be considered. However, vigorous dilution should be avoided since this may lead to nausea, vomiting, and aspiration. Emesis should be avoided to prevent re-exposure of the esophagus to the corrosive salts. In copper sulfate poisoning, vomiting is likely to appear suddenly. Therefore, the patient may need treatment with an antiemetic. Activated charcoal administration can be considered following potentially serious ingestion. It is most effective when given within 60 min after the ingestion of the Cu compound. The standard dose is 25–100 g for adults and adolescents; and 25–50 g for children aged 1–12 years or 0.5–1 g/kg body weight. Sucralfate may help to relieve the symptoms of mucosal injury. Studies have reported the use of steroid lowered mortality rate in copper sulfate poisoning.

Symptomatic methemoglobinemia patients should receive methylene blue. This normally happens at the levels of methemoglobin >20%–30%. Administer oxygen while preparing for methylene blue therapy. Methylene blue promotes the transformation of methemoglobin to hemoglobin by enhancing the enzymatic action of the methemoglobin reductase. The initial dose is 1–2 mg/kg/dose intravenously over 5 min. The dose may be repeated if cyanosis does not disappear within 60 min. Failure of methylene blue treatment indicates an inadequate dose of methylene blue, G-6-PD deficiency, or NADPH-dependent methemoglobin reductase deficiency. Hyperbaric oxygen may be beneficial if methylene blue is ineffective. Hyperbaric oxygen increases the dissolved oxygen that can protect the patient while the body reduces methemoglobin. Another alternative to methylene blue is ascorbic acid, a reducing agent, which can be given 100–500 mg twice daily either orally or intravenously. The hypotensive episodes can be treated with fluids, dopamine, and noradrenaline. For rhabdomyolysis, initial replacement of 4–6 L/day with close monitoring for fluid overload, mannitol (100 mg/day), and urine alkalinization are considerable in the early course of the disease, however, there is no clear evidence for these strategies.

Hemodialysis to remove Cu is ineffective, however, it may be indicated in patients with renal failure secondary to Cu poisoning. Perineal dialysis with albumin resulted in the extraction of more Cu than dialysate without albumin. On the other hand, the quantity of Cu removed by peritoneal dialysis was very small. Meanwhile, there is insufficient evidence regarding any role of hemoperfusion for Cu elimination.

13.8.2 Treatment

13.8.2.1 Acute

In the overdose, induction of emesis or gastric lavage is generally not required after acute ingestion of Cu compounds, due to persistent vomiting. Activated charcoal with or without a cathartic may be useful, however, no adequate data on the efficacy of decontamination methods after acute ingestion of Cu. Chelating agents are recommended in severe Cu poisoning, however, clinical pharmacokinetic data are lacking to guide their application. Intravenous calcium disodium versenate (CaNa$_2$ EDTA) or intramuscular dimercaprol (British anti-Lewisite; BAL) is the agent of choice in severe ingestion. D-Penicillamine (DPA) may be administered orally if tolerated. The adult dose is 1,000–1,500 mg/day divided every 6–12 h, before food. Whereas, child dose is starting with 10 mg/kg/day gradually increases to 30 mg/kg/day divided into two or three doses. Intramuscular BAL is usually appropriate in patients with vomiting and GI injury, which prevents oral DPA use. BAL-Cu complex will be primarily eliminated through bile; therefore, it is beneficial in renal failure patients. Dosing of BAL is a deep intramuscular injection of 3–5 mg/kg/dose every 4 h for 2 days, every 4–6 h for the next 2 days, thereafter, every 4–12 h till 7 days. On the other hand, DPA is more effective than BAL, therefore, wherever possible, DPA should be initiated concurrently or immediately following the starting of BAL. The dose of CaNa$_2$ EDTA is 75 mg/kg/day either deep intramuscularly or slow intravenous infusion in three to six divided doses for till 5 days. If required a second course, repeat after a minimum of 2 days. However, each course should not be exceeded by 500 mg/kg (Ashish et al. 2013).

13.8.2.2 Chronic

Chelation therapy with DPA is preferred for symptomatic neurological, psychiatric, and hepatic disorders. Few patients exhibit worsening neurological symptoms at the time of initial few weeks with chelation treatment. Normally, the symptoms alleviate, however, deterioration in certain patients indicates the need for alternative treatment such as BAL or triethylenetetramine dihydrochloride (trientine). Daily adult dose for DPA is 1–2 g orally in divided doses, for trientine 0.75–1.5 g orally and for Zn salts are 150–200 mg metallic Zn orally in divided doses. An investigational drug, tetrathiomolybdate has successfully demonstrated the ability to chelate Cu in sheep; however, human data are limited.

Wilson's disease: The therapy of Wilson's disease involves avoidance of diet rich in Cu and any of its supplements. Zinc acetate with a dose of 50 mg three times daily is an effective maintenance treatment. Zn promotes the synthesis of intestinal metallothionein that prevents the absorption of Cu by hindering the transport of Cu from the intestinal mucosa to the systemic circulation (Hoogenraad et al. 1987).

13.9 CONCLUSION

In recent years, nutritionists have been more concerned regarding Cu toxicity than Cu deficiency. The increased quantity of Cu was reported in the drinking water due to the wide use of Cu water pipes. Cooking with Cu cookware can also increase the Cu content of foods. Industrial exposure to Cu fumes, dust, or mists may result in metal fume fever with atrophic changes in nasal mucous membranes. High uptakes of Cu may cause liver and kidney damage and even death. Large quantities of various Cu salts have been taken in suicide attempts and produced Cu toxicity in humans. The toxicity effects of Cu are possibly due to redox cycling and the generation of reactive oxygen species that damage DNA. Chelation therapy may be used for both acute and chronic toxicity with Cu poisoning. For the treatment of Wilson's disease, a diet rich in Cu and supplements containing Cu should be avoided. In addition, chelating agents such as DPA, BAL, Zn acetate, etc., can be used.

REFERENCES

Abe, A., Yamashita, S., Noma, A. 1989. Sensitive, direct colorimetric assay for copper in serum. *Clinical Chemistry* 35:552–554.

Adelstein, S.J., Vallee, B.L. 1961. Copper metabolism in man. *New England Journal of Medicine* 265:892–897.

Agarwal, S.K., Tiwari, S.C., Dash, S.C. 1993. Spectrum of poisoning requiring hemodialysis in a tertiary care hospital in India. *International Journal of Artificial Organs* 16(1):20–22.

Agency for Toxic Substances and Disease Registry, US Public Health Service. 1997. *ATSDR's Toxicological Profiles: Copper.* Boca Raton, FL: Lewis Publishers, CRC Press Inc.

Akintonwa, A. 1989. Fatal poisoning by copper sulfate ingested from "spiritual water". *Veterinary and Human Toxicology* 31:453–454.

Andrews, P.L., Davis, C.J., Bingham, S., Davidson, H.I., Hawthorn, J., Maskell, L. 1990. The abdominal visceral innervation and the emetic reflex: pathways, pharmacology, and plasticity. *Canadian Journal of Physiology and Pharmacology* 68:325–345.

Araya, M., Pizarro, F., Olivares, M., et al. 2006. Understanding copper homeostasis in humans and copper effects on health. *Biological Research* 39:183–187.

Ashish, B., Neeti, K., Himanshu, K. 2013. Copper toxicity: a comprehensive study. *Research Journal of Recent Sciences* 2:58–67.

Askergren, A., Mellgren, M. 1975. Changes in the nasal mucosa after exposure to copper salt dust. a preliminary report. *Scandinavian Journal of Work, Environment & Health* 1:45–49.

Barceloux, D.G. 1999. Copper. *Clinical Toxicology* 37(2):217–230.

Barkoff, J.R. 1976. Urticaria secondary to a copper intrauterine device. *International Journal of Dermatology* 15:594–595.

Bentur, Y., Koren, G., McGuigan, M., Spielberg, S.P. 1988. An unusual skin exposure to copper; clinical and pharmacokinetic evaluation. *Clinical Toxicology* 29:371–380.

Bonham, M., Jacqueline, M., Bernadette, M.H., Strain, J.J. 2002. The immune system as aphysiological indicator of marginal copperstatus? *British Journal of Nutrition* 87:393–403.

Burnett, J.W. 1989. Copper. *Cutis* 43:322.

California Department of Health Services. 1991. Fact Sheets on Chemical Contaminants in Drinking Water Copper (Cu). Berkeley, CA: Water Toxicology Unit, Pesticideand Environmental Toxicology Section, California Department of Health Services.

Camakaris, J., Voskoboinik, I., Mercer, J.F. 1999. Molecular mechanisms of copper homeostasis. *Biochemical and Biophysical Research Communications* 261:225–232.

Carpenter, T.O., Pendrak, M.L., Anast, C.S. 1988. Metabolism of 25-hydroxyvitamin D in copper-laden rat: a model of Wilson's disease. *American Journal of Physiology* 254(2 Pt 1):E150–E154.

Cartwright, G.E., Wintrobe, M.M. 1964. Copper metabolism in normal subjects. *American Journal of Clinical Nutrition* 14:224–232.

Chantler, E.N., Scott, K., Filho, C.I., et al. 1984. Degradation of the copper-releasing intrauterine contraceptive device and its significance. *British Journal of Obstetrics and Gynaecology* 92:172–181.

Chuttani, H.K., Gupta, P.S., Gulati, S., Gupta, D.N. 1965. Acute copper sulfate poisoning. *American Journal of Medicine* 39:849–854.

Cohen, S.R. 1974. A review of the health hazards from copper exposure. *Journal of Occupational Medicine* 16:621–624.

Committee on Medical and Biological Effects of Environmental Pollutants. 1977. *Copper.* Washington, DC: National Academy of Sciences, 29–58.

Cousins, R.J. 1985. Absorption, transport, and hepatic metabolism of copper and zinc: special reference to metallothionein and ceruloplasmin. *Physiological Reviews* 65:238–309.

Cox, D.W. 1999. Disorders of copper transport. *British Medical Bulletin* 55(3):544–555.

Crayton, J.W., Walsh, W.J. 2007. Elevated serum copper levels in women with a history of post-partum depression. *Journal of Trace Elements in Medicine and Biology* 21:17–21.

Dash, S.C. 1989. Copper sulphate poisoning and acute renal failure. *International Journal of Artificial Organs* 12:610.

Davies, D.J.A., Bennett, B.G. 1985. Exposure of man to environmental copper: an exposure commitment assessment. *Science of the Total Environment* 46:215–227.

Davis, C.D. 2003. Low dietary copper increases fecal free radical production, fecal water alkaline phosphatase activity and cytotoxicity in healthy men. *Journal of Nutrition* 133:522–527.

Ellenhorn, M.J., Barceloux, D.G. 1988. *Medical Toxicology: Diagnosis and Treatment of Human Poisoning.* New York, NY: Elsevier Science Publishing Company, pp. 54, 84, 1022.

Encyclopaedia Britannica. 2019. Pierre-Marie-Alexis Millardet. https://www.britannica.com/biography/Pierre-Marie-Alexis-Millardet.

Evans, G.W. 1973. Copper homeostasis and metabolism in the mammalian system. *Physiological Reviews* 53:535–569.

Gonzalez-Revalderia, J., Garcia-Bermejo, S., Menchen- Herreros, S., Fernandez-Rodriguez, E. 1990. Biological variation of Zn, Cu, and Mg in serum of healthy subjects. *Clinical Chemistry* 36:2140–2141.

Groff, J.L., Gropper, S.S., Hunt, S.M. 1995. *Advanced Nutrition and Human Metabolism.* New York, NY: West Publishing Company.

Harper, M. 1987. Possible toxic metal exposure of prehistoric bronze workers. *British Journal of Industrial Medicine* 44(10):652–656.

Harris, E.D. 2001. Copper homeostasis: the role of cellular transporters. *Nutrition Reviews* 59:281–285.

Harris, E.D., Qian, Y., Tiffany-Castiglioni, E., Lacy, A.R., Reddy, M.C. 1998. Functional analysis of copper homeostasis in cell culture models: a new perspective on internal copper transport. *American Journal of Clinical Nutrition* 67(5 Suppl):988S–995S.

Haywood, S., Loughran, M. 1985. Copper toxicosis and tolerance in the rat. II. Tolerance–a liver protective adaptation. *Liver* 5:267–275.

Hebert, C.D., Elwell, M.R., Travlos, G.S., Fitz, C.J., Bucher, J.R. 1993. Subchronic toxicity of cupric sulfate administered in drinking water and feed to rats and mice. *Fundamental and Applied Toxicology* 21:461–475.

von Herbay, A., de Groot, H., Hegi, U., Stremmel, W., Strohmeyer, G., Sies, H. 1994. Low vitamin E content in plasma of patients with alcoholic liver disease, hemochromatosis and Wilson's disease. *Journal of Hepatology* 20(1):41–46.

Hopper, S.H., Adams, J.H.S. 1958. Copper poisoning from vending machines. *Public Health Reports* 1:910–914.

Hoogenraad, T.U., Van Hattum, J., Van den Hamer, C.J.A. 1987. Management of Wilson's disease with zinc sulphate experience in a series of 27 patients. *Journal of the Neurological Sciences* 77:137–146.

Institute of Medicine. 2001. Dietary reference intakes for vitamin A, vitamin K, arsenic, boron, chromium, copper, iodine, iron, manganese, molybdenum, nickel, silicon, vanadium and zinc. A report of the Panel on Micronutrients, Subcommittees on Upper Reference Levels of Nutrients and of Interpretation and Use of Dietary Reference Intakes, and the Standing Committee on the Scientific Evaluation of Dietary Reference Intakes. Food and Nutrition Board, Institute of Medicine. Washington, DC, National Academy Press.

Ishmael, J., Gopinath, C., Howell, J.M. 1971. Experimental copper toxicity in sheep. Histological and histochemical changes during the development of lesions in the liver. *Research in Veterinary Science* 12:358–366.

Jantsch, W., Kulig, K., Rumack, B.H. 1984–1985. Massive copper sulfate ingestion resulting in hepatotoxicity. *Journal of Toxicology - Clinical Toxicology* 22(6):585–588.

Johnson, P.E. 1989. Factors affecting copper absorption in humans and animals. *Advances in Experimental Medicine and Biology* 258:71–79.

Johnson, P.E., Milne, D.B, Lykken, G.I. 1992. Effects of age and sex on copper absorption, biological half-life, and status in humans. *American Journal of Clinical Nutrition* 56:917–925.

Klein, W.J. Jr, Metz, E.N., Price, A.R. 1972. Acute copper intoxication a hazard of hemodialysis. *Archives of Internal Medicine* 129:578–582.

Knobeloch, L., Schubert, C., Hayes, J., et al. 1998. Gastrointestinal upsets and new copper plumbing–is there a connection? *Wisconsin Medical Journal* 97:49–53.

Kurisaki, E., Kuroda, Y., Sato, M. 1988. Copper-binding protein in acute copper poisoning. *Forensic Science International* 38:3–11.

Landner, L., Lindestrom, L. 1999. Copper in society and in the environment. Vasteras, Swedish Environmental Research Group (MFG) (SCDA S-721 88).

Lauwerys, R.R., Hoet, P. 1992. *Industrial Chemical Exposure Guidelines for Biological Monitoring*, 2nd edn. Boca Raton, FL: Lewis Publishers, 51–52.

Linder, M.C., Hazehg-Azam, M. 1996. Copper biochemistry and molecular biology. *American Journal of Clinical Nutrition* 63:797S–811S.

Lönnerdal, B. 1996. Bioavailability of copper. *American Journal of Clinical Nutrition* 63:821S–829S.

Lucas, L.C., Lemons, J.E. 1992. Biodegradation of restorativemetallic systems. *Advances in Dental Research* 6:32–37.

Luza, S.C., Speisky, H.C. 1996. Liver copper storage and transport during development: implications for cytotoxicity. *American Journal of Clinical Nutrition* 63:812S–820S.

Martins da Costa, C., Baldwin, D., Portmann, B., et al. 1992. Value of urinary copper excretion after penicillamine challenge in the diagnosis of Wilson's disease. *Hepatology* 15:609–615.

Meret, S., Henkin, R.I. 1971. Simultaneous direct estimation by atomic absorption spectrophotometry of copper and zinc in serum, urine, and cerebrospinal fluid. *Clinical Chemistry* 17:360–373.

Mittal, S.R. 1972. Oxyhaemoglobinuria following copper sulphate poisoning: a case report and a review of the literature. *Forensic Science* 1:245–248.

Müller, T., Feichtinger, H., Berger, H., Muller, W. 1996. Endemic Tyrolean infantile cirrhosis: an ecogenetic disorder. *Lancet* 347:877–880.

National Institutes of Health. 2019. Copper: Fact Sheet for Health Professionals. Office of Dietary Supplements. https://ods.od.nih.gov/factsheets/Copper-HealthProfessional/#en3.

NSF. 2000. Drinking water treatment chemicals - health effects. NSF international standard for drinking water additives. Ann Arbor, MI, NSF International, p. 28 (ANSI/NSF 60-2000).

Okereke, T., Sternlieb, I., Morell, A.G., Scheinberg, I.H. 1972. Systemic absorption of intrauterine copper. *Science* 177:358–360.

Olivares, M., Uauy, R. 1996. Copper as an essential nutrient. *American Journal of Clinical Nutrition* 63:791S–796S.

Osredkar, J., Sustar, N. 2011. Copper and zinc, biological role and significance of copper/zinc imbalance. *Journal of Clinical Toxicology* S:3.

Peña, M.M.O., Lee, J., Thiele, D.J. 1999. A delicate balance: homeostatic control of copper uptake and distribution. *Journal of Nutrition* 129:1251–1260.

Popević, M.B., Kisić, G., Dukić, M., Bulat, P. 2011. Work ability assessment in a patient with Wilson's disease. *Arhiv za Higijenu Rada i Toksikologiju* 62(2):163–167.

Prasad, R., Kaur, G., Nath, R., Walia, B.N. 1996. Molecular basis of pathophysiology of Indian childhood cirrhosis: role of nuclear copper accumulation in liver. *Molecular and Cellular Biochemistry* 156:25–30.

Schaefer, M., Gitlin, J.D. 1999. Genetic disorders of membrane transport IV. Wilson's disease and Menkes disease. *American Journal of Physiology* 276(2):G311–G314.

Scheinberg, I.H., Gitlin, D. 1952. Deficiency of ceruloplasmin in patients with hepatolenticular degeneration (Wilson's disease). *Science* 116:484–485.

Scheinberg, I.H., Jaffe, M.E., Sternlieb, I. 1987. The use of trientine in preventing the effects of interrupting penicillamine therapy in Wilson's disease. *New England Journal of Medicine* 317:209–213.

Schroeder, W.H., Dobson, M., Kane, D.M., Johnson, N.D. 1987. Toxic trace elements associated with airborne particulate matter: a review. *Journal of the Air Pollution Control Association* 37:1267–1285.

Schwartz, E., Schmidt, E. 1986. Refractory shock secondary to copper sulfate ingestion. *Annals of Emergency Medicine* 15:952–954.

Semple, A.B., Parry, W.H., Phillips, D.E. 1960. Acute copper poisoning: an outbreak traced to contaminated water from a corroded geyser. *Lancet* 2:700–701.

Sharpe, W.E., DeWalle, D.R. 1985. Potential health implications for acid precipitation, corrosion, and metals contamination of drinking water. *Environmental Health Perspectives* 63:71–78.

Smith, C.H., Bidlack, W.R. 1980. Interrelationships of dietary ascorbic acid and iron on the tissue distribution of ascorbic acid, iron, and copper in female guinea pigs. *Journal of Nutrition* 110:1398–1408.

Sontz, E., Schwieger, J. 1995. The "green water" syndrome: copper-induced hemolysis and subsequent acute renal failure a consequence of a religious ritual. *American Journal of Medicine* 98:311–315.

Spitalny, K.C., Brondum, J., Vogt, R.L., Sargent, H.E., Kappel, S. 1984. Drinking-water-induced copper intoxication in a Vermont family. *Pediatrics* 74:1103–1106.

Stein, R.S., Jenkins, D., Korns, M.E. 1976. Death after use of cupric sulfate as emetic. *JAMA* 235:801.

Stephenson, T., Lester, J.N. 1987. Heavy metal behavior during the activated sludge process. I. extent of soluble and insoluble metal removal. *Science of the Total Environment* 63:199–214.

Strausak, D, Mercer, J.F., Dieter, H.H., Stremmel, W., Multhaup, G. 2001. Copper in disorders with neurological symptoms: Alzheimer's, Menkes, and Wilson diseases. *Brain Research Bulletin* 55(2):175–185.

Tamura, T., Turnlund, J.R. 2004. Effect of long-term, high-copper intake on the concentrations of plasma homocysteine and B vitamins in young men. *Nutrition* 20:757–759.

Tanner, M.S. 1999. Indian childhood cirrhosis and Tyrolean childhood cirrhosis. In: Leone, A., Mercer, J.F.B., eds. *Copper Transport and Its Disorders*. New York, NY: Plenum Press, pp. 127–137.

Thiery, M., Kosonen, A. 1987. The multi-sleeved copper T model TCu220C: effect of long-term use on corrosion and dissolution of copper. *Contraception* 35:163–170.

Timonen, H. 1976. Copper release from Copper-T intrauterine devices. *Contraception* 14:25–38.

Turnlund, J.R. 1988. Copper nutriture, bioavailability, and the influence of dietary factors. *Journal of the American Dietetic Association* 88:303–310.

Turnlund, J.R., Keyes, W.R., Kim, S.K., Domek, J.M. 2005. Long-term high copper intake: effects on copper absorption, retention, and homeostasis in men. *American Journal of Clinical Nutrition* 81(4):822–828.

Turnlund, J.R., Keyes, W.R., Peiffer, G.L., Scott, K.C. 1998. Copper absorption, excretion and retention by young men consuming low dietary copper determined by using the stable isotope ^{65}Cu. *American Journal of Clinical Nutrition* 67:1219–1225.

Twedt, D.C., Sternlieb, I., Gilbertson, S.R. 1979. Clinical, morphologic, and chemical studies on copper toxicosis of Bedlington terriers. *Journal of the American Veterinary Medical Association* 175:269–275.

US EPA. 1995. Effect of pH, DIC, orthophosphate and sulfate on drinking water cuprosolvency. Washington, DC, US Environmental Protection Agency, Office of Research and Development (EPA/600/R-95/085).

US FDA. 1994. Code of Federal Regulations. Vol. 21. Food and Drugs Part 182, Substances Generally Recognized as Safe, pp. 399–418, and Part 184, Direct Food Substances Affirmed as Generally Regarded as Safe, pp. 418–432. https://www.accessdata.fda.gov/scripts/cdrh/cfdocs/cfcfr/CFRSearch.cfm?CFRPart=182.

US NRC. 2000. *Copper in Drinking Water.* Washington, DC: National Research Council, National Academy Press.

Walker, W.R., Beveridge, S.J., Whitehouse, M.W. 1981. Dermal copper drugs: the copper bracelet and Cu(II) salicylate complexes. *Agents Action* 8(suppl):359–367.

Walsh, M., Crosson, F.J., Bayley, M. 1977. Acute copper intoxication. *American Journal of Diseases of Children* 131:149–151.

Weant, G.E. 1985. Sources of copper air emissions. Research Triangle Park, North Carolina: Air and Energy Engineering Research Laboratory US Environmental Protection Agency, EPA 600/2-85-046.

Weber, P.M., O'Reilly, S., Pollycove, M., Shepley, L. 1969. Gastrointestinal absorption of copper: studies with ^{64}Cu, ^{95}Zn, a whole body counter and the scintillation camera. *Journal of Nuclear Medicine* 10:591–596.

Whitehouse, M.W, Walker, W.R. 1977. The copper bracelet for arthritis. *The Medical Journal of Australia* 1:938.

Wilhelm, M., Ohnesorge, F.K., Hotzel, D. 1990. Cadmium, copper, lead, and zinc concentrations in human scalp and pubic hair. *Science of the Total Environment* 92:199–206.

Willis, M.S., Monaghan, S.A., Miller, M.L., et al. 2005. Zinc-induced copper deficiency: a report of three cases initially recognized on bone marrow examination. *American Journal of Clinical Pathology* 123(1):125–131.

World Health Organization. 2004. Copper in Drinking-Water: Background Document for Development of WHO Guidelines for Drinking-water Quality. https://www.who.int/water_sanitation_health/publications/copper/en/.

Yelin, G., Taff, M.L., Sadowski, G.E. 1987. Copper toxicity following massive ingestion of coins. *American Journal of Forensic Medicine and Pathology* 8:78–85.

Zumkley, H., Bertram, H.P., Lison, A., Knoll, O., Losse, H. 1979. Aluminum, zinc and copper concentrations in plasma in chronic renal insufficiency. *Clinical Nephrology* 12:18–21.

CHAPTER 14

Cadmium Exposure and Toxicity

Soisungwan Satarug
The University of Queensland

Kenneth R. Phelps
Albany Medical College

CONTENTS

14.1 INTRODUCTION

Cadmium (Cd), atomic weight 112, is found with zinc (Zn) and mercury (Hg) in group IIB of the periodic table. Cd is not a transition metal, but it has a similar ionic radius to that of calcium (Ca) and electronegativity similar to that of Zn. These chemical propensities allow Cd to enter cells and confer the potential to interfere with the physiological functions of both Ca and Zn. Thus, Cd is considered as a "high health risk" metal (Satarug et al., 2010, 2017a,b,c). It is present in zinc, copper, and lead ores and in phosphate rocks from which fertilizer is produced for agricultural use (IPCS, 1992; Satarug et al., 2003; Garrett, 2010; ATSDR, 2012; Lamb et al., 2016). Cadmium oxide (CdO) is released into the atmosphere by volcanic emissions and burning of fossil fuels, tobacco, and other biomass (IPCS, 1992; Memon and Schröder, 2009). CdO, which enters the human body through the lungs, is more bioavailable than Cd-containing compounds that enter through the gut (Satarug et al., 2002; Takenaka et al., 2004; Blum et al., 2015; Dumkova et al., 2016). Like other metals, Cd is highly persistent in the environment because it is not biodegradable. No normal excretory mechanism exists for Cd, and virtually all acquired Cd is consequently retained. Even if environmental exposure is low, long-term intake results in substantial accumulation of the metal in various organs and tissues (Satarug et al., 2003; Satarug, 2018). Ultimately, most retained Cd is sequestered in kidneys, where its half-life may range from 9 to 44 years (Elinder et al., 1976, 1978; Suwazono et al., 2009; Fransson et al., 2014).

Industrial production and use have made Cd ubiquitous in the environment, and agricultural application of phosphate fertilizers has transferred Cd to the food chain (Schroeder and Balassa 1963; Satarug et al., 2003; Clemens, 2006; Bandara et al., 2010; Clemens et al., 2013; Lamb et al., 2016). From the soil, plants take up more Cd than other toxic heavy metals such as lead and mercury (McLaughlin and Singh, 1999). Advanced technologies have reduced exposure due to smelting and industrial applications, but industrial recycling of Cd is limited (IPCS, 1992; Järup, 2003). An outbreak of "itai-itai" disease, a severe form of Cd poisoning, serves as a reminder of the health threat from Cd contamination of a staple food crop (Aoshima, 1987; Horiguchi et al., 2010). Similarly, livestock that graze on contaminated pasture can accumulate levels of renal and hepatic Cd that make the organs unsafe for human consumption (Wilkinson et al., 2003).

Because of phylogenic characteristics, tobacco, rice, other cereal grains, potatoes, salad vegetables, spinach, and Romaine lettuce accumulate Cd more efficiently than other plants (Clemens et al., 2013). Multiple detoxification mechanisms enable plants to tolerate high Cd concentrations (Cobbett, 2000; Cobbett and Goldsbrough 2002). These mechanisms involve an array of metal-binding ligands, including phytochelatins (PCs), metallothioneins (MTs), other low-molecular-weight thiols, glutathione, cysteine, γ-glutamylcysteine, and cysteinylglycine (Memon and Schröder, 2009; Pivato et al., 2014). Consequently, concentrations of Cd that are toxic to animals and humans are not toxic to plants. Children, women of reproductive age, and patients with low body iron stores, hypertension, and diabetes are at increased risk for Cd toxicity (Buchet et al., 1990; Satarug et al., 2000a; 2018a; Haswell-Elkins et al., 2008; Madrigal et al., 2019).

Cd and its compounds are classified as class I carcinogens by a world cancer authority (International Agency for Research on Cancer, IARC, 1993). In workers who inhaled CdO, the lungs, bones, and kidneys were the principal targets of Cd toxicity (Lauwerys et al., 1994; Roels et al., 1999). However, Cd affects many more tissues, organs, and systems than those associated with occupational exposure.

The goal of this chapter is to provide an update of knowledge on adverse effects of environmental exposure to Cd that were observed in epidemiologic studies worldwide together with the levels at which such exposure produced toxicity. Consideration is given to at-risk subpopulation groups. This information is relevant to public health policy regarding advisable exposure limits and tolerable body burden of Cd that are much lower than previously estimated, arguing strongly for a modification of the criteria by which toxicity of Cd is judged.

In Sections 14.2 and 14.3, we discuss current standards for tolerable intake and the urinary threshold of Cd. We also assess the utility of urinary Cd excretion as a measure of body burden of Cd. Dietary factors influencing the absorption rate of Cd are discussed together with nutritional and genetic factors that affect Cd toxicity. The pathogenesis of Cd-induced nephropathy is detailed in Section 14.4, and Cd-induced bone diseases are discussed in Section 14.5. Section 14.6 focuses on environmental exposure to Cd and its association with chronic kidney disease (CKD). Effects on blood pressure of environmental exposure to Cd and the pathogenesis of Cd-induced hypertension are explored in Section 14.7. Section 14.8 reviews Cd toxicity in organs other than the kidneys and bone and examines the implications of these effects in assessments of health risks related to Cd.

14.2 EXPOSURE SOURCES, LIMITS, AND BODY BURDEN ASSESSMENT

Food crops grown in Cd-rich soils constitute the major source of non-workplace exposure to the metal. Accordingly, dietary assessment methods such as food frequency questionnaires, duplicate diets, and total diet studies (TDSs) have been used to evaluate environmental exposure to Cd. TDS is a food safety monitoring program, conducted by food authority agencies such as the U.S. Food and Drug Administration (FDA), the European Food Safety Agency (EFSA), and the Food Standards of Australia and New Zealand (FSANZ), formerly known as the Australia and New Zealand Food Authority (ANZFA). It is known also as the "market basket survey" because it involves collection of samples of foodstuffs from supermarkets and retail stores for quantitation of various food additives, pesticide residues, contaminants, and nutrients (Callan et al., 2014). TDS is a reasonable method to identify sources of Cd in the human diet.

14.2.1 Dietary Sources of Cadmium

The average dietary Cd intake in the United States has been reported to be as low as 4.63 µg/day (Kim et al., 2018). This figure was based on data from 24-h dietary recalls in the U.S. National Health and Nutrition Examination Survey (NHANES) 2007–2012 cycle and Cd concentrations in foodstuffs determined in the U.S. FDA 2006–2013 TDS. The Women's Health Initiative study in the United States reported a higher dietary Cd intake of 10.9 µg/day (Adams et al., 2014). In the TDS report, principal sources of Cd and their percentage contributions to total dietary Cd intake were cereals and bread (34%), leafy vegetables (20%), potatoes (11%), legumes and nuts (7%), stem/root vegetables (6%), and fruits (5%) (Kim et al., 2019). As in TDS from Sweden and France, the bulk of dietary Cd (40%–60%) came from foods that are frequently consumed in large quantities such as staples and vegetables (Sand and Becker, 2012; Arnich et al., 2012).

The average dietary Cd intake in the United States of 4.63 µg/day was lower than values of 10.6 µg/day and 11.2 µg/day intake levels in Sweden and France, respectively (Sand and Becker, 2012; Arnich et al., 2012). The Swedish TDS also reported dietary intake for the high consumer of 23 µg/day, with additional Cd coming from seafood (shellfish) and spinach (Sand and Becker, 2012). For the high consumer, the French TDS reported dietary Cd intake of 18.9 µg/day with additional Cd coming from mollusks and crustaceans (Arnich et al., 2012).

14.2.2 Use and Misuse of TDS Data

TDS provides a reasonable method to gauge the relative contribution of each food item to total dietary Cd intake. It can also provide the basis to define a maximally permissible concentration (MPC) for Cd in a specific food group, as discussed in Section 14.2.3. However, the utility of dietary intake estimates in evaluating body burden is questionable. TDS data did not report the variability of dietary habits nor did it report the body status of essential metals, notably iron and zinc, that are known to govern Cd absorption and accumulation in the body (further discussed in Section 14.3). The utility of dietary estimates in risk assessments is not demonstrable, as studies summarized below indicate.

An association between breast cancer risk and environmental exposure became evident only when such exposure levels were evaluated with urinary Cd excretion (Gallagher et al., 2010; Larsson et al., 2015; Lin et al., 2016). In stark contrast, no association was observed when exposure levels were evaluated with estimates of dietary Cd (Van Maele-Fabry et al., 2016).

Only a weak correlation was seen between estimates of dietary Cd and urinary Cd excretion in an analysis of subgroup in the Women's Health Initiative study ($N = 1,002$ women, mean age 63.4). The mean dietary Cd of this cohort was 10.4 µg/day, while the mean urinary Cd excretion was 0.62 µg/g creatinine (Quraishi et al., 2016). Dietary Cd estimates, based on the data from the U.S. FDA 2006–2011 TDS, did not correlate with blood Cd levels in a study of an Asian subpopulation in NHANES 2011–2012 (Awata et al., 2017).

A study of Danish women reported mean dietary Cd of 14 µg/day (5th–95th percentiles: 8–22 µg/day), with leafy vegetables and soy-based products being the main sources (Eriksen et al., 2014; Vacchi-Suzzi et al., 2015). Only a marginal correlation was seen between dietary Cd intake estimates and urinary Cd levels in a study of 1,764 post-menopausal Danish women (Eriksen et al., 2014; Vacchi-Suzzi et al., 2015). Thus, dietary intake estimates are of limited utility in body burden assessment.

14.2.3 Dietary Exposure and the Urinary Cadmium Threshold

A dietary intake limit, known as the Provisional Tolerable Weekly Intake (PTWI), was first established for Cd in 1989 by the Joint Expert Committee on Food Additives and Contaminants (JECFA) of the Food and Agriculture Organization (FAO) and the World Health Organization (WHO) of the United Nations (FAO/WHO, 1989). By definition, the PTWI for a chemical with no intended function is an estimate of the amount of the chemical that can be ingested weekly over a lifetime without appreciable health risk (FAO/WHO, 1989). The first PTWI figures allocated to Cd were 400–500 µg/week (FAO/WHO, 1989). These figures were based on a critical Cd concentration of 200 µg/g kidney cortex, attainable after dietary Cd intake of 140–260 µg/day for over 50 years or 2,000 mg of Cd over a lifetime. In 1993, the previous guideline was retained but expressed in terms of the intake/kg body weight, corresponding to 7 µg/kg body weight/week or 1 µg/kg body weight/day (FAO/WHO, 1993). However, it was recognized that the model on which the PTWI was based did not include a safety factor, and it had a very modest safety margin between normal dietary intake and intake that would produce deleterious health effects.

In 2010, the tolerable weekly intake level for Cd was amended to tolerable monthly intake (TMI) level of 25 µg/kg body weight/month, equivalent to 0.83 µg/kg body weight/day or 58 µg/day for a 70-kg person (FAO/WHO, 2010). In addition, urinary Cd of 5.24 µg/g creatinine was adopted as a provisional urinary Cd threshold limit. While most countries rely on the FAO/WHO guidelines, the European Food Safety Authority (EFSA) established a tolerable dietary Cd intake level of 2.5 µg/kg body weight per week (25 µg per day for a 70-kg person) and a urinary Cd threshold of 1 µg/g creatinine (EFSA 2011, 2012). The tolerable dietary Cd intake and urinary Cd threshold levels established by EFSA are lower than the figures of FAO/WHO.

Both FAO/WHO and EFSA used kidney as the sole target of Cd toxicity, and both groups used population dietary Cd intake data in deriving an estimate of tolerable intake. Grains plus grain products, vegetables plus vegetable products and starchy roots plus tubers contributed, respectively, to 26.9%, 16.0%, and 13.2% of Cd in the European population diet (EFSA, 2011, 2012). FAO/WHO and EFSA also used urinary excretion of the low-molecular-weight protein β2-microglobulin (β_2MG) levels above 300 µg/g creatinine as a sign of toxicity from an excessive Cd intake. In the derivation of threshold limit, the FAO/WHO considered kidney Cd concentration of 180–200 µg/g kidney wet weight to be critical, while the EFSA considered kidney Cd concentration of 50 µg/g as a critical level.

Numerous epidemiologic studies showing renal toxicity at tissue concentrations lower than 50 µg/g and urinary Cd as low as one-tenth of 5.24 µg/g creatinine, the threshold established by FAO/WHO (Sections 14.6–14.8). Some longitudinal studies have linked all levels of lifetime environmental exposure to increased mortality across multiple populations. Notably, many adverse effects of Cd have observed in the absence of evidence for kidney effects (Section 14.8). These findings challenge current kidney-based risk assessment practice and suggest there are no acceptable levels of dietary Cd. Intake of this pervasive toxicant should thus be kept to a minimum.

14.2.4 Other Environmental Guidelines

Safe limits also were established for Cd in the environment and in foodstuffs [International Programme on Chemical Safety (IPCS) 1992; Codex, 2014]. Maximal acceptable concentrations are 3 mg/kg in soil that is for producing crops for humans and 3 µg/L in drinking water (IPCS, 1992). MPCs have also been established for Cd in staple foods and shellfish known to be hyper-accumulators of Cd. At present, the MPC for potatoes is 0.1 mg/kg, while the MPC for rice is 0.4 mg/kg dry grain weight (Codex, 2014). However, the MPC for Cd in rice has been called into question.

A study in Vietnam reported that staple rice alone constituted 93%–95% of total dietary Cd intake, and water spinach constituted another 2.4%–4% (Minh et al., 2012). In a low-exposure area, dietary Cd intake was calculated from consumption of 426 g of cooked rice per day and mean (range) rice Cd content of 0.076 (0.030–0.16) mg/kg dry rice grain weight. The dietary Cd intake was found to be 29 and 33 µg/day in women and men, respectively. In a high exposure area, dietary intake was 109 and 122 µg/day in women and men, respectively. These figures were calculated from consumption of 461 g of cooked rice per day and mean rice Cd content of 0.31 mg/kg dry grain weight. Dietary Cd intake of 109–122 µg/day in a high-exposure area exceeded the established tolerable exposure level.

In support of a lower MPC, rice Cd concentrations above 0.1 mg/kg have been linked to increased mortality from all causes, especially in women (Nogawa et al., 2018). In another study, rice Cd concentration as low as 0.27 mg/kg was associated with increased risk of itai-itai disease (Nogawa et al., 2017). The MPC for rice, currently at 0.4 mg/kg, should thus be reduced to the lowest achievable level. It can be inferred from the Vietnamese experience described above that rice Cd content should not be higher than 0.1 mg/kg.

14.2.5 Assessment of Cadmium Exposure and Body Burden

Dietary assessments and quantification of Cd in scalp hair, toenails, fingernails, blood, and urine have been proposed for evaluation of body burden. Dietary assessment has limited utility for this purpose (see Section 14.2.2). Biologic specimens such as scalp hair and toenails show promise, but reproducibility of assays across populations has yet to be confirmed. Of the potential parameters, urinary Cd has been widely used, especially in environmental exposure monitoring and surveillance of health effects in polluted areas.

Measurement of urinary and/or blood Cd concentrations has been included in the NHANES. This survey has become a rich data source for investigating adverse effects of environmental exposure to Cd. In most epidemiologic studies, urinary Cd excretion (E_{Cd}) has been normalized to urinary creatinine excretion (E_{cr}) as $[Cd]_u/[cr]_u$, where $[Cd]_u$ = urine concentration of Cd (µg/L), and $[cr]_u$ = urine creatinine concentration (mg/dL). The ratio $[Cd]_u/[cr]_u$ is expressed as µg/g creatinine, and it corrects for urinary dilution. Normalization of urinary Cd excretion is discussed further in Section 14.4.5.

Blood Cd has been proposed as a better estimate of intake for people with substantial nephron loss. Urinary Cd levels in these subjects may be lower than in similarly aged persons with optimal kidney function. In one study, persons dying from kidney disease had lower kidney Cd levels than those who died from other diseases (Lyon et al., 1999). Confounding effects of nephron loss on excretion of Cd (E_{Cd}) can be addressed by normalizing urinary Cd excretion (E_{Cd}) to creatinine clearance (C_{cr}) instead of E_{cr} (Section 14.4.5).

14.2.6 Urinary Cadmium as a Body Burden Indicator

The conclusion that urinary Cd reflects total body content of the metal is based on a direct relationship between excretion of Cd (E_{Cd}) and the amounts of Cd found in human organs, notably the liver and the kidneys (Orlowski et al., 1998; Satarug et al., 2002; Akerstrom et al., 2013; Wallin et al., 2014). The greatest concentration of Cd occurs in the kidneys, and kidney and liver Cd comprised two-thirds of the body burden in chronically exposed subjects (reviewed in Satarug, 2018). Urinary Cd levels correlated substantially with kidney Cd content and secondarily with determinants of kidney Cd such as age and gender. Consequently, urinary Cd serves as a measure of kidney burden.

An autopsy study in Poland reported that a urinary Cd level of 1.7 µg/g creatinine corresponded to kidney Cd level of 50 µg/g (Orlowski et al., 1998). In an Australian study of accident victims, the mean Cd levels in lung, liver, and kidney cortex were 0.13, 0.95, and 15.45 µg/g wet tissue weight (Satarug et al., 2002). In a subgroup, mean urinary Cd was 0.62 µg/L, ranging between 0.05 and 2.88 µg/L, while mean Cd levels in lung, liver, and kidney cortex were 0.12, 0.99, and 20.5 µg/g wet tissue weight, respectively. These urinary Cd concentrations (µg/L) correlated strongly with kidney Cd levels and age, but not with lung or liver. Thus, urinary Cd appeared to be indicative of kidney Cd burden. Intriguingly, these data on urinary Cd concentration (µg/L) are in line with data from kidney transplant donors.

In a study of Swedish donors ($N = 109$), mean kidney Cd was 12.9 µg/g, and the "urine-to-kidney Cd ratio" was 1:60. Accordingly, urinary Cd of 0.42 µg/g creatinine corresponded to kidney Cd level of 25 µg/g wet kidney weight (Akerstrom et al., 2013). In another study of transplant donors, means for Cd in urine, blood, and kidney in women were 0.34 µg/g creatinine, 0.54 µg/L and 17.1 µg/g kidney wet weight, respectively (Wallin et al., 2014). The corresponding figures in men were 0.23 µg/g creatinine, 0.46 µg/L and 12.5 µg/g, all of which were lower than women (Wallin et al., 2014). Like Australian study, a strong correlation was observed between kidney Cd and urinary Cd.

14.2.7 Ecological and Other Studies Demonstrating the Utility of Urinary Cadmium

14.2.7.1 Sweden

An ecological study in Sweden found the highest levels of Cd in garden soils and in carrots and potatoes grown in the area closest to an old battery factory (Hellström et al., 2007). Potatoes and vegetables caused dietary intake of Cd to increase by 18%–38%. Persons with urinary levels higher than 1 µg/g creatinine were those over 30 years of age, who lived in close proximity to a factory and consumed home-grown food items.

14.2.7.2 The Torres Strait, Australia

High Cd levels (7–76 mg/kg wet weight) were found in the offal of dugongs and turtles that constituted the diet of islanders living in the Torres Strait, Australia (Satarug et al., 2000a; Haswell-Elkins et al., 2007a,b; 2008). Of 182 residents, 12% had urinary Cd levels ≥2 µg/g creatinine, a value that exceeded the EFSA urinary threshold limit, while the mean urinary Cd was 0.83 µg/g creatinine (Haswell-Elkins et al., 2007a). Consumption of turtle livers and kidneys and locally gathered clams, peanuts, and coconuts accounted for 46% of total variation in the measured levels of urinary

Cd excretion (Haswell-Elkins et al. 2007b). These two studies provided data linking habitual consumption of high-Cd foods such as offal (turtle liver and kidney) to high Cd body burdens.

14.2.7.3 British Columbia, Canada

Bivalve mollusks and crustaceans are filter feeders that accumulate metals from the aquatic environment independent of environmental pollution. Contaminated waters further increase the accumulation (Whyte et al., 2009; Guéguen et al., 2011). It has been argued that Cd in these organisms is not bioavailable, and high MPC levels for Cd in oysters have been proposed (Kruzynski, 2004; Bendell, 2010). However, a study of non-smoking oyster growers (33 men, 28 women), aged 33–64 years (mean 47.3) in British Columbia, Canada, showed that elevated urinary Cd excretion levels were associated with a duration of oyster farming of at least 12 years together with an average consumption of 18 oysters/week (87 g/week) (Copes et al., 2008). The estimated Cd intake from oysters was 174 μg/week (24.8 μg/day), and the mean urinary Cd was 0.76 μg/g creatinine (range: 0.16–4.04 μg/g creatinine). This mean urinary Cd was approximately twofold higher than the mean urinary Cd of 0.35 μg/g creatinine recorded for non-smokers, aged 20–75 years, in the Canada Health Measure Survey (Garner and Levallois, 2016).

14.2.8 Summary

Cd is present in virtually all foodstuffs. Thus, foods that are frequently consumed in large quantities such as rice, potatoes, wheat, leafy salad vegetables, and other cereal crops are the most significant dietary Cd sources. TDS is a reasonable method to identify sources of Cd in the diet. TDS can also provide the basis to define a MPC for Cd in a specific food group. In contrast, it has limited utility in evaluating Cd body burden, which is best estimated with urinary Cd excretion (E_{Cd}) or a derivative thereof. The adoption of Cd excretion for this purpose is based on two facts: the tissue content of Cd is highest in the kidneys, and it varies directly with urinary Cd. Despite its utility, normalization of $[Cd]_u$ to $[cr]_u$ is problematic. A solution is proposed in Section 14.4.

14.3 CADMIUM ABSORPTION AND TOXICITY: THE INTERPLAY OF NUTRITION AND GENETICS

Cd has no known physiological function in the body, but it has high toxicity potential because of its propensity to bind sulfur-containing ligands with a high affinity. This propensity leads to an inactivation and loss of function of thiol (–SH) groups in glutathione and various proteins (Jomova and Valko, 2011; Rubino, 2015; Valko et al., 2016). A wide range of diversity of toxic effects of Cd can thus be predictable.

Although no mechanisms have been evolved to absorb or eliminate Cd from the body, Cd has a similar ionic radius to that of Ca and electronegativity similar to that of Zn. Consequently, Cd can be taken up by the same transporters and receptors that cells use to acquire Ca and Zn. Moreover, Cd also has the potential to interfere with physiological functions of both of these elements. This section examines MT, a carrier and reservoir of Cd, and transporters and receptors that are involved in intestinal absorption of Cd. Dietary and nutritional factors affecting absorption and body burden of Cd are discussed. The effects of zinc and iron on body burden and toxicity of Cd are summarized.

14.3.1 Metallothionein as Cadmium Reservoir

MTs, a group of low molecular weight (6–7 kDa) metal-binding proteins, are highly conserved and ubiquitously expressed in many tissues (Sabolić et al., 2010; Vašák and Meloni, 2011; Hennigar et al., 2016). There are many human MT isoforms, but MT-1 and MT-2 are expressed in most tissues, leucocytes included, while MT-3 is expressed in kidneys and neurons but not in leucocytes (Yoshida et al., 1998; Boonprasert et al., 2012; Hennigar et al., 2016; Sabolić et al., 2018). MT contains an unusually high molar content of cysteine (30%), and the sulfur atoms in this amino acid participate in metal sequestration. In theory, up to 7 atoms of zinc (Zn^{2+}) or cadmium (Cd^{2+}) or 12 atoms of copper (Cu^{2+}) can be sequestered per molecule of MT. MT-3 exhibits higher affinity for copper than MT-1 and MT-2 (Vašák and Meloni, 2011).

The synthesis of MT is induced by Zn, as evidenced by a study showing that hepatic MT expression was not detectable in zinc-deficient rats (Sato et al., 1984). Increased expression of MT was detected in erythrocytes and monocytes from men, aged 19–35 years, who received zinc supplements at a dose of 50 mg/day for 18 days (Sullivan et al., 1998). Like Zn, Cd is an inducer of cellular MT synthesis (Takeda et al., 1995; Yoshida et al., 1998; Katakai et al., 2001; Boonprasert et al., 2012). Because Cd exerts toxicity in the unbound state, i.e., as Cd^{2+} ions, complexes of Cd and MT (CdMT) are viewed as detoxified forms. Although expression of MT in hepatocytes prevents toxicity following Cd uptake, work with the normal rat liver cell line TRL1215 demonstrated the propensity for nitric oxide (NO) to displace Zn^{2+} ions or Cd^{2+} ions that were bound to MT (Misra et al. 1996; Katakai et al., 2001). The release of Cd^{2+} ions previously sequestered in MT implies that Cd toxicity may occur long after exposure cessation (Satarug et al., 2000b; Hoet et al., 2012).

14.3.2 Metal Transporters, Carriers, and Receptors

Absorption of dietary Cd is mediated by multiple transporters, carriers and channels that are used to acquire a range of nutritionally essential elements, notably iron (Fe^{2+}), zinc (Zn^{2+}), calcium (Ca^{2+}), and manganese (Mn^{2+}) (Vesey, 2010; Thévenod and Wolff, 2016; McDermott et al., 2016; Thevenod et al. 2019). Examples of metal transporters, carriers and receptors that have been implicated in intestinal Cd absorption include DMT1, a Zrt- and Irt-related protein (ZIP) of zinc transporter family (ZIP14), the Ca^{2+}-selective channel TRPV6, and the human neutrophil gelatinase-associated lipocalin (hNGAL) receptor (Vesey, 2010; Thévenod and Wolff, 2016; Thevenod et al. 2019). ZIP14 and TRPV6 are highly expressed by intestinal enterocytes (Fujishiro et al., 2017; Jorge-Nebert et al., 2015; Kovacs et al., 2011, 2013).

DMT1 was once thought to be the principal transporter for Cd in the enterocyte because this iron-import transporter has the same high affinity for Cd^{2+} ions as it does for iron (Km 0.5–1 µM) (Garrick et al., 2003). However, ferroportin 1 (FPN1) exports iron but not Cd from enterocytes into the portal blood circulation (Mitchell et al., 2014). Although basolateral export of Cd is not understood, calbindin, a calcium-binding protein, may be involved in cytoplasmic transport Cd to the basolateral cell surface and egress from the enterocyte into the portal circulation.

There is increasing evidence that zinc transporters, notably ZIP8 and ZIP14, mediate Cd uptake by various cell types (Fujishiro et al., 2012; Jenkitkasemwong et al., 2012; Schneider et al., 2014; Zhang et al., 2016; Aydemir and Cousins, 2018). In one study, ZIP8 mediated cellular uptake of not only zinc (Zn^{2+}) but also manganese (Mn^{2+}), iron (Fe^{2+}), cobalt (Co^{2+}), and selenite ($HSeO3^-$), one form of selenium. In this ZIP8-mediated metal uptake, selenite was taken up with Zn and bicarbonate $[(Zn^{2+})/ (HCO_3^-) (HSeO_3^-)]$ (McDermott et al., 2016). Like Zn and Mn, the presence of Se may reduce Cd uptake which may explain protective effect of Se against Cd toxicity (Section 14.3.6).

14.3.3 Factors Affecting the Cadmium Absorption Rate

The absorption rate of Cd in human is estimated to be 3%–7% (WHO, 1989; IPCS, 1992). However, a study of Japanese women suggested that up to 40% of ingested Cd was absorbed (Kikuchi et al., 2002, 2003; Horiguchi et al., 2004). An extremely small amount of Cd, 0.001%–0.005% of the total body content, is excreted in urine each day (Elinder et al., 1976, 1978). Because of the miniscule elimination rate, the body burden of Cd is essentially determined by the absorption rate.

Experimental studies showed that intestinal expression of metal transporters was influenced by age, physiological requirement, nutritional status, and dietary factors (Satarug et al., 2000a; Min et al., 2008, 2015). Consequently, Cd absorption rates may vary widely among people. In theory, the absorption rate of Cd will rise when the body is in short supply of the elements that share absorption and transport mechanisms with Cd, such as iron, zinc, and calcium (Flanagan et al., 1978; Finley, 1999). It will also rise in subjects whose diets are deficient in these elements.

14.3.3.1 Dietary Factors

Intestinal absorption of Cd appears to depend on whether the metal is presented in a bound or unbound form such as Cd salts; $CdCl_2$, $CdSO_4$ (Scheuhammer, 1988; Hiratsuka et al., 1999). In most experimental work, Cd salts are often used either in drinking water or in mixtures with animal food pellets. In the human diet, however, Cd is bound to metal-binding proteins and ligands such as MT and PC. These MT- and PC-bound forms of Cd, denoted, respectively, as CdMT and CdPC, can possibly be absorbed intact via transcytosis (Fujita et al., 1993) and receptor (hNGAL)-mediated endocytosis (Langelueddecke et al., 2014). Most cells, hepatocytes included, lack a protein-internalization mechanism for taking up CdMT; renal tubular cells are an exception because they are equipped for reabsorption of virtually all filtered proteins (Nielsen et al., 2016; Langelueddecke et al., 2012, 2013).

Figure 14.1 depicts the kidney as the site where most assimilated Cd is deposited. Cd absorbed from the diet is transported to liver via the hepatic portal system. Cd ions (Cd^{2+}) are taken up by hepatocytes and induce synthesis of MT. Cd is then tightly bound to MT in complexes of CdMT (Sabolić et al., 2010). Liver cells do not take up CdMT and CdPC (Sabolić et al., 2010). Consequently, newly absorbed CdMT and CdPC are distributed around the body. These complexes are ultimately

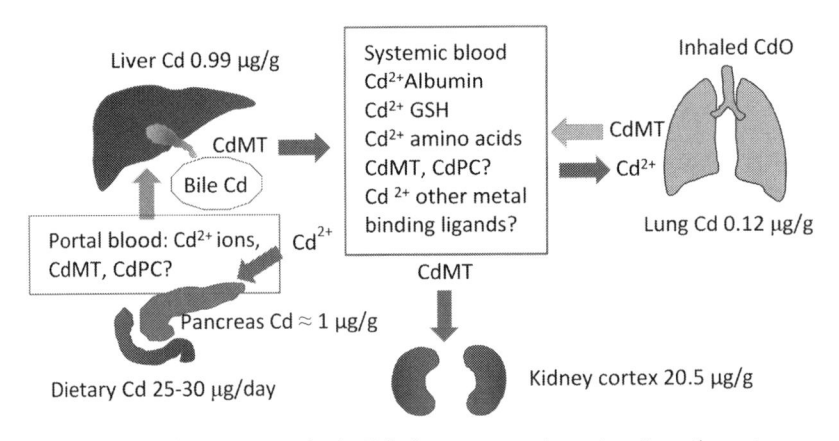

Figure 14.1 The kidney as the site of cadmium accumulation. Dietary Cd is absorbed and transported to liver via the hepatic portal system. The Cd^{2+} ions, taken up by hepatocytes, induce the synthesis of MT that forms complexes with Cd^{2+} ions as CdMT. Most cells, hepatocytes included, do not take up CdMT and CdPC. Consequently, newly absorbed CdMT and CdPC plus CdMT released from liver and lungs are delivered to kidneys, where they are taken up by kidney tubules. Cd that is secreted into bile may be excreted in feces or reabsorbed and returned to the liver. The lung, liver, and kidney cortex Cd concentrations of 0.12, 0.99, and 20.5 µg/g wet tissue weight were mean values recorded for Australian subjects, aged 2–70 years (mean 39.9). The mean urinary Cd was 0.62 µg/L, range; 0.05–2.88 µg/L (Satarug et al., 2002). Abbreviations: CdMT, cadmium-metallothionein complex; CdPC, cadmium-phytochelatin complex.

passed into glomerular filtrate by virtue of their small molecular weights. Inhaled CdO induces MT in lungs, and CdMT is formed in situ, and CdMT may be released to systemic circulation.

Most filtered Cd is selectively taken up and retained by proximal tubular cells. Consequently, the highest Cd concentrations are found in kidney cortex (Section 14.4.2). The liver serves as a reservoir that continuously releases CdMT into the systemic circulation. Cd that is secreted into bile may be excreted in feces or reabsorbed and returned to the liver (Kikuchi et al., 2002, 2003; Horiguchi et al., 2004).

14.3.3.2 Intestinal Metallothionein

Because CdMT is taken up largely by kidneys (Dorian et al., 1992; Sudo et al., 1994; Liu et al., 1998), dietary Cd is targeted to the kidney if it is absorbed as CdMT. Higher Cd levels in kidney than liver seen in human autopsy studies are consistent with this view of Cd absorption (reviewed in Satarug, 2018). In environmental exposure to low-level Cd conditions, initial uptake by the liver is also compatible with the autopsy observation if enough time has passed. In contrast, inhalation of CdO in dust and fume, especially in workplace exposure settings produced high Cd levels in both liver and kidneys; hepatic and renal Cd levels of workers were 42.3 and 110 µg/g, respectively (Ellis et al., 1984, 1985).

A direct accumulation of exogenous CdMT in the kidneys was evidenced from studies showing a relatively higher Cd accumulation in the kidneys in rats fed diets containing CdMT than in those fed with $CdCl_2$ in an equal molar of Cd as CdMT (Groten et al., 1991a, 1994). However, renal accumulation via redistribution of Cd from the liver was lower in CdMT fed rats than $CdCl_2$ fed rats (Groten et al., 1991a). In another rat study, a higher concentration of Cd in the proximal convoluted tubules was observed after CdMT than after $CdCl_2$ injection (Dorian et al. 1995).

If intestinal MT plays an important role in dietary Cd absorption and renal accumulation, then variability in Cd absorption may be due to variable capacities of different foods to promote expression of MT (Elsenhans et al. 1992, 1997). In rats, intestinal MT expression was induced by Cd^{2+}, Zn^{2+}, and Cu^{2+} but not by nickel or lead (Tandon et al., 1993). Dietary flavonoids interacted with copper and iron to influence expression of MT (Kuo et al., 1998). Quercetin decreased MT levels in cultured human intestinal cells, and genistein and biochanin increased those levels (Kuo et al., 1998).

14.3.3.3 Metals and Other Elements in the Diet

In mice, a diet deficient in iron or calcium led to much greater renal Cd accumulation than a diet deficient in copper, zinc, or manganese (Min et al., 2015). Renal Cd accumulation levels strongly correlated with intestinal expression of calcium transporter 1 (CaT1) and metallothionein isoform1 (MT1). Younger mice accumulated more Cd in kidneys than older mice, when fed with a calcium-deficient diet. In addition, an iron-deficient diet caused a greater increase in hepatic Cd accumulation than any other diet tested.

Calcium, zinc, iron, and fiber levels in the diet had a strong influence on Cd absorption and renal accumulation (Wing, 1993; Rimbach and Pallauf, 1997; Brzoska and Moniuszko-Jakoniuk, 1998; Lind et al., 1998). A 70%–80% reduction in Cd accumulation in the liver and kidneys was observed in rats after 8 weeks of Cd-containing food supplemented with calcium, phosphorus iron, and zinc (Groten et al., 1991b). The reduction of Cd accumulation was attributed to the presence of iron (Fe^{2+}). Addition of vitamin C improved iron uptake but did not decrease Cd accumulation. Iron in combinations with calcium, phosphorus, and zinc gave the most pronounced effect.

In experimental animals, a diet high in phytate (fiber) enhanced absorption and renal retention of Cd (Rimbach and Pallauf, 1997). Such effects require further study because they could endanger humans with high fiber intake. Supplementation of food commodities with citric acid, which is a common practice, enhances the bioavailability of essential metals such as zinc and may also result in increased absorption of Cd (Walter et al., 1998).

The presence of flavonoids and curcumin in the diet may reduce Cd absorption by the formation of complexes with Cd (Kukongviriyapan et al., 2016; Li et al., 2017). A study in mice showed that Cd administration with curcumin resulted in lower Cd levels in blood and organs (liver, kidney) than in mice that received Cd only (Kukongviriyapan et al., 2014; Li et al., 2017). Regular consumption of curcumin was linked to lowered blood Cd and lowered risk of hypertension in a Korean study (Choi et al., 2018).

14.3.4 Determinants of Body Burden of Cadmium

14.3.4.1 Low Body Iron Stores

Low body iron stores and iron deficiency are common globally, especially in children and women of child-bearing age. The prevalence of iron deficiency among Australian women aged 25–49 years was 20.3% (Ahmed et al., 2008). In a Korean study, the percentages of iron deficiency in adolescents and in women aged 19–49 years were 36.5% and 32.7%, respectively (Suh et al., 2016). Women with iron deficiency had higher blood Cd (1.53 µg/L) than those with normal iron status (1.03 µg/L).

In a Thai study, women with low body iron stores, defined by serum ferritin <20 µg/L, had threefold higher urinary Cd than women of the same age with adequate body iron status (Satarug et al., 2004). However, an effect of body iron stores on body burden of Cd was not observed in Japanese women (Horiguchi et al., 2004). In this population, exogenous Cd exposure may have been so high that intestinal absorption was not affected by iron status. Swedish women had higher blood Cd than men (Olsen et al., 2012). Non-smoking Norwegian women with low body iron stores had 1.42-fold higher blood Cd (0.37 µg/L) than similarly aged women with normal body iron stores; 26% of those with iron deficiency had high blood Cd levels (Meltzer et al., 2010).

In a Swedish study, women with low iron stores who consumed the diet high in shellfish (a high-Cd food) had 63% higher blood Cd and 24% higher urinary Cd than those with low body iron stores and mixed diets. These findings indicated that Cd in shellfish was bioavailable (Vahter et al., 1996).

Data from a study of Swedish kidney transplant donors showed that the rate of kidney Cd accumulation in non-smokers was 3.9 µg/g kidney wet weight for every 10-year increase in age (Barregard et al., 2010). This rate of kidney Cd accumulation rose to 4.5 µg/g kidney wet weight per 10 years in non-smoking women with low body iron stores. In a Swedish autopsy study, it was noted that most of the individuals with kidney Cd concentrations over 50 µg/g were women (Elinder et al., 1976).

14.3.4.2 The H63D Variant of the HFE Hemochromatosis Gene

The genetic disease hereditary hemochromatosis (HHC) is caused by the H63D variant of the hemochromatosis (HFE) gene. The phenotype is characterized by excessively enhanced absorption of dietary iron and the long-term consequences thereof (Anderson and Frazer, 2017). Hence, people with HHC may absorb also more Cd in the diet. In patients with HHC or iron deficiency, increased expression of the iron-importing protein divalent metal transporter 1 (DMT1) creates greater capacity for absorption of iron and (presumably) Cd. Increased expression of the iron-importing protein DMT1 in people with hemochromatosis or iron deficiency, in general, would provide them with a greater capacity to absorb iron and possibly Cd.

A longitudinal cohort of the U.S. men, aged 51–97 years, known as the Normative Aging Study, linked high toenail Cd levels to homozygous H63D variant alleles of the hemochromatosis (HFE) gene and to low hemoglobin levels (Ciesielski et al., 2018). In addition, lower vitamin C intake levels were associated with higher toenail Cd levels and a higher copy number of H63D variant alleles. These findings suggest that people with the H63D variant alleles or low hemoglobin may absorb more Cd if vitamin C intake levels were low.

14.3.4.3 ZIP8 and ZIP14 Genetic Variants

Blood Cd variation was linked to genetic variants of the zinc transporters ZIP8 and ZIP14 (Rentschler et al., 2014). Argentine Andean women carrying the GT or TT genotype of the rs4872479 ZIP14 variant had 1.25-fold higher blood Cd levels than GG genotype carriers. Also the Andean women carrying the AG or GG genotype of the rs10014145 ZIP8 variant had 1.18-fold higher blood Cd levels than those carrying the CC genotype. These findings implicated ZIP8 and ZIP14 in cellular Cd uptake. Tissues and organs in which these ZIP8 and ZIP14 are highly expressed, including ocular tissue, appear to be at increased risk of Cd toxicity. Cd exposure raises the risk of macular degeneration (MD), and zinc supplementation reduces that risk (Section 14.3.7).

14.3.5 Zinc Intake and Cadmium Toxicity

Data from NHANES 1988–1994 participants aged ≥50 years showed that women had a higher body burden of Cd than men. Median values for urinary Cd in women and men were 0.77 and 0.58 µg/g creatinine, respectively (Lin et al., 2013). Higher urinary Cd was associated with Zn intake levels below recommended dietary allowance (RDA) in both men and women. For both men and women, urinary Cd was associated with an increased mortality from cancer, but Zn intake below RDA was associated with an elevated cancer mortality risk in women only. There was a 1.55-fold increase in cancer mortality risk in women who consumed Zn below RDA, compared with women who met the RDA (Lin et al., 2013).

In another analysis of NHANES 1988–1994 data, environmental Cd exposure and inadequate Zn intake were associated with obstructive lung disease independent of smoking (Lin et al., 2010). The lowest tertile of Zn intake (<8.35 mg/day) was associated with a 1.89-fold increase in risk of obstructive lung disease, compared with the highest Zn intake tertile (>14.4 mg/day). The highest tertile of urinary Cd (>0.79 µg/g creatinine) was associated with increased risk of obstructive lung disease (OR 3.48), compared with the lowest tertile (urinary Cd <0.39 µg/g creatinine) in a model, adjusted for smoking. Thus, an approximately twofold increase in body burden of Cd was associated with an increased risk of obstructive lung disease. Dietary Zn intake of 15 mg/day, which was higher than RDA of 11 mg/day for men and 8 mg/day for women, was required to reduce Cd absorption and limit the body burden of Cd to levels not associated with obstructive lung disease.

An effect of Zn on estimated glomerular filtration rate (eGFR) was observed in an analysis of data from 1545 participants aged ≥20 years in NHANES 2011–2012. In this cohort, 7.5% had eGFR below 60 mL/min/1.73 m^2 (Lin et al., 2014). Blood Cd levels >0.53 µg/L were associated with a 2.04-fold increase in risk of low eGFR, compared with blood Cd levels <0.18 µg/L. For any given blood Cd level, serum Zn levels below 74 µg/dL were associated with a 3.38-fold increase in risk of low eGFR.

14.3.6 Selenium and Cadmium Toxicity

An effect of selenium (Se) on Cd toxicity was observed in a study of Bangladeshi preschool children, aged 4.4–5.4 years (Skröder et al., 2015). The measured Cd effects were kidney volume, determined by ultrasonography, and eGFR calculated from serum cystatin C levels. Urinary Cd levels were inversely associated with eGFR, especially in girls. Among the girls, an increment of 0.5 µg/L in urinary Cd was associated with a decrease in eGFR of 2.6 mL/min/1.73 m^2. The association between lower eGFR values and higher urinary Cd levels was stronger in those with urinary Se below 12.6 µg/L. Kidney volume showed also an inverse relationship to urinary Cd excretion

(Skröder et al., 2015). These findings linked Se to Cd toxicity in kidneys in children. As discussed in Section 14.3.2, if uptake of Cd and Se both were mediated by the zinc transporter ZIP8, reduced Cd toxicity in girls with urinary Se levels >12.6 µg/L was probably due to an inhibitory effect of Se on uptake of Cd.

A beneficial effect of Se was suggested in a Chinese case-control study that included 240 invasive breast cancer cases and 246 age-matched non-cancer controls (Wei et al., 2015). There was a 2.83-fold increase in breast cancer risk in women with urinary Cd in the highest tertile and urinary Se in the lowest tertile (Wei et al., 2015). The risk of breast cancer was also reduced in women with urinary Se in the middle tertile (Wei et al., 2015).

14.3.7 Zinc and Cadmium Toxicity in Ocular Tissues

Zinc (Zn) is present in relatively high concentrations in ocular tissues, especially in the retinal pigment epithelium (RPE) cells (Wills et al. 2008a; Ugarte et al., 2013; Ugarte and Osborne, 2014). Zinc transporters, notably ZIP8 and ZIP14, prominently expressed in RPE cells, mediate Zn uptake by RPE cells (Leung et al., 2008). However, these ZIP8 and ZIP14 are known to also mediate Cd uptake (Fujishiro et al., 2012; Schneider et al., 2014; Zhang et al., 2016), thereby accounting for Cd accumulation in RPE (Erie et al. 2005; Wills et al. 2008b).

Evidence that Cd may be involved in the pathogenesis of eye disease, such as MD comes from autopsy and donor studies showing accumulation of Cd in eye tissues. In an autopsy study, Cd was found in RPE and choroids in all eyes examined (Erie et al., 2005). A donor study observed that Cd in eyes from older donors (aged ≥50 years) was higher than those from younger donors (aged <55 years) (Wills et al., 2008b).

Data from NHANES 2005–2008 participants aged ≥60 years showed that blood Cd levels ≥0.66 µg/L were associated with a 1.56-fold increase in the risk of MD, compared with blood Cd levels <0.25 µg/L (Wu et al., 2014). Urinary Cd, as low as 0.35 µg/L, was associated with 3.31-fold increase in risk of MD in non-Hispanic whites. Thus, a slight elevation of Cd body burden was associated with an increased risk of MD. Non-Hispanic whites may be at increased risk for this complication. The implications of ocular toxicity for Cd health risk assessment are discussed in Section 14.8.

Ocular Cd toxicity was observed also in Koreans, but at blood Cd levels 2.7-fold higher than in U.S. NHANES. Blood Cd levels ≥1.80 µg/L were associated with an increased risk of MD in Korean men aged >40 years (OR 2.11), compared with blood Cd levels <0.79 µg/L (Kim et al., 2014). In another study of Korean subjects aged ≥40 years, the highest quartile of blood Cd levels was associated with an elevated risk of macular generation (OR 1.92) compared with the lowest quartile (Kim et al., 2016).

Work with the human RPE cell line, ARPE-19, showed that co-exposure with either Zn or Mn reduced Cd uptake and accumulation (Satarug et al., 2008). These findings were consistent with the fact that cellualr uptake of Cd, Zn and Mn was mediated by the same set of metals transporters (Fujishiro et al., 2012; Jenkitkasemwong et al., 2012; Schneider et al., 2014; Zhang et al., 2016; Aydemir and Cousins, 2018). The reduction of Cd uptake by Zn co-exposure provide a plausible explanation for the positive outcomes of high-dose zinc supplement in the Age-Related Eye Disease Study in the U.S. (AREDS, 2001; Chew et al., 2013). The AREDS supplement formulations included high-dose antioxidants [vitamin C (500 mg), vitamin E (400 IU), β-carotene (25 mg, equivalent to vitamin A 25,000 IU)], high-dose zinc (80 mg) with additional 2 mg Cu to prevent Cu deficiency anemia from high-dose zinc (AREDS, 2001).

In a 10-year follow-up of participants in AREDS, no adverse effects were associated with the AREDS formulation (Chew et al., 2013). Notably, there was a reduction in the risk of developing advanced MD (OR 0.66) and the development of moderate vision loss (OR 0.71). Reduced mortality from circulatory diseases was seen in participants, assigned to high-dose Zn supplement.

14.3.8 Summary of Dietary, Genetic, and Nutritional Influences of Cadmium Toxicity

Experimental studies indicated that the Zn transporters, notable ZIP8, and ZIP14, mediates cellular uptake of Se as selenite ($HSeO_3^-$), Fe^{2+}, Co^{2+}, Zn^{2+}, Mn^{2+}, and Cd^{2+}. In theory, cellular Cd uptake is reduced by any of these metals. Increased Cd levels in retinal pigment epithelial cells from eyes of patients with MD were noted, and epidemiologic studies linked elevated urinary and blood Cd levels to increased risk of MD. High-dose Zn supplementation in studies of age-related eye disease delayed progression of MD while reducing vision loss and mortality.

Adequate dietary Zn intake was associated with reduced risk of Cd-related obstructive lung disease and Cd-related CKD. Increased breast cancer risk was associated with elevated urinary Cd in combination with low urinary Se. A preventive effect of Se was noted in a study showing an inverse association between estimated glomerular filtration and urinary Cd. Indicators of low body iron stores were associated with elevations of urinary and blood Cd levels. Enhanced Cd absorption and low vitamin C intake accounted for high toenail Cd in men who carried the hemochromatosis (HFE) gene H63D variant.

These findings suggest that genetic considerations and the homeostasis of several other metals affect the propensity to Cd toxicity.

14.4 RENAL TOXICITY OF CADMIUM

14.4.1 Entry of Cd into the Circulation

Inorganic Cd enters the bloodstream through the lungs and duodenum. After assimilation, most circulating Cd adheres to red blood cells and albumin, but it may also bind to sulfhydryl groups in glutathione (GSH), sulfur-containing amino acids, and MT synthesized in absorbing tissues (Sabolić et al., 2010). In the colon, bacterial degradation of plant matter may liberate CdMT and Cd-phytochelatin complexes (CdPCs) for transcytosis by epithelial cells (Langelueddecke et al., 2014).

Most exogenous Cd traverses the gut and the hepatic portal system before reaching the systemic circulation. If hepatocytes do not remove Cd in the first pass, their exposure to the metal continues because albumin and red blood cells are not filtered by normal glomeruli. Hepatocytes respond to uptake of Cd by synthesizing MT and storing complexes of CdMT. Eventually, cell death liberates the complexes, which reach the kidneys through the circulation (Dudley et al., 1985; Chan et al., 1993). CdMT originating outside the liver is processed exclusively in kidneys (Dorian et al., 1995; Sudo et al., 1994).

14.4.2 Entry of Cd into Tubular Cells

Glomeruli filter Cd if it is bound to MT, GSH, amino acids, or (presumably) PC (Fujita et al., 1993; Sabolić et al., 2010). Segments S1–S3 of the proximal tubule reabsorb most or all filtered Cd (Nomiyama and Foulkes, 1977; Dorian et al., 1992, 1995; Felley-Bosco and Diezi, 1989; Sudo et al., 1994). At the apical brush border membrane (BBM), CdMT undergoes receptor-mediated endocytosis (RME), but the role of megalin:cubulin in this process has been questioned because of the low affinity of the complex for CdMT (Thevenod et al., 2019).

The capacity of the proximal tubule for reabsorption of CdMT is high (Nomiyama and Foulkes, 1977; Dorian et al., 1992, 1995). Other forms of Cd are taken up by multiple channels and transporters in the BBM, and by organic cation transporters in the basolateral membrane (BLM) (Zalups, 2000;

Soodvilai et al., 2011), but CdMT appears to be the principal species reabsorbed in the proximal tubule. A second receptor with greater affinity for CdMT, lipocalin 2, may mediate reabsorption in the distal tubule (Thevenod et al., 2019).

After RME of CdMT, the resulting endosome merges with a lysosome to which CdMT is transferred. MT is degraded within the lysosome, and Cd is transported to the cytoplasm. There, it stimulates synthesis of MT *in situ* and forms complexes of CdMT (Sabolić et al., 2010). The half-life of these complexes is estimated to be 10–30 years (Jarup and Akesson, 2009). In most chronically exposed subjects, the kidneys harbor a substantial majority of total body Cd (reviewed in Satarug, 2018).

14.4.3 Mechanisms of Cd-Induced Cytotoxicity

Because Cd is not a transition metal, it is believed to induce the creation of reactive oxygen species (ROS) by displacing copper (Cu), which is a transition metal, from intracellular MT (Thevenod, 1999, 2003). Electron transfer by Cu creates ROS that injure cell membranes, mitochondria, and transporters, including Na/K ATPase. Cytochrome c from damaged mitochondria amplifies production of ROS (Thevenod, 1999, 2003).

The amount of free Cd in cytoplasm is presumed to determine the toxicity of the metal to tubular cells (Goyer et al., 1989; Prozialeck and Edwards, 2012). A critical concentration is apparently reached more readily with acute and intense than with chronic and gradual exposure (Prozialeck et al., 2007); this difference probably reflects greater availability of MT when Cd accumulates slowly (Min et al. 1987; Liu et al. 1999). Within tubular cells, free Cd rises and falls with total Cd and determines the severity of microanatomic lesions (Goyer et al., 1989). In animals given daily intraperitoneal injections of $CdCl_2$, microscopic tissue injury was evident weeks before functional abnormalities appeared (Goyer et al., 1989). In similar experiments, autophagosomes, apoptosis, and distortion of mitochondrial anatomy became increasingly evident with time (Tanimoto et al., 1993).

Intracellular Cd also undermines the action of cadherins to bind tubular cells to one another and to basement membranes, and thereby disrupts the polarity of cells lining the proximal nephron. If Na/K ATPase is displaced from its normal basolateral location, the lack of access to interstitial fluid destroys the electrochemical gradient for sodium that drives apical co-transport of multiple substances (Prozialeck and Edwards, 2012).

14.4.4 Functional Expressions of Cd-Induced Nephrotoxicity

In humans, functional expressions of Cd tubulopathy include impaired reabsorption of filtered proteins and substances co-transported with sodium; release of intracellular proteins into filtrate; altered homeostasis of trace metals; and nephron loss resulting in reduced GFR.

14.4.4.1 Impaired Reabsorption of Small Filtered Proteins

Small, unbound plasma proteins are filtered freely by glomeruli and reabsorbed by proximal tubules. For several decades, investigators have measured excretion rates or urine concentrations of such proteins to document Cd-induced reabsorptive dysfunction (Peterson et al., 1969; Honda et al., 2010; Kim et al., 2015a). β_2-microglobulin (β_2MG) and retinol-binding protein (RBP) have been assayed most frequently.

β_2MG, molecular weight 11,000, is synthesized and shed by all nucleated cells (Argyropoulos et al., 2017). The protein is not bound to other carriers. Its glomerular sieving coefficient (GSC) is approximately 0.9 (Norden et al., 2001), and proximal tubular cells reabsorb >99% of the filtered protein (Portman et al., 1986). Because β_2MG is unstable in acid urine, specimens must be brought to neutral pH for testing. A tubular maximum (T_m) for β_2MG is demonstrable with sufficiently rapid intravenous administration (Gauthier et al., 1984). Cellular intoxication by Cd appears to reduce the T_m.

β_2MG was discovered in the urine of a patient with Cd nephropathy, and its utility for identification of tubulointerstitial (TI) disease was quickly confirmed (Peterson et al., 1969). In the study of Peterson and colleagues, the excretion rate of β_2MG $\left(E_{\beta_2 MG}\right)$ and the ratio of $[\beta_2 MG]_u$ to $[\text{albumin}]_u$ were much higher in TI than in glomerular disease. In animals exposed to Cd, Piscator and colleagues documented the coincidence of excessive $E_{\beta_2 MG}$ with normal $[\beta_2 MG]_p$ (Piscator et al., 1981); because excretion of β_2MG increased while filtration remained stable, reabsorption necessarily declined.

Retinol-binding protein 4 (RBP4) is synthesized in the liver. Its molecular weight is approximately 21,000, and most circulating RBP4 is bound to transthyretin. This complex is minimally filterable because its molecular weight exceeds 70,000. In contrast, unbound RPB4 is filtered, reabsorbed, and metabolized by proximal tubular cells, and its excessive appearance in urine is an early indicator of reabsorptive dysfunction. RBP4 is more stable in urine than β_2MG, and it survives freezing and thawing. Both RBP4 and metabolic products are measured with the urine assay (Norden et al., 2014).

Although $E_{\beta_2 MG}$ and E_{RBP4} identify tubulopathies, the parameters share attributes that compromise their suitability for this purpose. β_2MG production rises in response to many inflammatory and neoplastic conditions (Forman, 1982); conversely, $[RBP4]_p$ may fall in such conditions because the protein is a negative acute-phase reactant (Norden et al., 2014). If reabsorption rates of β_2MG and RBP4 *per nephron* remain constant as production rates of these substances change, excretion will vary directly with production. If production and reabsorption per nephron remain constant as nephrons are lost, excretion of the indicators will rise. Several studies of Cd-intoxicated subjects have demonstrated inverse relationships between GFR and $E_{\beta_2 MG}$ or E_{RBP}, but to the best of our knowledge, none quantified the individual contributions of low GFR and reduced biomarker reabsorption to increased $E_{\beta_2 MG}$ or E_{RBP} (Elinder et al., 1985; Jarup et al., 1995; Portman et al., 1986; Bernard et al., 1988; Satarug et al., 2019a).

14.4.4.2 *Release of Intracellular Proteins into Glomerular Filtrate*

A second category of markers for investigating Cd nephropathy consists of proteins synthesized in tubular cells and released to filtrate. Because their high molecular weights preclude filtration by glomeruli, their excessive appearance in urine reflects cellular injury exclusively (Price, 1992; Prozialeck and Edwards, 2010).

The most commonly employed marker of this type is N-acetyl-β-D-glucosaminidase (NAG). It exists in lysosomes in two isoforms. The first, NAG-A (acidic), is excreted by exocytosis into tubular filtrate. The second, membrane-bound NAG-B (basic), is released into filtrate after tubular cell injury. In a group of Cd-exposed subjects, Bernard and colleagues found a continuous linear relationship between $[NAG-B]_u/[cr]_u$ and $[Cd]_u/[cr]_u$. The authors concluded that NAG was a sensitive indicator of Cd-induced injury and suggested that any cellular content of Cd might inflict injury (Bernard et al., 1995). Many investigators have demonstrated correlations among $[NAG]_u/[cr]_u$, $[Cd]_u/[cr]_u$, and $[\beta_2 MG]_u/[cr]_u$ (Kawada et al., 1989, 1992; Koyama et al., 1992; Jarup et al., 1995; Honda et al., 2010; Swaddiwudhipong et al., 2010; Zhang et al., 2015).

Kidney injury molecule 1 (KIM-1) is a tubule-specific trans-membrane glycoprotein that appears exclusively in response to cell damage. It is not normally detectable. The protein may play a role in reconstitution of cell-to-cell adhesion after an insult (Prozialeck and Edwards, 2012). Appearance of this marker accompanies apoptosis and does not require necrosis (Prozialeck et al., 2009). In animal studies of Cd toxicity, KIM-1 was detectable in urine before MT and CC-16 (a plasma protein handled similarly to β_2MG) (Prozialeck et al., 2007). In humans exposed environmentally to Cd, $[KIM-1]_u/[cr]_u$ was increased before $[\beta_2 MG]_u/[cr]_u$, and $\log([KIM-1]_u/[cr]_u)$ correlated strongly with $\log([Cd]_u/[cr]_u)$ (Pennemans et al., 2011). KIM-1 excretion may provide the clearest early evidence of Cd-induced cell injury.

14.4.4.3 Impaired Reabsorption of Substances Co-Transported with Sodium

Proximal tubular cells reabsorb an array of filtered substances. Several, including glucose, phosphate, urate, and amino acids, are transported with sodium. Another substance, bicarbonate, is reclaimed from filtrate through the mediation of sodium–hydrogen exchanger 3. In the BLM, Na/K ATPase moves 2 K molecules inward as 3 Na molecules are extruded, and the resulting electrochemical Na gradient across the BBM promotes co-transport of Na with other filtered substances (Gonick, 1982; Palmer and Schnermann, 2015). ROS induced by Cd sequestration may undermine this process by compromising cell-to-cell adhesion and secondarily disrupting basolateral-apical polarity (Prozialeck and Edwards, 2012). ROS may also damage Na/K ATPase and apical Na-co-transporters directly (Gonick et al., 1975; Kim et al., 1990; Herak-Kramberger et al., 1996).

In the 1970s, studies of Cd-intoxicated patients in Japan showed reduced reabsorption of substances co-transported with Na (Nogawa et al., 1975; Saito et al., 1977). Collectively, these aberrations constitute the Fanconi syndrome. Cases were concentrated in areas where smokestack emissions or mining detritus had contaminated water and soil. In afflicted subjects, glycosuria and proteinuria were universal. Reduced reabsorption of amino acids, phosphate, and urate occurred in smaller subsets, but tubular bicarbonate reclamation was rarely affected. In periodic examinations after cessation or mitigation of Cd exposure, elements of the Fanconi syndrome tended to persist (Aoshima, 1987). Symptomatic osteomalacia developed in the most severely afflicted patients (Section 14.5.1; Nogawa et al., 1975; Saito et al., 1977; Takebayashi et al., 1987; Noda and Kitigawa, 1990; Noda et al., 1991; Baba et al., 2014).

14.4.4.4 Altered Homeostasis of Zinc and Copper

Proximal tubular cells reabsorb zinc (Zn) and copper (Cu) with specific transporters in the BBM (Moulis, 2010). Because Cd can occupy these transporters, one might anticipate that reabsorption of unbound Cd would reduce reabsorption and increase excretion of Zn and Cu. Many empiric observations suggest otherwise.

In micropuncture/microinjection studies in rats, Cd did not interfere with proximal tubular reabsorption of Zn (Barbier et al., 2004). In animals given a regimen of subcutaneous $CdCl_2$, the renal cortical content of Zn at 12 weeks was no different from that of controls (Prozialeck et al., 2016). In Japanese women severely intoxicated with Cd, $[Zn]_u$ was *lower* than that of unexposed women, and $[Zn]_p$ was normal (Nogawa et al., 1984). In autopsy studies of accident victims, the content of Zn in livers and kidneys rose with that of Cd (Satarug et al., 2001). In at least two clinical studies, $[Zn]_p$ was inversely related to E_{Cd} even though Zn was presumably sequestered in one or more organs (Pizent et al., 2003; Thijs et al., 1992). Taken together, these studies suggest that Cd promotes storage of exogenous Zn in the liver and possibly the kidneys, and thereby reduces endogenous fecal excretion of Zn. Sequestration of Zn appears to normalize or even reduce $[Zn]_p$ even though net intestinal assimilation of the metal is increased.

In the rats given $CdCl_2$ subcutaneously for 12 weeks, renal Cu consistently exceeded that of controls (Prozialeck et al., 2016). In the autopsy study of accident victims, hepatic Cu and Cd rose in tandem, but renal Cu and Cd were unrelated (Satarug et al., 2001). In the aforementioned Japanese women, increased $[Cu]_p$ and $[Cu]_u$ accompanied severe Cd intoxication (Nogawa et al., 1984). Taken together, these observations suggest that acquisition of Cd enhances hepatic storage of Cu; the renal content of Cu may or may not rise, but it does not fall.

None of the foregoing evidence suggests that Cd toxicity reduces the total body content of Zn or Cu. Instead, Zn and Cu accumulate with Cd in one or more organs. To the best of our knowledge, clinically evident deficiencies of Zn or Cu have not been attributed to Cd.

14.4.4.5 *Urinary Excretion of Cd*

The correct interpretation of E_{Cd} is debatable. In theory, Cd appears in urine for one or both of the following reasons: Cd is filtered by glomeruli, and a small, unreabsorbed fraction is excreted; alternatively, Cd is released into filtrate by tubular cells and subsequently excreted (Chaumont et al., 2013). The pathophysiologic significance of E_{Cd} depends on the principal source of the metal.

In chronically exposed subjects, we believe that most or all excreted Cd emanates from tubular cells. In rabbits with established Cd nephropathy, the reabsorptive maximum for filtered CdMT was much greater than any foreseeable filtration rate of Cd in humans (Nomiyama and Foulkes, 1977). In rodents given $CdCl_2$ over several weeks, microscopic tubular cell injury and cortical accumulation of Cd preceded the appearance of Cd in urine, and CdMT was identified in sloughed tubular cells (Tanimoto et al., 1993; Goyer et al., 1989). Intact proximal tubular cells were found to extrude MT-containing vesicles into the lumen (Sabolić et al., 2018). E_{Cd} correlated directly with tissue Cd in autopsied accident victims and human kidney donors (Satarug et al., 2002; Akerstrom et al., 2013). In some populations, E_{Cd} *rose* with GFR in exposed subjects as though the number of nephrons had determined the amount of excreted Cd (Weaver et al., 2011; Jin et al., 2018; Buser et al., 2016). In multiple surveys, E_{NAG} varied directly with E_{Cd} as if both substances had emanated from the intracellular space (Kawada et al., 1989, 1992; Koyama et al., 1992; Bernard et al., 1995; Zhang et al., 2015). In the aggregate, these observations suggest that in chronically intoxicated subjects, most or all excreted Cd is released from tubular cells.

14.4.5 Normalization of Excretion Rates

For decades, it has been customary to normalize the urine concentration of Cd and other biomarkers to that of creatinine ($[x]_u/[cr]_u$). This practice nullifies the effect of urine flow rate (V_u) on concentrations, but it introduces another significant confounder. Although creatinine metabolism is complex, muscle mass is the principal determinant of E_{cr} at any GFR (Heymsfield et al., 1983). Consequently, E_{cr} varies by as much as fourfold among subjects with a given E_x, and E_x/E_{cr} varies commensurately. Because $E_{cr} = [cr]_u V_u$, and $E_x = [x]_u V_u$, the ratio $[x]_u/[cr]_u$ incorporates the same imprecision (Heymsfield et al., 1983; Barr et al., 2004; Chaumont et al., 2013; Jenny-Burri et al., 2015).

To address this issue, we introduced the practice of normalizing E_x to creatinine clearance (C_{cr}), an approximation of GFR, in studies of Cd nephrotoxicity (Satarug et al., 2018b). Because E_x/C_{cr} quantifies the amount of x excreted per volume of filtrate, it also depicts the amount excreted per surviving nephron. Since $C_{cr} = E_{cr}/[cr]_p$, E_{cr} and $[cr]_p$ rise or fall by the same factor at a given C_{cr}; consequently, normalization of E_x to C_{cr} circumvents the effect of muscle mass on E_x/E_{cr} and $[x]_u/[cr]_u$. For any x, $E_x/C_{cr} = [x]_u[cr]_p/[cr]_u$ (see Appendix); in practice, therefore, calculation of E_x/C_{cr} consists of multiplying $[x]_u/[cr]_u$, the customary expression, by $[cr]_p$. Because E_x/C_{cr} retains the ratio $[x]_u/[cr]_u$, it preserves the desired nullification of V_u as a confounder of $[x]_u$; simultaneously, E_x/C_{cr} circumvents the effect of muscle mass on $[x]_u/[cr]_u$.

In addition to these benefits, E_x/C_{cr} addresses the scenario in which E_x is proportionately more reduced than E_{cr} as GFR declines. In theory, if substance x is released into filtrate from tubular cells, then E_x may fall in tandem with the number of nephrons. If E_{cr} does not fall proportionately, then E_x/E_{cr} and $[x]_u/[cr]_u$ understate the degree of injury per surviving nephron. Since C_{cr} varies directly with nephron number, E_x/C_{cr} relates injury more precisely to surviving nephron mass.

We recently used calculations of E_x/C_{cr} to conduct a quantitative analysis of Cd-induced nephron loss (Satarug et al., 2019b). In three subsets of Thai subjects defined by severity of Cd exposure, estimated GFR (eGFR) was inversely related to both E_{Cd}/C_{cr} and E_{NAG}/C_{cr}, and the two ratios were directly related to each other. We could not reconcile the evident biologic link between E_{Cd}/C_{cr} and E_{NAG}/C_{cr} with the premise that Cd had been filtered but not reabsorbed. In contrast, the inference that both Cd and NAG had emanated from tubular cells seemed logical and consistent with our data. We argued that eGFR, E_{Cd}/C_{cr}, and E_{NAG}/C_{cr} were interrelated because sequestered Cd was their common determinant.

14.4.6 Cd Excretion and Glomerular Filtration Rate

A paradox is evident in reported relationships of GFR to environmental Cd exposure. Some investigators found that E_{Cd} *rose* with GFR when exposure was low (Weaver et al., 2011; Jin et al., 2018; Buser et al., 2016). However, many investigators associated tubular dysfunction with low environmental exposure (Roels et al., 1993; Bernard et al., 1995; Jarup et al., 2000; Wallin et al., 2014), and at least two groups found that GFR fell from normal values as E_{Cd} rose minimally (Akesson et al., 2005; Satarug et al., 2018a,b). To reconcile these observations, we speculate that Cd nephropathy begins with a transitory phase in which cell injury is releasing Cd to filtrate but has not yet led to cell death; during that phase, the number of nephrons determines E_{Cd}. As Cd begins to destroy cells, E_{Cd} increases further even though nephrons drop out and GFR begins to decline.

A large body of work shows that GFR fell as a consequence of intense occupational or environmental exposure to Cd. Nephron loss was most extreme in polluted regions of Japan (Saito et al., 1977), but it was also documented in other Asian countries and in Europe. Progression of CKD often continued after cessation of exogenous exposure (Roels et al., 1989; Jarup et al., 1995; Swaddiwudhipong et al., 2012); this evolution may have resulted from the continued release of hepatic CdMT to the kidneys, or more ominously, it may have been due to continuous injury by a stable burden of intracellular Cd (Satarug and Moore, 2004). It is possible that nitric oxide generated within cells displaces Cd and Cu from MT and thereby facilitates ongoing injury (Misra et al., 1996; Satarug et al., 2000b).

Another point warrants emphasis. Reductions in GFR due to Cd nephropathy are sometimes attributed to glomerular injury. Although this inference may be at least partially correct, it is not necessary. Sufficient tubular injury disables glomerular filtration and ultimately leads to nephron atrophy, glomerulosclerosis, and interstitial inflammation and fibrosis (Schnaper, 2017).

14.4.7 Summary of the Renal Toxicity of Cd

To summarize this section on the kidney, we propose the following evolution of Cd nephropathy. Cd enters the circulation through the lungs and gut. Ionized Cd is absorbed by metal transporters in the duodenum, and protein-bound Cd may be absorbed by transcytosis in the colon. Most assimilated Cd passes through the liver, where hepatocytes create and store complexes of CdMT. These complexes are released over time to the kidneys, where they are filtered and internalized by proximal tubular cells. Cd is separated from MT in lysosomes and rejoined to synthesized MT in cytoplasm.

Cd that eludes intracellular complexation promotes synthesis of ROS that inflict injury. That injury induces autophagy, apoptosis, and necrosis of tubular cells, and it undermines adhesion of cells to one another. Cellular injury also leads to the release of proteins and CdMT into filtrate; compromises reabsorption of filtered proteins and substances co-transported with sodium; and ultimately reduces GFR through destruction of nephrons. At present, no treatments exist for mitigation of renal Cd toxicity or effective removal of Cd from tubular cells. Commonsense therapeutic measures include cessation of environmental exposure and control of hypertension and diabetes.

14.5 SKELETAL TOXICITY OF CD

The skeleton remodels itself in isolated bone multicellular units (BMUs) through sequential resorption by osteoclasts, production of osteoid matrix by osteoblasts, and mineralization of matrix with hydroxyapatite (Raisz, 2005). Although remodeling occurs in both cortices and trabecules, it is conventionally analyzed with histomorphometry of undecalcified sections of trabecular bone (Kulak and Dempster, 2010). Disorders of remodeling, the so-called metabolic bone diseases, include osteomalacia, osteoporosis, and osteitis fibrosa (OF). In osteomalacia, normal matrix production is followed by defective mineralization. Sequelae include weakness, generalized aching, and painful unicortical fractures (Looser's zones) (Fukumoto et al., 2015).

Osteoporosis results from remodeling in which resorption consistently exceeds formation. Newly formed bone is normally mineralized, but skeletal mass declines progressively. Trabecules are eventually interrupted, cortices are thinned, and bone strength suffers. Vertebral crush factors cause loss of height, and trauma leads to appendicular fractures, especially of the wrist and hip (Lorentzon and Cummings, 2015).

OF is the skeletal lesion of primary or secondary hyperparathyroidism. Because the activity of all bone cells is increased, the microscopic picture is dominated by osteoclastic lacunae and woven (not lamellar), incompletely mineralized bone (Kulak and Dempster, 2010). OF is not typically associated with Cd intoxication, and we do not consider it further in this chapter.

14.5.1 Cd-Induced Osteomalacia

In broad categories, osteomalacia results from vitamin D deficiency (insufficient intestinal calcium absorption), hypophosphatemia, metabolic acidosis, and toxic skeletal effects of metals and drugs. Each of these insults interferes with normal mineralization of available osteoid (Fukumoto et al., 2015).

In the 1970s and 1980s, aluminum (Al) caused osteomalacia in hemodialysis patients through direct skeletal toxicity (Ott et al., 1982), and its presence in bone was documented with positive Aluminon staining at mineralization fronts. Less commonly, iron was demonstrated in addition to or instead of aluminum (Pierides, 1984; Phelps et al., 1988). Although visible metal was commonly viewed as the inhibitor of mineralization, a significant fraction of thickened osteoid seams failed to show Al or Fe, and chelation therapy with deferoxamine occasionally restored mineralization despite the persistence of stainable metal in reconstituted bone (Ott et al., 1986; Rapoport et al., 1987).

Several groups found diffuse osteomalacia in bone from autopsied victims of itai-itai disease ("ouch-ouch" disease, IID), a syndrome associated in selected regions of Japan with extreme Cd toxicity (Takebayashi et al., 1987; Noda and Kitagawa, 1990; Noda et al., 1991; Baba et al., 2014). In the study of Noda et al., approximately half of mineralization fronts showed Fe by histochemical staining and X-ray microanalysis. Al was not found with either technique. The skeletal content of Cd was 4.5× higher than in control subjects, but Cd did not co-locate with Fe. Lesions of hemochromatosis were absent, and some subjects were also free of hemosiderin deposits. Other investigators demonstrated Fe at mineralization fronts in rats and monkeys given Cd (Hiratsuka et al., 1997; Kurata et al., 2014).

Although this finding is unexplained, we offer the following speculation. Through the creation of ROS and damage to mitochondria, Cd may ultimately liberate Fe from heme compounds. In turn, Fe may amplify the injury induced by ROS. In osteocytes, that injury may disrupt normal calcium and phosphate transport to mineralization fronts and allow Fe to accumulate at those sites. Because half of mineralization fronts did not show Fe staining in IID (Noda et al., 1991), osteoblasts and osteocytes are the likely foci of Cd toxicity.

The tubulopathy of Cd intoxication also leads to aberrations that may cause osteomalacia (Fukumoto et al., 2015). Proximal renal tubular acidosis compromises skeletal mineralization (Phelps et al., 1986), but in patients with IID, the plasma bicarbonate concentration was typically normal when reabsorption of other substances was compromised (Saito et al., 1977). Hypophosphatemia causes osteomalacia in many disorders of tubular phosphate handling (Wagner et al., 2014); Cd-induced osteomalacia has also been attributed to hypophosphatemia (Takebayashi et al., 2000), but the lesion also coincided with normophosphatemia in some patients with IID (Saito et al., 1977).

In theory, Cd toxicity could disrupt vitamin D metabolism and compromise skeletal mineralization secondarily. The proximal tubule synthesizes the active metabolite of vitamin D, 1,25-dihydroxyvitamin D (1,25(OH)$_2$D), from the hepatic precursor 25-hydroxyvitamin D (25OHD), which tubular cells acquire through megalin:cubilin-mediated endocytosis of vitamin D-binding protein (VDBP) (Kaseda et al., 2011). Although Cd interferes with reabsorption of VDBP (Uchida et al., 2007), [25OHD]$_p$ was nevertheless normal in a simian model of Cd nephropathy (Kurata et al., 2014). [1,25(OH)$_2$D]$_p$ was reduced in patients with Cd intoxication (Nogawa et al., 1987), but nephron loss may produce this result, and [1,25(OH)$_2$D]$_p$ correlated with GFR in a second study (Aoshima and Kasuya, 1991). In the third examination of this issue, [1,25(OH)2D]$_p$ was not low enough to cause osteomalacia in subjects with Cd tubulopathy (Kaewnate et al. 2012). In general, it is not clear that the metal inhibits 1α-hydroxylation of vitamin D. The pathogenesis of Cd-induced osteomalacia is multifactorial, and the principal cause may vary from patient to patient.

14.5.2 Cd-Induced Osteoporosis

In clinical practice, skeletal mineral is quantified by measurement of bone mineral density (BMD) with dual-energy X-ray absorptiometry (DEXA). BMD is reported in units of g/cm^2. For purposes of interpretation, BMD is most commonly related to the mean value of young, normal volunteers (t-score). A BMD between 1 and 2.5 standard deviations below the reference mean (t-score between −1 and −2.5) is labeled osteopenia, and a value >2.5 SD below that mean (t-score < −2.5) is labeled osteoporosis. In the absence of a recognizable cause of osteomalacia, a reduction in BMD is assumed to reflect resorption in excess of formation of normally mineralized bone (Lorentzon and Cummings, 2015).

Osteopenia and osteoporosis are highly prevalent among populations with all degrees of Cd exposure (Staessen et al., 1999; Alfven et al., 2000; Wang et al., 2003; Jarup and Alfven, 2004; Akesson et al., 2006; Gallagher et al., 2004; Narwot et al., 2010; Wu et al., 2010). Simultaneous renal manifestations are typically limited to modest tubular proteinuria. All of these studies have demonstrated statistical relationships between BMD and E_{Cd}, an increased prevalence of osteoporosis as E_{Cd} rises, or both findings. In multiple studies, calciuria also rose with E_{Cd}, and [PTH] simultaneously fell (Akesson et al., 2006; Schutte et al., 2008; Ibrahim et al., 2016). Calciuria was therefore a direct result of skeletal toxicity and not a secondary consequence of tubular toxicity of Cd. Reported cases in which hypercalciuria and stones accompanied other manifestations of Cd tubulopathy may be exceptions to this general rule (Kazantzis, 1979; Kaewnate et al., 2012).

The biologic basis for Cd-induced osteoporosis is well understood. In osteoblast cultures, either Ca or Mg blocked cellular uptake of Cd, and the Ca-Mg channel TRPM7 was identified as the likely avenue of Cd uptake (Levesque et al., 2008). In a separate study, exposure of osteoblasts to Cd induced apoptosis (Coonse et al., 2007). In cultures of marrow constituents, Cd enhanced the formation of multinucleated osteoclasts from macrophage precursors, and the extent of excavation of trabecular surfaces increased accordingly (Wilson et al., 1996). In neonatal mouse calvaria, Cd stimulated Ca release from bone by a mechanism that required carbonic anhydrase and prostaglandin E2 (Carlsson and Lundholm, 1996). Taken together, this work suggests that Cd promotes resorption and impedes formation of bone.

14.5.3 Summary of the Skeletal Toxicity of Cd

Cd causes two clinically significant skeletal lesions. Osteomalacia, a disorder of mineralization of newly formed bone, is an expression of severe Cd toxicity and is now uncommon. It can be explained as a consequence of altered bone cell function, but it may also be abetted by multiple consequences of Cd-induced tubulopathy. The osteopenia–osteoporosis continuum results from a failure to replace resorbed bone completely in the process of remodeling. It complicates all degrees of environmental Cd exposure and is highly prevalent wherever that exposure occurs. It is readily diagnosed with bone densitometry.

14.6 CADMIUM EXPOSURE, CHRONIC KIDNEY DISEASE, AND MORTALITY

Impaired tubular reabsorption, indicated by elevated urinary β_2MG excretion levels, is the most frequently reported effect of environmental exposure to Cd (see Section 14.2.3). This sign of Cd toxicity has long been dismissed and deemed not to be of clinical relevance. However, increasing evidence links Cd-induced tubulopathy to the pathogenesis and progression of CKD (Swaddiwudhipong et al., 2012, 2015; Satarug et al., 2018a,b, 2019a; see Section 14.4.6). CKD is defined as estimated glomerular filtration rate (eGFR) ≤ 60 mL/min/1.73 m^2 or urinary albumin to creatinine ratio above 30 mg/g for at least 3 months (Glassock et al., 2017).

CKD is a cause of morbidity and mortality and its prevalence continues to rise globally (De Nicola and Zoccali, 2016; Glassock et al., 2017). The relationship between Cd-induced tubulopathy and GFR has been discussed in Section 14.4.6. The present section summarizes data linking urinary Cd excretion to CKD and mortality.

14.6.1 Environmental Cadmium Exposure and CKD

14.6.1.1 Data from the United States

In NHANES 1999–2006, the prevalence of CKD in adult participants with a normal blood pressure range was 13.4% (Crews et al., 2010). The prevalence of CKD rose to 17.3%, 22%, and 27.5% in those with prehypertension, undiagnosed hypertension, and diagnosed hypertension, respectively. Table 14.1 provides a summary of data from the U.S. NHANES that link chronic Cd exposure to an increased risk of CKD even when exposure levels are relatively low. In NHANES 1999–2006, urinary Cd levels ≥ 1 µg/L were associated with 1.41- and 1.48-fold increases in risk of albuminuria

Table 14.1   Urinary and Blood Cadmium Levels Associated with Adverse Effects on Kidneys

NHANES Cycle Reference	Findings
1999–2006, n 5,426, aged ≥20 years, Ferraro et al. (2010).	Urinary Cd levels ≥1 µg/L were associated with increased risk of albuminuria[a] (OR 1.41) and low GFR[b] (OR 1.48).
1999–2006 n 14,778, aged ≥20 years, Navas-Acien et al. (2009).	Blood Cd levels ≥0.6 µg/L were associated with increased risk of albuminuria[a] (OR 1.92), low GFR[b] (OR 1.32) and low GFR plus albuminuria (OR 2.91).
2007–2012 n 2,926, aged ≥20 years, Madrigal et al. (2019).	Blood Cd levels >0.61 µg/L were associated with increased risk of low GFR (OR 1.80) and albuminuria (OR 1.60). Women with diabetes and hypertension plus blood Cd levels >0.61 had 4.9 mL/min/1.73 m² lower GFR than non-diabetic, normotensive women who had blood Cd <0.21 µg/L. Women with hypertension and blood Cd levels >0.61 µg/L had 5.77 mL/min/1.73 m² lower GFR than those with normal blood pressure and blood Cd <0.21 µg/L.
2011–2012 n 1,545, aged ≥20 years, Lin et al. (2014).	Blood Cd levels >0.53 µg/L were associated with increased risk of albuminuria (OR 2.04) and low GFR (OR 2.21).
2009–2012, n 2,926, aged ≥20 years, Zhu et al. (2019).	Urinary Cd levels >0.220 µg/L were associated with elevated urinary albumin levels, compared with urinary Cd <0.126 µg/L. Blood Cd levels >0.349 µg/L associated with elevated urinary albumin excretion, compared with blood Cd <0.243 µg/L.

NHANES, National Health and Nutrition Examination Survey; n, sample size; OR, odds ratio; HR, hazard ratio.
[a]   Urinary albumin to creatinine ratio ≥30 mg/g creatinine.
[b]   Estimated glomerular filtration rate (eGFR) <60 mL/min/1.73 m².

and GFR less than 60 mL/min/1.73m², respectively (Ferraro et al., 2010). Blood Cd levels ≥0.6 µg/L were associated with 1.92-, 1.53-, and 2.91-fold increases in risk of albuminuria, low GFR and albuminuria plus low GFR, respectively (Navas-Acien et al., 2009).

Associations of environmental Cd exposure with CKD were replicated in other NHANES datasets. In NHANES 2009–2012, elevated urinary albumin was associated with urinary Cd >0.220 µg/L, compared with urinary Cd <0.126 µg/L, but albuminuria was not associated with Pb or Hg excretion (Zhu et al., 2019). In NHANES 2011–2012, blood Cd levels >0.53 µg/L were associated with 2.04- and 2.21-fold increases in risk of albuminuria and low GFR, respectively (Lin et al., 2014).

In NHANES 2007–2012, blood Cd levels >0.61 µg/L were associated, respectively, with 1.60- and 1.80-fold increases in risk of albuminuria and low GFR when compared to blood Cd levels ≤0.11 µg/L (Madrigal et al., 2019). In this NHANES, women with diabetes and hypertension were more susceptible than normotensive, non-diabetic subjects to effects of Cd on GFR. On average, women with diabetes, hypertension, and blood Cd quartile 4 (0.61–9.3 µg/L) had eGFR 4.9 mL/min/1.73 m² lower than non-diabetic, normotensive women, who had the lowest blood Cd (0.11–0.21 µg/L). In those women with hypertension and the highest blood Cd quartile, the mean eGFR was 5.77 mL/min/1.73 m² lower than in normotensive with the lowest blood Cd quartile.

14.6.1.2 Data from Other Populations

Urinary Cd >0.8 µg/g creatinine was associated with a significant GFR reduction in Swedish women 53–64 years of age, and urinary Cd >0.60 µg/g creatinine was associated with tubular injury (Akesson et al., 2005). In a Chinese population study (n = 8,429), the highest quartile of estimated cumulative Cd intake was associated with a 4.05-fold increase in the prevalence of CKD when compared to the lowest quartile (Shi et al., 2017).

In a Korean study, an inverse association between blood Cd and GFR was observed in women only; blood Cd levels in the highest tertile were associated with 1.85 mL/min/1.73 m² lower GFR values, compared with the lowest tertile (Hwangbo et al., 2011). In another Korean study, elevated blood Cd levels were associated in increased CKD prevalence, especially in those with diabetes or hypertension (Kim et al., 2015b). Blood Pb and Hg levels did not show the same relationship. The highest quartile of blood Cd was associated with 1.97-fold increase in risk for low GFR in women, compared with the lowest quartile. The same relationship was not evident in men (Myong et al., 2012).

An inverse association of urinary Cd and eGFR was observed together with a positive association between urinary Cd and urinary β_2MG in residents of an area of Thailand affected by Cd pollution (Satarug et al., 2018a,b; 2019). In line with the U.S. and Korean data, Thai data indicated subjects with hypertension were more susceptible than normotensive subjects to effects of Cd on GFR; subjects with hypertension and the highest quartile of urinary Cd had 12 and 17 mL/min/1.73 m² lower eGFR values than hypertensive and normotensive subjects with the lowest urinary Cd quartile, respectively (Satarug et al., 2018b).

14.6.2 Increased Susceptibility to Renal Cadmium Toxicity in Diabetes

The Cadmibel study showed that people with diabetes were more susceptible than those with no diabetes to renal Cd toxicity (Buchet et al., 1990). A similar observation was subsequently made in women from Sweden (Akesson et al., 2005; Barregard et al., 2014), the Torres Strait, Australia (Haswell-Elkins et al., 2008), Korea (Hwangbo et al., 2011), and the United States (Madrigal et al., 2019). Similarly, experimental studies have shown that renal manifestations of diabetes and Cd toxicity are enhanced when both the metal and the disease are present. Injection of CdMT resulted in increased urinary excretion of proteins and calcium in both diabetic obese mice and non-obese littermate controls (Jin et al. 1994). However, in the diabetic mice, the dose of CdMT required to induce proteinuria and calciuria was one-fourth of that required in controls. CdMT induced renal glycosuria in both groups.

Hereditary diabetic Chinese hamsters also appeared to be more susceptible to Cd renal toxicity (Jin and Frankel, 1996). In type I diabetic rats, kidney Cd concentrations of 10–40 µg Cd/g wet weight were found to increase excretion rates of albumin, transferrin, and immunoglobulin (Bernard et al., 1991). However, none of the usual manifestations of tubular toxicity were seen at these renal concentrations of the metal. Hence, a renal Cd burden lower than the recognized threshold for toxicity in humans (less than 50 µg/g wet weight) may enhance the development of diabetic renal complications without concurrent tubular toxicity. The mechanisms underlying the involvement of Cd in the development and progression of diabetic nephropathy needs further study.

Renal morphological and histological changes such as dilatation of interstitial veins with apparent paravenous lymphatic infiltrates were more pronounced in vitamin C-deficient guinea pigs compared with those fed sufficiently with the vitamin (Nagyova et al., 1994). This suggests a role for vitamin C in the prevention of Cd toxicity. People with diabetes are generally known to have marginally reduced vitamin C status (Will et al., 1999; Shim et al., 2010). In NHANES 1988–1994, mean serum vitamin C concentration was lower in persons with newly diagnosed diabetes, compared with those without diabetes. NHANES 1988–1994 showed that vitamin C-deficiency prevalence was 13% and that mean serum vitamin C concentrations in persons with newly diagnosed diabetes were lower than those without diabetes (Will et al., 1999; Schleicher et al. 2009). Likewise, in a study of 2,048 Koreans, aged ≥30 years, lower serum vitamin C levels were observed in both non-smokers and smokers who had diabetes, compared with non-smokers and smokers without diabetes (Shim et al., 2010). In another study, diabetes patients with clinical nephropathy were found to have lower mean plasma vitamin C due to higher renal clearance of vitamin C compared to those with microalbuminuria (Hirsch et al., 1998). Low vitamin C levels may be one of the factors explaining increased susceptibility to renal Cd toxicity among people with diabetes (Buchet et al., 1990; Akesson et al., 2005; Barregard et al., 2014; Haswell-Elkins et al., 2008; Hwangbo et al., 2011; Madrigal et al., 2019).

14.6.3 Environmental Cadmium Exposure and Mortality

This section highlights longitudinal studies that linked lifetime Cd exposure to mortality.

14.6.3.1 The U.S. Experience

The geometric mean (GM) urinary Cd reported for a representative U.S. population was 0.210 µg/g creatinine. The 50th, 75th, 90th, and 95th percentile levels of urinary Cd were 0.208, 0.412, 0.678, 0.949 µg/g creatinine, respectively (Crinnion, 2010). NHANES 1999–2008 showed 94–98% of non-smokers and 96%–99% of smokers, aged 20–85 years, were environmentally exposed to Cd, reflected by their urinary Cd excretion levels (Riederer et al., 2013). Among non-smoking adults without CKD, enrolled in NHANES 1999–2006, the prevalences of urinary Cd >1, >0.7, and >0.5 µg/g creatinine were 1.7%, 4.8%, and 10.8% in men, and 2.5%, 7.1%, and 16% in women (Mortensen et al., 2011). In NHANES 2007–2012, urinary Cd levels suggested that 91.9% of adult participants had been exposed to environmental Cd (Buser et al., 2016). In an analysis of temporal trend over 18 years for environmental exposure to Cd in the United States (NHANES 1988–2006), the mean urinary Cd fell by 29% in men (0.58 vs. 0.41 µg/g creatinine, $P < 0.001$), but not in women (0.71 vs. 0.63 µg/g creatinine, $P = 0.66$) (Ferraro et al., 2012). In addition, this study noted an association between blood Cd and increased cardiovascular mortality, especially in women. The reduction of urinary Cd in men was attributable to lowered smoking prevalence (Tellez-Plaza et al., 2012).

In a longitudinal study of 1988–1994 NHANES participants, urinary Cd levels >0.48 µg/g were associated with a 4.29-fold increase in cancer mortality when compared to urinary levels <0.21 µg/g creatinine (Menke et al., 2009). In men only, a twofold increase in urinary Cd levels was, respectively, associated with 1.28-, 1.55-, 1.21-, and 1.36-fold increases in deaths from all causes, cancer, cardiovascular disease, and coronary heart disease, after adjustment for exposure from cigarette smoking. In this prospective study, the GM for urinary Cd was 0.28 µg/g creatinine in men and 0.40 µg/g creatinine in women (Menke et al., 2009).

Other longitudinal studies of women enrolled on the NHANES 1999–2008 linked urinary Cd levels ≥0.37 to ≥0.65 µg/g to increased risk of breast cancer (Gallagher et al., 2010), increased deaths from heart disease (Tellez-Plaza et al., 2012), increased mortality from lung cancer (Adams et al., 2012; Lin et al., 2013), and increased mortality from liver-related diseases (Hyder et al., 2013).

A meta-analysis of six prospective cohort studies conducted in populations with mean urinary Cd levels ≤1 µg/g demonstrated 1.38-, 1.56-, and 1.50-fold increases in deaths from all causes, cancer, and cardiovascular disease, respectively (Larsson and Wolk, 2016). Increased mortality has been linked to urinary Cd as low as 1 µg/g creatinine. This value is one-fifth of the threshold of 5.24 µg/g creatinine established by the FAO/WHO (Section 14.2.2).

14.6.3.2 The Japan Experience

The Japan nationwide study reported that the mean dietary Cd intake in 30 locations ranged between 12.5 and 70.5 µg/day (Ikeda et al., 2015). Urinary Cd ranged from 1.16 to 11.02 µg/g creatinine, and blood Cd from 0.46 to 3.98 µg/L (Ikeda et al., 2015). The mean urinary Cd excretion recorded in a large study (N = 10,753 women, 35–60 years) was 1.3 µg/g creatinine (range: 0.8–3.2 µg/g creatinine) (Ezaki et al., 2003). The Japan national mean urinary Cd was higher than 1 µg/g creatinine.

A 19-year follow-up of residents in three locations with no known Cd pollution was performed. There were 1.35- and 1.64-fold increases in deaths from all causes in men with urinary Cd 1.96-3.22 and ≥3.23 µg/g creatinine, respectively, when compared to men with urinary Cd <1.14 µg/g creatinine (Suwazono et al., 2014). Likewise, urinary Cd levels ≥4.66 µg/g were associated with 1.49-fold increases in mortality from all causes in women when compared with urinary Cd <1.46 µg/g creatinine (Suwazono et al., 2014).

In an analysis of mortality in an area polluted with Cd, there were 1.57 and 2.40-fold increases in deaths from all causes in men and women who showed evidence of kidney pathologies, notably proteinuria and glycosuria (Maruzeni et al., 2014). In both men and women, there were increased deaths from ischemic heart disease together with an increase in the incidence of diabetes and kidney disease. In a cancer mortality analysis, a 1.49-fold increase in deaths from cancer of all sites was observed, especially in women with the signs of Cd-linked kidney pathologies (proteinuria and glycosuria) at the baseline test 26 years earlier (Nishijo et al., 2018). The specific cancer types were uterus (HR 3.85), kidney and urinary tract (HR 10.1), and kidney (HR 7.71). At the baseline, the median urinary Cd levels in women and men with proteinuria and glycosuria were 8.3 and 10 µg/L, respectively. Paradoxically, in men, the risk of lung cancer and the risk of dying from cancer at any site were reduced (HR 0.53 and 0.79)

14.6.3.3 Other Populations

In a 20-year cohort study of 956 Belgian subjects, selected randomly from low- and high-exposure areas, the mean blood Cd in cohort participants was below <2 µg/L, and the mean urinary Cd excreted was below 2 µg/day (Nawrot et al. 2008). A twofold increase in urinary Cd was associated with 20% and 44% increases in mortality in the low- and high-exposure areas, respectively. Likewise, a twofold increase in blood Cd was associated with 25% and 33% increases in mortality in the low- and high-exposure areas, respectively.

In a 20.2-year follow-up of 2,920 middle-aged Swedish women, blood Cd levels ≥0.69 µg/L were associated with increased mortality risk (HR 2.06), compared with blood Cd levels ≤0.18 µg/L

(Moberg et al., 2017). Another prospective cohort study in Western Australia included 1,359 women, mean age 75.2 years, who had atherosclerotic vascular disease for 14.5 years (Deering et al., 2018). The median urinary Cd among cohort participants was 0.18 μg/L with the 25th and 75th percentile urinary Cd levels being 0.09 and 0.32 μg/L, respectively. A 2.7-fold difference in urinary Cd in the lowest and the highest quartile was associated with a 1.36-fold increase in risk of dying from heart failure in those who never smoked and a 1.17-fold increase in risk of having a heart failure event (Deering et al., 2018).

14.6.4 Summary of Cadmium-Linked CKD and Increased Mortality Risk

Epidemiologic studies in the United States show that a modest kidney burden of Cd, indicated by $[Cd]_u$ as low as 1 μg/L, was associated with an increased prevalence of albuminuria and eGFR less than 60 mL/min/1.73 m^2. Increased albumin excretion was not associated with Pb or Hg excretion, but it was associated with $[Cd]_u$ as low as 0.22 μg/L. Elevated β_2MG excretion has also proven to be a sensitive indicator of Cd-induced reabsorptive dysfunction, but it may also result from loss of nephrons for any reason. Increased susceptibility to Cd toxicity was observed in diabetic people in the United States, Belgium, Sweden, Australia, and Korea. Moreover, increased mortality risk was linked to environmental Cd exposure in the same countries. It is possible that in some parts of the world, environmental exposure to Cd inevitably eventuates in a toxic kidney burden. Public measures are required to reduce such exposure to the lowest achievable level.

14.7 CADMIUM AND HYPERTENSION

Hypertension affects 25%–30% of the adult population in most economically developed countries (Fryar et al., 2017). It is a risk factor for cardiovascular disease and both a cause and a consequence of CKD (Horowitz et al., 2015). This section summarizes effects of environmental Cd exposure on blood pressure increases and the risk of hypertension. Insights from Cd-induced hypertension in experimental animals are highlighted. We also discuss evidence that a second messenger produced in kidneys, 20-hydroxyeicosatetraenoic acid (20-HETE), regulates salt reabsorption, vascular tone, and volume homeostasis in response to renal Cd intoxication.

14.7.1 Epidemiologic Studies on Cadmium-Associated Hypertension

14.7.1.1 Studies in the United States and Canada

Evidence for a role of environmental Cd exposure in the pathogenesis of hypertension comes from a longitudinal study of Native Americans in the Strong Heart Study. In this study, higher urinary Cd excretion at baseline was associated with higher rates of increased systolic and diastolic blood pressure. A 10% increase in risk for hypertension accompanied each one-unit increment in log-transformed urinary Cd (Oliver-Williams et al., 2018).

NHANES 1999–2004 showed a positive association between blood Cd and blood pressure (Tellez-Plaza et al., 2008). The strength of the association was strong in non-smokers, moderate in former smokers, and weak or negligible in current smokers. A similar observation was made in the Canadian Health Measures Survey of 2007–2013, in which blood Cd levels were positively associated with both systolic and diastolic blood pressure levels, but the risk of hypertension was reduced by 52% in women who were current smokers (Garner and Levallois, 2017). This paradox is reminiscent of associated normotension and high-dose Cd exposure in Japanese population studies (Nakagawa and Nishijo, 1996; Kurihara et al., 2004). A similar finding was observed in animals given drinking water with high Cd levels (Section 14.7.3.2). Cd-induced hypertension may be mitigated in smokers by substances that accompany Cd. Tin, for example, has blood pressure-lowering effects (Sacerdoti et al., 1989; Escalante et al., 1991).

A positive association between blood Cd and blood pressure was replicated in NHANES 1999–2006 (Scinicariello et al., 2011). In addition, white and Mexican-American women in NHANES 1999–2006 were found to be more susceptible to apparent effects of Cd on blood pressure than black women; an increased risk of hypertension was seen in Caucasian (OR 1.54) and Mexican-American women (OR 2.38) who had blood Cd levels ≥0.4 µg/L, but not in black women or white, black, or Mexican-American men.

14.7.1.2 Studies in Asia

In a low-exposure area of Thailand, a threefold increase in urinary Cd levels, from 0.39 to 1.12 µg/L, was associated with an 11% increase in risk of hypertension (Satarug et al., 2005). That risk was further increased from 11% to 20% in those with increased excretion of NAG, an indicator of tubular injury (Section 14.4.4.2). All subjects in this study had normal eGFR. This study therefore suggested that Cd exposure may cause hypertension without a concurrent GFR reduction.

In a Chinese case-control study, urinary Cd levels >1.07 µg/L were associated with a 1.33-fold increase in risk of hypertension, compared with urinary Cd levels <0.54 µg/L (Wu et al., 2019). In another case-control study of non-smoking Chinese women, higher average blood pressure was noted in women with untreated essential hypertension who had higher blood and urinary Cd, compared with the age-matched normotensive control and untreated non-essential hypertensive groups who had lower blood and urinary Cd concentrations (Lin et al., 1995).

In a study of residents (71 men and 110 women) of Cd-polluted areas of Southeast China, blood Cd was associated with the prevalence of hypertension, especially in women with GM blood Cd of 3.84 µg/L (Chen et al., 2013). Blood Cd levels of 1–1.7 µg/L were associated with increased risk of hypertension in a study of residents (276 women and 165 men) in other polluted areas of China (Chen et al., 2015).

In the Korean population survey of 2008–2010, blood Cd levels were associated with increases in both systolic blood pressure (SBP) and diastolic blood pressure (DBP) (Lee and Kim, 2012). Doubling blood of Cd from 0.62 to 1.33 µg/L in men and from 0.73 to 1.57 µg/L in women was associated with increased prevalence of pre-hypertension and hypertension.

Under occupational exposure settings, blood Cd, but not blood Pb, showed a positive association with SBP and DBP in men who worked in copper smelting industries (An et al., 2017). Doubling of blood Cd was associated with an average 2.310 and 2.067 mmHg increases in SBP and DBP, respectively. The GM blood Cd and blood Pb in Korean workers were 1.05 µg/L and 5.84 µg/dL, respectively (An et al., 2017). These mean blood Cd and Pb levels were higher than the Korean national mean for blood Cd in men of 0.83 µg/dL and blood lead of 2.44 µg/dL, respectively.

14.7.1.3 Summary of Epidemiologic Findings on Cadmium-Related Hypertension

Most epidemiologic studies relied on blood Cd levels to examine effects of environmental Cd exposure on blood pressure. Like the U.S. data, blood Cd levels in Asian populations were associated with blood pressure increases although blood Cd levels in Asian populations were higher than the U.S. population. The increased risk of hypertension in Chinese subjects with urinary Cd levels >1.07 µg/L was in line with Thai data, in which urinary Cd of 1.12 µg/L was associated with increased hypertension risk (Satarug et al., 2005). The same range of urinary Cd (urinary Cd levels ≥1 µg/L) was associated, respectively, with 41% and 48% increases in risk of albuminuria and low eGFR in the U.S. population (Section 14.6; Ferraro et al., 2010). In the U.S., Canadian, and Chinese studies, a preponderance of Cd-associated hypertension in non-smoking women was noted. There is dearth of studies of the mechanism by which renal Cd raises blood pressure. Experimental evidence suggests that Cd induces hypertension by increasing tubular avidity for filtered sodium (Section 14.7.3).

14.7.2 The Kidney as the Regulator of Blood Pressure

An indispensable role of the kidneys in long-term regulation of systemic blood pressure levels was demonstrated in the 1970s, when the transplant of a kidney from a hypertensive rat to a normotensive host raised blood pressure, and the transplant of a normal kidney to hypertensive host lowered blood pressure (Dahl et al. 1972, 1974; Dahl and Heine 1975; Crowley and Coffman, 2014). A close relationship between renal perfusion pressure, urine flow, and sodium excretion was later described (Guyton, 1991) and led to the proposal that hypertension develops when the pressure-natriuresis response is shifted to higher pressure levels.

Numerous studies with various experimental models of hypertension have linked renal handling of sodium and water to blood pressure levels (McDonough et al., 2003; Yang et al., 2003; Morrison and Mindel, 2004; McDonough, 2010; Rossier et al., 2013). The regulation of sodium transport in kidneys by blood pressure is highly complex, and current knowledge about this phenomenon is fragmentary (Feraille and Dizin, 2016; McDonough and Nguyen, 2015; Gonzalez-Vicente et al., 2019). The eicosanoid 20-HETE has been proposed as one mediator of pressure natriuresis (Williams et al., 2010; Fan and Roman, 2017; Zhang et al., 2018). Nitric oxide and factors that suppress the renin–angiotensin–aldosterone system (RAS) are other postulated mediators (McDonough and Nguyen, 2015; Fan and Roman, 2017).

14.7.3 Insights from Cadmium-Induced Hypertension in Experimental Animals

14.7.3.1 Low-Dose, Lifelong Exposure Experiments

Under well-controlled conditions resembling low-dose, lifelong human exposure, male and female Long–Evans rats were given 5 ppm of cadmium (Cd), lead (Pb), or chromium(III) in drinking water from the time of weaning throughout life. Hypertension developed only in Cd-exposed rats (Schroeder, 1964). Hypertension occurred more frequently in female rats than in males and in those receiving both Cd and sodium chloride (0.1% NaCl). Female rats ingested more Cd than males.

Rats with Cd-induced hypertension showed increased mortality and marked renal vascular changes (Schroeder, 1964), aortic atherosclerosis (Revis et al., 1981), increased sodium retention (Perry and Erlanger, 1981), and reduced sodium excretion and urinary volume (Pena and Iturri, 1993). These findings suggested that lifelong intake of low-level Cd (5 ppm in drinking water) may enhance tubular Na reabsorption. The pathologies seen in the kidney and vascular system of Cd-induced hypertensive animals resembled those seen in some forms of human "essential" hypertension.

The Cd content in kidneys of hypertensive rats ranged between 5 and 50 µg/g wet weight (Perry et al., 1977, 1978, 1979). The average cadmium content of kidneys from cadmium-induced hypertensive ICR mice was 19.75 µg/g kidney wet weight (Kukongviriyapan et al., 2014). These levels were in the same range found in kidneys from non-occupationally exposed Australian subjects in an autopsy study (Satarug et al., 2002), and in biopsies from Swedish kidney transplant donors (Barregard et al., 2010; Wallin et al., 2014).

14.7.3.2 High-Dose, Short-Term Exposure Experiments

High-dose levels of Cd were used in short-term hypertension induction experiments. Results from these short-term, high-dose experiments, detailed below, suggest that hypertension developed when renal Cd accumulation and tubular damage reached a certain level, and that hypertension was sustained if exposure to Cd and tubular destruction continued. In addition, hypertension was intensified and more persistent when Cd was administered with Ni or Pb. Because persistent tubular damage eventually reduces GFR through nephron destruction, data from these high-dose experiments suggest that Cd may induce hypertension by reducing GFR. In contrast, lifelong intake of low-level Cd appeared to increase tubular sodium reabsorption and decreased sodium excretion (Section 14.7.3.1).

In female rats of the Long–Evans strain (Perry et al., 1977), hypertension developed only in the groups exposed to Cd at 1, 2.5, 5, and 10 ppm dose levels. Exposure to Cd at 25 and 50 ppm concentrations appeared to be toxic, as evidenced by substantial weight loss, but hypertension was nevertheless seen in the 25-ppm group. Hypertension occurred after 12 months of exposure in the 1 ppm group and after 6 months of exposure in the 2.5, 5, and 10 ppm groups (Perry et al., 1977).

In male Sprague-Dawley rats, a single-dose intravenous injection of Cd acetate at 0.32, 1.0, or 3.2 mg/kg body weight caused a rise in blood pressure after an acute transient fall (Puri, 1999). Repeated administration of Cd for 5 days at a dose of 1 mg/kg (i.p.) caused a rise in blood pressure detectable 2, 4, and 10 days after treatment (Puri, 1999).

Synergy between Cd and lead (Pb) or nickel (Ni) has been noted. Exposure to low-dose Cd and Pb in drinking water for 18 months (Perry and Erlanger, 1978; Perry et al., 1979) caused a greater increase in systolic blood pressure (by 43 mmHg) than exposure to Cd alone (15 mmHg). Repeated intraperitoneal administration of Cd or Ni at 0.1 to 1.0 mg/kg body weight for 7 days raised systolic blood pressure by 50 mmHg in rats exposed to Cd or Cd plus nickel. There was no increase in systolic blood pressure in the rats exposed to nickel alone (Wang et al., 2002). In the group of rats exposed to Cd plus nickel, high blood pressure persisted for 30 days, but for only 20 days in the group exposed to Cd alone.

14.7.4 Cd-Induced Hypertension and 20-Hydroxyeicosatetraenoic Acid

Formation and urinary excretion of the eicosanoid 20-HETE by human kidneys were documented in the 1990s (Schwartzman et al. 1990; Prakash et al. 1992; Satarug et al., 2006). Later, the enzymes CYP4F2 and CYP4A11 were found to be responsible for renal production of 20-HETE (Lasker et al., 2000). Immunostaining for expression MT1/2, HO-1, CYP2B6, CYP2E1, CYP3A4, CYP4A11 and CYP4F2 in human kidneys are shown in Figure 14.2a-g, respectively. Immunostaining for renal expression of α-SMA, CD68 and a negative control are shown in Figure 14.2h-j, respectively. In human kidneys, CYP4F2 was more prominently expressed in glomeruli and proximal tubular cells than the distal tubular cells (Figure 14.2g). In contrast, CYP4A11 was weakly expressed in the proximal and distal tubules, but not in glomeruli (Figure 14.2f). The zinc transporter, ZIP8, known to mediate Cd uptake, was more prominently expressed in the distal tubules than the proximal tubules, but not in the glomeruli (Ajjimaporn et al., 2012).

In the kidney, 20-HETE reduces tubular Na transport and enhances the response of afferent arteries to constrictor stimuli (Ge et al., 2013; Fan et al., 2015; Elshenawy et al., 2017). 20-HETE contributes to the natriuretic effects of parathyroid hormone, dopamine, endothelin, and angiotensin II (ANG II) in the proximal tubule. Elevations in renal perfusion pressure increase 20-HETE production, and 20-HETE partially mediates the pressure natriuretic response (Fan et al. 2015; Elshenawy et al., 2017). Some investigators have expressed a conflicting belief that 20-HETE contributes to the development of Cd-induced hypertension.

All Na transport in the nephron depends on the activity of Na/K-ATPase in the BLM (Rossier et al. 2013; Feraille and Dizin, 2016). In the thick ascending limb of Henle's loop, Cd also blocks apical egress of K through a specialized channel and thereby impedes the NaK2Cl co-transporter. In the proximal tubule, 20-HETE promotes internalization of sodium–hydrogen exchanger 3. All of these actions interfere with reabsorption of filtered Na.

In a cross-sectional study of 225 women, aged 33–55 years, who were residents of a Cd-polluted area of Thailand, urinary 20-HETE excretion showed a positive correlation with systolic blood pressure. Urinary 20-HETE increments from tertile 1 to tertile 3 were associated with a systolic blood pressure increase of 6 mmHg in normotensive women (Boonprasert et al., 2018). Urinary 20-HETE levels ≥469 pg/mL were associated with a 1.9-, 4.36-, and 1.53-fold increases in the prevalence of hypertension, higher urinary Cd levels, and higher urinary excretion of β_2MG, respectively. It is not

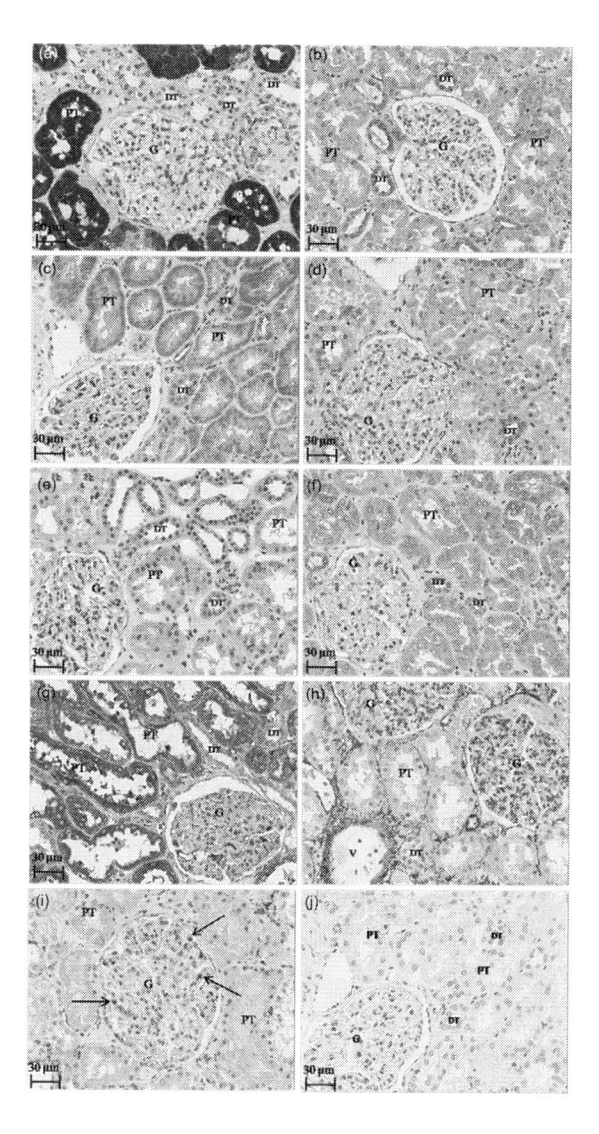

Figure 14.2 Localization of CYP4A11 and CYP4F2 in human kidneys. Immunostaining for renal expression of (a) MT1/2, (b) HO-1, (c) CYP2B6, (d) CYP2E1, (e) CYP3A4, (f) CYP4A11, (g) CYP4F2, (h) α-SMA, an indicator of fibrosis (i) CD68 (macrophage/monocyte specific marker, an indicator of inflammation), and (j) a negative control. Arrows indicate positive CD68 cells. Abbreviations: α-SMA, alpha smooth muscle actin; PT, proximal tubule; DT, distal tubule; G, glomerulus, and V, vessel. (From Boonprasert et al., *Toxicol. Lett.* 249:5–14, 2016. With permission.)

possible to discern whether 20-HETE was a mitigator or a cause of hypertension. It is noteworthy, however, that the mean blood Cd of 3.6 μg/L was nearly ten times higher than a level associated with an increased risk of hypertension in population with low-low-level environmental Cd exposure. Hypertension in the Thai women may have resulted from a reduction in GFR.

14.7.5 Summary of Cadmium and Hypertension

Data from a longitudinal study in the United States implicated the kidney burden of Cd, assessed with urinary Cd levels, in the pathogenesis of blood pressure increases and hypertension. A urinary Cd level of approximately 1 μg/L was associated with an increased prevalence of hypertension in Chinese and Thai studies. A positive association of blood Cd and blood pressure was observed in U.S., Canadian, Chinese, and Korean population studies. In U.S. Canadian, and Chinese studies, the preponderance of Cd-associated hypertension was seen in non-smoking women. Blood Cd levels ≥0.4 µg/L were associated with an increased risk of hypertension in Caucasian and Mexican-American women, but not in black women or white, black, or Mexican-American men.

Rats with Cd-induced hypertension showed increased sodium retention and reduced sodium excretion. Thus increased tubular avidity for filtered sodium appeared to be a possible mechanism by which lifelong, low-dose Cd intake causes hypertension. Whether 20-HETE is a mediator or a mitigator of Cd-induced hypertension is unresolved.

14.8 OTHER END-ORGAN EFFECTS AND IMPLICATIONS FOR RISK ASSESSMENT

In this section, findings from the U.S. NHANES that examined adverse effects of Cd in tissues and organs other kidneys and bone are summarized. Tables 14.2 and 14.3 present the urinary and blood Cd levels that were associated with the effects measured.

Table 14.2 Urinary Cadmium Levels Associated with Adverse Effects in Multiple Organs

Targets	NHANES Cycle Reference	Findings
Lung	1988–1994, n 7,455, 13.4-year follow-up, Adams et al. (2012).	Urinary Cd levels ≥0.58 µg/g creatinine were associated with an increased mortality from lung cancer (HR 3.22).
Liver	1988–1994, n 12,732, aged ≥20 years, Hyder et al. (2013).	In women, urinary Cd levels ≥0.83 µg/g creatinine were associated with liver inflammation in women (OR 1.26). In men, urinary Cd levels ≥0.65 µg/g creatinine were associated with liver inflammation (OR 2.21), NAFLD (OR 1.30) and NASH (OR 1.95).
Pancreas	1988–1994, n 8,722, aged ≥40 years, Schwartz et al. (2003).	Urinary Cd levels 1–2 µg/g creatinine were associated with increases in risk of pre-diabetes (OR 1.48) and diabetes (OR 1.24).
Pancreas	2005–2010, n 2,398, aged ≥40 years, Wallia et al. (2014).	Urinary Cd levels >1.4 µg/g creatinine were associated with an increased risk of pre-diabetes in non-smokers.
Breast	1999–2008, 92 cases, 2,884 controls, Gallagher et al. (2010).	Urinary Cd levels ≥0.37 µg/g creatinine were associated with an increased breast cancer risk (OR 2.50).
Brain	1988–1994, n 5,572, aged 20–59 years, Ciesielski et al. (2013).	Per 1 µg/L increase in urinary Cd was associated with a reduction in cognitive function (attention/perception domain) test scores by 1.93%.
Brain	1999–2006, n 2,023, aged 60–85, mean 7.5-year follow-up, Peng et al. (2017).	An interquartile range increase in urinary Cd of 0.51 µg/L was associated with an increased mortality from Alzheimer's disease (HR 1.58).
Blood vessels	1999–2004, n 6,456, aged ≥40 years, Tellez-Plaza et al. (2010).	Urinary Cd levels ≥0.69 µg/g creatinine were associated with an increased risk of peripheral arterial disease in men (OR 4.90).
Heart	1999–2004, n 8,989, mean 4.8-year follow-up, Tellez-Plaza et al. (2012).	Urinary Cd levels ≥0.57 µg/g creatinine were associated with an increased mortality from ischemic heart disease (HR 2.53).
Eye	2005–2008, n 5,390, aged ≥40 years, Wu et al. (2014).	Urinary Cd levels ≥0.35 µg/L were associated with increased risk of MD in non-Hispanic whites (OR 3.31).
Auditory system	2005–2008, n 878, aged 12–19 years, Shargorodsky et al. (2011).	Urinary Cd levels >0.15 µg/g creatinine were associated with low-frequency hearing loss (OR 3.08).

Geometric mean, the 50th, 75th, 90th, and 95th percentile values for urinary Cd levels in the representative U.S. general population are 0.210, 0.208, 0.412, 0.678, 0.949 µg/g creatinine, respectively (Crinnion, 2010).
NHANES, National Health and Nutrition Examination Survey; n, sample size; OR, odds ratio; HR, hazard ratio; NAFLD, non-alcoholic fatty liver disease; NASH, non-alcoholic steatohepatitis; MD, macular degeneration.
Urinary albumin to creatinine ratio ≥30 mg/g creatinine.
Estimated glomerular filtration rate (eGFR) <60 mL/min/1.73 m².

Table 14.3 Blood Cadmium Levels Associated with Adverse Effects in Multiple Organs

Targets	NHANES Cycle Reference	Findings
Lung	2007–2010 n 9,575, aged ≥20 years, Rokadia and Agarwal (2013).	Blood Cd levels ≥0.73 µg/L were associated with increases risk of obstructive lung disease[a] (OR 2.52).
Heart	2003–2012, n 12,511, aged 45–79 years, Hecht et al. (2016).	For non-smokers, blood Cd levels >0.49 µg/L were associated with increased risk of having myocardial infarction (PR 1.54).
Heart	2003–2014, n 14,832, aged 40–79 years, Hecht et al. (2019).	Blood Cd levels >0.52 µg/L were associated with increased risk of having angina (OR 1.45), compared with blood Cd levels <0.28 µg/L.
Brain	2007–2010, n 2,892, aged 20–39 years, Scinicariello and Buser (2015).	Blood Cd levels ≥0.54 µg/L were associated with depressive symptoms in non-smokers (OR 2.91), and smokers (OR 2.69).
Brain	2011–2014, n 2,068, aged 60–80 years, Li et al. (2018).	Blood Cd was inversely associated with cognitive function test scores in participants with blood Cd levels ≥0.63 µg/L.
Brain	1999–2004, n 4,060, aged ≥60 years, follow-up for 7–13 years, Min and Min (2016).	Blood Cd levels >0.6 µg/L were associated with mortality from Alzheimer's disease (HR 3.83).
Brain	1999–2006, n 6,141, aged 60–85, mean 7.5-year follow-up, Peng et al. (2017).	An interquartile increase in blood Cd of 0.36 µg/L associated with mortality from Alzheimer's disease (HR 1.22). Mean blood Cd 0.51 µg/L.
Pituitary/ thyroid axis	2007–2008, n 1,587, median age 45 years, Yorita Christensen (2013).	Blood Cd levels >0.6 µg/L and urinary Cd levels >0.4 µg/L were associated with decreased thyroid stimulating hormone levels.
Vestibular system	1999–2004, n 5,574, aged ≥40 years, Min et al. (2012).	Blood Cd levels >0.9 µg/L were associated with impaired balance control (OR 1.27).
Auditory system	1999–2004, n 3,698, aged 20–69 years, Choi et al. (2012).	Blood Cd levels >0.80 µg/L associated with an increase in pure-tone averages by 13.8%.
Eye	2005–2008, n 5,390, aged ≥40 years, Wu et al. (2014).	Blood Cd levels ≥0.66 µg/L associated with increased risk of having macular degeneration (OR 1.56), compared with blood Cd levels <0.25 µg/L.

Geometric mean, the 50th, 75th, 90th, and 95th percentile values for blood Cd levels in the representative U.S. general population are 0.304, 0.300, 0.500, 1.10, 1.60 µg/L, respectively (Crinnion, 2010).
NHANES, National Health and Nutrition Examination Survey; n, sample size; OR, odds ratio; HR, hazard ratio; PR, prevalence ratio.
Urinary albumin to creatinine ratio ≥30 mg/g.
Estimated glomerular filtration rate (eGFR) <60 mL/min/1.73 m².
[a] A ratio between forced expiratory volume in 1 s and forced vital capacity <0.7.

14.8.1 Liver and Pancreas

In NHANES 1988–1994, urinary Cd levels >0.83 µg/g creatinine were associated with 1.26-fold increase in risk of liver inflammation in women, while urinary Cd levels >0.65 µg/g creatinine were associated with 2.21-fold increase in risk of liver inflammation in men (Hyder et al., 2013). In men only, Cd levels >0.65 µg/g creatinine were associated with 1.30- and 1.95-fold increases in risk of having non-alcoholic fatty liver disease (NAFLD) and non-alcoholic steatohepatitis (NASH), respectively. A median 14.6-year follow-up study linked urinary Cd levels >0.65 creatinine to a 3.42-fold increase in mortality from non-malignant liver-related diseases (Hyder et al., 2013).

In NHANES 1988–1994, urinary Cd 1–1.99 µg/g creatinine levels were associated with 1.48- and 1.24-fold increases in risk of having pre-diabetes (abnormal fasting blood glucose) and diabetes, respectively (Schwartz et al., 2003). The fold increases in pre-diabetic and diabetic risks rose respectively to 2.05 and 1.45 in those with urinary Cd >2 µg/g creatinine levels.

In NHANES 2005–2010, urinary Cd levels above 1.375 µg/g creatinine were associated with an increased risk of pre-diabetic among non-smokers (Wallia et al., 2014). Cd has been shown to accumulate in the pancreatic islet β-cells (El Muayed et al., 2012; Wong et al., 2017). Loss of pancreatic β-cells due to Cd-induced apoptosis could be one of the many mechanisms by which Cd exposure increase pre-diabetic risk (Satarug and Moore, 2012; Treviño et al., 2015; Edwards and Ackerman, 2016). A particularly strong association of body burden indicator (urinary Cd) and risk of pre-diabetes seen, especially in the elderly, suggest that Cd levels in β-cells have had exceeded a threshold limit due to a prolonged exposure duration (Wallia et al., 2014). Notably, urinary Cd levels found to be associated with increased risk of pre-diabetes in moderate smokers and heavy smokers were approximately half of the level associated with an increase in pre-diabetic risk in non-smokers (Wallia et al., 2014). This was likely due to extensive kidney damage and loss of nephron among those who smoked, leading to lower urinary Cd levels than non-smokers of same age.

14.8.2 Blood Vessels, the Heart, and the Brain

In NHANES 1999–2004, urinary Cd levels ≥0.69 µg/g creatinine were associated with 4.90-fold increase in risk of having peripheral arterial disease (PAD) in men (Tellez-Plaza et al., 2010). In women, urinary Cd was not associated with PAD risk. However, U-shape curve was seen between blood Cd and the prevalence of PAD in women, which reflected an effect of Cd exposure on PAD risk at blood Cd levels below <0.3 µg/L. Women were thus particularly sensitive to vascular effects of Cd.

In NHANES 1988–1994, high urinary Cd levels were associated with risk of abnormal frontal T-wave, subclinical marker of ventricular arrhythmias, in participants, aged ≥40 years (Faramawi et al., 2012). In NHANES 2003–2012, blood Cd above 0.49 µg/L levels were associated with increased risk of having myocardial infarction in non-smokers, (PR 1.54) and smokers (PR 1.57), aged 45–79 years (Hecht et al., 2016). In NHANES 2003–2012, blood Cd levels >0.52 µg/L were associated with a 1.45-fold increase in risk of having angina in participants, aged 40–79 years (Hecht et al., 2019).

In a study of NHANES 1988–1994 participants, aged 20–59 years, 1 µg/L increment of urinary Cd, there was a 1.93% reduction in a neurocognitive function, assessed by attention/perception domain test (Ciesielski et al., 2013). In the NHANES 2005–2010, blood Cd was associated with a 1.48-fold increase in risk of depression in participants, aged ≥18 years (Berk et al., 2014). In NHANES 2007–2010, blood Cd levels ≥0.54 µg/L levels were respectively associated with 2.91- and 2.69-fold increases in risk of having depressive symptoms in non-smokers and smokers, aged 20–39 (Scinicariello and Buser, 2015).

In a follow-up of NHANES 1999–2004 participants, aged ≥60 years), blood Cd levels >0.6 µg/L were associated 3.83-fold increase in risk of dying from Alzheimer's disease, compared with blood Cd levels <0.3 µg/L (Min and Min, 2016). Elevated urinary Cd levels were also associated with increased risk of dying from Alzheimer's disease in participants in NHANES 1988–1994 and NHANES 1999–2006. In the mean 7.5-year follow-up, an interquartile increase in urinary Cd of 0.51 µg/L was associated with increased deaths from Alzheimer's disease (HR 1.58) in the the NHANES 1999–2006 participants, aged 60–85 years (Peng et al., 2017).

14.8.3 Eyes and Ears

In NHANES 2005–2008, the prevalence of MD was 2.8% in the 40–59-year age group and it rose to 13.4% in those aged 60 years or older (Wu et al., 2014). In this NHANES, blood Cd levels ≥0.66 µg/L were associated with evidence of AMD (OR 1.56), compared with blood Cd levels of 0.14–0.25 µg/L. Non-Hispanic Whites were particularly sensitive, given that urinary Cd levels ≥0.35 µg/L were associated with a 3.31-fold rise in risk of MD.

In NHANES 1999–2004, blood Cd levels ≥0.9 µg/L were associated with impaired balance control (OR 1.27) in participants, aged ≥40 years (Min et al., 2012). In the same NHANES 1999–2004, blood Cd levels ≥0.80 µg/L were associated with a rise of pure-tone averages (PTAs) by

13.8%, indicative of hearing disability, while blood Pb levels ≥2.80 µg/dL were associated with a rise in PTA by 18.6% (Choi et al., 2012). In the NHANES 2005–2008, urinary Cd levels >0.15 µg/g creatinine levels were associated with 3.08-fold increase in risk of low-frequency hearing loss in adolescents, aged 12–19 years, while blood Pb levels ≥2 µg/dL were associated with 2.22-fold increase in risk of high-frequency hearing loss (Shargorodsky et al., 2011).

14.8.4 Immunity and Pituitary-Thyroid Axis

In NHANES 1999–2012, blood Cd levels >0.606 µg/L were associated with increased risk of infections with hepatitis B virus (OR 1.72), and *Helicobacter pylori* (OR 1.5) in participants, aged ≥3 years (Krueger and Wade, 2016). These findings are consistent with immunosuppressive effects of Cd. In NHANES 2007–2010, blood Cd showed a positive association with serum thyroglobulin and free thyroxin or tetraiodothyronine in women, while showing a positive association with serum total triiodothyronine and thyroglobulin in men (Luo and Hendryx, 2014). In NHANES 2007–2008, blood Cd levels >0.6 µg/L and urinary Cd levels >0.4 µg/L were associated with a fall of thyroid-stimulating hormone, while urinary Cd was associated with increased total triiodothyronine and active or free triiodothyronine (Yorita Christensen, 2013).

14.8.5 Summary of Multiple Toxicity Targets and Implications for Risk Assessment

Evidence for Cd effects in organs other than kidneys and bone has emerged from NHANES database. In longitudinal studies, elevated Cd body burden was linked to increases in deaths from lung cancer, ischemic heart disease, and Alzheimer's disease. In cross-sectional studies, an elevated burden was linked to an increased prevalence of liver inflammation, non-alcoholic fatty liver disease, non-alcoholic steatohepatitis, breast cancer, pre-diabetes, diabetes, PAD, MD, hearing disability, and reduced neurocognitive function.

The multiplicity of Cd targets calls for a modification of the criteria by which toxicity of the metal is judged. Risk assessment should be based on the organ most sensitive to Cd. The kidney has been the conventional target for this assessment because it contains the highest Cd concentration. Urinary Cd levels ≥1 µg/L were associated with increased prevalence of CKD, but levels lower than 1 µg/L were associated increases in prevalence of MD, breast cancer, and mortality from Alzheimer's disease. To afford sufficient health protection, subpopulations with susceptible to Cd toxicity should be included in Cd health risk estimation.

ABBREVIATIONS

AMD	age-related macular degeneration
AREDS	age-related eye disease study
BBM	brush border membrane
β_2MG	beta$_2$-microglobulin
Ca	calcium
Cd	cadmium
CKD	chronic kidney disease
CdO	cadmium oxide
CdMT	cadmium-metallothionein complex
CdPC	cadmium-phytochelatin complex
C_{cr}	creatinine clearance, units of volume/time
Cu	copper

CYP4	cytochrome P450 family 4
DMT1	divalent metal transporter 1
E_x/C_{cr}	excretion rate of x per volume of filtrate, units of mass/volume
EFSA	European Food Safety Agency
FAO/WHO	Food and Agriculture Organization/World Health Organization
FDA	Food and Drugs Administration
FPN1	feroportin1
FFQ	food frequency questionnaire
GFR	glomerular filtration rate, units of volume/time
eGFR	estimated glomerular filtration rate, units of mL/min/1.73 m^2
GSH	glutathione
GSC	glomerular sieving coefficient
hNGAL	human neutrophil gelatinase-associated lipocalin
20-HETE	20-hydroxyeicosatetraenoic acid
KIM-1	kidney injury molecule-1
Mn	manganese
MT	metallothionein
NAG	N-acetyl-β-D-glucosaminidase
NAFLD	non-alcoholic fatty liver disease
NASH	non-alcoholic steatohepatitis
NHANES	National Health and Nutrition Examination Survey
NO	nitric oxide
Pb	lead
PC	phytochelatin
PTWI	provisional tolerable weekly intake
RBP	retinol-binding protein
RME	receptor-mediated endocytosis
ROS	reactive oxygen species
Sn	tin
Se	selenium
TALH	thick ascending limb of the loop of Henle
TDS	total diet study
V_u	urine flow rate, units of volume/time
ZIP	zrt- and irt-related protein (ZIP) of zinc transporter family
Zn	zinc

ACKNOWLEDGMENTS

We thank Dr Patharee Oungsakul, UQ School of Veterinary Sciences, for her assistance with literature acquisition. This work was supported with resources of the Stratton Veterans' Affairs Medical Center, Albany, NY, USA, and was made possible by facilities at that institution. Opinions expressed in this chapter are those of the authors and do not represent the official position of the United States Department of Veterans' Affairs.

APPENDIX

Demonstration that $E_x/C_{cr} = [x]_u[cr]_p/[cr]_u$.
Let V_u = urine flow rate, units of volume/time;
E_x = urinary excretion rate of substance x, units of mass/time;

$[x]_u$ = urinary concentration of substance x, units of mass/volume;
E_{cr} = urinary excretion rate of creatinine, units of mass/time;
$[cr]_p$ = plasma concentration of creatinine, units of mass/volume;
$[cr]_u$ = urine concentration of creatinine, units of mass/volume;
C_{cr} = renal creatinine clearance (an approximation of GFR) = $E_{cr}/[cr]_p$, units of volume/time;
E_x/C_{cr} = amount of x excreted per volume of filtrate, units of mass/volume.
$E_x/C_{cr} = [x]_u V_u/([cr]_u V_u/[cr]_p)$; cancelling V_u and rearranging,
$E_x/C_{cr} = [x]_u[cr]_p/[cr]_u$.

REFERENCES

Adams, S.V., Passarelli. M.N., Newcomb, P.A. 2012. Cadmium exposure and cancer mortality in the Third National Health and Nutrition Examination Survey cohort. *Occup. Environ. Med.* 69:153–156.

Adams, S.V., Quraishi, S.M., Shafer, M.M., Passarelli, M.N., Freney, E.P., Chlebowski, R.T., et al. 2014. Dietary cadmium exposure and risk of breast, endometrial, and ovarian cancer in the Women's Health Initiative. *Environ. Health Perspect.* 122:594–600.

Age-Related Eye Disease Study Research Group, 2001. A randomized, placebo controlled, clinical trial of high-dose supplementation with vitamins C and E, beta carotene, and zinc for age-related macular degeneration and vision loss: AREDS report no. 8. *Arch. Ophthalmol.* 119:1417–1436.

Ahmed F., Coyne T., Dobson A., McClintock, C. 2008. Iron status among Australian adults: Findings of a population based study in Queensland, Australia. *Asia Pac. J. Clin. Nutr.* 17:40–47.

Ajjimaporn, A., Botsford, T., Garrett, S.H., Sens, M.A., Zhou, X.D., Dunlevy, J.R., et al. 2012. ZIP8 expression in human proximal tubule cells, human urothelial cells transformed by Cd+2 and As+3 and in specimens of normal human urothelium and urothelial cancer. *Cancer Cell Int.* 12:16. doi: 10.1186/1475-2867-12-16.

Akerstrom, M., Barregard, L., Lundh, T., Sallsten, G. 2013. The relationship between cadmium in kidney and cadmium in urine and blood in an environmentally exposed population. *Toxicol. Appl. Pharmacol.* 268: 286–293.

Akesson, A., Bjellerup, P., Lundh, T., Lidfeldt, J., Nerbrand, C., Samisioe, G., et al. 2006. Cadmium-induced effects on bone in a population-based study of women. *Environ. Health Perspect.* 114: 830–834.

Akesson, A., Lundh. T., Vahter. M., Bjellerup, P., Lidfeldt, J., Nerbrand, C., et al. 2005. Tubular and glomerular kidney effects in Swedish women with low environmental cadmium exposure. *Environ. Health Perspect.* 113:1627–1631.

Alfven, T.Y., Elinder, C.-G., Carlsson, M.D., Grubb, A., Hellstrom, L., Persson. B., et al. 2000. Low-level cadmium exposure and osteoporosis. *J. Bone Miner. Res.* 15:1579–1586.

An, H.C., Sung, J.H., Lee, J., Sim, C.S., Kim, S.H., Kim, Y. 2017. The association between cadmium and lead exposure and blood pressure among workers of a smelting industry: A cross-sectional study. *Ann. Occup. Environ. Med.* 29:47. doi: 10.1186/s40557-017-0202-z.

Anderson, A.J., Frazer, D.M. 2017. Current understanding of iron homeostasis. *Am. J. Clin. Nutr.* 106:1559S–1566S.

Aoshima, K. 1987. Epidemiology and tubular dysfunction in the inhabitants of a cadmium-polluted area in the Jinzu River basin in Toyama Prefecture. *Tohoku J. Exp. Med.* 152: 151–172.

Aoshima, K., Kasuya, M. 1991. Preliminary study on serum levels of 1, 25-dihydroxyvitamin D and 25-hydroxyvitamin D in cadmium-induced renal tubular dysfunction. *Toxicol. Lett.* 57:91–99.

Argyropoulos, C.P., Chen, S.S., Ng Y.-H., Roumelioti, M.-E., Shaffi, K., Singh, P.P., et al. 2017. Rediscovering beta-2 microglobulin as a biomarker across the spectrum of kidney diseases. *Front. Med.* 4:73. doi: 10.3389/fmed.2017.00073.

Arnich, N., Sirot, V., Rivière, G., Jean, J., Noël, L., Guérin, T., et al. 2012. Dietary exposure to trace elements and health risk assessment in the 2nd French Total Diet Study. *Food Chem. Toxicol.* 50:2432–2449.

ATSDR, 2012. *Agency for Toxic Substances and Disease Registry, Toxicological Profile for Cadmium.* Department of Health and Humans Services, Public Health Service, Centers for Disease Control and Prevention: Atlanta, GA.

Awata, H., Linder, S., Mitchell, L.E., Delclos, G.L. 2017. Association of dietary intake and biomarker levels of arsenic, cadmium, lead, and mercury among Asian populations in the U.S.: NHANES 2011–2012. *Environ. Health Perspect.* 125: 314–323.

Aydemir, T.B., Cousins, R.J. 2018. The multiple faces of the metal transporter ZIP14 (SLC39A14). *J. Nutr.* 148:174–184.

Baba, H., Tsuneyama, K., Kumada, T., Aoshima, T., Imura, J. 2014. Histopathological analysis for osteomalacia and tubulopathy in itai-itai disease. *J. Toxicol. Sci.* 39: 91–96.

Bandara, J.M., Wijewardena, H.V., Liyanege, J., Upul, M.A., Bandara, J.M. 2010. Chronic renal failure in Sri Lanka caused by elevated dietary cadmium: Trojan horse of the green revolution. *Toxicol. Lett.* 198:33–39.

Barbier, O., Jacquillet, G., Tauc, M., Poujeol, P., Cougnon, M. 2004. Acute study of interaction among cadmium, calcium, and zinc transport along rat nephron in vivo. *Am. J. Physiol.* 287: F1067–F1075.

Barr, D.B., Wilder, L.C., Caudill, S.P., Gonzalez, A.J., Needham, L.L., Pirkle, J.L. 2004. Urinary creatinine concentrations in the U.S. population: Implications for urinary biologic monitoring measurements. *Environ. Health Perspect.* 113: 192–200.

Barregard, L., Fabricius-Lagging, E., Lundh, T., Mölne, J., Wallin, M., Olausson, M., et al. 2010. Cadmium, mercury, and lead in kidney cortex of living kidney donors: Impact of different exposure sources. *Environ. Res.* 110:47–54.

Barregard, L., Bergstrom, G., Fagerberg, B. 2014. Cadmium, type 2 diabetes, and kidney damage in a cohort of middle-aged women. *Environ. Res.* 135: 311–316.

Bendell, L.I. 2010. Cadmium in shellfish: The British Columbia, Canada experience: A mini-review. *Toxicol Lett.* 198:7–12.

Berk, M., Williams, L.J., Andreazza, A.C., Pasco, J.A., Dodd, S., Jacka, F.N., et al. 2014. Pop, heavy metal and the blues: Secondary analysis of persistent organic pollutants (POP), heavy metals and depressive symptoms in the NHANES National Epidemiological Survey. *BMJ* 4:e005142. doi: 10.1136/bmjopen-2014-005142.

Bernard, A., Vyskocyl, A., Mahieu, P., Lauwerys, R. 1988. Effect of renal insufficiency on the concentration of free retinol-binding protein in urine and serum. *Clin. Chim. Acta* 171:85–93.

Bernard, A., Schdeck, C., Cardenas, A., Buchet, J.-P., Lauwerys, R. 1991. Potentiation of diabetic nephropathy in uniephrectomized rats subchronically exposed to cadmium. *Toxicol. Lett.* 58:51–57.

Bernard, A., Thjioelemans, N., Roels, H., Lauwerys, R. 1995. Association between NAG-B and cadmium in urine with no evidence of a threshold. *Occup. Environ. Med.* 52:177–180.

Blum, J.L., Edwards, J.R., Prozialeck, W.C., Xiong, J.Q., Zelikoff, J.T. 2015. Effects of maternal exposure to cadmium oxide nanoparticles during pregnancy on maternal and offspring kidney injury markers using a murine model. *J. Toxicol. Environ. Health A* 78:711–724.

Boonprasert, K., Ruengweerayut, R., Aunpad, R., Satarug, S., Na-Bangchang, K. 2012. Expression of metallothionein isoforms in peripheral blood leukocytes from Thai population residing in cadmium-contaminated areas. *Environ. Toxicol. Pharmacol.* 34:935–944.

Boonprasert, K., Satarug, S., Morais, C., Gobe, G.C., Johnson, D.W., Na-Bangchang, K., et al. 2016. The stress response of human proximal tubule cells to cadmium involves up-regulation of haemoxygenase 1 and metallothionein but not cytochrome P450 enzymes. *Toxicol Lett.* 249:5–14.

Boonprasert, K., Vesey, D.A., Gobe, G.C., Ruenweerayut, R., Johnson, D.W., Na-Bangchang, K., et al. 2018. Is renal tubular cadmium toxicity clinically relevant? *Clin. Kidney J.* 11:681–687.

Brzoska, M.M., Moniuszko-Jakoniuk, J. 1998. The influence of calcium content in diet on cumulation and toxicity of cadmium in the organism. *Arch.Toxicol.* 72:63–73.

Buchet, J.P., Lauwerys, R., Roels, H., Bernard, A., Bruaux, P., Claeys, F., et al. 1990. Renal effects of cadmium body burden of the general population. *Lancet* 336:699–702.

Buser, M.C., Ingber, S.Z., Raines, N., Fowler, D.A., Scinicariello, F. 2016. Urinary and blood cadmium and lead and kidney function: NHANES 2007–2012. *Int. J. Hyg. Environ. Health* 219:261–267.

Callan, A., Hinwood, A., Devine A., 2014. Metals in commonly eaten groceries in Western Australia: A market basket survey and dietary assessment. *Food Addit. Contam. A* 31:1968–1981.

Carlsson, L., Lundholm, C.E. 1996. Characterisation of the effects of cadmium on the release of calcium and on the activity of some enzymes from neonatal mouse calvaria in culture. *Comp. Biochem. Physiol.* 115C: 251–256.

Chan, H.M., Zhu, L.F., Zhong, R., Grant, D., Goyer, R.A., Cherian, M.G. 1993. Nephrotoxicity in rats following liver transplantation from cadmium-exposed rats. *Toxicol. Appl. Pharmacol.* 123:89–96.

Chaumont, A., Voisin, C., Deumer, G., Haufroid, V., Annesi-Maesano, I., Roels, H., et al. 2013. Associations of urinary cadmium with age and urinary proteins: Further evidence of physiological variations unrelated to metal accumulation and toxicity. *Environ. Health Perspect.* 121:1047–1053.

Chen, X., Zhu, G., Lei L, Jin T. 2013.The association between blood pressure and blood cadmium in a Chinese population living in cadmium polluted area. *Environ. Toxicol. Pharmacol.* 36:595–599.

Chen, X., Wang, Z., Zhu, G., Liang, Y., Jin, T. 2015. Benchmark dose estimation of cadmium reference level for hypertension in a Chinese population. *Environ. Toxicol. Pharmacol.* 39:208–212.

Chew, E.Y., Clemons, T.E., Agrón, E., Sperduto, R.D., Sangiovanni, J.P., Kurinij, N., et al. 2013. Long-term effects of vitamins C and E, β-carotene, and zinc on age-related macular degeneration: AREDS report no. 35. *Ophthalmol.* 120:1604–1611.

Choi, Y.H., Hu, H., Mukherjee, B., Miller, J., Park. S.K. 2012. Environmental cadmium and lead exposures and hearing loss in U.S. adults: The National Health and Nutrition Examination Survey, 1999 to 2004. *Environ. Health Perspect.* 120:1544–1550.

Choi, J.W., Oh, C., Shim, S.Y., Jeong, S., Kim, H.S., Kim, M.S. 2018. Reduction in prevalence of hypertension and blood heavy metals among curry-consumed Korean. *Tohoku J. Exp. Med.* 244:219–229.

Ciesielski, T., Bellinger, D.C., Schwartz, J., Hauser, R., Wright, R.O. 2013. Associations between cadmium exposure and neurocognitive test scores in a cross-sectional study of US adults. *Environ. Health* 12:13. doi: 10.1186/1476-069X-12-13.

Ciesielski, T.H., Schwartz, J., Bellinger, D.C., Hauser, R., Amarasiriwardena, C., Sparrow, D., et al. 2018. Iron-processing genotypes, nutrient intakes, and cadmium levels in the Normative Aging Study: Evidence of sensitive subpopulations in cadmium risk assessment. *Environ Int.* 119:527–535.

Clemens, S. 2006. Toxic metal accumulation, responses to exposure and mechanisms of tolerance in plants. *Biochimie* 88:1707–1719.

Clemens, S., Aarts, M.G., Thomine, S., Verbruggen, N. 2013. Plant science: The key to preventing slow cadmium poisoning. *Trends Plant Sci.* 18:92–99.

Cobbett, C.S. 2000. Phytochelatins and their roles in heavy metal detoxification. *Plant Physiol.* 123:825–832.

Cobbett. C., Goldsbrough, P. 2002. Phytochelatins and metallothioneins: Roles in heavy metal detoxification and homeostasis. *Annu. Rev. Plant Biol.* 53:159–182.

Codex Alimentarius, 2014. CODEX STAN 193-1995, General Standard for Contaminants and Toxins in Food and Feed. www.fao.org/fileadmin/user_upload/livestockgov/documents/1_CXS_193e.pdf.

Coonse, K.G., Coonts, A.J., Morrison, E.V., Heggland, S.J. 2007. Cadmium induces apoptosis in the human osteoblast-like cell line Saos-2. *J. Toxicol. Environ. Health* 70:575–581.

Copes, R., Clark, N.A., Rideout, K., Palaty, J., Teschke, K., 2008. Uptake of cadmium from Pacific oysters (Crassostrea gigas) in British Columbia oyster growers. *Environ. Res.* 107: 160–169.

Crews, D.C., Plantinga, L.C., Miller, E.R., Saran, R, Hedgeman, E., Saydah, S.H., et al. 2010. Prevalence of chronic kidney disease in persons with undiagnosed or prehypertension in the United States. *Hypertens.* 55:1102–1109.

Crinnion, W.J. 2010. The CDC fourth national report on human exposure to environmental chemicals: What it tells us about our toxic burden and how it assists environmental medicine physicians. *Altern. Med. Rev.* 15:101–108.

Crowley, S.D., Coffman, T.M. 2014. The inextricable role of the kidney in hypertension. *J. Clin. Invest.* 124:2341–2347.

Dahl, L.K., Heine, M., Thompson, K. 1972. Genetic influence of renal homografts on the blood pressure of rats from different strains. *Proc. Soc. Exp. Biol. Med.* 140: 852–856.

Dahl, L.K., Heine, M., Thompson, K. 1974. Genetic influence of the kidneys on blood pressure: Evidence from chronic renal homografts in rats with opposite predispositions to hypertension. *Circ. Res.* 34:94–101.

Dahl, L.K., Heine, M. 1975. Primary role of renal homografts in setting chronic blood pressure levels in rats. *Circ. Res.* 36:692–696.

Deering, K.E., Callan, A.C., Prince, R.L., Lim, W.H., Thompson, P.L., Lewis, J.R., et al. 2018. Low-level cadmium exposure and cardiovascular outcomes in elderly Australian women: A cohort study. *Int. J. Hyg. Environ. Health* 221:347–354.

De Nicola, L., Zoccali, C. 2016. Chronic kidney disease prevalence in the general population: Heterogeneity and concerns. *Nephrol. Dial. Transplant.* 31:331–335.

Dorian, C., Gattone II, V.H., Klaassen, C.D. 1992. Renal cadmium deposition and injury as a result of accumulation of cadmium-metallothionein (CdMT) by the proximal convoluted tubules: A light microscopic autoradiography study with [109]CdMT. *Toxicol. Appl. Pharmacol.* 114:173–181.

Dorian, C., Gattone II, V.H., Klaassen, C.D. 1995. Discrepancy between the nephrotoxic potencies of cadmium-metallothionein and cadmium chloride and the renal concentration of cadmium in the proximal convoluted tubules. *Toxicol. Appl. Pharmacol.* 130:161–168.

Dudley, R.E., Gammal, L.M., Klaassen, C.D. 1985. Cadmium-induced hepatic and renal injury in chronically exposed rats: Likely role of hepatic cadmium-metallothionein in nephrotoxicity. *Toxicol. Appl. Pharmacol.* 77:414–426.

Dumkova, J., Vrlikova, L., Vecera, Z., Putnova, B., Docekal, B., Mikuska, P., et al. 2016. Inhaled cadmium oxide nanoparticles: Their in vivo fate and effect on target organs. *Int. J. Mol. Sci.* 17. pii: E874. doi:10.3390/ijms17060874.

Edwards, J., Ackerman, C. 2016. A review of diabetes mellitus and exposure to the environmental toxicant cadmium with an emphasis on likely mechanisms of action. *Curr. Diabetes Rev.* 12:252–258.

EFSA, 2011. Statement on tolerable weekly intake for cadmium *EFSA J.* 9:1975 (1–19).

EFSA, 2012. Cadmium dietary exposure in the European population *EFSA J.* 10:2551 (1–36).

Elinder, C.G., Lind, B., Kjellstorm, T., Linnman, L., Friberg, L. 1976. Cadmium in kidney cortex, liver and pancreas from Swedish autopsies: Estimation of biological half time in kidney cortex, considering calorie intake and smoking habits. *Arch. Environ. Health* 31:292–301.

Elinder, C.G., Kjellstorm, T., Lind, B., Molander, M.L., Silander, T. 1978. Cadmium concentrations in human liver, blood, and bile: Comparison with a metabolic model. *Environ. Res.* 17:236–241.

Elinder, C.G., Edling, C., Lindberg, E., Kagedal, B., Vesterberg, O. 1985. Assessment of renal function in workers previously exposed to cadmium. *Br. J. Indust. Med.* 42:754–760.

Ellis, K.J., Yuen, K., Yasumura, S. and Cohn, S.H. 1984. Dose-response analysis of cadmium in man: Body burden vs kidney dysfunction. *Environ. Res.* 33:216–226.

Ellis, K.J., Cohn, S.H. and Smith, T.J. 1985. Cadmium inhalation exposure estimates: Their significance with respect to kidney and liver burden. *J. Toxicol. Environ. Health* 15:173–187.

El Muayed, M., Raja, M.R., Zhang, X., MacRenaris, K.W., Bhatt, S., Chen, X., et al. 2012. Accumulation of cadmium in insulin-producing β cells. *Islets* 4:405–416.

Elsenhans, B., Kolb, K., Schumann, K., Forth, W. 1992. Endogenous intestinal metallothionein possibly contributes to the renal accumulation of cadmium. *IARC Sci. Publ.* 118:225–230.

Elsenhans, B., Strugala, G.J., Schafer, S.G. 1997. Small-intestinal absorption of cadmium and the significance of mucosal metallothionein. *Hum. Exp. Toxicol.* 16:429–434.

Elshenawy, O.H., Shoieb, S.M., Mohamed, A., El-Kadi, A.O. 2017. Clinical implications of 20-hydroxyeicosatetraenoic acid in the kidney, liver, lung and brain: An emerging therapeutic target. *Pharmaceutics* 9. pii: E9. doi:10.3390/pharmaceutics9010009.

Erie, J.C., Butz, J.A., Good, J.A., Erie, E.A., Burritt, M.F., Cameron, J.D. 2005. Heavy metal concentrations in human eyes. *Am. J. Ophthalmol.* 139:888–893.

Eriksen, K.T., Halkjaer, J., Sorensen, M., Meliker, J.R., McElroy, J.A., Tjonneland, A., et al. 2014. Dietary cadmium intake and risk of breast, endometrial and ovarian cancer in Danish postmenopausal women: A prospective cohort study. *PLoS One* 9:e100815.

Escalante, B., Sacerdoti, D., Davidian, M.M., Laniado-Schwartzman, M., McGiff, J.C. 1991. Chronic treatment with tin normalizes blood pressure in spontaneously hypertensive rats. *Hypertens.* 17:776–779.

Ezaki, T., Tsukahara, T., Moriguchi, J., Furuki, K., Fukui, Y., Ukai, H., et al. 2003. No clear-cut evidence for cadmium-induced renal tubular dysfunction among over 10,000 women in the Japanese general population: A nationwide large-scale survey. *Int. Arch. Occup. Environ. Health* 76:186–196.

Fan, F., Muroya, Y., Roman, R.J. 2015. Cytochrome P450 eicosanoids in hypertension and renal disease. *Curr. Opin. Nephrol. Hypertens.* 24:37–46.

Fan, F., Roman, R.J. 2017. Effect of cytochrome P450 metabolites of arachidonic acid in Nephrology. *J. Am. Soc. Nephrol.* 28:2845–2855.

FAO/WHO. 1989. *Evaluation of Certain Food Additives and Contaminants (Thirty-Third Report of the Joint FAO/WHO Expert Committee on Food Additives).* WHO Technical Report Series No. 776. World Health Organization: Geneva.

FAO/WHO. 1993. *Evaluation of Certain Food Additives and Contaminants (Forty-First Report of the Joint FAO/WHO Expert Committee on Food Additives).* WHO Technical Report Series No. 837. World Health Organization: Geneva.

FAO/WHO. 2010. *Joint FAO/WHO Expert Committee on Food Additives, Seventy-third Meeting, Geneva, 8–17 June 2010. Summary and Conclusions. JECFA/73/SC*. Food and Agriculture Organization of the United Nations; World Health Organization: Geneva.

Faramawi, M.F., Liu, Y., Caffrey, J.L., Lin, Y.S., Gandhi, S., Singh, K.P. 2012.The association between urinary cadmium and frontal T wave axis deviation in the US adults. *Int. J. Hyg. Environ. Health* 215:406–410.

Felley-Bosco, E., Diezi, J. 1989. Fate of cadmium in rat renal tubules: a micropuncture study. *Toxicol. Appl. Pharmacol.* 98:243–251.

Feraille, E., Dizin, E. 2016. Coordinated control of ENaC and Na+, K+-ATPase in renal collecting duct. *J. Am. Soc. Nephrol.* 27:2554–2563.

Ferraro, P.M., Costanzi, S., Naticchia, A., Sturniolo, A., Gambaro, G. 2010. Low level exposure to cadmium increases the risk of chronic kidney disease: Analysis of the NHANES 1999–2006. *BMC Public Health.* 10:304. doi: 10.1186/1471-2458-10-304.

Ferraro, P.M., Sturniolo, A., Naticchia, A., D'Alonzo, S., Gambaro, G. 2012. Temporal trend of cadmium exposure in the United States population suggests gender specificities. *Intern. Med. J.* 42:691–67.

Finley, J.W. 1999. Manganese absorption and retention by young women is associated with serum ferritin concentration. *Am. J. Clin. Nutr.* 70:37–43.

Flanagan, P.R., McLellan, J.S., Haist, J., Cherian, M.G., Chamberlain, M.J., Valberg, L.S. 1978. Increased dietary cadmium absorption in mice and human subjects with iron deficiency. *Gastroenterol.* 46:609–623.

Forman, D.T. 1982. Beta-2 microglobulin: An immunogenetic marker of inflammatory and malignant origin. *Ann. Clin. Lab. Sci.* 12:447–451.

Fransson, M.N., Barregard, L., Sallsten, G., Akerstrom, M., Johanson, G. 2014. Physiologically-based toxico-kinetic model for cadmium using Markov-Chain Monte Carlo analysis of concentrations in blood, urine, and kidney cortex from living kidney donors. *Toxiol. Sci.* 141:365–376.

Fryar, C.D., Ostchega, Y., Hales, C.M., Zhang, G., Kruszon-Moran, D. 2017. Hypertension prevalence and control among adults: United States, 2015–2016. *NCHS Data Brief* 289:1–8.

Fujishiro, H., Yano, Y., Takada, Y., Tanihara, M., Himeno, S. 2012. Roles of ZIP8, ZIP14, and DMT1 in transport of cadmium and manganese in mouse kidney proximal tubule cells. *Metallomics* 4:700–708.

Fujishiro, H., Hamao, S., Tanaka, R., Kambe, T., Himeno, S. 2017. Concentration-dependent roles of DMT1 and ZIP14 in cadmium absorption in Caco-2 cells. *J. Toxicol. Sci.* 42:559–567.

Fujita, Y., ElBelbasi, H.I., Min, K.-S., Onosaka, S., Okada, Y., Matsumoto, Y., et al. 1993. Fate of cadmium bound to phytochelatin in rats. *Res. Commun. Chem. Pathol. Pharmacol.* 82:357–365.

Fukumoto, S., Ozono, K., Michigami, T., Minagawa, M., Okazaki, R., Sugimoto, T., et al. 2015. Pathogenesis and diagnostic criteria for rickets and osteomalacia: Proposal by an expert panel supported by Ministry of Health, Labour and Welfare, Japan, The Japanese Society for Bone and Mineral Research and the Japan Endocrine Society. *Endocrine J.* 62:665–671.

Gallagher, C.M., Chen, J.J., Kovach, J.S. 2010. Environmental cadmium and breast cancer risk. *Aging (Albany NY)* 2:804–814.

Gallagher, C.M., Kovach, J.S., Meliker, J.R. 2004. Urinary cadmium and osteoporosis in U.S. women ≥ 50 years of age: NHANES 1988–1994 and 1999–2004. *Environ. Health Perspect.* 116:1338–1343.

Garner, R., Levallois, P. 2016. Cadmium levels and sources of exposure among Canadian adults. *Health Rep.* 27:10–18.

Garner, R.E., Levallois, P. 2017. Associations between cadmium levels in blood and urine, blood pressure and hypertension among Canadian adults. *Environ. Res.* 155:64–72.

Garrett, R.G. 2010. Natural sources of metals to the environment. *Human Ecological Risk Assessment* 6:945–963.

Garrett, S.H., Sens, M.A., Todd, J.H., Somji, S., Sens, D.A. 1999. Expression of MT-3 protein in the human kidney. *Toxicol. Lett.* 105:207–214.

Garrick M.D., Dolan K.G., Horbinski C., Ghio A.J., Higgins D., Porubcin M., et al. 2003. DMT1: A mammalian transporter for multiple metals. *Biometals* 16:41–54.

Glassock, R.J., Warnock, D.G., Delanaye P. 2017. The global burden of chronic kidney disease: Estimates, variability and pitfalls. *Nat. Rev. Nephrol.* 13:104–114.

Gauthier, C., Nguyen-Simonnet, H., Vincent, C., Revillard, J.-P., Pellet, M.V. 1984. Renal tubular absorption of β2 microglobulin. *Kidney Int.* 26:170–175.

Ge, Y., Murphy, S.R., Lu. Y., Falck, J., Liu, R., Roman, R.J. 2013. Endogenously produced 20-HETE modulates myogenic and TGF response in microperfused afferent arterioles. *Prostaglandins Other Lipid Mediat.* 102–103:42–48.

Gonick, H., Indraprasit, S., Neustein, H., Rosen, V. 1975. Cadmium-induced experimental Fanconi syndrome. *Curr. Prob. Clin. Biochem.* 4:111–118.

Gonick, H.C. 1982. Pathophysiology of human proximal tubular transport defects. *Klin. Wochenschr* 60:1201–1211.

Gonzalez-Vicente, A., Saez, F., Monzon, C.M., Asirwatham, J., Garvin, J.L. 2019. Thick ascending limb sodium transport in the pathogenesis of hypertension. *Physiol. Rev.* 99:235–309.

Goyer, R.A., Miller, C.R., Zhu, S.-Y., Victery, W. 1989. Non-metallothionein-bound cadmium in the pathogenesis of cadmium nephrotoxicity in the rat. *Toxicol. Appl. Pharmacol.* 101:232–244.

Groten, J.P., Sinkeldam, E.J., Luten, J.B., van Bladeren, P.J. 1991a. Cadmium accumulation and metallothionein concentrations after 4-week dietary exposure to cadmium chloride or cadmium-metallothionein in rats. *Toxicol. Appl. Pharmacol.* 111:504–513.

Groten, J.P., Sinkeldam, E.J., Muys, T., Luten, J.B., van Bladeren, P.J. 1991b. Interaction of dietary Ca, P, Mn, Cu, Fe, Zn and Se with accumulation and oral toxicity of cadmium in rats. *Food Chem. Toxicol.* 29:249–258.

Groten, J.P., Koeman, J.H., van Nesselrooij, J.H., Luten, J.B., Fentener van Vlissingen, J.M., Stenhuis, W.S., et al. 1994. Comparison of renal toxicity after long-term oral administration of cadmium chloride and cadmium-metallothionein in rats. *Fundam. Appl. Toxicol.* 23:544–552.

Guéguen, M., Amiard, J.C., Arnich, N., Badot, P.M., Claisse, D., Guérin, T., et al. 2011. Shellfish and residual chemical contaminants: Hazards, monitoring, and health risk assessment along French coasts. *Rev. Environ. Contam. Toxicol.* 213:55–111.

Guyton, A.C. 1991. Blood pressure control-special role of the kidneys and body fluids. *Science* 252:1813–1816.

Haswell-Elkins, M., McGrath, V., Moore. M., Satarug. S., Walmby, M., Ng, J. 2007a. Exploring potential dietary contributions including traditional seafood and other determinants of urinary cadmium levels among indigenous women of a Torres Strait Island (Australia). *J. Expo. Sci. Environ. Epidemiol.* 17:298–306.

Haswell-Elkins, M., Imray, P., Satarug, S., Moore, M.R., O'dea, K. 2007b. Urinary excretion of cadmium among Torres Strait Islanders (Australia) at risk of elevated dietary exposure through traditional foods. *J. Expo. Sci. Environ. Epidemiol.* 17:372–377.

Haswell-Elkins, M., Satarug, S., O'Rourke, P., Moore, M., Ng, J., McGrath, V., et al. 2008. Striking association between urinary cadmium level and albuminuria among Torres Strait Islander people with diabetes. *Environ. Res.* 106:379–383.

Hecht, E.M., Arheart, K.L., Lee, D.J., Hennekens, C.H., Hlaing, W.M. 2016. Interrelation of cadmium, smoking, and cardiovascular disease (from the National Health and Nutrition Examination Survey). *Am. J. Cardiol.* 118:204–209.

Hecht, E.M., Arheart, K.L., Lee, D.J., Hennekens, C.H., Hlaing, W.M. 2019. Interrelationships of cadmium, smoking, and angina in the National Health and Nutrition Examination Survey: A Cross-Sectional Study. *Cardiol.* 141:177–182.

Hellström, L., Persson, B., Brudin, L., Grawé, K.P., Oborn, I., Järup, L. 2007. Cadmium exposure pathways in a population living near a battery plant. *Sci. Total Environ.* 373:447–455.

Hennigar, S.R., Kelley, A.M., McClung, J.P. 2016. Metallothionein and zinc transporter expression in circulating human blood cells as biomarkers of zinc status: A systematic review. *Adv. Nutr.* 7:735–746.

Herak-Kramberger, C.M., Spindler, B., Biber, J., Murer, H., Sabolic, I. 1996. Renal type II Na/P_i-cotransporter is strongly impaired whereas the Na/sulphate-cotransporter and aquaporin 1 are unchanged in cadmium-treated rats. *Pflugers Arch.* 432:336–344.

Heymsfield, S.B., Arteaga, C., McManus, C., Smith, J., Moffitt, S. 1983. Measurement of muscle mass in humans: Validity of the 24-hour urinary creatinine method. *Am. J. Clin. Nutr.* 37:478–494.

Hiratsuka, H., Katsuta, O., Toyota, N., Tsuchitani, M., Akiba, T., Marumo, F., et al. 1997. Iron deposition at mineralization fronts and osteoid formation following chronic cadmium exposure in ovarietomized rats. *Toxicol. Appl. Pharmacol.* 143:348–356.

Hiratsuka, H., Satoh, S., Satoh, M., Nishijima, M., Katsuki, Y., Suzuki, J., et al. 1999. Tissue distribution of cadmium in rats given minimum amounts of cadmium-polluted rice or cadmium chloride for 8 months. *Toxicol. Appl. Pharmacol.* 160:183–191.

Hirsch, I.B., Atchley, D.H., Tsai, E., Labbe, R.F., Chait, A. 1998. Ascorbic acid clearance in diabetic nephropathy. *J. Diabetes Complications* 12:259–263.

Hoet, P., Haufroid, V., Deumer, G., Dumont, X., Lison, D., Hantson, P. 2012. Acute kidney injury following acute liver failure: Potential role of systemic cadmium mobilization? *Intens. Care Med.* 38:467–473.

Honda, R., Swaddiwudhipong, W., Nishijo, M., Mahasakpan, P., Teeyakasem, W., Ruangyuttikarn, W., et al. 2010. Cadmium induced renal dysfunction among residents of rice farming area downstream from a zinc-mineralized belt in Thailand. *Toxicol. Lett.* 198:26–32.

Horiguchi, H., Oguma, E., Sasaki, S., Miyamoto, K., Ikeda, Y., Machida, M., et al. 2004. Comprehensive study of the effects of age, iron deficiency, diabetes mellitus, and cadmium burden on dietary cadmium absorption in cadmium-exposed female Japanese farmers. *Toxicol. Appl. Pharmacol.* 196:114–123.

Horiguchi, H., Aoshima, K., Oguma, E., Sasaki, S., Miyamoto, K., Hosoi, Y., et al. 2010. Latest status of cadmium accumulation and its effects on kidneys, bone, and erythropoiesis in inhabitants of the formerly cadmium-polluted Jinzu River Basin in Toyama, Japan, after restoration of rice paddies. *Int. Arch. Occup. Environ. Health* 83:953–970.

Horowitz, B., Miskulin, D., Zager, P. 2015. Epidemiology of hypertension in CKD. *Adv. Chronic Kidney Dis.* 22:88–95.

Hwangbo, Y., Weaver, V.M., Tellez-Plaza, M., Guallar, E., Lee, B.K., Navas-Acien, A. 2011. Blood cadmium and estimated glomerular filtration rate in Korean adults. *Environ. Health Perspect.* 119:1800–1805.

Hyder, O., Chung, M., Cosgrove, D., Herman, J.M., Li, Z., Firoozmand, A., Gurakar, A., Koteish, A., Pawlik, T.M. 2013. Cadmium exposure and liver disease among US adults. *J. Gastrointest. Surg.* 17:1265–1273.

IARC, 1993. *IARC Monographs on the Evaluation of Carcinogenic Risks to Humans. Beryllium, Cadmium, Mercury, and Exposures in the Glass Manufacturing Industry.* International Agency for Research on Cancer: Lyons.

Ibrahim, K.S., Beshir, S., Shahy, E.M., Shaheen, W. 2016. Effect of occupational cadmium exposure on parathyroid gland. *Maced J. Med. Sci.* 4:302–306.

Ikeda, M., Nakatsuka, H., Watanabe, T., Shimbo, S. 2015. Estimation of daily cadmium intake from cadmium in blood or cadmium in urine. *Environ. Health Prev. Med.* 20:455–459.

IPCS, 1992. *International Programme on Chemical Safety, Environmental Health Criteria 134, Cadmium.* WHO: Geneva.

Järup, L. 2003. Hazards of heavy metal contamination. *Br. Med. Bull.* 68:167–182.

Jarup, L., Persson, B., Elinder, C.G. 1995. Decreased glomerular filtration rate in solderers exposed to cadmium. *Occup. Environ. Med.* 52:818–822.

Jarup, L., Hellstrom, L., Alfven, T., Carlsson, M.D., Grubb, A., Persson, B., et al. 2000. Low level exposure to cadmium and early kidney damage: The OSCAR study. *Occup. Environ. Med.* 57:668–672.

Jarup, L., Alfven, T. 2004. Low level cadmium exposure, renal and bone effects: The OSCAR study. *Biometals* 17:505–509.

Jarup, L., Akesson, A. 2009. Current status of cadmium as an environmental health problem. *Toxicol. Appl. Pharmacol.* 238:201–208.

Jenkitkasemwong, S., Wang, C.Y., Mackenzie, B., Knutson, M.D. 2012. Physiologic implications of metal-ion transport by ZIP14 and ZIP8. *Biometals* 25:643–655

Jenny-Burri, J., Haldiman, M., Bruschweiler, B.J., Bochud, M., Burnier, M., Paccaud, F., et al. 2015. Cadmium body burden of the Swiss population. *Food Addit. Contam. Part A Chem. Control Expo. Risk Assess.* 32:1265–1272.

Jin, T., Nordberg, G.F., Sehlin, J., Leffler, P., Wu, J. 1994. The susceptibility of spontaneously diabetic mice to cadmium metallothionein nephrotoxicity. *Toxicol.* 89:81–90.

Jin, T., Frankel, B.J. 1996. Cadmium-metallothionein nephrotoxicity is increased in genetically diabetic as compared with normal Chinese hamsters. *Pharmacol. Toxicol.* 79:105–108.

Jin, T., Nordberg, G., Sehlin, J., Wallin, H., Sandberg, S. 1999. The susceptibility to nephrotoxicity of streptozotocin-induced diabetic rats subchronically exposed to cadmium chloride in drinking water. *Toxicology.* 142:69–75.

Jin, R., Zhu, X., Shrubsole, M.J., Yu, C., Xia, Z., Dai, Q. 2018. Associations of renal function with urinary excretion of metals: evidence from NHANES 2003–2012. *Environ. Int.* 121:1355–1362.

Jomova, K., Valko, M. 2011. Advances in metal-induced oxidative stress and human disease. *Toxicology.* 283:65–87.

Jorge-Nebert L.F., Gálvez-Peralta M., Landero Figueroa J., Somarathna M., Hojyo S., Fukada T., et al. 2015. Comparing gene expression during cadmium uptake and distribution: Untreated versus oral Cd-treated wild-type and ZIP14 knockout mice. *Toxicol. Sci.* 143:26–35.

Kaewnate, Y., Niyomtam, S., Tangvarasittichai, O., Meemark, S., Pingmuagkaew, P., Tangvarasittichai, S. 2012. Association of elevated urinary cadmium with urinary stone, hypercalciuria and renal tubular dysfunction in the population of cadmium-contaminated area. *Bull. Environ. Contam. Toxicol.* 89:1120–1124.

Kaseda, R., Hosojima, M., Sato, H., Saito, A. 2011. Role of megalin and cubulin in the metabolism of vitamin D (D3). *Ther. Apher. Dial.* 15 (Suppl 1):14–17.

Katakai, K., Liu, J., Nakajima, K., Keefer, L.K., Waalkes, M.P. 2001. Nitric oxide induces metallothionein (MT) gene expression apparently by displacing zinc bound to MT. *Toxicol. Lett.* 119:103–108.

Kawada, T., Koyama, H., Suzuki, S. 1989. Cadmium, NAG activity, and β_2-microglobulin in the urine of cadmium pigment workers. *Br. J. Ind. Med.* 46:52–55.

Kawada, T., Shinmyo, R.R., Suzuki, S. 1992. Urinary cadmium and N-acetyl-β-D-glucosaminidase excretion of inhabitants living in a cadmium-polluted area. *Int. Arch. Occup. Environ. Health* 63:541–546.

Kazantzis, G. 1979. Renal tubular dysfunction and abnormalities of calcium metabolism in cadmium workers. *Environ. Health Perspect.* 28:155–159.

Kikuchi, Y., Nomiyama, T., Kumagai, N., Uemura, T., Omae, K. 2002. Cadmium concentration in current Japanese foods and beverages. *J. Occup. Health* 44:240–247.

Kikuchi, Y., Nomiyama, T., Kumagai, N., Dekio, F., Uemura, T., Takebayashi, T., et al. 2003. Uptake of cadmium in meals from the digestive tract of young non-smoking Japanese female volunteers. *J. Occup. Health* 45:43–52.

Kim, K.R., Lee, H.Y., Kim, C.K., Park, Y.S. 1990. Alteration of renal amino acid transport system in cadmium-intoxicated rats. *Toxicol. Appl. Pharmacol.* 106:102–111.

Kim, E.C., Cho, E., Jee, D. 2014. Association between blood cadmium level and age-related macular degeneration in a representative Korean population. *Invest. Ophthalmol. Vis. Sci.* 55:5702–5710.

Kim, Y.-D., Yim, D.-H., Eom S.-Y., Moon, S.-I., Park, C.-H., et al. 2015a. Temporal changes in urinary levels of cadmium, N-acetyl-β-D-gluosaminidase and β2-microglobuilin in individuals in a cadmium-contaminated area. *Environ. Toxicol. Pharmacol.* 39:35–41.

Kim, N.H., Hyun, Y.Y., Lee, K.B., Chang, Y., Ryu, S., Oh, K.H., Ahn, C. 2015b. Environmental heavy metal exposure and chronic kidney disease in the general population. *J. Korean Med. Sci.* 30:272–277.

Kim, M.H., Zhao, D., Cho, J., Guallar, E. 2016. Cadmium exposure and age-related macular degeneration. *J. Expo. Sci. Environ. Epidemiol.* 26:214–218.

Kim, K., Melough, M.M., Vance, T.M., Noh, H., Koo, S.I., Chun, O.K. 2018. Dietary cadmium intake and sources in the US. *Nutrients* 11. pii: E2. doi:10.3390/nu11010002.

Kovacs, G., Danko, T., Bergeron, M.J., Balazs, B., Suzuki, Y., Zsembery, A., et al. 2011. Heavy metal cations permeate the TRPV6 epithelial cation channel. *Cell Calcium.* 49:43–55.

Kovacs, G., Montalbetti, N., Franz, M.C., Graeter, S., Simonin, A., Hediger, M.A. 2013. Human TRPV5 and TRPV6: Key players in cadmium and zinc toxicity. *Cell Calcium.* 54:276–286.

Koyama, H., Satoh, H., Suzuki, S., Tohyama, C. 1992. Increased urinary cadmium excretion and its relationship to urinary N-acetyl-β-D-glucosaminidase activity in smokers. *Arch. Toxicol.* 66:598–601.

Krueger, W.S., Wade, T.J. 2016. Elevated blood lead and cadmium levels associated with chronic infections among non-smokers in a cross-sectional analysis of NHANES data. *Environ. Health* 15:16. doi: 10.1186/s12940-016-0113-4.

Kruzynski, G.M. 2004. Cadmium in oysters and scallops: The BC experience. *Toxicol. Lett.* 148:159–169.

Kukongviriyapan, U., Pannangpetch, P., Kukongviriyapan, V., Donpunha, W., Sompamit, K., Surawattanawan, P. 2014. Curcumin protects against cadmium-induced vascular dysfunction, hypertension and tissue cadmium accumulation in mice. *Nutrients* 6:1194–1208.

Kukongviriyapan, U., Apaijit, K., Kukongviriyapan. V. 2016. Oxidative stress and cardiovascular dysfunction associated with cadmium exposure: Beneficial effects of curcumin and tetrahydrocurcumin. *Tohoku J. Exp. Med.* 239:25–38.

Kulak, C.A., Dempster, D.W. 2010. Bone histomorphometry: A concise review for endocrinologists and clinicians. *Arq. Bras. Endocrinol. Metab.* 54:87–98.

Kuo, S.M., Leaviti, P.S. and Lin, C.P. 1998. Dietary flavonoids interact with trace metals and affect metallothionein levels in human intestinal cells. *Biol. Trace Elem. Res.* 62:135–153.

Kurata, Y., Katsuta, O., Doi, T., Kawasuso, T., Hiratsuka, H., Tsuchitani, M., Umemura, T. 2014. Chronic cadmium treatment induces tubular nephropathy and osteomalacic osteopenia in ovariectomized cynomolgus monkeys. *Vet. Pathol.* 51:919–931.

Kurihara, I., Kobayashi, E., Suwazono, Y., Uetani, M., Inaba, T., Oishiz, M., et al. 2004. Association between exposure to cadmium and blood pressure in Japanese peoples. *Arch. Environ. Health* 59:711–716.

Lamb, D.T., Kader, M., Ming, H., Wang, L., Abbasi, S., Megharaj, M., et al. 2016. Predicting plant uptake of cadmium: validated with long-term contaminated soils. *Ecotoxicol.* 25:1563–1574.

Langelueddecke, C., Roussa, E., Fenton, R.A., Wolff, N.A., Lee, W.K., Thévenod, F. 2012. Lipocalin-2 (24p3/ neutrophil gelatinase-associated lipocalin (NGAL)) receptor is expressed in distal nephron and mediates protein endocytosis. *J. Biol. Chem.* 287:159–169.

Langelueddecke, C., Roussa, E., Fenton, R.A., Thévenod, F. 2013. Expression and function of the lipocalin-2 (24p3/NGAL) receptor in rodent and human intestinal epithelia. *PLoS One* 8:e71586. doi: 10.1371/journal.pone.0071586.

Langelueddecke, C., Lee, W.-K., Thevenod, F. 2014. Differential transcytosis and toxicity of the hNGAL receptor ligands cadmium-metallothionein and cadmium-phytochelatin in colon-like Caco-2 cells: Implications for cadmium toxicity. *Toxicol. Lett.* 226:228–235.

Larsson, S.C., Orsini, N., Wolk, A. 2015. Urinary cadmium concentration and risk of breast cancer: a systematic review and dose-response meta-analysis. *Am. J. Epidemiol.* 182:375–380.

Larsson, S.C., Wolk, A. 2016. Urinary cadmium and mortality from all causes, cancer and cardiovascular disease in the general population: Systematic review and meta-analysis of cohort studies. *Int. J. Epidemiol.* 45:782–791.

Lasker, J.M., Chen, W.B., Wolf, I., Bloswick, B.P., Wilson, P.D., Powell, P.K. 2000. Formation of 20-hydroxyeicosatetraenoic acid, a vasoactive and natriuretic eicosanoid, in human kidney. Role of Cyp4F2 and Cyp4A11. *J. Biol. Chem.* 275:4118–4126.

Lauwerys, R.R., Bernard, A.M., Roels, H.A., Buchet, J.P. 1994. Cadmium: Exposure markers as predictors of nephrotoxic effects. *Clin. Chem.* 40(7):1391–1394.

Lee, B.K., Kim, Y. 2012. Association of blood cadmium with hypertension in the Korean general population: Analysis of the 2008–2010 Korean National Health and Nutrition Examination Survey data. *Am. J. Ind. Med.* 55:1060–1067.

Leung, K.W., Liu, M., Xu, X., Seiler, M.J., Barnstable, C.J., Tombran-Tink, J., 2008. Expression of ZnT and ZIP zinc transporters in the human RPE and their regulation by neurotrophic factors. *Investig. Ophthalmol. Vis. Sci.* 49:1221–1231.

Levesque, M., Martineau, C., Jumarie, C., Moreau, R. 2008. Characterization of cadmium uptake and cytotoxicity in human osteoblast-like MG-63 cells. *Toxicol. Appl. Pharmacol.* 231:308–317.

Li, X., Jiang, X., Sun, J., Zhu, C., Li, X., Tian, L., et al. 2017. Cytoprotective effects of dietary flavonoids against cadmium-induced toxicity. *Ann. NY Acad. Sci.* 1398:5–19.

Li, H., Wang, Z., Fu, Z., Yan, M., Wu, N., Wu, H., et al. 2018. Associations between blood cadmium levels and cognitive function in a cross-sectional study of US adults aged 60 years or older. *BMJ* 8:e020533. doi: 10.1136/bmjopen-2017-020533.

Lin, J.-L., Lu, F.-H., Yeh, K.-H. 1995. Increased body cadmium burden in Chinese women without smoking and occupational exposure. *Clin. Toxicol.* 33:639–644.

Lin, Y.S., Caffrey, J.L., Chang, M.H., Dowling, N., Lin. J.W. 2010. Cigarette smoking, cadmium exposure, and zinc intake on obstructive lung disorder. *Respir. Res.* 11:53. doi: 10.1186/1465-9921-11-53.

Lin, Y.S., Caffrey, J.L., Lin, J.W., Bayliss, D., Faramawi, M.F., Bateson, T.F., et al. 2013. Increased risk of cancer mortality associated with cadmium exposures in older Americans with low zinc intake. *J. Toxicol. Environ. Health A* 76:1–15.

Lin, Y.S., Ho, W.C., Caffrey, J.L., Sonawane, B. 2014. Low serum zinc is associated with elevated risk of cadmium nephrotoxicity. *Environ. Res.* 134:33–38.

Lin, J., Zhang, F., Lei, Y., 2016. Dietary intake and urinary level of cadmium and breast cancer risk: A meta-analysis. *Cancer Epidemiol.* 42:101–107.

Lind, Y., Engman, J., Jorhem, L., Glynm, A.W. 1998. Accumulation of cadmium from wheat bran, sugar-beet fibre, carrots and cadmium chloride in the liver and kidneys of mice. *Br. J. Nutr.* 80:205–211.

Liu, J., Habeecu, S.S, Liu, Y. and Klaassen, C.D. 1998. Acute cadmium injection is not a good model to study chronic Cd nephropathy: Comparison of chronic $CdCl_2$ and CdMT exposure with acute CdMT injection in rats. *Toxicol. Appl. Pharmacol.* 153:48–58.

Liu, Y., Liu, J., Habeedu, S.S.M., Klaassen, C.D. 1999. Metallothionein protects against the nephrotoxicity produced by chronic CdMT exposure. *Toxicol. Sci.* 50:221–227.

Lorentzon, M., Cummings, S.R. 2015. Osteoporosis: The evolution of a diagnosis. *J. Intern. Med.* 277:650–661.

Luo, J., Hendryx, M. 2014. Relationship between blood cadmium, lead, and serum thyroid measures in US adults: The National Health and Nutrition Examination Survey (NHANES) 2007–2010. *Int. J. Environ. Health Res.* 24:125–136.

Lyon, T.D., Aughey, E., Scott, R., and Fell, G.S. 1999. Cadmium concentrations in human kidney in the UK: 1978-1993. *J. Environ. Monit.* 1:227–231.

Madrigal, J.M., Ricardo, A.C., Persky, V., Turyk, M. 2019. Associations between blood cadmium concentration and kidney function in the U.S. population: Impact of sex, diabetes and hypertension. *Environ. Res.* 169:180–188.

Maruzeni, S., Nishijo, M., Nakamura, K., Morikawa, Y., Sakurai, M., Nakashima, M., et al. 2014. Mortality and causes of deaths of inhabitants with renal dysfunction induced by cadmium exposure of the polluted Jinzu River basin, Toyama, Japan: A 26-year follow-up. *Environ. Health* 13:18. doi: 10.1186/1476-069X-13-18.

McDermott, J.R., Geng, X., Jiang, L., Gálvez-Peralta, M., Chen, F., Nebert, D.W., et al. 2016. Zinc- and bicarbonate-dependent ZIP8 transporter mediates selenite uptake. *Oncotarget* 7:35327–35340.

McDonough, A.A., Leong, P.K., Yang, L.E. 2003. Mechanisms of pressure natriuresis: How blood pressure regulates renal sodium transport. *Ann. NY Acad. Sci.* 986:669–677.

McDonough, A.A. 2010. Mechanisms of proximal tubule sodium transport regulation that link extracellular fluid volume and blood pressure. *Am. J. Physiol. Regul. Integr. Comp. Physiol.* 298:R851–861.

McDonough, A.A., Nguyen, M.T.X. 2015. Maintaining balance under pressure: Integrated regulation of renal transporters during hypertension. *Hypertens.* 66:450–455.

McLaughlin, M.J. and Singh, B.R. 1999. Cadmium in soils and plants. In: *Developments in Plant and Soil Sciences*, Volume 85, eds. M. J. McLaughlin and B. R. Singh. Kluwer Academic Publishers: Dorddrecht/Boston/London, pp. 1–7.

Meltzer, H.M., Brantsaeter, A.L., Borch-Iohnsen, B., Ellingsen, D.G., Alexander, J., Thomassen, Y. et al. 2010. Low iron stores are related to higher blood concentrations of manganese, cobalt and cadmium in non-smoking, Norwegian women in the HUNT 2 study. *Environ. Res.* 110:497–504.

Menke, A., Muntner, P., Silbergeld, E.K., Platz, E.A., Guallar, E. 2009. Cadmium levels in urine and mortality among U.S. adults. *Environ. Health Perspect.* 117:190–196.

Memon, A.R., Schröder, P. 2009. Implications of metal accumulation mechanisms to phytoremediation. *Environ. Sci. Pollut. Res. Int.* 16:162–175.

Min, K.S., Hatta, A., Onosaka, S., Ohta, N., Okada, Y., Tanaka, K. 1987. Protective role of renal metallothionein against Cd nephropathy in rats. *Toxicol. Appl. Pharmacol.* 88:294–301.

Min, K.S., Ueda, H., Kihara, T., Tanaka, K. 2008. Increased hepatic accumulation of ingested Cd is associated with upregulation of several intestinal transporters in mice fed diets deficient in essential metals. *Toxicol Sci.* 106:284–289.

Min, K.B., Lee, K.J., Park, J.B., Min, J.Y. 2012. Lead and cadmium levels and balance and vestibular dysfunction among adult participants in the National Health and Nutrition Examination Survey (NHANES) 1999–2004. *Environ. Health Perspect.* 120, 413–417.

Min, K.S., Sano, E., Ueda, H., Sakazaki, F., Yamada, K., Takano, M., et al. 2015. Dietary deficiency of calcium and/or iron, an age-related risk factor for renal accumulation of cadmium in mice. *Biol. Pharm. Bull.* 38:1557–1563.

Min, J.Y., Min, K.B. 2016. Blood cadmium levels and Alzheimer's disease mortality risk in older US adults. *Environ. Health* 15: 69. doi: 10.1186/s12940-016-0155–7.

Minh, N.G., Hough, R.L., Thuy, L.T., Nyberg, Y., Mai, L.B., Vinh, N.C., et al. 2012. Assessing dietary exposure to cadmium in a metal recycling community in Vietnam: Age and gender aspects. *Sci. Total Environ.* 416:164–171.

Misra, R.R., Hochadel, J.F., Smith, G.T., Cook, J.C., Waalkes M.P., Wink, D.A. 1996. Evidence that nitric oxide enhances cadmium toxicity by displacing the metal from metallothionein. *Chem. Res. Toxicol.* 9:326–332.

Mitchell, C.J., Shawki, A., Ganz, T., Nemeth, E., Mackenzie, B. 2014. Functional properties of human ferroportin, a cellular iron exporter reactive also with cobalt and zinc. *Am. J. Physiol. Cell Physiol.* 306:C450–C459.

Moberg, L., Nilsson, P.M., Samsioe, G., Sallsten, G., Barregard, L., Engström, G., et al. 2017. Increased blood cadmium levels were not associated with increased fracture risk but with increased total mortality in women: The Malmö Diet and Cancer Study. *Osteoporos. Int.* 28:2401–2408.

Morrison, A.R., Mindel, G. 2004. Hypertension: A disorder of volume control? What is the evidence? *Adv. Chronic Kidney Dis.* 11:197–201.

Mortensen, M.E., Wong, L.Y., Osterloh, J.D. 2011. Smoking status and urine cadmium above levels associated with subclinical renal effects in U.S. adults without chronic kidney disease. *Int. J. Hyg. Environ. Health* 214:305–310.

Moulis, J-M. 2010. Cellular mechanisms of cadmium toxicity related to the homeostasis of essential metals. *Biometals* 23:877–896.

Myong, J.P., Kim, H.R., Baker, D., Choi, B. 2012. Blood cadmium and moderate-to-severe glomerular dysfunction in Korean adults: Analysis of KNHANES 2005–2008 data. *Int. Arch. Occup. Environ. Health* 85:885–893.

Nagyova, A., Galbavy, S., Ginter, E. 1994. Histopathological evidence of vitamin C protection against Cd-nephrotoxicity in guinea pigs. Exp.Toxicol. Pathol. 46:11–14.

Nakagawa, H., Nishijo, M. 1996. Environmental cadmium exposure, hypertension and cardiovascular risk. *J. Cardiovascular Risk* 3:11–17.

Nawrot, T.S., Van Hecke, E., Thijs, L., Richart, T., Kuznetsova, T., Jin, Y., et al. 2008. Cadmium-related mortality and long-term secular trends in the cadmium body burden of an environmentally exposed population. *Environ. Health Perspect.* 116:1620–1628.

Narwot, T., Geusens, P., Nulens, T.S., Nemery, B. 2010. Occupational cadmium exposure and calcium excretion, bone density, and osteoporosis in men. *J. Bone Miner. Res.* 15:1441–1445.

Navas-Acien, A., Tellez-Plaza, M., Guallar, E., Muntner, P., Silbergeld, E., Jaar, B., et al. 2009. Blood cadmium and lead and chronic kidney disease in US adults: A joint analysis. *Am. J. Epidemiol.* 170:1156–1164.

Nielsen, R., Christensen, E.I., Birn, H. 2016. Megalin and cubilin in proximal tubule protein reabsorption: From experimental models to human disease. *Kidney Int.* 89:58–67.

Nishijo, M., Nakagawa, H., Suwazono, Y., Nogawa, K., Sakurai, M., Ishizaki, M., et al. 2018. Cancer mortality in residents of the cadmium-polluted Jinzu river basin in Toyama, Japan. *Toxics* 6. pii: E23. doi:10.3390/toxics6020023.

Noda, M., Kitagawa, M. 1990. A quantitative study of iliac bone histopathology on 62 cases with itai-itai disease. *Calcif. Tiss. Int.* 47:66–74.

Noda, M., Yasuda, M., Kitagawa, M. 1991. Iron as a possible aggravating factor for osteopathy in itai-itai disease, a disease associated with chronic cadmium intoxication. *J. Bone Miner. Res.* 6:245–255.

Nogawa, K., Ishizaki, A., Fukushima, J., Shibata, I., Hagino, N. 1975. Studies on the women with acquired Fanconi syndrome observed in the Ichi River basin polluted by cadmium. *Environ. Res.* 10:280–307.

Nogawa, K., Yamada, Y., Honda, R., Tsuritani, I., Kobayashi, E., Ishizaki, M. 1984. Copper and zinc levels in serum and urine of cadmium-exposed people with special reference to renal tubular damage. *Environ. Res.* 33:29–38.

Nogawa, K., Tsuritani, I., Kido, T., Honda, R., Yamada, Y, Ishizaki, M. 1987. Mechanism for bone disease found in inhabitants environmentally exposed to cadmium: Decreased serum 1 alpha, 25-dihydroxyvitamin D level. *Int. Arch. Occup. Environ. Health* 59:21–30.

Nogawa, K., Sakurai, M., Ishizaki, M., Kido, T., Nakagawa, H., Suwazono, Y. 2017. Threshold limit values of the cadmium concentration in rice in the development of itai-itai disease using benchmark dose analysis. *J. Appl. Toxicol.* 37:962–966.

Nogawa, K., Suwazono, Y., Nishijo, M., Sakurai, M., Ishizaki, M., Morikawa, Y., et al. 2018. Relationship between mortality and rice cadmium concentration in inhabitants of the polluted Jinzu River basin, Toyama, Japan: A 26 year follow-up. *J. Appl. Toxicol.* 38:855–861.

Nomiyama, K., Foulkes, E.C. 1977. Reabsorption of filtered cadmium-metallothionein in the rabbit kidney. *Proc. Soc. Exp. Biol. Med.* 156:97–99.

Norden, A.G.W., Lapsley, M., Lee. P.J., Pusey, C.D., Scheinman, S.J., Tam, F.W.K., et al. 2001. Glomerular protein sieving and implications for renal failure in Fanconi syndrome. *Kidney Int.* 60:1885–1892.

Norden, A.G.W., Lapsley, M., Unwin, R.H.J. 2014. Urine retinol-binding protein 4: A functional biomarker of the proximal renal tubule. *Adv. Clin. Chem.* 63:85–122.

Oliver-Williams, C., Howard, A.G., Navas-Acien, A., Howard, B.V., Tellez-Plaza, M., Franceschini, N. 2018. Cadmium body burden, hypertension, and changes in blood pressure over time: Results from a prospective cohort study in American Indians. *J. Am. Soc. Hypertens.* 12:426–437.

Olsen, L., Lind, P.M., and Lind, L. 2012. Gender differences for associations between circulating levels of metals and coronary risk in the elderly. *Int. J. Hyg. Environ. Health* 215:411–417.

Orlowski, C., Piotrowski, J.K., Subdys, J.K., Gross, A. 1998. Urinary cadmium as indicator of renal cadmium in humans: An autopsy study. *Hum. Exp. Toxicol.* 17:302–306.

Ott, S.M., Maloney, N.A., Coburn, J.W., Alfrey, A.C., Sherrard, D.J. 1982. The prevalence of bone aluminum deposition in renal osteodystrophy and its relation to the response to calcitriol therapy. *N. Engl. J. Med.* 307:709–713.

Ott, S.M., Andress, D.L., Nebeker, H.G., Milliner, D.S., Maloney, N.A., Coburn, J.W., et al. 1986. Changes in bone histology after treatment with deferoxamine. *Kidney Int. Suppl.* 18:S108–S113.

Palmer, L.G., Schnermann, J. 2015. Integrated control of Na transport along the nephron. *Clin. J. Am. Soc. Nephrol.* 10:676–687.

Pena, A., Iturri, S.J. 1993. Cadmium as hypertensive agent. Effect on ion excretion in rats. *Comp. Biochem. Physiol. C* 106:315–319.

Peng, Q., Bakulski, K.M., Nan, B., Park, S.K. 2017. Cadmium and Alzheimer's disease mortality in U.S. adults: Updated evidence with a urinary biomarker and extended follow-up time. *Environ. Res.* 157:44–51.

Pennemans, V., DeWinter, L.M., Munters, E., Nawrot, T.S., Vamn Kerkhove, E., Rigo, J.-M., et al. 2011. The association between urinary kidney injury molecule 1 and urinary cadmium in elderly during long-term, low-dose cadmium exposure: A pilot study. *Environ. Health* 10:77.

Perry, H.M. Jr., Erlanger, M.W. 1981. Sodium retention in rats with cadmium-induced hypertension. *Sci. Total Environ.* 22:31–38.

Perry, H.M. Jr., Erlanger, M., Perry, E.F. 1977. Elevated systolic pressure following chronic low-level cadmiun feeding. *Am. J. Physiol.* 232:H114–H121.

Perry, H.M., Erlanger, M.W. 1978. Pressor effect of chronically feeding cadmium and lead together. *Trace Subst. Environ. Health* 12:268–275.

Perry, H.M., Erlanger, M., and Perry, E.F. 1978. Increase in the systolic pressure of rats chronically fed cadmium. *Am. J. Physiol.* 235:H385–H391.

Perry, H.M., Erlanger, M., Perry, E.F. 1979. Increase in the systolic pressure of rats chronically fed cadmium. *Environ. Health Perspect.* 28:251–260.

Perry, H.M., Erlanger, M., Perry, E.F. 1979. Increase in the systolic pressure of rats chronically fed cadmium. *Environ. Health Perspect.* 28:251–260.

Peterson, P.A., Evrin, P.-E., Berggard, I. 1969. Differentiation of glomerular, tubular, and normal proteinuria: Determination of urinary excretion of β_2-microglobulin, albumin, and total protein. *J. Clin. Invest.* 48:1189–1198.

Pivato, M., Fabrega-Prats, M., Masi, A. 2014. Low-molecular-weight thiols in plants: Functional and analytical implications. *Arch. Biochem. Biophys.* 560:83–99.

Phelps, K.R., Einhorn, T.A., Vigorita, V.J., Lieberman, R.L., Uribarri, J. 1986. Acidosis-induced osteomalacia: Metabolic studies and skeletal histomorphometry. *Bone* 7:171–179.

Phelps, K.R., Vigorita, V.J., Bansal, M., Einhorn, T.A. 1988. Histochemical demonstration of iron but not aluminum in a case of dialysis-associated osteomalacia. *Am. J. Med.* 84:775–780.

Pierides, A.M. 1984. Iron and aluminum osteomalacia in hemodialysis patients (letter). *N. Engl. J. Med.* 310:323.

Piscator, M., Bjorck, L., Nordberg, M. 1981. β_2-microglobulin levels in serum and urine of cadmium exposed rabbits. *Acta Pharmacol. Toxicol.* 49:1–7.

Pizent, A., Jurasovic, J., Telisman, S. 2003. Serum calcium, zinc, and copper in relation to biomarkers of lead and cadmium in men. *J. Trace Elem. Med. Biol.* 17:199–205.

Portman, R.J., Kissane, J.M., Robson, A.M. 1986. Use of β_2 microglobulin to diagnose tubulo-interstitial renal lesions in children. *Kidney Int.* 30:91–98.

Prakash, C., Zhang, J.Y., Falck, J.R. Chauman, K., Blair, A. 1992. 20-Hydroxyeicosatetraenoic acid is excreted as a glucuronide conjugate in human urine. *Biochem. Biophys. Res. Comm.* 158:728–733.

Price, R.G. 1992. Measurement of N-acetyl-beta-glucosaminidase and its isoenzymes in urine: Methods and clinical applications. *Eur J. Clin. Chem. Clin. Biochem.* 30:693–705.

Prozialeck, W.C., Vaidya, V.S., Liu, J., Waalkes, M.P., Edwards, J.R., Lamar, P.C., et al. 2007. Kidney injury molecule-1 is an early biomarker of cadmium nephrotoxicity. *Kidney Int.* 72:985–993.

Prozialeck, W.C., Edwards, J.R., Lamar, P.C., Liu, J., Vaidya, V.S., Bonventre, J.V. 2009. Expression of kidney injury molecule-1 (Kim-1) in relation to necrosis and apoptosis during the early stages of Cd-induced proximal tubular injury. *Toxicol. Appl. Pharmacol.* 238:306–314.

Prozialeck, W.C., Edwards, J.R. 2010. Early biomarkers of cadmium exposure and nephrotoxicity. *Biometals* 23:793–809.

Prozialeck, W.C., Edwards, J.R. 2012. Mechanisms of cadmium-induced proximal tubule injury: New insights with implications for biomonitoring and therapeutic interventions. *J. Pharmacol. Exp. Ther.* 343, 2–12.

Prozialeck, W.C., Lamar, P.C., Edwards, J.R. 2016. Effects of sub-chronic Cd exposure on levels of copper, selenium, zinc, iron and other essential metals in rat renal cortex. *Toxicol. Rep.* 3:740–746.

Puri, V.N. 1999. Cadmium induced hypertension. *Clin. Exp. Hypertens.* 21:79–84.

Quraishi, S.M., Adams, S.M., Meliker, J.R., Li, W., Luo, J., Neuhouser, M.L., et al. 2016. Urinary cadmium and estimated dietary cadmium in the Women's Health Initiative. *J. Expo. Sci. Environ. Epidemiol.* 26:303–308.

Raisz, L.G. 2005. Pathogenesis of osteoporosis: Concepts, conflicts, and prospects. *J. Clin. Invest.* 115:3318–3325.

Rapoport, J., Chaimovitz, C., Abulfil, A., Mostovlavsky, M., Gazit, D., Bab, I. 1987. Aluminum-related osteomalacia: Clinical and histologial improvement following treatment with desferrioxamine (DFO). *Isr. J. Med. Sci.* 23:1242–1246.

Rentschler, G., Kippler, M., Axmon, A., Raqib, R., Skerfving, S., Vahter, M., et al. 2014. Cadmium concentrations in human blood and urine are associated with polymorphisms in zinc transporter genes. *Metallomics* 6:885–891.

Revis, N.W., Zinsmeister, A.R., Bull, R. 1981. Atherosclerosis and hypertension induction by lead and cadmium ions: An effect prevented by calcium ion. *Proc. Natl. Acad. Sci. USA* 78:6494–6498.

Riederer, A.M., Belova, A., George, B.J., Anastas, P.T. 2013. Urinary cadmium in the 1999–2008 U.S. National Health and Nutrition Examination Survey (NHANES). *Environ. Sci. Technol.* 47:1137–1147.

Rimbach, G., Pallauf, J. 1997. Cadmium accumulation, zinc status and mineral bioavailability of growing rats fed diets high in zinc with increasing amounts of phytic acid. *Biol. Trace Elem. Res.* 57:59–70.

Roels, H.A., Lauwerys, R.R., Buchet, J.P., Bernard, A.M., Vos, A., Oversteyns, M. (1989). Health significance of cadmium induced renal dysfunction: A five year follow up. *Br. J. Indust. Med.* 46, 755–764.

Roels, H.A., Bernard, A.M., Buchet, J.P., Lauwerys, R.R., Hotter G., Ramis, I.U., et al. 1993. Markers of early renal changes induced by industrial pollutants. III Application to workers exposed to cadmium. *Br. J. Indust. Med.* 50:37–48.

Roels, H.A., Hoet, P., Lison, D. 1999. Usefulness of biomarkers of exposure to inorganic mercury, lead, or cadmium in controlling occupational and environmental risks of nephrotoxicity. *Ren. Fail.* 21:251–262.

Rokadia, H.K., Agarwal, S. 2013. Serum heavy metals and obstructive lung disease: Results from the National Health and Nutrition Examination Survey. *Chest* 143:388–397.

Rossier, B.C., Staub, O., Hummler, E. 2013. Genetic dissection of sodium and potassium transport along the aldosterone-sensitive distal nephron: Importance in the control of blood pressure and hypertension. *FEBS Lett.* 587:1929–1941.

Rubino, F.M. 2015. Toxicity of glutathione-binding metals: A Review of targets and mechanisms. *Toxics* 3:20–62. doi:10.3390/toxics3010020.

Sabolić, I., Breljak, D., Skarica, M., Herak-Kramberger, C.M. 2010. Role of metallothionein in cadmium traffic and toxicity in kidneys and other mammalian organs. *Biometals* 23:897–926.

Sabolić, I., Skarica, M., Ljubojevic, M., Breljak, D., Herak-Kramberger, C.M., Crljen, V., et al. 2018. Expression and immunolocalization of metallothioneins MT1, MT2 and MT3 in rat nephron. *J. Trace Elem. Med. Biol.* 46:62–75.

Sacerdoti, D., Escalante, B., Abraham, N.G., McGiff, J.C., Levere, R.D., Schwartzman, M.L. 1989. Treatment with tin prevents the development of hypertension in spontaneously hypertensive rats. *Science* 243:388–390.

Saito, H., Shioji, R., Hurukawa, Y., Nagai, K., Arikawa, T., Saito, T., et al. 1977. Cadmium-induced proximal tubular dysfunction in a cadmium-polluted area. *Contr. Nephrol.* 6:1–12.

Sand, S., Becker, W. 2012. Assessment of dietary cadmium exposure in Sweden and population health concern including scenario analysis. *Food Chem. Toxicol.* 50:536–544.

Satarug, S., Haswell-Elkins, M.R., Moore, M.R. 2000a. Safe levels of cadmium intake to prevent renal toxicity in human subjects. *Br. J. Nutr.* 84:791–802.

Satarug, S., Baker, J.R., Reilly, P.E.B., Esumi, H., Moore, M.R. 2000b. Evidence for a synergistic interaction between cadmium and endotoxin toxicity and for nitric oxide and cadmium displacement of metals in the kidney. *Nitric Oxide* 4:431–440.

Satarug, S., Baker, J.R., Reilly, P.E.B., Moore M.R., Williams, D.J. 2001. Changes in zinc and copper homeostasis in human livers and kidneys associated with exposure to environmental cadmium. *Hum. Exp. Toxicol.* 20:205–213.

Satarug, S., Baker, J.R., Reilly, P.E.B., Moore, M.R., Williams, D.J. 2002. Cadmium levels in the lung, liver, kidney cortex, and urine samples from Australians without occupational exposure to metals. *Arch. Environ. Health* 57:69–77.

Satarug, S., Baker, J.R., Urbenjapol, S., Haswell-Elkins, M., Reilly, P.E., Williams, D.J. et al. 2003. A global perspective on cadmium pollution and toxicity in non-occupationally exposed population. *Toxicol. Lett.* 31(137):65–83.

Satarug, S., Moore, M.R. 2004. Adverse health effects of chronic exposure to low-level cadmium in foodstuffs and cigarette smoke. *Environ. Health Perspect.* 112:1099–1103.

Satarug, S., Ujjin, P., Vanavanitkun, Y., Baker, J.R., Moore, M.R. 2004. Influence of body iron store status and cigarette smoking on cadmium body burden of healthy Thai women and men. *Toxicol. Lett.* 148:177–185.

Satarug, S., Nishijo, M., Ujjin, P., Vanavanitkun, Y., Moore, M.R., 2005. Cadmium-induced nephropathy in the development of high blood pressure. *Toxicol. Lett.* 157:57–68.

Satarug, S., Nishijo, M., Lasker, J.M., Edwards, R.J., Moore, M.R. 2006. Kidney dysfunction and hypertension: Role for cadmium, p450 and heme oxygenases? *Tohoku J. Exp. Med.* 208:179–202.

Satarug, S., Kikuchi, M., Wisedpanichkij, R., Li, B., Takeda, K., Na-Bangchang, K., et al. 2008. Prevention of cadmium accumulation in retinal pigment epithelium with manganese and zinc. *Exp. Eye Res.* 87:587–593.

Satarug, S., Garrett, S.H., Sens, M.A., Sens, D.A. 2010. Cadmium, environmental exposure, and health outcomes. *Environ. Health Perspect.* 118:182–190.

Satarug, S., Moore, M.R. 2012. Emerging roles of cadmium and heme oxygenase in type-2 diabetes and cancer susceptibility. *Tohoku J. Exp. Med.* 228:267–288.

Satarug, S., Vesey, D.A., Gobe, G.C. 2017a. Kidney cadmium toxicity, diabetes and high blood pressure: The perfect storm. *Tohoku J. Exp. Med.* 241:65–87.

Satarug, S., Vesey, D.A., Gobe, G.C. 2017b. Current health risk assessment practice for dietary cadmium: Data from different countries. *Food Chem. Toxicol.* 106:430–445.

Satarug, S., Vesey, D.A., Gobe, G.C. 2017c. Health risk assessment of dietary cadmium intake: Do current guidelines indicate how much is safe? *Environ. Health Perspect.* 125:284–288.

Satarug, S. 2018. Dietary cadmium intake and its effects on kidneys. *Toxics* 6, pii: E15. doi:10.3390/toxics6010015.

Satarug, S., Ruangyuttikarn, W., Nishijo, M., Ruiz, P. 2018a. Urinary cadmium threshold to prevent kidney disease development. *Toxics* 6. pii: E26. doi:10.3390/toxics6020026.

Satarug, S., Boonprasert, K., Gobe, G., Ruenweerayut, R., Johnson, D., Na-Bangchang, K., et al. 2018b. Chronic exposure to cadmium is associated with a marked reduction in glomerular filtration rate. *Clin. Kidney J.* doi:10.1093/ckj/sfy113.

Satarug, S., Vesey, D.A., Nishijo, M., Ruangyuttikarn, W., Gobe, G.C. 2019a. The inverse association of glomerular function and urinary β2-MG excretion and its implications for cadmium health risk assessment. *Environ. Res.* 173:40–47.

Satarug, S.S., Vesey, D.A., Ruangyuttikarn, W., Nishijo, M., Gobe, G.C., Phelps, K.R. 2019b. The source and pathophysiologic significance of excreted cadmium. *Toxics* 7:55.

Sato, M., Mehra, R.K., and Bremner, I. 1984. Measurement of plasma metallothionein-I in the assessment of the zinc status of zinc-deficient and stressed rats. *J. Nutr.* 114:1683–1689.

Scheuhammer, A.M. 1988. The dose-dependent deposition of cadmium in organs of Japanese quails following oral administration. *Toxicol. Appl. Pharmacol.* 95:153–161.

Schleicher, R.L., Carroll, M.D., Ford. E.S., Lacher, D.A. 2009. Serum vitamin C and the prevalence of vitamin C deficiency in the United States: 2003–2004 National Health and Nutrition Examination Survey (NHANES). *Am. J. Clin. Nutr.* 90:1252–1263.

Schnaper, H.W. 2017. The tubulointerstitial pathophysiology of progressive kidney disease. *Adv. Chron. Kidney Dis.* 24:107–116.

Scinicariello, F., Abadin, H.G., and Murray, H.E. 2011. Association of low-level blood lead and blood pressure in NHANES 1999-2006. *Environ. Res.* 111:1249–1257.

Shargorodsky, J., Curhan, S.G., Henderson, E., Eavey, R., Curhan, G.C. 2011. Heavy metals exposure and hearing loss in US adolescents. *Arch. Otolaryngol. Head Neck Surg.* 137:1183–1189.

Shi, Z., Taylor, A.W., Riley, M., Byles, J., Liu, J., Noakes, M. 2017. Association between dietary patterns, cadmium intake and chronic kidney disease among adults. *Clin. Nutr.* 5614:31366–31368.

Schneider, S.N., Liu, Z., Wang, B., Miller, M.L., Afton. S.E., Soleimani, M., et al. 2014. Oral cadmium in mice carrying 5 versus 2 copies of the Slc39a8 gene: Comparison of uptake, distribution, metal content, and toxicity. *Int. J. Toxicol.* 33:14–20.

Schroeder, H.A., Balassa, J.J. 1963. Cadmium: Uptake by vegetables from superphosphate in soil. *Science* 140:819–820.

Schroeder, H.A. 1964. Cadmium hypertension in rats. *Am. J. Physiol.* 207:62–66.

Schutte, R., Nawrot, T.S., Richart, T., Thijs, L., Vanderschueren, D., Kuznetsova, T., Van Hecke, E., Roels, H.A., Staessen, J.A. 2008. Bone resorption and environmental exposure to cadmium in women: A population study. *Environ. Health Perspect.* 116:777–783.

Schwartz, G.G., Il'yasova, D., Ivanova, A. 2003. Urinary cadmium, impaired fasting glucose, and diabetes in the NHANES III. *Diabetes Care* 26:468–470.

Schwartzman, M.L., Martasek, P., Rios, A.R., Levere, R.D., Solangi, K., Goodman, A.L., et al. 1990. Cytochrome P450-dependent arachidonic acid metabolism in human kidney. *Kidney Int.* 37:94–99.

Scinicariello, F., Buser, M.C. 2015. Blood cadmium and depressive symptoms in young adults (aged 20–39 years). *Psychol. Med.* 45:807–815.

Shim, J.E., Paik, H.Y., Shin, C.S., Park, K.S., Lee, H.K. 2010. Vitamin C nutriture in newly diagnosed diabetes. *J. Nutr. Sci. Vitaminol.* 56:217–221.

Skröder, H., Hawkesworth, S., Kippler, M., El Arifeen, S., Wagatsuma, Y., Moore, S.E., et al. 2015. Kidney function and blood pressure in preschool-aged children exposed to cadmium and arsenic—potential alleviation by selenium. *Environ. Res.* 140:205–213

Soodvilai, S., Nantavishit, J., Muanprasat, C., Chatsudthipong, V. 2011. Renal organic cation transporters mediated cadmium-induced nephrotoxicity. *Toxicol. Lett.* 204:38–42.

Staessen, J.A., Roels, H.A., Emelianjov, D., Kuznetsova, T., Thijs, L., Vangronsveld, J., et al. 1999. Environmental exposure to cadmium, forearm bone density, and risk of fractures: prospective population study. *Lancet* 353:1140–1144.

Sudo, J.-I., Hayashi, T., Soyama, M., Fukata, M., Kakuino, K. 1994. Kinetics of Cd^{2+} in plasma, liver and kidneys after single intravenous injection of Cd-metallothionein-II. *Eur. J. Pharmacol.* 270:229–235.

Suh, Y.J., Lee, J.E., Lee, D.H., Yi, H.G., Lee, M.H., Kim, C.S., et al. 2016. Prevalence and relationships of iron deficiency anemia with blood cadmium and vitamin D levels in Korean women. *J. Korean Med. Sci.* 31:25–32.

Sullivan, V.K., Burnett, F.R., Cousins, R.J. 1998. Metallothionein expression is increased in monocytes and erythrocytes of young men during zinc supplementation. *J. Nutr.* 28:707–713.

Suwazono, Y., Kido, T., Nakagawa, H., Nishijo, M., Honda, R., Kobayashi, E., et al. 2009. Biological half-life of cadmium in the urine in the habitats after cessation of exposure. *Biomarkers* 14:77–81.

Suwazono, Y., Nogawa, K., Morikawa, Y., Nishijo, M., Kobayashi, E., Kido, T., et al. 2014. Impact of urinary cadmium on mortality in the Japanese general population in cadmium non-polluted areas. *Int. J. Hyg. Environ. Health* 217:807–812.

Swaddiwudhipong, W., Limpatanachote, P., Nishijo, M., Honda, R., Mahasakpan, P., Krintratun, S. 2010. Cadmium-exposed population in Mae Sot District, Tak Province: 3. Associations between urinary cadmium and renal dysfunction, hypertension, diabetes, and urinary stones. *J. Med. Assoc. Thai.* 93:231–238.

Swaddiwudhipong, W., Limpatanachote, P., Mahasakpan, P., Krintratun, S., Punta, B., Funkhiew, T. 2012. Progress in cadmium-related health effects in persons with high environmental exposure in northwestern Thailand: A five-year follow-up. *Environ. Res.* 112:194–198.

Swaddiwudhipong, W., Nguntra, P., Kaewnate, Y., Mahasakpan, P., Limpatanachote, P., Aunjai, T., et al. 2015. Human health effects from cadmium exposure: Comparison between persons living in cadmium-contaminated and non-contaminated areas in northwestern Thailand. *Southeast Asian J. Trop. Med. Publ. Health* 46:133–142.

Takebayashi, S., Harada, T., Kamura, S., Satoh, T., Segawa, M., Yajima, K. 1987. Cadmium-induced osteopathy: Clinical and autopsy findings of four patients. *Bone Pathol. Appl. Pathol.* 5:190–197.

Takebayashi, S., Jimi, S., Seegawa, M., Kiyoshi, Y. 2000. Cadmium induces osteomalacia mediated by proximal tubular atrophy and disturbances of phosphate reabsorption. A study of 11 autopsies. *Pathol. Res. Pract.* 196:653–663.

Takeda, K., Fujita, H., Shibahara, S. 1995. Differential control of the metal-mediated activation of the human heme oxygenase-1 and metallothionein IIA genes. *Biochem. Biophys. Res. Commun.* 207:160–167.

Takenaka, S., Karg, E., Kreyling, W.G., Lentner, B., Schulz, H., Ziesenis, A., et al. 2004. Fate and toxic effects of inhaled ultrafine cadmium oxide particles in the rat lung. *Inhal. Toxicol.* 16 (Suppl 1):83–92.

Tandon, S.K., Khandelwal, S., Jain, V.K., Mathur, N. 1993. Influence of dietary iron deficiency on acute metal intoxication. *Biometals* 6:133–138.

Tanimoto, A., Hamada, T., Koide, O. 1993. Cell death and regeneration of renal proximal tubular cells in rats with subchronic cadmium intoxication. *Toxicol. Pathol.* 21:341–352.

Tellez-Plaza, M., Navas-Acien, A., Crainiceanu, C.M., Guallar, E. 2008. Cadmium exposure and hypertension in the 1999–2004 National Health and Nutrition Examination Survey (NHANES). *Environ. Health Perspect.* 116:51–56.

Tellez-Plaza, M., Navas-Acien, A., Crainiceanu, C.M., Sharrett, A.R., Guallar, E. 2010. Cadmium and peripheral arterial disease: Gender differences in the 1999–2004 US National Health and Nutrition Examination Survey. *Am. J. Epidemiol.* 172:671–681.

Tellez-Plaza M, Navas-Acien A, Menke A, Crainiceanu CM, Pastor-Barriuso R, Guallar, E. 2012. Cadmium exposure and all-cause and cardiovascular mortality in the U.S. general population. *Environ. Health Perspect.* 120:1017–1022.

Thevenod, F., Friedmann, J.M. 1999. Cadmium-mediated oxidative stress in kidney proximal tubule cells induces degradation of Na^+/K^+-ATPase through proteasomal and endo-/lysosomal proteolytic pathways. *FASEB J.* 13:1751–1761.

Thevenod, F. 2003. Nephrotoxicity and the proximal tubule. Insights from cadmium. *Nephron Physiol* 93:87–93.

Thévenod, F., Wolff, N.A. 2016. Iron transport in the kidney: Implications for physiology and cadmium nephrotoxicity. *Metallomics* 8:17–42.

Thevenod, F., Fels, J., Lee, W.-K., Zarbock, R. 2019. Channels, transporters and receptors for cadmium and cadmium complexes in eukaryotic cells: Myths and facts. *Biometals* doi: 10.1007/s10534-019-00176-6.

Thijs, L., Staessen, J., Amery. A., Bruaux. P., Buchet, J.-P., Claeys, F., et al. 1992. Determinants of serum zinc in a random population sample of four Belgian towns with different degrees of environmental exposure to cadmium. *Environ. Health Perspect.* 98:251–258.

Treviño, S., Waalkes, M.P., Flores Hernández, J.A., León-Chavez, B.A., Aguilar-Alonso, P., Brambila, E. 2015. Chronic cadmium exposure in rats produces pancreatic impairment and insulin resistance in multiple peripheral tissues. *Arch. Biochem. Biophys.* 583:27–35.

Uchida, M., Teranishi, H., Aoshima, K., Katoh, T., Kasuya, M., Inadera, H. 2007. Elevated urinary levels of vitamin D-binding protein in the inhabitants of a cadmium polluted area, Jinzu River Basin, Japan. *Tohoku J. Exp. Med.* 211:269–274.

Ugarte, M., Osborne, N.N., Brown, L.A., Bishop, P.N. 2013. Iron, zinc, and copper in retinal physiology and disease. *Surv. Ophthalmol.* 58:585–609.

Ugarte, M., Osborne, N.N. 2014. Recent advances in the understanding of the role of zinc in ocular tissues. *Metallomics* 6:189–200.

Vacchi-Suzzi, C., Eriksen, K.T., Levine, K., McElroy, J., Tjonneland, A., Raaschou-Nielsen, O., et al. 2015. Dietary intake estimates and urinary cadmium levels in Danish postmenopausal women. *PLoS One* 10, e0138784. doi: 10.1371/journal.pone.0138784.

Vahter, M., Berglund, M., Nermell, B., Akesson, A. 1996. Bioavailability of cadmium from shellfish and mixed diet in women. *Toxicol. Appl. Pharmacol.* 136:332–341.

Valko, M., Jomova, K., Rhodes, C.J., Kuča, K., Musílek, K. 2016. Redox- and non-redox-metal-induced formation of free radicals and their role in human disease. *Arch. Toxicol.* 90:1–37.

Van Maele-Fabry, G., Lombaert, N., Lison, D., 2016. Dietary exposure to cadmium and risk of breast cancer in postmenopausal women: A systematic review and meta-analysis. *Environ. Int.* 86:1–13.

Vašák, M., Meloni, G. 2017. Mammalian metallothionein-3: New functional and structural insights. *Int. J. Mol. Sci.* 18. pii: E1117.

Vesey, D.A. 2010. Transport pathways for cadmium in the intestine and kidney proximal tubule: Focus on the interaction with essential metals. *Toxicol. Lett.*198:13–19.

Wagner, C.A., Rubio-Aliaga, I., Biber, J., Hernando, N. 2014. Genetic diseases of renal phosphate handling. *Nephrol. Dialys. Transplant.* 29 (Suppl 4):iv45–iv54.

Wallia, A., Allen, N.B., Badon, S., El Muayed, M. 2014. Association between urinary cadmium levels and prediabetes in the NHANES 2005–2010 population. *Int. J. Hyg. Environ. Health* 217:854–860.

Wallin, M., Sallsten, G., Lundh, T., Barregard, L. 2014. Low-level cadmium exposure and effects on kidney function. *Occup. Environ. Med.* 71:848–854.

Wang, S.J., Paek, D.M., Kim, R.H., Cha, B.S. 2002. Variation of systemic blood pressure in rats exposed to cadmium and nickel. *Environ. Res.* 88:116–119.

Wang, H., Zhu, G.H., Shi, Y., Weng, S., Jin, T., Kong, Q., Nordberg, G.F. 2003. Influence of environmental cadmium exposure on forearm bone density. *J. Bone Miner. Res.* 18:553–560.

Walter, A., Rimbach, G, Most, E., Pallauf, J. 1998. Effect of citric acid supplements to a maize-soya diet on the in vitro availability of minerals, trace elements and heavy metals. *J. Vet. Med.* 45:517–524.

Weaver, V.M., Kim N.-S., Jaar, B.G., Schwartz, B.S., Parsons, P.J., Steuerwald, A.J., et al. 2011. Associations of low-level urine cadmium with kidney function in lead workers. *Occup. Environ. Med.* 68:250–256.

Wei, X.L., He, J.R., Cen, Y.L., Su, Y., Chen, L.J., Lin, Y., et al. 2015. Modified effect of urinary cadmium on breast cancer risk by selenium. *Clin. Chim. Acta* 438:80–85.

Whyte, A.L., Hook, G.R., Greening, G.E., Gibbs-Smith, E., Gardner, J.P. 2009. Human dietary exposure to heavy metals via the consumption of greenshell mussels (Perna canaliculus Gmelin 1791) from the Bay of Islands, northern New Zealand. *Sci. Total Environ.* 407:4348–4355.

Wilkinson, J.M., Hill, J., Phillips, C.J. 2003. The accumulation of potentially-toxic metals by grazing ruminants. *Proc. Nutr. Soc.* 62:267–277.

Will, J.C., Ford, E.S., Bowman, B.A. 1999. Serum vitamin C concentrations and diabetes: findings from the Third National Health and Nutrition Examination. *Am. J. Clin. Nutr.* 70:49–52.

Wills, N.K., Ramanujam, V.M., Chang, J., Kalariya, N., Lewis, J.R., Weng, T.X., et al. 2008b. Cadmium accumulation in the human retina: Effects of age, gender, and cellular toxicity. *Exp. Eye Res.* 86:41–51.

Wills, N.K., Ramanujam, V.M., Kalariya, N., Lewis, J.R., van Kuijk, F.J. 2008a. Copper and zinc distribution in the human retina: Relationship to cadmium accumulation, age, and gender. Exp. Eye Res. 87:80–88.

Wilson, A.K., Cerny, E.A., Smith, B.D., Wagh, A., Bhattacharyka, M.H. 1996. Effects of cadmium on osteoclast formation and activity *in vitro. Toxicol. Appl. Pharmacol.* 140:451–460.

Williams, J.M., Murphy, S., Burke, M., Roman, R.J. 2010. 20-hydroxyeicosatetraeonic acid: a new target for the treatment of hypertension. *J. Cardiovasc. Pharmacol.* 56:336–344.

Wing, A.M. 1993. The effect of whole wheat, wheat bran and zinc in the diet on the absorption and accumulation of cadmium in rats. *Br. J. Nutr.* 69:199–209.

Wong, W.P., Allen. N.B., Meyers. M.S., Link. E.O., Zhang. X., MacRenaris, K.W., et al. 2017.Exploring the association between demographics, SLC30A8 genotype, and human islet content of zinc, cadmium, copper, iron, manganese and nickel. *Sci Rep.* 2017 7:473. doi: 10.1038/s41598-017-00394-3.

Wu, Q., Magnus. J.H., Hentz, J.G. 2010. Urinary cadmium, osteopenia, and osteoporosis. *Osteoporosis Int.* 21:1449–1454.

Wu, E.W., Schaumberg, D.A., Park, S.K. 2014. Environmental cadmium and lead exposures and age-related macular degeneration in U.S. adults: The National Health and Nutrition Examination Survey 2005 to 2008. *Environ. Res.* 133:178–184.

Wu, W., Liu, D., Jiang, S., Zhang, K., Zhou, H., Lu, Q. 2019. Polymorphisms in gene MMP-2 modify the association of cadmium exposure with hypertension risk. *Environ. Int.* 124:441–447.

Yang, L.E., Zhong, H., Leong, P.K., Perianayagam, A., Campese, V.M., McDonough, A.A. 2003. Chronic renal injury-induced hypertension alters renal NHE3 distribution and abundance. *Am. J. Physiol. Renal. Physiol.* 284:F1056–F1065.

Yorita Christensen, K.L. 2013. Metals in blood and urine, and thyroid function among adults in the United States 2007–2008. *Int. J. Hyg. Environ. Health* 216:624–632.

Yoshida, M., Ohta, H., Yamauchi, Y., Seki, Y., Sagi, M., Yamazaki, K., et al. 1998. Age-dependent changes in metallothionein levels in liver and kidney of the Japanese. *Biol. Trace Elem. Res.* 63:167–175.

Zalups, R.K. 2000. Evidence for basolateral uptake of cadmium in the kidneys of rats. *Toxicol. Appl. Pharmacol.* 164:15–23.

Zhang, Y.R., Wang, P., Liang, X.X., Tan, C.S., Tan, J.B., Wang, J., et al. 2015. Associations between urinary excretion of cadmium and renal biomarkers in non-smoking females: A cross-sectional study in rural areas of South China. *Int. J. Environ. Res. Public Health* 12:11988–12001.

Zhang, R., Witkowska, K., Afonso Guerra-Assunção, J., Ren, M., Ng, F.L., Mauro, C., et al. 2016. A blood pressure-associated variant of the SLC39A8 gene influences cellular cadmium accumulation and toxicity. *Hum. Mol. Genet.* 25:4117–4126.

Zhang, C., Booz, G.W., Yu, Q., He, X., Wang, S., Fan, F. 2018. Conflicting roles of 20-HETE in hypertension and renal end organ damage. *Eur. J. Pharmacol.* 833:190–200.

Zhu, X.J., Wang, J.J., Mao, J.H., Shu, Q., Du, L.Z. 2019. Relationships between cadmium, lead and mercury levels and albuminuria: Results from the National Health and Nutrition Examination Survey Database 2009–2012. *Am. J. Epidemiol.* pii: kwz070. doi: 10.1093/aje/kwz070.

CHAPTER 15

Cobalt Toxicity and Human Health

Md. Hafiz Uddin and Marufa Rumman
Wayne State University

CONTENTS

15.1 INTRODUCTION

In the process of evolution, many inorganic elements became crucial for fundamental functions of the organism which are considered as essential. In cells, H, C, N, O represent 96% of the bulk of organic matter; Na, Mg, P, S, Cl, K, Ca represent about 3%; and B, Cr, Co, Cu, F, I, Fe, Mn, Mo, Ni, Se, Si, Sn, V, Zn represent about 0.5% [1]. Among these components, cobalt (Co) is an essential dietary trace metal which plays an important role in enzymes activation such as cholinesterases and carboxylases and an essential constituent of vitamin B_{12} (cyanocobalamin) [2]. Cobalt naturally exists in the environment, occurring in the earth's crust at about 0.002% primarily in the form of sulfides, oxides, and arsenides [3]. It is also present in the human diet, primarily in vegetables and fish, and in drinking water. Although cobalt is an essential micronutrient in the form of vitamin B_{12}, inorganic cobalt is not required in human diets [4]. Cobalt metal is usually mixed with other metals to form alloys, to make it harder and resistant to wear and corrosion. These alloys are used in a number of industrial applications, e.g., aircraft engines, magnets, and cutting and grinding tools. Artificial hip and knee joint prosthesis is another important use of it. Furthermore, cobalt compounds are used as colorants in ceramics, glass, and paints, and as paint driers [3]. Cobalt deficiency has never been documented in normal subjects who are fed a balanced diet. However, recent data suggest that obese children might have low levels of cobalt in their blood and may need cobalt supplementation [1,4], cobalt is acutely toxic in higher doses, and long-term, cumulative exposure even at a low level such as occupational exposure can give rise to adverse health effects involving various organs including liver, kidney, pancreas, heart, skeleton, and skeletal muscle. The cobalt toxicity affects cardiovascular system, brain, thyroid gland, lung, skin, and immune system, and may include a possible carcinogenic potential [1,4,5].

15.2 COBALT EXPOSURE

Humans can be exposed to cobalt by inhalation, ingestion, and through dermal contact with cobalt-containing materials. The occupational exposure to cobalt mostly occurs via inhalation usually in cobalt-processing plants, hard-metal industry, diamond polishing, and ceramic industry as part of the refining, welding, grinding, alloy production, and via the use of cobalt blue dyes [3,4,6]. In addition, exposure to cobalt radio-isotopes can occur to the workers of the nuclear industry particularly to ^{60}Co, a neutron activation product. Dermal exposure to stable cobalt can happen to workers exposed to specific occupational conditions, and dermal absorption of radio-cobalt is the major risk for workers in the nuclear industry [6,7]. Environmental exposure to cobalt oxide particles is associated with inhalation of particulate matter circulating in the air due to the combustion of cobalt sources, which generate primarily cobalt oxides [7]. Cobalt exposure is increasing due to the rising cobalt demand worldwide for its use in enhancing rechargeable battery efficiency, super-alloys, magnetic products, and nanoparticles. As the manufacturing industries are increasingly using cobalt, the potential exposure to both industrial workers and the general population is also increasing [8].

The largest source of cobalt exposure is through consumption of food and drinks containing cobalt such as vitamin B_{12} supplement, meat, and dairy products [3]. Cobalt is absorbed through the gastrointestinal tract and accumulated in different organs of the body and excreted mostly through kidney in the form of urine; however, efficiency decreases with time and may take years to expel if high amounts of cobalt is consumed. Dietary cobalt intake is limited to a few µg/day and highest concentrations found in fish, green leafy vegetables, and fresh cereals. The reported cobalt intake varied from 0.17 to 33 µg/kg/day in different countries including Canada, France, Italy, Australia, and New Zealand. In France, coffee intake by adults and chocolate by children are considered as the main contributors to cobalt consumption [1,9]. Numerous over-the-counter cobalt-containing supplements are available in the United States (e.g., Mother Earth Minerals, Ogden, UT), and recommended daily doses ranging from 200 to 1,000 µg cobalt/day. The United States Food and Drug Administration (FDA) has not recommended any dose for cobalt supplements; however, European Food Safety Authority (EFSA), Expert Group on Vitamins and Minerals (EGVM) in the United Kingdom suggested 600–1,400 µg cobalt/day. These doses are much higher than the typical dietary intake of cobalt (5–40 µg/day) in the United States [9]. Because of the lack of epidemiological data, dietary exposure to cobalt is not considered as a public health issue [1,9].

Over the last decades, the use of cobalt–chromium metal-on-metal (MoM) prosthesis in orthopedic joint replacements, especially in the young people, has created a new obvious source of cobalt exposure at levels far more above the general population [3,4]. Furthermore, suspicion has been raised that oral cobalt salt may be misused by athletes as an attractive alternative to traditional blood doping techniques for enhancing aerobic performance, despite harmful health risks. However, the frequency of this kind of practice among the athletes is unknown [1,4]. A concern has also been raised for dental restoration and implant receiving subjects for systemic effects of released cobalt by corrosion of hard metals containing cobalt [3,4].

15.3 EFFECT ON HUMAN HEALTH

The toxic effects of cobalt are notable and of significant interest for human health, and different forms of cobalt can have different toxicological effects [3]. In the 1950s, anemic patients were treated with oral cobalt, including children and pregnant women, at doses 6–150 mg/day. Although cobalt effectively stimulated the production of hemoglobin, toxic symptoms were often described as cobaltism which included neurological, cardiac, and endocrine effects, like goiters and thyroid diseases (inhibition by cobalt of tyrosine iodinase and myxedema) [1,4]. Eventually, synthetic erythropoietin and other drugs replaced cobalt in the treatment of anemia. In the past, beer industries used cobalt as a foam stabilizer. Therefore, in the 1960s, a small cohort of heavy beer drinkers who consumed cobalt doses of 5–10 mg/day with drinks for up to 12 months developed severe cardiomyopathy. To stimulate fat and carbohydrate metabolism, protein synthesis, erythrocyte production, and central nervous system repair, cobalt supplements are used at doses up to 1 mg/day [1].

High tissue levels of cobalt can cause toxicity as it competes with calcium thereby interfere with intracellular-binding proteins and cell signaling (Figure 15.1). Cobalt usually as ions (Co^{2+}) enters in to the cells utilizing similar mechanism of calcium (Ca^{2+}), nonetheless trapped in the cytoplasm because of its bonding tendency with biomolecules. The trapped cobalt ions are unsuccessful to export themselves through Ca pump [4]. Recently, it has been shown that cobalt modulates gene expressions and influences various pathways involving hypoxic response, oxidative stress, glycolysis, gluconeogenesis, energy metabolism, etc. Previous studies have showed that cobalt salts affect pancreatic cells, liver P-450 cytochrome activities, and platelet aggregation [1,10]. Allergic contact dermatitis is a common effect of dermal exposure to cobalt in humans [4]. In addition to cobalt metal particles, cobalt ions cause cytotoxicity in alveolar macrophages and alveolar type II cells. The harmful effects of health among workers employed in refinery industries associated with the production of alloys from a mixture of tungsten carbide, cobalt, and/or other metals are well documented [11,12]. Cobalt-induced (exerted by Co^{2+} and/or Co^{3+} ions) interstitial lung diseases, e.g., interstitial alveolitis, fibrosis, and asthma occurred due to the generation of deleterious reactive oxygen species (ROS) and subsequent oxidative damage to critical biomolecules within lung tissue [2,4]. International Agency for Research on Cancer (IARC, 2006) in their last review considered cobalt(II) salts as possibly carcinogen to humans (Group 2B) [2,5] and cobalt metal with tungsten carbide as probable carcinogen to humans (Group 2A) (all cobalt combinations and risk category by IARC and Report on Carcinogens (ROC) are shown in Table 15.1) [4,13]. In fact, a study showed that exposure to tungsten–carbide–cobalt alloy among the workers of hard metal industry leads to increased lung cancer risk; however, epidemiologic data for workers exposed to cobalt metal have not observed sufficient evidence to confirm the link. Recently, ROC lists "cobalt sulfate," "cobalt ions in vivo," and "cobalt–tungsten carbide: powders and hard metals" as reasonably anticipated to be human carcinogens [3].

The respiratory, hematological, cardiac, dermal, and neurological effects due to cobalt toxicity is well documented and reviewed extensively in various literature [4,14,15]. Occupational and public exposure, through inhalation, oral ingestion, dermal absorption, or by injection, of manufactured nanoparticles and implantation of MoM prosthetics is currently in increase and will probably continue to increase in the near future. However, there is still a lack of information about the impact of manufactured nanoparticles on human health as well as reliable data on risk assessment [16]. Here we have discussed more on the health effect that is associated with recent cobalt metal implant use and extensive use of cobalt nanoparticles. A schematic diagram of the effect of cobalt shown in Figure 15.1.

15.3.1 Metal-on-Metal Prosthesis

15.3.1.1 Hip Arthroplasty

The total hip arthroplasty (THA) has improved the quality of life of patients with different joint diseases. The first generation of polyethylene prostheses was problematic because of the production of significant amounts of wear particles through friction at the bearing interface. However, the ultra-high-molecular-weight polyethylene particles can cause periprosthetic osteolysis leading to arthroplasty failure. To overcome this problem, the orthopedic industry has developed four major kinds of prosthesis, namely MoM, ceramic-on-ceramic, metal-on-highly cross-linked polyethylene, and

Table 15.1 Classification of Carcinogenic Agents by International Agency for Research on Cancer (IARC) and Report on Carcinogens (ROC)

Agent	Group	Year
Implanted foreign bodies of metallic chromium or titanium and of cobalt-based, chromium-based, and titanium-based alloys, stainless steel and depleted uranium	3, the agent (mixture) is unclassifiable as to carcinogenicity in humans	1999
Implanted foreign bodies of metallic cobalt, metallic nickel, and an alloy powder containing 66%–67% nickel, 13%–16% chromium, and 7% iron	2B, possibly carcinogenic to humans	1999
Cobalt sulfate and other soluble cobalt(II) salts	2B, possibly carcinogenic to humans	2006
Cobalt and cobalt compounds	2B, possibly carcinogenic to humans	1991
Cobalt metal without tungsten carbide	2B, possibly carcinogenic to humans	2006
Cobalt metal with tungsten carbide	2A, probably carcinogenic to humans	2006
Cobalt sulfate	Reasonably anticipated to be human carcinogens	2014
Cobalt ions	Reasonably anticipated to be human carcinogens	2016
Cobalt–tungsten carbide: powders and hard metals	Reasonably anticipated to be human carcinogens	2014

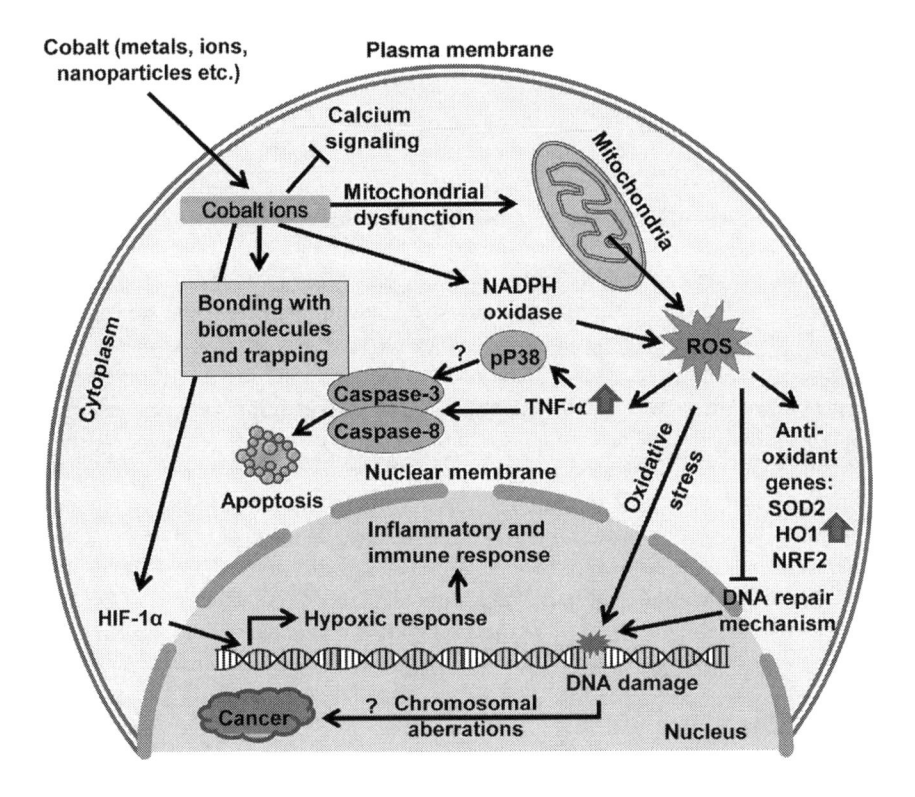

Figure 15.1 **Effect of cobalt on the biological systems.** A generalized schematic diagram showing the effect of cobalt on normal human cells. Cobalt mainly exerts its functions as ions (Co^{2+}) inside the cells, competes with calcium signaling, and generates a large amount of ROS by mitochondrial dysfunction. ROS influences a number of pathways leading to DNA damage, impairment of DNA repair machinery, and activation of antioxidant systems. Cobalt gets trapped by making bonds with various biomolecules in the cell cytoplasm. Cobalt also induces hypoxic response via HIF-1α to activate inflammatory and immune responses. ROS-induced TNF-α can phosphorylate P38 which, in turn, initiates apoptosis by an unknown mechanism. Unrepaired DNA may cause cancer through chromosomal aberration and currently unknown pathways.

ceramic-on-highly cross-linked polyethylene. Among these new devices, MoM is usually preferred because of low volumetric wear in hip simulators. Although ceramic-on-ceramic bearings are wear resistant, it offers low friction. Because of bearing noise, chipping during insertion, and unantici-pated fracture is less popular among patients [17].

Present-day MoM total hip resurfacings were introduced in the 1990s, representing about 10% of all hip arthroplasties in developed countries between 1990 and 2010. Bearings of MoM hip resurfacing are usually made from a high-carbon cobalt–chromium (Co-Cr) alloy [18]. MoM pros-theses such as cobalt–chromium, however, still generate metal particles by wear-tear. Cobalt ions (Co^{2+}) are generated from the corrosion of cobalt particles in periprosthetic tissue. Thus cobalt ions' concentration is found to be increased in the blood (>20 nM) of MoM THA patients. One study of 6 patients at 20 months' post-implantation showed 1,000-fold higher (30 vs. 0.03 mg/g) cobalt concentration in the periprosthetic tissues compared to control patients [17]. Recent, unexpectedly high failure rates of MoM hip implants have reinstated the issue of cobalt toxicity [19]. Although cobalt toxicity by MoM THA is a cause for concern, its pathophysiology has not been well studied [17]. Currently, adverse reaction occurs to cobalt debris during wear and tear of MoM hip implants of which the underlying mechanism is not fully understood [19].

Increased risk of cobalt toxicity occurs following orthopedic joint replacement including hip, knee, and spine with cobalt–alloy implants. Cobalt exposure from joint replacement is predicted to cost the orthopedic industry and healthcare providers billions of dollars because of early implant rejection due to inflammation and other complications. The patients on MoM bearings are signifi-cantly important and particularly prone to cobalt toxicity-induced early implant failures because of wear and tear [19].

Over the last two decades, MoM resurfacing hip replacement surgery has become increasingly common for osteoarthritis management. Younger or more active patients are particularly interested to use this as an alternative to total hip replacement. Despite the promising clinical reports, there is a concern regarding high levels of metal ions released from cobalt–chromium (Co-Cr) alloy MOM articulations. These implants can release metal ions, metal complexes, or metal particu-late after electrochemical or mechanical corrosion [20]. THA early failure is commonly caused by aseptic loosening of the implant due to adverse tissue responses around the prostheses by wear particles [18]. Cobalt toxicity has been related with the implant of chrome-cobalt bone prosthesis. In fact, a number of carriers of cobalt–chromium prosthesis have high serum cobalt concentra-tions (i.e. 41,000 nmo/L). Cobalt–chromium overload is associated with neurosensory, endocrine, respiratory, neurocognitive, cardiac, and neurological symptoms. Some reported complaints are tinnitus, deafness, vertigo, visual disturbances, skin rashes, hypothyroidism, tremor, dyspnea, mood disorders, heart failure, and peripheral neuropathy [1].

15.3.1.1.1 Effect of Metal-on-Metal Prosthetics on Human Health

Hip replacement surgery using implantable metal components has been performed in North America for over 50 years. During the past 6 years, cases of systemic symptoms associated with cobalt-containing metal alloy hip prosthetics (Co-HP) have been reported. Most of cobalt toxicity comes from inhalation and ingestion of cobalt. Patients with suspected prosthetic hip-associated cobalt toxicity (PHACT) usually showed neurologic, thyroid, and cardiac dysfunction; however, due to insufficient control subjects, it failed to establish a definitive link. Signs and symptoms appeared between 3 and 72 months after arthroplasty (median 19 months). Patients usually show elevated cobalt levels in one or more matrices [15,21].

It is evident that patients with a high cobalt serum concentration experience neurological and cardiovascular abnormalities. Patients who undergo surgery with ceramic-plastic revisions showed reduced cobalt concentrations in the body and improvement in neurological and cardiovascular func-tion [22,23]. Excess chromium–cobalt in the body is connected with neurosensory, neurocognitive, endocrine, respiratory, cardiac symptoms. Other complaints involve vertigo, tinnitus, deafness, visual disturbances, skin rashes, hypothyroidism, tremor, dyspnea, mood disorders, etc. [1]. Cobalt toxicity has been related to the implant of chrome-cobalt bone prosthesis. Although patients showed elevated levels of both chromium and cobalt, it is believed that cobalt is the most important contributor to toxicity [24]. In some patients, high serum cobalt concentrations (i.e. 41,000 nmo/L) have been detected [1,25].

Historically, neurotoxic symptoms such as hearing loss, visual impairment, and polyneuropathy were credited to cobaltous chloride infusions [21,26]. Most of the neurologic symptoms are optic nerve atrophy, audiometry documented hearing loss, and abnormal electrodiagnostic studies. Cobalt neurotoxicity is controversial and its mechanism is not clearly understood. Anticipated mechanisms include direct neuron cytotoxicity, neurotransmitter modulation, and interruption of mitochondrial oxidative phosphorylation [21,27]. Thyroid dysfunction has been associated with cobalt toxicity, particularly hypothyroidism attributed to cobalt chloride treatment for refractory anemia. The workers who occupationally exposed cobalt dye showed low-level chronic cobalt exposures and altered thyroid hormone metabolism in the absence of clinical disease [21].

Cobalt toxicity is classically associated with beer drinkers' cardiomyopathy. Drinking beer regularly containing excess cobalt is responsible for this type of cardiovascular dysfunction. In one case series, 28 men developed severe congestive heart failure secondary to cardiomyopathy. Five other men developed atrial fibrillation or flutter and eleven died [28]. Likewise, a Finnish study of cobalt workers showed that cumulative cobalt exposure was connected with subclinical echocardiographic changes [29]. Cobalt-induced myocardial injury has also been found in hard metal workers [21].

15.3.1.1.2 Effects on the Immune System

Metal ions leaching from metallic articulations may have adverse effects on the immune system. In vitro study showed that cobalt ions (Co^{2+}) at $100\,\mu M$ dose for 24–48 h resulted in significant decreases in cell viability, cytokine release, and induces apoptosis in resting human lymphocytes. Exposure at ten times lower dose led to significant decreases in cell proliferation and cytokine release but not apoptosis. The results indicate that cobalt ions can decrease lymphocyte proliferation and may contribute to altered adaptive immune system function [20]. A number of investigations showed that prolonged exposure for 5 days to cobalt metal debris can induce proliferation of resting lymphocytes. Cobalt toxicity may affect cytokine productions (Figure 15.1) (inhibits IL-6, IFN-γ, and TNF-α) and can modulate IL-2 secretion. Importantly, cobalt ion concentrations below the Medicines and Healthcare products Regulatory Agency (MHRA, ref: MDA/2012/036, issued: 25 June 2012) guideline levels (7 ppb) may cause impairment of immune regulation in patients with metal implants [18].

Implant debris causes excessive inflammation via danger signaling. Study indicates that cobalt-alloy particle induced NLRP3 inflammasome danger signals in vitro in human macrophage cell line (THP-1) and primary human macrophages as well as in in vivo murine model of inflammatory osteolysis. Blocking experiment confirms that cobalt alloy particles induce macrophage inflammation through danger signal, not through TLR4. Mice devoid of histidine that binds cobalt confirmed TLR4-independent inflammatory bone loss. The study suggests that orthopedic implant failures owing to an extreme innate immune reaction to cobalt debris are inflammasome mediated and require specific measures to prevent it [30].

Patients with total joint arthroplasty (TJA) can be exposed to a number of implant debris such as cobalt–chromium–molybdenum (CoCrMo) alloy, titanium (Ti) alloy, zirconium (Zr) oxide, and Zr alloy. Study evaluated human peri-implant cells such as osteoblasts, fibroblasts, and macrophages exposed to metal-based particles. The highest toxicity observed for CoCrMo alloy at a dose of only 50 particles per cell. Induction of interleukin (IL)-6, tumor necrosis factor-alpha (TNF- α), and IL-8 is a common phenomenon among all particles; however, the greatest response is observed in CoCrMo alloy, suggesting an inflammatory response surrounding TJA [31].

15.3.1.1.3 Immunosuppression and Bacterial Infection

Staphylococcus epidermidis infection occur in patients with MoM-THA and they suffer serious complications. Studies showed that innate immune cells such as neutrophils are exposed to cobalt ions (Co^{2+}), which is released by wear of MoM THA in the periprosthetic tissue. The concentration of Co^{2+} can be increased to 53 mM in periprosthetic tissue of patients which is sufficient to inhibit Hv1 proton channel of neutrophils. In fact, in vitro experiments showed that Co^{2+} inhibits proton currents even in sub-mM concentrations. Inhibition of proton channel impairs the cytosolic acid extrusion and decreases the production of superoxide in human neutrophils which, in turn, reduces the ability of human neutrophils to kill *S. epidermidis* by up to seven folds. Therefore, metal prostheses containing cobalt can promote bacterial infections in patients by immunosuppression which needs proper attention [17].

15.3.1.1.4 Alloy Shrapnel and Risk of Cancer

It has been shown that metal alloy containing 91% tungsten, 6% nickel, and 3% cobalt (WNC-91-6-3) when implanted as fragments in rat muscle causes rhabdomyosarcomas. This raises a concern that shrapnel may induce similar kinds of tumors in humans if not removed surgically. In vitro experiment with L6-C11 rat muscle cells showed that WNC-91-6-3 produces highly elevated ROS, causes DNA damage, and triggers an extensive hypoxic response within 24 h. Data confirmed the highest level of toxicity with WNC-91-6-3 among the alloys tested. The study suggests a synergistic combination effect among the metals in that specific proportion. Mechanistically, cobalt generates free radicals, binding of nickel to thiol and inhibitory effect to key enzymes could be associated with enhanced genotoxicity of WNC-91-6-3 [32]. Similar finding also observed in other studies such as Kalinich et al., 2005 [33]. Synergistic toxicity of cobalt and nickel reported in another study, where they treated H460 human lung epithelial cells with cobalt and nickel alone or in combination. The study reveals that co-treatment with cobalt and nickel is significantly more toxic than single treatment and the toxicity might be caused by ROS and double-strand breaks in DNA [34].

15.3.1.1.5 Others

The effect of cobalt in pregnant women bearing MoM implants and newborn is rare. A case study was performed on a 41-year-old, 12-gestational week pregnant patient with bilateral MoM hip arthroplasties. Blood levels of cobalt were extremely high (Co 138 µg/L). At 38 gestational week, a healthy male infant was delivered, and at 8 weeks the infant's cobalt level decreased without treatment. At the age of 14 weeks, the infant's development was normal and did not show any signs of toxicity. Although a case study is not confirmatory, the data can help while counseling patients with high cobalt ion levels whether to carry a pregnancy to term [35].

One of the major limitations in joint alloy arthroplasty is periprosthetic tissue reactions to wear debris, particularly polyethylene and cement. A study compared Birmingham hip resurfacing arthroplasty (BHR) and cementless total hip replacement with a 28-mm Metasul articulation (MTHR) after 2 years of implantation in 111 patients and 74 patients, respectively, with 130 healthy controls. Serum analysis observed a significantly higher concentration of chromium and cobalt in BHR compared to MTHR. Thus, wear debris reduction is an important necessity for the improvement of total hip replacement [36].

15.3.1.2 Oral Prosthetic

Nowadays, there is a growing level of utilization of non-precious metals and their alloys as a substitute for precious metal oral prosthetic devices. They are now commonly exploited as metal-ceramic or full-cast restorations, in addition to them serving as removable partial dentures. Metal alloy materials containing cobalt, chromium, and nickel are readily casted into the required thinner shapes for dental applications for the purpose of crowns, bridges, and fixed or removable partial dentures without any compromises concerning their firmness. The study showed that removable of partial dentures casted from cobalt, chromium, and/or nickel alloys accounts for about 90% of these medical devices. Among the alloys, cobalt–chromium (Co–Cr) alloys have been used as oral implants in dentistry for decades but a small-sized study was undertaken regarding their in vivo corrosion and biological impact [2].

Cobalt-containing metal alloy oral implants can leach cobalt ions from the corrosion which may form a complex by salivary biomolecules. A study found Co(II)-lactato, -formato, -histidinato, and/or -succinato complexes in the oral environment at sub-μM level either from saliva or local soft tissues. The toxic effect of such complexes is yet to be determined [2].

15.3.2 Cobalt Nanoparticles

Nanotechnology involves materials and systems in the submicron to nanometer range [16], and it is emerging at a fast pace with possibilities of an excellent socio-economic future. The apprehensions of bright future involvement makes nano objects ubiquitous [37]. It has been estimated that there are now more than 1,500 commercial nano products available on the world market [38]. Nanotechnology aids us to envision the new mysterious prospects in engineering, sophisticated electronics, environmental remediation, biosensing, and nanomedicine [37]. Metal-oxide nanoparticles have been commercially used in biosensors, water purification devices, diagnostics, cosmetics, and therapeutic agents [39]. Currently, nanoparticles have received much attention for their implications in cancer treatment [5]. For example, metallic cobalt as nanoparticles is used in biology and medicine in its simplest form cobalt oxide to complex organic compounds or biopolymers. This pervasive technology is likely to have a huge economic and social influence in almost all sectors of industrial and scientific activity [16].

Among nanomaterials, magnetic nanoparticles, in particular, iron, nickel, and cobalt, are one of the very vital and extensively exploited subcategories for its controlled use in the magnetic field. Recently, magnetic nanoparticles are widely used in sophisticated nano-based medicine, improved drug and gene delivery, nanoprobes, catalysts, and optical devices [5,37]. Moreover, these magnetic nanoparticles are used as contrast agents in magnetic resonance imaging (MRI) and as adjuvants for use in human vaccination [5,37,40]. More recently, cobalt nanoparticles have been suggested as an alternative to iron due to their greater effects on proton relaxation [5].

Although there is heavy use because of endless benefits, nanotechnology is also packed with negative side effects for human society and health [37]. There is still a lack of information about the potential risks of nanoparticles for the environment and for human health [10,16]. The unregulated use of nano products has become an accepted health problem because the inhalation, absorption, or ingestion leads to increased rates of chronic diseases [38]. A number of studies showed that cobalt nanoparticles (Co-NPs) are harmful to humans despite their usefulness [13] and concerns also arise because cobalt is also a skin sensitizer [40]. Most of the findings with Co-NPs toxicities came from human cells and animal studies.

15.3.2.1 Potential Impact of Cobalt Nanoparticles on Human Health

Co-NPs serve as efficient contrast agents for MRI among other biomedical use; however, the study observed significant neural damage in both hippocampus and cortex of the temporal lobe in Wistar rats when exposed to Co-NPs and Co^{2+} in 96–123 nM doses suggesting potential neurological risk in humans. Co^{2+} accumulated to a greater extent in brain and blood compared to Co-NPs, whereas Co-NPs amount was higher in the liver [41]. Experimental data indicate that Co-NPs can induce endothelial inflammatory responses not only via oxidative stress but also through various pathways. Internalization of Co-NPs by cells occurred rapidly and upregulation of adhesion molecules (ICAM-1, VCAM-1, E-selectin) at mRNA and protein levels takes place and subsequently releases monocyte chemoattractant protein-1 (MCP-1) and interleukin 8 (IL-8) in human aorta (HAECs) and HUVECs. In addition, Co-NPs induced time- and concentration-dependent metabolic impairment and oxidative stress compared with titanium oxide nanoparticles [38]. Mechanistically Co-NPs increase malondialdehyde and caspase-9 protein level leading to lipid peroxidation and apoptosis in PC12 cells [41]. Co^{2+} showed an association with enhanced weight gain in animals [41].

Tungsten carbide (WC) shows less toxicity to the cells; however, when attached with Co-NPs (WC-Co-NPs) its toxicity is enhanced almost similar to cobalt ions (Co^{2+}) in human keratinocytes (HaCaT). Gene expression analysis showed that WC exerted very little effects on the transcriptomic level after 3 and 72 h of exposure, whereas WC-Co-NPs caused significant transcriptional changes that were similar to Co^{2+}. Gene set enrichment analyses revealed that the differentially expressed genes were related to hypoxia response, carbohydrate metabolism, endocrine pathways, cell adhesion, and targets of several transcription factors (e.g., SOX2, YY1) [10]. Cobalt ferrite nanoparticles (Co-Fe-NPs) have great advantages compared to other contestants because of its physicochemical and magnetic properties with ease of synthesis resulting in extensive use in biomedical fields. Human body can be exposed to Co-Fe-NPs easily by ingestion, inhalation, adsorption, etc. which might induce oxidative stress, cytotoxicity, genotoxicity, inflammation, apoptosis, and developmental, metabolic, and hormonal abnormalities [37]. Research has shown the putative role of the hypoxia-inducible factor (HIF) pathway in the mechanism of Co-NPs toxicity (Figure 15.1) in human macrophage models using the U937 cell line, human alveolar macrophages, and monocyte-derived macrophages. Co-NPs induced HIF-1α stabilization which can be prevented by the addition of either ascorbic acid (100 mM) or glutathione (1 mM) suggesting the involvement of ROS-independent pathway in Co-NPs-induced cytotoxicity. Additionally, ascorbic acid causes the downregulation of IL-1b showing a possible link between HIF and the inflammatory response to Co-NPs [19].

Human lymphocytes play a foremost role in the immune system. Studies revealed that the Co_3O_4-NPs are more toxic to human lymphocytes when compared with other metal-oxide NPs (Fe_2O_3, SiO_2, and Al_2O_3 NPs). The Co_3O_4-NPs treatment causes reduction in cell viability and induces cell membrane damage in a dose-dependent manner. Further, Co_3O_4-NPs shown to deplete catalase, glutathione, and superoxide dismutase resulting in ROS-mediated oxidative stress. Chromosomal aberration was detected when exposed to Co_3O_4-NPs and Fe_2O_3-NPs at 100 mg/mL concentration. The oxidative stress leads to DNA damage and chromosomal aberrations in human lymphocytes [39].

15.3.2.1.1 Cobalt Nanoparticles Are Toxic to Cells and Tissues

In vitro studies showed that Co-NPs exposure can cause toxicity within 2–24 h of exposure in Balb/3T3 cells. The study observed nuclear and DNA damage by micronucleus test and comet assay; however, Co-NPs at <1 mM concentration did not show any detectable toxicity. By using radiolabeled compounds (^{60}Co-nano and ^{57}Co^{2+}), the authors also compared cobalt ions (cobalt chloride, $CoCl_2$) with Co-NPs and observed higher cytotoxicity and cellular uptake of Co-NPs [16]. A superparamagnetic zinc-cobalt ferrite nanoparticle (SFN) has been developed for early cancer diagnostics and targeted therapy. However, toxicity and biological evaluation showed that acute exposure (4 h) of SFN (100 µg/mL) decreases the viability of healthy HUVECs. The study also observed toxic effects in lungs, liver, and kidney tissues of New Zealand rabbits. Therefore, these NPs are not suitable for biomedical applications and should classify as toxic despite their interesting magnetic properties [42].

15.3.2.1.2 Cobalt Nanoparticles Largely Mediates Oxidative Damage

In vitro and in vivo studies showed toxicological harmful effects of Co-Fe-NPs on diverse organs and systems through the generation of ROS [37]. In vitro experiment in PC12 cells with Co-NPs showed downregulation of antioxidant response gene NRF2 along with reduced cell viability and increased apoptotic cell death [41]. In human macrophage cells (U937 cell line, human alveolar macrophages, and monocyte-derived macrophages), Co-NPs (5–20 mg/mL) induced enhanced cytotoxicity and elevated ROS, compared to cobalt ions (CoCl$_2$, up to 350 nM) [19]. Studies showed that bare cobalt oxide nanoparticles (CoO-NPs) provoked a significant amount of ROS in a dose-dependent manner within 24 h of exposure in primary human lymphocytes leading to cell death. To induce apoptotic cell death, the elevated ROS causes the elevation of TNF-α which, in turn, activates caspase-8, phosphorylate P38 mitogen-activated protein kinase (MAPK) followed by activation of caspase-3 (Figure 15.1). CoO-NPs can induce toxicity in vitro at 5 μg/mL (for 24 h culture) and in vivo at 200 μg/kg body weight (for 15 days). These findings suggested that bare CoO-NPs is toxic and fatal for human health [5].

15.3.2.1.3 Cobalt Nanoparticles Able to Penetrate Damage Skins

Co$_3$O$_4$-NPs are usually used in industry and in biomedicine, and skin absorption of these NPs is a deep concern. The workers and users can be exposed to Co$_3$O$_4$-NPs via powders or solutions. Studies demonstrated that Co$_3$O$_4$-NPs cannot penetrate Franz diffusion intact skin cells but able to penetrate when a damaged skin protocol is used. A long-term exposure to Co$_3$O$_4$-NPs (1,000 mg/L) induces cell damage and necrosis in cultured keratinocytes. The findings showed that a long-term exposure (7 days) can induce a dose-dependent cell viability reduction (effective concentration-50 (EC$_{50}$) values: 1.3×10^{-4} M; MTT essay; 3.7×10^5 M, AlamarBlue® assay) that seems to be associated to necrotic events (EC$_{50}$ value: 1.3×10^4 M, PI assay). It is recommended that study workers, and atopic subjects should use personal protective equipment to avoid contamination of the skin with Co$_3$O$_4$-NPs [40].

15.3.2.1.4 Genetic Background Can Affect Cobalt Nanoparticle Toxicity

Genetic background plays a role in Co-NPs toxicity. A study was also performed with long-term exposures of 12 weeks to sub-toxic doses (0.05 mg/mL) of Co-NPs to assess the oxidative DNA damage. Ogg1 deleted mouse embryonic fibroblast (MEF) failed to remove the 8-OH-dG lesions from DNA. Consequently, MEF accumulates ROS, cellular transformation, increases in metalloproteinases (MMPs), and anchorage-independent growth suggested potential cancer risk associated with Ogg1 genetic background. The study observed increase in expressions of antioxidant genes (Figure 15.1), Gstp1, Sod2, Ho1, and Keap1 in Ogg1 deleted MEF, which can be mediated by the activation of nuclear factor erythroid-2-related factor-2 (Nrf2). The conditioned media from MEF Ogg1 deleted cells promote the growth of HeLa cells. This further confirms that MEF Ogg1 deleted cells are prone to acquire oncogenic characteristics [13].

15.3.2.1.5 Modification of Cobalt Nanoparticles Surface to Overcome Toxicity

It has been shown in earlier studies that bare Co-NPs cause significant toxicity in diverse types of human cells as well as in animals. Surface coating of Co-NPs can reduce the toxicity. A study was conducted to coat cobalt ferrite nanoparticle (CoFe$_2$O$_4$-NPs) core with negatively charged polyacrylic acid (sodium salt) (PAA) and positively charged polyethylenimine (PEI). The effect of coated NPs on primary human myoblasts (MYO) and B16 mouse melanoma cell lines revealed that negatively charged PAA-coated NPs did not induce cytotoxicity, ROS, and did not activate the transcription factor NF-kB even at high concentrations (100 mg/mL) for 24 h. On the contrary, positively charged PEI NPs caused a dose-dependent necrotic cell death with the elevation of ROS at low concentrations (>4 mg/mL). Besides, PEI NPs-induced NF-kB activation 15–30 min after incubation in MYO cells might be through activation of TLR4 receptors. Thus Co-NPs toxicity can be reduced by the modification of surface coating materials [43].

15.4 CONCLUSION

Cobalt exerts both beneficial and harmful biological effects depending on the specific condition. Activation of HIF by cobalt can stimulate erythropoietin (Epo) production, leading to increased oxygen-carrying capacity of the blood; however, hypoxic adaptation increases the risk of carcinogenicity. Cobalt induces cytotoxicity by inducing apoptosis with immune response. The genotoxic effect of cobalt mediated by ROS governed the oxidative damage of DNA with inhibition of DNA repair mechanism. IARC classified cobalt as carcinogen for animals but possible carcinogen for humans due to the lack of sufficient evidence. Cobalt is accumulated primarily in liver, kidney, pancreas, heart, skeleton, and skeletal muscle, and excreted through urine. In blood, cobalt binds to albumin and inside cells with different cytosolic biomolecules. Cobalt (Co^{2+}) is imported into the cells similar to calcium (Ca^{2+}) but trapped in the cytosol due to its binding and fails to utilize Ca-pump for export (Figure 15.1). The novel use of cobalt in MoM bearings in hip joint arthroplasty and as nanoparticles brought up new problems associated with increased internal exposure. Patients with implant suffer from periprosthetic soft-tissue inflammation and adaptive immune response. Although cobalt-containing MoM articulations are well-documented clinical success, patients should be observed over a longer period to evaluate any evidence due to prolonged, elevated cobalt serum concentrations. Abuse of cobalt salt supplements by athletes as an attractive substitute to Epo doping for aerobic performance enhancement is another deep concern. Future epidemiological research with more subjects is necessary to confirm the risk of cancer associated with cobalt. The surface coating of cobalt nanoparticles used in the biomedical field can reduce harmful effect in patients. Finally, in our opinion, more in-depth research is necessary to investigate the impact of cobalt metal, ions, salts, and alloys on human health.

REFERENCES

1. Dolara P: Occurrence, exposure, effects, recommended intake and possible dietary use of selected trace compounds (aluminium, bismuth, cobalt, gold, lithium, nickel, silver). *International Journal of Food Sciences and Nutrition* 2014, 65(8):911–924.
2. Chang H, Tomoda S, Silwood CJ, Lynch E, Grootveld M: 1H NMR investigations of the molecular nature of cobalt(II) ions in human saliva. *Archives of Biochemistry and Biophysics* 2012, 520(1):51–65.
3. Suh M, Thompson CM, Brorby GP, Mittal L, Proctor DM: Inhalation cancer risk assessment of cobalt metal. *Regulatory Toxicology and Pharmacology (RTP)* 2016, 79:74–82.
4. Simonsen LO, Harbak H, Bennekou P: Cobalt metabolism and toxicology: a brief update. *The Science of the Total Environment* 2012, 432:210–215.
5. Chattopadhyay S, Dash SK, Tripathy S, Das B, Mandal D, Pramanik P, Roy S: Toxicity of cobalt oxide nanoparticles to normal cells; an in vitro and in vivo study. *Chemico-Biological Interactions* 2015, 226:58–71.
6. Ortega R, Bresson C, Fraysse A, Sandre C, Deves G, Gombert C, Tabarant M, Bleuet P, Seznec H, Simionovici A et al: Cobalt distribution in keratinocyte cells indicates nuclear and perinuclear accumulation and interaction with magnesium and zinc homeostasis. *Toxicology Letters* 2009, 188(1):26–32.
7. Ortega R, Bresson C, Darolles C, Gautier C, Roudeau S, Perrin L, Janin M, Floriani M, Aloin V, Carmona A et al: Low-solubility particles and a Trojan-horse type mechanism of toxicity: the case of cobalt oxide on human lung cells. *Particle and Fibre Toxicology* 2014, 11:14.
8. Smith LJ, Holmes AL, Kandpal SK, Mason MD, Zheng T, Wise JP, Sr.: The cytotoxicity and genotoxicity of soluble and particulate cobalt in human lung fibroblast cells. *Toxicology and Applied Pharmacology* 2014, 278(3):259–265.
9. Paustenbach DJ, Galbraith DA, Finley BL: Interpreting cobalt blood concentrations in hip implant patients. *Clinical Toxicology* 2014, 52(2):98–112.
10. Busch W, Kuhnel D, Schirmer K, Scholz S: Tungsten carbide cobalt nanoparticles exert hypoxia-like effects on the gene expression level in human keratinocytes. *BMC Genomics* 2010, 11:65.

11. Kazantzis G: Role of cobalt, iron, lead, manganese, mercury, platinum, selenium, and titanium in carcinogenesis. *Environmental Health Perspectives* 1981, 40:143–161.

12. Fedan JS, Cutler D: Hard metal-induced disease: effects of metal cations in vitro on guinea pig isolated airways. *Toxicology and Applied Pharmacology* 2001, 174(3):199–206.

13. Annangi B, Bach J, Vales G, Rubio L, Marcos R, Hernandez A: Long-term exposures to low doses of cobalt nanoparticles induce cell transformation enhanced by oxidative damage. *Nanotoxicology* 2015, 9(2):138–147.

14. Leggett RW: The biokinetics of inorganic cobalt in the human body. *The Science of the Total Environment* 2008, 389(2–3):259–269.

15. Barceloux DG: Cobalt. *Journal of Toxicology Clinical Toxicology* 1999, 37(2):201–206.

16. Ponti J, Sabbioni E, Munaro B, Broggi F, Marmorato P, Franchini F, Colognato R, Rossi F: Genotoxicity and morphological transformation induced by cobalt nanoparticles and cobalt chloride: an in vitro study in Balb/3T3 mouse fibroblasts. *Mutagenesis* 2009, 24(5):439–445.

17. Daou S, El Chemaly A, Christofilopoulos P, Bernard L, Hoffmeyer P, Demaurex N: The potential role of cobalt ions released from metal prosthesis on the inhibition of Hv1 proton channels and the decrease in Staphylococcus epidermidis killing by human neutrophils. *Biomaterials* 2011, 32(7):1769–1777.

18. Posada OM, Tate RJ, Grant MH: Toxicity of cobalt-chromium nanoparticles released from a resurfacing hip implant and cobalt ions on primary human lymphocytes in vitro. *Journal of Applied Toxicology (JAT)* 2015, 35(6):614–622.

19. Nyga A, Hart A, Tetley TD: Importance of the HIF pathway in cobalt nanoparticle-induced cytotoxicity and inflammation in human macrophages. *Nanotoxicology* 2015, 9(7):905–917.

20. Akbar M, Brewer JM, Grant MH: Effect of chromium and cobalt ions on primary human lymphocytes in vitro. *Journal of Immunotoxicology* 2011, 8(2):140–149.

21. Devlin JJ, Pomerleau AC, Brent J, Morgan BW, Deitchman S, Schwartz M: Clinical features, testing, and management of patients with suspected prosthetic hip-associated cobalt toxicity: a systematic review of cases. *Journal of Medical Toxicology: Official Journal of the American College of Medical Toxicology* 2013, 9(4):405–415.

22. Tower S: Arthroprosthetic cobaltism: identification of the at-risk patient. *Alaska Medicine* 2010, 52:28–32.

23. Tower SS: Arthroprosthetic cobaltism associated with metal on metal hip implants. *BMJ* 2012, 344:e430.

24. Rizzetti MC, Catalani S, Apostoli P, Padovani A: Cobalt toxicity after total hip replacement: a neglected adverse effect? *Muscle & Nerve* 2011, 43(1):146–147.

25. Mao X, Wong AA, Crawford RW: Cobalt toxicity: an emerging clinical problem in patients with metal-on-metal hip prostheses? *The Medical Journal of Australia* 2011, 194(12):649–651.

26. Kriss JP, Carnes WH, Gross RT: Hypothyroidism and thyroid hyperplasia in patients treated with cobalt. *Journal of the American Medical Association* 1955, 157(2):117–121.

27. Catalani S, Rizzetti MC, Padovani A, Apostoli P: Neurotoxicity of cobalt. *Human & Experimental Toxicology* 2012, 31(5):421–437.

28. McDermott PH, Delaney RL, Egan JD, Sullivan JF: Myocardosis and cardiac failure in men. *JAMA* 1966, 198(3):253–256.

29. Linna A, Oksa P, Groundstroem K, Halkosaari M, Palmroos P, Huikko S, Uitti J: Exposure to cobalt in the production of cobalt and cobalt compounds and its effect on the heart. *Occupational and Environmental Medicine* 2004, 61(11):877–885.

30. Samelko L, Landgraeber S, McAllister K, Jacobs J, Hallab NJ: Cobalt alloy implant debris induces inflammation and bone loss primarily through danger signaling, not TLR4 activation: implications for DAMP-ening implant related inflammation. *PloS One* 2016, 11(7):e0160141.

31. Dalal A, Pawar V, McAllister K, Weaver C, Hallab NJ: Orthopedic implant cobalt-alloy particles produce greater toxicity and inflammatory cytokines than titanium alloy and zirconium alloy-based particles in vitro, in human osteoblasts, fibroblasts, and macrophages. *Journal of Biomedical Materials Research Part A* 2012, 100(8):2147–2158.

32. Harris RM, Williams TD, Hodges NJ, Waring RH: Reactive oxygen species and oxidative DNA damage mediate the cytotoxicity of tungsten-nickel-cobalt alloys in vitro. *Toxicology and Applied Pharmacology* 2011, 250(1):19–28.

33. Kalinich JF, Emond CA, Dalton TK, Mog SR, Coleman GD, Kordell JE, Miller AC, McClain DE: Embedded weapons-grade tungsten alloy shrapnel rapidly induces metastatic high-grade rhabdomyosarcomas in F344 rats. *Environmental Health Perspectives* 2005, 113(6):729–734.

34. Patel E, Lynch C, Ruff V, Reynolds M: Co-exposure to nickel and cobalt chloride enhances cytotoxicity and oxidative stress in human lung epithelial cells. *Toxicology and Applied Pharmacology* 2012, 258(3):367–375.

35. Fritzsche J, Borisch C, Schaefer C: Case report: high chromium and cobalt levels in a pregnant patient with bilateral metal-on-metal hip arthroplasties. *Clinical Orthopaedics and Related Research* 2012, 470(8):2325–2331.

36. Witzleb WC, Ziegler J, Krummenauer F, Neumeister V, Guenther KP: Exposure to chromium, cobalt and molybdenum from metal-on-metal total hip replacement and hip resurfacing arthroplasty. *Acta Orthopaedica* 2006, 77(5):697–705.

37. Ahmad F, Zhou Y: Pitfalls and challenges in nanotoxicology: a case of cobalt ferrite ($CoFe_2O_4$) nanocomposites. *Chemical Research in Toxicology* 2017, 30(2):492–507.

38. Alinovi R, Goldoni M, Pinelli S, Campanini M, Aliatis I, Bersani D, Lottici PP, Iavicoli S, Petyx M, Mozzoni P et al: Oxidative and pro-inflammatory effects of cobalt and titanium oxide nanoparticles on aortic and venous endothelial cells. *Toxicology In Vitro: An International Journal Published in Association with BIBRA* 2015, 29(3):426–437.

39. Rajiv S, Jerobin J, Saranya V, Nainawat M, Sharma A, Makwana P, Gayathri C, Bharath L, Singh M, Kumar M et al: Comparative cytotoxicity and genotoxicity of cobalt (II, III) oxide, iron (III) oxide, silicon dioxide, and aluminum oxide nanoparticles on human lymphocytes in vitro. *Human & Experimental Toxicology* 2016, 35(2):170–183.

40. Mauro M, Crosera M, Pelin M, Florio C, Bellomo F, Adami G, Apostoli P, De Palma G, Bovenzi M, Campanini M et al: Cobalt oxide nanoparticles: behavior towards intact and impaired human skin and keratinocytes toxicity. *International Journal of Environmental Research and Public Health* 2015, 12(7):8263–8280.

41. Zheng F, Luo Z, Zheng C, Li J, Zeng J, Yang H, Chen J, Jin Y, Aschner M, Wu S et al: Comparison of the neurotoxicity associated with cobalt nanoparticles and cobalt chloride in Wistar rats. *Toxicology and Applied Pharmacology* 2019, 369:90–99.

42. Hanini A, Massoudi ME, Gavard J, Kacem K, Ammar S, Souilem O: Nanotoxicological study of polyol-made cobalt-zinc ferrite nanoparticles in rabbit. *Environmental Toxicology and Pharmacology* 2016, 45:321–327.

43. Lojk J, Strojan K, Mis K, Bregar BV, Hafner Bratkovic I, Bizjak M, Pirkmajer S, Pavlin M: Cell stress response to two different types of polymer coated cobalt ferrite nanoparticles. *Toxicology Letters* 2017, 270:108–118.

CHAPTER 16

Cobalt Toxicity

Muhammad Umar
Royal Liverpool University Hospital

Ayyaz Sultan
Royal Albert Edward Infirmary
Wrightington, Wigan and Leigh NHS Foundation Trust

Noman Jahangir
Leeds Children Hospital

Zobia Saeed
Royal Liverpool University Hospital

CONTENTS

16.1 INTRODUCTION

Cobalt is a silvery-gray metal, ductile heavy metal, naturally distributed as a trace element. Commonly, it is found as a stable isotope ^{59}Co (atomic number 27; atomic weight of 58.933) sharing many chemical and physical properties with iron and nickel. It is a ferromagnetic and transitional metal with high melting point (1,495°C), found in the earth's crust as cobaltous and cobaltic salt. Cobalt utilization is extensive, including the production of superalloys, lithium-based battery cathodes, oxidation catalysts, cemented carbides, and diamond tools. It is also used as radioisotopes, coloring/pigments and for electroplating purposes.

Cobalt, as a trace element, plays an essential role in maintaining biochemical processes such as regulation of gene expression, immune system, prevention of chronic diseases, and in antioxidant defense of body[1] for animals and humans. However, excessive cobalt exposure has been linked to several forms of adverse health-related issues. These effects can sometimes be seen at body-concentrations significantly less than lethal cobalt dose.[2]

In this chapter, we will discuss a brief history of cobalt use and its relation to humans.

16.2 HISTORY

As a coloring agent, cobalt has been used, since around 2000 BC, in pottery, jewelry, and glass in Egypt and Persia.[3] Cobalt was discovered in the ruins of Pompei which was destroyed in 79 AD. In the fifteenth century, cobalt was used as a blue pigment in Ming Dynasty (1368–1644 AD) pottery and in Venetian glass.[4] More recently, it has been used in Danish porcelain, for its distinct blue color. Leonardo Da Vinci used a bright blue pigment of cobalt in his oil paintings.

Cobalt was the first metal to be discovered since prehistoric period. Other known metals were in use (gold, silver, iron, copper, zinc, mercury, lead, and bismuth), but there are no known discoverers of those metals.[5] George Brandt, a Swedish chemist, first isolated cobalt in the eighteenth century, but it was T O Bregman, who identified it as an element in 1780.[6]

In the twentieth century, cobalt was extensively used in heavy metal industry. During World War II, cobalt alloys were used in manufacturing of jet engines and gas turbines. Since World War II, its use had increased significantly, reaching its peak in mid-1980s. At this time, the development of cobalt magnets leads to a significant size reduction of electrical and electronic equipment, playing an important role in our modern world.

In the 1960s, cobalt was used as a foam stabilizer in the beer industry which led to the high mortality due to cobalt-associated cardiomyopathy in heavy beer drinkers of Northern America.[7]

Over the last few decades, cobalt alloys have been used in healthcare sector in the manufacturing of arthroplasty implants. Cobalt-containing lithium-ion batteries have been increasingly used in recent years in smart portable devices and electric vehicles.

16.2.1 Cobalt Exposure

16.2.1.1 Environmental Exposure

Generally, low concentrations of different cobalt compounds are widely dispersed in nature. It is found in the earth's crust and atmosphere but rarely detected in drinking water. However, air, water, and soil pollution by cobalt can occur in areas near factories and heavily industrialized cities. In such regions, dietary (vegetable, cereals, and fish) cobalt intake and ingestion of contaminated dust are the main routes of exposure. The latter is particularly a significant route in children.[8]

Products such as eye pencil, eye shadow, lipstick, soaps are public sources of cobalt for humans.[9] Use of cobalt is forbidden in cosmetic products by the EU, but its presence is permissible as impurities when required. Although nickel is one of the most frequent causes of allergic contact dermatitis worldwide, concomitant nickel and cobalt allergy are frequently seen.[10] Cobalt may be absorbed from jewelry following oxidization to cobalt ions by sweat. This process is more applicable to jewelry which penetrates the skin.

Furthermore, leather goods and tattoo pigments containing cobalt may lead to contact dermatitis.[11]

16.2.1.2 Occupational Exposure

The hard metal industry is the primary source of occupational cobalt exposure, as it uses 15% of the worldwide production of cobalt.[12] Tungsten carbide is the crucial component of the mixture alloy in the hard metal industry. In this form, cobalt is mainly used as a binder. This combination of cobalt and tungsten carbide enhances cobalt's cellular uptake and toxicity.[2]

In the industrial environment, uptake of cobalt can result from inhalation of hard metal dust and direct contact to the skin. Smoking has also shown to increase cobalt intake via a dust-hand cigarette-mouth path.

In the hard metal industry, the highest levels of cobalt are measured in powder production areas, pressing department, and sintering workshop.[13] However, over the years, measures have been taken by the industry to improve hygiene and protection, leading to decreased levels of airborne cobalt and tungsten dust.

People working in the construction industry may also get cobalt exposure through skin contact with cement, leading to contact dermatitis. Similarly, employees in the e-waste recycling industry are also exposed to cobalt, as several electric and electronic devices contain cobalt. Formal recycling techniques used in advanced countries minimize the risk of this exposure by use of employee protection and appropriate equipment. In less developed countries, simple techniques are in use, such as open burning, acid bathing, and cutting. It leads to significantly high cobalt exposure to the workers.[14]

Diamond polishing is another industry, which poses an exposure risk to its workers. High-speed polishing disks made of micro diamonds are used, cemented in ultrafine cobalt metal powder. Use of these disks leads to cobalt dust formation, which can be inhaled by the workers. It has shown to cause respiratory problems.[2,15]

16.2.1.3 Medical Exposure

In the middle of the twentieth century, cobalt preparations were used in the treatment of anemia, as it showed a stimulant effect on the production of hemoglobin and red blood cells. Required daily dosage for these treatments was high, and it led to the development of severe adverse health effects. Due to a lack of significant positive response and treatment-related adverse effects, this is not in practice anymore.[16,17]

Cobalt supplements are also prescribed to some patients on post-menopausal hormone replacement therapy. It helps in reducing estrogen hyperexcretion by controlling overactivity of cytochrome enzymes.[18]

Over the last decades, the use of cobalt–chromium hard-metal alloys in orthopedic joint replacements has created a new source of cobalt exposure. The application of hard metals in metal-on-metal (MoM) bearings in hip joint arthroplasty has raised serious concern, especially given the high and an increasing number of hip replacements, in younger and more physically active patients. Approximately one million MoM articulations have been implanted since 1996. It became so popular that 10% of all hip arthroplasties performed in developed countries over the last two decades had MoM articulations.

There are two main types of hip replacements with MoM articulations: total hip replacement (THR) and hip resurfacing arthroplasty (HRA). The latter one is mainly used in the younger population, to preserve natural anatomy by just replacing the articular surfaces. This procedure is aimed to provide better activity levels and survivorship.

As two metal-articulating surfaces move against each other during activities of daily living, cobalt and chromium ions are released by friction, in the form of nano-sized wear particles. Further on, corrosion of metal surfaces and released metal particles can also contribute to raised metal ion levels in body.[19]

This source of cobalt exposure has extensively been studied. Certain factors have shown increased release of metal ions in patients with MoM hip implants. These factors include suboptimal implant position, modularity of implants, impaired renal function, bilateral hip replacements, and mixing components from different manufactures or different types.[20]

Madl et al.[21] studied wear particle generation by cobalt–chromium alloy MoM hip implants in patients and simulator systems. They concluded that under well-functioning conditions, MoM hip implants have a low wear rate and generate primarily oxidized chromium nanoparticles with minimal or no cobalt content. However, when there is an implant malpositioning, it leads to a significantly higher wear rate and creates larger-sized wear debris (up to 1,000 nm), containing higher concentrations of cobalt. Once released from the articular surfaces, the highly soluble cobalt ions bind with synovial proteins and adjacent tissue. Later these ions disseminate in blood and can lead to toxicity.[22]

16.2.2 Cobalt Uses

Contemporary use of cobalt can be broadly categorized into industrial utilization and biomedical applications.

16.3 APPLICATION OF COBALT IN MODERN INDUSTRIAL ERA

Contemporary demand for cobalt at an industrial level can be broadly categorized into metal and chemical.

1. **Cobalt Demand as Metal**:
   Soon after the discovery that cobalt bound tungsten carbide into a hard substance, this hard metal was used in many industrial applications, such as high-speed drills, cutting tools, metal rollers, and engine components used in automotive, mining, and general engineering industries. In these alloys, cobalt is used as a binder matrix metal, which results in a metal matrix with superior wear resistance and increased hardness, making it an ideal choice for these applications. Additionally, specific properties of excellent corrosion resistance and magnetic conductivity led to its increasing use in heavy industry.[16]

   Cobalt-containing superalloys have higher strength, high melting point, and superior resistance against corrosion, leading to its use in aerospace, petrochemical, and medical prosthetic manufacturing industry.

   Heavy industry use of cobalt puts its workers at a higher risk of cobalt exposure and may present as different forms of toxicity.
2. **Cobalt Demand as Chemical**:

Cobalt chemicals are used in the production of lithium-ion batteries, catalysts, pigments, polymers, and tires. Typically these are dry, granular, powdered salts, commonly in the form of cobalt oxide (CoO), cobalt hydroxide ($Co(OH)_2$), cobalt sulfate ($CoSO_4$), and cobalt acetate ($Co(CH_3COO)_2$), along with various niche compounds for other applications. Over the last few decades, the market for cobalt chemicals has spurred due to significant technological advancements. For example, lithium-ion batteries have become extremely popular, owing to their high energy, power density, and long lifetimes compared to other battery types. It makes them perfect not only for modern portable technologies (like smart devices) but also in the current automobile industry revolution of electric vehicles. As a result, demand for cobalt chemicals has increased exponentially. Recycling plants for these batteries have been a source of cobalt release in the environment.

16.4 COBALT ROLE IN BIOMEDICAL AND HEALTHCARE SYSTEM

1. **Cobalt Role in Healthcare System**:
   In common use, cobalt is in its stable form (^{59}Co); however, it is also available as unstable isotopes. ^{60}Co isotope is used as a source for gamma-ray for radiotherapy. Gamma-ray is used in several applications, such as external beam radiotherapy, medical supplies and waste sterilization, radiation treatment of foods for sterilization (cold pasteurization), industrial radiography for non-destructive detection of structural flaws in metal parts, pest insect sterilization, density measurements (e.g. concrete), and tank fill therapy.
   Another cobalt isotope (^{57}Co) is used for radiolabeling for vitamin B12 uptake and the Schilling test. This isotope has also been used as a source in X-ray fluorescence devices.
2. **Cobalt Role in Biochemical/Biomedical Reactions**:
   Diet is the primary source of cobalt exposure in the general population, although the human body requires only trace amounts of cobalt as a component of vitamin B12. The average daily dietary intake is highly variable and depends upon diet type and geographical location.
   Hundreds of food items contain cobalt in varying quantities. Certain everyday food items contain higher concentrations, such as chocolate, butter, nuts, fish, coffee, fresh cereals, and green leafy vegetables (broccoli and spinach).[23] Drinking water can also be a source of cobalt intake, but levels may vary significantly on a geographical basis. Environmental factors such as pollution and anthropogenic activities may contribute to the presence and variation of concentration of this metal in food. Food processing and packaging can also be a source of cobalt in diet.[24]
   "Total Diet Studies" (TDS) performed by several countries reported a mean dietary cobalt intake of 0.13–0.48 µg/kg bw/day in adults and 0.27–0.31 µg/kg bw/day in children. This amount is far below the tolerable daily intake of 1.6–8 µg/kg bw/day.[24]
   Role of cobalt has been established in protein synthesis, fat and carbohydrate metabolism, red blood cell production, and myelin sheath repair. These characteristics have been manipulated by manufacturers to promote cobalt-containing or vitamin B12 supplements. Cobalt has also been misused by athletes to enhance their anaerobic performance.[25] Along with other dietary intakes, these uses may contribute to unintentional ingestion of cobalt.

16.4.1 Cobalt Metabolism

Water-soluble cobalt salts are rapidly absorbed from the small intestine, though bioavailability is incomplete and quite variable. Another source of significant cobalt uptake is lungs, through inhalation, mainly metallic cobalt, and cobalt oxide in dust and welding fumes. Systemic dissemination of ultrafine particles happens through vascular and lymphatic systems and with the release of soluble cobalt ions.[26]

Following oral administration in humans, cobalt concentration in blood and serum is initially high, then decreases rapidly and gradually, to a low level in 24 hours. This decrease is due to tissue uptake in liver and kidney combined with excretion (urinary and fecal). The renal excretion is initially rapid but decreases within the first few days. A second slow phase follows, which lasts several weeks; however, cobalt retains in tissues for several years.[27]

During long-term exposure, cobalt accumulates in tissues, particularly liver and kidney, and its concentration is increased in whole-blood serum and urine.[28] This concentration decreases over 4 weeks after cessation of occupational exposure. Majority of cobalt ions bind to serum albumin, with only 5%–12% left as a free fraction. Total cobalt concentration in serum is thus much higher than its level in interstitial fluid, which is in equilibrium with free serum cobalt ions.[29]

16.4.2 Cobalt Pathogenesis

Cobalt ions and cobalt metal (nanoparticles) are cytotoxic and induce apoptosis. At higher concentrations, it can cause tissue necrosis and inflammatory response. The type of reaction depends upon the chemical form of cobalt exposure.

In occupational and environmental settings, predominant exposure is to cobalt metal particles. This exposure leads to an immune-mediated response and causes local adverse tissue reactions. It may present as adverse respiratory effects on the inhalation of cobalt dust.

This response can be of two forms: metal reactivity and metal allergy.[30]

16.4.2.1 Metal Reactivity

Metal reactivity is a normal innate immune response to a large amount of metal debris, leading to nonspecific foreign body reaction.

16.4.2.2 Metal Allergy

Metal allergy is an adaptive immune response to a small amount of metal debris in people with an allergic genetic predisposition. Allergy is typically associated with contact dermatitis.

Following MoM hip arthroplasty, corrosion and wear produce soluble metal ions and metal debris in the form of nanometric size huge number of wear particles with large area/volume ratio. Systemic dissemination of these particles happens through vascular and lymphatic systems. Cobalt ions are released from the surface of these nanoparticles to the biological environment. Nanoparticles are also phagocytosed by macrophages, leading to cobalt ion release by intra-cellular corrosion.[31] This release leads to increased blood concentration and systemic toxicity.

Cobalt toxicity related to metal hip implants may present as an adverse local reaction in peri-prosthetic soft tissues. It may lead to metallosis, tissue necrosis, osteolysis, pseudotumor formation, and aseptic lymphocyte-dominated vasculitis-associated lesions (ALVAL). Inflammatory and necrotic changes are caused by cell-mediated hypersensitivity and cytotoxicity due to metal ion release. Studies have shown that metal nanoparticles are more toxic than micrometer-sized particles leading to the concept of nanotoxicity.[32]

Free cobalt ions interact with various cellular receptors, ion channels, and biomolecules to initiate molecular mechanisms including generation of reactive oxygen species, lipid peroxidation, and alteration of calcium and iron homeostasis. These interactions create an interruption of mitochondrial function, triggering of erythropoiesis by interacting with body feedback systems, induction of genotoxic effects, deviation of DNA repair process, and interruption of thyroid iodine uptake.[33] Exposure to cobalt ions was also found to significantly inhibit osteoblast function by reducing alkaline phosphatase activity and calcium deposition. It also induces secretion of proteins such as IL-8 and MCP-1 in osteoblasts. The combination of these effects may play a role in osteolysis around the hip implants. Due to this multitude of local and systemic effects related to cobalt toxicity, it manifests as a clinical syndrome with symptoms related to multiple body systems. In the literature, it is also called arthroprosthetic cobaltism.[34]

A review study taking account of animal and human dose-response data for adverse health effects and biokinetic model discussed earlier has shown that severe effects including neurological and cardiac symptoms are seen with cobalt levels above 700 μg/L. In contrast, levels below 300 μg/L may lead to reversible hematological and endocrine symptoms.[35,36] However, published literature reporting systemic effects of cobalt suggests these effects can be seen in patients with lower cobalt levels as well, with a long-term exposure.[37,38]

So far, no consensus exists on a threshold value of cobalt concentration for the development of systemic health effects. In clinical practice, the threshold of suspicion for arthroprosthetic cobaltism should be kept low, mainly when dealing with a patient on risk of having raised cobalt levels, like the history of MoM hip surgery.

16.4.3 Measurement of Cobalt Levels

Mean cobalt blood levels in subjects without abnormal exposure is 0.3 μg/L, and 95% have a value of less than 0.6 μg/L.[39] According to the American Conference of Governmental Industrial Hygienists (ACGIH), the biologic exposure threshold for cobalt is 1 μg/L, value above this level indicates unsafe vocational exposure.

Various methods have been used to assess the level of cobalt exposure. In an occupational setting, target organ level is measured, in the form of skin patch testing (cutaneous intake) and analysis of exhaled breath condensate (inhaled cobalt concentration). However, exhaled breath condensate is not a reliable indicator of cobalt exposure in the workplace.[40]

A more reliable way is a systemic level measurement of ion concentration. It can be measured in urine, whole blood, or serum. Different units can be used to express the results of these measurements. For blood and serum, parts per billion (ppb), microgram per liter (μg/L), nanogram per milliliter (ng/mL), nanomole per liter (nm/L), and micromole per liter (μm/L) are used, whereas for urinary levels, microgram per gram creatinine (μg/g creatinine), microgram per liter (μg/L), and microgram per millimole creatinine (μg/mmol creatinine) are used.

Urinary cobalt concentration is commonly used in occupational exposure assessment. Usually, it shows a rapid increase in the first hours after cessation of exposure and reaches a peak at 3 hours.[41] Ion concentrations in urine are more variable and depend on the hydration of the individual as well.

Whole blood and serum are the preferred matrices in medical exposure settings, which are predominantly related to MoM hip implants. Both of these measurements are relatively similar and correlated. Paustenbach et al.[33] and Finley et al.[42] studied cobalt concentrations during dosing and following cobalt supplementation. Their studies suggested that the whole blood concentration might be the most appropriate measure to estimate the long-term average cobalt exposure,[22] whereas the serum concentration might give a better indication of recent or recently changed cobalt exposure. They also suggested that whole blood and serum concentration should not be used interchangeably, as there is no standard rate of conversion between the two.

16.4.3.1 Cobalt Level Measurement in Patients with Metal Hip Implants

In patients with metal hip implants, cobalt ion levels first increase to a maximum during the running-in phase of the implant. This phase can take approximately 9–12 months postoperatively. Further on, in a normally functioning hip, the ion concentration is expected to decrease to a steady state, whereas in a malfunctioning hip implant, it may increase further.[43] This concept of running-in phase is the basis of Medicines and Healthcare Products Regulatory Agency (MHRA) recommendation of checking the systemic cobalt and chromium levels after 12 months running-in period of MoM hip prosthesis, as a screening tool for in vivo performance.[44] Elevated metal ion concentrations have shown an association with an increased degree of wear, corrosion, and periprosthetic complications such as loosening.[45,46]

Multiple studies have discussed cobalt ion levels in patients with MoM hip implants. In a well-functioning implant, values range between 0.2 and 10 µg/L.[22] MHRA recommends a 7 µg/L threshold for the identification of patients with adverse local tissue reaction, which required further investigation and interventions.[44] However, the Mayo Clinic states values between 4 and 10 µg/L, which reflect the good condition of the MoM implant, whereas levels above 10 µg/L indicate significant implant wear. In a clinical scenario, cobalt ion levels should be interpreted carefully and serve as an adjunct to clinical and radiographic assessment.

Along with local effects related to the implant, elevated cobalt concentrations in patients with MoM articulation may show systemic effects of toxicity as well. Studies have shown an association between cobalt concentration and symptom severity. Van Der Straeten et al. concluded in their study that patients with repeated cobalt concentrations exceeding 20 µg/L are at risk of systemic cobalt toxicity.[46] However, there are no uniform criteria to guide clinicians in the detection and management of systemic cobalt toxicity concerning cobalt concentration levels.

A group of researchers developed a biokinetic model for cobalt, based on a series of novel oral dosing studies and human and animal toxicology data. Background cobalt blood concentration as a result of dietary intake was estimated to be 0.3 µg/L (0.04–0.9 µg/L). According to this model, systemic effects are unlikely to occur at cobalt levels below 300 µg/L in healthy individuals. It suggests that monitoring of patients with MoM implants might be useful from cobalt levels of 100 µg/L.

However, susceptible individuals may exhibit systemic toxicity even at lower cobalt concentration. This susceptibility has been attributed to the partitioning of cobalt in the serum. A large portion of cobalt (90%–95%) binds with albumin, whereas about 8% is free as ionic cobalt.[47] This free cobalt is the primary toxic form. Any condition which affects albumin levels in the body can affect the level of free ionic cobalt in the body. Conditions such as renal failure, iron deficiency, sepsis, alcoholism, malnutrition, and certain medications can reduce cobalt–albumin binding, ultimately leading to systemic toxicity at lower doses. So this subgroup of patients with MoM implants will need a closer follow-up.[33]

16.4.4 Cobalt Foe or Friend?

16.4.4.1 Cobalt Deficiency

Cobalt plays an essential role in the homeostasis of heme, DNA, amino acids, and fatty acids. It performs this crucial role as cyanocobalamin or vitamin B12, which is a co-factor for homocysteine methyltransferase. This enzyme converts homocysteine into methionine, an essential methyl donor for methylation reaction, needed for myelin maintenance and trans-methylation reactions. Its deficiency may lead to neurological diseases as peripheral neuropathy, subacute combined degeneration of spinal cord, ataxia, and irreversible dementia.

Reduced cyanocobalamin levels may also lead to an increase in plasma homocysteine levels, resulting in an increased risk of thrombosis and atherosclerosis, which may contribute to the development of myocardial infarction and stroke.

16.4.4.2 Cobalt Toxicity

Cobalt is acutely toxic in larger doses, and long-term exposure even at a low level can lead to a cumulative response, presenting as adverse health effects, affecting various organs and tissues. These effects may involve different parts and systems of the body, including skin, lungs, heart, nervous system, thyroid, blood, psychiatric, and possible carcinogenic potential.

16.4.4.3 Cardiac Toxicity

Cobalt-related cardiomyopathy was first reported in northern American heavy beer drinkers in the 1960s. It was attributed to the use of cobalt salts as a foam stabilizer in beers[48] in Quebec, Canada. It was known as beer drinkers cardiomyopathy, which showed signs similar to thiamine deficiency dilated cardiomyopathy (wet beriberi). The main difference between the two was cobalt cardiomyopathy had an acute onset. Beer-drinkers cardiomyopathy had very high mortality (40%–50%) and up to a third of the survivors had abnormal ECG findings.[49]

Further studies revealed that among this group, people with poor nutritional status were more prone to develop cobalt-related cardiac toxicity.[33] Since the cessation of cobalt stabilizer used in beer manufacturing, this type of cardiomyopathy is not seen anymore. However, cardiomyopathy has also been reported in hard metal workers[50] and in patients with hip implants (particularly MoM). Cobalt cardiomyopathy after hip implants is currently a topic of discussion in the literature, as a significant number of reports have been published over the last two decades.[51]

The exact mechanism for cobalt cardiomyopathy is not fully understood yet. It may be the result of cobalt accumulation in the cardiac tissue, leading to stimulation of carotid-body chemoceptors, mimicking the action of hypoxia. Molecular destruction of cardiac tissue has been studied in rat models. It has been attributed to a decrease in cardiac enzymes, including manganese superoxide dismutase, succinate cytochrome C oxidase, and also due to reduction of mitochondrial ATP production.[52]

Cobalt cardiomyopathy patients may present with exertional dyspnea, palpitations, and orthopnea. Investigations may reveal diastolic dysfunction, pericardial effusion, and reduced left ventricular ejection fraction (LVEF). An echocardiogram is recommended in patients with suspected cobaltism (MoM hip implants, hard metal industry) presenting with new cardiac symptoms, as these features can be picked with this simple non-invasive investigation.[51] Certain risk factors for cobalt cardiomyopathy have been reported in the literature, such as heavy alcohol intake, obesity, renal impairment, diabetes, and malnutrition.[51]

Histological review (biopsy or autopsy) of cobalt affected cardiac tissue has shown not only generic features of cardiomyopathy, such as myocardial hypertrophy and interstitial fibrosis but also some unique cobalt toxicity-related features such as increased vacuolation, lipofuscin, myofiber disarray, and abnormal mitochondrial forms with electron-dense deposits. A cardiac biopsy may play a role in the definitive diagnosis of cobalt toxicity in suspected cases.

Although cobalt cardiomyopathy can be fatal in some cases, the majority of patients notice the reversal of cardiac symptoms once cobalt levels are in control. It can be achieved by removal of the causative factor, like a revision of metal hip implant or change of environment. Chelation therapy can also play a role in reducing the system cobalt levels. The chelating agent is a chemical which reduces cobalt ion load by binding to it and aiding in its renal excretion. Different chelating agents have been reported in the literature, such as N-acetyl-cysteine, 2,3-dimercaptopropane-1-sulfonate, and ethylene diamine tetra-acetic acid, with variable results.[51] Although chelation therapy helps in reducing the metal ion levels, it is only recommended as an adjunct to treatment, whereas removal of the causative factor remains the treatment of choice. As chelation therapy relies on renal excretion, the patient's kidney functions should be checked regularly.

16.4.4.4 Respiratory Toxicity

The respiratory system was the first system found to be affected by hard metal dust containing cobalt exposure. It was reported as early as in 1940 when Jobs[53] published his study discussing the association between occupational hard metal exposure and respiratory changes, based upon chest radiographs showing pneumoconiosis. Initially, the term "occupational lung disease" was used, which was later replaced by "hard metal lung disease" to describe respiratory effects of cobalt-containing dust inhalation.[54] Cobalt dust inhalation may present as decreasing pulmonary function, increased frequency of cough due to irritation of respiratory cilia and sensitization of immune system, respiratory inflammation, and pulmonary fibrosis. These effects may be the result of the generation of oxidants, leading to free radical damage.

Predominantly, there are three distinct entities reported: occupational asthma, allergic alveolitis, or hypersensitivity pneumonitis and interstitial pneumonia.

Cobalt is a known allergen which produces asthma in workers exposed to cobalt alone (cobalt powder, cobalt salt) or in association with other metals (hard metal dust). Probably it is the result of type I allergic reaction, based on the presence of specific IgE antibodies against a complex of cobalt and albumin in workers with hard metal asthma[55]. Epidemiological study of hard metal workers demonstrated age and atopy as risk factors for hard metal asthma.[56] This asthma is characterized by wheezing, cough, and dyspnea, a positive cobalt-specific IgE test and a positive bronchoprovocation test with cobalt chloride.[57]

Allergic alveolitis or hypersensitivity pneumonitis occurs in the acute stage as an early inflammatory phase of fibrosis. It may present as a sudden occurrence of fever, anorexia, cough, and dyspnea with exertion. These symptoms may improve after removal from cobalt exposure (workplace) but may recur following re-exposure. If exposure persists for a long time, repeat episodes of alveolitis may progress to interstitial fibrosis.[58]

Interstitial pneumonia can be of two varieties: typical giant cell pneumonia or desquamative type. Giant cell pneumonia presents with weight loss, fatigue, exertional dyspnea, dry cough, wheezing, and chest pain. If the disease progresses to fibrosis stage, symptoms of cyanosis, digital clubbing, pulmonary hypertension, cor-pulmonale, and right heart failure may appear. Typically, interstitial lung disease appears after 10–12 years of exposure to hard metal dust, but symptoms may appear earlier (2–7 years), depending upon the duration and intensity of exposure.

Overall, cobalt toxicity related to pulmonary dysfunction can be obstructive or restrictive, presenting with relevant symptoms. Occasionally, it may also present with pneumothorax.

Unlike neurological symptoms, the respiratory disease may progress further even after cessation of the exposure, or return on re-exposure.[59]

16.4.4.5 Dermatological Toxicity

Cobalt is one of the three most common metallic sources of contact dermatitis, along with nickel and chromium. Cobalt exposure in an industry can lead to occupational skin diseases. Occupational contact dermatitis is the most common skin disease in the workplace.[2] It may be a result of long-term exposure or short but repetitive contact with hard metals.

Cobalt-related contact dermatitis may have an early onset in work, in a range of occupations, such as hairdressers/barbers, builders/building contractors, retail cash/checkout operators, machine operatives, and domestic cleaners. The hands are the most common part affected, for obvious reasons.[60] However, a study of allergic profile in construction workers has shown that dermatitis affected not only the exposed parts but also the covered parts.[61] In construction workers, chromate is the predominant allergen, followed by cobalt and nickel. However, isolated allergy to cobalt and nickel is unlikely. Cobalt allergy generally occurs secondary to a chromate allergy, which has already caused skin damage.[61] Cobalt and nickel commonly coexist in metal objects. Sensitization to nickel may increase the risk of developing cobalt allergy.[62]

In environmental setting, skin reactions may be caused by jewelry, cosmetics, and leather. There have been some reports of eczema with the use of cobalt-containing gel for fascial massage, cobalt-containing nail gel for nail art or by cobalt-releasing necklace. Skin symptoms resolved after removing further exposure to those products. Leather has also been found to be a source of cobalt-related contact dermatitis.[11] It may be associated with the use of leather containing clothing or furniture. Skin reactions have also been reported with oral cobalt intake, and it may present as acne and skin rashes.[63]

16.4.4.6 Neurological Toxicity

Cobalt toxicity may affect both peripheral and central nervous systems. It may present with a variety of symptoms related to hearing, balance, vision, sensory-motor function, and cognitive performance. It may also present as polyneuropathy.

Hearing loss is usually bilateral and of sensorineural nature. It is severe in high frequencies and can be progressive, dependent on exposure. Other associated symptoms can be of vertigo, dizziness, and tinnitus.

Visual impairment is usually the result of bilateral optic nerve atrophy. It may present with reduced visual acuity, blindness, decreased color vision, blurred vision, irregular cortical visual response, and retinal dysfunction.

Cobalt toxicity-related cognitive decline may present as poor concentration, memory loss, disorientation, attention deficit, inefficiency, and difficulty in registering new information.

Sensory and motor loss may present as incoordination, tremors, headaches, motor weakness, slow sensory neural conduction, dysesthesia, gait disturbance, numbness, and paresthesia.

Usually, neurological symptoms improve or resolve on the removal of cobalt exposure, but auditory and visual dysfunction may persist.

16.4.4.7 Endocrine Toxicity

Cobalt toxicity can lead to thyroid dysfunction, which may present as hypothyroidism or chronic thyroiditis. It may be a result of reduced iodine uptake and development of goitre. Histological analysis of cobalt affected thyroid tissue shows follicular distortion, cellular changes, and colloid depletion. Thyroid dysfunction is also considered as a risk factor for the development of cardiomyopathy related to cobalt toxicity.[64] This effect was first identified in the 1940s, following its use for the treatment of anemia.[65] The therapeutic use of cobalt for anemia treatment was associated with side effects such as thyroid dysfunction and reversible vision and hearing impairment.[66,67] Its use for anemia treatment ended in the 1970s when researchers developed more efficient drugs for this condition.

16.4.4.8 Hematological Toxicity

Cobalt toxicity may lead to inhibition of heme synthesis by acting upon two different sites in the biosynthetic pathway. It may lead to the formation of cobalt protoporphyrin rather than heme. However, it may also stimulate the oxidation of heme in many organs by induction of heme oxygenase. Oxidation of heme results in the stimulation of erythropoiesis.

Human studies have shown hematological stimulant effects, presenting as polycythemia, increase in hematocrit and hemoglobin levels.[68] It leads to the increased oxygen-carrying capacity of the blood, which is helpful under conditions of ischemia and hypoxia. Preconditioning with cobalt salts promotes tissue adaptation to hypoxia and enhances physical endurance performance. This effect of cobalt exposure has been misused by athletes to improve their aerobic performance in competitive sports, by use of cobalt-containing supplements.

16.4.4.9 Carcinogenic Potential

Cobalt metal and salts are genotoxic in mammalian in vitro test systems. Mainly it is caused by oxidative DNA damage by reactive oxygen species, along with inhibition of DNA repair.[32] Although evidence of carcinogenicity of cobalt is considered sufficient in experimental animals but is deemed to be inadequate in humans. However, cobalt, in combination with tungsten carbide, has been found to have carcinogenic potential.[69] International Agency for Research on Cancer

(IARC) classified the mixture of cobalt and tungsten carbide (Co/WA) as "probably carcinogenic to humans."[70] In an occupational exposure setting, cobalt-containing dust has been considered as a risk for lung cancer.

Numerous studies have been conducted to evaluate the risk of cancer in hip arthroplasty, particularly MoM articulation, due to increasing concerns. However, there is a lack of any significant evidence confirming the association of hip arthroplasty with a risk of cancer development.[71,72]

16.4.4.10 Psychiatric Toxicity

Certain psychological symptoms have been reported in patients with MoM hip implants. These symptoms include depression, lack of energy, fatigue, irritability, and anxiety.[34,73] However, there is lack any evidence suggesting cause-and-effect relationship between the development of these symptoms and cobalt.

16.4.4.11 Radioactive Toxicity

Typically, cobalt is present as a stable isotope, but radioactive isotopes can be produced by irradiating cobalt with thermal neutrons in a nuclear reactor. This form of cobalt is used as a source of gamma rays for different purposes such as sterilizing medical equipment and consumer products, food irradiation, manufacturing plastics, and as radiation therapy for treating cancer. Although the general population is not exposed to radioactive cobalt, patients on radiotherapy may be at risk of gamma rays from a cobalt source. Other people at risk are people working at nuclear facilities and nuclear waste storage sites. In 1984, one of the worst radiation contamination accidents occurred in North America, when a discarded radiotherapy unit was mistakenly disassembled in a junkyard in Juarez, Mexico.

Cobalt radiation exposure has been associated with nervous tissue damage, particularly peripheral nerves. In literature, variable complications have been reported following radiotherapy. Patients receiving radiotherapy for nasopharyngeal and cervical lymph node pathologies developed focal necrosis of frontal lobe.[74] In another study, patients receiving cobalt radiotherapy for all acquired severe visual disturbances, secondary to alterations in optic nerve including atrophy, terminal beading, lack of myelin and calcifications, confirmed on histopathology.[75] Cobalt radiotherapy-related myelopathy may also present as mild paralysis, brachial plexus neuropathy, and vocal cord paralysis.[76] Other effects of radioactive cobalt exposure may be temporary sterility due to impact on the reproductive system, increased infection risk secondary to reduced white cell count and dermatological changes such as skin blisters and burns. The amount of damage depends upon the amount of radiation and its duration of exposure.

REFERENCES

1. Strachan S. Trace elements. *Current Anaesthesia & Critical Care*; 21: 44–48.
2. Barceloux DG. Cobalt. *Journal of Toxicology: Clinical Toxicology*; 37: 201–206.
3. Planinsek F, Newkirk JB. *Kirk-Othmer Encyclopedia of Chemical Technology*. 3rd ed. New Jersey: John Wiley & Sons, 1979.
4. Britannica IE. *Encyclopaedia Britannica*. 1957. https://www.britannica.com/science/cobalt-chemical-element#ref1376
5. Weeks ME, Chemistry SLTJOP, 1946. Discovery of the elements. *ACS Publications*.
6. Hamilton EI. The geobiochemistry of cobalt. *Science of the Total Environment*; 150: 7–39.
7. Kesteloot H, Roelandt J, Willems J, et al. An enquiry into the role of cobalt in the heart disease of chronic beer drinkers. *Circulation*; 37: 854–864.

8. Cheyns K, Banza Lubaba Nkulu C, Ngombe LK, et al. Pathways of human exposure to cobalt in Katanga, a mining area of the D.R. Congo. *Science of the Total Environment*; 490: 313–321.

9. Bocca B, Pino A, Alimonti A, et al. Toxic metals contained in cosmetics: a status report. *Regulatory Toxicology Pharmacology*; 68: 447–467.

10. Hamann D, Thyssen JP, Hamann CR, et al. Jewellery: alloy composition and release of nickel, cobalt and lead assessed with the EU synthetic sweat method. *Contact Dermatitis*; 73: 231–238.

11. Bregnbak D, Thyssen JP, Zachariae C, et al. Association between cobalt allergy and dermatitis caused by leather articles: a questionnaire study. *Contact Dermatitis*; 72: 106–114.

12. Klasson M, Bryngelsson I-L, Pettersson C, et al. Occupational exposure to cobalt and tungsten in the Swedish hard metal industry: air concentrations of particle mass, number, and surface area. *Annals of Occupational Hygiene*; 60: 684–699.

13. Kraus T, Schramel P, Schaller KH, et al. Exposure assessment in the hard metal manufacturing industry with special regard to tungsten and its compounds. *Occupational and Environmental Medicine*; 58: 631–634.

14. Sthiannopkao S, Wong MH. Handling e-waste in developed and developing countries: initiatives, practices, and consequences. *Science of the Total Environment*; 463–464: 1147–1153.

15. Nemery B, Casier P, Roosels D, et al. Survey of cobalt exposure and respiratory health in diamond polishers. *The American Review of Respiratory Disease*; 145: 610–616.

16. Paustenbach DJ, Tvermoes BE, Unice KM, et al. A review of the health hazards posed by cobalt. *Critical Reviews in Toxicology*; 43: 316–362.

17. Domingo JL. Cobalt in the environment and its toxicological implications. *Reviews of Environmental Contamination and Toxicology*; 108: 105–132.

18. Wright JV. Bio-identical steroid hormone replacement: selected observations from 23 years of clinical and laboratory practice. *Annals of the New York Academy of Sciences*; 1057: 506–524.

19. Hart AJ, *et al.* Microfocus study of metal distribution and speciation in tissue extracted from revised metal on metal hip implants. *Journal of Physics: Conference Series*; 190: 012208. doi:10.1088/1742-6596/190/1/012208.

20. Bolland BJRF, Culliford DJ, Langton DJ, et al. High failure rates with a large-diameter hybrid metal-on-metal total hip replacement: clinical, radiological and retrieval analysis. *The Journal of Bone and Joint Surgery: British*; 93: 608–615.

21. Madl AK, Liong M, Kovochich M, et al. Toxicology of wear particles of cobalt-chromium alloy metal-on-metal hip implants Part I: physicochemical properties in patient and simulator studies. *Nanomedicine*; 11: 1201–1215.

22. Paustenbach DJ, Galbraith DA, Finley BL. Interpreting cobalt blood concentrations in hip implant patients. *Clinical Toxicology*; 52: 98–112.

23. Hokin B, Adams M, Ashton J, et al. Comparison of the dietary cobalt intake in three different Australian diets. *Asia Pacific Journal of Clinical Nutrition* 13: 289–291.

24. Arnich N, Sirot V, Rivière G, et al. Dietary exposure to trace elements and health risk assessment in the 2nd French Total Diet Study. *Food and Chemical Toxicology*; 50: 2432–2449.

25. Lippi G, Franchini M, Guidi GC. Blood doping by cobalt. Should we measure cobalt in athletes? *Journal of Occupational Medicine and Toxicology*; 1: 18.

26. Simonsen LO, Harbak H, Bennekou P. Cobalt metabolism and toxicology: a brief update. *Science of the Total Environment*; 432: 210–215.

27. Lauwerys R, Lison D. Health risks associated with cobalt exposure: an overview. *Science of the Total Environment*; 150: 1–6.

28. Leggett RW. The biokinetics of inorganic cobalt in the human body. *Science of the Total Environment*; 389: 259–269.

29. Jansen HM, Knollema S, van der Duin LV, et al. Pharmacokinetics and dosimetry of cobalt-55 and cobalt-57. *The Journal of Nuclear Medicine*; 37: 2082–2086.

30. Van Der Straeten C. The genesis and aftermath of metal ions and particles in metal-on-metal hip arthroplasty. 2013. http://hdl.handle.net/1854/LU-4337224

31. Kwon Y-M, Xia Z, Glyn-Jones S, et al. Dose-dependent cytotoxicity of clinically relevant cobalt nanoparticles and ions on macrophages in vitro. *Biomedical Materials*; 4: 025018.

32. Keegan GM, Learmonth ID, Case C. A systematic comparison of the actual, potential, and theoretical health effects of cobalt and chromium exposures from industry and surgical implants. *Critical Reviews in Toxicology*; 38: 645–674.

33. Paustenbach DJ, Tvermoes BE, Unice KM, et al. A review of the health hazards posed by cobalt. *Critical Reviews in Toxicology*; 43: 316–362.

34. Tower SS. Arthroprosthetic cobaltism: neurological and cardiac manifestations in two patients with metal-on-metal arthroplasty: a case report. *JBJS*; 92: 2847–2851.

35. Finley BL, Monnot AD, Gaffney SH, et al. Dose-response relationships for blood cobalt concentrations and health effects: a review of the literature and application of a biokinetic model. *Journal of Toxicology and Environmental Health, Part B*; 15: 493–523.

36. Unice KM, Monnot AD, Gaffney SH, et al. Inorganic cobalt supplementation: prediction of cobalt levels in whole blood and urine using a biokinetic model. *Food and Chemical Toxicology*; 50: 2456–2461.

37. Zywiel MG, Cherian JJ, Banerjee S, et al. Systemic cobalt toxicity from total hip arthroplasties. *The Bone & Joint Journal*; 98-B: 6–13.

38. Zywiel MG, Brandt J-M, Overgaard CB, et al. Fatal cardiomyopathy after revision total hip replacement for fracture of a ceramic liner. *The Bone & Joint Journal*; 95-B: 31–37.

39. Hodnett D, Wood DM, Raja K, et al. A healthy volunteer study to investigate trace element contamination of blood samples by stainless steel venepuncture needles. *Clinical Toxicology*; 50: 99–107.

40. Broding HC, Michalke B, Göen T, et al. Comparison between exhaled breath condensate analysis as a marker for cobalt and tungsten exposure and biomonitoring in workers of a hard metal alloy processing plant. *International Archives of Occupational and Environmental Health*; 82: 565–573.

41. Apostoli P, Porru S, Alessio L. Urinary cobalt excretion in short time occupational exposure to cobalt powders. *Science of the Total Environment*; 150: 129–132.

42. Finley BL, Unice KM, Kerger BD, et al. 31-day study of cobalt(II) chloride ingestion in humans: pharmacokinetics and clinical effects. *Journal of Toxicology and Environmental Health: Part A*; 76: 1210–1224.

43. MacDonald SJ. Can a safe level for metal ions in patients with metal-on-metal total hip arthroplasties be determined? *Journal of Arthroplasty*; 19: 71–77.

44. MHRA. MHRAhttps://assets.publishing.service.gov.uk/media/5954ca1ded915d0baa00009b/MDA-2017-018_Final.pdf (accessed 19 December 2018).

45. Sidaginamale R, Joyce JT, Lord JK, et al. Blood metal ion testing is an effective screening tool to identify poorly performing metal-on-metal bearing surfaces. *Bone & Joint Research*; 2: 84–95.

46. Van Der Straeten C, Van Quickenborne D, Seth P, et al. Systemic toxicity of metal ions in a metal-on-metal hip arthroplasty population. *Orthopaedic Proceedings*; 95-B, SUPP_34, 187–187.

47. Kerger BD, Gerads R, Gurleyuk H, et al. Cobalt speciation assay for human serum, Part I. Method for measuring large and small molecular cobalt and protein-binding capacity using size exclusion chromatography with inductively coupled plasma-mass spectroscopy detection. *Toxicological & Environmental Chemistry*; 95: 687–708.

48. Morin Y, Daniel P. Quebec beer-drinkers' cardiomyopathy: etiological considerations. *Canadian Medical Association Journal*; 97: 926–928.

49. Alexander CS. Cobalt-beer cardiomyopathy. A clinical and pathologic study of twenty-eight cases. *The American Journal of Medicine*; 53: 395–417.

50. Kennedy A, Dornan JD, King R. Fatal myocardial disease associated with industrial exposure to cobalt. *The Lancet*; 1: 412–414.

51. Umar M, Jahangir N, Faisal Khan M, et al. Cobalt cardiomyopathy in hip arthroplasty. *Arthroplast Today*; 5: 371–375.

52. Clyne N, Hofman-Bang C, Haga Y, et al. Chronic cobalt exposure affects antioxidants and ATP production in rat myocardium. *Scandinavian Journal of Clinical and Laboratory Investigation*; 61: 609–614.

53. Jobs H. Powder metallurgy as a source of dust from the medical and technical standpoint. *Vertravenargt*; 5: 142–148.

54. Bech AO, Kipling MD, Heather JC. Hard metal disease. *The British Journal of Industrial Medicine*; 19: 239–252.

55. Shirakawa T, Kusaka Y, Fujimura N, et al. The existence of specific antibodies to cobalt in hard metal asthma. *Clinical & Experimental Allergy*; 18: 451–460.

56. Kusaka Y, Iki M, Kumagai S, et al. Epidemiological study of hard metal asthma. *Occupational and Environmental Medicine*; 53: 188–193.

57. Shirakawa T, Kusaka Y, Fujimura N, et al. Occupational asthma from cobalt sensitivity in workers exposed to hard metal dust. *Chest*; 95: 29–37.

58. Dunlop P, Müller NL, Wilson J, et al. Hard metal lung disease: high resolution CT and histologic correlation of the initial findings and demonstration of interval improvement. *Journal of Thoracic Imaging*; 20: 301–304.

59. Ruediger HW. Hard metal particles and lung disease: coincidence or causality? *Respiration*; 67: 137–138.

60. Athavale P, Shum KW, Chen Y, et al. Occupational dermatitis related to chromium and cobalt: experience of dermatologists (EPIDERM) and occupational physicians (OPRA) in the U.K. over an 11-year period (1993–2004). *British Journal of Dermatology*; 157: 518–522.

61. Sarma N. Occupational allergic contact dermatitis among construction workers in India. *Indian Journal of Dermatology*; 54: 137–141.

62. Fischer T, Rystedt I. Cobalt allergy in hard metal workers. *Contact Dermatitis*; 9: 115–121.

63. Veien NK, Hattel T, Justesen O, et al. Oral challenge with nickel and cobalt in patients with positive patch tests to nickel and/or cobalt. *Acta Dermato-Venereologica*; 67: 321–325.

64. Roy PE, Bonenfant JL, Turcot L. Thyroid changes in cases of Quebec beer drinkers myocardosis. *American Journal of Clinical Pathology*; 50: 234–239.

65. Berk L, Burchenal JH, Castle WB. Erythropoietic effect of cobalt in patients with or without anemia. *The New England Journal of Medicine*; 240: 754–761.

66. Duckham JM, Lee HA. The treatment of refractory anaemia of chronic renal failure with cobalt chloride. *Quarterly Journal of Medicine*; 45: 277–294.

67. Licht A, Oliver M, Rachmilewitz EA. Optic atrophy following treatment with cobalt chloride in a patient with pancytopenia and hypercellular marrow. *Israel Journal of Medical Sciences*; 8: 61–66.

68. Toxicological Profile for Cobalt. 1–486. https://www.atsdr.cdc.gov/ToxProfiles/tp33.pdf

69. Wild P, Bourgkard E, Paris C. Lung cancer and exposure to metals: the epidemiological evidence. *Methods in Molecular Biology*; 472: 139–167.

70. IARC monographs on the evaluation of carcinogenic risks to humans. https://publications.iarc.fr/_publications/media/download/2705/29aacee6b89ff816188dcd990b61a16ad6486eec.pdf

71. Smith AJ, Dieppe P, Porter M, et al. Risk of cancer in first seven years after metal-on-metal hip replacement compared with other bearings and general population: linkage study between the National Joint Registry of England and Wales and hospital episode statistics. *BMJ*; 344: e2383–e2383.

72. Christian WV, Oliver LD, Paustenbach DJ, et al. Toxicology-based cancer causation analysis of CoCr-containing hip implants: a quantitative assessment of genotoxicity and tumorigenicity studies. *Journal of Applied Toxicology*; 34: 939–967.

73. Mao X, Wong AA, Crawford RW. Cobalt toxicity: an emerging clinical problem in patients with metal-on-metal hip prostheses? *Medical Journal of Australia*; 194: 649–651.

74. Llena JF, Céspedes G, Hirano A, et al. Vascular alterations in delayed radiation necrosis of the brain. An electron microscopical study. *Archives of Pathology & Laboratory Medicine*; 100: 531–534.

75. Fishman ML, Bean SC, Cogan DG. Optic atrophy following prophylactic chemotherapy and cranial radiation for acute lymphocytic leukemia. *American Journal of Ophthalmology*; 82: 571–576.

76. Johansson S, Svensson H, Denekamp J. Timescale of evolution of late radiation injury after postoperative radiotherapy of breast cancer patients. *International Journal of Radiation Oncology, Biology, Physics*; 48: 745–750.

CHAPTER 17

Iron-Siderophore and Tumorigenesis
Molecular Insight and Therapeutic Targets[1]

Sayantan Maitra
Institute of Pharmacy

Dibyendu Dutta
Darjeeling District Hospital

CONTENTS

17.1 INTRODUCTION

Iron (Fe) is a trace metal essential for the survival of almost all organisms including bacteria, plants, and human, as it participates in several metabolic processes including deoxyribonucleic acid (DNA) synthesis, oxygen transport, and electron transport. Though iron serves several beneficial activities in our body, its concentration should be closely monitored as in higher concentration it can generate free radicals which lead to tissue damage [1]. The body cannot synthesize iron and must acquire it. Dietary iron can be found mainly in two forms: heme and nonheme. All plant- and animal-derived food possess nonheme iron, while the heme iron is only found in animal-derived foods such as meat, fish, poultry, and eggs. Heme iron has greater bioavailability and absorbed more readily in comparison to nonheme iron [2]. Some good sources of iron are provided in Table 17.1.

Absorption of iron relies upon a number of dietary factors. Ascorbate and citrate act as a chelator and facilitate metal ion solubilization in duodenum thereby increase the absorption of iron. Iron is then readily transferred from these compounds into the basolateral cells of gastric endothelium.

---

[1] These authors contributed equally to this work.

Table 17.1　Good Sources of Iron

Food Item	Serving Size	Iron Content (mg)
Spleen, beef	3.5 oz.	39.4
Spinach, boiled	1/2 cup	3.2
Sweet potato, canned	1 cup	1.8
Wheat germ	1 oz.	2.6
Lentils	1/2 cup	3.3
Broccoli	1 medium stalk	2.1
Baked potato	1 medium	2.7
Oatmeal	1 cup	6.7

The potential absorption enhancing effect of ascorbic acid and citric acid is due to its ability to reduce ferric to ferrous iron [3]. In plant-based diets, the main inhibitor of iron absorption is phytate (myo-inositol hexokinasephosphate). Polyphenols present in fruits and vegetables also have inhibitory effects on iron absorption [2,4].

Iron deficiency is defined as the state when there are no mobilizable iron stores and in which there is a compromised supply of iron to tissues. This unavailability of iron leads to the occurrence of anemia, a condition in which the number of red blood cells (RBCs) in blood is low or blood cells have lower than normal level of hemoglobin [1].

Regulation of iron homeostasis is essential for all organisms. Free ferric iron can be lethal to cells hence its regulation is necessary. For human, the concentration of ferric ion is kept below 10^{-24}M [5]. The production of siderophores is one of the crucial strategies evolved by bacteria, fungi, and viruses to maintain iron homeostasis. Siderophores are the natural iron chelators, having low molecular weight (400–2,000 Da) and higher affinity for ferric iron. They are produced under the condition of iron deficiency with the aim to scavenge iron, but they also form complexes with other essential elements (Mn, Co, and Ni) available in the environment in order to make them available for microbes [6,7]. Fe^{3+} ion is a strong Lewis acid and is readily available to form complexes with the electron-pair donors. Siderophores utilize negatively charged oxygen groups as electron donor and thus form a stable complex with iron(III) [8]. Siderophores are mainly classified into three families depending upon the presence of functional group, i.e. hydroxymates, catecholates, and carboxylates. Most of the siderophores geometrically are octahedral, coordinating iron in a thermodynamically stable hexadentate conformation [9].

17.2　IRON METABOLISM IN NORMAL CELLS

Iron-containing proteins play pivotal roles in biological processes including erythropoiesis, cellular respiration, DNA synthesis, and detoxification. But, iron acts as a two-edged sword. Its higher concentration can catalyze the production of reactive oxygen species (ROS), which consequently leads to damage to nearby proteins, carbohydrates, lipids, and DNA. Therefore, to minimize this toxicity, iron homeostasis is very closely monitored by different mechanisms in human body as in Figure 17.1 [10].

Ferric iron (Fe^{3+}) is first converted into ferrous iron (Fe^{2+}) by duodenal cytochrome b (DCYTB) and then imported into enterocytes via divalent metal transporter 1 (DMT1) which is a symporter present on the apical membrane of enterocytes. Iron (Fe^{2+}) then exits the enterocyte by an efflux pump called ferroportin, integrated with oxidase hephaestin that helps to oxidize Fe^{2+} to Fe^{3+}. Two atoms of Fe^{3+} then gets loaded onto transferrin (TF) in order to travel through bloodstream to get delivered to bone marrow or peripheral tissues. The diferric-transferrin (TF-$[Fe^{3+}]_2$) complex gets

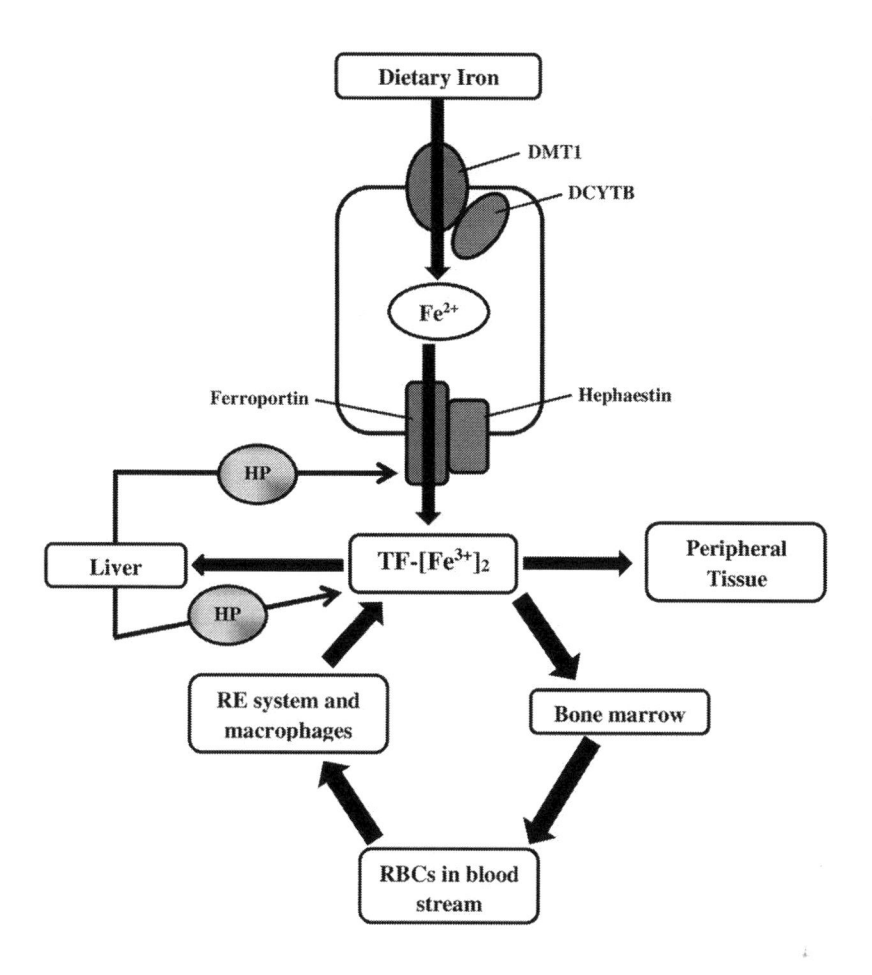

Figure 17.1 Dietary iron (Fe^{3+}) is absorbed in the apical membrane of duodenum in the form of Fe^{2+} by the action of duodenal cytochrome b(DCYTB) and brings into enterocytes via divalent metal transporter 1 (DMT1). Then, Fe^{2+} exits the enterocyte from basolateral surface, facilitated by the iron efflux pump ferroportin, which combines with the oxidase hephaestin to oxidize Fe^{2+} to form Fe^{3+} and the Fe^{3+} is loaded onto transferrin (TF). The diferric iron transferrin complex (TF-[Fe^{3+}]$_2$) circulates through the bloodstream to deliver iron to sites of utilization. Iron transported to bone marrow is used to synthesize hemoglobin and red blood cells (RBCs). The life span of RBCs is 90 days, after that they get catabolized by macrophages of the reticuloendothelial (RE) system. Followed by the release of iron from catabolized heme and exit of iron out of the macrophage by ferroportin, where it gets loaded onto TF in the bloodstream, in a process termed iron recycling. Excessive amount of iron into circulation triggers the synthesis of hepcidin (HP), a peptide hormone acts as a key regulator of systemic iron homeostasis. HP binds to ferroportin and induces its degradation followed by inhibition of both delivery of dietary iron through the enterocyte and iron recycling through the macrophage.

attached to transferrin receptor 1 (TFR1) present on the cell surface, and endocytosis of the complex takes place. Then the acidic condition of endosome triggers the dissociation of Fe^{3+} from TF, and the free Fe^{3+} is reduced to Fe^{2+} by six-transmembrane epithelial antigen of prostate (STEAP) reductase [10]. DMT1 then facilitates the transport of Fe^{2+} from endosome and forms a labile iron pool (LIP). Iron is then gets dispersed into multiple intracellular destinations and integrates into the active sites of enzymes that are involved in DNA synthesis, DNA repair, and cell cycle. Excess iron either stored in iron protein storage called ferritin or leaves the cell through ferroportin together with oxidase hephaestin [11,12].

17.3 IRON AND CANCER

The course of iron metabolism gets altered in cancer. The majority of studies have identified perturbations consistent with an enhanced requirement of iron for rapidly proliferating cancer cells. Expression of TFR1 is upregulated in several cancers including breast cancer, lung cancer, prostate cancer, lymphoma, glioma, leukemia, and others [12–14]. Reduction of endosomal ferric iron to ferrous iron is attributed to STEAP family metalloreductase and its expression also gets elevated in cancer including STEAP1, STEAP2, and STEAP3 [15].

17.3.1 Iron Uptake in Cancer

Overexpression of TFR1 and STEAP reductase in cancer cells gets accompanied by lipocalin 2 (LCN2), an overexpressed protein that is involved in alternative pathway of iron uptake. LCN2 binds with siderophores which were previously chelated with iron molecules. Thus, iron gets circulated in blood in the form of LCN2-siderophore complex and gets bind to cell surface receptor 24p3R and can serve as an alternative mechanism of iron delivery (Figure 17.2) [16].

17.3.2 Iron Storage in Cancer

Most cells, including cancer cells, store excess iron in ferritin. Ferritin is a 24-subunit protein that can hold up to 4,500 iron atoms in a ferrihydrite mineral core. Ferritin heavy chain and ferritin light chain are the two subunits of ferritin. Its activity is regulated by iron regulatory protein 1 (IRP1) and IRP2 [10]. It has been found that post-transcriptional IRP2 that represses activity of ferritin and elevates expression of TFR1, which leads to less storage of iron in cancer cells and makes them intracellularly available for metabolic and proliferative purpose [17].

17.3.3 Iron Efflux in Cancer

Cancer cells make metabolically available iron not only by increasing iron influx and decreasing iron storage but also by decreasing the iron efflux. Systemic iron levels are closely monitored by ferroportin–hepcidin regulatory axis. Ferroportin is expressed on the cell surface of enterocytes, and its expression is regulated by the circulatory peptide hormone hepcidin [18]. When levels of iron in storage and circulation are high, hepcidin is induced in hepatocytes via a bone morphogenetic protein (BMP)-mediated pathway and is released into the circulation. On the basolateral side of enterocytes, hepcidin binds to ferroportin and induces internalization of ferroportin into clathrin-coated pits and its subsequent lysosomal degradation (Figure 17.2), thus blocking the delivery of iron from the digestive tract to the blood [19,20].

17.4 THERAPEUTIC STRATEGIES BASED ON IRON DEPRIVATION

Elevated level of free iron may augment tumorigenesis; therefore decreasing the level of free, iron and targeting iron-containing proteins such as ribonucleotide reductase have proved to be an efficient strategy in chemotherapy [21]. Another approach to alter the increased level of free iron is chelation. There are numerous iron chelators readily available such as desferoxamine (DFO), triapine, deferiprone, and tachpyridine which are being preclinically or clinically used as chemotherapeutic agent [10]. Among these chelators, DFO significantly has both *in vitro* and *in vivo* antitumor activity which is mediated through depletion of iron pools [22]. Triapine also exhibits a broad spectrum of antiproliferative activity by inhibiting the enzyme ribonucleotide reductase as well as forming complexes with both free Fe^{2+} and Fe^{3+} [23,24].

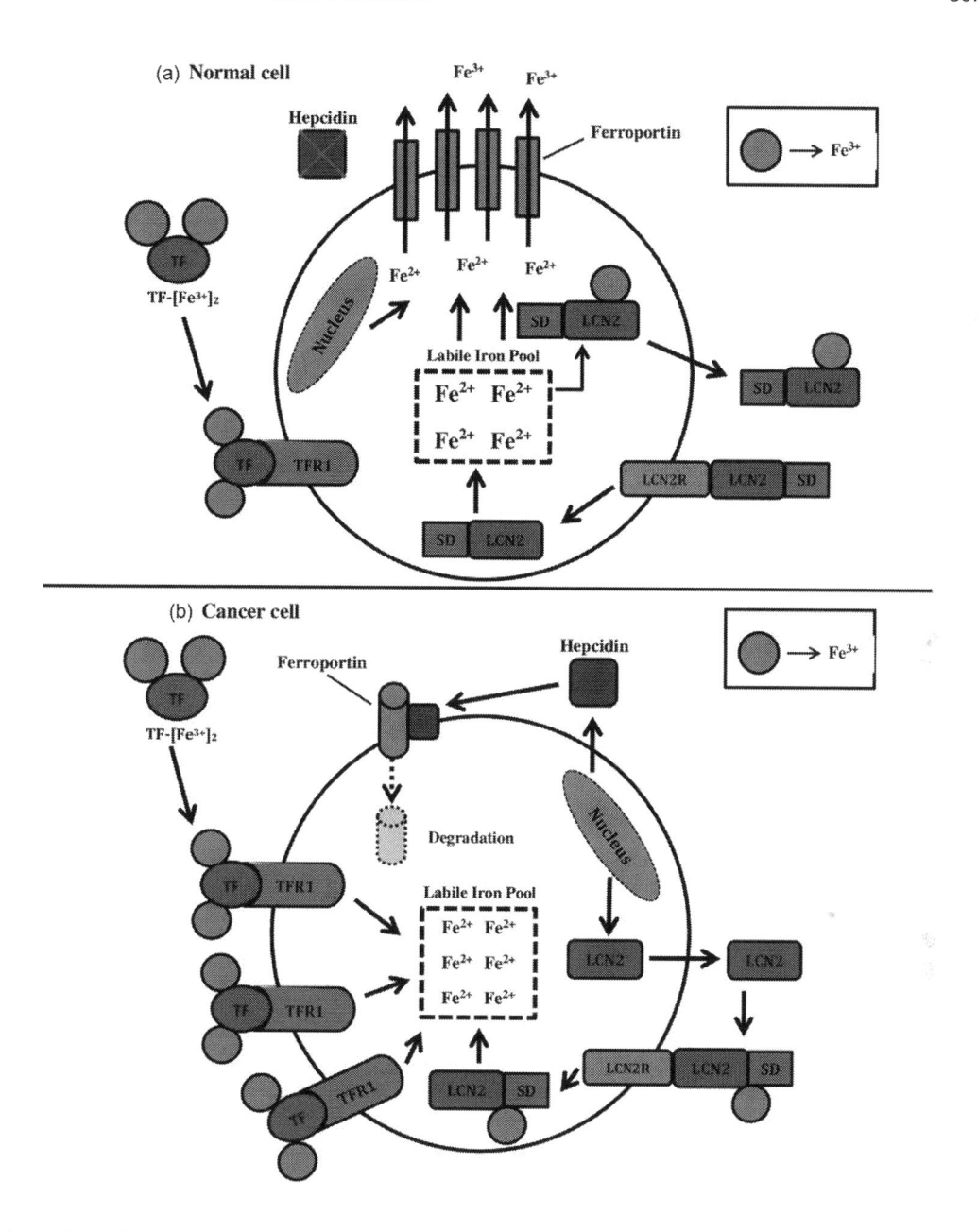

Figure 17.2 (a) Normal cells exhibit comparatively lower levels of transferrin receptor 1 (TFR1) and hepcidin and high levels of ferroportin, which collectively lead to a small pool of labile iron. In breast cells, lipocalin 2 (LCN2) forms a complex with a siderophore (SD) and may further reduce levels of intracellular iron by capturing and effluxing SD-bound iron from these cells. (b) On the contrary, cancer cells exhibit elevated expression of TFR1 and hepcidin and lower levels of ferroportin, which collectively lead to an increased labile iron pool. In breast cancer cells, LCN2, in a complex with SD-bound iron, may serve as a further source of iron. LCN2R: lipocalin 2 (LCN2) receptor.

Another strategy adapted to control the tumorigenesis via directly targeting TFR1. This can be achieved by two mechanisms, i.e. directly delivering the therapeutic molecules into malignant cells and blocking the normal functions of TFR1 which leads to death of malignant cells [25]. Many anticancer agents have been developed that mainly target TFR1, such as anti-TFR antibodies (HB21, 454A12, B3/25, OKT9, 7D3, 7579, and 42/6) [25]. Ligands that are responsible for delivering the anticancer agents directly into malignant cells are TF-doxorubicin, TF-cisplatin, TF-chlorambucil, etc. [10,21].

17.5 CONCLUSION

Collectively, alterations in the course of iron metabolism lead to tumor initiation, progression, and metastasis. In cancer cells, iron activates a wide variety of signal pathways and also the aberrant expression of numerous iron metabolism genes is also found in malignant tumors, suggesting the fundamental roles of iron in developing cancer. The therapeutic aim is to reduce the level of free iron in order to treat malignancy. To date, there are numerous preclinical and clinical studies that have been carried out to decrease the iron level by using a chelator. Still, more studies are warranted to further explore.

ABBREVIATIONS

BMP	bone morphogenetic protein
DCTYB	duodenal cytochrome b
DFO	desferoxamine
DMT1	divalent metal transporter 1
DNA	deoxyribonucleic acid
HP	hepcidin
IRP1	iron regulatory protein 1
IRP2	iron regulatory protein 2
Fe	iron
Fe^{2+}	ferrous ion
Fe^{3+}	ferric ion
LIP	labile iron pool
LCN2	lipocalin 2
RBC	red blood cell
RE	reticuloendothelial system
ROS	reactive oxygen species
SD	siderophores
STEAP	six-transmembrane epithelial antigen of prostate
TF	Transferrin
TF-[Fe^{3+}]$_2$	diferric-transferrin complex
TFR1	transferrin receptor 1

REFERENCES

1. Abbaspour N, Hurrell R, Kelishadi R. Review on iron and its importance for human health. *J Res Med Sci Off J Isfahan Univ Med Sci.* 2014 Feb;19(2):164–174.
2. Hurrell R, Egli I. Iron bioavailability and dietary reference values. *Am J Clin Nutr.* 2010 May;91(5):1461S–1467S.
3. Conrad ME, Schade SG. Ascorbic acid chelates in iron absorption: a role for hydrochloric acid and bile. *Gastroenterology.* 1968 Jul;55(1):35–45.
4. Cook JD, Monsen ER. Food iron absorption in human subjects. III. Comparison of the effect of animal proteins on nonheme iron absorption. *Am J Clin Nutr.* 1976 Aug;29(8):859–867.

5. Raymond KN, Dertz EA, Kim SS. Enterobactin: an archetype for microbial iron transport. *Proc Natl Acad Sci.* 2003 Apr 1;100(7):3584–3588.

6. Ahmed E, Holmström SJM. Siderophores in environmental research: roles and applications. *MicrobBiotechnol.* 2014 May;7(3):196–208.

7. Neilands JB. Siderophores: structure and function of microbial iron transport compounds. *J Biol Chem.* 1995 Nov 10;270(45):26723–26726.

8. Dhungana S, Crumbliss AL. Coordination chemistry and redox processes in siderophore-mediated iron transport. *Geomicrobiol J.* 2005 Apr;22(3–4):87–98.

9. Carroll CS, Moore MM. Ironing out siderophore biosynthesis: a review of non-ribosomal peptide synthetase (NRPS)-independent siderophore synthetases. *Crit Rev Biochem Mol Biol.* 2018;53(4):356–381.

10. Torti SV, Torti FM. Iron and cancer: more ore to be mined. *Nat Rev Cancer.* 2013 May;13(5):342–3555.

11. Gkouvatsos K, Papanikolaou G, Pantopoulos K. Regulation of iron transport and the role of transferrin. *BiochimBiophysActa.* 2012 Mar;1820(3):188–202.

12. Manz DH, Blanchette NL, Paul BT, Torti FM, Torti SV. Iron and cancer: recent insights. *Ann N Y Acad Sci.* 2016;1368(1):149–161.

13. Buss JL, Torti FM, Torti SV. The role of iron chelation in cancer therapy. *Curr Med Chem.* 2003 Jun;10(12):1021–1034.

14. Huang X. Iron overload and its association with cancer risk in humans: evidence for iron as a carcinogenic metal. *Mutat Res.* 2003 Dec 10;533(1–2):153–171.

15. Ohgami RS, Campagna DR, Greer EL, Antiochos B, McDonald A, Chen J, et al. Identification of a ferrireductase required for efficient transferrin-dependent iron uptake in erythroid cells. *Nat Genet.* 2005 Nov;37(11):1264–1269.

16. Bao G, Clifton M, Hoette TM, Mori K, Deng S-X, Qiu A, et al. Iron traffics in circulation bound to a siderocalin (Ngal)-catechol complex. *Nat Chem Biol.* 2010 Aug;6(8):602–629.

17. Wu KJ, Polack A, Dalla-Favera R. Coordinated regulation of iron-controlling genes, H-ferritin and IRP2, by c-MYC. *Science.* 1999 Jan 29;283(5402):676–679.

18. Ganz T, Nemeth E. Hepcidin and iron homeostasis. *Biochim Biophys Acta.* 2012 Sep;1823(9):1434–1443.

19. Ward DM, Kaplan J. Ferroportin-mediated iron transport: expression and regulation. *Biochim Biophys Acta.* 2012 Sep;1823(9):1426–1433.

20. Nemeth E, Tuttle MS, Powelson J, Vaughn MB, Donovan A, Ward DM, et al. Hepcidin regulates cellular iron efflux by binding to ferroportin and inducing its internalization. *Science.* 2004 Dec 17;306(5704):2090–2093.

21. Zhang C, Zhang F. Iron homeostasis and tumorigenesis: molecular mechanisms and therapeutic opportunities. *Protein Cell.* 2015 Feb;6(2):88–100.

22. Dayani PN, Bishop MC, Black K, Zeltzer PM. Desferoxamine (DFO)-mediated iron chelation: rationale for a novel approach to therapy for brain cancer. *J Neurooncol.* 2004 May;67(3):367–377.

23. Yu Y, Wong J, Lovejoy DB, Kalinowski DS, Richardson DR. Chelators at the cancer coalface: desferrioxamine to Triapine and beyond. *Clin Cancer Res Off J Am Assoc Cancer Res.* 2006 Dec 1;12(23):6876–6883.

24. Zhang C, Liu G, Huang M. Ribonucleotide reductase metallocofactor: assembly, maintenance and inhibition. *Front Biol.* 2014 Jan 2;9(2):104–113.

25. Daniels TR, Bernabeu E, Rodríguez JA, Patel S, Kozman M, Chiappetta DA, et al. The transferrin receptor and the targeted delivery of therapeutic agents against cancer. *Biochim Biophys Acta.* 2012 Mar;1820(3):291–317.

CHAPTER 18

Clinical Toxicology of Iron
Source, Toxidrome, Mechanism of Toxicity, and Management

Shilia Jacob Kurian, Sonal Sekhar Miraj, Ahmed Alshrief, Sreedharan Nair, and Mahadev Rao
Manipal College of Pharmaceutical Sciences

CONTENTS

18.1 INTRODUCTION

18.1.1 History

Iron (Fe), atomic number 26, has come from the Latin word *Ferrum*. Fe is the most abundant element on planet earth and the fourth most available element on the earth's crust and constitutes parts of the outer and inner core (Sheftel et al. 2012). Metallic form of Fe is rare in the earth's crust and is available in the meteorite form, compared to the Fe ores, which are the most abundant sources in the earth's crust. Smelting Fe ores at 1,500°C is the technique used to extract the metal (Fenton 2005). In the human body, Fe plays a vital role in growth, maintenance, and cell division. Approximately, 4 g of Fe will be present in a healthy adult, of which nearly 70% is in the hemoglobin, 25% in the Fe storage sites, and 5% in the myoglobin and few in tissue enzymes and plasma transferrin (Kang 2001).

In 2017, 4,400 cases of Fe exposure had been reported by the American Association of Poison Control Centers (AAPCC), resulting in eight major outcomes and two deaths. Of these, 82% were among children 6 years or younger (Spanierman 2019). Unintentional iron poisoning is common among children and results in minimal toxicity. This is because these Fe tablets seem to be appealing to the children, as they are brightly colored in appearance and sugar-coated. These chewable vitamin preparations are less toxic compared to adult preparations. Although some toxicities are observed in these patients, a decrease in fatalities has been observed due to effective preventive measures such as unit-dose packaging. On the other hand, intentional poisoning with Fe is not so common, but a majority of it is associated with mortality, approximately 10% in the intentional users versus 1% in the unintentional users (Liebelt 2019).

18.1.2 Properties (Physical and Chemical)

Fe is a lustrous, ductile, malleable, silver-gray metal. Fe is a transition metal, belonging to the vertical group in the periodic table. They possess properties between typical and less typical metals (Habashi 2010). The pure metal is less frequently seen and is found alloyed with carbon or other metals. Steel, an alloy of Fe and carbon, is the most popularly used form of Fe (Habashi 2013).

Fe compounds mainly exist in two oxidation states, i.e., +2 (ferrous) and +3 (ferric), while it also exists in higher states such as +4 and +6. Fe (+4) is an intermediate compound in several biochemical processes (Habashi 2013). Fe readily gets oxidized due to its affinity for oxygen, and in the presence of moisture, its surface gets converted to iron oxide hydrate. Hot air exposure to Fe causes the formation of intermediate iron oxide. Fe readily dissolves in dilute acids with the release of hydrogen and ferrous salts (Habashi 2013).

18.1.3 Uses (Commercial and Biological)

Commercial: Fe articles have become an indispensable part of today's world, owing to its low cost and high strength. Its application varies from food containers to family cars, from screwdrivers to machines. Steel is the most popular form of Fe owing to its low price and high tensile strength (Prawoto 2013).

Biological: Fe in the blood cells was first ascertained by Vincenzo Menghini (Sheftel et al. 2012). The red color of the blood is due to the presence of hemoglobin, which supplies oxygen to the body cells. Total hemoglobin contains nearly two-thirds of the Fe in the body, and its deficiency can cause anemia (Habashi 2013; Pillay 2017). Fe also plays a pivotal role in the production of ATP and the consumption of oxygen as a component of several mitochondrial enzymes, involved in the electron transport chain (Wessling-Resnick 2017).

18.2 SOURCES OF IRON

18.2.1 Dietary Sources

The daily requirement of Fe in adults is 10–20 mg, which increases to 23–30 mg during pregnancy (Pillay 2013). Two forms of Fe, available from dietary sources, are "heme" and "non-heme," with respect to the mechanism of absorption. The bioavailability of "non-heme" is less compared to "heme" form (Samaniego-Vaesken et al. 2017). This is because the "non-heme" Fe is available in the ferric form, which needs to be converted to the ferrous form to facilitate absorption (Hatcher et al. 2009). Nuts, vegetables, fish, eggs, meat, and cereals are the richest sources of iron. Currently, several Fe-fortified foods are available commercially (Samaniego-Vaesken et al. 2017).

18.3 TOXIC DOSE

The amount of elemental Fe ingested differs depending on the formulation (Pillay 2008). Table 18.1 gives the percentage of Fe content in various iron preparations available.

A single toxic dose of Fe has not been defined yet and therefore, doses <20 mg/kg of elemental Fe is considered not to be toxic. Doses between 20 and 60 mg/kg can cause moderate gastrointestinal symptoms, while doses >60 mg/kg can lead to severe morbidity and mortality. Approximately, 60–90 and 5–19 mg of elemental Fe are present in prenatal and children vitamin supplements, respectively (McGuigan 1996; Yuen and Becker 2019).

The AAPCC has set 40 mg/kg of elemental Fe as a threshold value to assess if a patient can be managed at home or needs to be evaluated at a healthcare facility (Chang and Rangan 2011). Serum levels >350 µg/dL between 2 and 6 hours of iron ingestion generally indicate significant intoxication. Likewise, the values >500 µg/dL suggest a severe risk of acute liver failure (Abhilash et al. 2013).

18.4 TOXICOKINETICS

Unlike toxicodynamics of Fe, toxicokinetic parameters are less well apprehended. Even though the plasma Fe binding mechanism is well perceived, the absorption, distribution, and storage remain less clear (Tenenbein 1998). The peak serum concentration of Fe occurs after 2–3 hours with therapeutic dosing and after 4–6 hours with overdose. Nevertheless, the peak values vary with the type and amount of Fe consumed. Following an overdose, the onset of symptoms usually begins within 4–6 hours (Manoguerra et al. 2005). Therefore, the toxicokinetic parameters are best estimated at

Table 18.1 Content of Elemental Iron in Different Iron Preparations

Iron Preparation	Route	Preparation Strength (Percentage of Elemental Iron, %)
Ferrous sulfate ($FeSO_4$)	PO	325 mg (20)
Ferrous fumarate ($C_4H_2FeO_4$)	PO	90–150 mg (33)
Ferrous gluconate ($C_{12}H_{24}FeO_{14}$)	PO	225–325 mg (12)
Ferrous succinate ($C_4H_4FeO_4$)	PO	100 mg (35)
[a]Iron polymaltose ($C_{12}H_{25}FeO_{14}$)	PO/IV/IM	—
Iron polysaccharide complex	PO	28–200 mg (100)
[a]Iron dextran (FeH_2O_4S)	IM/IV	—
Sodium ferric gluconate ($C_{66}H_{121}Fe_2NaO_{65}$)	IV	12.5 mg/mL up to 1 g (100)
Iron sorbitol citrate ($C_{12}H_{19}FeO_{13}$)	[a]IM	—

[a] Depends on product.

4–6 hours after iron ingestion. However, when patients present with Fe overdose, a single random measurement of Fe concentration may not be likely to give the true peak value (Robertson and Tenenbein 2005).

In the body, the disposal of Fe is a complex process, which is regulated to maintain homeostasis. Generally, 2%–15% of Fe gets absorbed from the gastrointestinal tract (GIT), particularly in the proximal small intestine. It occurs in two steps: first, ferrous ions get absorbed from the intestinal lumen into mucosal cells. Second, the absorbed ferrous form gets oxidized to the ferric form. Subsequently transferred to plasma and bound to the Fe-binding protein (transferrin) and thereby transported to the storage sites. The Fe within the liver gets utilized for the synthesis of Fe-containing proteins such as cytochromes or it is bound to transferrin. Consequently, it is transported to muscles and bone marrow for the synthesis of myoglobin and incorporation into hemoglobin, respectively, whereas the remaining gets stored in ferritin or hemosiderin. However, the senescent RBCs are phagocytosed by the reticuloendothelial cells of bone marrow and spleen, and the resultant hemoglobin catabolism upshots the plasma levels of Fe (Assi and Baz 2014).

The demand for Fe increases during blood loss, childhood, and pregnancy and so absorption increase. Only 10%–35% of Fe gets absorbed from a therapeutic dose, but 80%–95% gets absorbed during a Fe deficiency. The absorption amount decreases after an overdose as the capacity for absorption exceeded. A serious Fe overdose requires the consumption of elemental Fe in gram doses, yet the measured plasma concentrations are all in milligram amounts. The dose of its chelator-deferoxamine (DFOA or DFO) is determined by the amount of Fe absorbed. Although the maximum dose that can be administered is often less compared to the amount of Fe consumed, it is considered sufficient owing to the inefficient absorption of Fe at overdoses.

Transferrin is the primary Fe-binding protein present in plasma, which is produced in the liver. Albumin and other proteins too bind with Fe, but at lower affinity. There is approximately 3–5 g of Fe in the body. The exposure to Fe induces the release of apoferritin. Apoferritin is a protein commonly present in the intestinal mucosa membrane. The biological importance of apoferritin is its capability to bind as well as store Fe by combining with a ferric hydroxide–phosphate compound to form ferritin. The plasma ferrous gets oxidized in the presence of ferroxidase I and subsequently released from ferritin by the action of reducing agents such as ascorbic acid, cysteine, and reduced glutathione (Gurzau et al. 2003). At excess doses of Fe, transferrin becomes saturated, and the excess Fe will circulate as free Fe which is directly toxic (Yuen and Becker 2019)

Fe gets rapidly distributed and its entry into the tissues is majorly an active saturable process (for example, into the placenta), which involves transferrin receptors and endocytosis. The liver is a target organ in Fe poisoning since it passively absorbs Fe. The half-life of Fe after an overdose is approximately 6 hours which is similar to the therapeutic dose. Of the amount of Fe absorbed from the GIT (2%–15%), only 80% gets excreted from the body. Therefore, elimination after a therapeutic or an overdose is insignificant. Generally, the excess Fe gets excreted, but some remain in bile, shed intestinal cells and in urine and in trace amounts within nails, hair, and sweat (Tenenbein 1998).

18.5 MECHANISM OF TOXICITY

18.5.1 Acute

Fe toxicity occurs when the concentration of Fe exceeds the total iron-binding capacity (TIBC) and free iron circulates in the plasma. This effect can be local-corrosive or systemic-cellular effects (see Table 18.2). At the cellular level, metabolic acidosis is common. The unbound Fe from the plasma moves into the liver, heart, and endocrine cells; and the Fe participates in Fenton's reaction resulting in the formation of free radicals. These free radicals get involved in lipid peroxidation

Table 18.2   Local and Systemic Effect of Iron Toxicity

Local-Corrosive	Systemic-Cellular
Direct injury to the GIT, can lead to abdominal pain, nausea, vomiting, and diarrhea	Excessive post-arteriolar dilation resulting in venous pooling
Mucosal damage can result in blood as well as fluid loss; and result in hypovolemia	Higher capillary permeability resultant in decline in plasma volume
Hemorrhagic necrosis can result in hematemesis, perforation, and peritonitis	Oxidation of Fe^{+2} releases Fe^{+3} and $H^{+,}$ which attribute to metabolic acidosis
—	Inhibition of mitochondrial function can lead to liver damage, hypoglycemia, as well as hypoprothrombinemia

and cause organ damage by injuring the organelles (Kwiatkowski 2011; Yuen and Becker 2019; Robotham and Lietman 1980).

$$Fe^{2+} + H_2O_2 \rightarrow Fe^{3+} + OH^- + HO$$

18.5.2 Chronic

The amount of Fe absorbed, as well as excreted, is well regulated in the body. Any imbalances in this process can result in Fe accumulation. For instance, enhanced absorption can lead to Fe deposition in various tissues and further cause tissue damage and even end-organ failure.

Excess Fe usually gets absorbed as a result of inherited disorders affecting Fe metabolism or long-term therapeutic Fe intake. Fe overloading disorder classified as primary and secondary. Primary includes inherited diseases that enhance intestinal Fe absorption, such as HFE-related hemochromatosis, TFR2-related hemochromatosis, etc. Secondary accumulations are primarily due to blood transfusion to treat diseases originating from the erythroid system, such as ineffective erythropoiesis, chronic hemolytic anemias, or hypoplastic anemias. A diverse range of disorders that can cause Fe overload includes African Fe overload, porphyria cutanea tarda, neonatal hemochromatosis, etc.

In these conditions, the non-transferrin bound Fe reaches high concentrations, even up to 20 micromoles in some pathological conditions. The liver rapidly clears the excess Fe from the plasma, and subsequently iron gets accumulated in these tissues and results in toxicity. (Anderson 2007). Duodenal enterocytes, Fe-storing hepatocytes, and spleen macrophages release Fe into plasma through membrane ferroportin. Studies show that HFE mutations can impair the production of hepcidin, a molecule that inhibits ferroportin. Thereby, maintaining Fe homeostasis by preventing the secretion of Fe into the hepatic portal system and subsequently, decreasing Fe absorption and release of Fe from macrophages (Assi and Baz 2014).

18.6 CLINICAL MANIFESTATIONS

18.6.1 Acute

Fe toxicity has five clinical stages of manifestation (see Table 18.3). The first stage can be seen with doses as less as 10 mg/kg ingestion and is associated with damage to gastrointestinal (GI) mucosa. This phase exhibits vague GI symptoms such as diarrhea, vomiting, and GI blood loss. Additionally, early signs of shock and metabolic acidosis may also be seen. A lack of symptoms within 6 hours after ingestion predicts less chance to develop any further toxic sequelae. The second stage is referred to as the "latent phase" characterized by a marked improvement in GI symptoms. However, this phase has subtle evidence of cellular toxicity and metabolic acidosis. The latent

Table 18.3 Stages of Iron Poisoning

Stages	Time-to-Onset	Clinical Manifestations
Stage I	30 minutes to 6 hours	GI irritation
Stage II	6 to 24 hours	Recover from GI symptoms
Stage III	12 to 48 hours	Metabolic acidosis Dehydration Lactic acidosis
Stage IV	2 to 5 days	Hepatic disease
Stage V	3 to 6 weeks	Scarring following GI mucosa healing

phase usually occurs within 6–24 hours, although it may not be seen in all cases. The third stage occurs within 12–48 hours and accounts for most deaths associated with the toxicity. Apart from the oxidative phosphorylation pathway and generation of free radicals, this phase accounts for metabolic and cardiovascular symptoms. This has been associated with free Fe that inhibits thrombin and its related clotting pathway resulting in coagulopathy. In addition, metabolic acidosis develops due to the recurrence of vomiting and GI bleeding, which leads to hypovolemic shock. The fourth stage (although not always present) is characterized by hepatotoxicity and is seen within 48 hours. Hepatotoxicity is generally observed at serum Fe levels >1,000 mcg/dL and rarely at levels <700 mcg/dL. The fifth stage is usually observed in 3–6 weeks after the toxic ingestion, in which patients usually present with complaints of bowel obstruction due to stricture formations (Madiwale and Liebelt 2006; Chang and Rangan 2011).

18.6.2 Chronic

As a result of acquired and inherited conditions, chronic Fe overload occurs due to impaired Fe excretion mechanism in the body. The associated clinical implications include hepatic fibrosis, diabetes, cirrhosis, and cardiac disease. As mentioned earlier, Fe gets stored majorly in the liver, and therefore, it is the leading site for toxicity during Fe overload (Pietrangelo 2002). This can be explained by metabolic disorders such as hemochromatosis. This is characterized by normal Fe-driven erythropoiesis and accumulation of Fe in vital organs as a result of the gene mutation that alters Fe entry into the bloodstream (Kohgo et al. 2008). Hemochromatosis usually manifests as bronze skin color, abdominal and joints pain, fatigue, diabetes, and impotence. Vascular congestion in vital organs such as kidney, heart, GIT, and others may have occurred. Later, the cirrhotic changes may also lead to hepatocellular carcinoma.

18.7 DIAGNOSIS

Abdominal X-Ray: An abdominal radiograph could be helpful, as it would reveal the radiopaque pill fragments, though it cannot be directly correlated to the toxicity. However, it could generate a provisional diagnosis and suggest an initial line of treatment. A negative report should not be ignored since there is a possibility for Fe to be completed absorbed. (Madiwale and Liebelt 2006; Chang and Rangan 2011).

Serum Iron Level: Serum Fe level of 300 mcg/dL is considered as the cut off for clinical toxicity and should be obtained at the maximum by the sixth hour of ingestion. Evidence for clinical manifestations at levels <300 mcg/dL is less significant. Mild clinical presentations are seen in levels between 300 and 500 mcg/dL. Similarly, serious symptoms are observed between 500 and 1,000 mcg/dL (Madiwale and Liebelt 2006; Chang and Rangan 2011).

Total Iron-Binding Capacity Level: Previously, TIBC was considered as a chief parameter to predict clinical toxicity of Fe. However, current reports refute the same because of inaccurate TIBC values in the presence of Fe toxicity and DFOA, since this could give a false elevation or decrease. Moreover, in certain instances, TIBC levels have been reported to be greater than serum Fe values (Madiwale and Liebelt 2006; Chang and Rangan 2011).

Other Blood Tests: Elevated serum glucose and white blood cells (WBC) were once used as a specific marker of Fe toxicity. However, due to its less sensitivity and specificity, these have been less frequently used. Liver function tests can also be employed, as they depict the liver status and the extent of the injury. Renal function tests and electrolyte measurements help to calculate the anion gap and to identify electrolyte imbalances as well as prerenal azotemia (Spanierman 2019).

18.8 MANAGEMENT

Management starts with thorough patient evaluation, with the amount and formulation type of compound consumed. Initial management of toxicity will include inducing emesis or gastric lavage with a large-bore tube. In case a delay in obtaining serum Fe concentration is anticipated, evidence such as vomiting, diarrhea, elevated glucose, and WBC levels and a positive abdominal X-ray could be indicative of serum Fe concentration >300 mcg/dL. Subsequently, the individual can be given 50 mg/kg of DFOA. Urine "vin rose" color confirms serum Fe concentration greater than TIBC and initiates the chelation therapy and other management strategies. Patients with no evidence of Fe toxicity should be observed for at least 6 hours. Those with no symptoms could be discharged (Lacouture et al. 1981).

18.8.1 Supportive Care

Airway and breathing support should be provided as needed. Hypovolemic shock is the major cause of mortality in the initial phases, especially in patients presenting with severe GI symptoms. Hypovolemia and hypoperfusion can be managed by maintaining the circulatory volume using IV crystalloid solution. In the case of coagulopathy, vitamin K and fresh frozen plasma can be administered. Timely monitoring and supportive care should be provided throughout the therapy (Liebelt 2019; Yuen and Becker 2019).

18.8.2 Treatment

Decontamination: Gastric lavage could be performed with the largest tube available in a patient presenting with symptoms or an alleged history of Fe overdose. An abdominal radiograph should be obtained after the lavage to confirm that the tablets have been cleared if not, lavage should be repeated. The suitable lavage solutions are tap water or 0.9% saline. However, the use of gastric lavage in these patients is still under discussion, due to lack of evidence for its clinical benefit. Although it can be done in patients presenting within minutes of toxic amounts of Fe ingestion, the risk of perforations and aspirations are associated with this procedure (Madiwale and Liebelt 2006; Chang and Rangan 2011; Baranwal and Singhi 2003).

Whole bowel irrigations (WBI) may be performed in individuals whose radiograph reveals the tablets reached beyond the pylorus. It enhances the GI motility and decreases systemic Fe absorption, thereby reducing the transit time. WBI is considered the choice of method for decontamination of heavy metals and therefore, is recommended for iron toxicity. Polyethylene glycol, an osmotically neutral solution is considered for WBI (Chang and Rangan 2011).

Activated charcoal is used for GI decontamination and could be effective in ferrous sulfate overdose. But, the use of activated charcoal is not recommended in Fe toxicity due to its lack of efficacy in adsorbing Fe (Algren 2011; Manoguerra et al. 2005).

Chelation Therapy: Chelators in Fe toxicity are used to inhibit the hoarding of excess Fe and thereby avoid cardiac, hepatic, and endocrinological complications (Mobarra et al. 2016). The commonly employed Fe chelators include deferasirox, deferiprone, and DFOA. Deferasirox and deferiprone are classified as synthetic chelators, whereas DFOA is a siderophore (a natural compound obtained from microorganisms) and is the first approved chelator for Fe (Hatcher et al. 2009).

Deferoxamine: DFOA should be initiated in patients presenting with lethargy, vomiting, hypovolemia, shock, diarrhea, abdominal X-ray (showing large pills), metabolic acidosis, or serum Fe level > 500 mcg/dL (Madiwale and Liebelt 2006). DFOA is administered via parenteral route owing to its decreased absorption from the GIT and the plasma half-life 20 minutes. Therefore, DFOA is administered as a continuous infusion over 8–12 hours at a starting dose of 15 mg/kg/hour up to a maximum of 35 mg/kg/hour. Approximately 100 mg of DFOA binds to 8.5 mg of Fe. Higher doses and longer durations of DFOA therapy have been reported to cause pulmonary and neurotoxicity. The drug metabolites get excreted mainly via urine and partly in feces (Madiwale and Liebelt 2006; Kwiatkowski 2011).

DFOA works by forming a stable complex with ferric ions, which gets excreted via urine, giving it a "vin rose" color (Madiwale and Liebelt 2006). The entry of Fe-DFOA complex (ferrioxamine) from the extracellular into the intracellular space is restricted due to its diminished permeability, thereby limiting intracellular toxicity (Robotham and Lietman 1980). DFOA readily forms complexes with Fe from transferrin and ferritin, but not from hemoglobin (Tenenbein 1996). The commonly reported adverse reactions include injection site reactions, hypotension with rapid infusions, shock, tachycardia, hypersensitivity reactions, acute renal injury, and auditory dysfunction (Micromedex® 2019) (Figure 18.1).

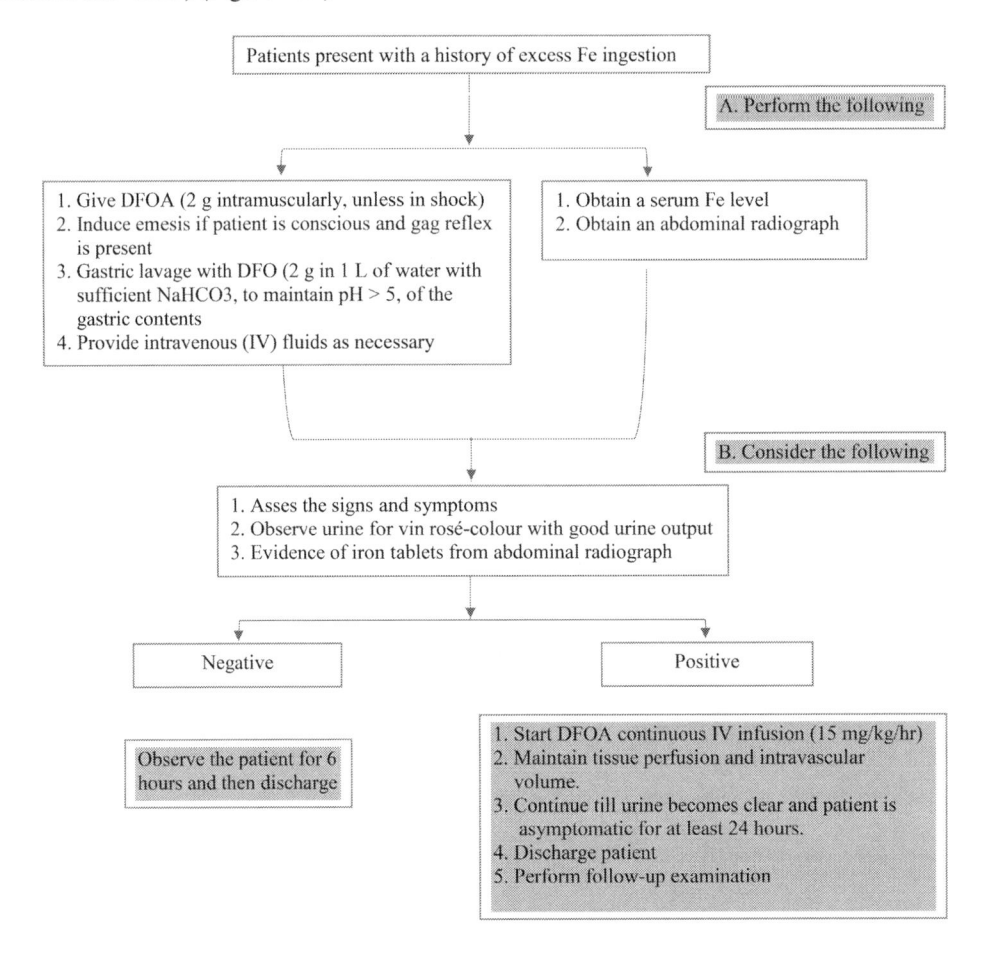

Figure 18.1   Treatment flowchart for patients with history of iron overdose.

Iron Removal Phlebotomy: Phlebotomy, the process of removing the blood from the body, is a frequently employed management strategy in patients with chronic Fe toxicity. This method is considered safer than the other methods including chelation and erythrocytopheresis. Removal of one unit of blood (approximately 500 ml) results in the removal of 200–250 mg of Fe (Barton et al. 1998). Additional care must be taken to ensure sufficient fluid is administered during the procedure in order to avoid hypovolemia. Phlebotomy should be performed once a week until serum ferritin levels reach approximately 50 mcg/L. Patients with hemoglobin <11 g/dL or hematocrit <0.33 may need a slight modification in the session schedules. It is believed that this method can improve chronic fatigue and cardiac function, stabilize liver disease, reverse hepatic fibrosis, and reduce skin pigmentation in patients with chronic Fe toxicity (Brissot et al. 2018).

Chelation therapy with DFOA is not recommended in patients with hemochromatosis, owing to its less efficacy and increased cost. Liver transplantation in hemochromatosis is mainly recommended in patients with hepatocellular carcinoma. Certain lifestyle and dietary modifications are advised to the patients. These include abstinence from alcohol and smoking and following a balanced diet, although no standard regimen exists (Brissot et al. 2018).

18.9 CONCLUSION

Accidental Fe overdose consumption is common in children, and intentional is seen more among adults. Although the outcome may rarely be fatal, substantial Fe ingestion can result in profound mental retardation or death (Robotham and Lietman 1980). Therefore, it is essential to appropriately manage individuals and conduct regular follow-ups. Management strategies include stabilizing vital functions, removing the unabsorbed Fe from GIT, and administering DFOA to chelate the circulating Fe and thereby prevent further complications (McGuigan 1996).

18.10 CASE STUDY

A 15-year-old girl was brought to the emergency department with an alleged history of consumption of 20 tablets (100 mg elemental Fe in each tablet, total ingested 2,000 mg of Fe). She was presented within 2 hours after ingestion with one episode of vomiting. Gastric lavage was performed. Arterial blood gas showed pH:7.03, pO_2:55.3 mmHg, pCO_2: 43.9 mmHg, HCO_3:22 mmol/L, and SPO_2:89%. In this view, the patient was shifted to the intensive care unit for the management of metabolic acidosis. The patient was conscious and oriented with stable vital parameters. Renal function test, liver function test, cardiac parameters, and complete blood count were performed to rule out systemic complications. Serum Fe levels on admission was 410 mcg/dL, TIBC was 12 mcg/dL, and ferritin was 15.23 ng/mL. DFOA was not initiated as the serum Fe levels were not >500 mcg/dL. Within 48 hours, serum Fe levels fell to 120 mcg/dL. TIBC (150 mcg/dL) and the metabolic acidosis had significantly improved. General conditions of the patient were improved and subsequently the patient was discharged.

REFERENCES

Abhilash, K.P., Arul, J.J., Bala, D. 2013. Fatal overdose of iron tablets in adults. *Indian Journal of Critical Care Medicine* 17(5):311–3.

Algren, D.A. 2011. Review of oral iron chelators (deferiprone and deferasirox) for the treatment of iron overload in pediatric patients. https://www.semanticscholar.org/paper/Review-of-Oral-Iron-Chelators-(-Deferiprone-and-)-Algren/15ed9e633efb40792b910a0c0a83f099f823b526

Anderson, G.J. 2007. Mechanisms of iron loading and toxicity. *American Journal of Hematology* 82(12 Suppl):1128–31.

Assi, T.B., Baz, E. 2014. Current applications of therapeutic phlebotomy. *Blood Transfusion* 12 (Suppl 1):s75–s83.

Baranwal, A.K., Singhi, S.C. 2003. Acute iron poisoning: management guidelines. *Indian Pediatrics* 40(6):534–40.

Barton, J.C., McDonnell, S.M., Adams, P.C., et al. 1998. Management of hemochromatosis. Hemochromatosis Management Working Group. *Annals of Internal Medicine*. 129(11):932–9.

Brissot, P., Pietrangelo, A., Adams, P.C., de Graaff, B., McLaren, C.E., Loréal, O. 2018. Haemochromatosis. *Nature Reviews Disease Primers* 4:18016.

Chang, T.P., Rangan, C. 2011. Iron poisoning: a literature-based review of epidemiology, diagnosis, and management. *Pediatric Emergency Care* 27(10):978–85.

Fenton, M.D. 2005. Mineral commodity profiles-iron and steel. Reston, VA: US Geological Survey. Open-File Report 2005-1254, https://pubs.usgs.gov/of/2005/1254/2005-1254.pdf.

Gurzau, E.S., Neagu, C., Gurzau, A.E. 2003. Essential metals-case study on iron. *Ecotoxicology and Environmental Safety* 56(1):190–200.

Habashi, F. 2010. Metals: typical and less typical, transition and inner transition. *Foundations of Chemistry* 12(1): 31–9.

Habashi F. 2013. Iron, physical and chemical properties. In: Kretsinger R.H., Uversky V.N., Permyakov E.A. (eds) *Encyclopedia of Metalloproteins*. Springer, New York.

Hatcher, H.C., Singh, R.N., Torti, F.M., Torti, S.V. 2009. Synthetic and natural iron chelators: therapeutic potential and clinical use. *Future Medicinal Chemistry* 1(9):1643–70.

Kang, J.O. 2001. Chronic iron overload and toxicity: clinical chemistry perspective. *Clinical Laboratory Science* 14(3):209–19.

Kohgo, Y., Ikuta, K., Ohtake, T., Torimoto, Y., Kato, J. 2008. Body iron metabolism and pathophysiology of iron overload. *International Journal of Hematology* 88(1):7–15.

Kwiatkowski, J.L. 2011. Management of transfusional iron overload: differential properties and efficacy of iron chelating agents. *Journal of Blood Medicine* 2:135–49.

Lacouture, P.G., Wason, S., Temple, A.R., Wallace, D.K., Lovejoy, F.H. Jr. 1981. Emergency assessment of severity in iron overdose by clinical and laboratory methods. *Journal of Pediatrics* 99(1):89–91.

Liebelt, E.L. 2019. Acute iron poisoning. UpToDate. https://www.uptodate.com/contents/acute-iron-poisoning.

Madiwale, T., Liebelt, E. 2006. Iron: not a benign therapeutic drug. *Current Opinion in Pediatrics* 18(2): 174–9.

Manoguerra, A.S., Erdman, A.R., Booze, L.L., et al. 2005. Iron ingestion: an evidence-based consensus guideline for out-of-hospital management. *Clinical Toxicology (Philadelphia, PA)*. 2005;43(6):553–70.

McGuigan, M.A. 1996. Acute iron poisoning. *Pediatric Annals* 25(1):33–8.

Micromedex®. 2019. Toxicology Management. Exposure Management [Internet]. Truvenhealth.com. http://truvenhealth.com/markets/provider/products/clinicalknowledge/toxicology-management.

Mobarra, N., Shanaki, M., Ehteram, H., et al. 2016. A review on iron chelators in treatment of iron overload syndromes. *International Journal of Hematology-Oncology and Stem Cell Research* 10(4):239–47.

Pietrangelo, A. 2002. Mechanism of iron toxicity. In: C. Hershko, eds, *Iron Chelation Therapy*, Springer, Boston, MA, pp. 19–43.

Pillay, V.V. 2008. *Comprehensive Medical Toxicology*. 2nd ed. Paras Medical Publisher. Hyderabad, India.

Pillay, V.V. 2013. *Modern Medical Toxicology*. 4th ed. Jaypee Brothers Medical Publishers (P) Ltd. New Delhi, India.

Pillay, V.V. 2017. *Textbook of Forensic Medicine & Toxicology*. 18th ed. Paras Medical Publisher. Hyderabad, India.

Prawoto, Y. 2013. Integration of mechanics into materials science research: A guide for material researchers in analytical, computational and experimental methods. Lulu.com.

Robertson, A., Tenenbein, M. 2005. Hepatotoxicity in acute iron poisoning. *Human & Experimental Toxicology* 24(11):559–62.

Robotham, J.L., Lietman, P.S. 1980. Acute iron poisoning. A review. *American Journal of Diseases of Children* 134(9):875–9.

Samaniego-Vaesken, M.L., Partearroyo, T., Olza, J., et al. 2017. Iron intake and dietary sources in the Spanish population: findings from the ANIBES study. *Nutrients* 9(3):203.

Sheftel, A.D., Mason, A.B., Ponka. P. 2012. The long history of iron in the Universe and in health and disease. *Biochimica et Biophysica Acta* 1820(3):161–87.

Spanierman, C.S. (2019). Iron toxicity. Emedicine. *Medscape.* https://emedicine.medscape.com/article/815213-overview#a6.

Tenenbein, M. 1996. Benefits of parenteral deferoxamine for acute iron poisoning. *Journal of Toxicology: Clinical Toxicology* 34(5):485–9.

Tenenbein, M. 1998. Toxicokinetics and toxicodynamics of iron poisoning. *Toxicology Letters* 102–103:653–6.

Wessling-Resnick, M. 2017. Excess iron: considerations related to development and early growth. *American Journal of Clinical Nutrition.* 106(Suppl 6):1600S–5S.

Yuen, H.W., Becker, W. 2019. Dimercaprol. [Updated 2019 Oct 30]. In: StatPearls [Internet]. Treasure Island (FL): StatPearls Publishing. https://www.ncbi.nlm.nih.gov/books/NBK459224/.

CHAPTER 19

Vanadium Toxicity Revisited
Striking the Right Balance between Potential New Generation Therapeutics and Adverse Side Effects

Rituparna Ghosh
Bhairab Ganguly College

Ahana Das
Maulana Azad College

Arnab Bandyopadhyay
CSIR-Indian Institute of Chemical Biology

Rajib Majumder
Adamas University

Samudra Prosad Banik
Maulana Azad College

CONTENTS

19.1 INTRODUCTION

Vanadium is a trace element belonging to group V and the fourth period of the periodic table. The name vanadium owes to Vanadis, the Norse goddess of love and beauty in Scandinavian mythology, since several vanadium compounds appear as bright multicolored chemical entities (Mukherjee et al. 2004). The metal was literally invented twice; initially by Andres Manuel Del Rio in 1801 which was mistakenly interpreted as chromium. Subsequently, in the French Academy of Science on February 28, 1831, Alexander Von Humboldt, a friend of Del Rio, along with Nils Sefstrom, a Swedish chemist, established the unique identity of vanadium as a new element (Trevino et al. 2019). Vanadium is the 22nd most abundant element on earth (0.013w/w), and it is widely distributed in all organisms. In humans,the vanadium content in blood plasma is around 200 nM, while in tissues is around 0.3 mg/kg and mainly found in bones, liver, and kidney. In vertebrates, vanadiumenters the organism principally via the digestive and respiratory tracts through food ingestion and air inhalation (Rehder 2008, 2015). It is found principally in rice, oats, beans, radishes, barley, buckwheat, lettuce, peas, potatoes, dill, parsley, black pepper, shellfish, meat, mushrooms, soy, wheat, and olives as well as in different food additives and nutritional supplement in trace amounts. Vanadium is also widely exploited as an industrial element owing to its hardness and ability to form alloys; subsequently, it has found indispensable use in the construction of machines and tools. Besides its use in iron and steel industry, vanadium is also used in photographic developer solutions, as a drying agent in various paints and varnishes, as a component of black dye, inks, and pigments that are employed by the ceramics, printing, and textile industries, as a reducing agent, as well as in the production of pesticides. Over the last two or three decades, a new application of vanadium has evolved in the form of redox flow batteries (Alotto et al. 2014). Probably, due to its versatility and widespread use (Yang et al. 2017), many toxic effects of the metal started to appear among people, especially in the workers of the industries directly or indirectly involving the use of vanadium ores and it soon became a potential threat to human health. In 1911, Dutton reported the first case of vanadium toxicity in the form of development of dry cough and irritation of eyes following occupational exposure to the fumes and dust of vanadium oxides (Dutton 1911). Since then, vanadium has been attributed to a plethora of environmental pollution such as soil and water pollution with associated cellular toxicity. This has, in turn, led to the discovery of a number of new generation medicines and therapies to combat the toxic effects of vanadium. However, the scientific community has also been able to better identify a handful of health augmenting effects associated with low concentrations of the metal which were known long before but are yet to cross the clinical trials. These observations are in perfect harmony with the justification of vanadium as an essential nutrient inside our body (Nriagu 1998). Therefore, vanadium presents an interesting model of a transition metal which is toxic to the body at higher concentration while at the same time presenting many health-boosting features at significantly lower concentrations.

In the subsequent sections, the chief cellular targets and consequences of vanadium-induced toxicity will be discussed along with the major scientific advances which have led to the identification of sites and mechanisms of vanadium uptake. Focus will also be given on effective therapeutic strategies, management of vanadium pollution in industries, and formulation of new vanadium-based drugs.

19.2 CHEMICAL AND PHYSICAL PROPERTIES OF VANADIUM

Pure vanadium (atomic number 23) is the 22nd most abundant element in the earth's crust with a silver-white, shiny, and ductile appearance. It can assume oxidation states from -5 to $+5$ and forms polymers frequently (Nechay 1984). However, vanadate (VO_3^-, V^{+5}) and vanadyl (VO^{+2}, V^{+4}) are the more frequently encountered forms in extracellular tissue fluids and cytosol, respectively. The interconversion between these two states is dependent upon pH, ligand availability, and susceptibility to aerial oxidation. A predominantly reducing intracellular environment maintained by glutathione and other agents is chiefly responsible for generation of vanadyl species. This is subsequently stabilized after binding to various cellular ligands such as proteins, amino acids, nucleic acids, phosphates, phospholipids, glutathione, citrate, oxalate, lactate, ascorbate, etc. (Barceloux 1999). Vanadyl binding has been shown to be especially strong in lysine-rich proteins which have been implicated in selective accumulation of vanadium in ascidians thriving in marine environment (Fukui et al. 2003). Therefore, the bioavailability of vanadium depends upon its state of conjugation to these various cellular molecules.

19.3 ENVIRONMENTAL DISTRIBUTION OF VANADIUM

19.3.1 Air

Anthropogenic sources contribute to about two-thirds of the atmospheric vanadium (Substances and Registry 1992). Vanadium is chiefly released into the air from discharges arising out of combustion of petroleum, coal, and heavy oils during the generation of electricity and heat. Very little vanadium is present in natural gas. Apart from this, discharges from metallurgy industry such as furnaces and crucibles employed in roasting of vanadium slag, smelting of vanadium pentoxide, and melting of ferrovanadium alloys also significantly add to the total vanadium load in the atmosphere. Substantial amounts of the metal are also released from boiler-cleaning operations arising out of dust laden with vanadium oxide. The rest of the vanadium release occurs naturally from continental dust, marine aerosols, and less significantly, from volcanic emission. Populations in areas with high levels of residual fuel oil consumption may also be exposed to above background levels of vanadium, both from increased particulate deposition upon food crops and soil in the vicinity of power plants and higher ambient air levels (Vanadium 1988). Vanadium enters air as an aerosol of simple or complex vanadium oxides from anthropogenic sources and as mineral particles of less soluble trivalent forms from natural sources. In earlier days, obtaining accurate measures of environmental vanadium release was very difficult due to the interference of other elements present, especially in high-polluted industrial belts. However, with the advent of sophisticated technology, such as neutron activation analysis, detection of vanadium in nanogram levels has been made possible (Galdino et al. 1987).

19.3.2 Soil

Vanadium has a broad distribution niche as a trace metal in the earth's crust; its concentration ranges between trace amounts to 400 µg/gm of dry soil, with an average of about 150 µg/gm (Vinogradov 1959). The chief source of vanadium in soil is the corresponding parent rock. Vanadium and chromium share similar chemical properties and exist in natural states in their corresponding anionic forms, as VO_4^{3-} and CrO_4^{2-}, respectively. In order to ensure accurate measurements of vanadium, it must be leached beforehand by alkaline solution of sodium carbonate. Vanadium is released naturally from soil by weathering of vanadium bearing rocks, atmospheric precipitation of vanadium-containing particulates as well as deposition of suspended particulate from water, and plant and animal wastes. Human beings also contribute to the load of vanadium in soil through the use of fertilizers containing materials with a high vanadium content such as rock phosphate, super-phosphate as well as dumping of industrial wastes such as slag heaps and mine tailings (Substances and Registry 1992). Vanadium is relatively mobile in neutral and alkaline soils as compared with other metals, which accounts for its subsequent leaching. However, the same is not true for acidic soils where the accumulated vanadium settles and stays for a much longer period.

19.3.3 Water

The highest source of vanadium in natural water supplies is from volcanic spring (Russo et al. 2014). Therefore, water bodies in the vicinity of a volcano can also contain substantial amounts of the element. The rest of the vanadium release into water bodies comes from erosion and subsequent leaching from rocks and soils as well as the discharge of industrial effluents especially from ash ponds and coal preparation (Vanadium 1988). In fresh water, vanadium exists as the vanadyl ion under reducing conditions and is oxidized to pentavalent vanadate when dissolved oxygen in water gets higher. Therefore, it is imperative that the vanadyl ion is the predominant form in water bodies with high Biological Oxygen Demand (BOD) value. Once in water, relatively small amounts are transported as suspended particles rather than a solution. Consequently, as vanadium from freshwater is finally discharged into the sea, it forms colloids. Ferric trioxide or by biochemical processes in organic matter. Marine plants and invertebrates accumulate much higher concentrations of vanadium compared with terrestrial plants and animals (Substances and Registry 1992).The maximum permissible limit of vanadium in water meant for human use is 140 µg/L according to European guidelines (Russo et al. 2014). Vanadium concentration in drinking water in regions not geographically or economically linked to vanadium exploitation stays typically between 1 and 6 µg/L(Davies and Bennett 1983) and those near industrial belts can go up to 100 µg/L (Vouk 1979). Therefore, vanadium contamination in drinking water does not generally pose much of a threat to the general population (Vanadium 1988).

19.3.4 Food

Food is the most significant point of exposure of the element for the general population (Byrne and Kosta 1978). Vanadium is required as a trace element in the body. Its dietary requirement of lesser than 10 µg/day is met chiefly from black pepper, dill seeds, mushrooms, parsley, shellfish, and spinach (Kohlmeier 2003, Ikebe and Tanaka 1979, Myron et al. 1977). It is also found in some proportions in beverages, fresh fruits and vegetables, cereals, liver, fats, and oil (Badmaev et al. 1999). It has been observed the food processing concentrates its vanadium content to some extent (French and Jones, 1993. Marine planktons and fishes have evolved specialized mechanisms to concentrate vanadyl from water. Therefore, seafoods are richer sources of vanadium as compared to others. High concentrations of vanadium exist in tobacco (Barceloux 1999).

19.4 PHYSIOLOGICAL RELEVANCE OF THE ELEMENT

The requirement of vanadium as a trace element in our body was established decades ago (Harland and Harden-Williams 1994, French and Jones 1993). Since then, scientists have come a long way in understanding the precise molecular functioning of vanadium in mimicking or modulating cellular processes. Several studies conducted in animal models deprived of vanadium in their diet have manifested clear evidence of reduced fertility, increased incidences of spontaneous abortion, and decreased functioning of the mammary gland. Other symptoms of vanadium deficiency included leg deformations, foot inflammation, and generalized musculoskeletal pain (Anke et al. 1986). It was subsequently found that vanadium in the form of pentavanadate is required to stimulate the phosphorylation and simultaneous activation of membrane receptor proteins (Nielsen 1991). Almost all of the physiological malfunctions and deformities are attributable to the role of vanadium in initiating phosphorylation-driven cellular signal transduction pathways. Vanadium is also needed for the functioning of several important metabolic enzymes such as isocitrate dehydrogenase and lactate dehydrogenase and is responsible for lowering elevated levels of serum creatine and β-lipoproteins and glucose in diabetic individuals (Yanardag et al. 2003). Vanadium deficiency affects thyroid metabolism, decreased thyroid weight, and thyroid weight/body weight ratio (Uthus and Nielsen 1990). Vanadium is known to promote release of Ca^{2+} via the production of inositol phosphate second messengers and prevent spasms in human bronchial smooth muscles (Cortijo et al. 1997). It is also implicated in lung inflammation and apoptosis through activation of a Reactive Oxygen Species (ROS)-dependent NADPH oxidase complex (Wang et al. 2003). Vanadium compounds are also known to regulate iron, glucose, and lipid metabolism; studies conducted in a range of species including chicken and rats have demonstrated deficiency symptoms of stunted growth, poor reproductive success, defective lipid metabolism and inhibition of Na^+-, K^+-ATPases in the kidney, brain, and heart cells (Nriagu 1998). However, these are of little concern to humans owing to the everyday requirement of vanadium of approximately 10 µg, which is more than catered for by any universal diet. Entry of vanadium through food sources is limited to 40 ng/g of body weight mostly through consumption of fishes, seafoods, and mushrooms which is considered to have no side effects. Under normal physiological conditions, less than 5% of the ingested vanadium in its amount is taken into the blood (Nielsen 1995a,b).However, bio absorption can be increased by an additional 5% through synthesized vanadium salts especially those used as insulin-mimetic (Bogden et al. 1982, Wiegmann et al. 1982). Inside the acidic environment of the vascular system most vanadium is transformed to the vanadyl (IV) state (Chasteen et al. 1986). Residual vanadium in the +5 state as vanadate is taken up by the intestinal epithelial cells through a poorly understood phosphate-based anionic transporter system (Cantley et al. 1978).

19.5 PORTALS OF VANADIUM ENTRY

19.5.1 Through Digestive Tract

In vertebrates, vanadium enters the organism principally via the digestive and respiratory tracts through food ingestion and air inhalation (Rehder 2008, 2015). The estimated daily dietary intake in the United States is 10–60 µg/day, and it is a little higher for processed foods. A little amount of vanadium also gets inside our body through drinking water (Health, Substances, and Registry 2012). Studies in animals have shown that less than 5% of the ingested vanadiumis absorbed while the rest is excreted via the feces. On the contrary, very little vanadium (0.2–1%) is absorbed in the human gut. Dietary composition, fasting, and speciation are the various factors on which extent of vanadium vary from individual to individual (Wilk et al. 2017, Ma et al. 2018, Greim 2009). Gastrointestinal problems such as abdominal pain, cramping, discomfort, irregular bowel discharge, nausea, and vomiting start showing only at dosage above 15 mg vanadium/day after atleast a period of 2weeks (Goldfine et al. 2000, Afkhami-Arekani et al. 2008). Most of the dietary vanadium is usually excreted in the feces, meaning that the vanadium accumulation in the body does not pose any significant threat (Korbecki et al. 2012, Costa Pessoa and Tomaz 2010).

19.5.2 Through Respiratory Tract

Inhalation of vanadium constitutes the chief portal of entry of the metal inside the human body. This has been established unequivocally through data from industry workers as well as studies in human and laboratory animals. The size and nature (organic/inorganic) of vanadium-containing particles and the solubility of vanadium compounds are the governing factors behind absorption rate of the metal in the respiratory tract. For example, insoluble vanadium pentoxide is relatively rapidly expelled by lungs in animals after sudden acute exposure, but the process occurs at a significantly slower rate when exposure to vanadium takes place on a regular basis. This occurs because over time the metal is slowly deposited in the lungs and tends to remain there. Soluble vanadium compounds are partially absorbed, but the extent of absorption is yet to be determined accurately. Adverse toxicity effects of vanadium inhalation have been reported in humans and animals at concentrations substantially higher than those found in environments. Workers regularly inhaling vanadium pentoxide dust have been shown to be affected from diverse respiratory obstructions and airway irritation (e.g., coughing, wheezing, and sore throat). Although the effects persist for days or weeks after the end of initial exposure, they do not generally affect lung function (Rehder 2013, Fallahi et al. 2018, Zhu et al. 2016, Yu et al. 2011, Wei et al. 2015). However, the effects are more pronounced in animal models with concomitant development of many lung lesions including alveolar/bronchiolar hyperplasia, inflammation, and fibrosis.

19.6 TOXICOKINETICS OF VANADIUM

19.6.1 Absorption

The absorption of vanadium compounds is affected by their solubility and the site of entry. In industrial workers, V_2O_5 is the chief species efficiently absorbed by the lungs. In contrast, vanadium salts have a poor absorption profile from the gastrointestinal tract and animal studies of lesser than 1–2 of the ingested vanadium (Roshchin et al. 1980). This also explains why the toxic effects of vanadium are more acute for those inhaling vanadium directly than others consuming significant amounts of vanadium through a biased vegetable or seafood-based diet. Vanadate (+5) from diet is absorbed about thrice more efficiently than the vanadyl (+4) species in the GI tract. The skin probably is a minor route of exposure, especially owing to low solubility of metallic vanadium. Finally, most forms of vanadium are converted into vanadyl (VO^{+2}) in the stomach and it remains in this form in the alkaline media of the duodenum (Substances and Registry 1992).

19.6.2 Distribution by Body Fluid

Once in the bloodstream, vanadium is selectively bound by four groups of serum proteins, namely transferrin >> hemoglobin ≈ immunoglobulin G > albumin in the order of affinity mentioned (Smith et al. 1995). In addition, vanadyl also shows a strong preference for binding to low-molecular-weight

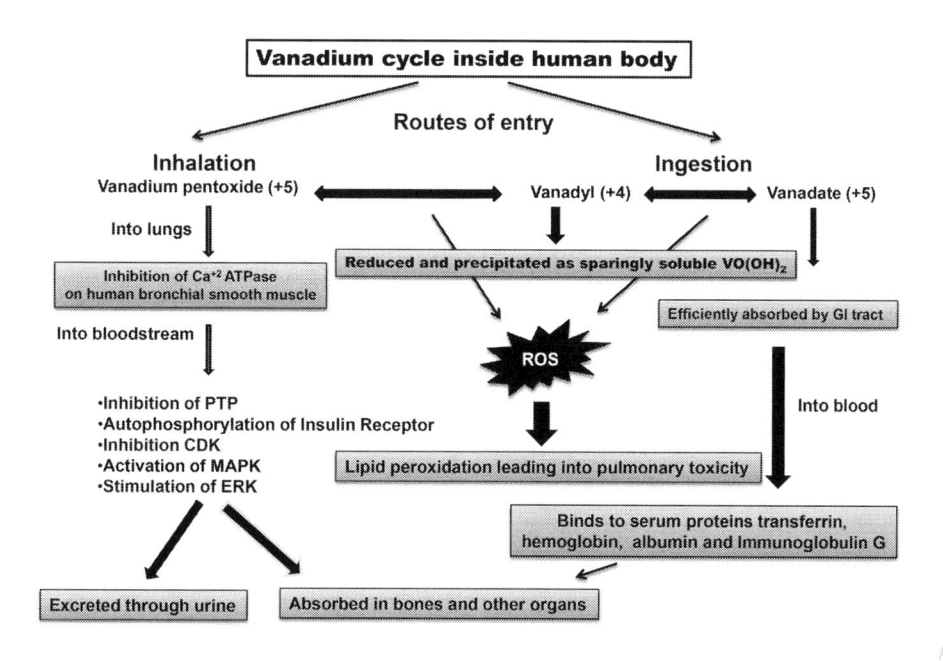

Figure 19.1 Vanadium cycle inside human body

negatively charged serum molecules such as citrate, oxalate, lactate, phosphate, glycine, and histidine (Sanna et al. 2009). V^{4+} ion shares the same binding site as Fe^{3+} ion in transferrin (Correia et al. 2017) and ferritin (Chasteen et al. 1986) and thus can displace iron from the protein. At sufficiently high concentrations, vanadyl also shows certain affinity to IgG. Subsequently, blood is distributed from blood to tissues rapidly and incorporated in various organs and tissues primarily in bones as well as in liver, kidney, brain, heart, and muscles (Figure 19.1).

19.6.3 Interconversion between Various Oxidation States of Vanadium

Vanadium alternates between the tetravalent form, vanadyl (V^{+4}), and the pentavalent form, vanadate (V^{+5}) inside the body. However, due its potential for affecting the signaling pathways, intracellular toxicity of vanadate is higher than the tetravalent form. Vanadate also reacts with a number of enzymes and is a potent inhibitor of the Na^{+}-, K^{+}-ATPase of plasma membranes (Mahmmoud et al. 2014). After intravenous administration of V^{+4} or V^{+5}, equilibrium is reached with respect to concentrations of both species.

19.6.4 Intracellular Transport of Vanadium

Vanadate compounds are transported inside the cells through anionic channels and in the process are converted into their corresponding vanadyl forms. Studies on this have demonstrated that passages of ionic forms of vanadium through intracellular causes changes in the cellular microvilli and the structure of the actinic filament network (Yang et al. 2004). Inside the cell, vanadium ions are subjected to oxidation-reduction processes mediated by reducers and metabolic oxidizers such as reduced glutathione, catechol, NADPH, NADH, cysteine, ascorbic acid, etc. and the form in which vanadium is found in abundance is dependent on the pool of these molecules (Goc 2006). Generally, more amounts of vanadium are found inside mitochondria and nucleus than in anywhere else.

19.6.5 Elimination/Excretion

Unabsorbed vanadium is expelled from the body through feces. Vanadium administered through parenteral route leads to excretion of about 10 of the metal through the fecal matters of humans and rats (Barceloux 1999). Vanadium has also been reported to be excreted by the body through bile and urine.

19.7 TOXIC EFFECTS OF VANADIUM

Vanadium is needed in a minute-scale amount as a trace element for proper execution of many physiological functions. However, at higher concentrations, and mostly in the form of inorganic compounds, it can be highly toxic to most of the organs and cellular locations of our body. Like any heavy metal, toxicity of vanadium also owes primarily to its elicitation of oxidative stress inside the cellular environment (Sharma et al. 2014) caused by an imbalance between generation of free radicals as a result of metabolic activities and antioxidants produced by the body in the form of defense responses. Generally, the magnitude of toxicity increases with the valence state of the metal with pentavalent vanadium being the most toxic. Vanadium toxicity is responsible for a multitude of ill effects such as respiratory dysfunction (Woodin et al. 2000), hematologic and biochemical alterations (Fortoul et al. 2011), renal toxicity (Zaporowska et al. 1993), immunotoxicity, mutagenicity (Avila-Costa et al. 2004), and neurotoxicity (Li et al. 2013). There has been substantial evidence that vanadium pentoxide causes development of carcinoma in lung cells of laboratory animals (Rondini et al. 2010). Also, the soluble organic compound vanadium acetylacetonate has been shown to induce progressively the formation of lung tumors (Greim 2009). The toxicity caused by organic and inorganic compounds of vanadium has been extensively reviewed by Ghosh and Banik (2016). In the following sections, the toxicity potential of vanadium and its compounds are discussed with respect to targets of action.

19.7.1 Mutagenic Potential

Vanadium compounds are effective inhibitors or stimulators of several enzymes involved in cellular replication, transcription, and DNA damage repair pathways. Therefore, they possess naturally a strong mutagenic potential (Stemmler and Burrows 2001). Altamirano-Lozano et al. (1996) showed single-strand DNA breaks in individual testis cells of mice following vanadium pentoxide exposure. Subsequently, it was shown that vanadium pentoxide and ammonium metavanadate can induce DNA strand breaks inhuman cell cultures (Ehrlich et al. 2009). Vanadium (IV)-mediated free radical causes hydroxylation of 2′-deoxyguanosine, which leads to DNA strand breaks. Vanadyl sulfate has been shown to damage DNA in human lymphocytes and HeLa cells (Wozniak and Blasiak 2004). Vanadium(III), vanadium(IV), and vanadium(V) compounds resulted in chromosome aberrations in Chinese hamster ovarian cell lines whereas vanadium(IV) and vanadium(V) compounds also induced aneuploidies in animal models (Greim 2009).

19.7.2 Toxicity Linked to Metabolism

Administration of vanadium in rabbits was associated with weight loss along with a decline in the albumin-globulin ratio, owing to both proteins. Vanadium also increases triglyceride concentrations in the liver and blood, decreases the serum cholesterol level, and increases levels of certain metabolic enzymes such as glutamate-oxaloacetate transaminase and glutamate-pyruvate transaminase activity. Additionally, exposure to the metal increases the catabolism of cysteine and reduces the synthesis of cholesterol and coenzyme A in rat liver (Roshchin et al. 1980).

19.7.3 Hematological Abnormalities

Vanadium causes peroxidative changes in the erythrocyte membrane leading to hemolysis and anemia (Ścibior et al. 2006, Zaporowska and Wasilewski 1992, Zaporowska et al. 1993). Additionally, administration of vanadium into the body is also known to increase the percentage of circulating reticulocytes and white blood cells (WBCs) in the peripheral blood. Increase in the number of WBCs is indicative of a vanadium-associated systemic inflammation (Al-Bayati et al. 1990). Vanadium also induces thrombocytosis and it might correlate with some thromboembolic diseases (González-Villalva et al. 2006).

19.7.4 Changes in Respiratory System

Since inhaled vanadium in the form of vanadium pentoxide is the most potent toxic form, it is imperative that the respiratory system of industrial plant workers will be the worst affected by vanadium toxicity. The cardinal signs of pentavalent vanadium associated changes are irritation, rhinitis, wheezing, nasal hemorrhage, cough, sore throat, and chest pain marked by tracheitis, pulmonary edema, bronchopneumonia, and enteritis (Sax 1989, Ress et al. 2003, Wörle-Knirsch et al. 2007). These are frequently observed symptoms in boilermakers in industries having repeated exposures to high concentration of vanadium pentoxide fumes (Substances and Registry 1992). The associated pathological changes following exposure to vanadium pentoxide dust have been extensively studied in animal models; chronic rhinitis, emphysema, patches of lung atelectasis, bronchopneumonia, and pyelonephritis have been visible in almost all cases (Roshchin et al. 1980) and in instances of severe poisoning, there is congestion in lungs with perivascular and focal edema, bronchitis, and focal interstitial pneumonia leading to death of the affected individual.

19.7.5 Cardiovascular Changes

Vanadium oxides also affect simultaneously the functioning of the cardiovascular system. Experiments in animal models exposed to vanadium fumes have indicated occurrence of arrhythmias and extrasystole, prolongation of the Q-RST interval, decrease in the height of the P and T waves of the ECG, perivascular swelling, as well as fatty changes in the myocardium (Roshchin et al. 1980). Alongside, intravenous injections of sodium ortho- and metavanadate have been shown to increase systolic blood pressure with concomitant vasoconstriction in the spleen, kidney, and intestines (Substances and Registry 1992).

19.7.6 Hepatic Changes

Vanadium results in partial necrosis of liver hepatocytes of rats and rabbits following secretion of cytokines and chemokines, a decrease in the rate of liver tissue respiration, and a fall in serum albumin levels (Substances and Registry 1992). The metal oxide also causes inhibition of cholesterol biosynthesis and induces lipid peroxidation in rat liver microsomes (Vanadium 2000). The visible histopathological changes following inhalation of vanadium include central vein congestion with scattered small hemorrhages and granular degeneration of hepatocytes (Vanadium 2000). Vanadium accumulation is also observed in the intermembrane space of liver mitochondria suggesting substantial damage caused by the metal (Hosseini et al. 2013).

19.7.7 Renal Changes

The excretory system is particularly vulnerable to damaging effects of vanadium toxicity. Acute vanadium exposure is known to cause glomerular hyperaemia, necrosis of convoluted tubules, albuminuria, granular degeneration of the epithelial cells of the convoluted tubules and fatty changes in the kidney as studied in animal models (Roshchin et al. 1980). Over the years, several important biomarkers have been identified to assess the extent of renal damage following vanadate exposure which has both diuretic and natriuretic effects on the kidney. Out of these, the urinary cystatin C ($CysC_u$) and kidney injury molecule-1 ($KIM-1_u$) are the most appropriate ones to evaluate renal function (Ścibior et al. 2014). The diuretic and natriuretic effects on kidney are believed to be due to the inhibition of Na^+-, K^+-ATPase thus blocking tubular reabsorption. Inhibition of renal brush border membrane vesicles by vanadium causes a time-dependent inhibition of citrate uptake in the membrane possibly contributing to nephrotoxicity (Sato et al. 2002). However, sufficient data on the effect on the human kidney is yet to arrive.

19.7.8 Immunological Changes

Vanadium may also affect the normal functioning of the immune system. It has been observed in animal models that administration of vanadium causes a sharp decline in the number of antibody-producing cells with decrease in peritoneal macrophages. Correspondingly, liver and spleen were enlarged with distinct formation of splenic megakaryocytes and red blood cell precursors (Substances and Registry 1992).

19.7.9 Neurological Changes

Vanadium is responsible for mediating oxidative damage in the central nervous system (CNS) (Garcia et al. 2005). Studies have revealed that vanadium toxicity in industry workers leads to fatigue, depression, and hostility (Li et al. 2013). Accumulation of vanadium caused damage in striatum leading to impairment in learning and memory (Sun et al. 2017). Imaging studies in animal models showed clear morphological alterations in the prefrontal cortex, loss of dendritic spines, and cell death in the hippocampal CA1 pyramidal cells and Purkinje cells of the cerebellum (Folarin et al. 2017). Additionally, the number of tyrosine hydroxylase-immuno reactive neurons in the substantia nigra pars compacta decreased substantially (Avila-Costa et al. 2004). Within the ependymal epithelium, there was cilia loss and detachment of cell layers following vanadium pentoxide inhalation (Avila-Costa et al. 2005). The damages cumulatively resulted in loss of permeability of the epithelium leading to neuronal death (Avila-Costa et al. 2005). Parallel studies in human revealed tremor in CNS with defective conditioned reflexes, time-dependent loss of dendritic spines with parallel loss of spatial memory (Avila-Costa et al. 2006), significant change in morphology of the hippocampus CA1 neurophile culminating to necrotic cell death (Vanadium 2000).

19.7.10 Effect on Reproductive System

Vanadium is a significant threat to global reproductive health. Vanadium derivatives exert their toxicities by releasing free radicals which disrupt testicular functions, spermatogenesis, and sperm maturation in rats. Exposure to vanadium(IV) has revealed ultra-structural changes and subsequent apoptosis (Aragon et al. 2005, Aragón and Altamirano-Lozano 2001). In all cases, the overdose of vanadium-borne free radicals causes a general imbalance between free radical load and corresponding scavenging agents such as the protective enzymes such as SOD, catalase, glutathione peroxidase,glucose-6-phosphate dehydrogenase, glutathione-S-transferase, and glutathione reductase. More evidence were obtained when levels of GSH, vitamins C and E were found to get depressed following vanadium exposure. There was also a significant increase in lipid peroxidation level, another hallmark of vanadium-induced stress (Chandra et al. 2007a–c). Vanadium exposure also brought about a drasticfall in epididymal sperm number, thus suggesting the role of metal toxicity in spermatogenesis.

19.8 THERAPEUTIC AND PHARMACOLOGICAL POTENTIALS OF VANADIUM

The therapeutic potential of many vanadium compounds has been discovered over the last few decades especially in the fields of diabetes and cancer. Since 2000, there have been more than 1,200 entries in Pubmed in this respect. The earliest known therapeutic use of vanadium dates back in 1899 by Lyonnet et al. where considerable reduction in blood glucose levels was attained in a sample of 60 diabetic individuals by application of sodium metavanadate (Thompson and Orvig 2006). The interest in anti-diabetic potential of vanadium was again renewed in the second half of the twentieth century chiefly because of the prospects of oral administration of the metal to substitute painful insulin injection (Heyliger et al. 1985). A substantial amount of pharmaceutical interest of vanadium owes to its ability of inhibiting the activity of Protein Tyrosine Phosphatases (PTPs), an intracellular enzyme implicated in cellular signal transduction pathways. PTPs occur in association with Protein Tyrosine Kinases (PTKs), which are mostly transmembrane receptor proteins responsible for downstream transmission of the signals through phosphorylation of keytyrosine residues. PTPs carry out subsequent dephosphorylation to switch off the cascade (Bhise et al. 2004). PTKs and PTPs have implicated in a plethora of cellular abnormalities in relation to proliferation, survival, motility, differentiation, and metabolism leading to ailments such as diabetes, cancer, chlorosis, anemia, and tuberculosis. This has led to the development of several kinase inhibitors and antibodies targeted against the kinases (Wu et al. 2016). However, the full therapeutic potential of this strategy had been far from realized chiefly due to the high sequence conservation in the PTP family of enzymes making it technically challenging to develop inhibitors targeting specific PTPs (Zhang 2017). Additionally, due to the polar nature of PTP active sites, it has been difficult to design inhibitor therapeutics which can cross the cell membrane.

Vanadium in the form of pentavalent orthovanadate acts as a reversible competitive inhibitor of PTP owing to its resemblance with phosphate group (Crans et al. 2004). Binding of vanadate into the PTP active site is facilitated by hydrogen bonds (Peters et al. 2003) which make it a strong inhibitor. Due to the sequence conservation of PTP active site, vanadium has evolved as a universal inhibitor of the pan PTP superfamily of enzymes (Gordon 1991). Some vanadium compounds, such as the peroxidovanadium complexes, can also bring about permanent irreversible inhibition of the active site by oxidizing critical cysteine residues (Evangelou 2002, Scrivens et al. 2003). This has enlarged the domain of metallotherapeutics and has led to the synthesis of several new oxovanadium-based compounds such as bis(maltolato)oxidovanadium (IV) (BMOV) as an inhibitor of class II LMW-PTP (Peters et al. 2003). Inhibition of PTPs is also implicated in treating diabetes, a long-recognized potential of vanadium-based compounds. Autophosphorylation of insulin receptors upon insulin binding causes docking and simultaneous activation of insulin receptor substrate 1 (IRS-1). This, in turn, initiates a signal cascade which culminates into deployment of GLUT-4 channels to the cell membrane resulting in increased uptake of glucose. In diabetic individuals, insulin doesn't function properly; therefore, subsequent glucose uptake ceases to occur efficiently. Vanadium-based compounds cause inhibition of PTPs thus resulting in sustained signal transmission by phospho-tyrosine residues and consequent activation of IRS-1 (Heyliger et al. 1985). In this way, vanadium mimics insulin and is able to lower blood glucose significantly. Structural analogy with phosphate has enabled vanadium to be used in other non-PTP inhibitor-based therapeutic formulations also.

The toxic effects of vanadium have also been cleverly redirected to destroy tumor cells. This, again, has been achieved through alteration of the levels of PTKs (Stern et al. 1993, Sabbioni et al. 1991). Vanadium complexes have shown promising results against L-1210 murine leukemia, rat liver tumors, fluid and solid Ehrlich ascites tumor, and TA3Ha murine mammary adenocarcinoma. They have been also deployed successfully to arrest the growth of human tumor colony formation and HEP-2 epidermoid carcinoma cells (Murthy et al. 1988) and improve the antioxidant status during the development of carcinogenesis (Mukherjee et al. 2004). A new organometallic compound, N,N′-ethylenebis (pyridoxylideneiminato) vanadium(IV) complex [Pyr2enV(IV)] has achieved 93 and 57 of cell mortality in A375 (human melanoma) and A549 (human lung carcinoma) cells, respectively (Strianese et al. 2013). Many new generation compounds are also in the pipeline for evaluation against chemical carcinogenesis through either of the two strategies: (a) inhibition of active carcinogenic metabolites (Evangelou 2002) and (b) arresting the activity of tyrosine phosphatases (Strianese et al. 2013).

Yet another success story of vanadium-based therapeutics had been the discovery of vanadium-containing enzymes (Winter and Moore 2009, Butler and Carter-Franklin 2004). These are superefficient catalysts equipped to carry out the activities of nitrogenase, haloperoxidase, and bromoperoxidase. Other health augmenting applications of the element include the use of vanadium sulfate to enhance muscle mass and improve the performance of athletes (Fawcett et al. 1997), increasing circulation of nitrate level in mice by peroxovanadium compounds, etc. Also, most of the cellular phosphate binding molecules such as ATPases are also subject to vanadium-mediated inhibition. For example, decavanadate can be used to inhibit actin-stimulated myosin ATPase and sarcoplasmic reticulum Ca^{2+} ATPase (Collauto et al. 2017). This presents yet another new avenue of utilizing the therapeutic potential of vanadium.

19.9 COPING WITH VANADIUM TOXICITY: THE CHALLENGES FOR THE FUTURE OF VANADIUM DRUGS

Among all the promises shown by vanadium as prospective therapeutic candidates, a majority of compounds are still to clear the clinical trial phases (Eisenberg et al. 1998) (Table 19.1) and quite a handful of them have been rejected altogether on account of severe side effects (Srivastava 2000, Domingo 1996). The compounds which have shown promising insulin-mimetic potential were found to be associated with a multitude of physiological abnormalities such as gastrointestinal toxicity, diarrhea, reduced thirst, loss of appetite, and reduced body weight gain (Heyliger et al. 1985, Gil et al. 1988, Brichard et al. 1988, Blondel et al. 1989, Ramanadham et al. 1989, Sakurai et al. 1990, Dai et al. 1994, Domingo et al. 1994). In fact, the toxicities associated with some of the therapeutics were so severe that they resulted in death of the laboratory animal during pilot studies. Once inside the cell, there is spontaneous interconversion between the vanadyl(IV) and vanadate(V) forms using Fenton-like reactions. This is a major source of vanadium-induced ROS generation inside the cell (Nechay 1984). These free radicals can bring about direct oxidation of cysteine residues in the active site of PTPs causing permanent irreversible damage to the enzymes (Barford 2004). Scientists have tried to circumvent this problem by generating big vanadium-containing organometallic complexes with reduced ROS load and enhanced bio absorption. However, many recent studies suggest that these complexes undergo speciation and subsequent decomposition releasing again free vanadium and other toxic ligands (Sanna et al. 2017, Levina and Lay 2017). The problem is even more complicated by the fact that chemistry speciation of vanadium at neutral pH is yet to be understood clearly. The most promising approach to reducing the *in vivo* speciation and non-target toxicity of ligands has been packaging of the oxidovanadaium compounds into nanocomposites. For example, development of a vanadate–chitosan complex packaged in a nanocomposite carrier has shown a lot

Table 19.1   Compounds of Vanadium Developed/Under Clinical Trial for Therapeutic Use

Type of Compound	Therapeutic Use	Whether in Clinical Trial	Reference
Vanadylsulfate (VOSO$_4$)	Type 2 Diabetes treatment	Yes	Cusi et al. (2001)
Sodium metavanadate (NaVO$_3$)	Diabetes treatment	Yes	Goldfine et al. (1995)
Sodium orthovanadate (Na$_3$VO$_4$)	Anti-cancer	No	Klein et al. (2008), Morita et al. (2010)
BMOV [bis(maltolato)oxovanadium]	Diabetes treatment	Yes	Scior et al. (2009), Nischwitz et al. (2013)
BEOV [bis(ethylmaltolato)oxovanadium]	Diabetes treatment	Yes	Thompson et al. (2009)
METVAN [bis(4,7-dimethyl-l, 10-phenanthroline) sulfatooxovanadium(IV)]	Anti-cancer	No	León et al. (2019)
BKOV [bis(kojato)oxovanadium(IV)]	Diabetes treatment	No	Korbecki et al. (2012)
Vanadium pentoxide(V$_2$O$_5$)	Anti-tumor	No	Öztürk et al. (2018)
Peroxovanadate	Anti-tumor	No	Djordjevitz et al. (1985)
Vanadocene dichloride	Anti-cancer	No	Ghosh et al. (2000)
Vanadocenes dithiocyanate (VCp$_2$(SCN)$_2$ × 0.5 H$_2$O)	Anti-cancer	No	Ghosh et al. (2000)
Vanadocene diselenocyanate [VCp$_2$(NCSe)$_2$]	Anti-cancer	No	Ghosh et al. (2000)
Ammoniummonovanadate (NH$_4$VO$_3$)	Anti-tumor	No	Bishayee et al. (2000), Chakraborty et al. (2007a)
Ammonium metavanadate (NH$_4$VO$_3$)	Anti-tumor	No	Bishayee et al. (1997), Chakraborty et al. (2007b)
Vanadyl-1,10-phenathroline complex [VO(phen)$_2$]	Anti-cancer	No	Sakurai et al. (1995)
Nicotinoyl hydrazine vanadium complexes	Anti-cancer	No	Nair et al. (2014)
Oxidovanadium flavonoid complexes	Anti-tumor	No	Leon et al. (2013, 2014)
Pyridinone ligated oxidovanadium complexes	Anti-melanoma	No	Rozzo et al. (2017)
Pyridoxylideneiminato vanadium	Anti-cancer	No	Strianese et al. (2013)
Vanadium(V)-peroxido-betaine	Anti-tumor	No	Petanidis et al. (2016)
Picolinato-bis(peroxido)oxovanadate	Anti-gliomas	No	Ajeawung et al. (2013)
VO-salen	Anti-leukemia	No	Meshkini et al. (2010)
Heteroleptic Schiff base vanadium complexes	Anti-cancer	No	Scalese et al. (2017)
Phenanthroline/quinolone ligated vanadium	Pancreas treatment	No	Kowalski et al. (2017)
Bis(acetylacetonato)-oxidovanadium (IV)	Pancreas treatment	No	Wu et al. (2016)
Schiff base vanadium complex	Anti-cancer	No	Sinha et al. (2017)
Bisperoxidovanadium compounds	Anti-cancer	No	Scrivens et al. (2003)

of promise as anti-diabetic therapeutic (Liu et al. 2016). Oxidovanadium complexes have also been successfully packaged with anionic polysaccharides (Kremer et al. 2015). Chen et al. reported packaging vanadium disulfide nanodots into PEG lipid micelles for effective delivery into tumor cells in mouse models (Chen et al. 2017). Stabilization of vanadium complexes by conjugating them with nano-particles has, therefore, been the only silver lining in the uphill task of developing vanadium-based therapeutics.

19.10 CONCLUDING REMARKS

Standing in the twenty-first century, the knowledge of toxicity associated with vanadium has been rendered outdated ever since man has realized that the consumption of vanadium by the general population is far below permissible levels. However, for the industry workers, who are at a far greater risk of vanadium-associated toxicity due to direct inhalation of vanadium pentoxide, several new dietary antioxidants have been developed (Zwolak 2020). Instead the same traditional knowledge has been exploited to discover new-age vanadium-based therapeutics for the treatment of deadly human ailments. However, often, the administration of vanadium in the form of therapeutic formulations has resulted in severe unwanted side effects in the lab animals leading to even death. Therefore, despite possessing promising properties most of these drugs have, the most persistent difficulty in designing vanadium-based drugs has been the unequal bioavailability of the metal in various tissue parts owing to the conjugation of vanadium to ferritin or transferring receptor inside the body. Vanadium is safe in adults if its consumption does not exceed 1.8 mg/day as set by the National Institute of Medicine. At higher doses, such as those used to treat diabetes, vanadium frequently causes unwanted side effects including abdominal discomfort, diarrhea, nausea, loss of energy, and problems with the nervous system. In parallel, the amount of elemental vanadium also varies in the different commonly available vanadium salts through which dietary supplementation is going to occur, for example, vanadyl sulfate contains 31 elemental vanadium; sodium metavanadate contains 42 elemental vanadium; and sodium orthovanadate contains 28 elemental vanadium. The biggest challenge therefore for the pharmaceutical industry is to find the optimal tissue-specific dosage of vanadium which will be minimally interfering with the normal physiology and functioning of other regions. Hopefully, with a fast advancing medical technology, the future is not too far where robust state of the art treatment procedures such as personalized medicine will be able to address these issues successfully to ensure a better disease-free life for the human society.

ACKNOWLEDGMENT

The authors express their deep gratitude to Prof. Amar K. Chandra, Department of Physiology, University of Calcutta, for his guidance in some of the studies reported and also to University Grants Commission, Govt. of India, for funding some of the works by the authors reported in this study.

CONFLICT OF INTEREST

The authors have no conflict of interest arising out of this chapter.

REFERENCES

Afkhami-Arekani, M, Mahdi Karimi, S MohammadiM, Forough, J. 2008. Effect of sodium metavanadate supplementation on lipid and glucose metabolism biomarkers in type 2 diabetic patients. *Malaysian Journal of Nutrition* 14 (1):113–119.

Ajeawung, NF, Faure, R, Jones, C, Kamnasaran, D. 2013. Preclinical evaluation of dipotassium bisperoxo (picolinato) oxovanadate V for the treatment of pediatric low-grade gliomas. *Future Oncology* 9 (8):1215–1229.

Al-Bayati, MA, Giri, SN, Rabbe, O, Rosenblatt, LS. 1990. Time and dose-response study of the effects of vanadate in rats: changes in blood cells, serum enzymes, protein, cholesterol, glucose, calcium, and inorganic phosphate. *Journal of Environmental Pathology Toxicology, Oncology* 10 (4–5):206–213.

Alotto, P, Guarnieri, M, Moro, F. 2014. Redox flow batteries for the storage of renewable energy: a review. *Renewable and Sustainable Energy Reviews* 29:325–335.

Altamirano-Lozano, M, Alvarez-Barrera, L, Basurto-Alcántara, F, Valverde, M. 1996. Reprotoxic and genotoxic studies of vanadium pentoxide in male mice. *Teratogenesis Carcinogenesis and Mutagenesis* 16 (1):7–17.

Anke, M, Groppel, B, Gruhn, K, Kosla, T, Szilagyi, M. 1986. New research on vanadium deficiency in ruminants. In: Anke M, Baumann W, Bräunlich H, Bruckner Chr, Groppel B (eds) 5. Spurenelementsymposium, Karl-Marx-Universität Leipzig, Friedrich-Schiller-Universität Jena, GDR, pp. 1266–1275.

Aragón, AM Altamirano-Lozano, MA. 2001. Sperm and testicular modifications induced by subchronic treatments with vanadium (IV) in CD-1 mice. *Reproductive Toxicology* 15(2):145–151.

Aragon, MA, Ayala, ME, Fortoul, TI, Bizarro, P, Altamirano-Lozano, MA. 2005. Vanadium induced ultrastructural changes and apoptosis in male germ cells. *Reproductive Toxicology* 20(1):127–134.

Avila-Costa, MR, Colín-Barenque, L, Zepeda-Rodríguez, A, Antuna, SB, Saldivar, L, Espejel-Maya, G, Mussali-Galante, P, Avila-Casado, MC, Reyes-Olivera, A, Anaya-Martinez, V. 2005. Ependymal epithelium disruption after vanadium pentoxide inhalation: a mice experimental model. *Neuroscience Letters* 381 (1–2):21–25.

Avila-Costa, MR, Fortoul, TI, Niño-Cabrera, G, Colín-Barenque, L, Bizarro-Nevares, P, Gutiérrez-Valdez, AL, Ordóñez-Librado, JL, Rodríguez-Lara, V, Mussali-Galante, P, and Díaz-Bech, P 2006. Hippocampal cell alterations induced by the inhalation of vanadium pentoxide (V_2O_5) promote memory deterioration. *Neurotoxicology* 27 (6):1007–1012.

Avila-Costa, MR, Montiel Flores, E, Colin-Barenque, L, Ordoñez, JL, Gutiérrez, AL, Niño-Cabrera, HG, Mussali-Galante, P, Fortoul, TI 2004. Nigrostriatal modifications after vanadium inhalation: an immunocytochemical and cytological approach. *Neurochemical Research* 29 (7):1365–1369.

Badmaev, V, Prakash, S, Majeed, M, 1999. Vanadium: a review of its potential role in the fight against diabetes. *The Journal of Alternative and Complementary Medicine* 5(3):273–291.

Barceloux, DG. 1999. Vanadium. *Journal of Toxicology: Clinical Toxicology* 37 (2):265–278.

Barford, D. 2004. The role of cysteine residues as redox-sensitive regulatory switches. *Current Opinion in Structural Biology* 14 (6):679–686.

Bhise, SB, Nalawade, AD, Wadhawa, H. 2004. Role of protein tyrosine kinase inhibitors in cancer therapeutics. *Indian Journal of Biochemistry & Biophysics* 41(6):273–280.

Bishayee, A, Karmakar, R, Mandal, A, Kundu, SN, Chattrjee M. 1997. Vanadium mediated chemoprevention against chemical hepatocarcinogenesis in rats: haematological and histological characterisitics. *European Journal of Cancer Prevention* 6(1):58–70.

Bishayee, A, Oinam, S, Basu, M, Chatterjee, M. 2000. Vanadium chemoprevention of 7,12-dimethybenz(α) anthracene-induced rat mammary carcinogenesis: probable involvement of representative hepatic phase I and II xenobiotic metabolizing enzymes. *Breast Cancer Research and Treatment* 63(2):133–145.

Blondel, O, Bailbe, D, Portha, B. 1989. *In vivo* insulin resistance in streptozotocin-diabetic rats—evidence for reversal following oral vanadate treatment. *Diabetologia* 32 (3):185–190.

Bogden, JD, Higashino, H, Lavenhar, MA, Bauman, JW, Kemp, FW, Aviv, A. 1982. Balance and tissue distribution of vanadium after short-term ingestion of vanadate. *The Journal of Nutrition* 112 (12):2279–2285.

Brichard, SM, Okitolonda, W, Henquin, JC. 1988. Long term improvement of glucose homeostasis by vanadate treatment in diabetic rats. *Endocrinology* 123 (4):2048–2053.

Butler, A Carter-Franklin, JN. 2004. The role of vanadium bromoperoxidase in the biosynthesis of halogenated marine natural products. *Natural Product Reports* 21 (1):180–188.

Byrne, AR, Kosta, L. 1978. Vanadium in foods and in human body fluids and tissues. *Science of the Total Environment* 10 (1):17–30.

Cantley, LC, Resh, MD, Guidotti, G. 1978. Vanadate inhibits the red cell (Na+, K+) ATPase from the cytoplasmic side. *Nature* 272 (5653):552–554.

Chakraborty, T, Chatterjee, A, Rana, A, Dhachinamoorthi, D, Kumar, PA, Chatterjee, M 2007. Carcinogen-induced early molecular events and its implication in the initiation of chemical hepatocarcinogenesis in rats: Chemopreventive role of vanadium on this process. *Biochimica et Biophysica Acta* 1772(1): 48–59.

Chakraborty, T, Chatterjee, A, Rana, A, Rana, B, Palanisamy, A, Madhappan, R, Chatterjee, M. 2007. Suppression of early stages of neoplastic transformation in a two-stage chemical hepatocarcinogenesis model: Supplementation of vanadium, a dietary micronutrient, limits cell proliferation and inhibits the formations of 8-hydroxy-20-deoxyguanosines and DNA strand-breaks in the liver of sprague-dawley rats. *Nutrition and Cancer* 59:228–247.

Chandra, AK, Ghosh, R, Chatterjee, A, Sarkar, M. 2007a. Effects of vanadate on male rat reproductive tract histology, oxidative stress markers and androgenic enzyme activities. *Journal of Inorganic Biochemistry* 101 (6):944–956.

Chandra, AK, Ghosh, R, Chatterjee, A, Sarkar, M. 2007b. Amelioration of vanadium-induced testicular toxicity and adrenocortical hyperactivity by vitamin E acetate in rats. *Molecular and Cellular Biochemistry* 306 (1–2):189–200.

Chandra, AK, Ghosh, R, Chatterjee, A, Sarkar,M 2007c. Vanadium-induced testicular toxicity and its prevention by oral supplementation of zinc sulphate. *Toxicology Mechanisms and Methods* 17 (4):175–187.

Chasteen, ND, Lord, EM, Thompson, HJ, and Grady, JK. 1986. Vanadium complexes of transferrin and ferritin in the rat. *Biochimica et Biophysica Acta -General Subjects* 884 (1):84–92.

Chen, Y, Cheng, L, Dong, Z, Chao, Y, Lei, H, Zhao, H, Wang, J, Liu, Z. 2017. Degradable vanadium disulfide nanostructures with unique optical and magnetic functions for cancer theranostics. *Angewandte Chemie International Edition* 56 (42):12991–12996.

Collauto, A, Mishra, S, Litvinov, A, Mchaourab, HS, Goldfarb, D. 2017. Direct spectroscopic detection of ATP turnover reveals mechanistic divergence of ABC exporters. *Structure* 25 (8):1264–1274.

Correia, I, Chorna, I, Cavaco, I, Roy, S, Kuznetsov, ML, Ribeiro, N, Justino, G, Marques, F, Santos-Silva, T, and Santos, MFA 2017. Interaction of [VIVO (acac) 2] with human serum transferrin and albumin. *Chemistry: An Asian Journal* 12 (16):2062–2084.

Cortijo, J, Villagrasa, V, Martí-Cabrera, M, Villar, V, Moreau, J, Advenier, C, Morcillo, EJ, and Small, RC 1997. The spasmogenic effects of vanadate in human isolated bronchus. *British Journal of Pharmacology* 121 (7):1339–1349.

Costa Pessoa, J and Tomaz, I.2010. Transport of therapeutic vanadium and ruthenium complexes by blood plasma components. *Current Medicinal Chemistry* 17 (31):3701–3738.

Crans, DC, Smee, JJ, Gaidamauskas, E, and Yang, L 2004. The chemistry and biochemistry of vanadium and the biological activities exerted by vanadium compounds. *Chemical Reviews* 104 (2):849–902.

Cusi, K, Cukier, S, DeFronzo, RA, Torres, MF, Puchulu, M, Pereira Redondo, JC 2001. Vanadyl Sulfate Improves Hepatic and Muscle Insulin Sensitivity in Type 2 Diabetes. *The Journal of Clinical Endocrinology & Metabolism* 86 (3):1410–1417.

Dai, S, Thompson, KH, Vera, E, McNeill, JH 1994. Toxicity studies on one-year treatment of non-diabetic and streptozotocin-diabetic rats with vanadyl sulphate. *Pharmacology and Toxicology* 75 (3–4):265–273.

Davies, DJA, Bennett, BG 1983. Exposure commitment assessments of environmental pollutants. *London University of London Monitoring Assessment, and Research Centre* 3 (30). https://www.scirp.org/ (S(35l jmbntvnsjt1aadkposzje))/reference/ReferencesPapers.aspx?ReferenceID=1223893

Djordjevitz, C, Wampler, GL 1985. Antitumor activity of peroxoheteroligand vanadates (V) in relation to biochemistry of vanadium. *J Inorg Biochem* 25(1):51–55.

Domingo, JL 1996. Vanadium: a review of the reproductive and developmental toxicity. *Reproductive Toxicology* 10 (3):175–182.

Domingo, JL, Gomez, M, Sanchez, DJ, Llobet, JM, and Keen, CL.1994. Relationship between reduction in food intake and amelioration of hyperglycemia by oral vanadate in STZ-induced diabetic rats. *Diabetes* 43 (10):1267–1267.

Dutton, WF 1911. Vanadiumism. *Journal of the American Medical Association* 56 (22):1648–1648.

Ehrlich, VA, Nersesyan, AK, Atefie, K, Hoelzl, C, Ferk, F, Bichler, J, Valic, E, Schaffer, A, Schulte-Hermann, R, and Fenech, M 2009. Inhalative exposure to vanadium pentoxide causes DNA damage in workers: results of a multiple end point study (vol. 116, pg 1689, 2008). *Environmental Health Perspectives* 117 (1):A15–A15.

Eisenberg, DM, Davis, RB, Ettner, SL, Appel, S, Wilkey, S, Van Rompay, M, and Kessler, RC 1998. Trends in alternative medicine use in the United States, 1990–1997: results of a follow-up national survey. *JAMA* 280 (18):1569–1575.

Ebru, Ö, Ayşe, K, Karaboğa, A, Alim, HD, Mükerrem BY 2018. Real-time Analysis of Impedance Alterations by the Effects of Vanadium Pentoxide on Several Carcinoma Cell Lines. *Turk J Pharm Sci* 15(1):1–6.

Evangelou, AM 2002. Vanadium in cancer treatment. *Critical Reviews in Oncology/Hematology* 42 (3):249–265.

Fallahi, P, Foddis, R, Elia, G, Ragusa, F, Patrizio, A, Guglielmi, G, Frenzilli, G, Benvenga, S, Cristaudo, A, and Antonelli, A 2018. Induction of Th1 chemokine secretion in dermal fibroblasts by vanadium pentoxide. *Molecular Medicine Reports* 17 (5):6914–6918.

Fawcett, JP, Farquhar, SJ, Thou, T, Shand, BI 1997Oral vanadyl sulphate does not affect blood cells, viscosity or biochemistry in humans. *Pharmacology and Toxicology* 80 (4):202–206.

Folarin, OR, Snyder, AM, Peters, DG, Olopade, F, Connor, JR and Olopade, JO 2017. Brain metal distribution and neuro-inflammatory profiles after chronic vanadium administration and withdrawal in mice. *Frontiers in Neuroanatomy* 11: 58.

Fortoul, TI, Rodriguez-Lara, V, Gonzalez-Villalva, A, Rojas-Lemus, M, Cano-Gutierrez, G, Ustarroz-Cano, M, Colin-Barenque, L, Montano, LF, García-Pelez, I, and Bizarro-Nevares, P 2011. Vanadium inhalation in a mouse model for the understanding of air-suspended particle systemic repercussion. *J Biomed Biotechnol.*. 2011:951043. doi:10.1155/2011/951043

French, RJ and Jones, PJH 1993. Role of vanadium in nutrition: metabolism, essentiality and dietary considerations. *Life Sciences* 52 (4):339–346.

Fukui, K, Ueki, T, Ohya, H, and Michibata, HJ 2003. Vanadium-binding protein in a vanadium-rich Ascidian ascidia s ydneiensis s amea: CW and pulsed EPR studies. *Journal of the American Chemical Society* 125 (21):6352–6353.

Galdino, S, Lins, M, Costa Dantas, C, and Van Grieken, RJ 1987. Radio-isotope neutron activation analysis for vanadium, manganese and tungsten in alloy steels. *Analytica Chimica Acta* 196: 337–343.

Garcia, GB, Biancardi, ME, Dario, Quiroga, A 2005. Vanadium (V)-induced neurotoxicity in the rat central nervous system: a histo-immunohistochemical study. *Drug and Chemical Toxicology* 28 (3):329–344.

Ghosh, R, and Banik, SP 2016Dual effects of vanadium: toxicity analysis in developing therapeutic lead-ups. In *Food Toxicology*. eds DBagchi and ASwaroop, 337–354CRC Press Boca Raton.

Ghosh, P, D'Cruz, OJ, Narla, RK, Uckum FM 2000. Apoptosois-inducing vanadocene compounds against human testicular cancer. *Clin Cancer Research* 6(4):1536–1543.

Gil, J, Miralpeix, M, Carreras, J, and Bartrons, R 1988. Insulin-like effects of vanadate on glucokinase activity and fructose 2, 6-bisphosphate levels in the liver of diabetic rats. *Journal of Biological Chemistry* 263 (4):1868–1871.

Goc, AJ 2006. Biological activity of vanadium compounds. *Central European Journal of Biology* 1 (3):314–332.

Goldfine, AB, Patti, M-E, Zuberi, L, Goldstein, BJ, LeBlanc, R, Landaker, EJ, Jiang, ZY, Willsky, GR, and Kahn, CR 2000. Metabolic effects of vanadyl sulfate in humans with non—insulin-dependent diabetes mellitus: in vivo and in vitro studies. *Metabolism* 49 (3):400–410.

Goldfine, AB, Simonson, DC, Folli, F, Patti, ME, Kahn, CR 1995. Metabolic effects of sodium metavanadate in humans with insulin-dependent and noninsulin-dependent diabetes mellitus in vivo and in vitro studies. *The Journal of Clinical Endocrinology & Metabolism* 80 (11):3311–3320.

González-Villalva, A, Fortoul, TI, Avila-Costa, MR, Piñón-Zarate, G, Rodriguez-Lara, V, Martínez-Levy, G, Rojas-Lemus, M, Bizarro-Nevarez, P, Díaz-Bech, P, Mussali-Galante, P 2006. Thrombocytosis induced in mice after subacute and subchronic V_2O_5 inhalation. *Toxicology and Industrial Health* 22 (3):113–116.

Gordon, JA 1991. Use of vanadate as protein-phosphotyrosine phosphatase inhibitor. In *Methods in Enzymology*, eds AM Pyle, DW Christianson Vol 201: 477–482. Elsevier, The Netherlands.

Greim, H 2009. *The MAK-Collection for Occupational Health and Safety*, Wiley-VCH, Belgium.

Harland, BF and Harden-Williams, BA 1994. Is vanadium of human nutritional importance yet? *Journal of the American Dietetic Association* 94 (8):891–894.

Heyliger, CE, Tahiliani, AG, and McNeill, JH 1985. Effect of vanadate on elevated blood glucose and depressed cardiac performance of diabetic rats. *Science* 227 (4693):1474–1477.

Hosseini, M-J, Shaki, F, Ghazi-Khansari, M, and Pourahmad, J 2013. Toxicity of vanadium on isolated rat liver mitochondria: a new mechanistic approach. *Metallomics* 5 (2):152–166.

Ignacio, EL, María, CR, Carlos, A, Franca, BS, Parajón-Costa, Enrique, JB 2019. Metvan, *bis*(4,7-Dimethyl-1,10-phenanthroline)sulfatooxidovanadium(IV): DFT and Spectroscopic Study—Antitumor Action on Human Bone and Colorectal Cancer Cell Lines. *Biological Trace Element Research* 191(1):81–87.

Ikebe, K, Tanaka, R 1979. Determination of vanadium and nickel in marine samples by flameless and flame atomic absorption spectrophotometry. *Bulletin of Environmental Contamination and Toxicology* 21 (4/5).

Jan, K, Irena, BB, Izabela, G, Dariusz, C 2012. Biochemical and medical importance of vanadium compounds. *Acta Biochim Pol.* 59(2):195–200.

Klein, A, Holko, P, Ligeza, J, Kordowiak, AM 2008. Sodium orthovanadate affects growth of some human epithelial cancer cells (A549, HTB44, DU145). *Folia Biol (Krakow)* 56(3–4):115–121.

Kohlmeier, M 2003. *Amino Acids and Nitrogen Compounds. Nutrient Metabolism.* Academic Press, London, 840p.

Korbecki, J, Baranowska-Bosiacka, I, Gutowska, I, Chlubek, D 2012. Biochemical and medical importance of vanadium compounds. *Acta Biochimica Polonica* 59 (2):195–200.

Kowalski, S, Hac, S, Wyrzykowski, D, Zauszkiewicz-Pawlak, A, Inkielewicz-Stepniak, I 2017. Selective cytotoxicity of vanadium complexes on human pancreatic ductal adenocarcinoma cell line by inducing necroptosis, apoptosis and mitotic catastrophe process. *Oncotarget* 8 (36):60324–60341.

Kremer, LE, McLeod, AI, Aitken, JB, Levina, A, Lay, PA 2015. Vanadium (V) and s-(IV) complexes of anionic polysaccharides: controlled release pharmaceutical formulations and models of vanadium bio-transformation products. *Journal of Inorganic Biochemistry* 147:227–234.

Leon, IE, DiVirgilio, AL, Porro,V, Muglia, CI, Naso, LG, Williams, PA, Bollati-Fogolin, M, Etcheverry, SB. 2013. Anti tumor properties of a vanadyl (IV) complex with the flavonoid chrysin [VO(chrysin)₂EtOH]₂ in a human osteosarcoma model: The role of oxidative stress and apoptosis. *Dalton Transactions* 42:11868–11880.

Leon, IE, Porro, V, Di Virgilio, AL, Naso, LG, Williams, PA, Bollati-Fogolín, M, Etcheverry, SB. 2014. Antiproliferative and apoptosis-inducing activity of an oxidovanadium(IV) complex with the flavonoid silibinin against osteosarcoma cells. *Journal of Biological Inorganic Chemistry* 19(1): 59–74.

Levina, A, Lay, PA 2017. Stabilities and biological activities of vanadium drugs: what is the nature of the active species? *Chemistry: An Asian Journal* 12 (14):1692–1699.

Li, H, Zhou, D, Zhang, Q, Feng, C, Zheng, W, He, K, Lan, YJ 2013. Vanadium exposure-induced neurobehav-ioral alterations among Chinese workers. *Neurotoxicology* 36:49–54.

Liu, Y, Jie, X, Guo, Y, Zhang, X, Wang, J, Xue, C 2016. Green synthesis of oxovanadium (IV)/chitosan nanocomposites and its ameliorative effect on hyperglycemia, insulin resistance, and oxidative stress. *Biological Trace Element Research* 169 (2):310–319.

Ma, J, Pan, L, Wang, Q, Lin, C, Duan, X, Hou, H 2018. Estimation of the daily soil/dust (SD) ingestion rate of children from Gansu Province, China via hand-to-mouth contact using tracer elements. *Environmental Geochemistry and Health* 40 (1):295–301.

Mahmmoud, YA, Shattock, M, Cornelius, F, and Pavlovic, D 2014. Inhibition of K+ transport through Na+, K+-ATPase by capsazepine: role of membrane span 10 of the α-subunit in the modulation of ion gating. *PLoS One* 9 (5) e96909.

Meshkini, A, Yazdanparast, R. 2010. Chemosensitization of human leukemia K562 cells to taxol by a vanadium-salen complex. *Experimental and Molecular Pathology* 89 (3):334–342.

Morita, A, Yamamoto, S, Wang, B, Tanaka, K, Suzuki, N, Aoki, S, Ito, A, Nanao, T, Ohya, S, Yoshino, M, Zhu, J, Enomoto, A, Matsumoto, Y, Funatsu, O, Hosoi, Y, Ikekita, M 2010. Sodium orthovanadate inhibits p53-mediated apoptosis. *Cancer Research* 70(1):257–265.

Mukherjee, B, Patra, B, Mahapatra, S, Banerjee, P, Tiwari, A, and Chatterjee, M 2004. Vanadium—an element of atypical biological significance. *Toxicology Letters* 150 (2):135–143.

Murthy, MS, Rao, LN, Kuo, LY, Toney, JH, and Marks, TJ 1988. Antitumor and toxicologic properties of the organometallic anticancer agent vanadocene dichloride. *Inorganica Chimica Acta* 152 (2):117–124.

Myron, DR, Givand, SH, Nielsen, FH 1977. Vanadium content of selected foods as determined by flameless atomic absorption spectroscopy. *Journal of Agricultural and Food Chemistry* 25 (2):297–300.

Nair, RS, Kuriakose, M, Somasundaram, V, Shenoi, V, Kurup, MR, Srinivas, P. 2014. The molecular response of vanadium complexes of nicotinoyl hydrazone in cervical cancers—A possible interference with hpv oncogenic markers. *Life Science* 116 (2):90–97.

Nechay, BR 1984. Mechanisms of action of vanadium. *Annual Review of Pharmacology and Toxicology* 24 (1):501–524.

Nielsen, FH 1991. Nutritional requirements for boron, silicon, vanadium, nickel, and arsenic: current knowledge and speculation. *The FASEB Journal* 5 (12):2661–2667.

Nielsen, FH 1995a. In Handbook of Metal-ligand interactions in biological fluids, bioinorganic medicine, eds G. Berthon and M Dekkar. 1:269–272 New York.

Nielsen, FH 1995b. Vanadium and its role in life. *Metal Ions in Biological Systems* 31:543–573.

Nischwitz, V, Davies, JT, Marshall, D, González, M, Gómez, Ariza, JL, Goenaga-Infante H 2013. Speciation studies of vanadium in human liver (HepG2) cells after *in vitro* exposure to bis(maltolato) oxovanadium(IV) using HPLC online with elemental and molecular mass spectrometry. *Metallomics* 5(12):1685–1697.

Nriagu, JO. 1998. History, occurrence, and uses of vanadium. In. *Vanadium in the Environment: Chemistry, and Biochemistry* ed Nriagu, JO. 1–36, Wiley, United States.

Petanidis, S, Kioseoglou, E, Domvri, K, Zarogoulidis, P, Carthy, JM, Anestakis, D, Moustakas, A, Salifoglou, A. 2016. *In vitro* and *ex vivo* vanadium antitumor activity in (TGF-β)-induced emt. Synergistic activity with carboplatin and correlation with tumor metastasis in cancer patients. *The International Journal of Biochemistry & Cell Biology* 74:121–134.

Peters, KG, Davis, MG, Howard, BW, Pokross, M, Rastogi, V, Diven, C, Greis, KD, Eby-Wilkens, E, Maier, M, Evdokimov, A. 2003. Mechanism of insulin sensitization by BMOV (bis maltolato oxo vanadium); unliganded vanadium (VO4) as the active component. *Journal of Inorganic Biochemistry* 96 (2–3):321–330.

Ramanadham, S, Mongold, JJ, Brownsey, RW, Cros, GH, McNeill, JH. 1989. Oral vanadyl sulfate in treatment of diabetes mellitus in rats. *American Journal of Physiology-Heart and Circulatory Physiology* 257 (3):H904–H911.

Rehder, D. 2008. *Bioinorganic Vanadium Chemistry.* Vol. 30, John Wiley & Sons, Germany.

Rehder, D. 2013. Vanadium. Its role for humans. In *Interrelations between Essential Metal Ions and Human Diseases*, eds Sigel A., Sigel H., Sigel R. 139–169. Springer, Dordrecht.

Rehder, D. 2015. The role of vanadium in biology. *Metallomics* 7 (5):730–742.

Ress, NB, Chou, BJ, Renne, RA, Dill, JA, Miller, RA, Roycroft, JH, Hailey, JR, Haseman, JK, Bucher, JR. 2003. Carcinogenicity of inhaled vanadium pentoxide in F344/N rats and B6C3F1 mice. *Toxicological Sciences* 74 (2):287–296.

Rondini, EA, Walters, DM, Bauer, AK. 2010. Vanadium pentoxide induces pulmonary inflammation and tumor promotion in a strain-dependent manner. *Particle and Fibre Toxicology* 7 (1):9.

Roshchin, AV, Ordzhonikidze, EK, Shalganova, IV. 1980. Vanadium–toxicity, metabolism, carrier state. *Journal of Hygiene, Epidemiology, Microbiology, and Immunology* 24 (4):377–383.

Rozzo, C, Sanna, D, Garribba, E, Serra, M, Cantara, A, Palmieri, G, Pisano, M. 2017. Antitumoral effect of vanadium compounds in malignant melanoma cell lines. *Journal of Inorganic Biochemistry* 174:14–24.

Russo, R, Sciacca, S, La Milia, DI, Poscia, A, Moscato, U. 2014. Vanadium in drinking water: toxic or therapeutic?! Systematic literature review and analysis of the population exposure in an Italian volcanic regionRoberta Russo. *European Journal of Public Health* 24 (suppl_2) doi:10.1093/eurpub/cku162.080.

Sabbioni, E, Pozzi, G, Pintar, A, Casella, L, Garattini, S. 1991. Cellular retention, cytotoxicity and morphological transformation by vanadium (IV) and vanadium (V) in BALB/3T3 cell lines. *Carcinogenesis* 12 (1):47–52.

Sakurai, H, Tamura, H, Okatani, K. 1995. Mechanism for a new vanadium complex: hydroxyl radical-dependent DNA cleavage by 1,10-phenanthroline-vanadyl complex in the presence of hydrogen peroxide. *Biochemical and Biophysical Research Communications* 206(1):133–137.

Sakurai, H, Tsuchiya, K, Nukatsuka, M, Sofue, M, Kawada, J. 1990. Insulin-like effect of vanadyl ion on streptozotocin-induced diabetic rats. *Journal of Endocrinology* 126 (3):451–459.

Sanna, D, Micera, G, Garribba, E. 2009. On the transport of vanadium in blood serum. *Inorganic Chemistry* 48 (13):5747–5757.

Sanna, D, Ugone, V, Micera, G, Buglyó, P, Bíró, L, Garribba, E. 2017. Speciation in human blood of Metvan, a vanadium based potential anti-tumor drug. *Dalton Transactions* 46 (28):8950–8967.

Sato, K, Kusaka, Y, Akino, H, Kanamaru, H, Okada, K. 2002. Direct effect of vanadium on citrate uptake by rat renal brush border membrane vesicles (BBMV). *Industrial Health* 40 (3):278–281.

Sax, Newton Irving. 1989. Dangerous properties of industrial materials. 7th Edition. New York: Van Nostrand Reinhold.

Scalese, G, Mosquillo, MF, Rostán, S, Castiglioni, J, Alho, I, Pérez, L, Correia, I, Marques, F, Costa Pessoa, J, Gambino, D. 2017. Heterolepticoxidovanadium(IV) complexes of 2-hydroxynaphtylaldimine and polypyridyl ligands against trypanosomacruzi and prostate cancer cells. *Journal of Inorganic Biochemistry* 175:154–166.

Ścibior, A, Gołębiowska, D, Adamczyk, A, Niedźwiecka, I, Fornal, E. 2014. The renal effects of vanadate exposure: potential biomarkers and oxidative stress as a mechanism of functional renal disorders—preliminary studies. *BioMed Research International* 2014. Article ID 740105 ll5 pages. doi:10.1155/2014/740105.

Ścibior, A, Zaporowska, H, Ostrowski, J. 2006. Selected haematological and biochemical parameters of blood in rats after subchronic administration of vanadium and/or magnesium in drinking water. *Archives of Environmental Contamination and Toxicology* 51 (2):287–295.

Scior, T, Mack, HG, García, JA, Koch, W 2009. Antidiabetic Bis-Maltolato-OxoVanadium(IV): conversion of inactive trans- to bioactive cis-BMOV for possible binding to target PTP-1B. *Drug Des Devel Ther* 2:221–231.

Scrivens, PJ, Alaoui-Jamali, MA, Giannini, G, Wang, T, Loignon, M, Batist, G, Sandor, VA 2003. Cdc25A-inhibitory properties and antineoplastic activity of bisperoxovanadium analogues. *Molecular Cancer Therapeutics* 2 (10):1053–1059.

Sharma, B, Singh, S, Siddiqi, N 2014. Biomedical implications of heavy metals induced imbalances in redox systems. *BioMed Research International* 2014: 640754. doi:10.1155/2014/640754

Sinha, A, Banerjee, K, Banerjee, A, Sarkar, A, Ahir, M, Adhikary, A, Chatterjee, M, Choudhuri, SK 2017. Induction of apoptosis in human colorectal cancer cell line, HCT-116 by a vanadium- schiff base complex. *Biomedicine & Pharmacotherapy* 92:509–518.

Smith, CA, Ainscough, EW, Brodie, AM 1995. Complexes of human lactoferrin with vanadium in oxidation states+ 3,+ 4 and+ 5. *Journal of the Chemical Society Dalton Transactions* 7:1121–1126.

Srivastava, AK 2000. Anti-diabetic and toxic effects of vanadium compounds. *Molecular and Cellular Biochemistry* 206 (1–2):177–182.

Stemmler, AJ Burrows, CJ 2001. Guanine versus deoxyribose damage in DNA oxidation mediated by vanadium (IV) and vanadium (V) complexes. *Journal of Biological Inorganic Chemistry* 6 (1):100–106.

Stern, A, Yin, X, Tsang, S-S, Davison, A, Moon, J 1993. Vanadium as a modulator of cellular regulatory cascades and oncogene expression. *Biochemistry and Cell Biology* 71 (3–4):103–112.

Strianese, M, Basile, A, Mazzone, A, Morello, S, Turco, MC, Pellecchia, C 2013. Therapeutic potential of a pyridoxal-based vanadium (IV) complex showing selective cytotoxicity for cancer versus healthy cells. *Journal of Cellular Physiology* 228 (11):2202–2209.

Substances and Disease Registry 1992. *Toxicological Profile for Vanadium*. US Department of Health and Human Services, Public Health Service, Atlanta.

Sun, L, Wang, K, Li, Y, Fan, Q, Zheng, W, Li, H 2017. Vanadium exposure-induced striatal learning and memory alterations in rats. *Neurotoxicology* 62:124–129.

Thompson, KH, Lichter, J, LeBel, C, Scaife, MC, McNeill, JH, Orvig, CJ. 2009. Vanadium treatment of type 2 diabetes: a view to the future. *Inorganic Biochemistry* 103 (4):554–558.

Thompson, KH, Orvig, CJ.2006. Vanadium in diabetes: 100 years from Phase 0 to Phase I. *Journal of Inorganic Biochemistry* 100 (12):1925–1935.

Treviño, S, Díaz, A, Sánchez-Lara, E, Sanchez-Gaytan, BL, Perez-Aguilar, JM, González-Vergara, E. (2019). Vanadium in biological action: Chemical, pharmacological aspects, and metabolic implications in diabetes mellitus. *Biological Trace Element Research* 188 (1):68–98. doi:10.1007/s12011-018-1540-6.

US Department of Health and Human Services 2012. *Toxicological Profile for Vanadium*. Public Health Service Agency for Toxic Substances, and Disease Registry. Atlanta, Georgia.

Uthus, EO, Nielsen, FH 1990. Effect of vanadium, iodine and their interaction on growth, blood variables, liver trace elements and thyroid status indices in rats. *Magnes Trace Elem.* 4:219–226.

Vanadium, 1988. *Environmental Health Criteria No. 81*. World Health Organization, Geneva.

Vanadium, 2000. *Chapter 6.12.9*. WHO Regional Office for Europe, Copenhagen, Denmark.

Vinogradov, AP. 1959. *The Geochemistry of Rare and Dispersed Elements in Soils*. Geochemical Society by Consultants Bureau, USA.

Vouk, VB. 1979. *Handbook on the Toxicology of Metals*. Elsevier-North-Holland Biomedical Press, The Netherlands.

Wang, L, Medan, D, Mercer, R, Overmiller, D, Leornard, S, Castranova, V, Shi, X, Ding, M, Huang, C, Rojanasakul, Y 2003. Vanadium-induced apoptosis and pulmonary inflammation in mice: role of reactive oxygen species. *Journal of Cellular Physiology* 195 (1):99–107.

Wei, TD, Li, SP, Liu, YX, Tan, CP, Li, J, Zhang, ZH, Lan, YJ, Zhang, Y 2015. Oxidative stress level of vanadium-exposed workers. *Journal of Sichuan University: Medical Science Edition* 46 (6):856–859.

Wiegmann, TB, Day, HD, Patak, RV 1982. Intestinal absorption and secretion of radioactive vanadium ($^{48}VO^{-3}$) in rats and effect of Al (OH)$_3$. *Journal of Toxicology: Part A Current Issues Environmental Health* 10 (2):233–245.

Wilk, A, Szypulska-Koziarska, D, Wiszniewska, B, 2017. The toxicity of vanadium on gastrointestinal, urinary and reproductive system, and its influence on fertility and fetuses malformations. *Advances in Hygiene and Experimental Medicine* 71: 850–859.

Winter, JM, Moore, BS 2009. Exploring the chemistry and biology of vanadium-dependent haloperoxidases. *Journal of Biological Chemistry* 284 (28):18577–18581.

Woodin, MA, Liu, Y, Neuberg, D, Hauser, R, Smith, TJ, Christiani, DC.2000. Acute respiratory symptoms in workers exposed to vanadium-rich fuel-oil ash. *American Journal of Industrial Medicine* 37 (4):353–363.

Wörle-Knirsch, JM, Kern, K, Schleh, C, Adelhelm, C, Feldmann, C, Krug, HF 2007. Nanoparticulate vanadium oxide potentiated vanadium toxicity in human lung cells. *Environmental Science and Technology* 41 (1):331–336.

Wozniak, K and Blasiak, J 2004. Vanadyl sulfate can differentially damage DNA in human lymphocytes and HeLa cells. *Archives of Toxicology* 78 (1):7–15.

Wu, P, Nielsen, TE, Clausen, MH 2016. Small-molecule kinase inhibitors: an analysis of FDA-approved drugs. *Drug Discovery Today* 21 (1):5–10.

Yanardag, R, Bolkent, S, Karabulut-Bulan, Ö, Tunali, S 2003. Effects of vanadyl sulfate on kidney in experimental diabetes. *Biological Trace Element Research* 95 (1):73–85.

Yang, J, Teng,Y,Wu,J,Chen, H,Wang,G,Song,L,Yue,W,Zuo,R, Zhai, Y.2017. Current status and associated human health risk of vanadium in soil in China. *Chemosphere* 171:635–643.

Yang, X-G, Yang, X-D, Yuan, L, Wang, K, Crans, DC 2004. The permeability and cytotoxicity of insulin-mimetic vanadium compounds. *Pharmaceutical Research* 21 (6):1026–1033.

Yu, D, Walters, DM, Zhu, L, Lee, P-K, Chen, Y 2011. Vanadium pentoxide (V_2O_5) induced mucin production by airway epithelium. *American Journal of Physiology: Lung Cellular and Molecular Physiology* 301 (1):L31–L39.

Zaporowska, H, Wasilewski, W 1992. Haematological effects of vanadium on living organisms. *Comparative Biochemistry and Physiology Part C: Comparative Pharmacology* 102 (2):223–231.

Zaporowska, H, Wasilewski, W, Słotwińska, M 1993. Effect of chronic vanadium administration in drinking water to rats. *Biometals* 6 (1):3–10.

Zhang, Z-Y 2017. Drugging the undruggable: therapeutic potential of targeting protein tyrosine phosphatases. *Accounts of Chemical Research* 50 (1):122–129.

Zhu, CW, Liu, YX, Huang, CJ, Gao, W, Hu, GL, Li, J, Zhang, Q, Lan, YJ 2016. Effect of vanadium exposure on neurobehavioral function in workers. *Chinese Journal of Industrial Hygiene and Occupational Diseases* 34 (2):103–106.

Zwolak, I 2020. Protective effects of dietary antioxidants against vanadium-induced toxicity: a review. *Oxidative Medicine and Cellular Longevity* 2020: Article ID 1490316 14 pages. doi:10.1155/2020/1490316

CHAPTER 20

The Ubiquitous Lead—Biological Effects, Toxicity, and Management

Sreejayan Nair
University of Wyoming

Debasis Bagchi
University of Houston College of Pharmacy

CONTENTS

20.1 INTRODUCTION

Lead is a naturally occurring mineral, and its use dates back to 10000 BC or earlier (Rich, 2014). Lead is purified from the extraction of lead ore (Galena, Plumbs, lead sulfide). Its softness, malleability, and resistance to corrosion had led to its wide use in a variety of products over the years, including jewelry, toys, vessels, pipes, glass-wares, solders, ammunitions, radiation-shield, cable, paint, pigments, cosmetics, and batteries. In the last century, tetraethyl lead was added to gasoline as a source of octane to prevent engine knockout. This practice was phased out in the mid-1970s following the U.S. Environmental Protection Agency's directive, which was based on the health concerns consequent to lead toxicity.

Being a ubiquitous metal and an environmental toxin, the majority of the population has some levels of lead in their blood. Recognizing the adverse effects of lead to humans, during recent years the regulatory agencies have reduced the reference values of lead in the blood. However, it should be noted that there are no acceptable blood levels of lead. The reference levels indicate the need for monitoring and or intervention of those subjects and do not suggest any "safe levels" of lead. For chronic exposure, however, measuring blood lead levels is futile as lead accumulates in the bones and soft tissues for decades.

Lead can cause a variety of biological effects. It can cause delayed memory loss, reduced IQ, and learning disabilities by affecting the brain. Lead also adversely affects the cardiovascular and hematopoietic system. Lead readily crosses the placental barrier, and studies have demonstrated that prenatal exposure to lead can result in learning disabilities in children. While it is a challenge to remove lead from the body following chronic exposure, acute lead toxicity can be treated with chelating agents, including ethylene-diamine tetra acetic acid.

20.2 BLOOD LEAD LEVELS

In 2012, the Centers for Disease Control and Protection (CDC) reduced the blood lead reference value for children to 5 µg/dL. Before that, blood lead levels of 10 µg/dL were considered elevated. Chelation therapy is indicated if the levels are greater than or equal to 45 µg/dL. In 2015, the National Institute for Occupational Safety and Health CDC's Adult Blood Lead Epidemiology and Surveillance (ABLES) program redefined elevated blood levels as 5 µg/dLfrom the previous 10 µg/dL for adults (2018). Additionally, it recommends the removal of subjects with a blood lead level of over 20 µg/dL from working around lead to reduce further exposure. However, as lead accumulates in bones and other body tissues, measuring lead blood levels will not be an accurate determinant of the total body burden of lead.

20.3 ABSORPTION, STORAGE, AND EXCRETION OF LEAD

Lead can be absorbed from the gastrointestinal tract, lungs (from lead dust), or to a very small extent from the skin. The lungs are the major site of absorption of lead—it can readily reach the systemic circulation via absorption from the bronchiolar epithelium. While organic lead is metabolized in the body, inorganic lead does not undergo biotransformation and is excreted unchanged by the urine (primarily) and through the biliary system into the feces. The biological half-life of lead in the adult human blood is approximately 30 days. About 90% of the lead in the blood is bound to the erythrocytes, and the remaining 1% is in the plasma. From the blood, lead is distributed to the body tissues, where it can accumulate for decades. About 95% of the total body stores of lead are found in the mineralized tissues, the bones, and teeth (Barry, 1975, 1981). The remaining 5% of lead is stored in the soft tissues, including liver, kidneys, lungs, muscle, and spleen, the liver being the primary site (Gerhardsson, Brune, Nordberg & Wester, 1986). In children, lead accumulates predominantly in the spongier, metabolically active trabecular bones. The lead is retained in the bones for decades and released from its body stores during bone remodeling, trauma, pregnancy, lactation, and in conditions where there is excessive bone resorption such as in case of osteoporosis. Lead can traverse the placental barrier and can cause toxicity in the fetus.

It has been proposed that lead competes with calcium for absorption from the gastrointestinal tract (Barltrop & Khoo, 1976). The competition with calcium for its absorption could also explain the deposition of lead onto osseous tissues such as the bones and the teeth. Animal studies have shown that feeding rodents with a low calcium diet can increase their susceptibility to lead toxicity (Six & Goyer, 1970). More recently, Rădulescu and Lundgren applied non-linear kinetic models to study the lead absorption into the brain tissues and observed that the effects of lead and calcium ingestion on the brain is rather specific to the brain and cannot be inferred from their levels or from their ratio within any of the other compartments (Radulescu & Lundgren, 2019). Their studies also suggest that the competition between calcium and lead transport does not mean that low lead levels in any tissue compartment can be associated with high calcium in that compartment or vice versa.

20.4 BIOLOGICAL EFFECTS OF LEAD

20.4.1 Central Nervous System

There is strong evidence to suggest that exposure to lead results in cognitive and neurobehavioral impairments in humans (Lanphear, Dietrich, Auinger & Cox, 2000; Needleman, Schell, Bellinger, Leviton & Allred, 1990; Toews, Kolber, Hayward, Krigman & Morell, 1978; Winneke, Lilienthal & Kramer, 1996). Elevated blood lead levels have been associated with lower IQ and cognition, and children were found to be at higher risk than adults. Studies by Canfield and coworkers demonstrated that blood lead level concentrations below 10 µg/dL are inversely and significantly associated with IQ scores in children (Canfield, Henderson, Cory-Slechta, Cox, Jusko & Lanphear, 2003). Using a nonlinear model, these authors postulate that IQ declined by 7.4 points as lifetime average blood lead concentrations increased from 1 to 10µg/dL. Furthermore, they show that declines in IQ are greater at concentrations lower than 10 µg/dL than at higher concentrations. Ruben and coworkers, in a recently published study, observed 565 New Zealanders for four decades and concluded that lead exposure during childhood was significantly associated with their lower cognitive function and socioeconomic status at the age of 38 years (Reuben et al., 2017).

Occupational exposure to lead can also result in impaired memory in adults. Khalil and coworkers evaluated a cohort of exposed and non-exposed workers approximately 22 years after they have been originally assessed for cognitive function using the Pittsburgh Occupational Exposures Test battery (Khalil, Morrow, Needleman, Talbott, Wilson & Cauley, 2009). They measured both the blood levels and tibia levels of lead and determined their cognitive function using a similar testing method and found the lead-exposed worker had lower cognitive performance and exhibited cognitive decline over 22 years. They also found a correlation between bone lead levels and cognitive scores in subjects 55 years of age or older. These studies suggest that cognitive impairment is more prevalent in older subjects who have been exposed to lead. Stewart and Schwartz performed a 15-year longitudinal study in U.S. organo-lead manufacturing workers (who were previously exposed to both inorganic and tetraethyl lead), inorganic lead workers in Korea, and Baltimore residents with environmental exposure to inorganic environmental lead (Stewart & Schwartz, 2007). These authors found that higher tibia lead was inversely correlated to cognitive outcome. These authors conclude that "a significant proportion of what is considered to be 'normal' age-related cognitive decline may, in fact, be due to past exposure to neurotoxicants such as lead." In addition to lower visual memory scores, exposure to lead resulted in the decline in both verbal and nonverbal memories and impaired visuospatial skills (Khalil, Morrow, Needleman, Talbott, Wilson & Cauley, 2009; Schwartz et al., 2000). Campbell and coworkers have reported that increased bone levels can be correlated to impaired language processing in children (Campbell, Needleman, Riess & Tobin, 2000).

In a recently published study, Lanphear and coworkers sought to examine the relationship between low blood lead concentrations <10 µg/dL with performance on tests of cognitive function (including arithmetic skills, reading skills, nonverbal reasoning, and short-term memory) in 4,853 children aged 6–16 years using the data from the Third National Health and Nutrition Examination Survey (NHANES III) (Lanphear, Dietrich, Auinger & Cox, 2000). After adjusting for confounding variables, the data obtained in this study showed an inverse relationship between blood lead concentration and the cognitive scores. It is of note that the geometric mean blood lead concentrations of the children in this study were 1.9µg/dL, indicating that deficits in cognitive and academic skills are associated with a blood lead concentration much lower than 5µg/dL.

20.4.2 Cardiovascular Disease

Cardiovascular disease is the leading cause of morbidity and mortality in the developed world, and the potential role of lead toxicity in the pathophysiology of cardiovascular disease has been explored (Navas-Acien, Guallar, Silbergeld & Rothenberg, 2007). Early studies by Hu and coworkers suggested a correlation between bone and blood lead levels of hypertension (Hu et al., 1996). Nash and coworkers studied lead levels in post- and peri-menopausal women and found a positive correlation between blood lead levels and both systolic and diastolic hypertension (Nash et al., 2003). Martin and coworkers conducted a cross-sectional analysis of 964 men and women aged 50–70 years and found a strong positive correlation of blood lead levels with both systolic and diastolic blood pressure and hypertension, after normalizing for potential confounding variables (Martin, Glass, Bandeen-Roche, Todd, Shi & Schwartz, 2006). Elevated bone lead levels were correlated with significant increases in QT and QRS intervals on the echocardiogram (Chen, Yen, Lo, Chu, Chiu & Chuang, 2013; Cheng, Schwartz, Vokonas, Weiss, Aro & Hu, 1998).

Obeng-Gyasi and coworkers hypothesized that lead exposure causes adverse cardiovascular function by altering blood pressure, inflammation, and lipid profile. To this end, these authors analyzed systolic and diastolic blood pressure, C-reactive proteins, and serum lipid levels. They used the data from the National Health and Nutrition Examination Survey (NHANES) 2007–2010 to examine the association between lead levels and the aforementioned endpoints. This study included adult subjects 20 years and above, who were classified into four quartiles based on the lead exposure—0–2, 2–5, 5–10, and 10 µg/dL. These authors found a positive association between lead exposure levels and systolic and diastolic blood pressure, C-reactive protein, triglycerides, low-density lipoprotein, cholesterol, and high-density lipoprotein cholesterol, showed differences between populations in the exposed and less-exposed occupations suggesting a profound effect of lead exposure on the cardiovascular system. In another recent study using nationally representative study samples of adults aged 20 years or older who were enrolled in NHANES-II study between 1988 and 1994 and followed up to 2011, which included over 14,000 adults, estimate that over 2,50,000 premature deaths from cardiovascular disease (Lanphear, Rauch, Auinger, Allen & Hornung, 2018).

20.4.3 Other Systems

Lead exposure has been shown to be associated with renal insufficiency, chronic kidney disease, nephropathy, loss of proximal tubules, and interstitial fibrosis (Benjelloun, Tarrass, Hachim, Medkouri, Benghanem & Ramdani, 2007; Wedeen et al., 1975). Toxic levels of lead have been associated with infertility, miscarriages, and developmental disabilities in children (Park et al., 2008). In the hematopoietic system, lead interferes with the activity of delta-aminolevulinic acid dehydratase, an enzyme that plays a critical role in the biosynthesis of heme, the cofactor found in hemoglobin, which can lead to anemia (Patrick, 2006).

20.5 MOLECULAR MECHANISMS LEADING TO LEAD TOXICITY

Molecular oxygen is a biradical, as its outermost shell contains two unpaired electrons in different orbitals. Because these two electrons are of similar spin and are in two different shells, the oxygen molecule is largely resistant to reduction. Any agent that reduces oxygen should provide a single electron in the opposite spin. This spin and orbital restriction can be countered by transition metal ions such as iron that can undergo redox reaction. A single electron reduction of molecular oxygen thus leads to the production of singlet oxygen. Singlet oxygen can accept another electron to form hydrogen peroxide, and hydrogen peroxide undergoes the classical Fenton reaction to produce hydroxyl radical. Singlet oxygen, hydrogen peroxide, and hydroxyl radicals (collectively referred to as reactive oxygen species) can oxidize biomolecules (proteins and carbohydrates), peroxidize lipid molecules (lipid peroxidation), crosslink DNA, and cause proinflammatory gene expression. Collectively this event is called "oxidative stress."

The human body has evolved mechanisms to counter oxidative stress. These mechanisms include sequestering transition metal ions with enzymes such as metallothionine, ceruloplasmin, and transferrin; scavenging oxygen free radicals with superoxide dismutase; and hydrogen peroxide by catalase. Endogenous antioxidants such as glutathione provide additional protection as a redox equivalent to maintain the redox balance in the cell. Glutathione in the cells is predominantly in its reduced (–SH) form. When it reacts with oxygen, it gets converted to the oxidized, disulfide form (S–S). The disulfide form is reduced back to the sulfydryl form by the glutathione reductase.

Studies have shown that lead toxicity is due to increased generation of reactive oxygen species (Flora, Gupta & Tiwari, 2012; Lopes, Peixe, Mesas & Paoliello, 2016). Lead primarily binds to the sulfydryl group of glutathione and inactivates the enzyme, which overwhelms the capacity of glutathione reductase. Lack of reduced glutathione results in the loss of a sink for reactive oxygen species, tipping the oxidative stress antioxidant balance toward oxidative stress. Studies have shown that lead also inactivates superoxide dismutase and catalase. This collectively leads to oxidative damage and lipid peroxidation. Oxidation of red blood cells, together with the inhibition of 5-aminolevulinic acid hydratase by lead, results in red blood cell hemolysis (Flora, Gupta & Tiwari, 2012).

In addition to these molecular mechanisms, as alluded to earlier, lead competes with calcium for absorption into various intracellular compartments and thereby interferes with the actions of calcium in several intracellular regulatory events. Studies have shown that lead activates the calcium-binding protein calmodulin which can affect a number of intracellular functions regulated by calmodulin, including the regulation of the second messenger protein kinase C (Goldstein, 1993). Other studies have suggested that lead levels can interfere with a number of neuronal functions including neurotransmitter release in a wide variety of neurons such as dopaminergic, GABergic, cholinergic, and N-methyl-D-aspartate (NMDA) neurons (Basha & Reddy, 2015; Duan, Peng, Shi & Jiang, 2017; Neal, Worley & Guilarte, 2011; NourEddine, Miloud & Abdelkader, 2005).

20.6 MANAGEMENT OF LEAD TOXICITY

Every patient with a blood level above 10 µg/dL should receive education about environmental lead exposure and should undergo follow-up monitoring. Chelation therapy for acute lead exposure is indicated for adult patients with a blood lead level above 80 µg/dL and for patients with levels above 50 µg/dL if they were to exhibit symptoms of lead toxicity. The commonly used chelating agents for adults are 2,3-dimercaptosuccinic acid (DMSA) succimer and calcium disodium ethylenediaminetetraacetic acid (EDTA). Dimercaprol and D-penicillamine are other options. Wherever possible, eliminating the source of lead exposure is more effective than using chelation therapy.

20.7 CONCLUSION

Lead is a ubiquitous heavy metal that can have a number of deleterious effects on the human body. Because it is retained in the bones for over decades, lead can exhibit chronic toxicity. With aggressive public health initiatives to reduce lead in the environment, the prevalence of lead toxicity continues to decline.

REFERENCES

Agency for Toxic Substances and Disease Registry (ATSDR). *Lead Toxicity: What is the Biological Fate of Lead in the Body?* [Online] Available from https://www.atsdr.cdc.gov/csem/csem.asp?csem=34&po=9.

Centers for Disease Control and Prevention. *CDC Response to Advisory Committee on Childhood Lead Poisoning Prevention Recommendations in "Low Level Lead Exposure Harms Children: A Renewed Call of Primary Prevention"*. [Online] Available from https://www.cdc.gov/nceh/lead/acclpp/CDC_Response_Lead_Exposure_Recs.pdf. [Accessed: 4/1/2020].

Centers for Disease Control and Prevention. *Adult Blood Lead Epidemiology and Surveillance (ABLES)*. [Online] Available from https://www.cdc.gov/niosh/topics/ables/description.html. [Accessed: 4/3/2020].

Barltrop D, & Khoo HE (1976). The influence of dietary minerals and fat on the absorption of lead. *Sci Total Environ* 6: 265–273.

Barry PS (1975). A comparison of concentrations of lead in human tissues. *Occup Environ Med* 32: 119–139.

Barry PS (1981). Concentrations of lead in the tissues of children. *Occup Environ Med* 38: 61–71.

Basha CD, & Reddy RG (2015). Long-term changes in brain cholinergic system and behavior in rats following gestational exposure to lead: protective effect of calcium supplement. *Interdisciplinary toxicology* 8: 159–168.

Benjelloun M, Tarrass F, Hachim K, Medkouri G, Benghanem MG, & Ramdani B (2007). Chronic lead poisoning: a "forgotten" cause of renal disease. *Saudi J Kidney Dis Transpl* 18: 83–86.

Campbell TF, Needleman HL, Riess JA, & Tobin MJ (2000). Bone lead levels and language processing performance. *Dev Neuropsychol* 18: 171–186.

Canfield RL, Henderson CR, Jr., Cory-Slechta DA, Cox C, Jusko TA, & Lanphear BP (2003). Intellectual impairment in children with blood lead concentrations below 10 microg per deciliter. *N Engl J Med* 348: 1517–1526.

Chen CC, Yen HW, Lo YH, Chu YH, Chiu YW, & Chuang HY (2013). The association of prolonged QT interval on electrocardiography and chronic lead exposure. *J Occup Environ Med* 55: 614–619.

Cheng Y, Schwartz J, Vokonas PS, Weiss ST, Aro A, & Hu H (1998). Electrocardiographic conduction disturbances in association with low-level lead exposure (the Normative Aging Study). *Am J Cardiol* 82: 594–599.

Duan Y, Peng L, Shi H, & Jiang Y (2017). The effects of lead on GABAergic interneurons in rodents. *Toxicol Ind Health* 33: 867–875.

Flora G, Gupta D, & Tiwari A (2012). Toxicity of lead: a review with recent updates. *Interdiscip Toxicol* 5: 47–58.

Gerhardsson L, Brune D, Nordberg GF, & Wester PO (1986). Distribution of cadmium, lead and zinc in lung, liver and kidney in long-term exposed smelter workers. *Sci Total Environ* 50: 65–85.

Goldstein GW (1993). Evidence that lead acts as a calcium substitute in second messenger metabolism. *Neurotoxicology* 14: 97–101.

Hu H, Aro A, Payton M, Korrick S, Sparrow D, Weiss ST, et al. (1996). The relationship of bone and blood lead to hypertension. *JAMA* 275: 1171–1176.

Khalil N, Morrow LA, Needleman H, Talbott EO, Wilson JW, & Cauley JA (2009). Association of cumulative lead and neurocognitive function in an occupational cohort. *Neuropsychology* 23: 10–19.

Lanphear BP, Dietrich K, Auinger P, & Cox C (2000). Cognitive deficits associated with blood lead concentrations <10 microg/dL in US children and adolescents. *Public Health Rep* 115: 521–529.

Lanphear BP, Rauch S, Auinger P, Allen RW, & Hornung RW (2018). Low-level lead exposure and mortality in US adults: a population-based cohort study. *Lancet Public Health* 3: e177–e184.

Lopes AC, Peixe TS, Mesas AE, & Paoliello MM (2016). Lead Exposure and Oxidative Stress: A Systematic Review. *Rev Environ Contam Toxicol* 236: 193–238.

Martin D, Glass TA, Bandeen-Roche K, Todd AC, Shi W, & Schwartz BS (2006). Association of blood lead and tibia lead with blood pressure and hypertension in a community sample of older adults. *Am J Epidemiol* 163: 467–478.

Nash D, Magder L, Lustberg M, Sherwin RW, Rubin RJ, Kaufmann RB, et al. (2003). Blood lead, blood pressure, and hypertension in perimenopausal and postmenopausal women. *JAMA* 289: 1523–1532.

Navas-Acien A, Guallar E, Silbergeld EK, & Rothenberg SJ (2007). Lead exposure and cardiovascular disease–a systematic review. *Environ Health Perspect* 115: 472–482.

Neal AP, Worley PF, & Guilarte TR (2011). Lead exposure during synaptogenesis alters NMDA receptor targeting via NMDA receptor inhibition. *Neurotoxicology* 32: 281–289.

Needleman HL, Schell A, Bellinger D, Leviton A, & Allred EN (1990). The long-term effects of exposure to low doses of lead in childhood. An 11-year follow-up report. *N Engl J Med* 322: 83–88.

NourEddine D, Miloud S, & Abdelkader A (2005). Effect of lead exposure on dopaminergic transmission in the rat brain. *Toxicology* 207: 363–368.

Park SK, O'Neill MS, Vokonas PS, Sparrow D, Wright RO, Coull B, et al. (2008). Air pollution and heart rate variability: effect modification by chronic lead exposure. *Epidemiology* 19: 111–120.

Patrick L (2006). Lead toxicity, a review of the literature. Part 1: Exposure, evaluation, and treatment. *Altern Med Rev* 11: 2–22.

Radulescu A, & Lundgren S (2019). A pharmacokinetic model of lead absorption and calcium competitive dynamics. *Sci Rep* 9: 14225.

Reuben A, Caspi A, Belsky DW, Broadbent J, Harrington H, Sugden K, et al. (2017). Association of childhood blood lead levels with cognitive function and socioeconomic status at age 38 years and with IQ change and socioeconomic mobility between childhood and adulthood. *Jama* 317: 1244–1251.

Rich V (2014) *International Lead Trade*. Woodhead Publishing Limited, Cambridge, England.

Schwartz BS, Stewart WF, Bolla KI, Simon PD, Bandeen-Roche K, Gordon PB, et al. (2000). Past adult lead exposure is associated with longitudinal decline in cognitive function. *Neurology* 55: 1144–1150.

Six KM, & Goyer RA (1970). Experimental enhancement of lead toxicity by low dietary calcium. *J Lab Clin Med* 76: 933–942.

Stewart WF, & Schwartz BS (2007). Effects of lead on the adult brain: a 15-year exploration. *Am J Ind Med* 50: 729–739.

Toews AD, Kolber A, Hayward J, Krigman MR, & Morell P (1978). Experimental lead encephalopathy in the suckling rat: concentration of lead in cellular fractions enriched in brain capillaries. *Brain Res* 147: 131–138.

Wedeen RP, Maesaka JK, Weiner B, Lipat GA, Lyons MM, Vitale LF, et al. (1975). Occupational lead nephropathy. *Am J Med* 59: 630–641.

Winneke G, Lilienthal H, & Kramer U (1996). The neurobehavioural toxicology and teratology of lead. *Arch Toxicol Suppl* 18: 57–70.

CHAPTER 21

Lead Toxicity
An Overview of Its Pathophysiology and Intervention Strategies

Rokeya Pervin and Md. Akil Hossain
Animal and Plant Quarantine Agency

Dipti Debnath
University of North Texas Health Science Center

Mohiuddin Ahmed Bhuiyan
University of Asia Pacific

CONTENTS

21.1 INTRODUCTION

Lead is the most ample heavy metal present in the environment. Not only the metallic lead but also its organic and inorganic salts have been known and extensively used since ancient civilizations. Lead and its alloys were used by the Romans on a large industrial scale. They utilized lead for the manufacturing of tableware, water pipes, and other things of daily use. Other than industrial applications, varieties of lead compounds were utilized as medicines for both internal and external applications [1]. The utilization of the "sugar of lead" as sweetening agent as well as a probable antibacterial agent in manufacturing wine was persisted till nineteenth century [2].

Widely available data of twentieth century indicated that lead is a highly toxic substance that is causing the prohibition of the use of lead-derived compounds in therapeutic applications, for instance internally used medications, and as additives of food. Compared to the levels in the 1970s, a reduced emission level (98%) of atmospheric lead is achieved by eliminating lead from the fuel of the vehicle [3]. Restraining emission from metal smelting and ore processing plants provided a further reduction in the emission level of lead [2]. Lead is still exploiting in various industries such as smelting, refining, battery manufacturing and recycling, car repair, etc., although numerous countries of the world have abolished its wide utilization. It is quite difficult for giving up its use completely due to its valuable characteristics like malleability, softness, poor conductibility, resistance to corrosion, and ductility. Environmental accumulations of lead as well as its hazards are increasing because of the continuous use and non-biodegradable nature of this compound [4].

In spite of the widespread occurrence and endurance of lead in the environment, it is still considered as one of the most dangerous toxins. It causes severe health threats in persons directly handling it, and in general people who live in environmentally degraded areas. It is estimated recently in the United States that a significant portion of children still has more than 5 µg/dL lead in their blood. Although, the Centers for Disease Control and Prevention (CDC) established 5 µg/dL of blood lead level as a reference value that requires therapeutic intervention [5]. A number of studies reported that occupational and environmental exposure of lead is associated with diseases of neurological, cardiovascular, renal, reproductive, skeletal, and immune systems in humans [6–8]. Both in adults and children, the nervous system is the most affected organ in lead toxicity. As the internal and external tissues of children are softer than adults, the toxicity in children has a greater impact than in adults. Infants and young children are especially sensitive to even low levels of lead [4]. Prevention of lead exposure and elimination of elevated blood lead levels both in children and adults can be achieved by knowing well about it. Therefore, this chapter reviews the common sources of lead exposure, the clinical consequences of lead toxicity, and recent recommendations for managing lead toxicity.

21.2 SOURCES AND ROUTES OF HUMAN EXPOSURE

Lead is a naturally occurring element in the earth's crust which is widely distributed throughout the environment in soil, water, and air by anthropogenic activities, elevating the natural levels. The lead content in the uppermost layer of soil can be characterized by atmospheric deposition from anthropogenic sources [9]. Much of the lead emitted into the atmosphere is in the form of inorganic salts. Drinking water may contain lead where historic lead soldering, piping or fittings are present at home or service connections. Tap water rarely contains lead due to the dissolution of lead from natural sources [10].

The main route of lead exposure in general people is ingestion. The European Food Safety Authority (EFSA) stated in a recent dietary assessment report that the consumption of certain foods in extremely high amounts has the biggest impact on dietary lead exposure. According to this report, the categories of food which have the greatest contribution in lead exposure are vegetables and vegetable products (8.4%), non-alcoholic beverages (10.2%), milk and dairy products (10.4%), and grains and grain products (16.1%) [11]. Lead exposure may also occur from glazed cookware, imported spices, and some traditional cosmetics and medicines. Exposure of lead in infants and young children possibly happens mostly from the ingestion of dust and soil. Exposure of lead in this age group can also be happened by chipped or flaking lead paint. Hand to mouth activities or direct ingestion may cause lead exposure from these sources. An estimated average amount of soil that a child may ingest per day is 100 mg [9].

Another principal route of lead exposure is inhalation which usually happens mostly in people near point emissions sources of lead; such as greater exposure of lead is prevalent for those people who live or work near hazardous waste sites [9]. Other lead exposure sources associated with air comprise smoking (environmental tobacco, mainstream and side stream smokes) and occupational exposure, such as, persons working with lead metal (e.g., casting, alloying, smelting, and refining), old leaded paints (e.g., burning or removal), leaded glass, lead-acid batteries, solders or pigments, scrap or waste, etc. [9]. Trace amount of lead is also present in coal; and burning of fuel for industrial, transportation, domestic, or other purposes can cause the release of this heavy metal in the environment. Exposure of lead in infants and children by inhalation route is comparatively insignificant [12].

21.3 GENERAL MECHANISM OF TOXICITY

Lead toxicity is usually occurred by generating an increased amount of reactive oxygen species (ROS) and by interfering with an antioxidant generation [13]. Lead is not a redox-active element and it cannot directly play a role in those reactions which initiate the ROS formation. It was found that the generation of ROS in erythrocytes is increased by the interaction of lead with oxyhemoglobin [14]. The most significant contribution of lead in the initiation and expansion of oxidative stress is arisen by its interference with the enzymes and other cellular components/mechanisms of the defensive system which are responsible for preventing oxidative damage [15]. Glutathione (GSH), a tri-peptide of cysteine, histidine, and glutamate, is one of the most significant elements that protects cell components from ROS damage [2]. In healthy cells and tissues, 90% of GSH exists in reduced form and 10% in oxidized form, and it usually functions as an antioxidant defense mechanism. GSH after being converted (oxidized) to glutathione disulfide is reduced back to GSH by glutathione reductase [16]. Lead inactivates glutathione by binding to GSH's sulphydryl group, which inhibits sulphydryl-dependent enzymes (e.g., glutathione reductase, superoxide dismutase, catalase, etc.) and causes GSH replenishment to become inefficient. Inhibition of these enzymes leads to the production of reactive oxygen species with resultant oxidative stress. The increase in oxidative stress leads to the damage of the cell membrane because of lipid peroxidation. Lead obstructs the activities of 5-aminolevulinic acid dehydratase and directs to hemoglobin oxidation, which together with the lipid peroxidation can cause hemolysis [17].

Lead enters the intravascular space and rapidly binds to red blood cells. The estimated half-life of lead in the blood is 30 days. Lead from the blood diffuses into the soft tissues, such as the brain, bone marrow, liver, kidneys, etc. Then this heavy metal diffuses into bone and remains deposited there for a longer period of time where its half-life is of several decades. Increased bone turnover with pregnancy, lactation, menopause, or immobilization causes the rise of blood lead levels [13] (Figure 21.1).

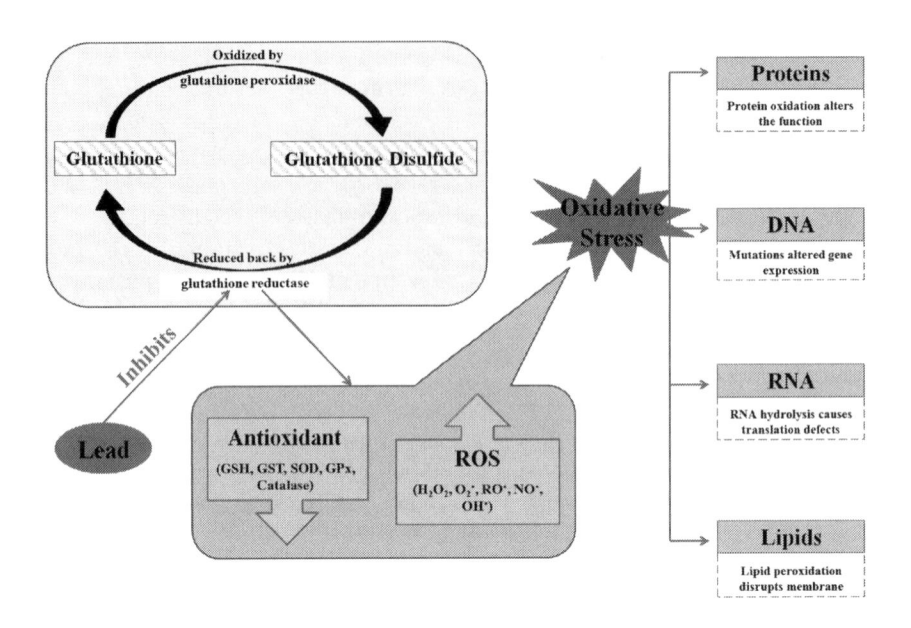

Figure 21.1 Induction of lead toxicity by promoting oxidative stress and increased levels of reactive oxygen species [2, 17].

21.4 KINETICS AND METABOLISM

Absorption of lead depends on the chemical and physical state of the metal and is influenced by the genetic factors, nutritional status, physiological status, and age of a person. The most common routes of lead absorption in children and adults are ingestion and inhalation, correspondingly. The absorption of lead in children may be 5–10 times greater than in adults [18–20]. Children, malnourished individuals, and pregnant women can absorb 40%–70% of ingested lead. About 5%–15% of ingested lead is absorbed in the gut of adults, which at fasting conditions may rise up to 45% [21]. Dietary deficiencies of zinc, calcium, iron, selenium or phosphate, copper, and ascorbic acid can cause the increase of gastrointestinal lead absorption, and currently, it is recommended that the supplementation is only applicable when a specific deficiency is proven. Lead competes with calcium for binding proteins involved with gastrointestinal absorption as they have a similar ionic size. Certain groups of people such as children and nursing mothers more efficiently absorb calcium than general people. It is hypothesized that these groups of people also more efficiently absorb lead due to their increased calcium absorption efficiency (particularly if an individual has calcium deficiency) [14]. Most inhaled lead is absorbed directly [21].

The absorption route of lead has a negligible effect on its distribution throughout the body. After the absorption, lead is mostly transported to the blood where it binds with erythrocyte proteins. Lead is distributed to soft tissues (e.g., brain, kidneys, liver) and bone with the circulation of blood, and gradually this metallic compound aggregates in those tissues and bone. The half-life of lead in the soft tissue and blood is 20–40 days whereas in bone is 10–30 years [9]. However, the half-life of lead in the blood differs contingent upon the exposure duration [22,23]. Several years of lead half-life in serum have been found in adults with occupational exposure [24,25]. The amount of lead deposited in the teeth and bone increases with the increase of age; roughly 70% of lead is deposited in children's bones and more than 95% of lead is deposited in the teeth and bones of adults [26]. The absorption and elimination of lead into bones are similar to those of calcium [27]. Accumulated lead can re-enter the blood and can re-distributed at times of stress or increased calcium use for instance menopause, pregnancy, chronic disease, lactation, chelation, and broken bones [28,29].

Adults excrete most of the lead within several weeks after an acute exposure, while children likely retain more lead for a longer time [20].

21.5 CLINICAL FEATURES

Acute lead poisoning is not so common and many lead exposures may appear asymptomatic [30]. But, it is reported in many studies that the normal functions of several organs started to hamper at a blood lead concentration ranging from <10 to 25 µg/dL [31,32]. Besides that, the symptoms and time to initiate symptoms after lead exposure may differ, and symptoms are non-specific. Typical symptoms of lead poisoning usually appear when the concentrations of lead in the blood of children and adults were 25–50 and 40–60 µg/dL, respectively. Severe lead toxicity is often reported when the lead concentration in the blood reaches to 70 µg/dL in children and 100 µg/dL in adults [21]. Due to the underdevelopment of skeletal systems and blood-brain barrier in infants and younger children, they are particularly vulnerable to the toxic effects of lead [33,34]. Moreover, <10 µg/dL of lead concentration in blood may cause abnormal differentiation of brain cell patterns ("pruning") in infants and younger children [34–36]. It has been widely described that increased levels of lead in blood negatively affects the intelligence quotient (IQ) level in children [37–42] (Table 21.1).

21.5.1 Effect on the Nervous System

Comparatively the nervous system appears to be the most sensitive and major target for lead-induced toxicity than other organ systems [44]. In the nervous system, lead mainly affects the neuronal signaling and calcium-based reactions. Both the central and peripheral nervous systems are affected by lead exposure. The effects of lead are more prominent in the central nervous system in children whereas this heavy metal remarkably affects the peripheral nervous system in adults [45,46]. An explicit result of lead exposure is encephalopathy, and the possibility of the incidence of this progressive degenerative brain disease increases with the increase of blood lead level and is more probable when the blood lead level exceeds 70 µg/dL [17,30]. The major clinical manifestations of lead poisoning are irritability, poor attention span, dullness, muscular tremor, headache, hallucinations, and loss of memory. Exposure with a very high amount of lead exhibits more severe symptoms comprising delirium, paralysis, lack of coordination, convulsions, ataxia, and coma [47]. Effects of lead exposure have been found also on the peripheral nervous system as peripheral neuropathy, relating attenuated motor activity caused by the destruction of the myelin sheath that insulates the nerves, therefore severely impairs the transduction of nerve impulse, resulting in muscular weakness, particularly of the exterior muscles, lack of muscular coordination, and fatigue [48].

Lead exposure-associated neurological effects are particularly devastating in developing offsprings and young children since the developing nervous system absorbs a high portion of the lead. In the brain of children, the accessibility of the proportion of systemically circulating lead is remarkably higher than in the brain of adults [49]. The formation of synapses in the developing brain of children is seriously impaired by lead in the cerebral cortex. Lead also interferes with the development of neurochemicals, including neurotransmitters, and the organization of ion channels. Lead poisoning also causes loss of neuron myelin sheath, reduction in the number of neurons; it interferes with neurotransmission and decreases neuronal growth. Children with more than 10 µg/dL of blood lead concentration are at greater risk of developmental disabilities [4]. The effect of lead on children's cognitive abilities takes place at very low levels [48,50,51]. It seems that there is no lower threshold level of lead which is considered safe for the nervous system [52]. The lead level in blood <5 µg/dL was reported to be related to decreased academic performance [46,53]. The lead level in blood <10 µg/dL was found to be linked with reduced IQ and behavior problems for instance aggression [54]. Increased level of lead in blood is also related to reduced cognitive performance

Table 21.1 Severity, Clinical Presentations and Eradication/Treatment Strategies of Lead Toxicity Depending on Blood-Lead Concentrations [21,30,31,43].

Blood Lead Concentration (μg/dL)	Toxicity Level	Clinical Presentations		Intervention Strategies
		Adults	Children	
<10	Asymptomatic or impaired abilities	Not applicable	Impairment of hearing and speech, decreased learning and memory, lower intelligence quotient, signs of attention-deficit/hyperactive disorder or hyperactivity, impairment of fine motor coordination, decreased verbal ability	Monitoring the level, minimizing exposure, discuss health risks
10–40	Mild	Not applicable	Mild lethargy/fatigue, irritability, paresthesia or myalgia, occasional abdominal discomfort	Removing exposure source, medical evaluation, monitoring the level twice a year
41–70	Moderate	Chronic hypertensive effects, impaired psychometrics, moodiness, somnolence, fatigue, reproductive effects, lessened leisure interest.	Constipation, vomiting, weight loss, tremor, muscular exhaustibility, headache, difficulty in concentrating, arthralgia, diffuse abdominal pain, general fatigue	Removing exposure source, medical evaluation, monitoring the level every month, administering EDTA or DMSA depending on symptoms
71–80	Severe	Insomnia, memory loss, headache, metallic taste, decreased libido, myalgia/arthralgia, constipation, abdominal pain, nephropathy	Bluish black pigmentation on gingival tissue, paresthesia or paralysis, colic (severe or intermittent cramps), encephalopathy	Treating with BAL when there is encephalopathy, otherwise, administer EDTA or DMSA depending on symptoms
100–150	Severe, acute	Encephalopathy, a number of CNS effects, nephropathy, anemia	Encephalopathy, nephropathy, anemia, seizures	Treating with BAL when there is encephalopathy, otherwise, administer EDTA or DMSA depending on symptoms

CNS, central nervous system; DMSA, dimercaptosuccinic acid; EDTA, ethylene diamine tetraacetic acid.

and with other psychiatric conditions such as anxiety and depression [55]. Lead exposure at early childhood and the prenatal stage was reported to correlate with violent crimes in adulthood [51]. Lead at higher level causes permanent damage of the brain and even death [46,56].

21.5.2 Effect on the Hematopoietic System

The hematopoietic system is directly affected by lead through the prevention of hemoglobin synthesis as a result of hindering several vital enzymes that are responsible for heme synthesis. In addition, by decreasing the strength of the cell membrane, lead minimizes the lifetime of circulating erythrocytes. The collective consequence of these two actions leads to anemia [57]. Anemia caused on account of lead poisoning can be of two types: hemolytic anemia, which is associated with acute high-level lead exposure, and frank anemia, which is caused only when the blood lead level is significantly elevated for prolonged periods [58]. Lead significantly affects the heme synthesis pathway in a dose-dependent manner by downregulating the three key enzymes, such as aminolevulinic acid synthetase (ALAS), δ-aminolevulinic acid dehydratase (ALAD), and ferrochelatase which are involved in the synthesis of heme [59]. Lead hinders these three fundamental enzymes of this pathway; however, the impact of lead on ALAD is extremely severe, and the level of lead toxicity is measured clinically by the degree of ALAD inhibition [17]. In addition, ALAS inhibition could be responsible for liver toxicity. Therefore, the production of heme is blocked by the combined inhibition of these three enzymes through the heme synthesis pathway [17]. The mechanisms involved in the reduction of the life span of erythrocytes are not yet sufficiently revealed.

Reticulocytosis and basophilic stippling are two other recorded hematological effects of lead exposure which are also potential biomarkers of lead toxicity diagnosis [17]. These aggregates are degradation products of ribonucleic acid [60]. Logistic regression analysis conducted in a recent study on children demonstrated that reduction of platelet count together with the reduction of RBC and hemoglobin counts are associated with an extreme lead concentration in blood [61].

21.5.3 Effect on Renal System

In occupational lead exposure, nephropathy is a well-documented disease [30]. Typically the dysfunction of the urinary system appears at a high lead level (>60 μg/dL), but the renal injury has also been reported at a low lead level (~10 μg/dL) [62]. Functional abnormality of the urinary system can be of two types, such as chronic nephropathy and acute nephropathy. Chronic nephropathy is extremely severe and can direct to morphological as well as functional changes that are irreversible. It is indicated by the tubule-interstitial and glomerular modifications that result in a renal breakdown, hyperuricemia, and hypertension. In contrast, acute nephropathy is comparatively less severe and is functionally defined by the impairment of the tubular transport system and morphologically characterized by the degenerative changes in the tubular epithelium accompanied by the development of nuclear inclusion bodies that contain lead-protein complexes. It does not allow the protein to excrete in the urine and may elevate the abnormal excretion of amino acids, phosphates, and glucose, a combination defined to as Fanconi's syndrome [63].

21.5.4 Effect on Cardiovascular System

Both chronic and acute lead poisoning causes cardiac and vascular damage with potentially lethal consequences including hypertension and cardiovascular disease [64]. Even the lower amount of lead exposure can pertain to the hypertension of both humans and animals. Other major adverse effects include cerebro-vascular accidents, peripheral vascular disease, and ischemic coronary heart disease [17]. Different studies reported that lead-induced renal dysfunction may play a role in the pathogenesis of hypertension. It is also strongly suspected that lead-induced hypertension may be operated by oxidative stress; however, the precise mechanism of lead-induced hypertension is not yet known. Even though renal dysfunction and hypertension are two proposed mechanisms for the effect of lead on cardiovascular disease, other mechanisms are likely to be involved [65].

A study was conducted in 14,000 participants who had <10 µg/dL lead in their blood [66]. Those participants were followed up for 12 years. After multivariate adjustment, the risk of cardiovascular events was remarkably higher in individuals who had high lead concentration (≥3.62 µg/dL) in their blood in comparison with those who had low lead concentration (<1.94 µg/dL). There was an increased pervasiveness of peripheral arterial disease in individuals who had higher levels of lead in their blood. Individuals with blood lead levels>40 µg/dL have also been reported to present delayed electrocardiogram conduction. The increased cardiovascular mortality seen in individuals with lead toxicity has been associated with increased lead levels both in the bone and blood, with a greater association with lead levels in bone [67]. Weisskopf et al studied bone lead levels and cardiovascular mortality among 850 male patients [68]. After multivariable analysis, it was revealed that participants with higher lead levels in bone were more likely to die due to cardiovascular disorders as well as other disorders.

21.5.5 Effect on the Reproductive System

Lead exposure adversely affects the reproductive systems of both men and women in a number of ways [17]. Most of the adverse effects have been seen in lead-exposed workers [30]. Abnormal spermatogenesis (reduced number and motility), reduced libido, infertility, chromosomal damage, changes in serum testosterone, and abnormal prostatic function are the most common adverse effects seen in men [17]. When the lead level in blood in a male person exceeds 40 µg/dL, sperm number is reduced and other changes in the sperm volume are occurred [4]. Lead exposure-associated harmful effects in the reproductive system of females are more severe than in males. Blood lead levels in mother and infant are typically similar since lead present in the mother's blood passes into the fetus through the placenta as well as via breast milk. Moreover, women are more vulnerable to infertility, pre-eclampsia, premature membrane rupture, miscarriage, premature delivery, and pregnancy hypertension [69]. Likewise, it has also been reported that lead has a direct impact on the developmental stages of the fetus [70]. Reduced head circumference, low birth weight, and low birth length have been seen due to lead exposure. The evidence associated with the probable teratogenic effect of lead is not well-documented [71].

21.5.6 Effect on Hepatic System

The evidence of the effects of chronic exposure of lead to the liver is inadequate. It has been shown in different studies that the inhibition of cytochrome P450 is directed by lead toxicity. But it tends to induce CYP51, a vital enzyme that is responsible for the synthesis of cholesterol, thus raises the level of cholesterol [72]. Lead nitrate also causes the proliferation of liver cells in rats without concomitant liver cell necrosis [30]. It has been suggested that the effects of lead on haem synthesis may affect the functional capacity of hepatic cytochrome P450 enzymes to metabolize drugs.

21.5.7 Effect on Other Organs

In severe cases of lead poisoning, children or adults may present with severe cramping abdominal pain (colic-like pain), which may be mistaken for an acute abdomen or appendicitis. Lead colic is a symptom of chronic lead poisoning and is associated with obstinate constipation [73]. The primary sites of lead storage in the human body are bones [74]. Lead has been linked to problems with the development and health of bones. At high levels, lead can result in slowed growth in children. There is a growing body of scientific literature that supports the role of lead exposure in the development of osteoporosis. Rat studies have found that lead exposure is associated with decreased bone mineral density (BMD). However, human studies are limited. Lead-exposed animals had decreased bone mass that resulted in bones that were more susceptible to fracture. Researchers are currently investigating the impacts of lead on dental health. One study found pre- and perinatal exposure to lead increased prevalence of caries in rat pups by almost 40% [73]. In a recent study in Iran, skin hyperpigmentation was reported in patients of lead poisoning [75]. Environmental Protection Agency of the United States has classified elemental lead and inorganic lead compounds as Group 2B: probable human carcinogens, revised in 2011 [76]. This classification is based in part on animal studies, which have been criticized because the doses of lead administered were extremely high. The National Toxicology Program (NTP) classified lead and lead compounds as "reasonably anticipated to be a carcinogen" [77].

21.6 DIAGNOSIS

In order to inhibit and cure lead poisoning and toxicity, proper diagnosis is a primary and rather important issue. In order to make a proper diagnosis, an inquiry about the possible routes of exposure is a must [78]. The inquiry should include medical history and the determination of clinical signs. Measurement of blood lead levels is the standard screening test for lead toxicity. There is no blood lead level below which no harm occurs. But for diagnostic purpose blood levels more than 10 µg/dL are considered abnormal [30]. Lead poisoning can also be evaluated by measuring erythrocyte protoporphyrin (EP) in blood samples [79]. EP is known to increase when the amount of lead in the blood is high, with a delay of a few weeks. However, the EP level alone is not sensitive enough to identify elevated blood lead levels below about 35 µg/dL [79]. Due to this higher threshold for detection and the fact that EP levels also increase in iron deficiency, the use of this method for detecting lead exposure has decreased [4]. Basophilic stripping is an important sign of lead poisoning. This stripping makes dots in red blood cells visible through the microscope [79]. Thus an examination of blood film for such signs could be effective in detecting lead poisoning.

The measurement of blood lead level does not give the actual account of lead stored in the body; it is just an indicator of the acute or recent lead exposure. Whole-body lead can be measured in bones noninvasively by X-ray fluorescence; this may be the best measure of cumulative exposure and total body burden. X-rays may also reveal lead-containing foreign materials, such as paint chips in the gastrointestinal tract [4]. Chronic lead nephropathy occurred due to years of lead exposure manifested in kidney biopsy by moderate focal atrophy, loss of proximal tubules, and interstitial fibrosis [80].

21.7 MANAGEMENT STRATEGIES

Management of lead toxicity requires extensive risk assessment and caregiver education. The most effective intervention for lead toxicity is early detection and removal of the lead source to prevent further exposure [20]. Other intervention strategies of lead toxicity are decontamination, supportive care, and chelation, while the medical interventions and treatments are variable depending on the confirmed blood lead level [30].

21.7.1 Prevention

Taking the toxic effects of lead into consideration, preventive measures are more preferable than the treatment options. Because complete removal of lead or reversing its harmful effects in the body is nearly impossible after it enters into the body [17]. The best approach is to avoid exposure to lead [81]. It is recommended to frequently wash children's hands and also to increase their intake of calcium and iron. It is also recommended to discourage children from putting their hands, which can be contaminated, in their mouth habitually, thus increasing the chances of getting poisoned by lead. Vacuuming frequently and eliminating the use or presence of lead-containing objects like blinds and jewelry in the house can also help to prevent exposures. House pipes containing lead or plumbing solder fitted in old houses should be replaced to avoid lead contamination through drinking water. It is believed that hot water contains higher lead levels than cold water, so it is recommended that for household uses cold water should be preferred to hot water [4]. Banning the setting up of those industries near to residential areas that deal with lead and fully prohibiting the use of lead where a suitable alternative is obtainable are the vital preventive measures recommended by the public health services for controlling lead [17]. Despite these preliminary measures, nutrition is considered to have a great role in preventing lead-induced toxicity. It is found in different studies that uptake of some nutrients, such as vitamins, flavonoids, and mineral elements can protect individuals to be affected by environmental lead and those who already contain lead in their body [17].

21.7.2 Antioxidants

It has revealed in different studies that antioxidants can prevent as well as cure the harmful effects that originated from the free radical generation in the body [17]. Non-enzymatic antioxidants, such as vitamins, flavonoids, minerals, carotenoids, etc. are components of various vegetables, fruits, grains, nuts, and meats, while enzymatic antioxidants, for instance, GPX, CAT, SOD are endogenously developed in the cell [82]. The amount of endogenous antioxidants generated at normal physiological condition is sufficient for quenching the free radicals which are produced at the normal physiological rate. Oxidative stress can occur if the balance between antioxidants and free radicals are hampered. Natural or environmental factors may increase the free radical concentrations and may lead to oxidative stress [83]. Lead exposure generates an impermissible amount of free radical and hence develops oxidative stress, at the same time it depletes the reserves of endogenous antioxidants and disables the biological system for reversing the harmful effects. A chain reaction starts with the generation of free radicals which leads to the disruption of the cell membrane, oxidation of protein, lipid peroxidation, and oxidation of nucleic acids, such as RNA and DNA leading to cancer [84]. In this case exogenous antioxidants can play an important role. It is suggested from various studies that a variety of toxic effects of lead and particularly the emergence of oxidative stress can be prevented or inhibited by the administration of different antioxidants [17]. Antioxidants are taken through the diet or as supplement form for maintaining the equilibrium between antioxidants and free radicals and consequently hinder a variety of harmful effects [85,86]. It is already revealed that persons who take antioxidant-containing diet can reap different health benefits more than other persons. Numerous antioxidants are naturally present in food whereas the supplementary form of antioxidants is usually rich with a single or few antioxidants, and hence food is always more preferential than supplements for boosting up the antioxidant levels [17].

There are a number of antioxidants that are believed to act against toxicity of chemicals like lead and its related compounds. In a recent study beta-carotene was found to have an antioxidant effect and some beneficial effects in lead poisoning, independent of chelation [87]. The authors also found notably reduced levels of homocysteine due to the application of beta-carotene in lead-exposed workers. Recently a study on a group of workers occupationally exposed to lead found that those treated with *N*-acetylcysteine (NAC) showed a considerable lowering in their blood lead levels. In addition, all groups receiving NAC were shown to have remarkably increased activity of glutamate dehydrogenase. It was further reported that treatment with NAC normalized the homocysteine level and reduced oxidative stress. It was thus concluded that NAC could be recommended as an alternative therapy for chronic lead toxicity in humans [88]. Other antioxidants reported to be effective against the toxic effect of lead are discussed below.

21.7.2.1 *Vitamins*

Vitamins (predominantly B, C, and E) play a very significant and competing role against the harmful effects of lead poisoning. Both the chelating of lead from tissues and the restoration of the pro/antioxidant balance can be obtained by the application of vitamins [17]. Lead-associated toxicity preventive effects of different eminent vitamins are discussed below.

21.7.2.1.1 *Vitamin B*

It is revealed that thiamine (vitamin B1) and pyridoxine (vitamin B6) have important properties of curing the harmful effects of lead poisoning. Vitamin B1 is found to have protective effects on the impact of short-term lead exposure [17]. Pyridoxine stimulates GSH production and thus provides antioxidant effects in curing poisonous effects of lead. It also has a moderate-level of lead chelating capability [89]. The interfering feature of pyridoxine in lead absorption and/or the existence of the ring in the nitrogen atom of vitamin B6 can be associated with the lead chelating effect of pyridoxine. Cysteine, a non-essential amino acid, can be synthesized from dietary methionine through the metabolic trans-sulfuration pathway, where pyridoxine acts as a vital co-factor [17].

21.7.2.1.2 *Vitamin C*

Perhaps, the effects of vitamin C in the prevention of lead-induced oxidative stress are studied most widely among all vitamins. Due to the metal chelating and ROS quenching properties of vitamin C, it became a very efficient detoxifying agent for lead [90,91]. Shan et al. (2009) studied the protective effects of thiamine and vitamin C against lead-associated harmful effects on the reproductive system of male mice [92]. Lead exposure demonstrated a substantial reduction of sperm motility and sperm count, and the stimulation of apoptosis by activating Bcl-2, Fas/Fas-L, and caspase-3. Concentration-dependently reversions of lead-induced oxidative stress, apoptosis, and DNA damage in liver cells of rats are achieved by the co-administration of thiamine and vitamin C [93]. It is also reported that the acute hepatotoxic effect of lead can be reduced by supplementing the combination of silymarin and vitamin C [94].

21.7.2.1.3 *Vitamin E*

Vitamin E has potent antioxidative effects, and this vitamin is effective in preventing lipid peroxidation in the membrane by blocking the free radical chain reaction. It is reported that the administration of vitamin E prevented the harmful effects of lead in rats by prohibiting oxidative stress as a result of free radicals scavenging [95]. It was found that the application of vitamin E reverses the lead-induced ALAD inhibition in the erythrocyte [96]. Thyroid dysfunction is restored by the application of vitamin E through the maintenance of the hepatic cell membrane architecture which is indirectly disrupted by lead-induced lipid peroxidation. It is also evident that vitamin E can exert its effect more profoundly in combination with other antioxidants than alone. Faster recovery of rats from lead-induced pathologic conditions is achieved by the application of vitamin E along with monoisoamyl derivative [97].

21.7.2.2 *Flavonoids*

Flavonoids are a group of natural polyphenolic substances which are widely available in fruits, vegetables, flowers, grains, stems, roots, bark, and certain beverages [98]. Flavonoids are comprehensively explored for their antioxidative properties. Similar to other anti-oxidants, flavonoids can prevent or inhibit oxidative stress by the cessation of free radical chain reaction and by the chelation of redox-active metal ions [99,100].

21.7.2.2.1 Quercetin

Quercetin is a widely explored and globally distributed bioflavonoid. Vegetables, fruits, and tea are the dietary sources of quercetin. The metal chelating and antioxidant properties of this flavonoid are considered because of the presence of many hydroxyl groups in its chemical structure and conjugated electrons [17]. These hydroxyl groups, including the carbonyl group, easily donate electrons by undergoing resonance and stabilize free radicals that can initiate lipid peroxidation [101]. Quercetin form a covalent bond with lead ions by the ortho-phenolic groups present in the quercetin B ring and consequently chelates lead [102]. It is reported in a study that the lead-induced histopathological injuries in the kidney of rats were protected by quercetin [103]. In that study it was found that quercetin noticeably reduced the level of ROS and the ratio of reduced glutathione (GSH) and oxidized glutathione (GSSG) in the kidney of lead-treated rats. Moreover, it repressed the increased level of 8-hydroxydeoxyguanosine, together with the restoration of GPx, Cu/Zn-SOD, and CAT activities in the kidney of lead-treated rats, and thus the inhibition of lead-induced apoptosis in the kidney of rats was confirmed [103].

The role of quercetin in curing the lead-induced impairment of synaptic plasticity in the rat model was investigated [104]. A range of parameters was considered in this study including paired-pulse reactions (PPR), population spike (PS) amplitude, excitatory postsynaptic potential (EPSP), and input/output (I/O) functions in the dentate gyrus area of different lead-treated rat groups. Treatment with quercetin demonstrated significant improvement in all of these parameters. The reduction of hippocampal lead concentration with the treatment of quercetin was also found. Thus, the therapeutic and medicinal effects of quercetin, together with its lower toxicity profile, have made it a very promising therapy in the field of heavy metal toxicity.

21.7.2.2.2 Alpha Lipoic Acid

Lipoic acid, also known as α-lipoic acid, is a naturally occurring organosulfur compound that is synthesized in animals, plants, and humans. Alpha lipoic acid is an antioxidant and found in many foods, such as carrots, potatoes, spinach, beets, and red meats [105]. This antioxidant is soluble both in water and fat which enables it to play its role in both aqueous and fatty regions of the body, and hence believed it as a universal antioxidant [106]. Alpha lipoic acid shows its antioxidative effect in two ways: first, it attacks ROS and prohibits lipid peroxides formation, and second, it regenerates and replenishes other antioxidants, such as vitamin C, vitamin E, etc. [107]. Alpha lipoic acid is usually used in combination with other chelating agents because of the inability of this antioxidant to chelate metal and the efficiency of this compound to persistently restrain the developed oxidative stress. It is found in a study that lipoic acid can significantly improve the lead-induced pathologic conditions on the liver, kidneys, and erythrocyte membranes by reverting the generated oxidative stress [108]. Alpha lipoic acid is reported to remove lead more effectively from the brain than from any other organs like kidneys, liver, and other soft tissues [109].

21.7.2.2.3 Curcumin

Curcumin is a biologically active yellow-colored polyphenolic compound widely found in turmeric. In metal poisoning curcumin is reported to show antioxidant, metal chelating, and radical scavenging effects [110–113]. Lead-induced neurotoxicity in rats was protected by the application of curcumin, where this antioxidant significantly improved the levels of different biomarkers of oxidative stress (SOD, CAT, and GSH) in various parts of the brain [114]. It is also reported that the level of lead in the brain of rats is remarkably reduced by chelation with curcumin [115,116].

21.7.2.2.4 Puerarin

Puerarin, an isoflavonoid compound, is widely distributed in the variety of plants and herbs, and this is the major bioactive compound in Pueraria. This compound is reported in a recent study for its GSK-3β and Akt phosphorylation promoting effects in lead acetate-exposed PC-12 cells [4]. It is concluded in this report that puerarin as a natural estrogen might be a promising candidate to prevent and treat lead-induced neurotoxicity associated chronic diseases.

21.7.2.3 Herbal Antioxidants

Lower cost and negligible side effects of herbal antioxidants made them attractive as clinical medicines. However, specific applications of herbal antioxidants in the form of potential medicines have been very limited. Longer treatment duration of herbal antioxidants makes them suitable for preventive instead of therapeutic purposes. Poor bioavailability and repetitive and extremely higher doses requirements for maintaining the therapeutic concentration in the body are other most serious limitations of herbal antioxidants [17]. Very few herbal antioxidants as discussed below are indicated for their protective effects against lead-induced oxidative stress.

21.7.2.3.1 Garlic

Garlic has been widely used both as a food and for medicinal purposes throughout ancient and modern history. This medicinal food has important pharmacological and therapeutic effects. Allicin is the major bioactive compound of garlic that provides pharmacological effects [117]. Garlic prevents oxidative stress by scavenging free radicals and chelating lead ions. It is reported that the lead burden in soft tissues, blood, bone, kidney, liver, etc. can be prevented by the consumption of garlic [118,119]. Similarly in a study, an aqueous extract of garlic is found to prevent lead-induced liver injury in rats [120].

21.7.2.3.2 Centella asiatica

Centella asiatica, commonly known as Indian pennywort and gotu kola, is an important medicinal herb which is widely used in the orient and is recently being popular in the West. It has a variety of therapeutic effects including healing metal toxicity [121]. Lead-induced oxidative stress in rats has significantly been controlled by the application of *C. asiatica* in combination with chelating agent dimercaptosuccinic acid (DMSA) [122]. Promising effects against lead-induced oxidative stress have also been found from the application of *C. asiatica* alone [123,124]. It is speculated that *C. asiatica* can cross the blood-brain barrier and restore the levels of modified neurotransmitters as well as recover the harmed antioxidant/prooxidant equilibrium which arises from lead exposure [125].

21.7.3 Chelation Therapy

Succimer, edetate calcium disodium (calcium-EDTA), and dimercaprol are the three agents mainly used for chelation [3]. Chelation therapy with dimercaprol (also known as "British anti-Lewisite" or BAL) and EDTA is recommended for a lead level of more than 70 µg/dL in children and more than 80 µg/dL in adults [30]. Dimercaprol is the most preferred chelator to treat severe symptoms of lead toxicity or encephalopathy. The typical dose of dimercaprol is 50–75 mg/m^2 by intra-muscular administration every 4 h for 3 to 5 days [21,30]. Peanut allergy should be evaluated in patients prior to the starting of treatment as the commercially available formulation contains peanut oil. For preventing the dissociation of dimercaprol-metal chelate as well as the subsequent lead re-absorption, the alkalinity of urine should be maintained during the treatment period. The majority of the undesirable effects of dimercaprol therapy are temporary and related to dose, which includes hypertension, nausea, headache, vomiting, hyperpyrexia, tachycardia, and mild-to-moderate leucopenia [21].

Calcium-EDTA, which is the calcium chelate of the disodium salt of ethylene-diamine-tetracetic acid (EDTA) is known to have a great affinity to the removing agent. The chelator for lead has a stronger affinity to lead than calcium and therefore the lead chelate is occurred by exchanging. This is then excreted in the urine, leaving behind harmless calcium [4]. Although calcium-EDTA is a potent lead chelator, it can elevate lead re-distribution to the central nervous system if dimercaprol is not given 4 h prior to the administration of calcium EDTA. It is recommended not to administer the calcium-EDTA as a single therapy. Currently, calcium-EDTA is applied only with dimercaprol in the case of severe encephalopathy. The typical dose per day is $1,000-1,500$ mg/m^2 as a continuous intravenous infusion for 5 days. Calcium-EDTA is recommended to start after 4 h of the administration of dimercaprol and with sufficient urinary output. The most severe adverse effect related to calcium-EDTA is renal toxicity which is possibly reduced by reducing the amount and frequency doses, at the same time maintaining sufficient hydration. The infusion of calcium EDTA should be discontinued for no less than 1 h prior to testing the lead level in the blood for allowing a precise measurement [21].

Succimer is the preferred chelating agent for asymptomatic or mild cases of lead toxicity and is accessible only in an oral capsule formulation. Succimer may be indicated as a chelating agent in children with a whole blood lead concentration of less than 25 μg/dL. The manufacturer recommends that succimer should be dosed at 350 mg/m^2 and administered thrice a day for 5 days, then decreased to twice a day for 14 days. The dosage for children and adults can also be expressed as 10 mg/kg per dose using the same regimen. The most frequently reported adverse effects of succimer are gastrointestinal disorders and a mild increase in the concentration of serum transaminase [21]. D penicillamine and DMSA are also used as chelating agents [30]. Comparatively, penicillamine is less usually used, partly due to the increased risk of interstitial nephritis in adults; this risk does not appear to be increased in children at a dosage of 20 mg/kg/day [21].

Chelation therapy successfully reduces high blood lead levels but may not prevent lead-induced cognitive defects associated with lower lead levels. This may be because of the inability to reverse pre-existing tissue damage or the inability to remove a sufficient amount of lead from the tissues. Chelation therapy is usually not indicated for adults with blood lead concentration less than 45 μg/dL owing to the possible risk of adverse drug reactions and concern regarding remobilized lead. Chelation therapy for children with blood lead concentrations less than 45 μg/dL is still controversial, particularly with two studies failing for demonstrating a change in cognitive outcomes with succimer administrations [21,126,127]. It is suggested to re-check lead levels in the blood after 1–3 weeks of chelation therapy, since lead levels may rebound because of the release of lead from storage sites; rebound levels are approximately 70% of the pretreatment whole blood lead level at 2 weeks after the last dose of succimer [21]. Rebound levels may also imply continued exposure and necessitate further investigations [3].

21.8 CONCLUSION

Lead is present everywhere in the environment and is recognized for not having any physiological function in the body. However, reports on its toxicity are well documented and lead toxicity appeared to be rather prominent among all the heavy metal toxicity. Moreover, a small amount of lead can cause toxicity. Lead toxicity hampers the functions of the nervous system, reproductive system, digestive system, respiratory system, etc. Additionally, lead obstructs the normal DNA transcription process, impedes enzymes from executing their usual activities, and leads to the disability of bones. A number of techniques are introduced nowadays to reduce the levels of lead from the body and to reverse the toxicity. Antioxidants and chelation therapy are the most prominent among those techniques. Prevention of lead accumulation by incorporation of various natural and synthetic antioxidants is regarded to be the best approach as chelation therapy can significantly reduce the lead level in the body but cannot remove all leads from the body. To prevent and treat the lead-induced toxicity and oxidative stress, a variety of naturally occurring antioxidants, such as flavonoids, vitamins, and herbal antioxidants have been documented. Because of the individual differences of genetic, dietary, and environmental factors, the treatment strategies may not be effective equally for all. Although there are some treatment options available at the moment, it is undoubtedly better to prevent direct exposure to lead and thus obstruct future consequences. One needs to be aware of the toxic effects of lead so that appropriate and timely therapy can be instituted in affected individuals. It is also recommended that parents should educate their children about how to prevent accidental lead poisoning.

REFERENCES

1. Nriagu JO. Saturnine drugs and medicinal exposure to lead: a historical outline. In: Needleman HL, editor. *Human Lead Exposure*. Boca Raton, FL: CRC Press; 1992. pp. 3–22.
2. Machiej S. Molecular mechanisms of lead toxicity. *BioTechnologia JBCBB* 2014;95(2):137–49.
3. Pirkle JL, Brody DJ, Gunter EW, Kramer RA, Paschal DC, Flegal KM, et al. The decline in blood lead levels in the United States. The National Health and Nutrition Examination Surveys (NHANES). *JAMA* 1994;272(4):284–91.
4. Wani AL, Ara A, Usmani JA. Lead toxicity: a review. *Interdiscip Toxicol* 2015;8(2):55–64.
5. Brink LL, Talbott EO, Sharma RK, Marsh GM, Wu WC, Rager JR, et al. Do US ambient air lead levels have a significant impact on childhood blood lead levels: results of a national study. *J Environ Public Health* 2013;2013:278042.
6. Silveira EA, Siman FD, de Oliveira Faria T, Vescovi MV, Furieri LB, Lizardo JH, et al. Low-dose chronic lead exposure increases systolic arterial pressure and vascular reactivity of rat aortas. *Free Radic Biol Med* 2014;67:366–76.
7. Zhang A, Hu H, Sánchez BN, Ettinger AS, Park SK, Cantonwine D, et al. Association between prenatal lead exposure and blood pressure in children. *Environ Health Perspect* 2012;120(3):445–50.
8. Xie X, Ding G, Cui C, Chen L, Gao Y, Zhou Y, et al. The effects of low-level prenatal lead exposure on birth outcomes. *Environ Pollut* 2013;175:30–4.
9. EFSA Panel on Contaminants in the Food Chain (CONTAM). Scientific opinion on lead in food. *EFSA J* 2010;8(4):1570. https://efsa.onlinelibrary.wiley.com/doi/pdf/10.2903/j.efsa.2010.1570 [accessed 30.12.2019].
10. World Health Organisation (WHO). Guidelines for drinking-water quality: fourth edition incorporating the first addendum. 2017. https://www.who.int/water_sanitation_health/publications/drinking-water-quality-guidelines-4-including-1st-addendum/en/ [accessed 30.12.2019].
11. European Food Safety Authority (EFSA). Lead dietary exposure in the European population. *EFSA J* 2012;10(7):2831. https://efsa.onlinelibrary.wiley.com/doi/pdf/10.2903/j.efsa.2012.2831 [accessed 30.12.2019].
12. Committee on Toxicity of Chemicals in Food Consumer Products and the Environment (COT). Statement on the potential risks from lead in the infant diet. 2013. https://cot.food.gov.uk/sites/default/files/cot/cotstatlead.pdf [accessed 30.12.2019].
13. Kathuria P, Rowden AK, O'Malley RK. Lead toxicity. In: Lorenzo N, Ramachandran TS, editors. *Hudson Street*. New York, NY: Medscape; 2018. pp. 1–22. https://emedicine.medscape.com/article/1174752-overview [accessed 30.12.2019].
14. Ribarov SR, Bochev PG. Lead-hemoglobin interaction as a possible source of reactive oxygen species – a chemiluminescent study. *Arch Biochem Biophys* 1982;213(1):288–92.
15. Gurer H, Ercal N. Can antioxidants be beneficial in the treatment of lead poisoning? *Free Radic Biol Med* 2000;29(10):927–45.

16. Franco R, Schoneveld OJ, Pappa A, Panayiotidis MI. The central role of glutathione in the pathophysiology of human diseases. *Arch Physiol Biochem* 2007;113(4–5):234–58.

17. Flora G, Gupta D, Tiwari A. Toxicity of lead: a review with recent updates. *Interdiscip Toxicol* 2012;5(2):47–58.

18. Alexander FW. The uptake of lead by children in differing environments. *Environ Health Perspect* 1974;7:155–9.

19. James HM, Hilburn ME, Blair JA. Effects of meals and meal times on uptake of lead from the gastrointestinal tract in humans. *Hum Toxicol* 1985;4(4):401–7.

20. Ziegler EE, Edwards BB, Jensen RL, Mahaffey KR, Fomon SJ. Absorption and retention of lead by infants. *Pediatr Res* 1978;12(1):29–34.

21. Gracia RC, Snodgrass WR. Lead toxicity and chelation therapy. *Am J Health-Syst Pharm* 2007;64(1):45–53.

22. Rabinowitz MB, Wetherill GW, Kopple JD. Kinetic analysis of lead metabolism in healthy humans. *J Clin Invest* 1976;58(2):260–70.

23. Griffin TB, Coulston F, Wills H, Russell JC. Clinical studies on men continuously exposed to airborne particulate lead. *Environ Qual Saf Suppl* 1975;2:221–40.

24. Schutz A, Skerfving S, Ranstam J, Christoffersson JO. Kinetics of lead in blood after the end of occupational exposure. *Scand J Work Environ Health* 1987;13(3):221–3.

25. Hryhorczuk DO, Rabinowitz MB, Hessl SM, Hoffman D, Hogan MM, Mallin K, et al. Elimination kinetics of blood lead in workers with chronic lead intoxication. *Am J Ind Med* 1985;8(1):33–42.

26. Barry PS. A comparison of concentrations of lead in human tissues. *Br J Ind Med* 1975;32(2):119–39.

27. Sargent JD. The role of nutrition in the prevention of lead poisoning in children. *Pediatr Ann* 1994;23(11):636–42.

28. Mahaffey KR. Biokinetics of lead during pregnancy. *Fundam Appl Toxicol* 1991;16(1):15–6.

29. Silbergeld EK. Lead in bone: implications for toxicology during pregnancy and lactation. *Environ Health Perspect* 1991;91:63–70.

30. Vaibhav S, Priyanka S, Avanish T. Lead poisoning. *Indian J Med Spec* 2018;9:146–9.

31. Canfield RL, Henderson CR Jr, Cory-Slechta DA, Cox C, Jusko TA, Lanphear BP. Intellectual impairment in children with blood lead concentrations below 10 microg per deciliter. *N Engl J Med* 2003;348(16):1517–26.

32. Pocock SJ, Smith M, Baghurst P. Environmental lead and children's intelligence: a systematic review of the epidemiological evidence. *BMJ* 1994;309(6963):1189–97.

33. Goldstein GW. Neurologic concepts of lead poisoning in children. *Pediatr Ann* 1992;21(6):384–8.

34. Goyer RA. Lead toxicity: current concerns. *Environ Health Perspect* 1993;100:177–87.

35. Silbergeld EK. Mechanisms of lead neurotoxicity, or looking beyond the lamppost. *FASEB J* 1992;6(13):3201–6.

36. Goldstein GW. Lead poisoning and brain cell function. *Environ Health Perspect* 1990;89:91–4.

37. Baghurst PA, McMichael AJ, Wigg NR, Vimpani GV, Robertson EF, Roberts RJ, et al. Environmental exposure to lead and children's intelligence at the age of seven years. The Port Pirie Cohort Study. *N Engl J Med* 1992;327(18):1279–84.

38. Faust D, Brown J. Moderately elevated blood lead levels: effects on neuropsychologic functioning in children. *Pediatrics* 1987;80(5):623–9.

39. McMichael AJ, Baghurst PA, Wigg NR, Vimpani GV, Robertson EF, Roberts RJ. Port Pirie Cohort Study: environmental exposure to lead and children's abilities at the age of four years. *N Engl J Med* 1988;319(8):468–75.

40. Hawk BA, Schroeder SR, Robinson G, Otto D, Mushak P, Kleinbaum D, et al. Relation of lead and social factors to IQ of low-SES children: a partial replication. *Am J Ment Defic* 1986;91(2):178–83.

41. Fulton M, Raab G, Thomson G, Laxen D, Hunter R, Hepburn W. Influence of blood lead on the ability and attainment of children in Edinburgh. *Lancet* 1987;1(8544):1221–6.

42. Yule Q, Lansdown R, Millar IB, Urbanowicz MA. The relationship between blood lead concentrations, intelligence and attainment in a school population: a pilot study. *Dev Med Child Neurol* 1981;23(5):567–76.

43. Bellinger DC, Needleman HL. Intellectual impairment and blood lead levels. *N Engl J Med* 2003;349(5):500–2.

44. Cory-Slechta DA. Legacy of lead exposure: consequences for the central nervous system. *Otolaryngol Head Neck Surg* 1996;114(2):224–6.

45. Brent J. A review of: "medical toxicology". *Clin Toxicol* 2006;44(3):355.

46. Bellinger DC. Lead. *Pediatrics* 2004;113(4 Suppl):1016–22.

47. Flora SJS, Flora G, Saxena G. Environmental occurrence, health effects and management of lead poisoning. In: Jose SC, Jose S, editors. *Lead: Chemistry, Analytical Aspects, Environmental Impact and Health Effects*. Amsterdam: Elsevier; 2006. pp. 158–228.

48. Sanders T, Liu Y, Buchner V, Tchounwou PB. Neurotoxic effects and biomarkers of lead exposure: a review. *Rev Environ Health* 2009;24(1):15–45.

49. Needleman H. Lead poisoning. *Annu Rev Med* 2004;55:209–22.

50. Xu J, Yan HC, Yang B, Tong LS, Zou YX, Tian Y. Effects of lead exposure on hippocampal metabotropic glutamate receptor subtype 3 and 7 in developmental rats. *J Negat Results Biomed* 2009;8:5.

51. Park SK, O'Neill MS, Vokonas PS, Sparrow D, Wright RO, Coull B, et al. Air pollution and heart rate variability: effect modification by chronic lead exposure. *Epidemiology* 2008;19(1):111–20.

52. Meyer PA, McGeehin MA, Falk H. A global approach to childhood lead poisoning prevention. *Int J Hyg Environ Health* 2003;206(4–5):363–9.

53. Needleman HL, Schell A, Bellinger D, Leviton A, Allred EN. The long-term effects of exposure to low doses of lead in childhood. An 11-year follow-up report. *N Engl J Med* 1990;322(2):83–8.

54. Guidotti TL, Ragain L. Protecting children from toxic exposure: three strategies. *Pediatr Clin North Am* 2007;54(2):227–35.

55. Jacobs DE, Clickner RP, Zhou JY, Viet SM, Marker DA, Rogers JW, et al. The prevalence of lead-based paint hazards in U.S. housing. *Environ Health Perspect* 2002;110(10):A599–606.

56. Cleveland LM, Minter ML, Cobb KA, Scott AA, German VF. Lead hazards for pregnant women and children: part 1: immigrants and the poor shoulder most of the burden of lead exposure in this country. Part 1 of a two-part article details how exposure happens, whom it affects, and the harm it can do. *Am J Nurs* 2008;108(10):40–9; quiz 50.

57. Guidotti TL, McNamara J, Moses MS. The interpretation of trace element analysis in body fluids. *Indian J Med Res* 2008;128(4):524–32.

58. Vij AG. Hemopoietic, hemostatic and mutagenic effects of lead and possible prevention by zinc and vitamin C. *Al Ameen J Med Sci* 2009;2(2):27–36.

59. Piomelli S. Childhood lead poisoning. *Pediatr Clin North Am* 2002;49(6):1285–304.

60. Patrick L. Lead toxicity part II: the role of free radical damage and the use of antioxidants in the pathology and treatment of lead toxicity. *Altern Med Rev* 2006;11(2):114–27.

61. Li C, Ni ZM, Ye LX, Chen JW, Wang Q, Zhou YK. Dose-response relationship between blood lead levels and hematological parameters in children from central China. *Environ Res* 2018;164:501–6.

62. Grant LD. Lead and compounds. In: Lippman M, editor. *Environmental Toxicants: Human Exposures and Their Health Effects*. Hoboken, NJ: John Wiley & Sons; 2008. pp. 757–809.

63. Rastogi SK. Renal effects of environmental and occupational lead exposure. *Indian J Occup Environ Med* 2008;12(3):103–6.

64. Navas-Acien A, Guallar E, Silbergeld EK, Rothenberg SJ. Lead exposure and cardiovascular disease – a systematic review. *Environ Health Perspect* 2007;115(3):472–82.

65. Alissa EM, Ferns GA. Heavy metal poisoning and cardiovascular disease. *J Toxicol* 2011;2011:870125.

66. Menke A, Muntner P, Batuman V, Silbergeld EK, Guallar E. Blood lead below 0.48 micromol/L (10 microg/dL) and mortality among US adults. *Circulation* 2006;114(13):1388–94.

67. Navas-Acien A, Selvin E, Sharrett AR, Calderon-Aranda E, Silbergeld E, Guallar E. Lead, cadmium, smoking, and increased risk of peripheral arterial disease. *Circulation* 2004;109(25):3196–201.

68. Weisskopf MG, Jain N, Nie H, Sparrow D, Vokonas P, Schwartz J, et al. A prospective study of bone lead concentration and death from all causes, cardiovascular diseases, and cancer in the Department of Veterans Affairs Normative Aging Study. *Circulation* 2009;120(12):1056–64.

69. Flora SJ. Arsenic-induced oxidative stress and its reversibility. *Free Radic Biol Med* 2011;51(2):257–81.

70. Saleh HA, El-Aziz GA, El-Fark MM, El-Gohary M. Effect of maternal lead exposure on craniofacial ossification in rat fetuses and the role of antioxidant therapy. *Anat Histol Embryol* 2009;38(5):392–9.

71. Winder C. Lead, reproduction and development. *Neurotoxicology* 1993;14(2–3):303–17.

72. Mudipalli A. Lead hepatotoxicity & potential health effects. *Indian J Med Res* 2007;126(6):518–27.

73. Agency for Toxic Substances and Disease Registry (ATSDR). Lead toxicity. 2017. https://www.atsdr.cdc.gov/csem/lead/docs/CSEM-Lead_toxicity_508.pdf [accessed 30.12.2019].

74. Renner R. Exposure on tap: drinking water as an overlooked source of lead. *Environ Health Perspect* 2010;118(2):A68–74.

75. Talaie H, Nasiri S, Gheisari M, Dadkhahfar S, Ahmadi S. Observational study of dermatological manifestations in patients admitted to a tertiary poison center in Iran. *Turk J Med Sci* 2018;48(1):136–41.

76. Agency for Toxic Substances and Disease Registry (ATSDR). Toxicological profile for lead. 2009. https://www.atsdr.cdc.gov/toxprofiles/tp13.pdf [accessed 30.12.2019].

77. National Toxicology Program (NTP). Lead and lead compounds. 2004. http://ntp.niehs.nih.gov/ntp/roc/content/profiles/lead.pdf [accessed 30.12.2019].

78. Nevin R. Understanding international crime trends: the legacy of preschool lead exposure. *Environ Res* 2007;104(3):315–36.

79. Patrick L. Lead toxicity, a review of the literature. Part 1: exposure, evaluation, and treatment. *Altern Med Rev* 2006;11(1):2–22.

80. Benjelloun M, Tarrass F, Hachim K, Medkouri G, Benghanem MG, Ramdani B. Chronic lead poisoning: a "forgotten" cause of renal disease. *Saudi J Kidney Dis Transpl* 2007;18(1):83–6.

81. Rossi E. Low level environmental lead exposure – a continuing challenge. *Clin Biochem Rev* 2008;29(2):63–70.

82. Flora SJ. Structural, chemical and biological aspects of antioxidants for strategies against metal and metalloid exposure. *Oxid Med Cell Longev* 2009;2(4):191–206.

83. Blokhina O, Virolainen E, Fagerstedt KV. Antioxidants, oxidative damage and oxygen deprivation stress: a review. *Ann Bot* 2003;91:179–94.

84. Gurer H, Ercal N. Can antioxidants be beneficial in the treatment of lead poisoning? *Free Radic Biol Med* 2000;29(10):927–45.

85. Pietta PG. Flavonoids as antioxidants. *J Nat Prod* 2000;63(7):1035–42.

86. Willcox JK, Ash SL, Catignani GL. Antioxidants and prevention of chronic disease. *Crit Rev Food Sci Nutr* 2004;44(4):275–95.

87. Kasperczyk S, Dobrakowski M, Kasperczyk J, Romuk E, Prokopowicz A, Birkner E. The influence of beta-carotene on homocysteine level and oxidative stress in lead-exposed workers. *Med Pr* 2014;65(3):309–16.

88. Kasperczyk S, Dobrakowski M, Kasperczyk A, Romuk E, Rykaczewska-Czerwińska M, Pawlas N, et al. Effect of N-acetylcysteine administration on homocysteine level, oxidative damage to proteins, and levels of iron (Fe) and Fe-related proteins in lead-exposed workers. *Toxicol Ind Health* 2016;32(9):1607–18.

89. Ahamed M, Siddiqui MK. Environmental lead toxicity and nutritional factors. *Clin Nut* 2007;26(4):400–8.

90. Das KK, Saha S. L-ascorbic acid and α tocopherol supplementation and antioxidant status in nickel- or lead-exposed rat brain tissue. *J Basic Clin Physiol Pharmacol* 2010;21(4):325–46.

91. Tariq SA. Role of ascorbic acid in scavenging free radicals and lead toxicity from biosystems. *Mol Biotechnol* 2007;37(1):62–5.

92. Shan G, Tang T, Zhang X. The protective effect of ascorbic acid and thiamine supplementation against damage caused by lead in the testes of mice. *J Huazhong Univ Sci Technolog Med Sci* 2009;29(1):68–72.

93. Wang C, Liang J, Zhang C, Bi Y, Shi X, Shi Q. Effect of ascorbic acid and thiamine supplementation at different concentrations on lead toxicity in liver. *Ann Occup Hyg* 2007;51(6):563–9.

94. Shalan MG, Mostafa MS, Hassouna MM, El-Nabi SE, El-Refaie A. Amelioration of lead toxicity on rat liver with vitamin C and silymarin supplements. *Toxicology* 2005;206(1):1–15.

95. Sajitha GR, Jose R, Andrews A, Ajantha KG, Augustine P, Augusti KT. Garlic oil and vitamin E prevent the adverse effects of lead acetate and ethanol separately as well as in combination in the drinking water of rats. *Indian J Clin Biochem* 2010;25(3):280–8.

96. Rendón-Ramirez A, Cerbon-Solórzano J, Maldonado-Vega M, Quintanar-Escorza MA, Calderón-Salinas JV. Vitamin-E reduces the oxidative damage on delta-aminolevulinic dehydratase induced by lead intoxication in rat erythrocytes. *Toxicol In Vitro* 2007;21(6):1121–6.

97. Flora SJ, Pande M, Mehta A. Beneficial effect of combined administration of some naturally occurring antioxidants (vitamins) and thiol chelators in the treatment of chronic lead intoxication. *Chem Biol Interact* 2003;145(3):267–80.

98. Youdim KA, Spencer JP, Schroeter H, Rice-Evans C. Dietary flavonoids as potential neuroprotectants. *Biol Chem* 2002;383(3–4):503–19.

99. Terao J. Dietary flavonoids as antioxidants. *Forum Nutr* 2009;61:87–94.

100. Rice-Evans C. Flavonoid antioxidants. *Curr Med Chem* 2001;8(7):797–807.

101. Beecher GR. Overview of dietary flavonoids: nomenclature, occurrence and intake. *J Nutr* 2003;133(10):3248S–54S.

102. Bravo A, Anacona JR. Metal complexes of the flavonoid quercetin: antibacterial properties. *Trans Met Chem* 2001;26:20–3.

103. Liu CM, Ma JQ, Sun YZ. Quercetin protects the rat kidney against oxidative stress-mediated DNA damage and apoptosis induced by lead. *Environ Toxicol Pharmacol* 2010;30(3):264–71.

104. Hu P, Wang M, Chen WH, Liu J, Chen L, Yin ST, et al. Quercetin relieves chronic lead exposure-induced impairment of synaptic plasticity in rat dentate gyrus in vivo. *Naunyn Schmiedebergs Arch Pharmacol* 2008;378(1):43–51.

105. Durrani AI, Schwartz H, Nagl M, Sontag G. Determination of free α-lipoic acid in foodstuffs by HPLC coupled with CEAD and ESI-MS. *Food Chem* 2010;120(4):1143–8.

106. De Araújo DP, Lobato Rde F, Cavalcanti JR, Sampaio LR, Araújo PV, Silva MC, et al. The contributions of antioxidant activity of lipoic acid in reducing neurogenerative progression of Parkinson's disease: a review. *Int J Neurosci* 2011;121(2):51–7.

107. Haleagrahara N, Jackie T, Chakravarthi S, Kulur AB. Protective effect of alpha-lipoic acid against lead acetate-induced oxidative stress in the bone marrow of rats. *Int J Pharmacol* 2011;7(2):217–27.

108. Sivaprasad R, Nagaraj M, Varalakshmi P. Combined efficacies of lipoic acid and 2,3-dimercaptosuccinic acid against lead-induced lipid peroxidationin in rat liver. *J Nutr Biochem* 2004;15(1):18–23.

109. Pande M, Flora SJ. Lead induced oxidative damage and its response to combined administration of alpha-lipoic acid and succimers in rats. *Toxicology* 2002;177(2–3):187–96.

110. Sethi P, Jyoti A, Hussain E, Sharma D. Curcumin attenuates aluminium-induced functional neurotoxicity in rats. *Pharmacol Biochem Behav* 2009;93(1):31–9.

111. Agarwal R, Goel SK, Behari JR. Detoxification and antioxidant effects of curcumin in rats experimentally exposed to mercury. *J Appl Toxicol* 2010;30(5):457–68.

112. Singh P, Sankhla V. In situ protective effect of curcumin on cadmium chloride induced genotoxicity in bone marrow chromosomes of Swiss albino mice. *J Cell Mol Biol* 2010;8(2):57–64.

113. Rao MV, Jhala DD, Patel A, Chettiar SS. Cytogenetic alteration induced by nickel and chromium in human blood cultures and its amelioration by curcumin. *Int J Hum Genet* 2008;8(3):301–5.

114. Shukla PK, Khanna VK, Khan MY, Srimal RC. Protective effect of curcumin against lead neurotoxicity in rat. *Hum Exp Toxicol* 2003;22(12):653–8.

115. Daniel S, Limson JL, Dairam A, Watkins GM, Daya S. Through metal binding, curcumin protects against lead- and cadmium-induced lipid peroxidation in rat brain homogenates and against lead-induced tissue damage in rat brain. *J Inorg Biochem* 2004;98(2):266–75.

116. Dairam A, Limson JL, Watkins GM, Antunes E, Daya S. Curcuminoids, curcumin, and demethoxycurcumin reduce lead-induced memory deficits in male Wistar rats. *J Agric Food Chem* 2007;55(3):1039–44.

117. Sharma V, Sharma A, Kansal L. The effect of oral administration of Allium sativum extracts on lead nitrate induced toxicity in male mice. *Food Chem Toxicol* 2010;48(3):928–36.

118. Senapati SK, Dey S, Dwivedi SK, Swarup D. Effect of garlic (*Allium sativum* L.) extract on tissue lead level in rats. *J Ethnopharmacol* 2001;76(3):229–32.

119. Pourjafar M, Aghbolaghi PA, Shakhse-Niaie M. Effect of garlic along with lead acetate administration on lead burden of some tissues in mice. *Pak J Biol Sci* 2007;10(16):2772–4.

120. Kilikdar D, Mukherjee D, Mitra E, Ghosh AK, Basu A, Chandra AM, et al. Protective effect of aqueous garlic extract against lead-induced hepatic injury in rats. *Indian J Exp Biol* 2011;49(7):498–510.

121. Flora SJ, Gupta R. Beneficial effects of *Centella asiatica* aqueous extract against arsenic-induced oxidative stress and essential metal status in rats. *Phytother Res* 2007;21(10):980–8.

122. Saxena G, Flora SJ. Changes in brain biogenic amines and haem biosynthesis and their response to combined administration of succimers and *Centella asiatica* in lead poisoned rats. *J Pharm Pharmacol* 2006;58(4):547–59.

123. Ponnusamy K, Mohan M, Nagaraja HS. Protective antioxidant effect of *Centella asiatica* bioflavonoids on lead acetate induced neurotoxicity. *Med J Malaysia* 2008;63(Suppl A):102.

124. Sainath SB, Meena R, Supriya Ch, Reddy KP, Reddy PS. Protective role of *Centella asiatica* on lead-induced oxidative stress and suppressed reproductive health in male rats. *Environ Toxicol Pharmacol* 2011;32(2):146–54.

125. Hussin M, Abdul-Hamid A, Mohamad S, Saari N, Ismail M, Bejo MH. Protective effect of *Centella asiatica* extract and powder on oxidative stress in rats. *Food Chem* 2007;100(2):535–41.
126. Dietrich KN, Ware JH, Salganik M, Radcliffe J, Rogan WJ, Rhoads GG, et al. Effect of chelation therapy on the neuropsychological and behavioral development of lead-exposed children after school entry. *Pediatrics* 2004;114(1):19–26.
127. Rogan WJ, Dietrich KN, Ware JH, Dockery DW, Salganik M, Radcliffe J, et al. The effect of chelation therapy with succimer on neuropsychological development in children exposed to lead. *N Engl J Med* 2001;344(19):1421–6.

CHAPTER 22

Arsenic Toxicity
Source, Mechanism, Global Health Complication, and Possible Way to Defeat

Dipti Debnath
University of North Texas Health Science Center

Biddut Deb Nath
Centre for the Rehabilitation of the Paralysed (CRP)

Rokeya Pervin and Md. Akil Hossain
Animal and Plant Quarantine Agency

CONTENTS

22.1 INTRODUCTION

The word "arsenic" originated from the Greek name arsenikon, meaning powerful which is a frequent element in the environment.[1–5] In the periodic table arsenic (As) is located as a 33rd chemical element, and as it shows both characteristics of metal and nonmetal, it is also categorized as a metalloid; often it is called metal and in terms of toxicology it is known as a heavy metal.[6] Arsenic is found in three allotropic modes: α (yellow), β (black), γ (gray), of the metallic state, and in several ionic forms.[7] As a natural element arsenic ranks 20th position for the most occurring trace element in the earth's surface and is widespread in the atmosphere.[8] Its combination with certain non-weathering resistant mineral sediments (for instance sulfide minerals) has contributed to its discharge into the environment in large quantities.[9] Arsenic is utilized in steel hardening and paint, semiconductor, pesticides, glass processing, fungicides, and rodenticides production.[10] It is often utilized to treat certain diseases (e.g., chronic myeloid leukemia, sleeping sickness) as a component of medicines. These effectiveness and exploitation of arsenic are also pertaining to pollution. Pollution of drinking water obtained from natural mineral sources is considered as the main cause of human arsenic toxicity instead of smelting, farming (fertilizers or pesticides), or mining sources.[11] Various countries, either more or less industrialized, have arsenic-contaminated drinking water.[12,13] The issue becomes a significant concern in the United States for instance, in Millard County drinking water arsenic content from private and public sources varies from 14 to 166 ppb.[14] In 2001, the Environment Protection Agency in the United States decreased the allowable amount of arsenic in drinking water from 50 to 10 ppb.[15] Higher levels up to 100–5,000 µg/L in surface water and groundwater can be detected in sulfide mineralization regions.[16] North and South America, some regions of Europe (Serbia, Romania, Hungary, and Croatia), and Southeast Asia (Philippines, China, India, Bangladesh, and Taiwan) have recorded high levels of arsenic in groundwater.[17–21] To the extent that toxic arsenic compounds can occur normally or anthropogenically in the atmosphere, comprehension of their toxicity is warranted.[22,23]

Arsenic ingestion can occur in several ways like inhalation, parenteral route, oral route, and dermal contact to some level. Arsenic air quantities in remote areas vary from 1 to 3 ng/m^3, however, quantities in metropolitans may vary from 20 to 100 ng/m^3. Arsenic levels are typically less than 10 µg/L in water, although there may be higher amounts near man-made sources or natural mineral deposits. Natural soil concentration of arsenic typically ranges from 1 to 40 mg/kg, but the introduction of fertilizers or the disposal of waste may generate much higher yields. Food intake is the biggest source of arsenic ingestion, with a typical intake of about 50 µg/day from food because the availability of arsenic in various foods varies from 20 to 140 ng/kg.[24] Ingestion from the water, soil, and the air is typically much lower, but in regions with arsenic pollution, exposure from these sources may become severe. Employees who manufacture or use arsenic chemicals in such workplaces as wineries, smelting, glass making, processing and implementation of pesticides, pharmaceuticals, ceramics, wood restoration or processing of semiconductors, metal ores refining, may be subjected to significantly greater scales of arsenic.[25] Another potential source of public exposure to arsenic is hazardous waste sites (HWS).[26] Such sites may be exposed through a number of mechanisms, including air pollution, polluted water or soil consumption, or via the food supply. Assessing the harmful effects of arsenic is not easy because hazard varies in several different organic and inorganic compounds depending on its solubility as well as its oxidation state.[27,28] Many reports have shown that arsenic toxicity differs on the various aspects like biological species, human vulnerability, amount of exposure, intensity and length, gender and age, genetic and dietary factors.[29] Many occurrences of arsenic human toxicity have been correlated with inorganic arsenic ingestion. Inorganic pentavalent arsenate (AsV) is two to ten times less poisonous than trivalent arsenite (AsIII).[30] Another inorganic arsenic element known as Gallium arsenide (GaAs) is a probable human health concern due to its extensive application in the microelectronics field.[31] Excessive levels of arsenic toxicity are of main concern because arsenic can lead to a variety of human health effects like cancer.[8] According to the latest report, interest in the arsenic poison has been increased in Thailand, Taiwan, Argentina, Bangladesh, Hungary, Finland, India, Chile, Mexico, China, and Inner Mongolia because of higher levels of arsenic consumption in their drinking water and emergence of several clinicopathological problems. The main effects are hematologic disorders like eosinophilia anemia and leukopenia, diabetes, carcinoma, developmental anomalies, neurobehavioral and neurologic disorders, portal fibrosis, hearing loss, skin cancer, cardiovascular and peripheral vascular disease.[32,33] Due to its abundance in the environment and its toxicity, serious environmental safety concerns are the potential for arsenic contamination of air, soil, and water from both natural and anthropogenic origins.[4,5]

22.2 CLINICAL UTILIZATION OF ARSENIC

22.2.1 Ancient Uses

Greek practitioners such as Galen and Hippocrates familiarized arsenic use and after that it was commonly used as a therapeutic agent. Different dosage forms of arsenic became available like tablets, solutions, injectable, and pastes. During nineteenth century, Fowler's solution, a 1% formulation of arsenic trioxide was extensively employed. The British *Pharmaceutical and Therapeutic Products* manual in 1958 which was edited by Martindale identified Fowler's solution indications as gingivitis and stomatitis in child, skin conditions (eczema, dermatitis herpetiformis, and psoriasis), Vincent's angina, and leukemia. This solution is often used as a health tonic. Lifelong use of Fowler's solution resulted in chronic arsenic poisoning which further caused angiosarcoma of the liver, haemangiosarcoma, and nasopharyngeal carcinoma. Before the Second World War arsenic was a preferred medication for syphilis.[34–37] Arsenic derivatives also have been used as chemical weapons throughout World War I. Lewisite and Adamsite are chemical agents of arsenic commonly known as battle gases. The chemical name of the Lewisite is 2-chlorovinyldichloroarsine which induces several health complications like severe blisters with slow-healing and respiratory inflammation. Its application in wartime led to the discovery of an antidote, leading in British anti-lewisite (BAL; 2, 3-dimercaptopropanol) production.[3,38]

22.2.2 Recent Uses

An arsenic compound called arsenic trioxide is now frequently used as an enhancer of apoptosis (programmed cell death) in patients with acute promyelocytic leukemia.[39–42] Arsenic releases apoptosis-inducing factors (AIF) from intermembrane space of the mitochondria and then it translocates to the nucleus of the cell, and by this mechanism arsenic causes apoptosis.[43] Then apoptosis-inducing factor triggers apoptosis, leading to impaired biochemistry, condensation of chromatin, degradation of DNA, and ultimately death of cells.[44] Arsenic remains an important component of many conventional non-western medicines. Most conventional Chinese medicines contain arsenic sulfide and can be used as tablets, pills, and other dosage forms and used for various treatment purposes like as an anti-inflammatory agent, some malignant tumors, asthma, psoriasis, hemorrhoids, syphilis, cough, rheumatism, an analgesic, and are also recommended as a health tonic.[45–47] Arsenic-containing naturopathic remedies are used in certain homeopathic formulations and hematological cancers in India and similarly, arsenic is used for hemorrhoids in naturopathic remedies in Korea.[48–50] Nevertheless, arsenic seems to be more a toxin than a desired component, and often with lead and mercury.[51,52] The California Department of Health Services tested 251 items in commercial herbal shops and found arsenic at concentrations of 20.4–114,000 parts per million (ppm) in 36 items (14%) with an mean of 145.53 ppm and an median of 180.5 ppm.[46]

22.3 CHEMISTRY OF ARSENIC

Arsenic primarily occurs in three states of valence such as −3, +3, and +5. In natural waters, pentavalent and trivalent arsenic forms are commonly found which are soluble in a wide variety of pH.[53,54] Trivalent species are more stable and prevalent in reducing atmospheric conditions, while pentavalent species are predominant in oxidizing atmospheric conditions. In the absence of oxygen, microorganisms in the soil can reduce arsenite to arsine (a poisonous gas).[55–57] Arsenic species can be methylated by animals, humans, and microorganisms as monomethylarsonic acid, dimethyl-arsinic acid, and trimethylarsine oxide.[58–61] Typically, pentavalent compounds are less toxic than trivalent compounds, and arsine gas is the most toxic of all. There has been a large range of arsenic poisoning.[62–64] There are organic arsenic compounds, but they are usually non-toxic, and typically inorganic arsenic compounds like monomethylarsonate and dimethylarsinate are even more poisonous to biological species than organic forms, whereas pentavalent arsenic is 60 times less poisonous to biological species than trivalent arsenic. In short, the toxicity of various arsenic compounds differs according to the following sequence: arsenite > arsenate > monomethylarsonate > dimethylarsinate.[65] Depending on the redox potential and pH, inorganic arsenic in water may be found in ionic pentavalent forms and non-ionic trivalent forms. The prevalent arsenic element is trivalent arsenite as H_3AsO_3 under reduced conditions, such as waters of almost neutral pH. The negatively charged form of pentavalent arsenate is prevalent in an oxidizing atmosphere and pH above neutral pH. It is a common fact that ionized arsenate forms always get completely eliminated and react more easily because of charged forms in all established arsenic extraction methods.[66,67] Depending on the water's atmospheric circumstances inorganic arsenic in water is found primarily in ionic pentavalent and non-ionic trivalent forms in varying percentages.[65,66] Trivalent arsenic is well established to be more poisonous than pentavalent forms and harder to eliminate from water.[15] Therefore, peroxidation of trivalent to pentavalent is needed before the operation of precipitation, adsorption, or ion exchange to maintain a high effect of elimination.[66,68,69]

22.4 SOURCES OF ARSENIC

22.4.1 Major Arsenic Minerals

It's very hard to find a pure form of arsenic metal in nature because in over 200 minerals which include arsenides, arsenates, sulfides, arsenites, oxides, and elemental it exists as a key ingredient. Arsenic is frequently associated with platinum deposits, molybdenum, bismuth, copper, antimony, cobalt, silver, phosphorus, zinc, tellurium, gold, nickel, mercury, iron, cadmium, selenium, uranium, tungsten, tin and lead, specially sulfosalts, and sulfides comprising compounds. Scorodite, arsenolite, realgar, pentoxide, claudetitis, arsenopyrite, and orpiment are generally known as arsenic-containing minerals.[70,71] Arsenic can also be oxidized to arsenate and arsenite through sulfide weathering. As a by-product of the smelting of lead, nickel, and copper, arsenic oxide is also produced. According to the environmental regulations it needs to be eliminated arsenic from ores to prevent entry to the atmosphere in solids, fluids or effluent gases. In mineralized regions, the highest proportions of these minerals occur. In the Earth's crust arsenopyrite is generated only under high-temperature conditions, along with other abundant orpiment, realgar, and As-sulfide minerals.[72]

Arsenopyrite is far less frequent than arsenic pyrite, which is arguably the most important origin of arsenic, although somewhat available in coal deposits. Arsenopyrite in sulfide ores combined with sediment-hosted gold deposits that are developed at about 100°C temperature and are collected from hydrothermal liquids known as the earliest developed minerals. Primarily, it is responsible for the indigenous arsenic, then the arsenic pyrite, and indeed afterward it is transformed into orpiment and realgar. At the recent phase of ore mineralization, oxides and sulfates are developed.[73,74]

22.4.2 Rock-Forming Minerals

Some common rock-forming minerals contain arsenic as different concentrations but not as a key element. It is 1.5 mg/kg for the crustal plentiful supply of arsenic and the element is highly chalcophile. Around 20% of natural minerals are elemental arsenic polymorphs, arsenides, oxides, arsenites, and alloys, another 20% are sulfides, and the rest of the major 60% are arsenates. Sulfur is the highest proportion that tends to occur in sulfur minerals in accordance with arsenic chemistry and the most prevalent of these is pyrite. Sulfide minerals have recorded arsenic quantities of more than 10^5 mg/kg, but usually, quantities are much smaller.[75,76] Consequently, arsenic quantities are usually 1 mg/kg or less in silicate minerals.[4] Average arsenic concentrations of 1–10 mg/kg are present in many metamorphic and igneous rocks and equal amounts of numerous carbonate rocks and minerals.[77] Galena, chalcopyrite, and pyrite can vary, even within a defined grain, however, in some scenarios achieve up to 10% by weight. Arsenic can be identified as a replacement for sulfur in the crystal lattice of many sulfide minerals. In addition to being a major component of ore deposits, pyrite is also produced under reduced environments (authigenic pyrite) in low-temperature sedimentary conditions. Throughout today's geochemical cycles, authigenic pyrite plays a crucial role. The wealthy sources of natural arsenic emergence include the many water sources, sediments of a number of rivers, oceans, and lakes. Typically, pyrite is selectively produced in areas of extreme reduction, such as across submerged plant roots or other organic decomposing nuclei. It can sometimes be available as a framboidal pyrite in a distinctive feature form. In several oxide minerals and hydrous metal oxides, elevated levels of arsenic can also be present whether as part of the mineral composition or as sorbed particles. Phosphate minerals, however, are far less frequent than oxide minerals, making a consequently minor contribution to most sediment's load. The most typical minerals of silicate comprise around 1 mg/kg or less and minerals of carbonate generally contain less than 10 mg/kg.[78–80]

22.4.3 Sedimentary Rocks

Usually, arsenic content in sedimentary rocks is 5–10 mg/kg, which is a fairly higher average terrestrial prevalence.[81] If a connection is made with igneous rocks, sedimentary rocks are similarly abundant in arsenic. Although minerals for instance feldspars and quartz abound principally in arsenic, its comparably inferior containers are sands and sandstone. The average density of arsenic in sandstone is about 4 mg/kg. In comparison to argillaceous deposits which have significantly higher levels and a wider variety of arsenic than sandstones, traditionally almost 13 mg/kg on average.[82] The larger values represent the greater percentage of clays, organic matter, minerals, and oxides in sulfide. Because of their intensified pyrite quantity, arsenic concentration exists in black shales mostly at the higher end of the scale. Marine shales generally have higher levels of sulfur. Relatively high densities of arsenic from mid-ocean settings (mid-Atlantic ridge, 174 mg/kg average) were assessed for shales and in this situation, Atlantic ridge gases can be a source of high arsenic. Although, arsenic quantities in bituminous and coal deposits are changeable but often high. Arsenic densities of 100–900 mg/kg are found in organic-rich shale samples from Germany. Incredibly high levels of up to 35,000 mg/kg of some coal samples were observed. Carbonate rocks usually have lower levels, indicating low mineral levels. The sources with the greatest reported arsenic densities are ironstones and iron-rich rocks, mostly several thousand mg/kg. Another arsenic-enriched source is phosphorites which contain arsenic levels of up to 400 mg/kg.[83]

22.4.4 Soils

The level of 5–10 mg/kg generally defines arsenic baseline concentrations in soils are normal. For American soils,[84–86] Shacklette et al. reported an average of 7.4 mg/kg (901 samples) and in world soils, Boyle and Jonasson reported an estimated baseline concentration of 7.2 mg/kg.[70,71] A very much higher mean value of 11.3 mg/kg was reported by Ure and Berrow. As a consequence of the historical usage of arsenic pesticides to fruit crops, they also noted quantities in the range 366–732 mg/kg in orchard soils.[81] Due to the high prevalence of sulfide mineral stages under reduced conditions alone peats and bog soils may have higher levels (average 13 mg/kg). Pyrite oxidation in sulfide-rich soils like mineral veins, dewatered mangrove marshes, and pyrite-rich shales developed acid sulfate soils may also be comparatively enriched in arsenic. In the B horizons of acid sulfate deposits resulting from weathering of pyrite-rich shales in Canada, Dudas identified arsenic levels up to 45 mg/kg. Due to the volatilization or leaching of arsenic to lower levels, levels in the overlying leached horizons were small (1.5–8.0 mg/kg).[87,88] Relatively high levels of up to 41 mg/kg in Vietnam's Mekong Delta acid sulfate soils were detected by Gusafsson and Tin. Despite the dominant sources of arsenic in soils being geological, and thus relying to some degree on the parent rock content composition. Additional supplies from industrial sources such as smelting and fossil-fuel gasoline vapors and agricultural supplies like phosphate fertilizers and pesticides are likely to be obtained nearby.[89,90]

22.4.5 Contaminated Surficial Deposits

Sediments and soils polluted by mining materials, particularly effluent and mine tailings, likely carry arsenic at levels far greater than the standard. Concentrations can jump up to a few thousand mg/kg in tailings-contaminated soils and tailings piles. In addition to representing an enhanced prevalence of primary arsenic-rich sulfide minerals, the elevated levels reflect secondary iron arsenates and iron oxides produced as initial ore mineral reaction products. The secondary sulfide minerals have different solubility in surface and groundwater oxidizing circumstances and the primary minerals in the tailings piles are vulnerable to oxidation. A widely known product of sulfide oxidation is called scorodite and its solubility is regarded in such oxidizing sulfide conditions to control arsenic concentrations. Under most groundwater conditions, scorodite is metastable and appears to dissolve incongruously, producing iron oxides and releasing arsenic into solution.[91] To practice, indeed, a broad range of interactions with Fe-As solubility are noticed, some of which related to the mineral type. Arsenic bound to iron oxide, especially under oxidizing conditions, is comparatively immobile, and secondary arsenolite is generally soluble.[92,93]

22.4.6 The Atmosphere

Even though the previous review has made it known that arsenic levels in the environment are small, contributions from smelting and other industrial activities, combustion of fossil fuel, and volcanic activity also boost them. Concentrations of approximately 10^{-5}–10^{-3} g/m^3 were observed in uncontaminated areas, rising to 0.003–0.18 g/m^3 in the city areas, and exceeding 1 g/m^3 near industrial sites.[16] A majority of the arsenic in the environment is particulate. Based on the relative values of wet and dry accumulation and proximity to sources of pollution, total arsenic accumulation levels were estimated in the range <1–1,000 µg/m^2/year. A mid-Atlantic coast value calculation ranged from 38 to 266 µg/m^2/year (29%–55% as a dry deposition).[94] In addition, water arsenic is airborne in nature, whether either deposition is dry or wet, and therefore this can increase slightly the concentration of aqueous substances. Nonetheless, it cannot be established that there is a real health hazard to drinking water supplies from toxic arsenic. In parts of China, atmospheric arsenic from coal-burning was cited as a significant cause of lung cancer. The hazard, however, lies in the direct inhalation of domestic coal-fire smoke and, in particular, in the ingestion of dried food over coal fires, instead of in the drinkable water influenced by environmental sources.[95]

22.4.7 Anthropogenic Sources

Arsenic has been utilized in various countries for a broad variety of applications. Some common anthropogenic sources of arsenic include pharmaceuticals, cosmetics, pesticides, mining operations and processing wastes, wood treatments, vitamin supplements, nutraceuticals, poultry and swine feed additives, dyes, herbicides, FGD plants, electronic manufacturing, paints, coal combustion, smelters, cigarettes, cattle dips, and highly soluble trioxide stockpiles are also not unusual. According to the record, 22% of the world's arsenic output is used in agricultural chemicals, 70% in timber treatment as copper chrome arsenate, and the rest in non-ferrous alloys, pharmaceutical, and glass. The main industrial activities that lead to anthropogenic arsenic pollution of soil, air, and water are burning of fossil fuels, smelting of non-ferrous metals, and mining. It is proven from the previous history that the degradation of large tracts of arable land is due to excessive arsenic-containing pesticides. Environmental pollution has also resulted by the use of arsenic in wood processing. The main uses in the United States were glass processing, wood preservation, and agricultural uses. A somewhat larger insight into the production of the different inorganic and organic sources of arsenic is gained with the emergence of new analytical techniques. Nonetheless, there are measurement of how much methylated arsenic derivatives lead to atmospheric quantities of arsenic in the atmosphere and measurements of volatilization at low temperatures. Overall arsenic releases from anthropogenic sources into the ecosystem are in the range of 30,000 tons As per year. The burning of coal and copper smelting alone leads to nearly 60% of arsenic pollution. There are also inconsistencies in the details of anthropogenic arsenic sources including wood and cow dung burning (Africa and India), forest burning and agricultural, smelting and gold mining (South Africa and *Union of Soviet Socialist Republics*), and herbicide use. Arsenic coal combustion at 6,240 tons As per year, a mix of steel and smelter production at 14,350 tons As per year, and herbicide use at 3,440 tons As per year were estimated by Chilvers and Peterson. Although global emissions to the air from anthropogenic sources, especially copper smelters contribute indirectly to land and terrestrial pollution, large quantities of arsenic are introduced to the land directly as a landfill from sludge disposal and slag and from refining and smelting operations.[96–98]

22.5 CLINICAL FEATURES OF ARSENIC

22.5.1 Acute Poisoning

Several acute arsenic poisoning incidents are caused by accidental exposure of pesticides or insecticides and less often by attempted suicide. Less than 5 mg lead in diarrhea and vomiting but fix within 12 h and it is also confirmed that medication is not needed.[99] In case of acute poisoning, the lethal dose of arsenic varies from 100 to 300 mg.[100] "The acute inorganic arsenic lethal dose to humans has been calculated to be around 0.6 mg/kg/day" according to the Risk Assessment Information System database statement.[101] A 23-year-old man who had swallowed 8 g of arsenic lived for 8 days.[102] Another case of a student, who absorbed 30 g arsenic, received aid after 15 h and lasted 48 h but died in spite of British arsenic antidote named anti-lewisite therapy and hemodialysis following gastric lavage.[103] Death normally occurs between 24 h to 4 days, based on the amount of arsenic ingested. At first, the clinical features refer typically to the gastrointestinal system and are frequent watery diarrhea, vomiting, nausea, and colicky abdominal pain. Extreme abdominal pain may occur and imitate an acute abdomen.[104] In the absence of other gastrointestinal symptoms,[105] prolonged salivation[106] occurs and maybe the reporting problem. Certain pathological symptoms are severe cardiomyopathy,[102,107] a systemic rash of the skin, seizures,[106] and acute psychosis. A prevalent trait is diarrhea due to higher blood vessel permeability. The substantial watery stools are referred to as "choleroid diarrhea." The stools are characterized as "bloody rice water" diarrhea in acute arsenic poisoning due to blood in the gastrointestinal tract but the word "rice water" is used in cholera. Due to gastrointestinal tract secretion a tremendous loss of fluid occur, which results in severe dehydration, decreased circulating amount of blood, and concomitant circulatory failure and these are the causes of death. Hepatic steatosis, gastritis, and oesophagitis are identified on postmortem inspection.[102]

Documented hematological complications are normocytic normochromic anemia, intravascular coagulation, basophilic stippling, bone marrow dysfunction, hemoglobinuria, and extreme pancytopenia. Four of the eight sailors were susceptible to arsine confirmed renal failure. Prominent features of acute toxicity are pulmonary edema and respiratory failure.[107–109] Peripheral neuropathy, which can last for as many as 2 years, is the most severe neurological form.[102,110,111] Rapid and extreme ascending weakness can result from peripheral neuropathy which is identical to the condition of Guillain-Barré, demanding mechanical ventilation.[107] A typical symptom is encephalopathy and if the encephalopathy etiology is unclear, the risk of arsenic poisoning must be acknowledged. After arsphenamines has given intravenously encephalopathy appeared.[112] It is assumed that the cause of encephalopathy is due to hemorrhage.[113] With acute arsenic poisoning metabolic alterations are also identified. Hypocalcaemia and hypoglycemia have developed in cattle and in a single patient acidosis developed. The urinary arsenic level is the best predictor of recent intake which is 1–2 days in acute toxicity.[102,114]

22.5.2 Chronic Poisoning

Lifelong exposure to arsenic progresses to multi-system illness, and malignancy is the most serious consequence. Arsenic poisoning clinical characteristics differ among geographic regions, racial groups, and persons. Which reasons decide the frequency of a specific clinical symptom or which body structure is affected is ambiguous. Therefore, a broad array of clinical characteristics is common among people subjected to chronic arsenic toxicity. The progression of sore throat, vomiting, and abdominal pain is gradual with un-specific symptoms.

22.5.2.1 Skin

There are dramatic changes in the skin with long-term access.[115] The primary indication is often dependent on solar keratosis, palmar, and hyperpigmentation, and a key trait is dermatological alterations. Keratosis may seem to be a consistent thickening or distinct nodule.[116] It is highlighted that a crucial diagnostic factor is both palmar and solar keratosis. Unusual darkening of the skin as a distinctive "raindrop" look or diffuse dark brown spots and less distinct diffuse darkening of the skin.[117] Bowen's disease may be caused by enhanced ultraviolet radiation penetration and excess skin melanin levels, which is arsenic-related skin cancer and rarely occurs in Asian people. In the case of non-melanin colored skin arsenic may induce a basal cell carcinoma. After contamination the dormant duration can be just as prolonged as 60 years. It has been documented that poison spread in vineyard laborers employing arsenical pesticides, patients administered with Fowler's solution, from drinking polluted wine and in sheep dip workers.[118] A further sign linked to arsenic accumulation is noticeable axial white lines in the fingernails and toenails termed Mee's lines in keratin-rich portions.[119]

22.5.2.2 Gastrointestinal System

Whereas diarrhea is an early-onset symptom and significant of acute arsenic poisoning, diarrhea results in repeated episodes of chronic poisoning and can be linked with vomiting. If some other characteristics such as neuropathy and skin alterations are indeed visible, skepticism of arsenic consumption should be provoked.[120] In West Bengal, India 248 patients with confirmed chronic arsenic poisoning who ingested arsenic polluted drinking water for 1–15 years, 69 of them were biopsied and from them, 63 showed non-cirrhotic portal fibrosis as well as 76.6% people were found with hepatomegaly.[121] From another analysis, for 5 of 42 patients with partial septal cirrhosis, arsenic was regarded as the etiological factor, an inert type of macronodular cirrhosis.[122]

22.5.2.3 Cardiovascular System

Due to arsenic exposure, a higher incidence of heart disease is noted in smelter employees.[123–125] Depending on a matrix for cumulative arsenic ingestion, research in Millard County, USA, showed a remarkable rise in morbidity from hypertensive cardiovascular disease in both females and males. Rahman et al. noted an elevated risk of high blood pressure in 1999 in a major survey of 1,481 people subjected to arsenic in well water in Bangladesh.[126] In "arseniasis-hyperendemic villages" 74 Taiwanese patients with ischaemic heart disease were analyzed, indicating a connection between long-term arsenic ingestion and ischaemic heart disease.[127,128] Arsenic is the main reason for cardiac arrhythmias, heart disease, and myocardial injury.[129,130] Due to long-term heavy arsenic interaction in artesian well water, black foot disease which is a distinctive peripheral vascular disease appears that makes foot gangrene exceptional to a confined area on Taiwan's southwest coast. Chile is another country that confirms peripheral vascular disease.[128,131]

22.5.2.4 Neurological System

Heavy metals like arsenic, lead, and mercury always threat the neurological system as the biggest target for harmful effects. There are several and varying neurological consequences. Peripheral neuropathy that imitates Guillain-Barré syndrome with identical electromyographic observations is the most widespread discovery.[132] Originally, neuropathy becomes tactile with a mask and anesthesia packaging. The poisoning consequences also include loss of memory, frustration, and behavioral changes.[133] Two staff confirmed cognitive deficits from 14 to 18 months of exposure and went back to normal mental function after termination from the source of arsenic.[134] In a big survey of 8,102 women and men who encountered long-term arsenic ingestion from well water, an enhanced frequency of cerebrovascular disease, notably cerebral infarction, was detected.[135]

22.5.2.5 Respiratory System

West Bengal, India's findings demonstrate both obstructive and restrictive lung disease.[116] For patients with the characteristic skin lesions of chronic arsenic poisoning, the respiratory disease was more severe.[136] Similar studies have been documented in Chilean children of connection between respiratory disease and skin symptoms.[131] Necropsy analysis in a restricted patient population promotes the possibility of enhanced deposition of arsenic in the lung while the reason is unknown.[137,138] A high incidence of bronchitis appears in a study of patients with Taiwan's black foot disease.[128]

22.5.2.6 *Malignant Disease*

As several millions of people are probable victims, the association between malignancy and arsenic is of serious concern. Arsenic is related to bladder, liver, skin, kidney, and lung cancers in India and Bangladesh.[139] Some other countries have proof that arsenic consumption induces kidney,[140] bladder,[141] liver, skin,[118] and lung[124,140] malignancies. Taiwanese information also records cancers of the stomach, kidney, colon, bladder, larynx, nasal cavity, skin, liver, bone, and lung as well as lymphoma.[128] Although not completely defined, the mechanisms may be a negative impact on enhanced protooncogene c-myc free radical induction and development, DNA methylation, and DNA repair. In certain conditions, arsenic may function as a tumor progressor, co-carcinogen, or tumor booster. Elevated arsenic concentrations in livestock are teratogenic.[142] In a group of people who drank arsenic from well water in Finland, structural chromosome abnormalities have been identified.[143]

22.6 MECHANISMS OF ARSENIC TOXICITY

Arsenic utilizes its damaging effect by impairing cellular respiration by inhibiting several mitochondrial enzymes and uncoupling oxidative phosphorylation; this is one of the mechanisms among others. Some arsenic poisoning commonly depends on its capacity to deal with enzyme and protein sulfhydryl groups as well as to replace phosphorus in a multitude of biochemical reactions.[144] Inactivation of enzymes like thiolase and dihydrolipoyl dehydrogenase occurs when arsenic and sulfhydryl groups of protein *in vitro* reaction takes place, as a result, inhibited oxidation of beta oxidation of fatty acids and pyruvate is generated.[145] Genotoxicity trials have shown that arsenic substances involve in the creation of micronuclei in both rodent and human cells in culture,[146–148] and exposed human cells,[149] trigger chromosomal distortions and prevent DNA repair and sister-chromatid transactions. The genotoxicity pathway is unclear but may be because of arsenate's power to act as an analog phosphate or fix enzymes or arsenate's capacity to stop DNA replication.[150] Without having any animal models there is another way, that is, experiments of *in vitro* cell conversion become a viable way to access information on arsenic poisoning's carcinogenic pathways. Arsenic derivatives and arsenic are detrimental to mouse cells and also hamster embryo cells of Syria and cause morphological alterations.[151–153] Arsenic trioxide has also been noted to cause damage to DNA in human lymphocytes[154] as observed by comet assay. It has also been demonstrated that arsenic substances trigger *c-fos* gene expression, in mammalian cells oxidative stress protein heme oxygenase, promote gene multiplication, block DNA repair, and stop mitotic cells.[155–158] Methylation is the key metabolic mechanism in humans for inorganic arsenic. Until elimination in the urine, most inorganic arsenic is metabolized into dimethylarsinic acid and monomethylarsonic acid. Arsenic methylation includes a two-electron conversion of pentavalent to trivalent arsenic derivatives supplemented by the transition from a methyl donor, including S-adenosylmethionine, of a methyl group.[8,159,160]

In the previous time, the common concept was that arsenite was the most likely source of arsenic carcinogenesis and that arsenic compound methylation was a mechanism of detoxification. Many scientists assumed that arsenic methylation was of great significance in reducing the carcinogenicity and/or poisoning.[161] The current belief of arsenic carcinogenesis is that arsenic methylation can be a toxication pathway rather than detoxification, and there are several chemical sources of arsenic that can be casual in carcinogenesis. A large part of the data has assembled on the toxication mechanism of methylated arsenic in a brief time frame.[162] Monomethylarsonic acid is recognized to have different observational mechanisms like genotoxicity,[163] enzyme inhibition,[164,165] and cell toxicity[166] in some of the biological activities. Monomethylarsonic as a trigger of arsenic carcinogenesis is an excellent candidate. Dimethylarsinic acid was observed arsenic exposed to human urine who has given a chelator.[167] There is presently limited information available on trivalent methylated compound tissue levels. Dimethylarsinic acid is regarded to also have clastogenicity[167] in varying observational systems such as cell toxicity,[166] enzyme inhibition,[164,165] and genotoxicity.[163] Trimethylarsinic acid can be generated from the oxide of trimethylarsine. One group finding indicates that only rats in their urine may contain significant quantities of trimethylarsine oxide. A Trimethylarsinic acid molecule does not have ionizable hydroxyl groups that inhibit trivalent arsenic molecules' capacity to communicate with DNA. Trimethylarsinic acid can be present in human tissues at lesser concentrations than rat tissues. Even though, arsenic is thought to be a serious public health issue, but there is still confusion about the molecular mechanism by which and the dosage at which it leads to cancer.[8,161]

22.7 PREVENTION, MANAGEMENT, AND SUBSEQUENT REGULATIONS

In developing countries, where millions of people's lives are damaged, human disaster due to arsenic poisoning is most serious. Several difficulties need to be specified in fixing the growing issue of arsenic toxicity and ill health. Data is needed to decide whether there is a limit for the manifestation of carcinogenic effects and to assess the exposure dose and duration.[168] More analyses are needed to connect noxious aspects with minerals, antioxidant, and vitamins preventive task as well as nutritional status, age, gender, and potential genetic polymorphism. In the same family, as is typically seen in Bangladesh, there is a defined variability in clinical characteristics among people. This could be identical to patients with irritable bowel syndrome because of "rapid or gradual" methylators of arsenic.[169] An aim is to provide secure drinking water. A number of different complexity strategies are present for eliminating arsenic from drinking water.[170] Especially in developing countries, the approach that is urgently needed should be price-effective, efficient, and community sustainable. Ion-exchange or precipitation techniques are among the available methods for separating arsenic from water. Arsenic filtration from tube wells has generated a variety of filters with different usability problems, cost, and sophistication; servicing and performance are associated with their use. Serious consideration is required for the system and the cost of a way to dispose arsenic stored after filtration. Employment of natural iron-treated items like iron oxide-coated sand, iron-treated gel beads, and iron-treated activated carbon are documented by fruitful research and the most powerful form of which was iron oxide-coated sand.[171] A widely accessible, interesting, and economical alternative is to collect rainwater and pump surface water. The cheapest option, indeed, would rely on society generosity to encourage the use of a non-contaminated neighbor's well.

Medication may include chelation treatment, disinfection, support measures, and separating patients from the origin of the contamination. Regarding current poisonings, lavage should be recommended, accompanied by activated charcoal, while activated charcoal may not adsorb large quantities of arsenic.[172] It is crucial to evaluate the intravascular volume and to prescribe adequate pressors, liquids, and electrolytes. A central venous catheter and Foley catheter should be inserted in chronic toxic patients; a large urine output should be sustained to improve excretion. Urine alkalinization at a pH of 7 may prohibit red cell metabolic byproducts from being stored in the renal tubules. Compounds that link to and improve the fecal and urinary elimination of toxic metals, also form complexes with toxic metal ions *in vivo* are called chelating agents. Metal complexes produced which are water soluble can be easily eliminated in feces and urine, reducing the levels of the body.[173] British anti-lewisite (BAL) has been the first chelating agent of preference to combat arsenic toxicity for several years. It is also confirmed that D-penicillamine is successful.[174] BAL has recently been shown to boost the material of brain arsenic.[173,175] BAL has been chemically synthesized to form water-soluble compounds, 2,3-dimercaptopropane-1-sulfonate (DMPS) and

2,3-dimercaptosucinic acid (DMSA), to fix this issue.[173] DMSA and DMPS are far more efficient than BAL depending on their therapeutic indexes, and DMSA has been utilized with great results to cure arsenic poisoning.[176,177] In certain cases where BAL and D-penicillamine did not have it, the DMSA has also found to be effective. People with chronic gastroenteritis, BAL should be recommended because DMSA can only be given orally, restricting its use to patients without intense gastroenteritis.[178] The medication of exposure to arsine gas varies as chelating agents may be inadequate. Without protecting hemolysis chelation treatment can enhance urinary arsenic elimination. To resolve the hemolytic reaction, exchange transfusions may be feasible, and hemodialysis may be appropriate for renal failure caused by arsine.

22.8 CONCLUSION

Exposure to arsenic harms millions of people around the world. Arsenic exposure usually takes place through dermal interaction, consumption, and inhalation. Due to such exposure considerable number of systemic health consequences occur in varieties of organs and tissue system like hematopoietic system, kidney, skin, gastrointestinal tract, respiratory system, bladder, and liver. New research has shown that arsenic use in drinking water has been tied to two non-cancer risk factors like diabetes mellitus and hypertension, which are considered as a leading source of death and morbidity. Some new investigation has shown in several countries, for instance, Bangladesh and Chile that arsenic raises infant mortality, spontaneous abortions, stillbirths, and pre-term births. Even at low doses of human inhalation, arsenic has cytotoxic, and biochemical impacts have been proved by *in vitro* and animal experiments. These are the consequences: protein-DNA cross-links induction, modified DNA-repair genes function, and changed output of colony formation, impaired methylation of DNA, oxidative damage initiation and gene expression, apoptosis activation, hindrance of pyruvate dehydrogenase, and abnormal glucocorticoid receptor function. In regards, the proof of human cancer is quite powerful, particularly for bladder, skin, kidney, liver, and lung cancer. In order to explore the dose-response link between non-cancerous endpoints and arsenic consumption, more epidemiological experiments are strongly advised. Such endpoints likely causes disease and death because of the very significant populations subjected and even slight increases in relative risk at low doses of arsenic exposure could be of considerable public health importance. In establishing a through arsenic risk analysis and administration scheme, such evidence is also crucial.

REFERENCES

1. Winship KA. Toxicity of inorganic arsenic salts. *Adverse Drug React Acute Poison Rev* 1984;3(3):129–60.
2. Wade MJ, Davis BK, Carlisle JS, Klein AK, Valoppi LM. Environmental transformation of toxic metals. *Occup Med* 1993;8(3):574–601.
3. Gorby MS. Arsenic poisoning. *West J Med* 1988;149(3):308–15.
4. Smedley PL, Kinniburgh DG. A review of the source, behaviour and distribution of arsenic in natural waters. *Appl Geochem* 2002;17(5):517–68.
5. Ng JC, Wang J, Shraim A. A global health problem caused by arsenic from natural sources. *Chemosphere* 2003;52(9):1353–9.
6. Mandal BK, Suzuki KT. Arsenic round the world: a review. *Talanta* 2002;58(1):201–35.
7. Orloff K, Mistry K, Metcalf S. Biomonitoring for environmental exposures to arsenic. *J Toxicol Environ Health B Crit Rev* 2009;12(7):509–24.
8. National Research Council. *Arsenic in Drinking Water*. National Academy Press, Washington, DC; 1999.
9. Mudroch A, Clair TA. Transport of arsenic and mercury from gold mining activities through an aquatic system. *Sci Total Environ* 1986;57:205–16.

10. Hathaway GJ, Proctor NH, Hughes JP, Fischman ML. Arsenic and arsine. In: Proctor NH, Hughes JP, editors. *Chemical Hazards of the Workplace*. 3rd ed. Van Nostrand Reinhold, New York, NY; 1991. pp. 92–96.

11. Matschullat J. Arsenic in the geosphere-a review. *Sci Total Environ* 2000;249(1–3):297–312.

12. Gebel T. Confounding variables in the environmental toxicology of arsenic. *Toxicology* 2000;144(1–3):155–62.

13. Zaw M, Emett MT. Arsenic removal from water using advanced oxidation processes. *Toxicol Lett* 2002;133(1):113–8.

14. Lewis DR, Southwick JW, Ouellet-Hellstrom R, Rench J, Calderon RL. Drinking water arsenic in Utah: a cohort mortality study. *Environ Health Perspect* 1999;107(5):359–65.

15. Chowdhury UK, Biswas BK, Chowdhury TR, Samanta G, Mandal BK, Basu GC, et al. Groundwater arsenic contamination in Bangladesh and West Bengal, India. *Environ Health Perspect* 2000;108(5):393–7.

16. World Health Organization. *Arsenic and Arsenic Compounds*. Environmental Health Criteria 224. 2nd ed. World Health Organization, Geneva; 2001.

17. World Health Organization. *Guidelines for Drinking-Water Quality*. 4th ed. World Health Organization, Geneva; 2011.

18. Gebel TW. Arsenic and drinking water contamination. *Science* 1999;283(5407):1455.

19. Ćavar S, Klapec T, Grubešić RJ, Valek M. High exposure to arsenic from drinking water at several localities in eastern Croatia. *Sci Total Environ* 2005;339(1–3):277–82.

20. Romić Ž, Habuda-Stanić M, Kalajdžić B, Kuleš M. Arsenic distribution, concentration and speciation in groundwater of the Osijek area, eastern Croatia. *Appl Geochem* 2011;26(1):37–44.

21. Rowland HA, Omoregie EO, Millot R, Jimenez C, Mertens J, Baciu C, et al. Geochemistry and arsenic behaviour in groundwater resources of the Pannonian Basin (Hungary and Romania). *Appl Geochem* 2011;26(1):1–7.

22. Nevens F, Fevery J, Van Steenbergen W, Sciot R, Desmet V, De Groote J. Arsenic and non-cirrhotic portal hypertension: a report of eight cases. *J Hepatol* 1990;11(1):80–5.

23. Luh MD, Baker RA, Henley DE. Arsenic analysis and toxicity-a review. *Sci Total Environ* 1973;2(1):1–2.

24. Agency for Toxic Substances and Disease Registry. *Toxicological Profile for Arsenic TP-92/09*. Center for Disease Control, Atlanta, GA; 2000.

25. Hartwig A, Groblinghoff UD, Beyersmann D, Natarajan AT, Filon R, Mullenders LH. Interaction of arsenic (III) with nucleotide excision repair in UV-irradiated human fibroblasts. *Carcinogenesis* 1997;18(2):399–405.

26. Agency for Toxic Substances and Disease Registry. *Hazardous Chemicals Data*. Center for Disease Control, Atlanta; 1992.

27. Marafante E, Vahter M. Solubility, retention, and metabolism of intratracheally and orally administered inorganic arsenic compounds in the hamster. *Environ Res* 1987;42(1):72–82.

28. Venugopal B, Lucky TD. Metal Toxicity in Mammals. Plenum Press, New York, NY; 1978.

29. Chen CJ, Lin LJ. Human carcinogenicity and atherogenicity induced by chronic exposure to inorganic arsenic. In: Nriagu JO, editor. *Arsenic in the Environment; Part II: Human Health and Ecosystem Effects*. John Wiley & Sons, New York, NY; 1994. pp. 109–131.

30. Kosnett MJ. Arsenic. In: Olson KK, editor. *Poisoning and Drug Overdose*. Appleton & Lange, Norwalk, CT; 1994. pp. 87–89.

31. National Toxicology Program (NTP). NTP toxicology and carcinogenesis studies of gallium arsenide (CAS No. 1303-00-0) in F344/N rats and B6C3F1 mice (inhalation studies). *Natl Toxicol Program Tech Rep* 2000;492:1–306.

32. Chappell WR, Beck BD, Brown KG, Chaney R, Cothern R, Cothern CR, et al. Inorganic arsenic: a need and an opportunity to improve risk assessment. *Environ Health Perspect* 1997;105(10):1060–7.

33. Mazumder DN, Haque R, Ghosh N, De BK, Santra A, Chakraborty D, et al. Arsenic levels in drinking water and the prevalence of skin lesions in West Bengal, India. *Int J Epidemiol* 1998;27(5):871–7.

34. Regelson W, Kim U, Ospina J, Holland JF. Hemangioendothelial sarcoma of liver from chronic arsenic intoxication by Fowler's solution. *Cancer* 1968;21(3):514–22.

35. Lander JJ, Stanley RJ, Sumner HW, Boswell DC, Aach RD. Angiosarcoma of the liver associated with Fowler's solution (potassium arsenite). *Gastroenterology* 1975;68(6):1582–6.

36. Neshiwat LF, Friedland ML, Schorr-Lesnick B, Feldman S, Glucksman WJ, Russo Jr RD. Hepatic angiosarcoma. *Am J Med* 1992;93(2):219–22.

37. Prystowsky SD, Elfenbein GJ, Lamberg SI. Nasopharyngeal carcinoma associated with long-term arsenic ingestion. *Arch Dermatol* 1978;114(4):602–3.
38. Malachowski ME. An update on arsenic. *Clin Lab Med* 1990;10:459–71.
39. Shen ZX, Chen GQ, Ni JH, Li XS, Xiong SM, Qiu QY, et al. Use of arsenic trioxide (As_2O_3) in the treatment of acute promyelocytic leukemia (APL): II. Clinical efficacy and pharmacokinetics in relapsed patients. *Blood* 1997;89(9):3354–60.
40. Bergstrom SK, Gillan E, Quinn JJ, Altman AJ. Arsenic trioxide in the treatment of a patient with multiply recurrent, ATRA-resistant promyelocytic leukemia: a case report. *J Pediatr Hematol Oncol* 1998;20(6):545–7.
41. Soignet SL, Maslak P, Wang ZG, Jhanwar S, Calleja E, Dardashti LJ, et al. Complete remission after treatment of acute promyelocytic leukemia with arsenic trioxide. *N Engl J Med* 1998;339(19):1341–8.
42. Zhu J, Chen Z, Lallemand-Breitenbach V, de Thé H. How acute promyelocytic leukaemia revived arsenic. *Nat Rev Cancer* 2002;2(9):705.
43. Lorenzo HK, Susin SA, Penninger J, Kroemer G. Apoptosis inducing factor (AIF): a phylogenetically old, caspase-independent effector of cell death. *Cell Death Differ* 1999;6(6):516.
44. Susin SA, Lorenzo HK, Zamzami N, Marzo I, Snow BE, Brothers GM, et al. Molecular characterization of mitochondrial apoptosis-inducing factor. *Nature* 1999;397(6718):441–6.
45. Wong SS, Tan KC, Goh CL. Cutaneous manifestations of chronic arsenicism: review of seventeen cases. *J Am Acad Dermatol* 1998;38(2):179–85.
46. Ko RJ, Ko R. Causes, epidemiology, and clinical evaluation of suspected herbal poisoning. *J Clin Toxicol* 1999;37(6):697–708.
47. Shen ZY, Tan LJ, Cai WJ, Shen J, Chen C, Tang XM, et al. Arsenic trioxide induces apoptosis of oesophageal carcinoma in vitro. *Int J Mol Med* 1999;4(1):33–40.
48. Kew J, Morris C, Aihie A, Fysh R, Jones S, Brooks D. Arsenic and mercury intoxication due to Indian ethnic remedies. *BMJ Br Med J* 1993;306(6876):506–7.
49. Treleaven J, Meller S, Farmer P, Birchall D, Goldman J, Piller G. Arsenic and ayurveda. *Leuk Lymphoma* 1993;10(4–5):343–5.
50. Mitchell-Heggs CA, Conway M, Cassar J. Herbal medicine as a cause of combined lead and arsenic poisoning. *Hum Exp Toxicol* 1990;9(3):195–6.
51. Ong ES, Yong YL, Woo SO. Determination of arsenic in traditional Chinese medicine by microwave digestion with flow injection-inductively coupled plasma mass spectrometry (FI-ICP-MS). *J AOAC Int* 1999;82(4):963–7.
52. Wong ST, Chan HL, Teo SK. The spectrum of cutaneous and internal malignancies in chronic arsenic toxicity. *Singapore Med J* 1998;39(4):171–3.
53. Feng ZU, Xia YA, Tian DE, Wu KE, Schmitt M, Kwok RK, et al. DNA damage in buccal epithelial cells from individuals chronically exposed to arsenic via drinking water in Inner Mongolia, China. *Anticancer Res* 2001;21(1A):51–7.
54. Bell FG. *Environmental Geology: Principles and Practice*. Blackwell Science, Oxford; 1998.
55. Bachofen R, Birch L, Buchs U, Ferloni P, Flynn I, Jud G, et al. Volatilization of arsenic compounds by microorganisms. Battelle Press, Columbus, OH; 1995.
56. Gao S, Burau RG. Environmental factors affecting rates of arsine evolution from and mineralization of arsenicals in soil. *J Environ Qual* 1997;26(3):753–63.
57. Cheng CN, Focht D. Production of arsine and methylarsines in soil and in culture. *Appl Environ Microbiol* 1979;38(3):494–8.
58. Ridley WP, Dizikes LJ, Wood JM. Biomethylation of toxic elements in the environment. *Science* 1977;197(4301):329–32.
59. Woolson EA. Generation of alkylarsines from soil. *Weed Sci* 1977;25(5):412–6.
60. Cullen WR, Reimer KJ. Arsenic speciation in the environment. *Chem Rev* 1989;89(4):713–64.
61. Gadd GM. Microbial formation and transformation of organometallic and organometalloid compounds. *FEMS Microbiol Rev* 1993;11(4):297–316.
62. Smedley PL, Edmunds WM, Pelig-Ba KB. Mobility of arsenic in groundwater in the Obuasi gold-mining area of Ghana: some implications for human health. *Geol Soc Spec Publ* 1996;113(1):163–81.
63. Cervantes C, Ji G, Ramírez JL, Silver S. Resistance to arsenic compounds in microorganisms. *FEMS Microbiol Rev* 1994;15(4):355–67.

64. Leonard A. Arsenic. In: Meriam E, editor. *Metals and Their Compounds in the Environment*. Wiley-VCH Verlag GmbH & Co. KGaA, Weinheim; 1991. pp. 751–72.

65. Jain CK, Ali I. Arsenic: occurrence, toxicity and speciation techniques. *Water Res* 2000;34(17):4304–12.

66. Sharma VK, Sohn M. Aquatic arsenic: toxicity, speciation, transformations, and remediation. *Environ Int* 2009;35(4):743–59.

67. Van Halem D. *Subsurface Iron and Arsenic Removal for Drinking Water Treatment in Bangladesh Water Management*. Academic Press, Delft; 2011.

68. Korte NE, Fernando Q. A review of arsenic (III) in groundwater. *Crit Rev Environ Sci Technol* 1991;21(1):1–39.

69. Guan X, Ma J, Dong H, Jiang L. Removal of arsenic from water: effect of calcium ions on As (III) removal in the $KMnO_4$–Fe (II) process. *Water Res* 2009;43(20):5119–28.

70. Boyle RW, Jonasson IR. The geochemistry of arsenic and its use as an indicator element in geochemical prospecting. *J Geochem Explor* 1973;2(3):251–96.

71. Welch AH, Lico MS, Hughes JL. Arsenic in ground water of the western United States. *Groundwater* 1988;26(3):333–47.

72. Rittle KA, Drever JI, Colberg PJ. Precipitation of arsenic during bacterial sulfate reduction. *Geomicrobiol J* 1995;13(1):1–11.

73. Arehart GB, Chryssoulis SL, Kesler SE. Gold and arsenic in iron sulfides from sediment-hosted disseminated gold deposits; implications for depositional processes. *Econ Geol* 1993;88(1):171–85.

74. Barker SL, Hickey KA, Cline JS, Dipple GM, Kilburn MR, Vaughan JR, et al. Uncloaking invisible gold: use of nanoSIMS to evaluate gold, trace elements, and sulfur isotopes in pyrite from Carlin-type gold deposits. *Econ Geol* 2009;104(7):897–904.

75. Smedley PL, Nicolli HB, Macdonald DM, Barros AJ, Tullio JO. Hydrogeochemistry of arsenic and other inorganic constituents in groundwaters from La Pampa, Argentina. *Appl Geochem* 2002;17(3):259–84.

76. O'Reilly J, Watts MJ, Shaw RA, Marcilla AL, Ward NI. Arsenic contamination of natural waters in San Juan and La Pampa, Argentina. *Environ Geochem Health* 2010;32(6):491–515.

77. Holland HD, Turekian KK. *Treatise on Geochemistry: Environmental Geochemistry*. Elsevier, Pergamon; 2004.

78. Goldberg S. Chemical modeling of arsenate adsorption on aluminum and Iron oxide minerals. *Soil Sci Soc Am J* 1986;50(5):1154–7.

79. Hiemstra T, Van Riemsdijk WH. A surface structural approach to ion adsorption: the charge distribution (CD) model. *J Colloid Interf Sci* 1996;179(2):488–508.

80. Goldberg S, Johnston CT. Mechanisms of arsenic adsorption on amorphous oxides evaluated using macroscopic measurements, vibrational spectroscopy, and surface complexation modeling. *J Colloid Interf Sci* 2001;234(1):204–16.

81. Marshall CP, Fairbridge RW, editors. *Encyclopedia of Geochemistry*. Springer Science & Business Media, Netherlands; 1999.

82. Bowen HJ, editor. *Environmental Chemistry*. Royal Society of Chemistry, United Kingdom; 2007.

83. Belkin HE, Zheng B, Finkelman RB. Geological sciences. In: 2nd World Chinese Conference. Standford University; 2000. p. 522.

84. Shacklette HT, Boerngen JG, Keith JR, US Geol. Survey Circular. 692. US Government Printing Office, Washington, DC; 1974.

85. Cannon HL, Petrie WL. A review of recent activity in the United States. *Philos Trans R Soc Lond B Biol Sci* 1979;288(1026):137–49.

86. Yang Q, Jung HB, Culbertson CW, Marvinney RG, Loiselle MC, Locke DB, et al. Spatial pattern of groundwater arsenic occurrence and association with bedrock geology in greater Augusta, Maine. *Environ Sci Technol* 2009;43(8):2714–9.

87. Dudas MJ. Enriched levels of arsenic in post-active acid sulfate soils in Alberta. *Soil Sci Soc Am J* 1984;48(6):1451–2.

88. Bennett B, Dudas MJ. Release of arsenic and molybdenum by reductive dissolution of iron oxides in a soil with enriched levels of native arsenic. *J Environ Eng Sci* 2003;2(4):265–72.

89. Nguyen KP, Itoi R. Source and release mechanism of arsenic in aquifers of the Mekong Delta, Vietnam. *J Contam Hydrol* 2009;103(1–2):58–69.

90. Gustafsson JP, Tin NT. Arsenic and selenium in some Vietnamese acid sulphate soils. *Sci Total Environ* 1994;151(2):153–8.

91. Robins RJ, Hamill JD, Parr AJ, Smith K, Walton NJ, Rhodes MJ. Potential for use of nicotinic acid as a selective agent for isolation of high nicotine-producing lines of *Nicotiana rustica* hairy root cultures. *Plant Cell Rep* 1987;6(2):122–6.

92. Krause E, Ettel VA. Solubilities and stabilities of ferric arsenate compounds. *Hydrometallurgy* 1989;22(3):311–37.

93. Pauwels H, Pettenati M, Greffié C. The combined effect of abandoned mines and agriculture on groundwater chemistry. *J Contam Hydrol* 2010;115(1–4):64–78.

94. Scudlark JR, Church TM. The atmospheric deposition of arsenic and association with acid precipitation. *Atmos Environ* 1988;22(5):937–43.

95. Finkelman RB, Belkin HE, Zheng B. Health impacts of domestic coal use in China. *Proc Natl Acad Sci* 1999;96(7):3427–31.

96. Andreae MO, Klumpp D. Biosynthesis and release of organoarsenic compounds by marine algae. *Environ Sci Technol* 1979;13(6):738–41.

97. Sullivan KA, Aller RC. Diagenetic cycling of arsenic in Amazon shelf sediments. *Geochim Cosmochim Acta* 1996;60(9):1465–77.

98. Winkel L, Berg M, Amini M, Hug SJ, Johnson CA. Predicting groundwater arsenic contamination in Southeast Asia from surface parameters. *Nat Geosci* 2008;1(8):536–42.

99. Kingston RL, Hall S, Sioris L. Clinical observations and medical outcome in 149 cases of arsenate ant killer ingestion. *J Clin Toxicol* 1993;31(4):581–91.

100. Schoolmeester WL, White DR. Arsenic poisoning. *South Med J* 1980;73(2):198–208.

101. Opresko DM. Risk Assessment Information System database. Oak ridge reservation environmental restoration program; 1992.

102. Ghariani M, Adrien ML, Raucoules M, Bayle J, Jacomet Y, Grimaud D. Subacute arsenic poisoning. *Ann Fr Anesth Reanim* 1991;10(3):304–7.

103. Logemann E, Krützfeldt B, Pollak S. Suicidal administration of elemental arsenic. *Arch Kriminol* 1990;185(3–4):80–8.

104. Mueller PD, Benowitz NL. Toxicologic causes of acute abdominal disorders. *Emerg Med Clin North Am* 1989;7(3):667–82.

105. Armstrong CW, Stroube RB, Rubio T, Siudyla EA, Miller GB. Outbreak of fatal arsenic poisoning caused by contaminated drinking water. *Int Arch Environ Health* 1984;39(4):276–9.

106. Campbell JP, Alvarez JA. Acute arsenic intoxication. *Am Fam Physician* 1989;40(6):93–7.

107. Greenberg C, Davies S, McGowan T, Schorer A, Drage C. Acute respiratory failure following severe arsenic poisoning. *Chest* 1979;76(5):596–8.

108. Wilkinson SP, McHugh P, Horsley S, Tubbs H, Lewis M, Thould A, et al. Arsine toxicity aboard the Asiafreighter. *Br Med J* 1975;3(5983):559–63.

109. Lerman BB, Ali N, Green D. Megaloblastic, dyserythropoietic anemia following arsenic ingestion. *Ann Clin Lab Sci* 1980;10(6):515–7.

110. Freeman JW, Couch JR. Prolonged encephalopathy with arsenic poisoning. *Neurology* 1978;28(8):853–5.

111. Le Quesne PM. Metal-induced diseases of the nervous system. *Br J Hosp Med* 1982;28(5):534–8.

112. Call RA, Gunn FD. Arsenical encephalopathy; report of a case. *Arch Pathol* 1949;48(2):119–28.

113. Beckett WS, Moore JL, Keogh JP, Bleecker ML. Acute encephalopathy due to occupational exposure to arsenic. *Br J Ind Med* 1986;43(1):66–7.

114. Breukink HJ, van Lieshout CG, van Buekelen P, Jansen HM. Arsenic poisoning through the roof. A case of arsenic poisoning in cattle. *Tijdschr Diergeneeskd* 1980;105(9):347–61.

115. Lien HC, Tsai TF, Lee YY, Hsiao CH. Merkel cell carcinoma and chronic arsenicism. *J Am Acad Dermatol* 1999;41(4):641–3.

116. Mazumder DN, Das JG, Santra A, Pal A, Ghose A, Sarkar S. Chronic arsenic toxicity in west Bengal-the worst calamity in the world. *J Indian Med Assoc* 1998;96(1):4–7.

117. Smith AH, Arroyo AP, Mazumder DN, Kosnett MJ, Hernandez AL, Beeris M, et al. Arsenic-induced skin lesions among Atacameno people in Northern Chile despite good nutrition and centuries of exposure. *Environ Health Perspect* 2000;108(7):617–20.

118. Everall JD, Dowd PM. Influence of environmental factors excluding ultra violet radiation on the incidence of skin cancer. *Bull Cancer* 1978;65(3):241–7.

119. Fincher RM, Koerker RM. Long-term survival in acute arsenic encephalopathy. Follow-up using newer measures of electrophysiologic parameters. *Am J Med* 1987;82(3):549–52.

120. Poklis A, Saady JJ. Arsenic poisoning: acute or chronic? Suicide or murder? *Am J Forensic Med Pathol* 1990;11(3):226–32.

121. Santra A, Das JG, De BK, Roy B, Guha DM. Hepatic manifestations in chronic arsenic toxicity. *Indian J Gastroenterol* 1999;18(4):152–5.

122. Nevens F, Staessen D, Sciot R, Van Damme B, Desmet V, Fevery J, et al. Clinical aspects of incomplete septal cirrhosis in comparison with macronodular cirrhosis. *Gastroenterology* 1994;106(2):459–63.

123. Pinto SS, Enterline PE, Henderson V, Varner MO. Mortality experience in relation to a measured arsenic trioxide exposure. *Environ Health Perspect* 1977;19:127–30.

124. Axelson O, Dahlgren E, Jansson CD, Rehnlund SO. Arsenic exposure and mortality: a case-referent study from a Swedish copper smelter. *Occup Environ Med* 1978;35(1):8–15.

125. Lee-Feldstein A. A comparison of several measures of exposure to arsenic: matched case-control study of copper smelter employees. *Am J Epidemiol* 1989;129(1):112–24.

126. Rahman M, Tondel M, Ahmad SA, Chowdhury IA, Faruquee MH, Axelson O. Hypertension and arsenic exposure in Bangladesh. *Hypertension* 1999;33(1):74–8.

127. Hsueh YM, Wu WL, Huang YL, Chiou HY, Tseng CH, Chen CJ. Low serum carotene level and increased risk of ischemic heart disease related to long-term arsenic exposure. *Atherosclerosis* 1998;141(2):249–57.

128. Tsai SM, Wang TN, Ko YC. Mortality for certain diseases in areas with high levels of arsenic in drinking water. *Int Arch Environ Health* 1999;54(3):186–93.

129. Benowitz NL. Cardiotoxicity in the workplace. *Occup Med* 1992;7(3):465–78.

130. Goldsmith S, From AH. Arsenic-induced atypical ventricular tachycardia. *N Engl J Med* 1980;303(19):1096–8.

131. Borgoño JM, Vicent P, Venturíno H, Infante A. Arsenic in the drinking water of the city of Antofagasta: epidemiological and clinical study before and after the installation of a treatment plant. *Environ Health Perspect* 1977;19:103–5.

132. Goddard MJ, Tanhehco JL, Dau PC. Chronic arsenic poisoning masquerading as Landry-Guillain-Barré syndrome. *Electromyogr Clin Neurophysiol* 1992;32(9):419–23.

133. Schenk VW, Stolk PJ. Psychosis following arsenic (possibly thalium) poisoning. *Psychiatr Neurol Neurochir* 1967;70(1):31–7.

134. Morton WE, Caron GA. Encephalopathy: an uncommon manifestation of workplace arsenic poisoning? *Am J Ind Med* 1989;15(1):1–5.

135. Chiou HY, Huang WI, Su CL, Chang SF, Hsu YH, Chen CJ. Dose-response relationship between prevalence of cerebrovascular disease and ingested inorganic arsenic. *Stroke* 1997;28(9):1717–23.

136. Mazumder DN, Haque R, Ghosh N, De BK, Santra A, Chakraborti D, et al. Arsenic in drinking water and the prevalence of respiratory effects in West Bengal, India. *Int J Epidemiol* 2000;29(6):1047–52.

137. Saady JJ, Blanke RV, Poklis A. Estimation of the body burden of arsenic in a child fatally poisoned by arsenite weedkiller. *J Anal Toxicol* 1989;13(5):310–2.

138. Quatrehomme G, Ricq O, Lapalus PH, Jacomet Y, Ollier A. Acute arsenic intoxication: forensic and toxicologic aspects (an observation). *J Forensic Sci* 1992;37(4):1163–71.

139. Rahman MM, Chowdhury UK, Mukherjee SC, Mondal BK, Paul K, Lodh D, et al. Chronic arsenic toxicity in Bangladesh and West Bengal, India-a review and commentary. *J Clin Toxicol* 2001;39(7):683–700.

140. Hopenhayn-Rich C, Biggs ML, Smith AH. Lung and kidney cancer mortality associated with arsenic in drinking water in Cordoba, Argentina. *Int J Epidemiol* 1998;27(4):561–9.

141. Guo HR, Chiang HS, Hu H, Lipsitz SR, Monson RR. Arsenic in drinking water and incidence of urinary cancers. *Epidemiology* 1997;8(5):545–50.

142. Hood RD, Vedel-Macrander GC. Evaluation of the effect of BAL (2, 3-dimercaptopropanol) on arsenite-induced teratogenesis in mice. *Toxicol Appl Pharmacol* 1984;73(1):1–7.

143. Mäki-Paakkanen J, Kurttio P, Paldy A, Pekkanen J. Association between the clastogenic effect in peripheral lymphocytes and human exposure to arsenic through drinking water. *Environ Mol Mutagen* 1998;32(4):301–13.

144. Goyer RA. Toxic effects of metals. In: Klaassen CD, editor. *Cassarett & Doull's Toxicology- The Basic Science of Poisons*. McGraw Hill, New York; 1996. pp. 691–736.

145. Belton JC, Benson NC, Hanna ML, Taylor RT. Growth inhibitory and cytotoxic effects of three arsenic compounds on cultured Chinese hamster ovary cells. *J Environ Sci Health A* 1985;20(1):37–72.

146. Barrett JC, Lamb PW, Wang TC, Te Lee C. Mechanisms of arsenic-induced cell transformation. *Biol Trace Elem Res* 1989;21(1):421–9.

147. Hartmann A, Speit G. Comparative investigations of the genotoxic effects of metals in the single cell gel (SCG) assay and the sister chromatid exchange (SCE) test. *Environ Mol Mutagen* 1994;23(4):299–305.

148. Jha AN, Noditi M, Nilsson R, Natarajan AT. Genotoxic effects of sodium arsenite on human cells. *Mutat Res* 1992;284(2):215–21.

149. Vega L, Gonsebatt ME, Ostrosky-Wegman P. Aneugenic effect of sodium arsenite on human lympho-cytes in vitro: an individual susceptibility effect detected. *Mutat Res* 1995;334(3):365–73.

150. Li JH, Rossman TG. Inhibition of DNA ligase activity by arsenite: a possible mechanism of its comutagenesis. *Mol Toxicol* 1989;2(1):1–9.

151. Landolph JR. Molecular and cellular mechanisms of transformation of C3H/10T 1/2 Cl 8 and diploid human fibroblasts by unique carcinogenic, nonmutagenic metal compounds. *Biol Trace Elem Res* 1989;21(1):459–67.

152. Lee TC, Oshimura M, Barrett JC. Comparison of arsenic-induced cell transformation, cytotoxic-ity, mutation and cytogenetic effects in Syrian hamster embryo cells in culture. *Carcinogenesis* 1985;6(10):1421–6.

153. Takahashi M, Barrett JC, Tsutsui T. Transformation by inorganic arsenic compounds of normal Syrian hamster embryo cells into a neoplastic state in which they become anchorage-independent and cause tumors in newborn hamsters. *Int J Cancer* 2002;99(5):629–34.

154. Schaumloffel N, Gebel T. Heterogeneity of the DNA damage provoked by antimony and arsenic. *Mutagenesis* 1998;13(3):281–6.

155. Gonsebatt ME, Vega L, Salazar AM, Montero R, Guzman P, Blas J, et al. Cytogenetic effects in human exposure to arsenic. *Mutat Res* 1997;386(3):219–28.

156. Nakamuro K, Sayato Y. Comparative studies of chromosomal aberration induced by trivalent and pentavalent arsenic. *Mutat Res* 1981;88(1):73–80.

157. Natarajan AT, Boei JJ, Darroudi F, Van Diemen PC, Dulout F, Hande MP, et al. Current cytogenetic methods for detecting exposure and effects of mutagens and carcinogens. *Environ Health Perspect* 1996;104(suppl 3):445–8.

158. Ramirez P, Eastmond DA, Laclette JP, Ostrosky-Wegman P. Disruption of microtubule assembly and spindle formation as a mechanism for the induction of aneuploid cells by sodium arsenite and vanadium pentoxide. *Mutat Res* 1997;386(3):291–8.

159. Thompson DJ. A chemical hypothesis for arsenic methylation in mammals. *Chem Biol Interact* 1993;88(2–3):89–114.

160. Cullen WR, McBride BC, Reglinski J. The reduction of trimethylarsine oxide to trimethylarsine by thiols: a mechanistic model for the biological reduction of arsenicals. *J Inorg Biochem* 1984;21(1):45–60.

161. Kitchin KT. Recent advances in arsenic carcinogenesis: modes of action, animal model systems, and methylated arsenic metabolites. *Toxicol Appl Pharmacol* 2001;172(3):249–61.

162. Aposhian HV, Gurzau ES, Le XC, Gurzau A, Healy SM, Lu X, et al. Occurrence of monomethylarson-ous acid in urine of humans exposed to inorganic arsenic. *Chem Res Toxicol* 2000;13(8):693–7.

163. Mass MJ, Tennant A, Roop B, Kundu K, Brock K, Kligerman A, et al. Methylated arsenic (III) species react directly with DNA and are potential proximate or ultimate genotoxic forms of arsenic. *Toxicologist* 2001;60:358.

164. Lin S, Cullen WR, Thomas DJ. Methylarsenicals and arsinothiols are potent inhibitors of mouse liver thioredoxin reductase. *Chem Res Toxicol* 1999;12(10):924–30.

165. Styblo M, Serves SV, Cullen WR, Thomas DJ. Comparative inhibition of yeast glutathione reductase by arsenicals and arsenothiols. *Chem Res Toxicol* 1997;10(1):27–33.

166. Petrick JS, Ayala-Fierro F, Cullen WR, Carter DE, Aposhian HV. Monomethylarsonous acid (MMAIII) is more toxic than arsenite in Chang human hepatocytes. *Toxicol Appl Pharmacol* 2000;163(2):203–7.

167. Le XC, Lu X, Ma M, Cullen WR, Aposhian HV, Zheng B. Speciation of key arsenic metabolic interme-diates in human urine. *Anal Chem* 2000;72(21):5172–7.

168. Abernathy CO, Liu YP, Longfellow D, Aposhian HV, Beck B, Fowler B, et al. Arsenic: health effects, mechanisms of actions, and research issues. *Environ Health Perspect* 1999;107(7):593–7.

169. Clark DW. Genetically determined variability in acetylation and oxidation. *Drugs* 1985;29(4):342–75.

170. Sutherland D, Swash PM, Macqueen AC, McWilliam LE, Ross DJ, Wood SC. A field based evaluation of household arsenic removal technologies for the treatment of drinking water. *Environ Tech* 2002;23(12):1385–404.

171. Yuan T, Yong HJ, Ong SL, Luo QF, Jun NW. Arsenic removal from household drinking water by adsorption. *J Environ Sci Health A* 2002;37(9):1721–36.

172. Al-Mahasneh QM, Rodgers GC. An in vitro study of efficacy of activated charcoal as an adsorbent for arsenical salts. In: International Congress on Clinical Toxicology, Societe des Sciences Medicales de Grande Duche de Luxembourg, Luxembourg; 1990.

173. Jones MM. New developments in therapeutic chelating agents as antidotes for metal poisoning. *Crit Rev Toxicol* 1991;21(3):209–33.

174. Peterson RG, Rumack BH. D-penicillamine therapy of acute arsenic poisoning. *J Pediatr* 1977;91(4):661–6.

175. Ding GS, Liang YY. Antidotal effects of dimercaptosuccinic acid. *J Appl Toxicol* 1991;11(1):7–14.

176. Kosnett MJ, Becker CE. Dimercaptosuccinic acid as a treatment for arsenic poisoning. *Vet Hum Toxicol* 1987;29(6):462.

177. Kosnett MJ, Becker CE. Dimercaptosuccinic acid-utility in acute and chronic arsenic poisoning. *Vet Hum Toxicol* 1988;30(4):369.

178. Mahieu P, Buchet JP, Lauwerys R. Clinical and biological evolution of acute oral intoxication with arsenious anhydride and considerations on the therapeutic attitude. *J Clin Exp Toxicol* 1987;7(4):273–8.

CHAPTER **23**

Clinical Toxicology of Mercury
Source, Toxidrome, Mechanism of Toxicity and Management

Sonal Sekhar Miraj, Pooja Gopal Poojari, Girish Thunga P, and Mahadev Rao
Manipal College of Pharmaceutical Sciences

CONTENTS

23.1 INTRODUCTION

23.1.1 History

Mercury (Hg), with atomic number 80, has come from the Greek term "hydrargyrias," indicating "water silver" (Broussard et al. 2002). Hg occurs in nature in three forms such as elemental, organic, and inorganic. All of these have harmful impacts on health (Rafati-Rahimzadeh et al. 2014). Exposure to Hg is the second-highest common reason for toxicity with metal poisoning (Patrick 2002). Metallic Hg poisoning was well aware at the time of Aristotle itself. A massive-scale occupational poisoning with Hg has happened while the sculpture of Buddha of Nara was built in Japan during the eighth century. In the last century, few major tragedies of Hg poisoning happened (Satoh 2000). One was Minamata disease in Japan, which reported the poisoning of 2,200 people due to the intake of fishes polluted with Hg. Similarly, another instance of Hg poisoning accounted in Niigata with 700 cases between the 1950s and 1960s (Clifton II 2007; Grandjean et al. 2010; Mostafalou and Abdollahi 2013; Watanabe and Satoh 1996). In Iraq, an Hg poisoning happened in 1971, when wheat grains were sprayed with organic Hg-based fungicides. This resulted in the death of more than 500 victims, who had taken bread made of tainted wheat (Bakir et al. 1973; Clarkson 1993). However, Hg has been useful for different medicinal and non-medical purposes. Prehistoric artists used cinnabar, a red metal (contains mercuric sulfide) for ancient cave drawing. Nowadays, Hg is synthesized as a byproduct from gold as well as bauxite mining. However, the largest source of Hg occurs naturally. Degassing of granite rock contributes for the majority of Hg (80%) present in the environment. Hg has been widely used in numerous industrial products, for example, battery, thermometer, and barometer (Broussard et al. 2002).

23.1.2 Properties

Hg is a very heavy, odorless, silver-colored liquid. On the other hand, in solid form, Hg is a ductile, malleable tin-white mass, which can be cut with a knife. Hg is liquid at 15°C and one atmosphere. Near to its boiling point of 356.7°C, Hg can be slowly oxidized to mercuric oxide (HgO) in the presence of oxygen (Micromedex® 2019).

23.1.3 Uses

Medical: These include diuretic, skin lotion, laxative, antiseptic, and for the therapy of syphilis (Broussard et al. 2002). The metallic form is useful in dentistry due to its peculiarity of easy amalgamation with other metals (Azevedo et al. 2012)

Commercial: Hg is found in numerous products, for example, battery, thermometer, and barometer. Certain items that have Hg are high-power discharge and fluorescent lights (Broussard et al. 2002). Other uses include electrolytic production of chlorine and caustic soda, especially in chlor-alkali industries. Additionally, elemental Hg is used in gold mining, coolant, mirror coating, and as a neutron absorber in nuclear power plants. Hg is also employed in laboratory, agricultural, and pharmaceutical chemicals (as an ingredient in many pharmaceuticals like diuretics, antiseptics, etc.) and for the manufacturing of Hg cells and Hg salts (Micromedex® 2019). Mercuric chloride ($HgCl_2$) makes use of different purposes such as photographic intensifier, wood additive, dry battery depolarizer, as a catalyst in the synthesis of compounds like disinfectants and vinyl chloride, and for isolating lead from gold. Recently, the organic Hg compounds have gained great attention since they are largely present in the food chain and often employed for the inhibition of growth of bacteria in medicinal preparations. In addition, natural Hg is present in fungicides as well as in industrial wastes (Broussard et al. 2002).

23.2 SOURCES OF MERCURY

Currently, Hg is viewed as a pollutant in the environment with a high degree of hazard to the health of public due to its potential toxic nature and fast mobility in the ecologies. Contacts with Hg can be possible from the natural as well as artificial sources.

23.2.1 Natural Sources

Natural sources of Hg are earthquakes, volcanic eruptions, and volatilization of Hg found in the marine eco-system as well as vegetation (Azevedo et al. 2012). Normally, inorganic forms of Hg are found in the outside layer of Earth. Cinnabar (mercuric sulfide; HgS) has been considered as the most significant ore of Hg. Outgassing process from the surface of Earth, including land and sea, discharges Hg fumes into the environment. A volcanic eruption is a significant natural source of Hg (Clarkson 1997).

23.2.2 Anthropogenic Sources

Activities of humans that can cause Hg exposure comprise of mining, fossil fuel and waste burning, use of coal, petroleum, and working in chlor-alkali industries (Azevedo et al. 2012). An outmost anthropogenic source of Hg comes from cinnabar mine. Power stations based on coal burning are a significant source of Hg to the environment. Waste incinerators also generate Hg fumes, which resembles the continuous working of Hg in fluorescent lamps and batteries of electronic gadgets and cameras. Additionally, crematoria are also sites of Hg emission because of cremation of bodies with amalgam tooth fillings and pacemakers containing Hg batteries (Clarkson 1997) (see Table 23.1).

23.2.2.1 Sources of Exposure

23.2.2.1.1 Inorganic Hg

Inorganic Hg (HgO) is used generally in various kinds of batteries. The explosion of these batteries in restrained places can cause acute exposures to Hg and other metals poisons. Inorganic Hg still continues to be a basic part of fluorescent lamps (Clarkson 1997).

23.2.2.1.2 Inhaled Hg Fumes

In spite of regulatory frameworks meant for controlling occupational exposures, metallic Hg is still commonly used and pollutes the environment. As discussed earlier, metallic Hg has been used in many instruments including thermometers, barometers, blood pressure apparatus. Inorganic Hg is present as an amalgam with metals such as silver and copper. These have been incorporated in a number of individuals as dental amalgam. Consequently, vanished Hg fumes from the amalgam surface are often subjected to be breathed. Therefore, dental amalgam has been recognized as a significant source of exposure in humans to Hg fumes (Clarkson 1997).

23.2.2.1.3 Methylmercury (CH_3Hg^+) Compounds

Fish and fish food products are a considerable source of human intake of methylmercury chemicals (Clarkson 1997). Dietary exposure to Hg as methylmercury (organic Hg exposure) happens majorly from the consumption of fish, shellfish, and marine products (Micromedex® 2019).

Table 23.1 Sources, Toxicokinetics, Mechanism of Toxicity of Mercury Poison

Hg Forms	Elemental	Inorganic	Organic
Sources	Thermometers, dental amalgams, old latex paints, incinerators, fossil fuels, and occupational	Batteries, disinfectants	Pesticides, fungicides, insecticides, certain vaccines, fish, and poultry
Absorption	75%–85% of fumes	7%–15% of the orally administered dose, 2%–3% of topically (dermal) dose in animals	≥95% in intestine, 100% of inhaled fumes
Distribution	Lipophilic, readily distribute to throughout the body Crosses blood-brain barrier (BBB) and placenta Primarily accumulates in brain and kidney	Does not crosses BBB and placenta Detected in neonatal brain Accumulates in kidney	Lipophilic, readily distribute to throughout the body Crosses BBB and placenta Primarily accumulates in brain and kidney
Metabolism	Converted to inorganic Hg by catalase and H_2O_2 after intracellular oxidation	Undergoes methylation by intestinal microbiota Binds to induce biosynthesis of metallothionein	Cysteine complex is required for absorption into intracellular Gradually methylated to inorganic Hg in the brain, fetal liver and generate free radicals
Elimination	Urine, feces, saliva, sweat	Urine, bile, feces, saliva, sweat	≈90% in bile and feces and ≈10% in urine
Half-life	60 days	40 days	65 days
Mechanism of toxicity	Oxidation to inorganic Hg (divalent)	Binds to sulfhydryl groups of enzymes and structural proteins	Demethylation to inorganic Hg (divalent) Generation of free radicals Binds to sulfhydryl groups of enzymes and structural proteins

H_2O_2, hydrogen peroxide; Hg, mercury.
Source: Modified from Patrick, L. 2002. Mercury toxicity and antioxidants: part I: role of glutathione and alpha-lipoic acid in the treatment of mercury toxicity. *Alternative Medicine Review* 7(6):456–71.

23.3 TOXIC DOSE

23.3.1 Toxicity

In 2004, the Joint Food and Agriculture Organization (FAO) of the United National/WHO Expert Committee on Food Additives (JECFA) put forward 1.6 mg/kg of body weight as a safe quantity of methylmercury, without showing any neurological problem. The US Environmental Protection Agency recommended 5.8 mcg/L as a reference blood concentration of Hg. WHO states below 6 mcg/g as the appropriate concentration of Hg in human hair (JECFA 2006). Thermometers contain around 500 mg of elemental Hg, which can cause symptomatic poisoning whenever vaporized by vacuuming. There have been reports of lethal cases to the children even after 14 days of contact to Hg, after the spillage of 0.5–1 ounce of Hg from a broken glass thermometer in a home (Micromedex® 2019).

23.3.2 Minimum Lethal Exposure

The estimated quantity of ingested Hg, which can be lethal to a man is considered as 100 g (1,429 mg/kg). A daily dose of 75 mg of Hg in drinking water is also thought to be lethal. Patients having Hg concentrations of 400–900 and 500–1,600 mcg/L in the blood and urine, respectively, have reported dead from acute Hg vapor poisoning (Micromedex® 2019).

23.4 TOXICOKINETICS

Similar to various metals, Hg is existing in numerous oxidative states, such as inorganic as well as organic compounds (see Table 23.1). The oxidative states involve elemental Hg, mercurous, or mercuric. Hg in any structural form is harmful. However, it can be distinguished by the way it is ingested, transformed to different Hg forms, the clinical presentations, and the response to various management strategies. Poisoning from Hg can appear from ingestion, inhalation, or skin contacts (Broussard et al. 2002). Elemental Hg can be easily evaporated at room temperature, therefore, the way of absorption will be regularly via lungs. In people, approximately 75%–85% of Hg is absorbed through this way, whereas below 3% only will be retained dermally. If elemental Hg is consumed orally, below 0.1% will be taken up from the GI tract and, hence, when orally ingested elemental Hg is somewhat not dangerous (Hursh et al. 1989).

Elemental Hg is highly lipophilic in nature. In the circulation, elemental Hg binds to various proteins, tissues, and erythrocytes. In erythrocytes, catalase can oxidize elemental Hg to an inorganic metabolite. If elemental Hg crosses the blood-brain barrier (BBB), it become ionized and subsequently trapped in the compartment, where, it is accessible to show its neurotoxicity. Elemental Hg has the longest ability to retain in the brain with noticeable levels available for a considerable length of time (Matsuo et al. 1989; Rothstein and Hayes 1960; Takahata et al. 1970). Half-life of elemental Hg in adults is around 60 days. Elemental Hg is also bioconverted to Hg^{2+} and CH_3Hg^{1+} in the gut by the activity of microorganisms (Broussard et al. 2002).

Inorganic Hg salts are available in two oxidation states, such as "mercurous" and "mercuric" forms. The main storage place of inorganic Hg is in the kidneys. Elimination of inorganic Hg is mainly via defecation. Mercuric particles don't cross the BBB or the placenta effectively. However, the slow elimination as well as the fact that exposure routinely happens over a prolonged time period, which allows for significant CNS build-up of mercuric ions and resultant toxicity. The half-life of inorganic Hg is roughly 40 days (Broussard et al. 2002).

Organic Hg occurs in three forms, which include aryl, short, and long chain alkyl compounds. Organic mercurial is absorbed more completely from the GI tract than inorganic salts, since former are more lipid-soluble and bind to sulfhydryl groups. Once absorbed in tissues, the aryl as well as long chain alkyl compounds are converted to divalent cations that exhibit properties of inorganic Hg toxicity. The short chain alkyl mercurials are readily absorbed in the GI tract (90%–95%) and stay stable in their initial structures. Alkyl organic Hg compounds have high lipid solvency and are dispersed consistently throughout the body, accumulating in the brain, kidney, liver, hair, and skin. Organic mercurials also cross the BBB, placenta and enter erythrocytes, attributing to neurological symptoms, teratogenic effects, and high blood-to-plasma ratio, respectively.

Methylmercury possess a greater affinity toward sulfhydryl groups, which accounts for its enzyme dysfunction activities. Such an enzyme, which undergo inhibition is choline acetyl transferase, which is related with the last step in synthesis of acetylcholine. This hindrance can prompt to deficiency of acetylcholine, adds to the symptoms and adverse effects related to motor dysfunction. Excretion of majority alkyl Hg happens via defecation (90%), secondary to enterohepatic circulation. Methylmercury has around 65 days of the biological half-life (Broussard et al. 2002).

23.5 MECHANISM OF TOXICITY

Hg ions depicts harmful effects by precipitating proteins, inhibiting enzyme, and by generalized corrosive activities. Hg ions bind to sulfhydryl groups and thereby modify the structure and capacity of key proteins and enzymes. Cellular metabolism might be weakened by Hg binding to amines and phosphoryl groups. Elemental Hg fumes are highly lipophilic that enable easy penetration through membrane of the cells and subsequently form mercuric state by oxidation. Mercuric salts form more dissolvable divalent compounds, therefore, this state has higher lethality compared to mercurous salts which produce monovalent Hg compounds. Consequently, after ingested, mercuric salts can be more quickly absorbed and lead to more harmful effects. Only an average of 10% of an inorganic salt will be absorbed in contrast to 90% of organic forms from GI tract. Consequently, inorganic salts attribute to corrosive effects on the GI mucosa, since it is available in the GI tract for prolonged time. Short-chained alkyl groups, like methylmercury are potentially hazards. These are largely absorbing at GI tract and distributing to brain, liver, and kidney. Finally, these are eliminated mainly through feces. Aryl compounds of Hg are eliminated in the form of mercuric ions (Broussard et al. 2002; Micromedex® 2019).

23.6 CLINICAL MANIFESTATIONS

The symptoms of an individual presented with Hg poisoning rely on the dose, length, and route of exposure. Acute toxicity is usually connected with the inhalation of elemental Hg or ingestion of inorganic Hg. Exposure to organic Hg may cause chronic toxicity (Broussard et al. 2002).

23.6.1 Acute (In Human)

Acute exposure caused due to inhalation of elemental Hg can affect primarily the lungs. Initial manifestations include weakness, chills, fever, shortness of breath, GI symptoms (like nausea, vomiting, diarrhea, abdominal pain, metallic taste), headache, tremor, visual disturbances, dyspnea, cough, and pleuritic chest pain. Additional clinical presentations are lethargy, confusion and oral cavity problems. Hg from dental amalgams can cause stomatitis. Pulmonary complications of inhaled toxicity comprise of interstitial emphysema and fibrosis, pneumothorax, pneumatocele, and pneumomediastinum.

Most the time, acute exposure to mercurous or mercuric salt happens orally. The destructive effects of these compounds represent the vast majority of the intense signs and indicate lethality. The acute presentation can include bloody stool, stomach ache, vomiting, and hypovolemic shock. Generally, systemic effects start few hours' after ingestion and may persist few days, that often comprise of oral cavity problems such as metallic taste, gingival irritation, bad breath, wobbly tooth, mucosal inflammation, and oliguria or anuria related to renal tubular necrosis (Broussard et al. 2002; Micromedex® 2019; Rafati-Rahimzadeh et al. 2014).

23.6.2 Chronic (In Human)

Chronic exposure is usually caused from long-term exposure at occupational places to elemental Hg (which will be transformed to the inorganic compounds), topical use of mercurial medicaments, or the chronic usage of cathartics or diuretics consisting Hg. Exposure to chronic and high-dose acute Hg leads to renal, neurological, psychological, and cutaneous clinical manifestations (Broussard et al. 2002).

Chronic inhalation leads to the classic triad of neuropsychiatric disturbances such as personality changes, hallucinations, delirium, insomnia, irritability, fatigue, memory loss, and erethism. In addition, tremor (fine intention tremor of fingers that progresses to choreiform movements of limbs) and gingivostomatitis can also often occur. Children and some adults may develop acrodynia associated with severe leg cramps, irritability, insomnia, diaphoresis, hypertension, miliarial rash, and peeling erythematous skin (on the fingers, hands, and feet). Renal dysfunction can also report (Micromedex® 2019). Acrodynia (pink disease), seen to be an Hg hypersensitivity, presents with dermatological manifestations, excessive sweating, elevated heart rate and blood pressure, photophobia, anorexia, sleep deprivation, poor muscle tone, irritation, and diarrhea. Organic Hg poisoning regularly occurs after consumption of contaminated foods, especially fish. The long chain alkyl as well as aryl type of organic Hg have same properties of inorganic Hg. Organic Hg targets specific locales in the brain, such as cerebral cortex, motor and sensory centers, auditory center, and cerebellum. The onset of adverse effects can occur days to weeks after exposure. Symptoms identified with the toxicity are mainly neurological, for example visual disturbance, ataxia, paresthesias, loss of hearing, dysarthria, mental deterioration, muscle tremor and with severe exposure, paralysis and death (Broussard et al. 2002; Porter and Moyer 1999).

23.6.3 Chronic (In Animals)

The breathing of harmful elemental Hg vapors which leads to severe dyspnea and compromised respiratory function are usually lethal at higher levels of exposure. Neurologic manifestations may develop even if exposure is not massive. Inorganic Hg attributes to its corrosive nature, causes mainly GI symptoms, including colic, anorexia, stomatitis, pharyngitis, vomiting, diarrhea, shock, difficulty breathing, and dehydration. Death usually occurs within hours of higher levels of exposure. Animals that remain alive may exhibit skin problems, anuria, polydypsia, blood in urine, or melena. Neurologic manifestations, such as CNS depression or excitation similar to organic Hg poisoning, may occur after chronic exposure. Based on the degree of exposure to organic Hg compounds, such as methylmercury, clinical symptoms may take days to occur. Neurologic manifestations that prevail include blindness, incoordination, tremors, abnormal behavior, hypermetria, nystagmus (cats), and tonic-clonic seizures. Advanced stages may be indicated by depression, anorexia, total blindness, paralysis, and death. The nervous system of young, developing animals is easily affected by organic Hg exposure, which is commonly manifested by cerebellar ataxia and death (Blakley 2019).

23.7 DIAGNOSIS

Exposure to Hg or its salts can be detected particularly in the samples, such as blood, urine, and hair. The estimated amount of Hg in the urine as well as blood can correlate with toxicity (Broussard et al. 2002). However, 24-h urinary Hg concentration is the best marker for chronic exposure (Micromedex® 2019).

23.7.1 Urine

Levels of Hg in urine are usually <10–20 mcg per 24 h. Urinary elimination of Hg is a good indicator of inorganic as well as elemental Hg exposure, however, it is not reliable for organic form (methylmercury) because excretion occurs predominately in the feces (Broussard et al. 2002). Urine is a more reliable sample for testing elemental and inorganic Hg. Amount >100 mcg/L, produce neurological signs while more than 800 mcg/L are often associated with mortality (Goldman et al. 2001).

23.7.2 Blood

In acute intoxication, concentration of methylmercury in erythrocytes becomes high, however, it differs in chronic toxicity. The whole blood Hg concentration is often <10 mcg/L, nevertheless it usually reaches upto 20 mcg/L. After long-term exposure to Hg fumes, the blood Hg concentration can rise to 35 mcg/L (Klaassen et al. 1996).

23.7.3 Hair

Sulfhydryl content of hair is high, therefore, Hg binds covalently with sulfur. Consequently, Hg can be seen in high levels in hair samples. But, hair sample analysis has high chance of false positive results. Therefore, it should not be considered alone for confirming toxicity of Hg. In hair, generally, concentration of Hg does not exceed 10 mg/kg. On the other hand, in moderate and severe methylmercury intoxications, these concentrations can rise to 200–800 and 2,400 mg/kg, respectively. Maternal hair Hg concentrations >10 parts per million can escalate the risk of neurological abnormalities in babies (Broussard et al. 2002).

23.8 MANAGEMENT

23.8.1 Supportive Care

Decision regarding treatment of Hg poisoning relies primarily on the type of Hg. Alike, any other poison, fundamental necessity is to acquire as much data on source, time, type, and way of Hg exposure. Supportive care starts with the ABCs (airway, breathe, circulation), particularly while dealing with the inhalation of elemental Hg and the ingestion of acidic inorganic Hg. In the event, patients exposed to Hg through the skin contacts, decontamination with copious irrigation of the exposed area should be performed. Vigorous hydration might be required for acute inorganic Hg ingestion. Gastric lavage is suggested for organic ingestion, particularly if the compound is seen on the stomach radiographs (Broussard et al. 2002).

Inhalation is the prime source of remarkable lethality, particularly after vaporization from warming or vacuuming spills. Ventilate the region for 15 min and get Hg globules with cardboard or pipe tape. If vacuuming is required, dispose the sack in two fixed plastic packs after the use. Thermometers contain around 500 mg of Hg and can cause massive damage whenever evaporated by vacuuming. Therefore, immediately shift the patient to natural air, followed by screen for respiratory distress as well as hypoxia. Thereafter provide 100% humidified oxygen, intubate, and assist ventilation as required. Administer beta-2 adrenergic agonists for bronchospasm (Micromedex® 2019).

23.8.2 Treatment

Chelation therapy can promote Hg expulsion from the body. However, there exist a paucity of clear data showing improved clinical response. Chelation ought to be performed in patients with serious manifestations after acute exposure. In addition, this can be considered in patients who are symptomatic after chronic presentation. Asymptomatic patients with raised urinary Hg levels presumably don't warrant chelation. Hg chelators basically used were succimer (Meso 2,3-dimercaptosuccinic acid; DMSA) given orally and frequently used for chronic poisoning and dimercaprol (British anti-Lewisite; BAL) administer intramuscular and used in patients with severe acute poisoning. Likewise, D-penicillamine (DPA) and N-acetyl penicillamine (NAP) can be also administered. Dosing of DPA is 250 mg orally four times daily for 1–2 weeks in adults and 20–30 mg/kg daily in four divided doses (maximum of 250 mg per dose) in children. DPA is useful only for elemental and inorganic Hg poisoning and is not effective for organic Hg poisoning (Ford et al._2001). Hypersensitivity reactions and renal toxicity are very common adverse reactions of DPA (Yilmaz et al. 2010) NAP is an analog of DPA, which is comparatively higher chelator of Hg. Currently, succimer is preferred over penicillamines, due its powerful metal-mobilizing capacity as well as fewer adverse effects (Katzung et al. 2009).

Parenteral chelation (intramuscular [IM] dimercaprol or intravenous[IV] unithiol) treatment ought to be initiated in patients with substantial acute exposures. If the victim can't take oral prescriptions, dimercaprol can be administered by titrating the dose downward over 10 days. The dosage regimen starts with 5 mg/kg, followed by 2.5 mg/kg once or multiple times every day for 10 days. At the point when the patient is improving and ready to endure oral medications, dimercaprol can be replaced with succimer with no holding up period between medicines. Succimer may be administered based on the following dosing plans, which involves 10 mg/kg orally three times every day for 5 days and thereafter, 10 mg/kg two times daily for 14 days. WHO recommends succimer should be initiated in children with urinary Hg levels $\geq$500 mcg/L, although patients are asymptomatic. Child dose calculation is based on the body surface area (BSA), 350 mg/m^2 thrice daily for the initial 5 days, followed by 350 mg/m^2 twice daily for next 14 days. If required, this can be repeated with an interval of 2 weeks' period between the regimens (Brent et al. 2017; Rafati-Rahimzadeh et al. 2014). Unithiol (2,3-dimercapto-1-propane sulfonic acid; DMPS) is another water-soluble analog of the dimercaprol, which replaced succimer in Europe. In the body, unithiol undergo oxidation ($\approx$80% of oxidation within first half of the hour) to form disulfide. Unithiol enters into the renal cells and clears the Hg accumulated in the tissues, thereby, eliminates Hg into the urine. Unithiol has the ability to remove deposits of mercuric compounds from almost all tissues except from brain tissues. Unithiol can be administered either orally or IV injection. Dosing for adult patients is 250 mg IV fourth hourly for the initial 2 days, followed by 250 mg IV sixth hourly for the next 2 days, and then 250 mg IV eight hourly. After IV therapy, route conversion to oral can be initiated with 300 mg thrice daily for 7 weeks. Duration of the therapy relies on the Hg concentration in blood as well as urine (Brent et al. 2017).

Currently, new succimer analogues such as DMSA esters found to be better antidotes for Hg. These are either mono or di esters of DMSA which may promote Hg elimination from the tissue. These compounds remove Hg from the kidneys as well as bile. Mono isoamyl ester of DMSA (MiADMSA) is a water-soluble lipophilic chelating agent, which can easily penetrate into intracellular space and largely distributed within the cells. MiADMSA can remove the heavy metals from intra as well as extra cellular sites (Flora et al. 2012). MiADMSA is usually given through oral or intraperitoneal route with doses of 25, 50, and 100 mg/kg. Other novel succimer analogous are Monomethyl DMSA (MmDMSA) and Monocyclohexyl DMSA (MchDMSA). Both MmDMSA and MchDMSA are oral chelating agents with lipophilic nature and can penetrate into tissues. Recently, combination chelation therapy has been considered important treatment strategies for the heavy metal poisoning. Co-administration of succimer with MiADMSA has shown greater effective compared to MiADMSA mono-therapy. This modality provides certain benefits, such as reduce dose of chelating agents, give better clinical outcomes, and reduce the possible adverse reactions (Flora and Pachauri 2010; Rafati-Rahimzadeh et al. 2014).

Alpha-lipoic acid (ALA) has showed beneficial effect in Hg poisoning by increasing levels of glutathione intracellularly to promote the mobilization and elimination of Hg (Patrick 2002). Additionally, ALA protects against cellular damage and neurotoxicity. Dihydrolipoic acid (DHLA), a reduced ALA form, seems to possess direct heavy metal-binding action. While comparing with dithiol-chelating agents (DMSA, DMPS, etc.), ALA has the ability to bind as well as mobilize heavy metals from tissue, however, with lesser effect. On the other hand, upturns in levels of glutathione with administration of ALA are not just from the conversion of oxidized glutathione to reduced form, but also from the glutathione synthesis (Packer et al. 1997). DHLA, a potent antioxidant has the ability to restore glutathione, vitamin E, oxidized ascorbate and coenzyme Q (Queiroz et al. 1998), and in addition to attribute the ability of ALA to promote levels of intracellular glutathione.

23.9 CONCLUSION

Hg has been used for industrial, religious, and medicinal purposes since ancient times, however, it can cause severe poisoning. Hg is a volatile fluid and, therefore, current valid concerns is its versatile environmental exposure from diverse sources. These include ingestion of fish with high Hg content. Similarly, other potential paths of exposure are use of dental amalgams and newborn child immunizations. Exposure to Hg can adversely affect any organ and body systems. Hg toxicity should be considered as a silent risk to environment as well as human life worldwide. Therefore, legislative and non-administrative associations need to distinguish high risk individuals to Hg poisoning. Also, they should ensure safe transport and handling of Hg compounds. Novel interventional treatment strategies such as access to new antitoxins, chelating specialists, combined treatment of various chelating agents, and certain nanosorbents can improve the management of Hg poisoning.

23.10 CASE STUDY

A 12-year_old boy brought to out-patient department of a tertiary healthcare facility with clinical presentations of abdominal pain, extremity pain, and dermal eruptions. The child was diagnosed with chronic Hg intoxication. His blood Hg level was 141 mcg/L. Clinician started D-penicillamine (DPA) as a management strategy. On tenth day after therapy, the skin lesions in both of the extremities as well as the gluteal area became worse. These were accompanied with itching, a burning sensation, pain, and elevated body temperature. Conversely, no rash, swelling, or elevated temperature were detected in the joints. The child's medical history revealed no particular features. However, his 10-year-old male sibling was also earlier diagnosed with chronic Hg intoxication. His mother informed that both children were fond of playing with battery buttons at their residence.

On the physical examination, his body weight was 48.5 kg and height was 151 cm. Vital signs showed body temperature was 37.6°C, pulse rate was 126 beats/min, and blood pressure was 112/82 mmHg. Overall, general state of the child was good. However, while palpating extremities a diffused sensitivity was present. But, no rash, elevated temperature, swelling, or movement restriction at the joints were observed. At the both lower extremities and the gluteal area, substantial linear erythematosus and ulcerated lesions were present. Lab parameters of the child were as follows. His urinary analysis indicated density: 1034, pH: 5.1, protein: >310, sulfosalicylic acid (SAA) test: +++, ketone: trace, and glucose: negative. Hematological parameters were Hb: 14.1 g/dL, WBC count: 8,200/mm^3, MCV:83 fL, MCH: 28.5 pg, MCHC: 35 g/dL, platelets: 438,000/mm^3, RDW: 13%, and ESR: 16 mm/h. Liver enzyme levels showed AST: 110 IU/L, ALT: 89 IU/L, and LDH: 1030 IU/L. CK: 3020 IU/dL, blood lead (Pb) level: 22.5 mcg/L, blood Hg level: 132 mcg/L (normal: 0–100 mcg/L), rheumatoid factor (RF), C-reactive protein (CRP) and anti-nuclear antibody (ANA) tested to be negative. Moreover, serum levels of both C3 and C4 were normal. Levels of 5-aminolevulinic acid (ALA) and porphobillinogen (PBG) in urine were normal. Electromyography (EMG) demonstrated myogenic involvement, however, the EMG became normal after 2 months.

Treatment with DPA children (20 mg/kg/daily in four divided doses) was immediately started in the child who was admitted in the hospital. Ibuprofen as well as gabapentin were also given to decrease the diffuse body pain. Halogenoderma was noted in connection with dermal eruptions. In the light of an elevated level of liver enzymes along with proteinuria, clinician halted DPA, because of the possibility of DPA related hepato- and nephrotoxicity. Dermal eruptions were disappeared after discontinuing DPA. The child was treated with dimercaprol (administered as deep intramuscular injection with 5 mg/kg every fourth hourly for 2 days, followed by 2.5 mg/kg twice daily for 10 days) as a replacement for D-penicillamine. Before initiating dimercaprol, glucose

6-phosphate dehydrogenase (G6PD) test was performed to rule out the deficiency, since dimercaprol has potential risk of causing hemolytic anemia (Dawn and Whited 2019). Subsequently after starting dimercaprol, hepatic and kidney function tests became normal. Thereafter treatment with DPA was re-started. After 3 weeks of the therapy, the child was discharged from the hospital since pain in the extremities as well as head was subsided and pain in abdomen was completely relieved. The blood level of Hg came down to 78 mcg/L at the time of discharge. In the second month of treatment, the complaints of the child were entirely resolved. The child was given a total course of 7-month DPA therapy. The blood level of Hg further reduced to 24 mcg/L. However, the levels were monitored for 1 year.

REFERENCES

Azevedo, B.F., Furieri, L.B., Peçanha, F.M., et al. 2012. Toxic effects of mercury on the cardiovascular and central nervous systems. *Journal of Biomedicine and Biotechnology* 1–11. doi:10.1155/2012/949048.

Bakir F., Damluji, S.F., Amin-Zaki, L., et al. 1973. Methylmercury poisoning in Iraq. *Science* 181(4096):230–41. doi:10.1126/science.181.4096.230.

Blakley, B.R. 2019. Overview of Mercury Poisoning. Merck Manual - *Veterinary Manual.* https://www.merckvetmanual.com/toxicology/mercury-poisoning/overview-of-mercury-poisoning.

Brent, J., Burkhart, K., Dargan, P., et al. 2017. *Critical Care Toxicology: Diagnosis and Management of the Critically Poisoned Patient.* Springer Nature, Switzerland AG

Broussard, L.A., Hammett-Stabler, C.A., Winecker, R.E., and Ropero-Miller, J.D. 2002. The toxicology of mercury. *Laboratory Medicine* 33(8):614–25.

Clarkson, T.W. 1993. Molecular and ionic mimicry of toxic metals. *Annual Review of Pharmacology and Toxicology* 33(1):545–71. doi:10.1146/annurev.pharmtox.33.1.545.

Clarkson, T.W. 1997. The toxicology of mercury. *Critical Reviews in Clinical Laboratory Sciences* 34(4):369–403. doi:10.3109/10408369708998098.

Clifton II, J.C. 2007. Mercury exposure and public health. *Pediatric Clinics of North America* 54(2):237–69. doi:10.1016/j.pcl.2007.02.005.

Dawn, L. and Whited, L. 2019. Dimercaprol. [Updated 2019 Oct 23]. In: *StatPearls [Internet].* Treasure Island, FL: StatPearls Publishing. https://www.ncbi.nlm.nih.gov/books/NBK549804/.

Flora, S.J.S., Bhadauria, S., Pachauri, V., and Yadav, A. 2012. Monoisoamyl 2,3 –dimercaptosuccinic acid (MiADMSA) demonstrates higher efficacy by oral route in reversing arsenic toxicity: a pharmacokinetic approach. *Basic and Clinical Pharmacology and Toxicology* 110(5):449–59.

Flora, S.J.S. and Pachauri, V. 2010. Chelation in metal intoxication. *International Journal of Environmental Research and Public Health* 7(7):2745–88.

Ford, M.D., Delaney, K.A., Ling, L.J., and Erickson, T. 2001. *Clinical Toxicology.* Philadelphia, PA: W.B. Saunders Company.

Goldman, L.R., Shannon, M.W., and American Academy of Pediatrics: Committee on Environmental Health. 2001. Technical report: mercury in the environment: implications for Pediatricians. *Pediatrics* 108(1):197–205. doi:10.1542/peds.108.1.197.

Grandjean, P., Satoh, H., Murata, K., and Eto, K. 2010. Adverse effects of methylmercury: environmental health research implications. *Environmental Health Perspectives* 118(8):1137–45. doi:10.1289/ehp.0901757.

Hursh, J.B., Clarkson, T.W., Miles, E.F., and Goldsmith, L.A. 1989. Percutaneous absorption of mercury vapor by man. *Archives of Environmental Health: An International Journal* 44(2):120–27. doi:10.1080/00039896.1989.9934385.

JECFA. 2006. Methylmercury. Summary and Conclusions of the 67th Joint FAO/WHO Expert Committee on Food Additives. Geneva, World Health Organization, International Programme on Chemical Safety. *WHO Technical Report Series* 940. http://www.who.int/foodsafety/chem/en/.

Katzung, B.G., Masters, S.B., and Trevor, A.J. 2009. *Basic and Clinical Pharmacology.* New York, NY: McGraw Hill.

Klaassen, C.D., Acosta, D., Amdur, M.O., et al. 1996. *Casarett and Doull's Toxicology: The Basic Science of Poisons.* New York, NY: McGraw-Hill.

Matsuo, N., Suzuki, T., and Akagi, H. 1989. Mercury concentration in organs of contemporary Japanese. *Archives of Environmental Health: An International Journal* 44(5):298–303. doi:10.1080/00039896.1 989.9935897.

Micromedex® Toxicology Management | Exposure Management [Internet]. Truvenhealth.com. 2019. http://truvenhealth.com/markets/provider/products/clinicalknowledge/toxicology-management.

Mostafalou, S. and Abdollahi, M. 2013. Environmental pollution by mercury and related health concerns: renotice of a silent threat. *Archives of Industrial Hygiene and Toxicology* 64(1):179–181. doi:10.2478/10004-1254-64-2013-2325.

Porter, W.H. and Moyer, T.P. 1999. Clinical Toxicology. In: C.A. Burtis and E.R. Ashwood, eds, *Tietz Textbook of Clinical Chemistry*. 3rd ed. Philadelphia, PA: WB Saunders. pp. 992–993.

Packer, L., Tritschler, H.J., and Wessel, K. 1997. Neuroprotection by the metabolic antioxidant alpha-lipoic acid. *Free Radical Biology and Medicine* 22(1–2):359–78.

Patrick, L. 2002. Mercury toxicity and antioxidants: part I: role of glutathione and alpha-lipoic acid in the treatment of mercury toxicity. *Alternative Medicine Review* 7(6):456–71.

Queiroz, M.L., Pena, S.C., Salles, T.S., et al. 1998. Abnormal antioxidant system in erythrocytes of mercury-exposed workers. *Human and Experimental Toxicology* 17(4):225–30.

Rafati-Rahimzadeh, M., Rafati-Rahimzadeh, M., Kazemi, S., and Moghadamnia, A.A. 2014. Current approaches of the management of mercury poisoning: need of the hour. *DARU Journal of Pharmaceutical Sciences* 22(1):46. doi:10.1186/2008-2231-22-46.

Rothstein, A. and Hayes, A.D. 1960. The metabolism of mercury in the rat studied by isotope techniques. *Journal of Pharmacology and Experimental Therapeutics* 130(2):166–76.

Satoh, H. 2000. Occupational and environmental toxicology of mercury and its compounds. *Industrial Health* 38(2):153–64. doi:10.2486/indhealth.38.153.

Takahata, N., Hayashi, H., Watanabe, S., and Anso, T. 1970. Accumulation of mercury in the brains of two autopsy cases with chronic inorganic mercury poisoning. *Folia psychiatrica et neurologica japonica* 24(1):59–69. doi:10.1111/j.1440-1819.1970.tb01457.x.

Watanabe, C. and Satoh, H. 1996. Evolution of our understanding of methylmercury as a health threat. *Environmental Health Perspectives* 104(Suppl 2):367–79. doi:10.2307/3432657.

Yilmaz, C., Okur, M., Geylani, H., Caksen, H., Tuncer, O., and Ataş, B. 2010. Chronic mercury poisoning: report of two siblings. *Indian Journal of Occupational and Environmental Medicine* 14(1):17–9.

CHAPTER 24

An Overview of Tungsten Toxicity

Ola Wasel and Jennifer L. Freeman
Purdue University

CONTENTS

24.1 INTRODUCTION

Tungsten (W) is a transition metal with an atomic number of 74, a molecular weight of 183.84, and belongs to Group VIB of the periodic table. Tungsten is present naturally in rocks and minerals. Tungsten is not present in a pure form, but it is naturally combined with other metals.[1] The most common forms of tungsten that are used in industrial applications are wolframite and scheelite.[1] Tungsten has the highest melting point and highest tensile strength at a temperature of over 1,665°C compared to all other metals.[2] Tungsten has several oxidation states: 0, +2, +3, +4, +5, and +6. The physical and chemical properties of tungsten compounds vary based on the oxidation state (Table 24.1). Tungsten is used in the forms of tungsten carbide, metallic tungsten, tungsten chemicals, and tungsten alloy in many different applications (Figure 24.1).[1,3]

Table 24.1 Physical and Chemical Properties of Tungsten Substances

Tungsten Substance	Molecular Formula	CAS Registry Number	Melting Point (°C)	Boiling Point (°C)	Solubility	References
Tungsten metal	W	7440-33-7	3,410	5,900 at 760 mm Hg	Insoluble in water Soluble in nitric acid/hydrogen fluoride	4
Tungsten carbide	WC	12070-12-1	2,785	6,000	Insoluble in water Soluble in nitric acid/hydrogen fluoride	4
Ditungsten carbide	W_2C	12070-13-2	2,800	N/A	Insoluble in water	5
Tungsten oxide	WO_2	12036-22-5	Decomposes at 1,500–1,700	N/A	Insoluble in water Insoluble in organic solvents	
Tungsten trioxide	WO_3	1314-35-8	1,472	N/A	Insoluble in water Slightly soluble in acids	5
Sodium tungstate dihydrate	$Na_2WO_4 \cdot 2H_2O$	10213-10-2	Decomposes at 100 and melts at 692	N/A	Water: very soluble 1×10^6 mg/L Insoluble in alcohol and acids	2, 5–7
Ammonium paratungstate	$(NH_4)_{10}H_2(W_2O_7)_6 \cdots 4H_2O$	11120-25-5	N/A	N/A	Soluble in water Insoluble in alcohol	8
Tungsten chloride	WCl_6	13283-01-7	275	346.75	Decomposes in water Soluble in ethanol, organic solvent, lingroin	5, 7, 8
Tungsten hexafluoride	WF_6	7783-82-6	2.3	17.5	Decomposes in water Dissolves in benzene, dioxane	5, 8

CAS, Chemical Abstracts Service.

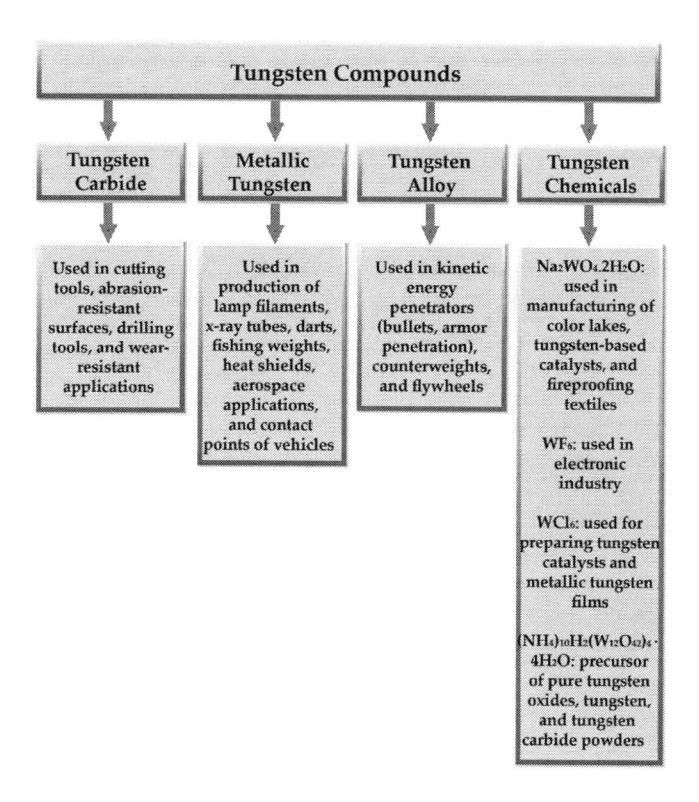

Figure 24.1 Forms and uses of tungsten substances.

24.2 ENVIRONMENTAL CONCENTRATIONS AND EXPOSURE

24.2.1 Environmental Exposure

Tungsten is a metal that can be present in the air, soil, and water through natural or anthropogenic activities. Tungsten can be released into the air as windblown dust or from ore processing, from hard-metal and tungsten carbide production, or from the combustion of municipal waste.[9] Tungsten can contaminate waterways via weathering of ores and soils, and contaminated effluents of mining and manufacturing processes. Dry deposition or precipitation of tungsten dust from the air, industrial waste landfills, sewage sludge application, and municipal solid waste ash that contain tungsten result in increased concentrations of tungsten in soil.[9]

Tungsten's concentration in air is associated with the presence of industrial facilities. For example, tungsten concentrations in particulate matter near a metal and processing facility in Sweet Home, Oregon, were between 1.5 and 3.0 ng/m^3. Concentrations within 20 km of the facility were <0.5 ng/m^3.[10] Sahle et al. (1994) showed that tungsten oxide was released from the hard metal industry as a byproduct during the reduction stage of the raw material.[11] Germani et al. (1981) measured tungsten concentrations in the ore, ore concentrate, and electrostatic precipitator dust in copper smelters and found concentrations were approximately 4, 4, and 44 mg/g, respectively.[12] In addition, tungsten can be released into the air via municipal waste incineration (MWI) plants.

Tungsten is not required by the United States Environmental Protection Agency (US EPA) to be monitored in surface or groundwater, so information about environmental water concentrations of tungsten is limited, but tungsten has been detected in surface and groundwater in areas that are naturally rich in tungsten, such as Northern Iceland. In this area, concentrations of tungsten in groundwater ranged between 0.03–11.5 and 0.015–0.49 ppb (μg/L) in lowland areas and in highland areas, respectively. In addition, tungsten concentrations in surface water were reported to be approximately 0.005–0.09, 0.005–0.34, and 0.005–0.33 ppb in lakes, rivers and streams, and peat soil water, respectively.[13] In the US, tungsten was found to contaminate groundwater at a Superfund National Priority List hazardous waste site.[9] Analysis of water from private wells in this area in Churchill County, Nevada revealed that the concentrations of tungsten ranged from 2 to 520 ppb[3]. These concentrations of tungsten in tap water were subsequently associated with an increase in tungsten concentrations in the urine of the residents of this county compared to the general population of the US.[14]

Tungsten is present in soil naturally, but applying sewage sludge and fertilizers that contain tungsten can increase tungsten concentrations in soil.[3] Another important anthropogenic source of tungsten in soil are military activities including the use of military tools containing tungsten such as ammunition. Soil samples collected from a depth up to 5 cm, from the Saudi Arabia- Kuwait border area, where there was heavy use of explosives during the Gulf War, revealed the presence of 126.5 mg tungsten/kg soil.[15] Upon moving to the use of tungsten-based bullets instead of lead, very high concentrations of tungsten were also detected in soil at Camp Edwards on the Massachusetts Military Reservation (2,800 mg/kg), but the concentration of tungsten decreased with increasing soil depth.[16] Tungsten particle size is also reported to decrease as you move away from the source. For example, surface dust samples from waste metal at a hard metal processing plant in Sweet Home, Oregon showed that the median tungsten particle size was 1.5 μm,[17] but smaller tungsten particle sizes were detected in areas 75 m away from the mill.

24.2.2 Occupational Exposure

Workers can be exposed to higher levels of tungsten compared to the general population during metal extraction from ores, manufacturing tungsten carbide, and in the tool-cutting industry. Exposure to tungsten dust occurs during different tasks throughout production processes. Tungsten dusts are produced during crushing, loading, grinding operations, and shaping of the products. Kraus et al. (2001) studied exposure to tungsten in different processes in a hard metal plant. This study showed that the maximum tungsten concentrations in ambient air were in the production of heavy metal alloys (125.0–417 μg/m^3).[18] Powder processing was the second highest (177.0–254.0 μg/m^3), while the lowest was in wet grinding (3 μg/m^3). Biological monitoring showed that the mean tungsten concentration in urine had the following trend classified by task: grinding > production of tungsten carbide > heavy alloy production > powder processing > sintering > forming > maintenance (94.4, 42.1, 24.9, 12.2, 12.5, 10.7, and 3.4 μg/g creatinine, respectively). In addition, the study showed that there was no correlation between the tungsten concentrations in air and urine on a task basis. The authors explained this by the differences in the bioavailability of tungsten forms in different processes.[18] Another study measured the personal air concentration of inhalable tungsten for 72 workers in a Swedish hard metal plant. It was found that the mean tungsten exposure concentrations to the inhalable fraction of tungsten ranged between 0.00027 and 0.570 mg/m^3 depending on the department. The highest concentration of tungsten was found in the powder production department.[19]

Occupational exposure to tungsten is regulated by the American Conference of Governmental Industrial Hygienists (ACGIH), and the National Institute of Occupational Safety and Health (NIOSH). The ACGIH set the threshold limit value (TLV)-time-weighted average (TWA) as 1 and 5 mg/m^3 for soluble and insoluble tungsten substances exposure, respectively. NIOSH's recommended exposure level (REL) 10-h TWA values for soluble and insoluble tungsten substances (as W) are 1 and 5 mg/m^3, respectively.[9]

24.3 TOXICOKINETICS

24.3.1 Inhalation Exposure

The presence of tungsten in urine and blood in workers exposed to tungsten dusts during the production of tungsten products confirms that inhaled tungsten is absorbed.[9] Results from hard metal cutting tool plants showed that tungsten-exposed workers have relatively high concentrations of tungsten in their urine compared to unexposed workers (7.06 and 0.08 µg/L, respectively).[20] Animal studies have been completed to understand the toxicokinetics of tungsten using radiolabeled tungsten compounds. The highest concentration of tungstic oxide ($^{181}WO_3$) was found in the lung after inhalation by Beagle dogs.[21] In a study designed to evaluate the pharmacokinetics of tungsten administered through inhalation, 16-week-old male rats were exposed to radiolabeled sodium tungstate ($Na_2^{188}WO_4$) as an aerosol with a concentration of 256 mg tungsten/m³ for 90 min. Concentrations of ^{188}W were detected by gamma spectrometry indicating the presence of tungsten in many organs such as the thyroid, kidney, spleen, and lymph nodes with the highest concentrations detected in the lung and thyroid gland.[22] The activity of ^{188}W decreased in all tissues, but some activity was observed in femur, lung, and spleen 21 days post-exposure, indicating retention of tungsten in the body. This study showed that the initial half-lives were 4–6 h and the long-term half-lives were 6–67 days.[22] Furthermore, the study revealed that urinary excretion is the main route of tungsten elimination.[22]

Rajendran et al. (2012) assessed the toxicokinetics of tungsten blue oxide ($WO_{2.9}$) for a longer exposure period. In this study, Sprague-Dawley rats were exposed to aerosols of tungsten blue oxide for 6 h/day for 28 days with concentrations of 15 or 124 mg/kg/day. This study showed that tungsten blue oxide was absorbed systemically with increased blood levels of tungsten and detection in the femur, kidney, and lung. Tungsten was the highest in lung, followed by femur and kidney, which were ranked as the second- and third-highest tungsten levels compared to other organs.[23] Tungsten concentrations reached a steady state at day 14 of exposure. The systemic elimination half-lives were approximately 23 and 154 h for 15 and 124 mg/kg/day, respectively. It is important to note that the urinary elimination of tungsten is three times lower than fecal elimination. The authors referred the high fecal elimination of tungsten to deposition in the lung, which consequently was cleared from the respiratory tract and then through the gastrointestinal tract. This result is consistent with a Beagle dog study, where 60% of the inhaled ^{188}W-labeled aerosol deposited in the lung and one-third of this deposited activity was cleared to systemic circulation after 10 days of exposure and the remaining lung deposition was cleared to the gastrointestinal tract.[21]

Another study evaluated the role of olfactory transport of inhaled tungsten to the central nervous system. In this study, male CD rats were exposed to aerosols of radiolabeled sodium tungstate for 90 min. Interestingly, this study showed that the main route for tungsten transport to the brain is systemic circulation, not via olfactory transport, which is only responsible for a minimal role in delivering tungsten to the brain.[24]

24.3.2 Oral Exposure

As mentioned earlier, humans can be exposed to tungsten via contaminated drinking water, allowing for oral exposure. An intoxication was reported for a healthy 19-year-old male who drank 250 mL of a mixture of beer and wine on a hot 155-mm gun-barrel.[25] Analysis using inductively coupled plasma emission-mass spectrometry (ICP-MS) showed very high concentrations of tungsten in gastric content, blood, and urine. Also, tungsten was found in nails and hair with signs of encephalopathy. Although the patient had six hemodialyses, the blood level remained high (0.005 mg/L) at day 13 post-ingestion and remained in urine until day 33.

In addition, animal studies showed the absorption and distribution of tungsten upon oral exposure. A study with male C57BL/6J mice exposed to drinking water containing sodium tungstate dihydrate ($Na_2WO_4 \cdot 2H_2O$) in a range of 15–1,000 mg tungsten/kg for 1 or 16 weeks, showed an increase in tungsten in bone marrow.[26] Another study with male C57BL/6J mice was performed where mice were exposed to sodium tungstate dihydrate in drinking water at concentrations of 0, 62.5, 125, or 200 mg/kg/day for 28 days.[27] ICP-MS analysis showed dose-dependent accumulation of tungsten in the kidney, liver, colon, bone, brain, and spleen. The highest concentration was found in the bones and the lowest concentration was found in brain tissue. Higher levels of tungsten were observed in female mice indicating sex differences in the uptake of tungsten.

McDonald et al. (2007) treated Sprague-Dawley rats and C57BL/6N mice with a range of sodium tungstate dihydrate concentrations by either oral gavage or intravenous administration.[28] Tungsten concentration in plasma reached a maximum level at 4 h in rats and 1 h in mice but significantly decreased within 24 h in both rodent models.[28] Also, tungsten was found in liver, kidney, femur, uterus, and intestine. In another study, rats and mice were repeatedly exposed to sodium tungstate dihydrate by gavage (10 mg/kg) or drinking water (560 mg/L) for 14 consecutive days.[29] Tungsten was detected in plasma, kidney, liver, femur, and intestine. In addition, transplacental transport of tungsten was evident with the observed elevated tungsten concentration in neonates. Accumulation was observed in the intestine, kidney, and femur. Tungsten concentration in mice fetuses was eight times higher than rat fetuses, confirming interspecies differences. McInturf et al. (2011) found that feeding rats with sodium tungstate (NaW) by gavage for 70 days lead to the presence of tungsten mainly in bone and spleen, followed by kidney, gastrointestinal tract, and with trace amounts in lung, liver, brain, blood, thymus, testes, and ovaries. Upon gestational exposure, tungsten was found in bone and gastrointestinal tract.[30] Overall, it can be concluded that the soluble forms of tungsten can be absorbed via inhalation or oral exposure and distributed to various organs, while tungsten is excreted mainly via urine and feces.

24.4 ADVERSE HEALTH EFFECTS

24.4.1 Cancer

Exposure to tungsten was a suspected cause of a cluster of leukemia cases in Churchill County, Nevada that included 16 cases between 1997 and 2002 due to the presence of military bases near this area. The concentrations of tungsten in tap water in this area ranged from not detectable to 336 µg/L, with a mean value of 4.66 µg/L, which was not significantly different compared to other counties in Nevada.[3] Rubin and colleagues investigated the potential causes of this childhood leukemia cluster by analyzing water, soil, blood, and urine samples from members of the community including the leukemia cases.[31] Although the concentrations of tungsten in urine and water were relatively high compared to the national comparison values, there were no significant differences between leukemia cases and comparison children.

Since that time many *in vitro* and animal studies have further investigated the potential carcinogenicity of tungsten. Laulicht et al. (2015) treated immortalized human bronchial epithelial cells (Beas-2B) with 0, 50, or 250 µM sodium tungstate (Na_2WO_4) for 6 weeks. Then, the cells were incubated in soft agar without any additional tungsten exposure. The results of this study showed that the number of colonies derived from tungsten-treated cells is significantly higher than untreated cells.[32] Also, there was an increase in anchorage-independent growth in soft agar in all tungsten-treated cells, which indicates tumorigenicity of the cells. The same study showed the ability of tungsten-treated cells to heal a wound produced by a scratch, indicating the ability of the cells to migrate. A month after injecting the transformed colonies derived from tungsten-treated cells into 12 female athymic nude mice, 100% of the tungsten-transformed cells formed visible tumors, which was absent in the control group. Also, RNA sequencing of the colonies derived from tungsten-treated cells and control cells revealed significant alterations in 535 genes. Among the altered genes were those related to the development of lung cancer. Pathway analysis showed that the top canonical pathways were "Role of Tissue Factor in Cancer," "Molecular Mechanisms of Cancer," "mTOR signaling," "Chronic Myeloid Leukemia Signaling," "p53 Signaling," and "PI3K/Akt signaling," which are involved in cancer, suggesting that tungsten is a carcinogen.[32] Although this study highlighted the potential carcinogenicity of tungsten, it is important to note that the concentrations used in the study weren't environmentally relevant, thus it is difficult to extrapolate the results for human effects. Bolt et al. (2015) showed that tungsten was detected in the urine of breast cancer patients who used a tungsten-based shield during intraoperative radiotherapy. The mean urinary tungsten concentration was 1.76 ng/mL. It is important to note that tungsten was detected in the urine of patients who did mastectomy, which indicates the mobilization of tungsten.[33] Furthermore, this study reported that tungsten induced lung cancer metastasis in female BALB/c mice that were treated with 15 ppm sodium tungstate dihydrate in their tap water and injected with 66Cl4 cells, a breast cancer cell line, in the fourth mammary fat pad.[33] This study showed that tungsten exposure caused alterations that enhanced cell invasion and metastasis such as matrix metalloproteinases and myeloid-derived suppressor cells. The concentrations of tungsten in this study were also relatively high compared to environmental levels. A 28-day repeated dosing experiment showed that oral exposure to tungsten oxide (WO_3) nanoparticles at 1,000 mg/kg led to a significant increase in DNA damage in Wistar rats compared to control and tungsten microparticles.[34]

24.4.2 Hard Metal Disease

Occupational exposure to inhaled dust of hard metal has been associated with pulmonary fibrosis, memory deficits, and increased mortality due to lung cancer (reviewed in Ref. 3). Metals such as cobalt (Co), yttrium, copper, nickel (Ni), iron (Fe), and molybdenum are added to tungsten to form tungsten alloy that has certain metallurgical properties. The adverse effects resulting from exposure to hard metal are believed to be driven by the metals combined with tungsten.[35] The International Agency for Research on Cancer (IARC) classified cobalt as possibly carcinogenic to humans (group B2), while cobalt with tungsten carbide (hard metal) is considered as probably carcinogenic to humans (group A2), indicating that a mixture of cobalt and tungsten may increase the toxicity of cobalt.[36] Lison et al. (1995) suggested that reactive oxygen species produced by cobalt particles increased in the tungsten carbide-cobalt particles.[37] In this study, casein-elicited mouse peritoneal macrophages were exposed to tungsten carbide-cobalt (WC-Co) particles at a range of 50–300 μg/mL in absence or presence of 1 mM butylated hydroxytoluene (BHT). Cytotoxicity was measured as lactate dehydrogenase (LDH) activity released in the cell-free fraction of the culture medium. Percentage of LDH increased with an increasing dose of WC-Co particles. The addition of BHT significantly decreased the percent of LDH released, indicating the presence of a protective effect against cytotoxicity caused by WC-Co particles.[37] It is important to note that the addition of Trolox, a hydrophilic antioxidant, didn't reduce the percent of LDH released. These results suggested that cytotoxicity induced by WC-Co particles was targeting lipid membrane via lipid peroxidation.[37] This study proposed that an interaction between WC and cobalt particles increased the production of ROS. They explained that electrons move from cobalt to the surface of WC particles, which can reduce oxygen and generate ROS and also increase the concentration of cobalt ions.[37] This proposed mechanism explains the differential toxicity of cobalt chloride ($CoCl_2$) and WC-Co nanoparticles on different types of human cells, where WC-Co nanoparticles significantly reduced cell viability compared to cobalt ion treatment.[38]

Many *in vitro* studies investigated the toxicity of tungsten carbide/cobalt and compared its toxicity to tungsten alone or cobalt alone. Lombaert et al. (2004) compared the apoptogenic effect of tungsten carbide (WC), cobalt particles (Co particles), $CoCl_2$, and tungsten carbide-cobalt

(WC-Co) particles using human peripheral blood mononucleated cells (PBMC). The cells were exposed to 2.0–6.0 µg/mL co-equivalent or 33.3–100.0 µg/mL WC. The apoptotic potential was examined by Annexin-V-positive/propidium iodide negative fluorescence after 15 min and 6 h post exposure.[39] Annexin-V-positive/propidium iodide negative fluorescence indicated early apoptosis. All compounds caused a significant increase in relative apoptotic activity compared to the control. Interestingly, there was a significant difference between WC-Co and WC. To understand the mechanism behind the observed apoptotic activity, apoptosis inhibitors were added to the cells treated with the four metal compounds and percent inhibition of apoptosis was measured. The results revealed differential sensitivity to apoptosis inhibitors.[39] A significant increase in apoptosis inhibition was observed upon adding caspase 9 inhibitor to WC or WC-Co particles-treated cells for 6 h, which wasn't observed with caspase 8 inhibitor treatment. On the other hand, a significant increase in apoptosis inhibition was observed upon adding caspase 8 or caspase 9 to Co particles-treated cells.[39] These results suggest that WC-Co particles and Co particles induce apoptosis via two different mechanisms. The sensitivity of WC-Co treated cells to caspase 9, which plays an important role in initiating the intrinsic apoptosis pathway, can be explained by increased production of ROS by cobalt in the presence of tungsten carbide.[37,39] Quantification of metal-induced apoptotic DNA fragmentation showed that at a longer exposure period (14 h), DNA fragmentation was significantly increased in Co particles and WC particles-treated cells. Furthermore, DNA fragmentation was significantly higher in WC-Co treated-cells compared to WC alone or Co alone, suggesting an additive apoptogenic effect.[39] In a follow-up study by Lombaert et al. (2013) exposure of human PBMC to WC-Co or Co particles for 15 min caused production of ROS, which induced an increase in expression of Heme oxygenase 1 (HMOX1), an oxidative stress response gene that protects cells from oxidative damage and stimulates a cascade of events including p38/mitogen-activated protein kinases (MAPK) activation, hypoxia inducible factor 1-α (HIF-1α) stabilization, and ATM-independent p53 stabilization. These changes weren't observed in $CoCl_2$-treated cells. HIF-1α activation is involved in alterations in cellular metabolism pathways in cancerous cells such as increased glucose uptake and lactate production.[40] Also, vascular endothelial growth factor (VEGF), responsible for angiogenesis, was induced by hypoxia and is a target gene of transcription factor HIF-1α.[40] Stabilization of HIF-1α by WC-Co particles may explain the carcinogenicity of hard metal dust.

On the other hand, Busch et al. (2010), showed that gene transcription alterations in a human keratinocyte cell line, HaCaT, treated with WC-Co weren't significantly different compared to cells treated with $CoCl_2$.[41] Although treatment of HaCaT cells with 33 µg/mL WC-Co or 3 µg/mL $CoCl_2$ for 3 days caused stimulation of transcription of genes involved in hypoxia pathways, these changes weren't observed in cells treated with 30 µg/mL tungsten carbide (WC) nanoparticles. Altogether, the results of this study indicated the importance of cobalt ions in the toxicity of WC-Co particles, regardless of the presence of WC in the nanoparticles.[41] Ding et al. (2009) showed that the size of WC-Co particles mediated its toxicity. Exposure of WC-Co nanoparticles (80 nm) in a mouse epidermal cell line (JB6 P+) caused an increase in activation of activator protein (Ap-1) and nuclear factor kappa-light-chain-enhancer of activated B cells (Nf-κP), MAPK signaling pathways, and a decrease in glutathione (GSH) levels. These changes were greater than the changes that occurred after treatment by WC-Co fine particles (4 µm).[42] These results indicated that the oxidative stress induced by WC-Co increased with the decreasing size of particles. The same group showed that exposure to WC-Co nanoparticles and fine particles resulted in the stimulation of proapoptotic factors such as Fas, Fas-associated protein with death domain (FADD), caspase 3, 8 and 9, BID and BAX in JB6 P+ cells. However, exposure to WC-Co nanoparticles caused higher apoptotic stimulation than the fine particles.[43] Contradictory to the previous studies, a study by Kühnel et al. (2012) showed that crystalline structure and surface area of WC nanoparticles was the determining factor for increased production of ROS, regardless of the absence or presence of Co. In this study, WC particles of a mean particle size of 113 ± 2 nm, WC-Co particles of a mean particle size (145 ± 5 nm), or WC particles of a mean particle size of (145 ± 5 nm) were added to HepG2 or HaCat cells.

The results showed that large particles of WC and WC-Co didn't cause a decrease in cell viability or an increase of ROS. The authors explained that the high surface area of WC (113 nm) may explain the increased ROS production. Also, they suggested that the crystalline structure of WC (113 nm), which contained carbon black is associated with the ROS production and genotoxicity of these particles.[44] This study emphasized the importance of considering different parameters while assessing the toxicity of WC nanoparticles. Armstead et al. (2014) showed that nano and micro WC-Co particles decreased the viability of Beas-2B cells compared to the control treatment.[45] The results showed that nanoparticles induced a more pronounced decrease in viability compared to microparticles. The authors explained the increased toxicity of nanoparticles by their ability to be internalized by lung epithelial cells compared to the microparticles.[45]

24.4.3 Interaction with Other Metals

As a replacement of lead-based ammunition the "green bullet" containing tungsten alloy was introduced as a "safer, less toxic" replacement. *In vitro* and *in vivo* studies have investigated the toxicity of different compositions of tungsten alloy and questions surfaced on the safety of tungsten alloy. Kalinich et al. (2005) implanted two doses (4 pellets or 20 pellets) of weapon-grade tungsten alloy, tungsten-nickel-cobalt (W/Ni/Co) intramuscularly into F344 rats that was composed of 91.1% W, 6% Ni, and 2.9% Co and compared it to 100% Ni pellets.[46] An increase in red blood cells, white blood cells, neutrophils, lymphocytes, and monocytes was detected after 1 month of exposure to the 20 pellets of tungsten alloy. Also, 100% of the rats developed high-grade pleomorphic rhabdomyosarcomas after 4–5 months post-implantation of the higher dose of tungsten alloy. The low dose of tungsten alloy and nickel pellets also induced tumor development, but at a slower rate compared to the higher dose of tungsten alloy. Unexpectedly, tungsten alloy that had a lower concentration of Ni resulted in lower toxicity compared to the 100% Ni pellets. This study emphasized that W, Co, and Ni may have a synergistic toxicity interaction. Implantation of tungsten alloy mimics exposure to shrapnel, suggesting that if the shrapnel isn't removed, tumors may develop in humans. To understand the mechanism behind the toxicity of tungsten alloy, Harris et al. (2011) compared the toxicity of W/Ni/Co alloy that had different ratios of the three metals [W/Ni/Co (91%:6%:3%) or W/Ni/Co (97%:2%:1%)] and tungsten-nickel-iron (W/Ni/Fe) alloy (97%:2%:1%) to the toxicity of each individual metal using a rat skeletal muscle cell line (L6-c11). Exposure to W/Ni/Co (91%:6%:3%), W/Ni/Co (97%:2%:1%), or W/Ni/Fe alloy (97%:2%:1%) caused a significant decrease in caspase 3 activity after 24 h of exposure, but the lowest activity was observed in W/Ni/Co (91%:6%:3%).[47] Moreover, the three types of tungsten alloy induced ROS production with the following trend W/Ni/Co (91%:6%:3%) > W/Ni/Co (97%:2%:1%) > W/Ni/Fe (97%:2%:1%).[47] Individual metal exposure revealed that only cobalt induced production of ROS, but with lesser extent compared to tungsten alloy, indicating a potential interaction between cobalt and tungsten as described by Lison et al. (1995)[37] who suggested the interaction between tungsten carbide, cobalt, and dissolved oxygen. Microarray analysis revealed that only the W/Ni/Co (91%:6%:3%) caused the upregulation of genes involved in DNA damage and stress response.[47] The authors suggested that W/Ni/Co (91%:6%:3%) induced HIF-1α stabilization via increased solubility of nickel and cobalt and the formation of free radicals.[47] These effects combined with inhibition of caspase 3 may explain the carcinogenicity of W/Ni/Co (91%:6%:3%).[47] These results are consistent with Schuster et al. who found that W/Ni/Fe (97.1%:1.7%:1.2%) alloy didn't induce tumor formation around implanted alloy pellets in F344 rats.[48] Schuster et al. explained this result by the low dissolution rate of W/Ni/Fe (97.1%:1.7%:1.2%) compared to W/Ni/Co (91.1%:6%:2.9%) alloy. These results highlight the role of mobilized ions in the toxicity of tungsten alloy. Roedel et al. (2012) showed that both tungsten/nickel/cobalt alloy (91.1%W/6%Ni/2.9%Co) and tungsten/nickel/iron alloy (92%W/5%Ni/3%Fe) particles were phagocytosed by lung macrophages, induced influx of neutrophils, production of inflammatory cytokines, and generation of ROS after 24 h post intratracheal instillation in rats.

In this study, the concentration of nickel was relatively comparable to tungsten alloy-containing cobalt and tungsten alloy-containing iron, which may explain the similar effects observed in both treatments.[49] These results emphasized that the ratio between components of the tungsten alloy plays an important role in toxicity.

Furthermore, Harris et al. (2015) compared the effects of W/Ni/Co (91%:6%:3%) in rat cells (L6-cl1) and human primary muscle cells (HSkMc) to reveal species-specific response to this type of alloy. Results showed that both W/Ni/Co (91%:6%:3%) and W/Ni/Co (97%:2%:1%) caused changes in the transcription of genes involved in glycolysis, hypoxia, stress responses, DNA damage, and cell death.[50] Adams et al. (2015) showed that soluble forms of nickel or cobalt caused genetic changes in nerve growth factor (NGF) differentiated pheochromocytoma cells (PC12), emphasizing that addition of tungsten had a minimal role on developing adverse effects.[51] These studies highlight that the form of the tested compound will affect the toxicity, thus experimental design should be very accurate to mimic the objectives of the research.

24.5 CONCLUSIONS

This chapter discussed the toxicity of tungsten compounds. It has been shown that different forms of tungsten vary in their properties, which affect their behavior in the environment and their potential adverse health effects. Based on the current literature, animal studies that investigated the toxicity of the soluble forms of tungsten (e.g., sodium tungstate) used relatively high concentrations compared to the environmental concentrations; therefore, it is extremely difficult to extrapolate the results of these studies for human risk assessment purposes. Results of toxicological studies of tungsten alloys showed that the composition of alloy affects the toxicological outcomes emphasizing the need of assessing the toxicity of each product. Also, it is important to consider adjusting ratios between components of the alloy to achieve the balance between getting desired characteristics and minimizing toxicity. Studies also confirmed that different structural forms of tungsten compounds change the toxicity. For example, nano-sized particles had higher toxicity compared to macro-sized particles of the same chemical structure. Thus, moving forward it is imperative to take all of these factors into consideration when designing toxicity studies to assess the new tungsten-based products, which will help investigators to achieve accurate results that can be applied to the real-world environment and relative exposure scenarios.

REFERENCES

1. van der Voet, G. B. et al. Metals and health: a clinical toxicological perspective on tungsten and review of the literature. *Mil. Med.* **172**, 1002–1005 (2007).
2. Ashford R. D. *Ashford's Dictionary of Industrial Chemicals: Properties, Production, Uses.* London: Wavelength Publications, Ltd (1994).
3. Keith, S. et al. Toxicological profile for tungsten. *Agency Toxic Subst. Dis. Regist.* 153 (2005).
4. HSDB. Tungsten. Hazardous Substances Data Bank. National Library of Medicine. http://toxnet.nlm. nih.gov (2004).
5. Lide, D. R. Tungsten. In: *CRC Handbook of Chemistry and Physics.* 81st ed. Boca Raton, FL: CRC Press LLC, pp. 3–207 (2000).
6. O'Neil, M. J. et al. *The Merck Index. An Encyclopedia of Chemicals, Drugs, and Biologicals.* Whitehouse Station, NJ: Merck Research Laboratories, p. 1748 (2001).
7. Penrice, T. W. Tungsten. In: Kirk Othmer (ed.), *Kirk-Othmer Encyclopedia of Chemical Technology.* New York, NY: John Wiley & Sons, p. 590 (1997b).
8. Lewis, R. J. *Hawley's Condensed Chemical Dictionary.* Chichester: John Wiley & Sons, Inc (1997).

9. Keith, S. and Wohlers, D. *Addendum to the Toxicological Profile for Tungsten* (2015). https://www.atsdr. cdc.gov/toxprofiles/Tungsten_Addendum_508.pdf

10. Sheppard, P. R., Speakman, R. J., Farris, C., and Witten, M. L. Multiple environmental monitoring techniques for assessing spatial patterns of airborne tungsten. *Environ. Sci. Technol.* **41**, 406–410 (2007).

11. Sahle, W., Laszlo, I., Krantz, S., and Christensson, B. Airborne tungsten oxide whiskers in a hard-metal industry. Preliminary findings. Hygiene. *Ann. Occup. Hyg.* **38**, 37–44 (1994).

12. Germani, M. S., Small, M., Zoller, W. H., and Moyers, J. L. Fractionation of elements during copper smelting. *Environ. Sci. Technol.* **15**, 299–305 (1981).

13. Arnórsson, S. and Lindvall, R. The distribution of arsenic, molybdenum and tungsten in natural waters in basaltic terrain, N-Iceland. In: Cidu, R., ed., Water-Rock Interact. Proceedings of the Tenth International Symposium on Water-Rock Interaction. WRI-10, Villasimius Italy 10–15 July 2001. Amsterdam, Netherlands: AA. Balkema, pp. 961–964 (2001).

14. CDC. Churchill County (Fallon), Nevada Exposure Assessment, 2003. Available at: http://www.cdc. gov/nceh/clusters/Fallon/study.htm (2003).

15. Sadiq, M., Mian, A. A., and AlThagafi, K. Inter-city comparison of metals in scalp hair collected after the Gulf War 1991. *J. Env. Sci Health Part A* **27**, 1415–1431 (1992).

16. Clausen, J. L. and Korte, N. Environmental fate of tungsten from military use. *Sci. Total Env.* **407**, 2887–2893 (2008).

17. Sheppard, P. R. et al. Comparison of size and geography of airborne tungsten particles in Fallon, Nevada, and Sweet Home, Oregon, with implications for public health. *J. Env. Public Health* 2012, 509458 (2012).

18. Kraus, T. et al. Exposure assessment in the hard metal manufacturing industry with special regard to tungsten and its compounds. *Occup. Environ. Med.* **58**, 631–634 (2001).

19. Klasson, M. et al. Occupational exposure to cobalt and tungsten in the Swedish hard metal industry: air concentrations of particle mass, number, and surface area. *Ann. Occup. Hyg.* **60**, 684–699 (2016).

20. De Palma, G., Manini, P., Sarnico, M., Molinari, S., and Apostoli, P. Biological monitoring of tungsten (and cobalt) in workers of a hard metal alloy industry. *Int. Arch. Occup. Environ. Health* **83**, 173–181 (2010).

21. Aamodt, R. L. Inhalation of 181 W labelled tungstic oxide by six beagle dogs. *Health Phys.* **28**, 733–743 (1975).

22. Radcliffe, P. M. et al. Pharmacokinetics of radiolabeled tungsten (188W) in male Sprague-Dawley rats following acute sodium tungstate inhalation. *Inhal. Toxicol.* **22**, 69–76 (2010).

23. Rajendran, N. et al. Toxicologic evaluation of tungsten: 28-day inhalation study of tungsten blue oxide in rats. *Inhal. Toxicol.* **24**, 985–994 (2012).

24. Radcliffe, P. M. et al. Acute sodium tungstate inhalation is associated with minimal olfactory transport of tungsten (188W) to the rat brain. *Neurotoxicology* **30**, 445–450 (2009).

25. Marquet, P. et al. Tungsten determination in biological fluids, hair and nails by plasma emission spectrometry in a case of severe acute intoxication in manTungsten determination in biological fluids, hair and nails by plasma emission spectrometry in a case of sever. *J. Forensic Sci.* **42**, 527–530 (1997).

26. Kelly, A. D. R. et al. In vivo tungsten exposure alters B-cell development and increases DNA damage in murine bone marrow. *Toxicol. Sci.* **131**, 434–446 (2013).

27. Guandalini, G. S. et al. Tissue distribution of tungsten in mice following oral exposure to sodium tungstate. *Chem. Res. Toxicol.* **24**, 488–493 (2011).

28. McDonald, J. D. et al. Disposition and clearance of tungsten after single-dose oral and intravenous exposure in rodents. *J. Toxicol. Env. Health A* **70**, 829–836 (2007).

29. Weber, W. M. et al. Disposition of tungsten in rodents after repeat oral and drinking water exposures. *Toxicol. Env. Chem.* **90**, 445–455 (2008).

30. McInturf, S. M. et al. The potential reproductive, neurobehavioral and systemic effects of soluble sodium tungstate exposure in Sprague-Dawley rats. *Toxicol. Appl. Pharmacol.* **254**, 133–137 (2011).

31. Rubin, C.S., Holmes, A.K., Belson, M.G., Jones, R.L., Flanders, W.D., Kieszak, S.M., Osterloh, J., Luber, G.E., Blount, B.C., Barr, D.B., Steinberg, K.K., Satten, G.A., McGeehin, M.A., and Todd, R.L. Investigating childhood leukemia in Churchill County, Nevada. *Enivron. Health Perspect.* **115**, 151–157 (2007).

32. Laulicht, B. et al. Tungsten-induced carcinogenesis in human bronchial epithelial cells. *Physiol. Appl. Pharmacol.* **288**, 33–39 (2015).

33. Bolt, A. M. et al. Tungsten targets the tumor microenvironment to enhance breast cancer metastasis. *Toxicol. Sci.* **143**, 165–177 (2015).

34. Chinde, S. and Grover, P. Toxicological assessment of nano and micron-sized tungsten oxide after 28 days repeated oral administration to Wistar rats. *Mutat. Res. Genet. Toxicol. Environ. Mutagen.* **819**, 1–13 (2017).

35. Davison, A. G. et al. Interstitial lung disease and asthma in hard-metal workers: bronchoalveolar lavage, ultrastructural, and analytical findings and results of bronchial provocation tests. *Thorax* **38**, 119–128 (1983).

36. Rousseau, M. C., Straif, K., and Siemiatycki, J. IARC carcinogen update. *Environ. Health Perspect.* **113**, 580–583 (2005).

37. Lison, D., Carbonnelle, P., Lauwerys, R., Mollo, L., and Fubini, B. Physicochemical mechanism of the interaction between cobalt metal and carbide particles to generate toxic activated oxygen species. *Chem. Res. Toxicol.* **8**, 600–606 (1995).

38. Bastian, S. et al. Toxicity of tungsten carbide and cobalt-doped tungsten carbide nanoparticles in mammalian cells in vitro. *Environ. Health Perspect.* **117**, 530–535 (2009).

39. Lombaert, N. et al. Evaluation of the apoptogenic potential of hard metal dust (WC-Co), tungsten carbide and metallic cobalt. *Toxicol. Lett.* **154**, 23–34 (2004).

40. Weidemann, A. and Johnson, R. S. Biology of HIF-1α. *Cell Death Differ.* **15**, 621–627 (2008).

41. Busch, W., Kühnel, D., Schirmer, K., and Scholz, S. Tungsten carbide cobalt nanoparticles exert hypoxia-like effects on the gene expression level in human keratinocytes. *BMC Genomics* **11**, 65 (2010).

42. Ding, M. et al. Size-dependent effects of tungsten carbide-cobalt particles on oxygen radical production and activation of cell signaling pathways in murine epidermal cells. *Toxicol. Appl. Pharmacol.* **241**, 260–268 (2009).

43. Zhao, J. et al. Apoptosis induced by tungsten carbide-cobalt nanoparticles in JB6 cells involves ROS generation through both extrinsic and intrinsic apoptosis pathways. *Int. J. Oncol.* **42**, 1349–1359 (2013).

44. Kühnel, D. et al. Comparative evaluation of particle properties, formation of reactive oxygen species and genotoxic potential of tungsten carbide based nanoparticles in vitro. *J. Hazard. Mater.* **227–228**, 418–426 (2012).

45. Armstead, A. L., Arena, C. B., and Li, B. Exploring the potential role of tungsten carbide cobalt (WC-Co) nanoparticle internalization in observed toxicity toward lung epithelial cells in vitro. *Toxicol. Appl. Pharmacol.* **278**, 1–8 (2014).

46. Kalinich, J. F. et al. Embedded weapons-grade tungsten alloy shrapnel rapidly induces metastatic high-grade rhabdomyosarcomas in F344 rats. *Environ. Health Perspect.* **113**, 729–734 (2005).

47. Harris, R. M., Williams, T. D., Hodges, N. J., and Waring, R. H. Reactive oxygen species and oxidative DNA damage mediate the cytotoxicity of tungsten-nickel-cobalt alloys in vitro. *Toxicol. Appl. Pharmacol.* **250**, 19–28 (2011).

48. Schuster, B.E., Roszell, L.E., Murr, L.E., Ramirez, D.A., Demaree, J.D., Klotz, B.R., Rosencrance, A.B., Dennis, W.E., Bao, W., Perkins, E.J., Dillman, J.F., and Bannon, D.I. In vivo corrosion, tumor outcome, and microarray gene expression for two types of muscle-implanted tungsten alloys. *Toxicol. Appl. Pharmacol.* **265**, 128–138 (2012).

49. Roedel, E. Q., Cafasso, D. E., Lee, K. W. M., and Pierce, L. M. Pulmonary toxicity after exposure to military-relevant heavy metal tungsten alloy particles. *Toxicol. Appl. Pharmacol.* **259**, 74–86 (2012).

50. Harris, R. M., Williams, T. D., Waring, R. H., and Hodges, N. J. Molecular basis of carcinogenicity of tungsten alloy particles. *Toxicol. Appl. Pharmacol.* **283**, 223–233 (2015).

51. Adams, V. H., Dennis, W. E., and Bannon, D. I. Toxic and transcriptional responses of PC12 cells to soluble tungsten alloy surrogates. *Toxicol. Rep.* **2**, 1437–1444 (2015).

CHAPTER 25

Boron Toxicity
An Insight on Its Influence on Wheat Growth

**Anamika Pandey, Mohd. Kamran Khan, Mehmet Hamurcu,
Fatma Gokmen Yilmaz, and Sait Gezgin**
Selcuk University

CONTENTS

25.1 INTRODUCTION

Boron toxicity is one of the major limiting factors for crop production in several arid and semi-arid areas of the world including Turkey, South Australia, Iraq, Chile, Spain, and USA where B level is high in irrigation water or soil (Cartwright et al. 1984; Nable et al. 1997; Kalayci et al. 1998; Jefferies et al. 2000; El-Shazoly et al. 2019). Even after 100 years of research on B in plants (Warington 1923), there are still several aspects that need to be focused in order to understand their role, especially in higher plants (Lewis 2019; González-Fontes 2019). Wheat is one of the major food crops consumed as the main component of the human diet around the world. However, it is sensitive to high B levels in the soil and wheat yields are affected under B toxic growth conditions. In this chapter, several facets of B toxicity in wheat and molecular advancement in understanding the physiological and biochemical mechanism and developing the B tolerant wheat genotypes have been discussed. The objectives that are required to be completed in this research area and the possible strategies toward them have also been mentioned.

25.2 B - AN ESSENTIAL ELEMENT FOR PLANTS

Boron is an essential micronutrient necessary for plant growth and development whose deficiency and toxicity both are common in plants (Landi et al. 2019). It is required by plants in a small amount ranging from 1 µg/g to 1 mg/g for normal growth. Moreover, the range between B deficiency and toxicity in plants is too narrow. It is the only microelement that enters the plant roots in the form of neutral solute that is boric acid that may convert into borate ions at high pH. Due to its small size and uncharged nature, it easily enters the phospholipid membrane and gets randomly distributed in the plant (Brown et al. 2002; Brdar-Jokanovic 2020). At high levels of B in the soil, this high permeability becomes more difficult to manage (Reid 2010). Though B deficiency can be handled using soil or foliar fertilizers, B toxicity is challenging to manage. As reclamation of B toxic soils is unrealistic, improvement of B tolerance in plants has been the target of researchers for several decades. Although different levels of B tolerance have been observed in different plant species, transferring them to conventional crops and increasing the yield has been a challenge.

25.3 DIFFERENT ROLES OF B IN PLANT METABOLISM

The interest in the role of boron in wheat growth and metabolism has been continued for almost a century (Morris 1931; Catav et al. 2018; Brdar-Jokanovic 2020). The main role of B is known for maintaining the integrity, structure, and stability of the cell wall. At cytoplasmic pH (7.0–7.5) or lower (around 5.0), most of the B in plants (more than 98%) exists in the form of boric acid. Both boric acid and borate anions can bind to molecules like sorbitol, ribose, apiose containing cis-hydroxyl groups (Matoh and Kobayashi 2002). In 1993, Matoh et al. isolated B-polysaccharide complex from cell walls of radish roots for the first time and further, Kobayashi et al. (1996) classified that polysaccharide as a pectic polysaccharide rhamnogalacturonan II (RGII). The cross linking of B with the RG II molecules forming ester bonds contributes to the cell wall strength (O'Neill et al. 1996; Funakawa and Miwa 2015). However, this strength depends on the fact that whether B binds with the apiose of side chain A or B of the pectic polysaccharide as only the one in chain A develops a firm borate bridge (Reuhs et al. 2004). The dicots with more pectin in their cell walls have higher B requirements and higher tolerance to B toxicity than monocots (Hu et al. 1996; O'Neill et al. 2004). B also forms complexes with other compounds including amino acids, phenolics, uronic acids, and polyalcohols (Hu et al. 1996; Brown et al. 2002).

Not only in the cell wall, but also B plays a crucial role in linking the plasma membrane with the cell wall via the formation of glycosyl inositol phosphoryl ceramides based complexes with B-RGII structures (Wang et al. 2015). Due to its role in membrane functioning, B participates in enzymatic reactions, hormones, and ions transport across the membrane (Brown et al. 2002; Goldbach and Wimmer 2007; Mosa et al. 2016). The involvement of B in maintaining the proton gradient across the membrane by controlling the proton-pumping ATPase activity has been observed (Goldbach and Wimmer 2007; Rossini Oliva et al. 2018). B regulates the activity of the water channels in the plasma membrane by the accumulation of hydroxyl radicals (Yu et al. 2002).

B stress is known to affect photosynthesis by hindering the thylakoid electron transport and the energy gradient across the membrane consequently affecting the activity of chloroplast membrane (Wang et al. 2015; Li et al. 2017; Shah et al. 2017; Shireen et al. 2018). Both B deficiency and toxicity cause oxidative damage by hydrogen peroxide and lipid peroxidation, further reducing the intercellular CO_2, leaf gas exchange, stomatal conductance, transpiration rate, and photosynthetic rate (Huang et al. 2014; Shah et al. 2017). However, starch and hexose may remain unconsumed due to the inhibition of growth, thus, their accumulation remains unaffected. Consequently, the accumulation of hexoses and starch feedback regulates the CO_2 assimilation (Han et al. 2008). B is found to be controlling nitrate accumulation and nitrate metabolism by the differential regulation of nitrate reductase (NR) activity (Eraslan et al. 2007; Cervilla et al. 2009; Seth and Aery 2017).

B stress regulates the expression of high molecular weight proteins associated with protein biosynthesis, cellular transport, proteolysis, detoxification, and antioxidant defense (Sang et al. 2015). B enhances the sugar content and its transport to the fast-growing organs like roots, fruits to address their sugar requirements (Pappin et al. 2012). The formation of B-polyol complexes and movement of B to roots for their discharge into soil is largely regulated by B-carbohydrate interactions (Pappin et al. 2012). B is associated with the metabolic pathways of auxin production especially indole-3-acetic acid, which in turn regulates the root length. The differential expression of IAA synthetic genes and rootward/shootward IAA transport genes seems to be regulating the root development at different stages of B treatment. B deficiency also inhibits root elongation due to a shortage in pyrimidine bases obstructing the synthesis of nucleic acids development (Reid et al. 2004; González-Fontes et al. 2008; Shireen et al. 2018; Brdar-Jokanovic 2020). In addition, an ethylene/auxin/ROS-dependent pathway is known to be regulating root cell elongation under B stress condition (Camacho-Cristobal et al. 2015). In addition, a constant supply of boron is necessary for cell division and meristematic activity in the root tissues (Tariq and Mott 2007).

B affects the germination of pollen and growth of pollen tube by regulating the even distribution of callose in the pollen tube wall, maintaining the cell wall plastic extensibility by association with pectin and less accumulation of phenolics in the cell wall (Wang et al. 2003). B deficiency negatively affects the number of grain set index and grains per spikelet in wheat (Cheng and Rerkasem 1993; Rerkasem et al. 2019). The effect of B on seed composition might be due to its role in carbon and nitrogen metabolism. The role of B in increasing the endogenous level of auxins and in transcriptional upregulation of proteins involved in embryogenesis has been confirmed in several studies (Camacho-Cristóbal et al. 2008; Pandey et al. 2012). In addition, B contributes to biological cell activities at both the transcription and translation levels (Dzondo-Gadet et al. 2002). B suppresses the Borate Exporter BOR1 translationally and contributes to the prevention of B toxicity (Aibara et al. 2018). B application has constructive effects on nodules and symbiotic nitrogen fixation in legumes (Bellaloui et al. 2014; Bellaloui and Mengistu 2015).

25.4 B UPTAKE AND TRANSLOCATION IN WHEAT

B is usually uptaken by plants via roots in the form of boric acid at normal soil pH of 5.5–7.5 (Camacho-Cristóbal et al. 2008; Landi et al. 2019). Initially, it was thought that passive diffusion of B across the plasma membrane in the form of boric acid due to the permeability of lipid bilayers is the only method of B adsorption from roots. However, with the discovery of several cell wall-related genes, B transporters, and signal transduction mechanisms, the entry of B in plants has been redefined (Wang et al. 2015). B enters into root cells through passive diffusion when B is present in the soil in normal to high concentrations. However, under low soil B concentrations, B entrance is facilitated by the borate exporters and boric acid channels (Stangoulis et al. 2001; Takano et al. 2008; Tanaka and Fujiwara 2008; Miwa and Fujiwara 2010). Not only for uptake, but B transporters are also involved in the exclusion of B from tissues under B toxic conditions (Miwa et al. 2007; Sutton et al. 2007; Schnurbusch et al. 2010). Where transporters and channels present in roots encourage the B uptake and loading into xylem, the ones in anthers and seeds elucidate the maintenance of intracellular B levels in these tissues under low transpiration rates (Tanaka et al. 2013; Yoshinari and Takano 2017).

B mostly moves in the xylem of plants based on the transpirational force that leads to the accumulation of B at the tips of the leaves and in mature parts leading to chlorotic and necrotic patches (Nable et al. 1997; Wimmer et al. 2003). That is why symptoms of B toxicity typically occurs on the margins of the mature leaves, however, B deficiency symptoms appear first in meristematic tissues like young leaves (Marschner 1995). Due to decreased mobility of B in the phloem, it is constantly required by the plant during its vegetative growth. In some plants like the ones with high sorbitol content and the ones with sucrose (Suc) as their photoassimilate, B forms complexes and passes via phloem causing more rapid accumulation in young leaves (Brown and Shelp 1997; Hu et al. 1997; Stangoulis et al. 2001, 2010). In such plants, the symptoms of B toxicity stress can be more on sink organs like fruit ailments, stem death, etc. Though B toxicity cannot be observed in roots in terms of patches, suppression of root length is an observable parameter (Wang et al. 2010). B uptake in plants is also found to be associated with stomatal conductance (Macho-Rivero et al. 2018). Under excess B supply, B concentrations may significantly vary between margins and tips of the leaves and midribs and petioles. In addition, B concentration in phloem exudates is lower than the leaves (Brown and Shelp 1997).

In wheat, as the ear of the wheat spike is completely covered by leaf sheath, the rate of water loss via transpiration is minimized. This leads to the minimal transport of B to the spikes via xylem (Rawson 1996). Then, B and sugars must be supplied through the phloem. Thus, in wheat, B passes through the phloem via the formation of bis-Suc borate complex, and thus, translocate Suc as their main photoassimilate. During the unloading of sucrose, B is also released from this complex and reaches the sink organs. Thus, the formation of bis-Suc borate complex in wheat could be of great significance in maintaining its reproductive growth (Rerkasem and Jamjod 2004; Stangoulis et al. 2010).

25.5 GENETIC VARIATION IN WHEAT TOWARD B TOXICITY TOLERANCE

B toxicity is a worldwide problem in agricultural soils especially in dry regions of the world, specifically with alkaline soils. It has been identified that the level of B causing toxicity varies in clayey, loamy, sandy, and loamy sand soils (Robertson et al. 1975). As a large number of factors such as soil pH, lime content, agro-physiological mechanisms regulate the B use efficiency of plants, it is crucial to identify potential genotypes with positive growth response in B toxic conditions and utilize them in breeding programs. Several reports on the genetic variability in B efficiency of wheat genotypes and differences in their response to high and low B supply are available (Karaman et al. 2012) (Table 25.1).

Bread wheat is found to be more tolerant toward B toxicity as compared to durum wheat (Kalayci et al. 1998; Karaman et al. 2012; Turan et al. 2018). This may be due to lower tissue B concentration and higher agronomic efficiency (dry matter yield) in bread wheat genotypes as compared to durum wheat (Yau et al. 1995; Taban and Erdal 2000). Boron toxicity tolerance in wheat is found to be strongly associated with the less B accumulation in the tissues (Nable 1988; Reid 2010). The limited B accumulation in resistant cultivars might be due to active efflux (Schnurbusch et al. 2010), controlled absorption, or active B exclusion from the tissues.

In addition, there are varieties with higher B concentration in tissues and enhanced B toxicity tolerance simultaneously which proposes the existence of mechanism other than exclusion (Nejad et al. 2015). Moreover, the absence of a consistent relationship between the dry weight and tissue B concentration suggests the involvement of other strategies in B toxicity tolerance (Torun et al. 2006). The mechanism of B redistribution in leaf tissues from more toxic regions (cytoplasm) to less toxic regions (cell wall) may account for B toxicity tolerance in such genotypes (Reid and Fitzpatrick 2009a, 2009b). In addition, the genotypes with higher yield in a specific high B expanse may have a lower yield in another B toxic region signifying the importance of genotype-environment in B toxicity tolerance (Pallotta et al. 2014). For example, the B-tolerant wheat genotypes showing a 16% greater yield in southern Australia revealed major yield loss in northern regions (Schnurbusch et al. 2010).

Table 25.1 Wheat Genotypes Determined to Be Tolerant to B Toxicity in Previously Conducted Studies

Name of the Genotype	Type	Origin	Given B Treatment	Growth Condition	References
Greek	*T. aestivum*	Greece	50 mg B per L	Nutrient solution	Nable (1988)
Halbred	*T. aestivum*	Australia	150 mg B per kg	Soil	Paull et al. (1988)
India 126	*T. aestivum*	India	150 mg B per L	Nutrient solution	Chantachume et al. (1995)
Klein Granador Lin Calel B. Inca		Argentina			
Aus 4903 Aus 4041		Australia			
Turkey 1473		Turkey			
G61450		Greece			
ICDW 7674	*Triticum durum*	Afghanistan	50 mg B per kg	Soil	Yau et al. (1997)
Bolal 2973	*T. aestivum*	Turkey	25 mg B per kg	Soil	Kalayci et al. (1998)
IAC 287 IAC24	*T. aestivum*	Brazil	2 mg B per L	Nutrient solution	Furlani et al. (2003)
Sabil-1 Stn "S"	*T. durum*	Syria	25 mg B per kg	Soil	Torun et al. (2006)
BDMM-98/11S	*T. aestivum*	Turkey	30 mg B per kg	Soil	Karaman et al. (2012)
KRL 99	*T. aestivum*	India	100 mM B per L	Greenhouse	Sharma et al. (2014)
BT-Schomburgk		Australia			
Kharchia-65		India			
AUS1473	*T. aestivum*	Turkey	–	–	Pallotta et al. (2014)
AUS10105	*T. durum*	India			
AUS10344	*T. durum*	Iraq			
AUS14010	*T. durum*	China			
AUS14740	*T. durum*	Afghanistan			
Benvenuto Inca	*T. aestivum*	Argentina			
Bolal-2973	*T. aestivum*	Turkey			
Bonza	*T. aestivum*	Colombia			
Etawah	*T. aestivum*	India			
G61450	*T. aestivum*	Greece/Italy			
Klein Granador	*T. aestivum*	Argentina			
Lerma 52	*T. aestivum*	CIMMYT/ Nepal			
Mentana	*T. aestivum*	Italy			
Carnamah	*T. aestivum*	Australia			
Correll	*T. aestivum*	Australia			
Currawa	*T. aestivum*	Australia			
Dagger	*T. aestivum*	Australia			
Espada	*T. aestivum*	Australia			
Frame	*T. aestivum*	Australia			
Gladius	*T. aestivum*	Australia			
Gurkha	*T. aestivum*	Australia			
Halberd	*T. aestivum*	Australia			
Heron	*T. aestivum*	Australia			
Insignia	*T. aestivum*	Australia			
Kalka	*T. durum*	Australia			

(Continued)

Table 25.1 (*Continued*)   Wheat Genotypes Determined to Be Tolerant to B Toxicity in Previously Conducted Studies

Name of the Genotype	Type	Origin	Given B Treatment	Growth Condition	References
Arg	*T. aestivum*	Iran	40 mg B per kg	Soil	Nejad et al. (2015)
Krichauff	*T. aestivum*	Australia			
Mace	*T. aestivum*	Australia			
Matong	*T. aestivum*	Australia			
Olympic	*T. aestivum*	Australia			
Quadrat	*T. aestivum*	Australia			
Spear	*T. aestivum*	Australia			
Tjilkuri	*T. durum*	Australia			
Wyuna	*T. aestivum*	Australia			
Yitpi	*T. aestivum*	Australia	Yitpi	*T. aestivum*	Australia
Kalyan Sona	*T. aestivum*	India	133 kg H_3BO_3 per ha	Field soil	Brdar et al. (2017)
Simonida		Serbia			
Teodora		Serbia			

Genetic diversity in the B tolerance level of cultivated and wild wheat genotypes has been assessed in several studies (Schnurbusch et al. 2008; Emon 2012; Emon et al. 2015). Schnurbusch et al. (2008) found that durum and bread wheat used in their experiment were more tolerant than *Triticum urartu* and *Triticum monococcum* accessions, may be due to the selection of diploid accessions for the study from an unsuitable region. In some studies, *Aegilops tauschii* showed the highest tolerance toward B toxicity in the screening of *Aegilops* sp. and *Triticum* sp. suggesting that D genome may also contribute toward B toxicity tolerance in wheat. This is in agreement with the concept that suggested 7B and 7D chromosomes and 5A, 5B, and 5D chromosomes of wheat that contains orthologous genes of B toxicity tolerant barley genes are linked with B toxicity tolerance (Sutton et al. 2007; Schnurbusch et al. 2008). Based on such studies, it could be proposed that a thorough screening of true genome progenitors of tetraploid and hexaploid wheat could provide more accessions with high B tolerance levels. Emon et al. (2012, 2015) conducted a series of B toxicity experiments on 12 wild wheat species including *Aegilops* and *Triticum* species with two tolerant and one susceptible bread wheat cultivars. They observed high tolerance in some of the *Aegilops species* and *T. dicoccoides* on the basis of root growth and suggested that B toxicity tolerant, *Ae. longissima* and *Ae. sharonensis* can be the potential candidates for wheat breeding programs. The results of such studies suggested that wild wheat and its relatives are the potential material for developing B toxicity tolerance in cultivated wheat genotypes.

25.6 PHYSIOLOGICAL AND BIOCHEMICAL CHANGES IN WHEAT UNDER B TOXICITY

The toxic B supply influences the enzymatic activities and biochemical constituents of wheat genotypes finally leading to decreased crop yield. The excess B supply enhances the proline, soluble sugar content, total phenol content, and soluble protein content of wheat crops. An understanding of enzymatic activities and biochemical analysis gives an opinion on why different cultivars with similar tissue B concentrations show different responses of B toxicity. High B causes accumulation of phenolic compounds in wheat leaves by increasing the level of pentose phosphate pathway (PPP) that supplies erythrose-4-phosphate for the biogenesis of phenolic compounds (Seth and Aery 2014; Pardossi et al. 2015; Mishra and Heckathorn 2016; Seth and Aery 2017). Higher B levels lead to the reduced proline breakdown and enhanced expression of proteins involved in amino acid metabolism, cellular transport, proteolysis, and antioxidants production (Sang et al. 2015; Seth and Aery 2017). Proline act as osmoprotectant and facilitates the plant tolerance toward B stress by maintaining the cellular balance (da Costa et al. 2011; Kayıhan et al. 2016). Wheat genotypes have been observed to be adapted to B stress conditions by the formation of B-polyol complexes and the movement of B from source to sink organs. This leads to the increase in the soluble sugar contents and increased expression of efflux transporters, finally pumping out excess B into the soil and minimizing the tissue B concentrations (Reid et al. 2004; Miwa et al. 2007; Martinez-Cuenca et al. 2015; Wakuta et al. 2016).

Excess B supply leads to the production of ROS such as hydrogen peroxide (H_2O_2) and superoxide anion ($\bullet O_2$) causing the oxidative damage of nucleic acids, proteins, and lipids (Ardic et al. 2009). This stimulates the scavenging system of ROS species that minimizes the cellular damage. The osmoprotective molecules such as proline and antioxidant enzymes such as POX, SOD, and CAT support the plant to deal with the ROS damage. B toxicity influences the peroxidase enzyme-based gene expression in wheat enhancing the peroxidase activity and the detoxification of hydrogen peroxide (Kayıhan et al. 2016). The lower production of ROS and enhanced antioxidant enzyme activities in tolerant wheat genotypes as compared to susceptible cultivars may contribute to competent scavenging of $\bullet O_2$ and H_2O_2 under boron toxicity and thus, greater tolerance toward B toxicity. Other than antioxidant enzyme activities, malonialdehyde (MDA) is a potential indicator of oxidative injury to membrane lipids. However, tolerant cultivars may have less increase in MDA as compared to sensitive cultivars under B toxicity (Erdal et al. 2014).

B toxic growth conditions not only have physiological and biochemical effects on wheat plants, but also have genotoxic consequences (Kekec et al. 2010; Catav et al. 2018). The genotoxic effects can be estimated by using the molecular markers and comparing the intensity or presence of band profiles for tolerant and susceptible genotypes (Cenkci et al. 2009). Moreover, B toxicity has different genotoxic influences on root and shoot samples because of the distinct rate of B accumulation in tissues (Catav et al. 2018).

25.7 WHEAT QTLs INVOLVED IN B TOXICITY TOLERANCE

As genetic diversity of wheat genotypes for boron toxicity tolerance has been well identified, it can be effectively utilized in the breeding program by the establishment of molecular markers linked with QTLs responsible for B tolerance in wheat (Moody et al. 1993; Jefferies et al. 2000). A number of studies have been conducted in this direction. The genetic control of B tolerance of wheat was initially found to be associated with three unlinked genes, Bo1, Bo2, and Bo3 (Paull et al. 1991) that were mapped on chromosomes 4A, 7B, and 7D (Paull et al. 1991, 1992; Jefferies et al. 2000). However, other locus controlling tolerance to boron was also suggested. The method of selecting B tolerant germplasm via treating plants with high B in a hydroponic system or field experiments is a labor-intensive method. Thus, the estimation of the chromosomal location containing the genes providing B tolerance would expedite the selection of the tolerant material using marker-assisted selection method.

The Bo1 gene location on Chromosome 7B providing B tolerance was believed to be derived from cultivar Halbred (that originally derived from cultivar Federation); while regions on 4A and 7D conferring B tolerance were proposed to be derived from an exotic Greek B tolerant line, G61450 (Paull 1990; Chantachume et al. 1995; Jefferies et al. 1999). However, the markers developed for these chromosomal locations can be useful only for screening a population where the B tolerant genes are derived from these sources. The breeding programs for developing tolerance can be more benefitted by marker-assisted selection methods by pyramiding the genes for B tolerance from different sources (Paull et al. 1991; Jefferies et al. 2000). Jefferies et al. (2000) constructed AFLP- and

RFLP-based linkage map and identified that locus on 7B and 7D chromosomes is linked with boron tolerance. The 7B chromosome was found to be involved in B uptake and reducing the negative effects of B toxicity on root growth (Jefferies et al. 2000).

Schnurbusch et al. (2007) put efforts to determine PCR marker that is closely associated with Bo1, a major QTL on chromosome 7B, and an important source of B tolerance in wheat genotypes. The major B tolerance locus, Bo1, on the long arm of chromosome 7B was mapped as a discrete Mendelian locus by the addition of 28 molecular markers close to the region reducing the size of the genetic interval (Schnurbusch et al. 2007). The STS marker, AWW5L7, that co-segregated with Bo1 was identified as a potential tool for B tolerance breeding in wheat (Schnurbusch et al. 2008) as it was more closely located to the region as compared to other SSR markers, GWM344 and WMC276, that were previously described by Martin et al. (2004). AWW5L7 is strongly associated with B tolerance in both durum and bread wheat genotypes and suppresses the negative effects of B toxicity on root length (Schnurbusch et al. 2007, 2008). It has been predicted that another major B tolerance locus different from Bo1 and unlinked to Xpsr121-7BL is present on chromosome 7B (Schnurbusch et al. 2008).

In another study, the expressed leaf symptoms were used to determine the genetic variation of durum wheat accessions under B toxicity (Torun et al. 2006). As the boron tolerance locus on 7DS chromosome was found to be associated with the effect of B toxicity on leaf symptoms, markers co-segregating with this trait locus can be used for the selection of B tolerant wheat germplasm. This may allow the selection of B tolerant wheat resources on the basis of different tolerance mechanisms other than root growth and reduced B uptake and accumulation (Figure 25.1).

Bo1 allele of 7B chromosome appeared to be efficiently utilized in breeding programs providing high B tolerance to a number of hexaploid and tetraploid cultivars of Australia, Iraq, India, and China (Jamjod 1996; Schnurbusch et al. 2008). However, the other high B tolerance locus recognized on chromosome 4AL has not been extensively utilized in breeding programs. Wheat material combining both the sources of B tolerance might be efficient in providing high yields in diverse environments (Torun et al. 2006; Yau and Ryan 2008; Schnurbusch et al. 2010).

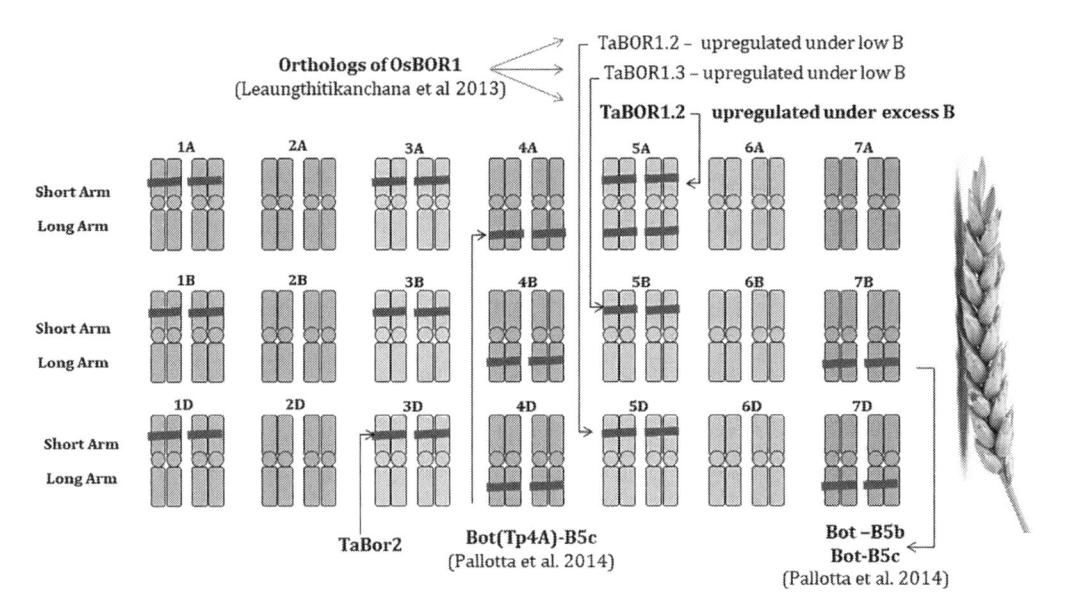

Figure 25.1 Different genes of wheat involved in B toxicity tolerance and their chromosomal location. The red marks (dark colored marks) represent the rough location of the boron transporters on different wheat chromosomes as per the information provided in Leaungthitikanchana et al. (2013) and Pallotta et al. (2014).

25.8 WHEAT GENES EXPRESSION: AN IMPORTANT FOOTSTEP IN B TOXICITY TOLERANCE

B moves in plants via (a) passive diffusion, (b) boric acid channel-facilitated diffusion, or (c) borate transport (Reid and Fitzpatrick 2009b). Under normal B growth conditions, plants easily take up B in the form of boric acid via passive diffusion (Stangoulis et al. 2001). However, under B toxicity, boric acid channels and B transporters at plasma membrane come into the role and contributes to less B accumulation in the tissues or greater B exclusion from the tissues, thus, reducing the effects of B toxicity (Nable 1988; Takano et al. 2008; Miwa and Fujiwara 2010).

Few studies suggested that reduction in plasma membrane intrinsic proteins (PIP) aquaporin levels might diminish the outcomes of boric acid toxicity by controlling the transport of boric acid (Martinez-Ballesta et al. 2008). Moreover, the second group of Nodulin 26-like intrinsic proteins (NIPs) is supposed to perform as permeases for boric acid and actively involved in B movement (Takano et al. 2006; Miwa and Fujiwara 2010; Roberts and Routray 2017; Yoshinari and Takano 2017). The NIP family gene, NIP5;1 was overexpressed in Arabidopsis roots under low B supply and was the first gene described as a boric acid channel (Takano et al. 2006). However, under high B supply, mRNA of NIP5;1 is degraded regulating the expression of NIP5;1 and inhibiting the excessive B transport for the plant adaptation to B toxic condition (Tanaka et al. 2011).

Bor borate exporters are known to have an important role in controlling B toxicity (Schnurbusch et al. 2007; Yoshinari and Takano 2017). To date, seven genes have been reported in Arabidopsis thaliana encoding BOR-type borate transporters (BOR1–BOR7) (Yoshinari and Takano 2017). Borate exporter BOR1 supports the movement of B into shoots under low B; however, under high B, BOR1 protein is selectively downregulated and BOR1 expression is transcriptionally suppressed. This post-transcriptional control of BOR1 supports the plants to escape B toxicity (Aibara et al. 2018). The clones of orthologs of BOR1 of Saccharomyces cerevisiae in wheat (Ta-BOR2) reduced the root B concentrations on their enhanced expression (Sutton et al. 2007; Takano et al. 2002, 2007; Reid 2007).

Though most of the BOR transporters regulate B uptake under low B supply, AtBOR4 is basically associated with B export from tissues providing enhanced B tolerance under high B supply (Miwa et al. 2007, 2014; Lv et al. 2017). Not only in roots, but also in leaves B transporters play a role in sending out the toxic B from cytoplasm to apoplasm where comparatively more B can be accumulated (Reid and Fitzpatrick 2009b). TaBor2 with higher sequence similarity with AtBOR4 expressed in the outer plasma membrane of root tissues lessen the B entry into xylem avoiding B toxicity (Nakagawa et al. 2007; Reid 2007; Miwa et al. 2007, 2014).

Several homologs of Bor transporter gene family of Arabidopsis have been reported in wheat (Sutton et al. 2007; Leaungthitikanchana et al. 2013; Pallotta et al. 2014; Reid 2014). Among the mentioned B transporters, Reid (2007) worked with the B efflux transporter TaBOR2 gene (D genome) of bread wheat, where the full-length cDNA was cloned into pGEM-T easy vectors and expression analysis confirmed that TaBOR2 gene expression in wheat was correlated with B toxicity tolerance. TaBOR1, the ortholog of AtBor1 in wheat regulates the movement of B into the shoot and its entrance into roots. Three *Triticum aestivum* BOR1 genes, TaBOR1.1, TaBOR1.2, and TaBOR1.3 (located on 5D, 5A, and 5B wheat chromosomes, respectively) orthologous to B toxicity tolerant gene of rice, OsBOR1 act as B efflux transporters (Leaungthitikanchana et al. 2013). Transient expression of TaBOR1-GFP fusions was observed in tobacco and Arabidopsis and differential accumulation of these genes under different B stress conditions. Among these genes, the overexpression of TaBOR1.2 under high B supply in root-shoot tissues demonstrated their role as B efflux transporters providing high B tolerance. Although the functional characterization of these three genes has been done, they were assigned to bread wheat only, and functional analysis of such genes is required in durum wheat genotypes as well.

Though QTLs for high B tolerance have been discovered in barley and wheat, only barley genes were shown to have significant roles in B tolerance till 2014 (Paull et al. 1991; Jefferies et al. 1999, 2000). The HvNIP2;1 (6H) and HvBot1 (4H) genes of barley encoding a member of the NIP aquaporin family and an anion-permeable transporter are involved in B stress responses (Schnurbusch et al. 2010). However, their orthologs were mapped on different locations in wheat and do not associate with the major B tolerance wheat QTLs, Bo1 (7BL) and Bo4 (4AL) (Chantachume et al. 1995; Jamjod 1996; Jefferies et al. 2000). In a study in 2014, Pallotta et al. reported the role of three root-specific B transporter genes (Bot genes) and confirmed the presence and function of their Bot-B5 alleles, (Bot-B5b), (Bot(Tp4A)-B5c) and (Bot-B5c) in durum and bread wheat by transforming them in yeast cells and their transient expression was observed in onion epidermal cells. The Bot genes were reported in durum and bread wheat cultivars only. However, Pallotta et al. (2014) formerly identified Bot-B5b allele via Ae. tauschii accessions and proposed that Aegilops can be the source of B tolerance in cultivated wheat and it should be explored in detail.

25.9 CONCLUSION

B toxicity is one of the major agricultural issues that have been actively explored for more than three decades. However, still its physiological and molecular aspects of the stress need detailed research. Dissimilar to deficiency, boron toxicity in the soil is challenging to amend, thus, the only solution to deal with the problem is developing B toxicity tolerant crops. The situation of the climate crisis is making it worse where B toxicity occurs simultaneously with other abiotic stresses. In such scenario, the interaction between B and other abiotic stresses like salinity (BorSal) (Wimmer et al. 2001; Pandey et al. 2019), drought (Bellaloui and Mengistu 2015; Naeem et al. 2018), heat (Eisvand et al. 2018) etc. should be studied both at the physiological and molecular level. Wheat being a majorly consumed crop can be focused for developing B toxicity tolerance and increasing the production yield. Though wheat genotypes have shown large genetic variation for B toxicity tolerance, wild accessions of the wheat genetic pool still need a thorough screening and may provide efficient resources for B tolerance. The utilization of contemporary gene editing, transformation, metabolomics, and proteomics-based approaches may facilitate the identification of novel B tolerant genes in unexplored wheat germplasm that can be further integrated into respective breeding programs to get enhanced yield in high B soils.

ACKNOWLEDGMENTS

The authors acknowledge TUBITAK 1001 (No 119O455) project for providing funding to perform research work in this direction.

REFERENCES

Aibara I, Hirai T, Kasai K, Takano J, Onouchi H, Naito S, Fujiwara T, Miwa K (2018) Boron-dependent translational suppression of the borate exporter BOR1 contributes to the avoidance of boron toxicity. *Plant Physiology* 177 (2):759–774.

Ardic M, Sekmen AH, Tokur S, Ozdemir F, Turkan I (2009) Antioxidant responses of chickpea plants subjected to boron toxicity. *Plant Biology (Stuttg)* 11 (3):328–338. doi:10.1111/j.1438-8677.2008.00132.x

Bellaloui N, Mengistu A (2015) Effects of boron nutrition and water stress on nitrogen fixation, seed δ15N and δ13C dynamics, and seed composition in soybean cultivars differing in maturities. *The Scientific World Journal* 2015:407872. doi:10.1155/2015/407872

Bellaloui N, Mengistu A, Kassem MA, Abel CA, Zobiole L (2014) Role of Boron Nutrient in Nodules Growth and Nitrogen Fixation in Soybean Genotypes Under Water Stress Conditions, *Advances in Biology and Ecology of Nitrogen Fixation*, Takuji Ohyama, IntechOpen, doi:10.5772/56994.

Brdar-Jokanovic M (2020) Boron toxicity and deficiency in agricultural plants. *International Journal of Molecular Sciences* 21 (4). doi:10.3390/ijms21041424

Brdar JM, Zoric M, Kondic SA, Maksimovic I, Kobiljski B, Kraljevic BM (2017) Boron tolerance in wheat accessions of different origin estimated in controlled and field conditions. *Journal of Agricultural Science and Technology* 19:345–356.

Brown P, Bellaloui N, Wimmer M, Bassil E, Ruiz J, Hu H, Pfeffer H, Dannel F, Römheld V (2002) Boron in plant biology. *Plant Biology* 4 (02):205–223.

Brown PH, Shelp BJ (1997) Boron mobility in plants. *Plant and Soil* 193 (1–2):85–101.

Camacho-Cristobal JJ, Martin-Rejano EM, Herrera-Rodriguez MB, Navarro-Gochicoa MT, Rexach J, Gonzalez-Fontes A (2015) Boron deficiency inhibits root cell elongation via an ethylene/auxin/ROS-dependent pathway in Arabidopsis seedlings. *Journal of Experimental Botany* 66 (13):3831–3840. doi:10.1093/jxb/erv186

Camacho-Cristóbal JJ, Rexach J, González-Fontes A (2008) Boron in plants: deficiency and toxicity. *Journal of Integrative Plant Biology* 50 (10):1247–1255.

Cartwright B, Zarcinas B, Mayfield A (1984) Toxic concentrations of boron in a red-brown earth at Gladstone, South Australia. *Soil Research* 22 (3):261–272.

Catav SS, Genc TO, Kesik Oktay M, Kucukakyuz K (2018) Effect of boron toxicity on oxidative stress and genotoxicity in wheat (Triticum aestivum L.). *Bulletin of Environmental Contamination and Toxicology* 100 (4):502–508. doi:10.1007/s00128-018-2292-x

Cenkci S, Yıldız M, Ciğerci İH, Konuk M, Bozdağ A (2009) Toxic chemicals-induced genotoxicity detected by random amplified polymorphic DNA (RAPD) in bean (Phaseolus vulgaris L.) seedlings. *Chemosphere* 76 (7):900–906.

Cervilla L, Blasco B, Ríos J, Rosales M, Rubio-Wilhelmi M, Sánchez-Rodríguez E, Romero L, Ruiz J (2009) Response of nitrogen metabolism to boron toxicity in tomato plants. *Plant Biology* 11 (5):671–677.

Chantachume Y, Smith D, Hollamby G, Paull J, Rathjen A (1995) Screening for boron tolerance in wheat (*T. aestivum*) by solution culture in filter paper. *Plant and Soil* 177 (2):249–254.

Cheng C, Rerkasem B (1993) Effects of boron on pollen viability in wheat. *Plant and Soil* 155 (1):313–315.

da Costa RCL, Lobato AKDS, da Silveira JAG, Laughinghouse IV HD (2011) ABA-mediated proline synthesis in cowpea leaves exposed to water deficiency and rehydration. *Turkish Journal of Agriculture and Forestry* 35 (3):309–317.

Dzondo-Gadet M, Mayap-Nzietchueng R, Hess K, Nabet P, Belleville F, Dousset B (2002) Action of boron at the molecular level. *Biological Trace Element Research* 85 (1):23–33.

Eisvand H, Kamaei H, Nazarian F (2018) Chlorophyll fluorescence, yield and yield components of bread wheat affected by phosphate bio-fertilizer, zinc and boron under late-season heat stress. *Photosynthetica* 56 (4):1287–1296.

El-Shazoly RM, Metwally AA, Hamada AM (2019) Salicylic acid or thiamin increases tolerance to boron toxicity stress in wheat. *Journal of Plant Nutrition* 42 (7):702–722. doi:10.1080/01904167.2018.1549670

Emon R, Gustafson K, Bebeli P, Jahiruddin M, Haque M, Ross K, Gustafson J (2012) Screening Aegilops-Triticum species for Boron tolerance. *African Journal of Agricultural Research* 7 (12):1631–1936.

Emon RM (2012) Screening Aegilops-Triticum species for Boron tolerance. *African Journal of Agricultural Research* 7 (12). doi:10.5897/ajar11.2084

Emon RM, Nevame AYM, Gustafson PJ, Haque MS, Jahiruddin M, Islam MM (2015) Morpho-genetic study and detection of boron toxicity tolerance of wild wheat genotypes. *Journal of Applied Biotechnology* 3 (2):41.

Eraslan F, Inal A, Gunes A, Alpaslan M (2007) Boron toxicity alters nitrate reductase activity, proline accumulation, membrane permeability, and mineral constituents of tomato and pepper plants. *Journal of Plant Nutrition* 30 (6):981–994. doi:10.1080/15226510701373221

Erdal S, Genc E, Karaman A, Khosroushahi F, Kizilkaya M, Demir Y, Yanmis D (2014) Differential responses of two wheat varieties to increasing boron toxicity. Changes on antioxidant activity, oxidative damage and DNA profile. *Journal of Environmental Protection and Ecology* 15:1217–1229.

Funakawa H, Miwa K (2015) Synthesis of borate cross-linked rhamnogalacturonan II. *Frontiers in Plant Science* 6:223.

Furlani ÂMC, Carvalho CP, Freitas JGd, Verdial MF (2003) Wheat cultivar tolerance to boron deficiency and toxicity in nutrient solution. *Scientia Agricola* 60 (2):359–370.

Goldbach HE, Wimmer MA (2007) Boron in plants and animals: is there a role beyond cell-wall structure? *Journal of Plant Nutrition and Soil Science* 170 (1):39–48. doi:10.1002/jpln.200625161

González-Fontes A (2019) Why boron is an essential element for vascular plants: a comment on Lewis (2019) 'Boron: the essential element for vascular plants that never was'. *New Phytologist* 226(5):1232–1237.

González-Fontes A, Rexach J, Navarro-Gochicoa MT, Herrera-Rodríguez MB, Beato VM, Maldonado JM, Camacho-Cristóbal JJ (2008) Is boron involved solely in structural roles in vascular plants? *Plant Signaling & Behavior* 3 (1):24–26.

Han S, Chen LS, Jiang HX, Smith BR, Yang LT, Xie CY (2008) Boron deficiency decreases growth and photosynthesis, and increases starch and hexoses in leaves of citrus seedlings. *Journal of Plant Physiology* 165 (13):1331–1341. doi:10.1016/j.jplph.2007.11.002

Hu H, Brown PH, Labavitch JM (1996) Species variability in boron requirement is correlated with cell wall pectin. *Journal of Experimental Botany* 47 (2):227–232.

Hu H, Penn SG, Lebrilla CB, Brown PH (1997) Isolation and characterization of soluble boron complexes in higher plants (the mechanism of phloem mobility of boron). *Plant Physiology* 113 (2):649–655.

Huang J-H, Cai Z-J, Wen S-X, Guo P, Ye X, Lin G-Z, Chen L-S (2014) Effects of boron toxicity on root and leaf anatomy in two Citrus species differing in boron tolerance. *Trees* 28 (6):1653–1666.

Jamjod S (1996) Genetics of boron tolerance in durum wheat. University of Adelaide, Department of Plant Science.

Jefferies SP, Barr AR, Karakousis A, Kretschmer JM, Manning S, Chalmers KJ, Nelson JC, Islam AKMR, Langridge P (1999) Mapping of chromosome regions conferring boron toxicity tolerance in barley (*Hordeum vulgare* L.). *Theoretical and Applied Genetics* 98 (8):1293–1303. doi:10.1007/s001220051195

Jefferies SP, Pallotta MA, Paull JG, Karakousis A, Kretschmer JM, Manning S, Islam AKMR, Langridge P, Chalmers KJ (2000) Mapping and validation of chromosome regions conferring boron toxicity tolerance in wheat (*Triticum aestivum*). *Theoretical and Applied Genetics* 101 (5):767–777. doi:10.1007/s001220051542

Kalayci M, Alkan A, Cakmak I, Bayramoğlu O, Yilmaz A, Aydin M, Ozbek V, Ekiz H, Ozberisoy F (1998) Studies on differential response of wheat cultivars to boron toxicity. *Euphytica* 100 (1–3):123–129.

Karaman MR, Zengin M, Horuz A (2012) Assessment of resistance of wheat genotypes (*T. aestivum* and *T. durum*) to boron toxicity. *Proceedings of World Academy of Science, Engineering and Technology*, World Academy of Science, Engineering and Technology Paris, France, pp. 809–812.

Kayıhan C, Öz MT, Eyidoğan F, Yücel M, Öktem HA (2016) Physiological, biochemical, and transcriptomic responses to boron toxicity in leaf and root tissues of contrasting wheat cultivars. *Plant Molecular Biology Reporter*. doi:10.1007/s11105-016-1008-9

Kekec G, Sakcali MS, Uzonur I (2010) Assessment of genotoxic effects of boron on wheat (*Triticum aestivum* L.) and bean (*Phaseolus vulgaris* L.) by using RAPD analysis. *Bulletin of Environmental Contamination and Toxicology* 84 (6):759–764.

Kobayashi M, Matoh T, Azuma J (1996) Two chains of rhamnogalacturonan II are cross-linked by borate-diol ester bonds in higher plant cell walls. *Plant Physiology* 110 (3):1017–1020. doi:10.1104/pp.110.3.1017

Landi M, Margaritopoulou T, Papadakis IE, Araniti F (2019) Boron toxicity in higher plants: an update. *Planta* 250 (4):1011–1032. doi:10.1007/s00425-019-03220-4

Leaungthitikanchana S, Fujibe T, Tanaka M, Wang S, Sotta N, Takano J, Fujiwara T (2013) Differential expression of three BOR1 genes corresponding to different genomes in response to boron conditions in hexaploid wheat (*Triticum aestivum* L.). *Plant and Cell Physiology* 54 (7):1056–1063. doi:10.1093/pcp/pct059

Lewis DH (2019) Boron: the essential element for vascular plants that never was. *New Phytologist* 221 (4):1685–1690.

Li XW, Liu JY, Fang J, Tao L, Shen RF, Li YL, Xiao HD, Feng YM, Wen HX, Guan JH (2017) Boron supply enhances aluminum tolerance in root border cells of pea (*Pisum sativum*) by interacting with cell wall pectins. *Frontiers in Plant Science* 8:742.

Lv Q, Wang L, Wang J-Z, Li P, Chen Y-L, Du J, He Y-K, Bao F (2017) SHB1/HY1 alleviates excess boron stress by increasing BOR4 expression level and maintaining boron homeostasis in Arabidopsis roots. *Frontiers in Plant Science* 8:790.

Macho-Rivero MA, Herrera-Rodríguez MB, Brejcha R, Schäffner AR, Tanaka N, Fujiwara T, González-Fontes A, Camacho-Cristóbal JJ (2018) Boron toxicity reduces water transport from root to shoot in Arabidopsis plants. Evidence for a reduced transpiration rate and expression of major pip aquaporin genes. *Plant and Cell Physiology* 59 (4):841–849.

Marschner H (1995) *Mineral Nutrition of Higher Plants*, 2nd edition. Academic, Great Britain.

Martin E, Eastwood R, Ogbonnaya F, Emebiri L (2004) Molecular marker for the boron tolerance of the wheat cultivar Yanac. *Proceedings of the 54th Australian Cereal Chemistry Conference and 11th Wheat Breeders Assembly*, Canberra, pp. 29–32.

Martinez-Ballesta MdC, Bastías E, Zhu C, Schäffner AR, González-Moro B, González-Murua C, Carvajal M (2008) Boric acid and salinity effects on maize roots. Response of aquaporins ZmPIP1 and ZmPIP2, and plasma membrane H+-ATPase, in relation to water and nutrient uptake. *Physiologia Plantarum* 132 (4):479–490.

Martinez-Cuenca M-R, Martinez-Alcantara B, Quiñones A, Ruiz M, Iglesias DJ, Primo-Millo E, Forner-Giner MA (2015) Physiological and molecular responses to excess boron in Citrus macrophylla W. *PloS One* 10 (7):e0134372.

Matoh T, Ishigaki K-i, Ohno K, Azuma J-i (1993) Isolation and characterization of a boron-polysaccharide complex from radish roots. *Plant and Cell Physiology* 34 (4):639–642.

Matoh T, Kobayashi M (2002) Boron function in plant cell walls. In: Goldbach HE, Brown PH, Rerkasem B, Thellier M, Wimmer MA, Bell RW (eds) *Boron in Plant and Animal Nutrition*. Springer, Boston, MA. doi:10.1007/978-1-4615-0607-2_13.

Mishra S, Heckathorn S (2016) Boron stress and plant carbon and nitrogen relations. In: *Progress in Botany 77.* Springer, pp. 333–355.

Miwa K, Aibara I, Fujiwara T (2014) Arabidopsis thaliana BOR4 is upregulated under high boron conditions and confers tolerance to high boron. *Soil Science and Plant Nutrition* 60 (3):349–355.

Miwa K, Fujiwara T (2010) Boron transport in plants: coordinated regulation of transporters. *Annals of Botany* 105 (7):1103–1108.

Miwa K, Takano J, Omori H, Seki M, Shinozaki K, Fujiwara T (2007) Plants tolerant of high boron levels. *Science* 318 (5855):1417–1417.

Moody DB, Rathjen AJ, Cartwright B (1993) Yield evaluation of a gene for boron tolerance using backcross-derived lines. In: Randall PJ, Delhaize E, Richards RA, Munns R (eds) *Genetic Aspects of Plant Mineral Nutrition: The Fourth International Symposium on Genetic Aspects of Plant Mineral Nutrition*, 30 September – 4 October 1991, Canberra, Australia. Springer Netherlands, Dordrecht, pp. 363–366. doi:10.1007/978-94-011-1650-3_45

Morris HS (1931) Physiological effects of boron on wheat. *Bulletin of the Torrey Botanical Club* 58:1–30.

Mosa KA, Kumar K, Chhikara S, Musante C, White JC, Dhankher OP (2016) Enhanced boron tolerance in plants mediated by bidirectional transport through plasma membrane intrinsic proteins. *Scientific Reports* 6:21640.

Nable RO (1988) Resistance to boron toxicity amongst several barley and wheat cultivars: a preliminary examination of the resistance mechanism. *Plant and Soil* 112 (1):45–52.

Nable RO, Bañuelos GS, Paull JG (1997) Boron toxicity. *Plant and Soil* 193 (1–2):181–198.

Naeem M, Naeem MS, Ahmad R, Ahmad R, Ashraf MY, Ihsan MZ, Nawaz F, Athar H-u-R, Ashraf M, Abbas HT (2018) Improving drought tolerance in maize by foliar application of boron: water status, antioxidative defense and photosynthetic capacity. *Archives of Agronomy and Soil Science* 64 (5):626–639.

Nakagawa Y, Hanaoka H, Kobayashi M, Miyoshi K, Miwa K, Fujiwara T (2007) Cell-type specificity of the expression of Os BOR1, a rice efflux boron transporter gene, is regulated in response to boron availability for efficient boron uptake and xylem loading. *The Plant Cell* 19 (8):2624–2635.

Nejad SG, Savaghebi G, Farahbakhsh M, Amiri RM, Rezaei H (2015) Tolerance of some wheat varieties to boron toxicity. *Cereal Research Communications* 43 (3):384–393.

O'Neill MA, Ishii T, Albersheim P, Darvill AG (2004) Rhamnogalacturonan II: structure and function of a borate cross-linked cell wall pectic polysaccharide. *Annual Review of Plant Biology* 55:109–139.

O'Neill MA, Warrenfeltz D, Kates K, Pellerin P, Doco T, Darvill AG, Albersheim P (1996) Rhamnogalacturonan-II, a pectic polysaccharide in the walls of growing plant cell, forms a dimer that is covalently cross-linked by a borate ester in vitro conditions for the formation and hydrolysis of the dimer. *Journal of Biological Chemistry* 271 (37):22923–22930.

Pallotta M, Schnurbusch T, Hayes J, Hay A, Baumann U, Paull J, Langridge P, Sutton T (2014) Molecular basis of adaptation to high soil boron in wheat landraces and elite cultivars. *Nature* 514 (7520):88–91. doi:10.1038/nature13538

Pandey A, Khan MK, Hakki EE, Gezgin S, Hamurcu M (2019) Combined boron toxicity and salinity stress-an insight into its interaction in plants. *Plants (Basel)* 8 (10). doi:10.3390/plants8100364

Pandey DK, Singh AK, Chaudhary B (2012) Boron-mediated plant somatic embryogenesis: a provocative model. *Journal of Botany* 2012:375829.

Pappin B, Kiefel MJ, Houston TA (2012) Boron-carbohydrate interactions. In: *Carbohydrates - Comprehensive Studies on Glycobiology and Glycotechnology*, Chuan-Fa Chang, IntechOpen. doi:10.5772/50630

Pardossi A, Romani M, Carmassi G, Guidi L, Landi M, Incrocci L, Maggini R, Puccinelli M, Vacca W, Ziliani M (2015) Boron accumulation and tolerance in sweet basil (*Ocimum basilicum* L.) with green or purple leaves. *Plant and Soil* 395 (1–2):375–389.

Paull JG (1990) Genetic studies on the tolerance of wheat to high concentrations of boron. PhD Thesis, University of Adelaide.

Paull JG, Cartwright B, Rathjen AJ (1988) Responses of wheat and barley genotypes to toxic concentrations of soil boron. *Euphytica* 39 (2):137–144. doi:10.1007/bf00039866

Paull J, Nable R, Rathjen A (1992) Physiological and genetic control of the tolerance of wheat to high concentrations of boron and implications for plant breeding. *Plant and Soil* 146 (1–2):251–260.

Paull J, Rathjen A, Cartwright B (1991) Major gene control of tolerance of bread wheat (*Triticum aestivum* L.) to high concentrations of soil boron. *Euphytica* 55 (3):217–228.

Rawson H (1996) The developmental stage during which boron limitation causes sterility in wheat genotypes and the recovery of fertility. *Functional Plant Biology* 23 (6):709–717.

Reid R (2007) Identification of boron transporter genes likely to be responsible for tolerance to boron toxicity in wheat and barley. *Plant and Cell Physiology* 48 (12):1673–1678. doi:10.1093/pcp/pcm159

Reid R (2010) Can we really increase yields by making crop plants tolerant to boron toxicity? *Plant Science* 178 (1):9–11. doi:10.1016/j.plantsci.2009.10.006

Reid R (2014) Understanding the boron transport network in plants. *Plant and Soil* 385 (1–2):1–13. doi:10.1007/s11104-014-2149-y

Reid R, Fitzpatrick K (2009a) Influence of leaf tolerance mechanisms and rain on boron toxicity in barley and wheat. *Plant Physiology* 151 (1):413–420. doi:10.1104/pp.109.141069

Reid R, Fitzpatrick KL (2009b) Redistribution of boron in leaves reduces boron toxicity. *Plant Signaling & Behavior* 4 (11):1091–1093.

Reid RJ, Hayes JE, Post A, Stangoulis JCR, Graham RD (2004) A critical analysis of the causes of boron toxicity in plants. *Plant, Cell & Environment* 27 (11):1405–1414. doi:10.1111/j.1365-3040.2004.01243.x

Rerkasem B, Jamjod S (2004) Boron deficiency in wheat: a review. *Field Crops Research* 89 (2–3):173–186. doi:10.1016/j.fcr.2004.01.022

Rerkasem B, Lordkaew S, Yimyam N, Jamjod S (2019) Evaluating Boron Efficiency in Heat Tolerant Wheat Germplasm. *International Journal of Agriculture and Biology* 21 (2):385–390.

Reuhs BL, Glenn J, Stephens SB, Kim JS, Christie DB, Glushka JG, Zablackis E, Albersheim P, Darvill AG, O'Neill MA (2004) L-Galactose replaces L-fucose in the pectic polysaccharide rhamnogalacturonan II synthesized by the L-fucose-deficient mur1 Arabidopsis mutant. *Planta* 219 (1):147–157.

Roberts DM, Routray P (2017) The nodulin 26 intrinsic protein subfamily. In: Chaumont F, Tyerman S (eds) *Plant Aquaporins. Signaling and Communication in Plants*. Springer, Cham. doi:10.1007/978-3-319-49395-4_13

Robertson L, Knezek B, Belo J (1975) A survey of Michigan soils as related to possible boron toxicities. *Communications in Soil Science and Plant Analysis* 6 (4):359–373.

Rossini Oliva S, Mingorance MD, Sanhueza D, Fry SC, Leidi EO (2018) Active proton efflux, nutrient retention and boron-bridging of pectin are related to greater tolerance of proton toxicity in the roots of two Erica species. *Plant Physiology and Biochemistry* 126:142–151. doi:10.1016/j.plaphy.2018.02.029

Sang W, Huang Z-R, Qi Y-P, Yang L-T, Guo P, Chen L-S (2015) An investigation of boron-toxicity in leaves of two citrus species differing in boron-tolerance using comparative proteomics. *Journal of Proteomics* 123:128–146.

Schnurbusch T, Collins NC, Eastwood RF, Sutton T, Jefferies SP, Langridge P (2007) Fine mapping and targeted SNP survey using rice-wheat gene colinearity in the region of the Bo1 boron toxicity tolerance locus of bread wheat. *Theoretical and Applied Genetics* 115 (4):451–461. doi:10.1007/s00122-007-0579-0

Schnurbusch T, Hayes J, Sutton T (2010) Boron toxicity tolerance in wheat and barley: Australian perspectives. *Breeding Science* 60 (4):297–304. doi:10.1270/jsbbs.60.297

Schnurbusch T, Langridge P, Sutton T (2008) The Bo1-specific PCR marker AWW5L7 is predictive of boron tolerance status in a range of exotic durum and bread wheats. *Genome* 51 (12):963–971. doi:10.1139/G08-084

Seth K, Aery NC (2014) Effect of boron on the contents of chlorophyll, carotenoid, phenol and soluble leaf protein in mung bean, *Vigna radiata* (L.) Wilczek. *Proceedings of the National Academy of Sciences, India Section B: Biological Sciences* 84 (3):713–719.

Seth K, Aery NC (2017) Boron induced changes in biochemical constituents, enzymatic activities, and growth performance of wheat. *Acta Physiologiae Plantarum* 39 (11). doi:10.1007/s11738-017-2541-3

Shah A, Wu X, Ullah A, Fahad S, Muhammad R, Yan L, Jiang C (2017) Deficiency and toxicity of boron: alterations in growth, oxidative damage and uptake by citrange orange plants. *Ecotoxicology and Environmental Safety* 145:575–582.

Sharma S, Kumar A, Setter T, Singh M, Lata C, Prasad K, Kulshrestha N (2014) Boron tolerance in wheat varieties. *Vegetos* 27 (2):322–328.

Shireen F, Nawaz MA, Chen C, Zhang Q, Zheng Z, Sohail H, Sun J, Cao H, Huang Y, Bie Z (2018) Boron: functions and approaches to enhance its availability in plants for sustainable agriculture. *International Journal of Molecular Sciences* 19 (7). doi:10.3390/ijms19071856

Stangoulis JC, Reid RJ, Brown PH, Graham RD (2001) Kinetic analysis of boron transport in Chara. *Planta* 213 (1):142–146.

Stangoulis J, Tate M, Graham R, Bucknall M, Palmer L, Boughton B, Reid R (2010) The mechanism of boron mobility in wheat and canola phloem. *Plant Physiology* 153 (2):876–881. doi:10.1104/pp.110.155655

Sutton T, Baumann U, Hayes J, Collins NC, Shi BJ, Schnurbusch T, Hay A, Mayo G, Pallotta M, Tester M, Langridge P (2007) Boron-toxicity tolerance in barley arising from efflux transporter amplification. *Science* 318 (5855):1446–1449. doi:10.1126/science.1146853

Taban S, Erdal ☒ (2000) Effects of boron on growth of various wheat varieties and distribution of boron in aerial part. *Turkish Journal of Agriculture and Forestry* 24 (2):255–262.

Takano J, Kobayashi M, Noda Y, Fujiwara T (2007) Saccharomyces cerevisiae Bor1p is a boron exporter and a key determinant of boron tolerance. *FEMS Microbiology Letters* 267 (2):230–235.

Takano J, Miwa K, Fujiwara T (2008) Boron transport mechanisms: collaboration of channels and transporters. *Trends in Plant Science* 13 (8):451–457. doi:10.1016/j.tplants.2008.05.007

Takano J, Noguchi K, Yasumori M, Kobayashi M, Gajdos Z, Miwa K, Hayashi H, Yoneyama T, Fujiwara T (2002) Arabidopsis boron transporter for xylem loading. *Nature* 420 (6913):337.

Takano J, Wada M, Ludewig U, Schaaf G, von Wiren N, Fujiwara T (2006) The Arabidopsis major intrinsic protein NIP5;1 is essential for efficient boron uptake and plant development under boron limitation. *Plant Cell* 18 (6):1498–1509. doi:10.1105/tpc.106.041640

Tanaka M, Fujiwara T (2008) Physiological roles and transport mechanisms of boron: perspectives from plants. *Pflügers Archiv-European Journal of Physiology* 456 (4):671–677.

Tanaka M, Takano J, Chiba Y, Lombardo F, Ogasawara Y, Onouchi H, Naito S, Fujiwara T (2011) Boron-dependent degradation of NIP5;1 mRNA for acclimation to excess boron conditions in Arabidopsis. *Plant Cell* 23 (9):3547–3559. doi:10.1105/tpc.111.088351

Tanaka N, Uraguchi S, Saito A, Kajikawa M, Kasai K, Sato Y, Nagamura Y, Fujiwara T (2013) Roles of pollen-specific boron efflux transporter, OsBOR4, in the rice fertilization process. *Plant and Cell Physiology* 54 (12):2011–2019.

Tariq M, Mott C (2007) The significance of boron in plant nutrition and environment - a review. *Journal of Agronomy* 6 (1):1.

Torun AA, Yazici A, Erdem H, Çakmak İ (2006) Genotypic variation in tolerance to boron toxicity in 70 durum wheat genotypes. *Turkish Journal of Agriculture and Forestry* 30 (1):49–58.

Turan MA, Taban S, Kayin GB, Taban N (2018) Effect of boron application on calcium and boron concentrations in cell wall of durum (*Triticum durum*) and bread (*Triticum aestivum*) wheat. *Journal of Plant Nutrition* 41 (11):1351–1357. doi:10.1080/01904167.2018.1450424

Wakuta S, Fujikawa T, Naito S, Takano J (2016) Tolerance to excess-boron conditions acquired by stabilization of a BOR1 variant with weak polarity in Arabidopsis. *Frontiers in Cell and Developmental Biology* 4:4.

Wang Q, Lu L, Wu X, Li Y, Lin J (2003) Boron influences pollen germination and pollen tube growth in *Picea meyeri*. *Tree Physiology* 23 (5):345–351.

Wang B-L, Shi L, Li Y-X, Zhang W-H (2010) Boron toxicity is alleviated by hydrogen sulfide in cucumber (*Cucumis sativus* L.) seedlings. *Planta* 231 (6):1301–1309.

Wang N, Yang C, Pan Z, Liu Y, Peng Sa (2015) Boron deficiency in woody plants: various responses and tolerance mechanisms. *Frontiers in Plant Science* 6:916.

Warington K (1923) The effect of boric acid and borax on the broad bean and certain other plants. *Annals of Botany* 37 (148):629–672.

Wimmer MA, Muehling KH, Läuchli A, Brown PH, Goldbach HE (2001) Interaction of salinity and boron toxicity in wheat (*Triticum aestivum* L.). *Plant Nutrition* 426–427. doi:10.1007/0-306-47624-x_206

Wimmer MA, Muhling KH, Lauchli A, Brown PH, Goldbach HE (2003) The interaction between salinity and boron toxicity affects the subcellular distribution of ions and proteins in wheat leaves. *Plant, Cell and Environment* 26 (8):1267–1274. doi:10.1046/j.0016-8025.2003.01051.x

Yau S-k, Nachit MM, Ryan J (1997) Variation in boron-toxicity tolerance in a durum wheat core collection. In: Bell RW, Rerkasem B (eds) *Boron in Soils and Plants: Proceedings of the International Symposium on Boron in Soils and Plants Held at Chiang Mai*, Thailand, 7–11 September, 1997. Springer Netherlands, Dordrecht, pp. 117–120. doi:10.1007/978-94-011-5564-9_22

Yau S, Nachit M, Ryan J, Hamblin J (1995) Phenotypic variation in boron-toxicity tolerance at seedling stage in durum wheat (*Triticum durum*). *Euphytica* 83 (3):185–191.

Yau SK, Ryan J (2008) Boron toxicity tolerance in crops: a viable alternative to soil amelioration. *Crop Science* 48 (3):854. doi:10.2135/cropsci2007.10.0539

Yoshinari A, Takano J (2017) Insights into the mechanisms underlying boron homeostasis in plants. *Frontiers in Plant Science* 8:1951. doi:10.3389/fpls.2017.01951

Yu Q, Hlavacka A, Matoh T, Volkmann D, Menzel D, Goldbach HE, Baluška F (2002) Short-term boron deprivation inhibits endocytosis of cell wall pectins in meristematic cells of maize and wheat root apices. *Plant Physiology* 130 (1):415–421.

CHAPTER 26

Diversity of Mechanisms for Boron Toxicity in Mammals

Diana Rodríguez-Vera, Antonio Abad-García, Mónica Barrón-González, Julia J. Segura-Uribe, Eunice D. Farfán-García, and Marvin A. Soriano-Ursúa
Instituto Politécnico Nacional

CONTENTS

26.1 INTRODUCTION

Due to the discovery of new compounds in nature, the number of boron-containing compounds (BCCs) is increasing. Also, the group of BCCs is expanding due to the synthesis of new molecules of potential boron-based drugs or bio-materials (Fernandes, Denny, and Dos Santos 2019; Soriano-Ursúa, Farfán-García, and Geninatti-Crich 2019; Yinghuai et al. 2019).

Reports on the toxicity of BCCs are scarce. Additionally, data are centered on the analysis of a few BCCs such as boric acid or borax. Therefore, the reason of the delay in BCCs exploration in the biomedical field (compared with the exploration and inclusion of other elements to design new molecules for biomedical purposes, such as chloride or fluoride in the last decades) could be some reports suggesting high toxicity of these BCCs (Farfán-García et al. 2016).

In the past five decades, the toxicology of boron has been further explored due to the increasing interest in the specific effects of BCCs known from ancient investigations (Mogoşanu et al. 2016), as well as the introduction of new boron-containing molecules for diagnosis and therapy in medicine for mammal species.

Currently, available data allow discarding high toxicity of molecules, including boron in their structure (Turkez et al. 2012). Moreover, these data have triggered the interest to develop and study the pharmacological and toxicological profile of BCCs (Lu et al. 2019; Soriano-Ursúa, Farfán-García, and Geninatti-Crich 2019).

Several recent reports involve most of the systems and organization levels regarding the biological effects of boric acid in mammals (Pizzorno 2015; Abdelnour et al. 2018; Khaliq, Juming, and Ke-Mei 2018). Furthermore, some attractive molecular toxicity results, which could be applied to therapy purposes, have been reported (Scott and Walmsley 2015). However, some of the possible mechanisms are being clarified, and the transportation, action, and depuration of boric acid are being elucidated (Soriano-Ursúa, Farfán-García, and Geninatti-Crich 2019).

Recently, data of BCCs effects, such as bortezomib or tavaborole, on mammals, which have been studied for some years, were reported (Farfán-García et al. 2016). Furthermore, new data on natural compounds with bioactive effects on humans, such as fructoborates and other sugar-borates found in vegetables included in the standard diet, have been investigated (Militaru et al. 2013; Mogoşanu et al. 2016).

In the present chapter, the toxicology of natural and synthetic BCCs reported in mammals in the last years is explored. Notably, the diversity of mechanisms is suggested or demonstrated for the reported biological effects.

26.2 TOXICITY RELATED TO KINETICS

Boron plays many beneficial functions in biological, metabolic, and physiological processes in plants and animals, as well as in preventing some nutritional disorders in mammals (Liu et al. 2015; Khaliq, Juming, and Ke-Mei 2018). As boron deficiency has been correlated with low immune function and high incidence of osteoporosis, which considerably increases mortality, boron supplementation has been suggested for livestock (Wang et al. 2019). On the contrary, excessive boron levels are related to cell damage and toxicity on different animal species (Figure 26.1) (Abdelnour et al. 2018). However, boron is considered non-toxic, as it is known that the acute oral lethal dose 50 (LD50) values for boric acid in small mammal species range from 2,500 to 6,000 mg/kg of body weight.

Although boron supplementation has been suggested, the mechanisms of BCCs used as boron sources are unclear, for which efforts have focused on controlling the adequate doses to obtain the beneficial effects while limiting the adverse effects.

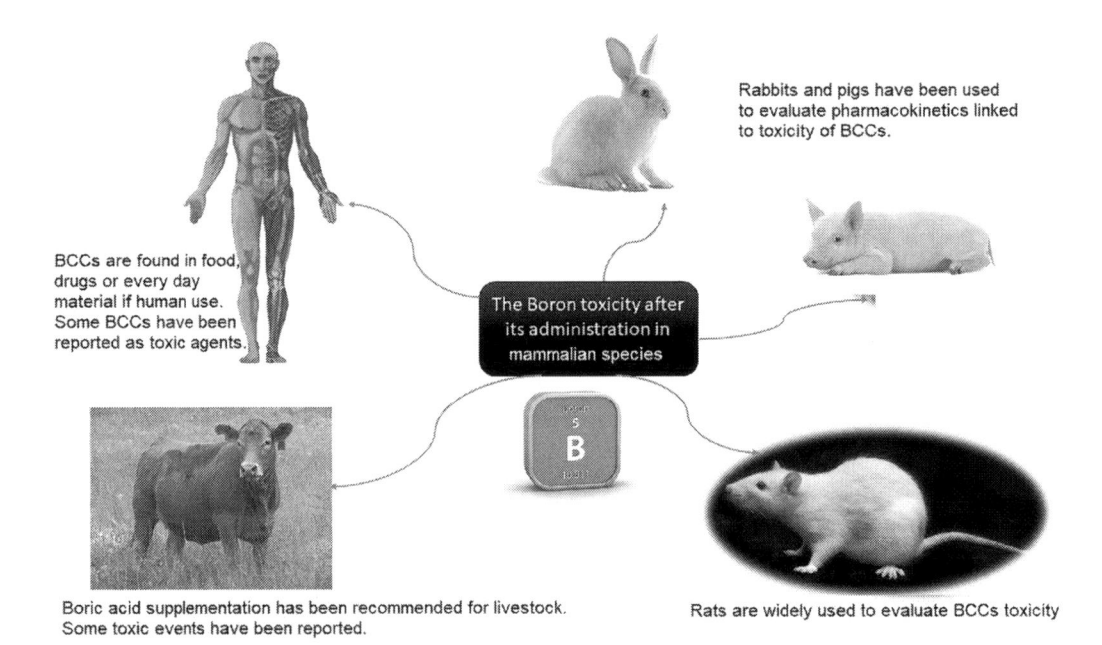

Rabbits and pigs have been used to evaluate pharmacokinetics linked to toxicity of BCCs.

BCCs are found in food, drugs or every day material if human use. Some BCCs have been reported as toxic agents.

The Boron toxicity after its administration in mammalian species

Boric acid supplementation has been recommended for livestock. Some toxic events have been reported.

Rats are widely used to evaluate BCCs toxicity

Figure 26.1 Mammal species that show some toxic effects after BCCs administration.

The average concentrations of boron and the most common BCCs to which mammals are exposed should be considered. On the one hand, the abundance of boron and BCCs in the Earth's crust is around 10 mg/kg (from 5 mg/kg in basalts to 100 mg/kg in shales) and about 4.5 mg/L in the oceans. Surface water levels range from 0.01 to 2 mg/L. On the other hand, borax pentahydrate, borax, sodium perborate, boric acid, colemanite, and ulexite are recognized as the most important commercial BCCs (Ocampo-Néstor et al. 2017). Moreover, sugar-borate esters are common compounds found in vegetables (Hunter et al. 2019). Mammal exposures to boron, generally as boric acid or borates, may occur through food and water ingestion or food contamination with pesticides containing BCCs, as well as by inhalation of boron-containing dust or powders, or BCCs from cosmetics or pharmaceutical preparations. Conversely, the most common inorganic BCCs, boric acid and borax (sodium tetraborates), are used in industrial and medical applications (Soriano-Ursúa, Farfán-García, and Geninatti-Crich 2019). Therefore, BCCs toxicity from natural sources in mammals is not easily identified.

Although poorly described, the existence of one or a few mechanisms related to each reported adverse effect seems to play a critical role in the kinetics (absorption, distribution, metabolism, and excretion) of BCCs. Blood concentrations of boric acid and other BCCs are associated with the presence and intensity of adverse effects. Therefore, physiological and toxicological approaches are recently centered on the understanding of the mechanisms by which BCCs enter the cell and the physical and biological processes that determine their distribution in plants and animals (Stangoulis and Reid 2002).

The toxicological profile for boron follows the ADME acronym (absorption, distribution, metabolism, and elimination).

Regarding absorption, orally administered inorganic borates can be converted to boric acid and its derivatives at physiological pH in the aqueous layer covering the mucosal levels before absorption, while it is suggested that organic borates could be more stable (Hunter et al. 2019). Boric acid is readily and entirely absorbed in both animals and humans and is evenly distributed through the body fluids by passive or facilitated diffusion (Ocampo-Néstor et al. 2017). In mammals such as sheep and cattle, borates have shown to be readily absorbed from the digestive tract, while the percutaneous absorption of boric acid in mammals is minimal with full skin, but it increased in rabbits with severe burned or partially denuded skin. The increased absorption of BCCs is linked to increased toxicity.

As boric acid can be easily diluted in water, it is expected to be rapidly distributed throughout the body and the mammalian placenta. Distribution appears to occur by passive diffusion through body fluids. In contrast to blood and soft tissues, bones show a selective uptake of boron and a remarkably long time of retention, which were at least three-fold higher than other tissues; however, the mechanisms for this phenomenon are unclear. Numerous experimental and epidemiological investigations show the beneficial effects of both nutritional and pharmacological intakes of BCCs on metabolic balance, as well as bone formation and mineralization, bone mechanical properties, and maintenance in mammals (Sizmaz, Koksal, and Yildiz 2017). However, boron retention has not been described. Consequently, boron could be distributed to kidneys, liver, muscle, brain, adrenals, epididymis, testes, seminal vesicles, and blood.

Boron distribution appears to be similar and consistent with high membrane permeability and passive distribution. Therefore, some features of the boron distribution in mammalian cells are the following:

1. It is quickly distributed through body water.
2. It is not accumulated in soft tissues.
3. In soft tissues, its concentrations are equivalent to plasmatic concentrations.
4. Boron levels are higher in the bone than other tissues.
5. Boron (as boric acid) is rapidly excreted in the urine, with a half-life of less than 24 h.

Moreover, boron toxicity is uncommon in mammals and appears to occur mainly through excessive accumulation of BCCs (Ince, Filazi, and Yurdakok-dikmen 2017). Signs of toxicity could be related to the distribution stage and might include nausea, poor appetite, weight loss, decreased sexual activity, seminal volume and sperm count, and mortality for boric acid.

As an inorganic element, boron is not metabolized in mammals, apparently because these organisms do not have the required energy to break the bond between boron and oxygen atoms. However, some researchers have found borates in blood, urine, and tissues, for which they should be studied as bioactive agents.

The liver is known as the largest body gland, which is related to many biotransformation processes, and more prone to the risk of toxic substances exposure. However, BCCs have induced positive effects on the development and protection of the liver. As an example, the overall liver metabolism was enhanced by BCCs in pregnant cattle and induced a significant reduction in serum very-low-density lipoprotein and triglycerides. Furthermore, the incidence of liver damage during early lactation was reduced (Goszczyński, Fink, and Boratyński 2018). The modulation of effects on oxidative stress is suggested as mechanisms of action. Dietary sodium borate (30 g B per day) improves the metabolic status of cattle during the periparturient period by stimulating glucose metabolism, limiting lipolysis intensity, and increasing serum concentrations of P, Mg, and Ca, hence preventing metabolic ailments (Kabu and Akosman 2013; Sizmaz, Koksal, and Yildiz 2017). In rabbits, boron did not affect any hematological parameters, which suggested positive effects of boron on fatty liver and visceral fat by reducing oxidative stress at the mitochondria level—affecting the Krebs cycle, glucose-alanine cycle, and methionine metabolism. Consequently, boron reduced oxidative stress and positively affected the liver lipid profile. Thus, at optimal supplementation, boron modifies hepatocyte functions, including the storage and metabolism of glycogen (Watabe et al. 2017).

The strong affinity of boric acid and borates for cis-hydroxy groups (Stangoulis and Reid 2002; Brustad et al. 2008) is a feature of interest for analyzing the ability to form bioactive and toxic molecules. This phenomena has been supported by *in silico* studies, which suggest the ability of boron to disrupt interactions on a wide group of molecules, including macromolecules such as receptors, transporters, and enzymes (Yang, Gao, and Wang 2003), and could reasonably explain some possible mechanisms of action for biological effects (Cie et al. 2016).

Biotransformation processes have been described poorly for BCCs. It is known that boric acid is rapidly eliminated unchanged by glomerular filtration in the urine (Ince, Filazi, and Yurdakok-dikmen 2017; Watabe et al. 2017).

As BCCs toxicity is related to tissue accumulation, researchers have approached their toxicity due to the positive results of boron in cancer therapy. Evidence has revealed that boron cluster-containing compounds do not cause high toxicity *in vivo* in boron neutron capture in cancer therapy (BNCT) (Barth et al. 2012). In contrast, some studies *in vitro* indicate a complicated interaction pattern between metallacarboranes and various cell types (Goszczyński, Fink, and Boratyński 2018). The physicochemical studies have revealed that metallacarboranes can be assumed to be non-classical amphiles, which explains their self-assembly in aqueous environments and their ability to interact with lipid bilayers. Regarding pharmacokinetics, the influx of metallacarboranes to cells is faster than their efflux, which in turn leads to boron cluster accumulation to high concentrations inside the cells, while limiting their clearance.

26.3 TOXICITY IN MURINE SPECIES

Based on cellular and metabolic processes, several studies indicate the anticancer effects of BCCs (Doğan et al. 2017), whereas other results suggest their ability as antimicrobials. Therefore, due to the antineoplastic and antimicrobial activity of BCCs, toxicity is evaluated for their use in humans. For this purpose, mainly murine species are used for testing. Unsurprisingly, data from these species are abundant. From observations in these animals, it is often stated that these compounds tested at high doses are toxic, but not carcinogenic or mutagenic. The principal toxicities reported are reproductive and developmental (Fail et al. 1998).

Accordingly, boric acid and borates have been widely administered to mice (Table 26.1). A detailed description of BCCs toxicity and kinetics in mice is beyond the aim of this chapter. However, the high variability of these features is observed depending on the analyzed BCCs. For example, the determination of boron concentrations in blood showed that in the first group of mice administered intravenously with 0.5 mg of sodium pentaborate, boron blood levels were 40 ppm after 1 min, which decreased to 18 ppm after 2 h; in the second group administered with 2.1 mg of sodium pentaborate, the boron blood levels were 158 ppm after 1 min, which decreased to 40 ppm after 2 h. These results suggest a two-order clearance with a rapid phase in the beginning. In both groups, boron was mainly eliminated in the urine, and half of the administered boron was excreted within an hour for both doses (Konikowski and Farr 1965). The kinetics and toxicity of tavaborole and other BCCs have been described as well. Toxicity for neoplasia is minimal, even after a 2-year exposition (Ciaravino, Plattner, and Chanda 2013). Moreover, higher molecular weight compounds have been administered to mice, such as carboranes and boron-containing complexes, in doses up to 78 mg/kg body weight, which were well-tolerated and no morbidity or mortality was observed (Vicente et al. 2003).

Similarly, BCCs have been studied in rats (Table 26.2). It has been observed that a high dose or a chronic exposition to sodium borate is required to induce notable toxicity (Umbetov et al. 2016). No carcinogenic potential was found after a 2-year administration of tavaborole in rats (Ciaravino, Plattner, and Chanda 2013). Moreover, boron-containing polyhedral complexes showed similar toxicity profiles to other molecules used in cancer therapy (Koo et al. 2007).

Table 26.1 BCCs Tested for Toxicity Profiles in Mice

Compound or Group	
Pentaborane	2,4-divinyl-nido-o-carboranyldeuteroporphyrin IX
L-(+)-2-beta-Tropanyl diphenylborinate	Boron oxide
Carboranes	Boron containing peptides on L1210
Pentaerythritol Di-(p-methylbenzeneboronate)	Cytotoxicity of ribo- and arabinoside boron nucleosides
Boron fluoride	5-o-Carboranyl-1-(2-deoxy-2-fluoro-beta-D-arabinofuranosyl)uracil
Decaborane	Boron-containing phenylalanine and tyrosine methyl esters
4,4,8,8-tetraethyl-3,3a,4,8-tetrahydro 3a,4a,4-diazabora-S-indacene	Boric acid
Meta-carboranes	Neodymium-iron-boron magnets
Boron tribromide	(Guanidine biboric acid adduct)
Boron trifluoride	Dodecahydro-closo-dodecaborate
Polyhedral carboranes.	Decaborane
10B(n, alpha)7Li reaction	Boron nitride
Diazaborine derivative (Sa 84.474)	3-Thienylboronic acid
Sulfhydryl borane	PEG-coated boron nitride
Nido-carboranylporphyrin	Amine-boranes
Boron-containing thiouracil	Borates
Trimethylamine-carbomethoxyborane, tetrakis-mu-(trimethylamine-boranecarboxylato)-bis(trimethylamine- carboxyborane)-dicopper(II), and N, N-dimethyl-n-octadecylamine borane	Copper tetracarbonyltetraphenylporphyrin

Deprivation of boron in rats seems to be damaging for bones and the immune system. In contrast, signs of boron toxicity include decreased bodyweight, as well as low weight of some organs, such as the spleen, kidneys, and pancreas. Also, boron deficiency impairs the early development of rodents. These effects have been described in pigs or ruminants (Kabu and Akosman 2013).

Regarding reproduction, high doses of boron or BCCs are required to induce disruption of fertility (data ranging from 100- to 1,000-fold greater than physiological). The No Observed Adverse Effect Level (NOAEL) was 55 mg B/kg per day, and the Lowest Observed Adverse Effect Level (LOAEL) was 76 mg B/kg per day. Therefore, rats receiving 58.5 mg B/kg per day were found sterile: the absence of spermatozoa in semen and testicular atrophy in males and decreased ovulation in most of the examined ovaries in females. An attempt to obtain litters by mating the treated females with the males fed control diet was not successful. Apparently, female rats are not at significant risk of reproductive failure due to borates at a dose <50 mg/kg. In multigenerational studies of boron toxicity (as borax or boric acid), no adverse effects on reproduction or gross pathology were found in Sprague-Dawley rats treated with a sub-chronic or chronic dose up to 17.5 mg B/kg per day examined until the third generation.

In contrast, male rats showed adverse reproductive effects induced by borates administration, such as a decreasing reproductive capacity and visible testicular lesions. However, these adverse effects were not the result of boron accumulation in the reproductive organs because boron levels were no higher than those found in blood and other soft tissues (Ince, Filazi, and Yurdakok-dikmen 2017). Testicular testosterone decreased with an increasing boron dose in diets, although Leydig cells, responsible for the biosynthesis of testosterone, are intact despite testicular atrophy. Therefore, impaired fertility in male rats is dose-dependent by targeting the high proliferative cells, germ cells, through decreasing DNA synthetic rate rather than the induction of DNA damage. In a previous study, the administration of boron at a 125 mg/kg dose showed no adverse effects on fertility, sperm characteristics, or prenatal development of the impregnated females. In contrast, none of the male rats treated with 500 mg/kg could impregnate untreated females, suggesting a definitive loss of fertility (El-Dakdoky and El-Wahab 2013).

In terms of biochemistry, lactate and pyruvate production and DNA synthesis are disrupted, although just a few molecular details of this mechanism are known. These molecules are significantly reduced, which could be related to germ epithelial sloughing and testicular atrophy resulting from damaged energy production and mitosis/meiosis in the Sertoli cells. Also, increased activity of the superoxide dismutase and glutathione peroxidase has been reported.

Boron interaction with hormones (Pecenin et al. 2018), transmembrane signaling, and transmembrane movement of regulatory ions has been demonstrated. Boron may increase plasma estradiol levels, which tends to increase plasma testosterone concentrations in males, or an additional mechanism could alter sex steroids concentration, disrupting fertility (Bello et al. 2018).

Concerning development, boron improves the metabolic status of mammals during the periparturient period by stimulating glucose metabolism, limiting lipolysis intensity, increasing bone condensation, and serum concentrations of P, Mg, and Ca, hence preventing metabolic ailments. The required doses are species-dependent (Kabu et al. 2015; Sizmaz, Koksal, and Yildiz 2017). On this basis, the number of implantation sites and fetuses was slightly higher in rats fed 2 µg B/g diet than in rats fed a low boron diet (~0.04 ppm B). Consistently, recent studies documented that boron supplementation plays a vital role in embryogenesis, immunity, and psychomotor functions (Pizzorno 2015; Dessordi et al. 2017).

Contrastingly, the excessive administration of boron seems to induce undesirable developmental changes, mainly bone malformation or unexplained intrauterine death. The mean blood boron levels in pregnant rats on gestation day 20 in the pivotal developmental toxicity study were reported to be 1.27 and 1.53 µg B/g at the NOAEL and LOAEL, respectively. The apparent toxicity of boron in the developmental stages has prompted several research groups to examine the effect of pregnancy on the ability to excrete boron. Scientists have mentioned that liver metabolism could be key to the

Table 26.2 BCCs Tested for Toxicity Profiles in Rats

Compound or Group	
Nano-hexagonal boron nitride-hydroxy apatite	Pro-soft Val-boropro
Tavaborole	Borocaptate
L-(4-[10]Boronophenyl)alanine	Boron hydrides
Boric acid	Sodium borocaptate and boronophenylalanine
Borax	L-arginine oxoborolidinone
Phenylboronic acid	Borates and borides
Boron oxide	Disodium mercaptoundecahydro-closo-dodecaborate
4-Borono-2-(18)F-fluoro-phenylalanine	Decaborane-14 (B10H14) and pentaborane-9 (B5H9)
Fructoborate	Boronated porphyrin
Bortezomib	Boron trifluoride
Boron-containing protoporphyrin IX derivatives	Aminoethoxydiphenyl borate
Prochelator BSIH	Decaborane
Boronated porphyrin BOPP	5-o-carboranyl pyrimidine nucleosides
BODIPY	Boronate-substituted stilbenes
Boronate prochelator BHAPI	Diphenylboroxazolidones of L-α-amino acids
Phenylboronic acid	Dodecahydro-closo-dodecaborate
Boronic acid-containing CXCR1/2	p-Methylbenzeneboronate
Boron trifluoride	Carbide whiskers
10B-paraboronophenylalanine	5-o-Carboranyl-1-(2-deoxy-2-fluoro-beta-D-arabinofuranosyl) uracil
AN6414	Meta-carboranes
Boronophenylalanine (BPA) and decahydrodecaborate (GB-10)	Boronated protoporphyrin (BOPP)
P-boronophenylalanine	2-Aminoethoxydiphenyl borate
Boron-containing nucleosides	Oxaborole 6-carboxamides
Boron trifluoride	Borocaptate sodium
Pentaerythritol di-(p-methylbenzeneboronate)	Decaborane
Bortezomib	Sodium mercaptoundecahydrododecaborate (borocaptate)
Boron hydride	Dipeptidyl boronic acid
Boronated chlorin e6-based photosensitizers	Boranes
Boron-containing polyamines	Amine cyanoboranes, amine carboxyboranes
Boron estrogens	Dust of borides and carbides
P-boronophenylalanine	Meso-tetra[(nido-carboranylmethyl)phenyl]porphyrins
Carboranyl uridines	Heptyl cyclohexylboronate
Boronated porphyrin EC032	Sulfhydryl borane monomer and dimer
Boronated porphyrin TABP-1	Base-boronated nucleosides and phosphate-boronated nucleotides
Beta-5-o-carboranyl-2′-deoxyuridine	Decarborane
2-Aminoethoxydiphenyl borate	3′-Aminocyanoborane-2′, 3′-dideoxypyrimidines
Boron-rich building blocks	Boronated protohaemins
BODIPY	Ribo- and arabinoside boron nucleoside
Novel boron-containing, nonclassical antifolates	Decarborane
Glutamine-functionalized carborane	Polyhedral borane anion-substituted tetraphenyl porphyrins
Boron-containing ether lipid B-Et-11-ome	Carborane
Borophenylalanine	Pentaborane
Bis(2-chloroethyl)amino group joined to boron	4-Dihydroxy boryl phenylalanine
Tetraphenylborate	

toxicity of BCCs in development. Consequently, boron ingestion can enhance or ameliorate bone density and embryonic development (Abdelnour et al. 2018). Evidence also indicates the influence of boron compounds on the response to steroid hormones involved in the embryonic bone turnover such as estrogen and vitamin D3 (Nielsen 2014; Bozkurt and Küçükyilmaz 2015).

Although many BCCs have shown to be safe for mammals, some compounds among them should be considered as highly toxic. Pentaborane concentrations of 7.8 ppm for mice and 10.4 ppm for rats caused 50% death in rats and mice during a single 60-min exposure period (Weir et al. 1964). Some five-membered cyclic boronic acids induced neurotoxicity with a micromolar intraperitoneal single dose. Acute signs of such neurotoxicity in animals are hypnosis, ataxia, convulsions, and central nervous system depression. For this reason, each new BCCs and the new boron-free compounds should be evaluated for feeding supplements for livestock, poultry, or humans, as well as drug development (Pérez-Rodríguez et al. 2017).

26.4 TOXICITY OF BCC USED AS DRUGS IN HUMANS

Increasingly scientific data showing the importance of boron in humans is being continuously generated for establishing the benefits and toxicity of boron consumption (Pizzorno 2015; Soriano-Ursúa, Farfán-García, and Geninatti-Crich 2019). It is also important to know all the regulations for boron use, its physiological effects, health uses, and the currently defined limits of toxicity (Scorei 2013).

Efficient boron uptake in human cells and the borate ability to inhibit a variety of enzyme systems, as well as the discovery of boron chemistry, are the basis of new drugs with boron (Dembitsky, Al Quntar, and Srebnik 2011).

It must be pointed out that the main source of human consumption of natural BCCs is through food and water (Pizzorno 2015; Farfán-García et al. 2016). In food, boron is present as fructoborates in fruits, vegetables such as potatoes and avocadoes, legumes, nuts, eggs, and dried food, from which some studies show that boron content on most diets is around 1.5–3 mg/day (Nieves 2013). However, various everyday items, such as medicines, cosmetics, toys, detergents, adhesives, antibiotics, insecticides antiseptics, among others, contain BCCs (Nieves 2013; Farfán-García et al. 2016).

Boron was discovered in 1870 as sodium borate and boric acid, which were used to preserve foods (in previous human applications, these compounds probably were used as a mix of inorganic BCCs). During both World War I and II, BCCs performed a vital role in preventing food crises. After these wars, boron was considered relatively as non-toxic because no deaths were attributed to using it as a preservative (Soriano-Ursúa, Farfán-García, and Geninatti-Crich 2019).

From 1930 to 1950, several reports described that boron posed a risk to health, probably because of some fatal cases related to high doses of boric acid intoxication, which is an ingredient of insecticides (Baker et al. 2009). Since that period, boron is considered an essential nutrient in humans. However, its recommendation intake has not determined yet. The World Health Organization has established a tolerance intake, defined as the highest level of daily nutrient intake with no risk of adverse health effects, which is around 0.4 mg/B/kg body weight/day. However, even with this information, the actual boron requirement levels are still unclear in many countries (Scorei 2013). Although some negative effects have been related to toxicity, in which the 50% lethal dose as boric acid for one-time administration is 2.6 g/kg body weight (Nieves 2013), only a few case reports have described boron intoxication in humans. Ingestion of >25–76 mg of boron/kg of body weight can cause a variety of symptoms, such as gastrointestinal symptoms, vomiting, diarrhea, abdominal pain, headache, lightheadedness, and rash (Craan, Myres, and Green 1997; Nieves 2013). Furthermore, since 1983, the Expert Panel of the American College of Toxicology concluded that boric acid is safe in concentrations ≤5% (50 mg/g). As an ingredient in many products for adults, European countries adopted this quantity. Additionally, the

European Food Safety Authority established a safe consumption of 10 mg of boron per day for adults (Craan, Myres, and Green 1997; Scorei 2013).

BCCs from water and diet is limited. Although boron is found in aquatic and terrestrial plants, no bioaccumulation occurs through the food chain. Boron levels are relatively high in hazelnut, butter, avocado, peanuts, and other nuts with shells, wine, beer, and in fruits, vegetables with leaves, and legumes (Barth et al. 2012; Nieves 2013). However, any toxicity of these foods has not been attributed to boron.

As a consequence of the relationship to toxicity, the use of boron in the medical field is limited. For example, natural BCCs have been used therapeutically in humans with excellent results, even if clear regulations regarding their use in drugs are lacking (Scorei 2013). No association was found between fructoborates administration and several signs or symptoms, such as clinical or ophthalmologic signs, body weight gain, food consumption, food efficiency, motor activity findings, or even death in rats. Neither adverse changes in hematology, coagulation, clinical chemistry, or urinalysis parameters in male or female rats. At necropsy, no macroscopic, histopathological findings, or organ weight changes were related to fructoborates administration. Under the conditions of this study, based on the toxicological endpoints evaluated, the NOAEL for fructoborates in the diet was 1,161 and 1,171 mg/kg bodyweight/day in male and female rats, respectively (Hunter et al. 2019).

The design and use of potent boron-containing drugs are attractive since boron addition seems to increase the affinity and selectivity of targeted proteins. However, some toxicological data found in experimental models has limited the therapeutic use of these boron-containing drugs (Farfán-García et al. 2016).

Concerning clinical cases of boric acid poisoning, data indicate that the systemic toxicity of BCCs is related to some particularly affected organs and depends on the dose, duration, and time of exposure, similar to boron-free compounds (Craan, Myres, and Green 1997). Furthermore, only a few BCCs have been studied and used in humans, from which their toxicological data has been published (Figure 26.2). Therefore, additional studies are required to support the hypotheses regarding BCCs (Bolt, Başaran, and Duydu 2012; Farfán-García et al. 2016).

Due to the low toxicity and high affinity to targets, an increasing number of BCCs are being identified as potential drugs. Data accumulation of the physical, chemical, and biological properties of boron are increasing BCCs synthesis by medicinal chemists (Baker et al. 2009).

One of the main reasons why BCCs are considered toxic is associated with the initial use of these compounds as antibiotics by damaging DNA (Hansen, Jolly, and Linder 2015) and their use as therapy prescribed by oncologists by inhibiting the proteasome and inducing cell damage in the human body (Adams 2004; Baker et al. 2009). Consequently, boric acid, and later, para-boronophenylalanine were used as boron neutron-capture therapy agents, which were found safe for humans (Baker et al. 2009).

Due to the diversity of biological effects and toxicological profiles, the main problem is not what happens to the boron atom but what happens to the rest of the molecule in the organism (Baker et al. 2009). On this basis, organoboron chemistry is allowing medicinal chemists to build many drug-like BCCs to explore their use in chemotherapeutics (Yang, Gao, and Wang 2003; Baker et al. 2009). For medical purposes, the application of many BCCs as therapeutic agents is now developing with a great diversification, suggesting general low toxicity of BCCs which is not associated with the presence of the boron atom in the structure of a given compound (Price et al. 1997; Farfán-García et al. 2016). These therapeutic actions could be achieved by using the same targets for the desired effect or by interaction in a new system (Table 26.3). BCCs, which act as potent drugs, are expected to show high toxicity to some other species, although the information available for humans is scarce (Scorei and Popa 2010; Farfán-García et al. 2016).

Simple boronic acids have not shown to induce fatal toxicity in humans. However, the limitations for their use as therapeutic agents also have limited the studies in humans. Some molecules, including boronic or structurally related moieties, such as benzoxaboroles or the proteasome inhibitors

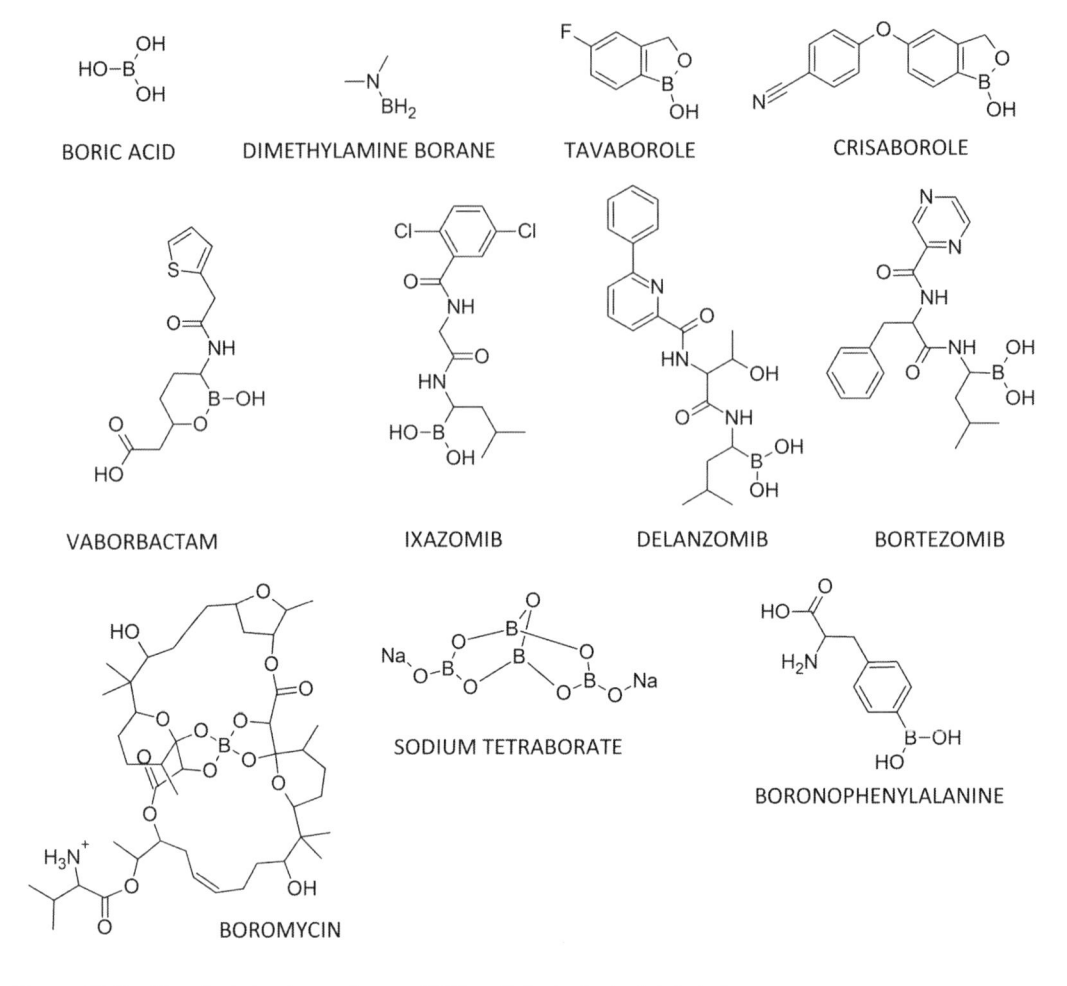

BORIC ACID   DIMETHYLAMINE BORANE   TAVABOROLE   CRISABOROLE

VABORBACTAM   IXAZOMIB   DELANZOMIB   BORTEZOMIB

SODIUM TETRABORATE

BOROMYCIN

BORONOPHENYLALANINE

Figure 26.2 Chemical structures of some BCCs with toxicological data in humans.

(Table 26.3), have shown similar toxicity as their free-boron analogs. Boronic acid compounds have shown effects as potent enzyme inhibitors, as boron neutron capture agents in cancer therapy, and as antibody mimics that also recognize important saccharides. The interaction of the boronic moiety with diols seems to play a key role in these mechanisms. Some examples of these BCCs are boronic acid, bonzoxaborole compounds, boron-containing anticoagulants, and phenylboronic acid derivatives, which have been developed and considered for different clinical treatments with no available information about toxicity (Priestley et al. 2002; Baker et al. 2006, 2009; Venkatraman et al. 2009). Also, boronic acid inhibitors of DPP4 have contributed substantially to the efficacy of post-meal blood glucose levels (Baker et al. 2009). The ability of these compounds to interact with diols could involve a high affinity of enzymatic or non-enzymatic systems triggering the molecular mechanism of toxicity.

Other groups, as aminoboranes or carboranes, have been poorly studied in humans. In the case of diazaborines, they have shown an antimicrobial activity, and other classes of diazaborines have not mentioned any reports of toxicity (Baldock et al. 1998; Baker et al. 2009).

Undeniably, the development of BCCs is appealing. At present, only a limited number of cases of boron intoxication involving humans have been described. Oral exposure of adults to high levels of boric acid has resulted in little or no observable toxicity, as seen in accidental poisonings up to

Table 26.3   Some Clinical Findings after BCC Administration

BCC	Posology	Beneficial Effects	Adverse Effects	Suggested Mechanisms of Action
Boric acid	10 mg/day Topical solution 2.5% 600 mg 3–6 g infants 15–20 g adults	Diminishes pain in osteoarthritis (Newnham 1994) Disrupts membrane permeability (Farfán-García et al. 2016) Avoids eye infection (Aviñó-Martínez, España-Gregori, and Peris-Martínez 2008) Inhibits the growth of *Candida albicans* (Mullins and Trouton 2015) Inhibits key enzymes of microorganisms (Duydu, Başaran, and Bolt 2012)	Produces nausea, vomiting, greenish diarrhea, and dehydration shortly after administration (Weir and Fisher 1972; Restuccio, Mortensen, and Kelley 1992; Ishi, Fujizuka, and Takahashi 1993; Corradi et al. 2010) In the longer term, it can lead to hypotension, metabolic acidosis, oliguric renal failure (cloudy swelling, and granular degeneration of renal tubular cells) (Ishi, Fujizuka, and Takahashi 1993; Hamilton and Wolf 2007), a generalized erythematous rash, several superficial skin abrasions, congestion, and exfoliation of the mucosa (Locatelli et al. 1987; Heindel et al. 1992; Restuccio, Mortensen, and Kelley 1992; Corradi et al. 2010)	Interactions with enzymes, membrane receptors, and transporters. No particular mechanism has been established for specific adverse effects. The treatment for boric acid poisoning is limited to alleviates signs and symptoms, albeit some authors suggest hemodialysis (Ishi, Fujizuka, and Takahashi 1993; Teshima, Taniyama, and Oishi 2001; Hamilton and Wolf 2007)
Sodium borate	10 mg/day >15 mg B/kg	Produces a significant decline in the level of sex hormones globulin binding, highly sensitive CRP, and TNF-α (Naghii et al. 2011; Farfán-García et al. 2016)	Toxicity has been found related to reproduction and development (Hakki, Bozkurt, and Hakki 2010). Evidence of irritability, disturbed sleeping, vomiting, severe diarrhea, seizures, anemia, and death (Ishi, Fujizuka, and Takahashi 1993)	Borates act in the same way as boric acid, by modulating the expression of proteins and by directly interacting with some membrane receptors (Farfán-García et al. 2016) Molecular mechanisms are unknown, but effects are strongly related to those found for boric acid (Naghii et al. 2011)
Bortezomib	1–2 mg/m²/day intravenously >1.3 mg/m²/day intravenously	Limits the proliferation of multiple myeloma. A proteasome inhibitor (Farfán-García et al. 2016; PubChem Database n.d.) In large clinical trials of bortezomib, elevations in serum aminotransferase levels were common, occurring in ~10% of patients. However, values higher than five times the upper limit of normal were rare (PubChem Database n.d.)	Stimulates peripheral neurotoxicity and lung toxicity (Simone et al. 2013) Patients show an increased risk of thrombocytopenia, neutropenia, gastrointestinal toxicity, peripheral neuropathy, infection, and fatigue, although the quality of evidence is highly variable (Argyriou et al. 2014). Bortezomib is typically given with other chemotherapeutic agents, including cyclophosphamide and dexamethasone, which can cause hepatitis B reactivation. However, no reports on hepatitis B reactivation attributable to bortezomib alone were found (Wester et al. 1998a)	No specific mechanisms have been associated with any adverse effect (PubChem Database n.d.)

(*Continued*)

Table 26.3 (*Continued*) Some Clinical Findings after BCC Administration

BCC	Posology	Beneficial Effects	Adverse Effects	Suggested Mechanisms of Action
Ixazomib	Oral, 4 mg on days 1, 8 and 15 in 28-day cycles		Common toxicities are more commonly observed with IRd versus placebo-Rd were thrombocytoperia, nausea, vomiting, diarrhea, constipation, rash, peripheral neuropathy, peripheral edema, and back pain (Moreau et al. 2016; PubChem Database n.d.)	
Delanzomib	2.4 mg/m², on days 1, 8, and 15 in 28-day cycles		Adverse effects were rash, thrombocytopenia, neutropenia, nausea, vomiting, anorexia, fatigue, pyrexia, and peripheral neuropathy (Vogl et al. 2017)	
Tavaborole	Topic application 5%	Favorable benefit-risk profile in the treatment of onychomycosis (Elewski et al. 2015)	Adverse effects found in <3% of patients at the application site include exfoliation and erythema (Farfán-García et al. 2016). Treatment-related adverse events include application site exfoliation, application site erythema, and application site dermatitis, and ingrown toenail (Elewski et al. 2015)	While inhibition of enzymes in pathogenic agents has been demonstrated, the mechanisms for adverse effects have not yet been established (Farfán-García et al. 2016)
Crisaborole		Improvement was found for atopic dermatitis and onychomycosis (Farfán-García et al. 2016; Tom et al. 2016). No evidence of mutagenic or clastogenic potential, as well as altered effects on fertility, has been observed. Oral LD50 value for rats is >500 mg/kg (PubChem Database n.d.)	In all studies, most adverse effects were minimal, including gastrointestinal alterations, local mild pain, and nasopharyngitis (Farfán-García et al. 2016; Tom et al. 2016). Hypersensitivity reactions such as contact urticaria may occur, and discontinuation of the treatment is advised (PubChem Database n.d.)	While inhibition of enzymes in pathogenic agents has been demonstrated, the mechanisms for adverse effects have not been established (Farfán-García et al. 2016)

(Continued)

Table 26.3 (*Continued*)   Some Clinical Findings after BCC Administration

BCC	Posology	Beneficial Effects	Adverse Effects	Suggested Mechanisms of Action
Vaborbactam		It has been used in trials investigating the treatment of bacterial infections. It is indicated for the treatment of complicated urinary tract infection, including pyelonephritis, complicated intra-abdominal infection, hospital-acquired pneumonia, including ventilator-associated pneumonia, treatment of patients with bacteremia in association with these infections. It is also indicated for the treatment of infections due to aerobic Gram-negative organisms in adults with limited treatment options (Griffith et al. 2016)	In a pharmacokinetic study, mild lethargy was observed in the highest dose group (PubChem Database n.d.)	
Dimethylamine borane			It causes dizziness, nausea, diarrhea, cognitive dysfunction, slurred speech, irritability, ataxia, peripheral neuropathy, cerebellar damage, and parkinsonism (Tsan et al. 2005; Kuo et al. 2006; Farfán-García et al. 2016)	Although myelin and axonal degeneration have been reported,[a] the molecular mechanisms are unknown (Liu et al. 2011; Farfán-García et al. 2016)
Boromycin			Hemolysis (Farfán-García et al. 2016; Moreira, Aziz, and Dick 2016)	An ionophore for potassium ions,[b] probably through direct interaction on membrane receptors (Farfán-García et al. 2016; Moreira, Aziz, and Dick 2016)

[a] Data from occupational exposure.
[b] Tests carried out in vitro by using human cells.

88 g, of which 90% of the cases were asymptomatic (Uluisik, Karakaya, and Koc 2018). Moreover, recent studies have indicated that boron has positive effects on human health, emphasizing its beneficial roles in bone development, the antioxidant defense system, mineral and hormone metabolism, wound healing, energy metabolism, and the immune system (Kuru et al. 2019). In light of the evidence, boron could be considered as a promising option to make positive contributions to human health (Bakirdere, Orenay, and Korkmaz 2010).

26.5 TOXICITY MECHANISMS AND THEIR IMPLICATIONS

Paracelsus's statement *"The dose makes the poison"* is often found in BCCs-toxicity exploration, suggesting dose-dependent toxicity. BCCs-toxic properties are closely tied to their microenvironment concentration (Farfán-García et al. 2016). In this sense, boron compounds are toxic to all species tested at high doses. However, evidence regarding BCCs being mutagenic or carcinogenic on eukaryotic cells is weak (Scott and Walmsley 2015). Moreover, the major toxicity related to chronic exposition in rats is developmental and reproductive impairment, whereas no clear evidence was found to suggest that boron, as it is in nature, interferes with human fertility and reproduction (Bakirdere, Orenay, and Korkmaz 2010).

Mechanisms for BCCs-toxicity in mammals can be divided into those related to pharmacodynamics and those related to pharmacokinetics.

Among the mechanisms related to pharmacodynamics, the effects of BCCs have been demonstrated or suggested on all types of known receptors, including toxic effects (Figure 26.3). Numerous active natural organic BCCs exist, comprising boric acid/borates esters presented as cis-diol biological molecules, such as plant-based organic BCCs, a few polyketide antibiotics (as borophycin and boromycin), boron siderophore complexes (as vibrioferrin and rhizoferrin), and the bacteria signaling molecule AI-2 (furanosyl borate diester). From these compounds, plant-based organic BCCs such as sugars and polyalcohol borate esters (as fructose borate and glucose borate esters) are significant in the nutrition of mammals (Donoiu et al. 2018; Hunter et al. 2019). From the studied synthetic BCCs group, boronic acids and other bioactive boronoxygenated compounds share the ability to interact with receptors (enzymes, membrane or cytosolic receptors) through hydrogen bond formation, particularly hydroxyl groups linked to boron. However, the boron atom could also be key in those interactions (Soriano-Ursúa, Das, and Trujillo-Ferrara 2014; Soriano-Ursúa et al. 2019).

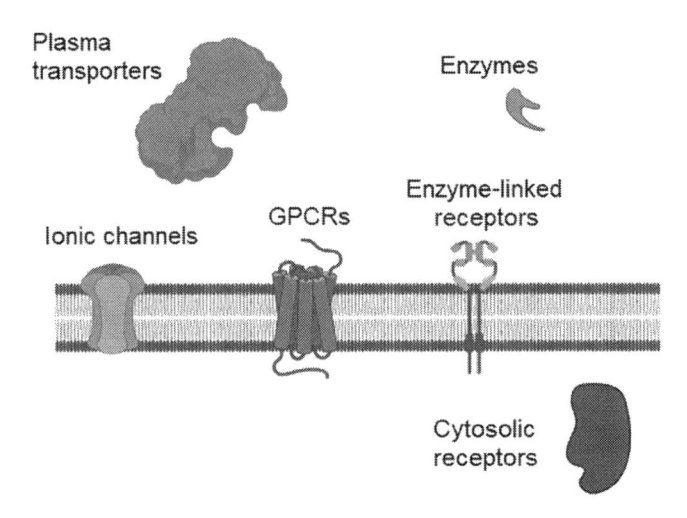

Figure 26.3 Some pharmacodynamics mechanisms for BCCs linked to toxicity.

The chemical details about other mechanisms involving structures, including several aromatic rings or complex polyhedral forms (carboranes), which interact with steroid hormone receptors, are less known.

Toxicity related to ionic channels is well described (Rosalez et al. 2020); for example, in the heart of rats perfused with 2-aminoethoxydiphenyl borate, it provoked a period of ectopic tachycardia. As perfusion continued, the rate of spontaneous ventricular depolarization increased and became disorganized inducing heart fibrillation. All these changes are related to the ability of 2-aminoethoxydiphenyl borate to activate transient calcium channels and act on myocardial voltage-independent calcium channels (Wang et al. 2012). Also, toxicity is related to the activation of cytoplasmic or nuclear receptors, which could explain some toxic effects limiting reproduction or modifying the morphology and functions of reproductive organs in mammals (Wang, Zhao, and Chen 2008). In this regard, boric acid could interact with membrane or cytoplasmic steroid receptors and reach steroid transporters, increasing the availability of endogenous steroids, as well as inducing biological effects cataloged as undesirable effects (Bello et al. 2018). Carboranes also have been considered as BCCs with high potential to induce cell-damage in mammals for which have been proposed as a treatment for steroid-dependent cancer (Ohta et al. 2015).

Data related to toxicity through action on G-protein coupled receptors or enzyme-like receptors is scarce. Some reports have studied the interaction of BCCs on these receptors, some as markers and others by labeling the targeted receptors. In the latter case, boron-dypirromethene-derivatives are widely used (Soriano-Ursúa, Farfán-García, and Geninatti-Crich 2019).

Among data related to toxicity and pharmacokinetics, some features seem to be shared for BCCs (Figure 26.4). Regardless of the source, boron is rapidly found in tissues and body fluids in humans, mostly as boric acid or borate anions. Also, it is known that urinary boron excretion changes rapidly depending on boron intake, conferring the kidney a key role in homeostasis and BCCs-concentration regulator (Nieves 2013), whereas the liver or the lungs show a weak ability for biotransformation.

Lower toxicity of some BCCs has been observed in adult humans in comparison with other mammals. For example, blood boron concentration in humans with high exposure to boric acid is at least six times lower than blood levels causing toxic effects for development and reproduction in other species. Some workgroups targeting humans exposed to boron have concluded that many BCCs should not be considered toxic for human reproduction (Duydu, Başaran, and Bolt 2012; Duydu et al. 2012; Farfán-García et al. 2016) as a consequence of differences in absorption and clearance rate (Teshima et al. 1992; Farfán-García et al. 2016). These differences depend on age,

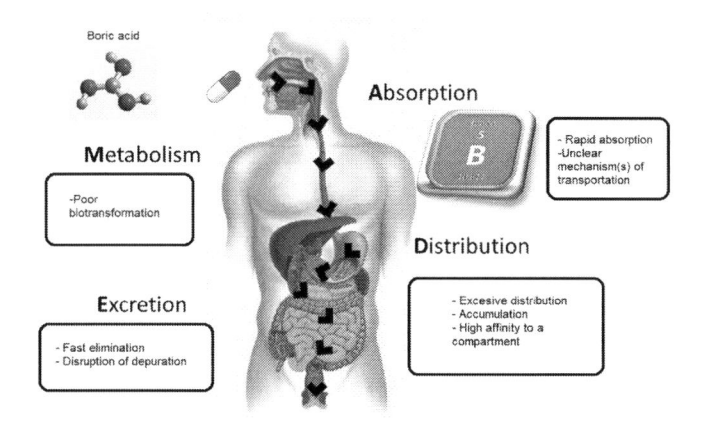

Figure 26.4 Pharmacokinetics mechanisms for BCCs linked to toxicity.

species, vehicle, and route of administration. As an example, the percutaneous absorption of boron as boric acid, borax, and disodium octaborate tetrahydrate through intact human skin is low and significantly less than the average daily dietary intake, but the absorption increases on damaged skin or with hydrophilic gels as a vehicle (Wester et al. 1998b).

Other BCCs are distributed in lipophilic tissues. SCYX-7158, a benzoxaborole agent for trypanosomiasis, demonstrates a high bioavailability, low intravenous plasma clearance, and a 24-h elimination half-life with high tissue distribution in mammals. Outstandingly, this compound is found in brain tissue and spreads through the cerebrospinal fluid at high concentrations following a 25 mg/kg oral dose in rodents (Jacobs et al. 2011). This fact highlights the relevance of studying individual compounds since the presumption of a hydrophilic BCC-group could not be directly associated with the kinetic behavior of mammals. Additionally, some data support selective phylogenetic or species-dependent cellular kinetics of these compounds (Murray 1998; Farfán-García et al. 2016).

Researchers are trying to build models with innovative strategies (Raies and Bajic 2016) to predict several categories of BCCs toxicity for their therapeutic use (Bakri et al. 2014), considering both pharmacodynamics and pharmacokinetics, based on acute toxicity, mutagenicity, tumorigenicity, skin and eye irritation, reproductive effects and multiple-dose effects (Chen et al. 2013). However, data limitations complicate possible theoretical inferences (Andrade-Jorge et al. 2018).

In conclusion, the advances in boron chemistry have expanded the applications of this metalloid in the biomedical field. BCCs have a wide range of applications in chemistry, life sciences, and energy research, which lead to a higher exposition in mammals. However, due to the limited knowledge of boron-toxicity, the use of boron as a versatile element to develop new therapeutic agents for several diseases or molecules with other applications is limited. Despite this, it is well known that an increasing number of BCCs show limited toxicity based on the doses required to induce undesirable effects. In contrast, carefully pharmacological characterization is required to propose new potential agents or to apply those recently developed as the information regarding structure-toxicity relationship is scarcely found in open databases for most of these compounds.

REFERENCES

Abdelnour, S. A., M. E. Abd El-Hack, A. A. Swelum, A. Perillo, and C. Losacco. 2018. The Vital Roles of Boron in Animal Health and Production: A Comprehensive Review. *Journal of Trace Elements in Medicine and Biology* 50: 296–304. https://doi.org/10.1016/j.jtemb.2018.07.018.

Adams, J. 2004. The Development of Proteasome Inhibitors as Anticancer Drugs. *Cancer Cell* 5 (5): 417–21. https://doi.org/10.1016/S1535-6108(04)00120-5.

Andrade-Jorge, E., A. K. García-Ávila, A. L. Ocampo-Néstor, J. G. Trujillo-Ferrara, and M. A. Soriano-Ursúa. 2018. Advances of Bioinformatics Applied to Development and Evaluation of Boron-Containing Compounds. Current Organic Chemistry 22 (3): 298–306. https://doi.org/10.2174/13852728216661704 27124336.

Argyriou, A. A., G. Cavaletti, J. Bruna, A. P. Kyritsis, and H. P. Kalofonos. 2014. Bortezomib-Induced Peripheral Neurotoxicity: An Update. *Archives of Toxicology* 88 (9): 1669–79. https://doi.org/10.1007/s00204-014-1316-5.

Aviñó-Martínez J. A., E. España-Gregori, and C. Peris-Martínez. 2008. Successful Boric Acid Treatment of Aspergillus Niger Infection in an Exenterated Orbit. *Environmental Health Perspectives* 24 (1): 79–81. https://doi.org/10.2307/3431968.

Baker, S. J., Y. K. Zhang, T. Akama, et al. 2006. Discovery of a New Boron-Containing Antifungal Agent, 5-Fluoro-1,3-Dihydro-1-Hydroxy-2,1-Benzoxaborole (AN2690), for the Potential Treatment of Onychomycosis. *Journal of Medicinal Chemistry* 49 (15): 4447–50. https://doi.org/10.1021/jm0603724.

Baker, S. J., C. Z. Ding, T. Akama, et al. 2009. Therapeutic Potential of Boron-Containing Compounds. *Future Medicinal Chemistry* 1 (7): 1275–88. https://doi.org/10.4155/fmc.09.71.

Bakirdere, S., S. Orenay, and M. Korkmaz. 2010. Effect of Boron on Human Health. *The Open Mineral Processing Journal* 3 (1): 54–9. https://doi.org/10.2174/1874841401003010054.

Bakri, R., A. A. Parikesit, C. P. Satriyanto, D. Kerami, and U. S. F. Tambunan. 2014. Utilization of Boron Compounds for the Modification of Suberoyl Anilide Hydroxamic Acid as Inhibitor of Histone Deacetylase Class II Homo Sapiens. *Advances in Bioinformatics* 2014: 104823. https://doi.org/10.1155/2014/104823.

Baldock, C., G. J. De Boer, J. B. Rafferty, A. R. Stuitje, and D. W. Rice. 1998. Mechanism of Action of Diazaborines. *Biochemical Pharmacology* 55 (10): 1541–9. https://doi.org/10.1016/S0006-2952(97)00684-9.

Barth, R. F., M.G.H. Vicente, O.K. Harling, et al. 2012. Current Status of Boron Neutron Capture Therapy of High-Grade Gliomas and Recurrent Head and Neck Cancer. *Radiation Oncology* 7 (1): 146. https://doi.org/10.1186/1748-717X-7-146.

Bello, M., C. Guadarrama-García, L. M. Velasco-Silveyra, E. D. Farfán-García, and M. A. Soriano-Ursúa. 2018. Several Effects of Boron Are Induced by Uncoupling Steroid Hormones from Their Transporters in Blood. *Medical Hypotheses* 118: 78–83. https://doi.org/10.1016/j.mehy.2018.06.024.

Bolt, H. M., N. Başaran, and Y. Duydu. 2012. Human Environmental and Occupational Exposures to Boric Acid: Reconciliation with Experimental Reproductive Toxicity Data. *Journal of Toxicology and Environmental Health - Part A: Current Issues* 75 (8–10): 508–14. https://doi.org/10.1080/15287394.2012.675301.

Bozkurt, M., and K. Küçükyilmaz. 2015. An Evaluation on the Potential Role of Boron in Poultry Nutrition. Part I: Production performance. *World's Poultry Science Journal* 71 (2): 327–38. https://doi.org/10.1017/S0043933915000331.

Brustad, E., M. L. Bushey, J. Wook Lee, D. Groff, W. Liu, and P. G. Schultz. 2008. A Genetically Encoded Boronate-Containing Amino Acid. *Angewandte Chemie - International Edition* 47 (43): 8220–3. https://doi.org/10.1002/anie.200803240.

Chen, L., J. Lu, J. Zhang, K. R. Feng, M. Y. Zheng, and Y. D. Cai. 2013. Predicting Chemical Toxicity Effects Based on Chemical-Chemical Interactions. *PLoS One* 8 (2): 4–12. https://doi.org/10.1371/journal.pone.0056517.

Ciaravino, V., J. Plattner, and S. Chanda. 2013. An Assessment of the Genetic Toxicology of Novel Boron-Containing Therapeutic Agents. *Environmental and Molecular Mutagenesis* 54 (5): 338–46. https://doi.org/10.1002/em.21779.

Cie, M., K. Labus, M. Lewa, T. Ko, J. Liesiene, and J. Bryjak. 2016. Effective L-Tyrosine Hydroxylation by Native and Immobilized Tyrosinase. *PLoS One* 11: 10. https://doi.org/10.1371/journal.pone.0164213.

Corradi, F., C. Brusasco, S. Palermo, and G. Belvederi. 2010. A Case Report of Massive Acute Boric Acid Poisoning. *European Journal of Emergency Medicine* 17 (1): 48–51. https://doi.org/10.1097/MEJ.0b013e32832d8516.

Craan, A. G., A. W. Myres, and D. W. Green. 1997. Hazard Assessment of Boric Acid in Toys. *Regulatory Toxicology and Pharmacology* 26 (3): 271–80. https://doi.org/10.1006/rtph.1997.1155.

Dembitsky, V. M., A. A. A. Al Quntar, and M. Srebnik. 2011. Natural and Synthetic Small Boron-Containing Molecules as Potential Inhibitors of Bacterial and Fungal Quorum Sensing. *Chemical Reviews* 111 (1): 209–37. https://doi.org/10.1021/cr100093b.

Dessordi, R., A. L. Spirlandeli, A. Zamarioli, J. B. Volpon, and A. Navarro. 2017. Boron Supplementation Improves Bone Health of Non-obese Diabetic Mice. *Journal of Trace Elements in Medicine and Biology* 39: 169–75. https://doi.org/10.1016/j.jtemb.2016.09.011.

Doğan, A., S. Demirci, H. Apdik, et al. 2017. A New Hope for Obesity Management: Boron Inhibits Adipogenesis in Progenitor Cells through the Wnt/β-Catenin Pathway. *Metabolism* 69: 130–42. https://doi.org/10.1016/j.metabol.2017.01.021.

Donoiu, I., C. Militaru, O. Obleagă, et al. 2018. Effects of Boron-Containing Compounds on Cardiovascular Disease Risk Factors – A Review. *Journal of Trace Elements in Medicine and Biology* 50: 47–56. https://doi.org/10.1016/j.jtemb.2018.06.003.

Duydu, Y., N. Başaran, and H. M. Bolt. 2012. Exposure Assessment of Boron in Bandirma Boric Acid Production Plant. *Journal of Trace Elements in Medicine and Biology* 26 (2–3): 161–4. https://doi.org/10.1016/j.jtemb.2012.03.008.

Duydu, Y., N. Başaran, A. Üstünda, et al. 2012. Assessment of DNA Integrity (COMET Assay) in Sperm Cells of Boron-Exposed Workers. *Archives of Toxicology* 86 (1): 27–35. https://doi.org/10.1007/s00204-011-0743-9.

El-Dakdoky, M. H., and H. M. F. Abd El-Wahab. 2013. Impact of Boric Acid Exposure at Different Concentrations on Testicular DNA and Male Rats Fertility. *Toxicology Mechanisms and Methods* 23 (5): 360–7. https://doi.org/10.3109/15376516.2013.764951.

Elewski, B. E., R. Aly, S. L. Baldwin, et al. 2015. Efficacy and Safety of Tavaborole Topical Solution, 5%, a Novel Boron-Based Antifungal Agent, for the Treatment of Toenail Onychomycosis: Results from 2 Randomized Phase-III Studies. *Journal of the American Academy of Dermatology* 73 (1): 62–9. https://doi.org/10.1016/j.jaad.2015.04.010.

Fail, P. A., R. E. Chapin, C. J. Price, and J. J. Heindel. 1998. General, Reproductive, Developmental, and Endocrine Toxicity of Boronated Compounds. *Reproductive Toxicology* 12 (1): 1–18. https://doi.org/10.1016/S0890-6238(97)00095-6.

Farfán-García, E. D., N. T. Castillo-Mendieta, F. J. Ciprés-Flores, I. I. Padilla-Martínez, J. G. Trujillo-Ferrara, and M. A. Soriano-Ursúa. 2016. Current Data Regarding the Structure-Toxicity Relationship of Boron-Containing Compounds. *Toxicology Letters* 258: 115–25. https://doi.org/10.1016/j.toxlet.2016.06.018.

Fernandes, G. F. S., W. A. Denny, and J. L. Dos Santos. 2019. Boron in Drug Design: Recent Advances in the Development of New Therapeutic Agents. *European Journal of Medicinal Chemistry* 179: 791–804. https://doi.org/10.1016/j.ejmech.2019.06.092.

Griffith, D. C., J. S. Loutit, E. E. Morgan, S. Durso, and M. N. Dudley. 2016. Phase 1 Study of the Safety, Tolerability, and Pharmacokinetics of the β-Lactamase Inhibitor Vaborbactam (RPX7009) in Healthy Adult Subjects. *Antimicrobial Agents and Chemotherapy* 60 (10): 6326–32.

Goszczyński, T. M., K. Fink, and J. Boratyński. 2018. Icosahedral Boron Clusters as Modifying Entities for Biomolecules. *Expert Opinion on Biological Therapy* 18 (1): 205–13. https://doi.org/10.1080/14712598.2018.1473369.

Hakki, S. S., B. S. Bozkurt, and E. E. Hakki. 2010. Boron Regulates Mineralized Tissue-Associated Proteins in Osteoblasts (MC3T3-E1). *Journal of Trace Elements in Medicine and Biology* 24 (4): 243–50. https://doi.org/10.1016/j.jtemb.2010.03.003.

Hamilton, R. A., and B. C. Wolf. 2007. Accidental Boric Acid Poisoning Following the Ingestion of Household Pesticide. *Journal of Forensic Sciences* 52 (3): 706–8. https://doi.org/10.1111/j.1556-4029.2007.00420.x.

Hansen, M. M., R. A. Jolly, and R. J. Linder. 2015. Boronic Acids and Derivatives - Probing the Structure-Activity Relationships for Mutagenicity. *Organic Process Research and Development* 19 (11): 1507–16. https://doi.org/10.1021/acs.oprd.5b00150.

Heindel, J. J., C. J. Price, E. A. Field, et al. 1992. Developmental Toxicity of Boric Acid in Mice and Rats. *Toxicological Sciences* 18 (2): 266–77. https://doi.org/10.1093/toxsci/18.2.266.

Hunter, J. M., B. V. Nemzer, N. Rangavajla, et al. 2019. The Fructoborates: Part of a Family of Naturally Occurring Sugar–Borate Complexes—Biochemistry, Physiology, and Impact on Human Health: A Review. *Biological Trace Element Research* 188 (1): 11–25. https://doi.org/10.1007/s12011-018-1550-4.

Ince, S., A. Filazi, and B. Yurdakok-dikmen. 2017. Boron. In *Reproductive and Developmental Toxicology*, 521–35. Elsevier Inc. https://doi.org/10.1016/B978-0-12-804239-7.00030-5.

Ishi, Y., N. Fujizuka, and T. Takahashi. 1993. A fatal case of acute boric acid poisoning. *Journal of Toxicology: Clinical Toxicology* 31 (2): 345–52.

Jacobs, R. T., B. Nare, S. A. Wring, et al. 2011. SCYX-7158, an Orally-Active Benzoxaborole for the Treatment of Stage 2 Human African Trypanosomiasis. *PLoS Neglected Tropical Diseases* 5: 6. https://doi.org/10.1371/journal.pntd.0001151.

Kabu, M., and M. S. Akosman. 2013. Biological Effects of Boron. In *Reviews of Environmental Contamination and Toxicology*, 57–75. New York, NY: Springer. https://doi.org/10.1007/978-1-4614-6470-9_2.

Kabu, M., C. Uyarlar, K. Zarczynska, W. Milewska, and P. Sobiech. 2015. The Role of Boron in Animal Health. *Journal of Elementology* 20 (2): 535–41. https://doi.org/10.5601/jelem.2014.19.3.706.

Khaliq, H., Z. Juming, and P. Ke-Mei. 2018. The Physiological Role of Boron on Health. *Biological Trace Element Research* 186 (1): 31–51. https://doi.org/10.1007/s12011-018-1284-3.

Konikowski, T., and L. E. Farr. 1965. Determination of Microgram Quantities of Inorganic Boron in Mammalian Specimens. *Clinical Chemistry* 11 (3): 378–85. https://doi.org/10.1093/clinchem/11.3.378.

Koo, M. S., T. Ozawa, R. A. Santos, et al. 2007. Synthesis and Comparative Toxicity of a Series of Polyhedral Borane Anion-Substituted Tetraphenyl Porphyrins. *Journal of Medicinal Chemistry* 50 (4): 820–7. https://doi.org/10.1021/jm060895b.

Kuo, H. C., C. C. Huang, C. C. Chu, and N. S. Chu. 2006. Axonal Polyneuropathy after Acute Dimethylamine Borane Intoxication. *Archives of Neurology* 63 (7): 1009–12. https://doi.org/10.1001/archneur.63.7.1009.

Kuru, R., S. Yilmaz, G. Balan, et al. 2019. Boron-Rich Diet May Regulate Blood Lipid Profile and Prevent Obesity: A Non-Drug and Self-Controlled Clinical Trial. *Journal of Trace Elements in Medicine and Biology* 54: 191–8. https://doi.org/10.1016/j.jtemb.2019.04.021.

Liu, J., H. Zhang, Z. Xiao, F. Wang, X. Wang, and Y. Wang. 2011. Combined 3D-QSAR, Molecular Docking and Molecular Dynamics Study on Derivatives of Peptide Epoxyketone and Tyropeptin-Boronic Acid as Inhibitors against the B5 Subunit of Human 20S Proteasome. *International Journal of Molecular Sciences* 12 (3): 1807–35. https://doi.org/10.3390/ijms12031807.

Liu, Z., H. Chen, K. Chen, Y. Shao, D. O. Kiesewetter, G. Niu, and X. Chen. 2015. Boramino Acid as a Marker for Amino Acid Transporters. *Science Advances* 1 (8): 1–7. https://doi.org/10.1126/sciadv.1500694.

Locatelli, C., C. Minoia, M. Tonini, and L. Manzo. 1987. Human Toxicology of Boron with Special Reference to Boric Acid Poisoning. *Giornale Italiano Di Medicina Del Lavoro* 9 (3–4): 141–6.

Lu, L., Q. Zhang, M. Ren, et al. 2019. Effects of Boron on Cytotoxicity, Apoptosis, and Cell Cycle of Cultured Rat Sertoli Cells In Vitro. *Biological Trace Element Research* 1–8. https://doi.org/10.1007/s12011-019-01911-3.

Militaru, C., I. Donoiu, A. Craciun, I. D. Scorei, A. M. Bulearca, and R. I. Scorei. 2013. Oral Resveratrol and Calcium Fructoborate Supplementation in Subjects with Stable Angina Pectoris: Effects on Lipid Profiles, Inflammation Markers, and Quality of Life. *Nutrition* 29 (1): 178–83. https://doi.org/10.1016/j.nut.2012.07.006.

Mogoşanu, G. D., A. Biţă, L. E. Bejenaru, et al. 2016. Calcium Fructoborate for Bone and Cardiovascular Health. *Biological Trace Element Research* 172 (2): 277–81. https://doi.org/10.1007/s12011-015-0590-2.

Moreau, P., T. Masszi, N. Grzasko, et al. 2016. Oral Ixazomib, Lenalidomide, and Dexamethasone for Multiple Myeloma. *New England Journal of Medicine* 374 (17): 1621–34. https://doi.org/10.1056/NEJMoa1516282.

Moreira, W., D. B. Aziz, and T. Dick. 2016. Boromycin Kills Mycobacterial Persisters without Detectable Resistance. *Frontiers in Microbiology* 7: 99. https://doi.org/10.3389/fmicb.2016.00199.

Mullins, Z. M., and K. M. Trouton. 2015. Basic Study: Is Intravaginal Boric Acid Non-Inferior to Metronidazole in Symptomatic Bacterial Vaginosis? Study Protocol for a Randomized Controlled Trial. *Trials* 16 (1): 1–7. https://doi.org/10.1186/s13063-015-0852-5.

Murray, F. J. 1998. A Comparative Review of the Pharmacokinetics of Boric Acid in Rodents and Humans. *Biological Trace Element Research* 66: 331–41. https://doi.org/10.1007/BF02783146.

Naghii, M. R., M. Mofid, A. R. Asgari, M. Hedayati, and M. S. Daneshpour. 2011. Comparative Effects of Daily and Weekly Boron Supplementation on Plasma Steroid Hormones and Proinflammatory Cytokines. *Journal of Trace Elements in Medicine and Biology* 25 (1): 54–8. https://doi.org/10.1016/j.jtemb.2010.10.001.

Newnham, R. E. 1994. Essentiality of Boron for Healthy Bones and Joints. *Environmental Health Perspectives* 102 (s7): 83–5. https://doi.org/10.1289/ehp.94102s783.

Nielsen, F. H. 2014. Update on the Human Health Effects of Boron. *Journal of Trace Elements in Medicine and Biology* 28 (4): 383–7. https://doi.org/10.1016/j.jtemb.2014.06.023.

Nieves, J. W. 2013. Skeletal Effects of Nutrients and Nutraceuticals, beyond Calcium and Vitamin D. *Osteoporosis International* 24 (3): 771–86. https://doi.org/10.1007/s00198-012-2214-4.

Ocampo-Néstor, A. L., J. G. Trujillo-Ferrara, A. Abad-García, C. Reyes-López, S. Geninatti-Crich, and M. A. Soriano-Ursúa. 2017. Boron's Journey: Advances in the Study and Application of Pharmacokinetics. *Expert Opinion on Therapeutic Patents* 27 (2): 203–15. https://doi.org/10.1080/13543776.2017.1252750.

Ohta, K., T. Ogawa, A. Oda, A. Kaise, and Y. Endo. 2015. Design and Synthesis of Carborane-Containing Estrogen Receptor-Beta (ERβ)-Selective Ligands. *Bioorganic and Medicinal Chemistry Letters* 25 (19): 4174–8. https://doi.org/10.1016/j.bmcl.2015.08.007.

Pecenin, M. F., L. Borges-Pereira, J. Levano-Garcia, et al. 2018. Blocking IP3 Signal Transduction Pathways Inhibits Melatonin-Induced Ca2+ Signals and Impairs P. Falciparum Development and Proliferation in Erythrocytes. *Cell Calcium* 72: 81–90. https://doi.org/10.1016/j.ceca.2018.02.004.

Pérez-Rodríguez, M., E. García-Mendoza, E. D. Farfán-García, et al. 2017. Not All Boronic Acids with a Five-Membered Cycle Induce Tremor, Neuronal Damage and Decreased Dopamine. *NeuroToxicology* 62: 92–9. https://doi.org/10.1016/j.neuro.2017.06.004.

Pizzorno, L. 2015. Nothing Boring about Boron. *Integrative Medicine (Boulder)* 14 (4): 35–48.

Price, C. J., P. L. Strong, F. J. Murray, and M. M. Goldberg. 1997. Blood Boron Concentrations in Pregnant Rats Fed Boric Acid throughout Gestation. *Reproductive Toxicology* 11 (6): 833–42. https://doi.org/10.1016/S0890-6238(97)00067-1.

Priestley, E. S., I. De Lucca, B. Ghavimi, S. Erickson-Viitanen, and C. P. Decicco. 2002. P1 Phenethyl Peptide Boronic Acid Inhibitors of HCV NS3 Protease. *Bioorganic and Medicinal Chemistry Letters* 12 (21): 3199–202. https://doi.org/10.1016/S0960-894X(02)00682-0.

PubChem Database. n.d. Crisaborole. National Center for Biotechnology Information.

Raies, A. B., and V. B. Bajic. 2016. In Silico Toxicology: Computational Methods for the Prediction of Chemical Toxicity. *Wiley Interdisciplinary Reviews: Computational Molecular Science* 6: 147–72. https://doi.org/10.1002/wcms.1240.

Restuccio, A., M. E. Mortensen, and M. T. Kelley. 1992. Fatal Ingestion of Boric Acid in an Adult. *American Journal of Emergency Medicine* 10 (6): 545–7. https://doi.org/10.1016/0735-6757(92)90180-6.

Rosalez, M.N, E. Estevez-Fregoso, A. Alatorre, A. Abad-García, and M. A. Soriano-Ursúa. 2020. 2-Aminoethyldiphenyl Borinate: A Multitarget Compound with Potential as a Drug Precursor. *Current Molecular Pharmacology* 13 (1): 57–75. https://doi.org/10.2174/1874467212666191025145429.

Scorei, R. I. 2013. Boron in Human Nutrition and Its Regulations Use. *Journal of Nutritional Therapeutics* 2: 22–9. https://doi.org/10.6000/1929-5634.2013.02.01.3.

Scorei, R. I., and R. Popa. 2010. Boron-Containing Compounds as Preventive and Chemotherapeutic Agents for Cancer. *Anti-Cancer Agents in Medicinal Chemistry* 10 (4): 346–51. https://doi.org/10.2174/187152010791162289.

Scott, H., and R. M. Walmsley. 2015. Ames Positive Boronic Acids Are Not All Eukaryotic Genotoxins. *Mutation Research. Genetic Toxicology and Environmental Mutagenesis* 777: 68–72. https://doi.org/10.1016/j.mrgentox.2014.12.002.

Simone, U. De, L. Manzo, C. Ferrari, J. Bakeine, C. Locatelli, and T. Coccini. 2013. Short and Long-Term Exposure of CNS Cell Lines to BPA-f a Radiosensitizer for Boron Neutron Capture Therapy: Safety Dose Evaluation by a Battery of Cytotoxicity Tests. *Neurotoxicology* 35: 84–90. https://doi.org/10.1016/j.neuro.2012.12.006.

Sizmaz, O., B. H. Koksal, and G. Yildiz. 2017. Rumen Microbial Fermentation, Protozoan Abundance and Boron Availability in Yearling Rams Fed Diets with Different Boron Concentrations. *Journal of Animal and Feed Sciences* 26 (1): 59–64. https://doi.org/10.22358/jafs/69038/2017h.

Soriano-Ursúa, M. A., B. C. Das, and J. G. Trujillo-Ferrara. 2014. Boron-Containing Compounds: Chemico-Biological Properties and Expanding Medicinal Potential in Prevention, Diagnosis and Therapy. *Expert Opinion on Therapeutic Patents* 24 (5): 485–500. https://doi.org/10.1517/13543776.2014.881472.

Soriano-Ursúa, M. A., E. D. Farfán-García, and S. Geninatti-Crich. 2019. Turning Fear of Boron Toxicity into Boron-Containing Drug Design. *Current Medicinal Chemistry* 26 (26): 5005–18. https://doi.org/10.2174/0929867326666190327154954.

Soriano-Ursúa, M. A., M. Bello, C. F. Hernández-Martínez, et al. 2019. Cell-Based Assays and Molecular Dynamics Analysis of a Boron-Containing Agonist with Different Profiles of Binding to Human and Guinea Pig Beta2 Adrenoceptors. *European Biophysics Journal* 48 (1): 83–97. https://doi.org/10.1007/s00249-018-1336-9.

Stangoulis, J. C. R., and R. J. Reid. 2002. Boron Toxicity in Plants and Animals. In Goldbach H. E., P. H. Brown, B. Rerkasem, et al. (eds), *Boron in Plant and Animal Nutrition*, 227–40. Boston, MA: Springer.

Teshima, D., K. Morishita, Y. Ueda, et al. 1992. Clinical Management of Boric Acid Ingestion: Pharmacokinetic Assessment of Efficacy of Hemodialysis for Treatment of Acute Boric Acid Poisoning. *Journal of Pharmacobio-Dynamics* 15: 287–94. https://doi.org/10.1248/bpb1978.15.287.

Teshima, D., T. Taniyama, and R. Oishi. 2001. Usefulness of Forced Diuresis for Acute Boric Acid Poisoning in an Adult. *Journal of Clinical Pharmacy and Therapeutics* 26 (5): 387–90. https://doi.org/10.1046/j.1365-2710.2001.00365.x.

Tom, W. L., M. V. Syoc, S. Chanda, and L. T. Zane. 2016. Pharmacokinetic Profile, Safety, and Tolerability of Crisaborole Topical Ointment, 2% in Adolescents with Atopic Dermatitis: An Open-Label Phase 2a Study. *Pediatric Dermatology* 33 (2): 150–9. https://doi.org/10.1111/pde.12780.

Tsan, Y. T., K. Y. Peng, D. Z. Hung, W. H. Hu, and D. Y. Yang. 2005. Case Report: The Clinical Toxicity of Dimethylamine Borane. *Environmental Health Perspectives* 113 (12): 1784–6. https://doi.org/10.1289/ehp.8287.

Turkez, H., F. Geyikoglu, A. Tatar, M. S. Keles, and I. Kaplan. 2012. The Effects of Some Boron Compounds against Heavy Metal Toxicity in Human Blood. *Experimental and Toxicologic Pathology* 64 (1–2): 93–101. https://doi.org/10.1016/j.etp.2010.06.011.

Uluisik, I., H. C. Karakaya, and A. Koc. 2018. The Importance of Boron in Biological Systems. *Journal of Trace Elements in Medicine and Biology* 45: 156–62. https://doi.org/10.1016/j.jtemb.2017.10.008.

Umbetov, T., A. Berdalinova, A. Koyshybayev, K. Umbetova, and G. Sultanova. 2016. Structure of the Spleen at Chronic Intoxication of the Organism by Sodium Tetraborate and after Intoxication. *Georgian Medical News* 5(254): 81–7.

Venkatraman, S., W. Wu, A. Prongay, V. Girijavallabhan, and F. G. Njoroge. 2009. Potent Inhibitors of HCV-NS3 Protease Derived from Boronic Acids. *Bioorganic and Medicinal Chemistry Letters* 19 (1): 180–3. https://doi.org/10.1016/j.bmcl.2008.10.124.

Vicente, M. G. H., A. Wickramasinghe, D. J. Nurco, et al. 2003. Synthesis, Toxicity and Biodistribution of Two 5,15-Di[3,5-(Nido-Carboranylmethyl)Phenyl]Porphyrins in EMT-6 Tumor Bearing Mice. *Bioorganic and Medicinal Chemistry* 11 (14): 3101–8. https://doi.org/10.1016/S0968-0896(03)00240-2.

Vogl, D. T., T. G. Martin, R. Vij, et al. 2017. Phase I/II Study of the Novel Proteasome Inhibitor Delanzomib (CEP-18770) for Relapsed and Refractory Multiple Myeloma. *Leukemia & Lymphoma* 58 (8): 1872–9. https://doi.org/10.1080/10428194.2016.1263842.

Wang, Y., Y. Zhao, and X. Chen. 2008. Experimental Study on the Estrogen-like Effect of Boric Acid. *Biological Trace Element Research* 121 (2): 160–70. https://doi.org/10.1007/s12011-007-8041-3.

Wang, P., P. K. Umeda, O. F. Sharifov, et al. 2012. Evidence That 2-Aminoethoxydiphenyl Borate Provokes Fibrillation in Perfused Rat Hearts via Voltage-Independent Calcium Channels. *European Journal of Pharmacology* 681 (1–3): 60–7. https://doi.org/10.1016/j.ejphar.2012.01.045.

Wang, Z., G. Gao, C. Duan, and H. Yang. 2019. Biomedicine & Pharmacotherapy Progress of Immunotherapy of Anti-α-Synuclein in Parkinson's Disease. *Biomedicine & Pharmacotherapy* 115: 108843. https://doi.org/10.1016/j.biopha.2019.108843.

Watabe, T., K. Hanaoka, S. Naka, et al. 2017. Practical Calculation Method to Estimate the Absolute Boron Concentration in Tissues Using 18F-FBPA PET. *Annals of Nuclear Medicine* 31 (6): 481–5. https://doi.org/10.1007/s12149-017-1172-5.

Weir, R. J., and R. S. Fisher. 1972. Toxicologic Studies on Borax and Boric Acid. *Toxicology and Applied Pharmacology* 23 (3): 351–64. https://doi.org/10.1016/0041-008X(72)90037-3.

Weir, F. W., V. M. Seabaugh, M. M. Mershon, D. G. Burke, and M. H. Weeks. 1964. Short Exposure Inhalation Toxicity of Pentaborane in Animals. *Toxicology and Applied Pharmacology* 6 (1): 121–31. https://doi.org/10.1016/0041-008X(64)90029-8.

Wester, R. C., T. Hartway, H. I. Maibach, et al. 1998a. In Vitro Percutaneous Absorption of Boron as Boric Acid, Borax, and Disodium Octaborate Tetrahydrate in Human Skin. *Biological Trace Element Research* 66 (1–3): 111–20. https://doi.org/10.1007/bf02783131.

Wester, R. C., X. Hui, T. Hartway, et al. 1998b. In Vivo Percutaneous Absorption of Boric Acid, Borax, and Disodium Octaborate Tetrahydrate in Humans Compared to in Vitro Absorption in Human Skin from Infinite and Finite Doses. *Toxicological Sciences* 45 (1): 42–51. https://doi.org/10.1093/toxsci/45.1.42.

Yang, W., X. Gao, and B. Wang. 2003. Boronic Acid Compounds as Potential Pharmaceutical Agents. *Medicinal Research Reviews* 23 (3): 346–68. https://doi.org/10.1002/med.10043.

Yinghuai, Z., X. Lin, H. Xie, J. Li, N. S. Hosmane, and Y. Zhang. 2019. The Current Status and Perspectives of Delivery Strategy for Boron-Based Drugs. *Current Medicinal Chemistry* 26 (26): 5019–35. https://doi.org/10.2174/0929867325666180904105212.

Lithium
Is It Good or Bad for Plant Growth?

Mohsin Tanveer
University of Tasmania and Chinese Academy of Sciences

Lei Wang
Chinese Academy of Sciences

CONTENTS

27.1 INTRODUCTION

27.1.1 Sources of Li Contamination in Agri-Environment

Lithium (Li) is widely distributed in the earth crust and is present in small amounts (Kashin, 2019). Lithium is a highly reactive element and mostly exists in the form of salts or carbonates (Choi et al., 2019). It is reported that average values for Li compounds are 30 mg/kg in the earth's crust, 25 mg/kg in soil, 2 mg/kg in drinking water, 170–190 µg/L in seawater, and 2 ng/m^3 in the atmosphere (Birch, 1988; Ribas, 1991). Lithium minerals are being mined around the globe including China, Russia, Canada, and Australia as major contributing regions worldwide to produce different Li-based items (Moore, 2007). Lithium is being added in our ecosystem especially in agri-environment via various means (Table 27.1) and Li brines and Li shaley rocks are other major sources of Li contamination (Vine and Dooley, 1980; Shahzad et al., 2016). Industrial wastewater or water resources near to Li mines also contribute toward the addition of Li in our agriculture

Table 27.1   Different Sources of Lithium Entry in Agri-Environment

Li Source	Li Concentration	References
Li brine	20 mg/L in dead sea 1,500 mg/L in Salar de Atacama	Moore (2007)
Li shaley rocks	20–100 mg/kg	Vine and Dooley (1980)
Authigenic clays	200–500 ppm in detrital clays	Chan et al. (1997)
Land disposal of Li based material	–	Kszoz and Stewart (2003) and Choi et al. (2019)
Li based industrial effluent	–	Choi et al. (2019)
Felsic rocks	80 mg/kg	Grigor'ev (2003) and Kashin (2019)
Intermediate rocks	20 mg/kg	
Mafic rocks	20 mg/kg	
Sedimentary rocks	33 mg/kg	
Clay deposits	55 mg/kg	
Sandstones	30 mg/kg	

Table 27.2   Sensitivity of Different Plant Species to Li Stress

Plant Species	Sensitivity to Li Stress	Li Concentration at Which 25% Growth Reduced (ppm)	Li Concentration at Which Growth Completely Reduced (ppm)
Avocado (*Persea americana*) Soybean (*Glycine max*) Sour orange (*Citrus auruntiuam*)	Very sensitive	6–8	100–300
Grape (*Vitis vinifera*) Red kidney bean (*Phaseolus vulgaris*)	Sensitive	8–12	300–600
Cotton (*Gossypium hirsutum*) Dallis grass (*Paspalum dilatatum*) Red beat (*Beta vulgaris*)	Moderate sensitive	20–35	500–2,000
Rhodes grass (*Chloris gayana*) Sweet corn (*Zea mays*)	Tolerant	65–70	2,500–3,000

Source: Conceived from Bingham et al. (1964).

environment (Hull et al., 2014). Li is also being added in our soil by Li leaching from Li-based batteries disposal sites (Kszos et al., 2003; Aral and Vecchio-Sadus, 2008). According to a report, the industrial use of Li includes industries producing colorings, batteries, and metal alloys manufacturing industries, and Li-based pool-cleaning chemicals are major sources of Li contamination for water bodies and soil, thus polluting environments and providing a direct source of Li toxicity for plants (Schrauzer, 2002).

27.1.2 Controversial Role of Li in Plant Physiology

Controversial results have been reported so far relating to the role of Li in plant physiology. It is yet to be discovered whether Li is an essential or non-essential for plant growth while some studies reported different essential characteristics of Li in different living organisms. Numerous plant species can absorb various amounts of Li and showed growth improvement regardless of this that Li seems not to be essential for the plant growth (Jiang et al., 2014; Shahzad et al., 2016, 2017); however, some contrasting evidence revealed opposite trend and showed that Li is highly toxic to plants and disturbs numerous physiological and biochemical processes in plants (Kabata-Pendias and Mukherjee, 2007).

Table 27.3 Response of Different Plant Species under Low and High Lithium Concentration

Plant Species	Toxicity Level	Effects	References
Maize	Low (16 mg/dm³)	Increases maize yield significantly	Antonkiewicz et al. (2017)
	High (64 mg/dm³)	Reduces maize yield significantly	
Apocynum venetum	Low (50 mg/kg)	No significant decline root and shoot biomass, photosynthetic pigment, and CO_2 accumulation	Jiang et al. (2014)
	High (>200 mg/kg)	Significant decline in biomass production, associated with a reduction in photosynthetic pigment accumulation	
Helianthus annuus (Sunflower)	Low (20–40 mM)	Hypocotyl length was improved	Stolarz et al. (2015)
	High (60 and 80 mM)	Reduction of up to 55% was observed in hypocotyl length	
Lettuce	Low (2.5 mg/dm³)	Significant increase in root growth	Kalinowska et al. (2013)
	High (50 or 100 mg/dm³)	Root growth was reduced significantly	
Sunflower	Low (5 mg/dm³)	Overall shoot growth was improved by 10%	Hawrylak-Nowak et al. (2012)
	High (50 mg/dm³)	Shoot growth was decreased by 27%	
Soybean	Low (<40 mg/dm³)	Seed germination and plant growth was improved	Ribeiro et al. (2019)
	High (>60 mg/dm³)	Seed germination and plant growth was significantly reduced	
Soybean	Low (<40 mg/dm³)	Seed germination and plant growth was improved	dos Santos et al. (2019)
	High (40–60 mg/dm³)	No significant effects on plant growth	
	Very high (>60 mg/dm³)	Seed germination and plant yield was significantly reduced	

The sensitivity of different plant species to Li toxicity also described the extent of growth reduction in response to Lithium stress (Table 27.2). Therefore, in the following section a comparison has been made, which shows the reported positive effects of Li in plants and the negative effects of Li in plants. In the end, some key factors are discussed to further explore the role of Li in plants.

27.1.3 Positive Effects of Li on Plant Metabolism

Lithium is a non-essential element and has not been observed directly involved in plant metabolism (Shahzad et al., 2016), however, a number of studies reported that Li at low concentration improves overall plant growth (Table 27.3, Shahzad et al., 2017; Tanveer et al., 2019). In a study, it was observed that Li application at low concentration (20 mg/kg) increased the shoot biomass and root biomass production up to 98% and 64%, respectively as compared to control (Bakhat et al., 2019). Such Li-induced growth improvement was due to a higher accumulation of photosynthetic pigments and leaf gas exchange (Jiang et al., 2014). Contrarily, Bakhat et al. (2019) observed that chlorophyll contents were decreased at low concentration, while increased under high Li concentration as compared with control, indicating the chlorophyll dilution effect due to higher growth of spinach shoot. Under high Li concentration, a higher rate of ROS production and lipid peroxidation was observed in different species, however, Li at low concentration activates the ROS detoxification system and reduces ROS production (Hawrylak-Nowak et al., 2012; Shahzad et al., 2017). Moreover, Li also improves the accumulation of osmolytes and ascorbic acid (Makus et al., 2006). In another study it was observed that Li improves plant growth by increasing root length and relative water contents in roots, suggesting Li may play an important role in water retention in the root (Kalinowska et al., 2013). Further studies are required to further explore the role of Li in plant metabolism.

27.2 TOXIC EFFECTS OF LI ON PLANT METABOLISM

27.2.1 Li Toxicity Limits Plant Growth

Optimal plant growth and development is determined by numerous factors such as genetic factors, climatic conditions, soil fertility, and topography. As mentioned above, Li in low concentration is beneficial for plant growth, however, a high concentration of Li results in significant growth and yield reduction (Shahzad et al., 2017). Lithium reduces plant growth and development by altering or disrupting numerous physiological mechanisms in plants (Table 27.3, Shahzad et al., 2016). In tobacco, the Li application significantly reduces plant growth by inducing the development of necrotic spots on leaves and thus reduces total chlorophyll contents (Naranjo et al., 2003). Similar results are also reported in citrus plants (Aral and Vecchio Sadus, 2008). Li-induced growth reduction was also associated with Li-induced disrupted pollen development and altered rhythmic movement of petals (Birch, 1991; Zonia and Tupy, 1995a). In lettuce, a high concentration of Li resulted in a significant reduction in stem diameter, specific leaf area, root and shoot length (da Silva et al., 2019). In brassica, Li toxicity reduced seed germination and reduced chlorophyll contents and phenolic compound production (Li et al., 2009).

The growth of the aboveground plant part strongly depends on the below-ground plant part. Plant roots are the first organ that comes in contact with Li in soil and Li in excess alters root gravitropic growth of roots (Mulkey, 2005). Moreover, Li toxicity also induced necrosis in the root meristematic zone and the induction of necrotic spots in Li-treated plants seems to be mediated by ethylene (Naranjo et al., 2003). Tissue turgidity is also very important for plant growth and Magalhães et al. (1990) reported that Li reduced dry weight of the roots in radish and leaves of watercress due to reduced water retention in roots. Likewise, in *Arabidopsis*, Li reduced relative water contents in plant roots by reducing either water retention in roots or by reducing water uptake (Duff et al., 2014). The role of aquaporins in response to Li should be examined to underpin the mechanisms behind above-mentioned responses to Li. In wheat, Li reduced net water content and dry weight; however, it was not clear whether tissue desiccation under Li stress was due to reduced water uptake or elevated transpiration rate (Kent, 1941).

Some other reported toxic effects of Li toxicity in plants include abnormal pollen development and inhibits pollen germination (Zonia and Tupy, 1995a), deformation of pollen, inducing symmetrical mitoses in the microspores (Zonia and Tupy, 1995b), inhibition of the rhythmic movements of pulvini and petals (Liskow, 1993; Birch, 2012). Moreover, inhibition of glycogen synthase kinase-3β has been proposed as a possible mechanism for the effect of lithium on development (Klein and Melton, 1996), and the most widely accepted action of lithium is the inhibition of inositol monophosphatases (Hanson, 1991; Berridge, 2009). Moreover, Li toxicity induces the depletion of cellular inositol and inhibits the inositol cycle and calcium signaling (Gillaspy et al., 1995; Saxena et al., 2013). In short, Li reduces plant growth and yield by reducing the development of various agronomical traits and altering numerous physiological processes.

27.2.2 Li Toxicity Reduces Photosynthesis in Plants

Lithium reduces plant growth by reducing the carbon assimilation process and chlorophyll contents. Photosynthesis is a complex mechanism in plants and depends on the operation of numerous physiological mechanisms simultaneously. Photosynthesis can be reduced largely via two ways: one via stomatal limitation and second via non-stomatal limitation (Figure 27.1). Stomatal limitation includes the stomatal conductance and rate of CO_2 influx and accumulation in cells, while non-stromal limitation includes the reduction in photosynthetic pigment in photosynthetically active cells It has been revealed in different studies that Li toxicity reduces chlorophyll contents For example, in sunflower Li toxicity reduced carotenoid contents in leaves by 16%, and in maize

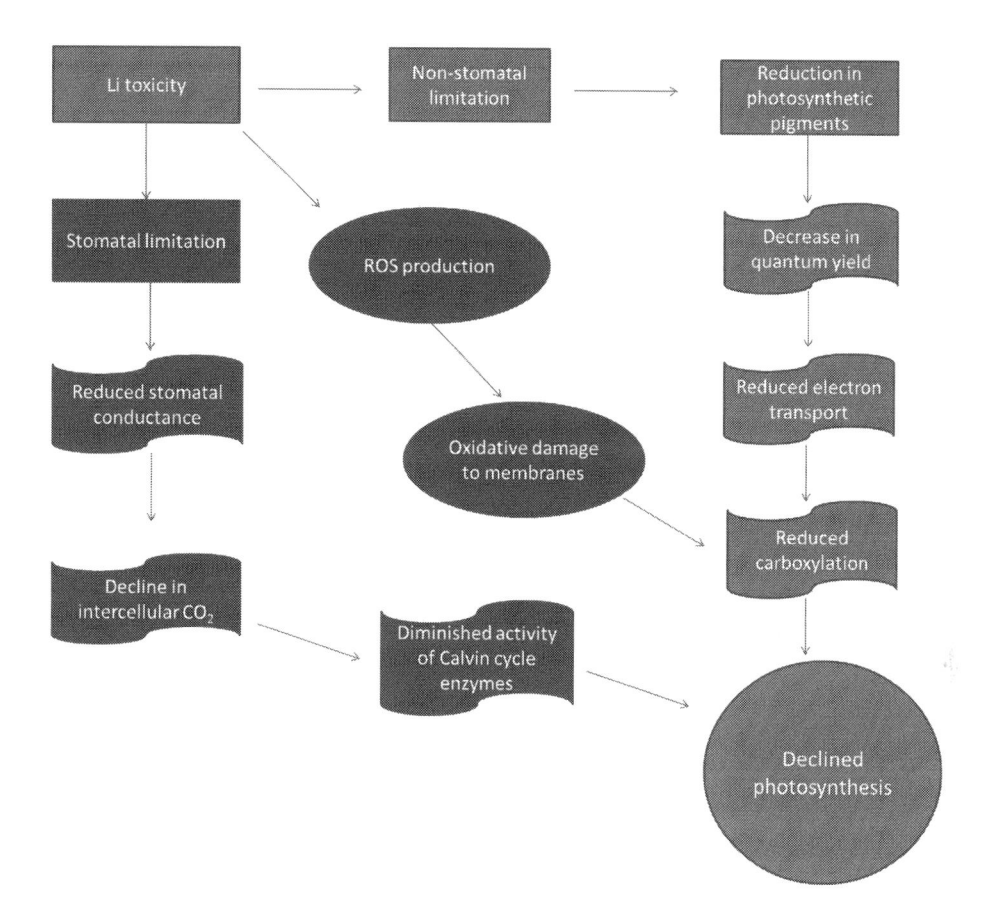

Figure 27.1 Effect of Li toxicity on photosynthesis. Li toxicity reduces photosynthesis by inducing stomatal and non-stomatal limitations. In stomatal limitation, a decline in stomatal conductance and intercellular CO_2 accumulation results in declined photosynthesis, while in non-stomatal limitation Li toxicity may reduce the biosynthesis of photosynthetic pigment or degrade the ultra-structure of the chloroplast, which ultimately reduce photosynthesis. Li toxicity also induces ROS production, which causes oxidative damage to chloroplast membrane and concomitantly reduce photosynthesis.

carotenoid and chlorophyll contents were decreased by 67% and 45%, respectively (Hawrylak-Nowak et al., 2012). Likewise, in brassica, Li induced reduction in total chlorophyll contents was suggested to be associated with the disintegration of chloroplast fine structure, instability of pigment-protein complexes, or changes in the quantity and composition of metabolites (Dubey, 2005; Li et al., 2009). Li toxicity causes development on necrotic spots in leaves, as shown in maize (Hawrylak-Nowak et al., 2012), red bean, tomato (Bingham et al., 1964), and lettuce (Kalinowska et al., 2013). The development of necrotic spots could be due to the degradation of chlorophyll contents (Shahzad et al., 2016). Several other reasons may also explain such response, including higher catalytic action of chlorophyllase, pheophorbide oxygenase, red chlorophyll catabolite reductase, and Mg-dechelatase, which are responsible for chlorophyll bleaching (Harpaz-Saad et al., 2007).

27.2.3 Li Toxicity Results in Oxidative Damage

Reactive oxygen species (ROS) production is an inevitable mechanism in plants under stress conditions and Li-induced growth reduction was also associated with ROS production under Li stress (Oktem et al., 2005; Shahzad et al., 2016, 2017). Studies revealed that Li toxicity increased the level of lipid peroxidation and malonaldehyde production (Hawrylak-Nowak et al., 2012); however, its exact mechanism is not known. It was suggested earlier that, Li toxicity may lead to higher intracellular levels of O^{2-} or OH^- radical through a Fenton-type reaction known to initiate lipid peroxidation (Shahzad et al., 2016). ROS react with different biological membranes and biomolecules such as lipids, photosynthetic pigments, and nucleic acids and the ultimate effect of ROs would be related to the disruption in the normal functioning of these biomolecules (Singh et al., 2010). ROS production due to Li toxicity can oxidize aldehydes and replace hydrogen from unsaturated fatty acid, which concomitantly can alter the ultrastructure of biological membranes.

27.2.4 Li Toxicity Impairs Ionic Homeostasis in Plants

Lithium belongs to group IA of the periodic table. It is an alkali metal with similar chemical and physical properties to other members of group IA and some members of group IIA (Shahzad et al., 2016). Li interacts with other members of group IA and IIA such as Na^+, K^+, Ca^{2+}, and Mg^{2+} and impairs ionic homeostasis in plants (Kabata-Pendias and Mukherjee, 2007). For example, Li toxicity reduces K^+ uptake in tobacco leaves (Naranjo et al., 2003). Li-induced growth reduction in different plant species was observed to decrease in K^+ uptake under Li stress (Li et al., 2009; Hawrylak-Nowak et al., 2012). Li has also been observed as a downstream element for Ca^{2+} uptake and Ca^{2+} homeostasis (Bakhat et al., 2019). Other than K^+ or Ca^{2+}, Li also interact with other elements, for example, Li increased Fe uptake and plant growth (Magalhães et al., 1990). No exact mechanism has been identified; however, we proposed a model deciphering Li-induced disruption in normal ionic homeostasis in plants under Li stress (Figure 27.2). According to this model, Li enters plant cells via non-selective cation channels and depolarizes the plasma membrane, which concomitantly results in the activation of depolarization activated K^+ outward rectifying channels to activate K^+ efflux. Moreover, cytosolic Li toxicity causes ROS production, which further activates ROS activated NSCC channels. In Li-tolerant plant species, a decline in cytosolic K^+ may compensate by releasing K^+ from the vacuolar pool, mediated by tonoplast K^+-permeable channels (TPK in the model), while Li sensitive plant species may lack or inefficient in replenishing cytosolic K from the vacuole. Future studies are required to test this model.

27.3 FUTURE RESEARCH PERSPECTIVE

Lithium stress tolerance mechanism is not explored yet; however, the Li detoxification process should be considered to improve Li toxicity in plants (Tanveer et al., 2019). As mentioned above, Li may induce ROS production by generating free oxygen radicals from Fenton type reactions (Shahzad et al., 2016). These ROS can cause oxidative damage and reduce plant growth (Kalinowska et al., 2013; Shahzad et al., 2017). Nonetheless, improving Li tolerance in plants by reducing ROS production is imperative, and this can be achieved by increasing antioxidant production. However, it is not always essential to increase antioxidant activity in order to improve stress tolerance in plants (Tanveer and Shabala, 2018). The possible explanations for such discrepancy lay in some facts that some ROS such as H_2O_2 plays an important role in stress signaling and plant metabolism, so tempering with it may result in pleiotropic effects (Tanveer and Ahmed, 2020). It becomes increasingly evident that substantial variations exist in the production of the antioxidants in various plant tissues and at various time points (Tanveer et al., 2019). For instance, GPX activity was inhibited at 1 mM Li, but at the same time SOD activity was increased, showing the differential response of antioxidants to Li stress (Nciri et al., 2012). Hence, the inter-specific or intra-specific aspects of ROS production and scavenging should be considered.

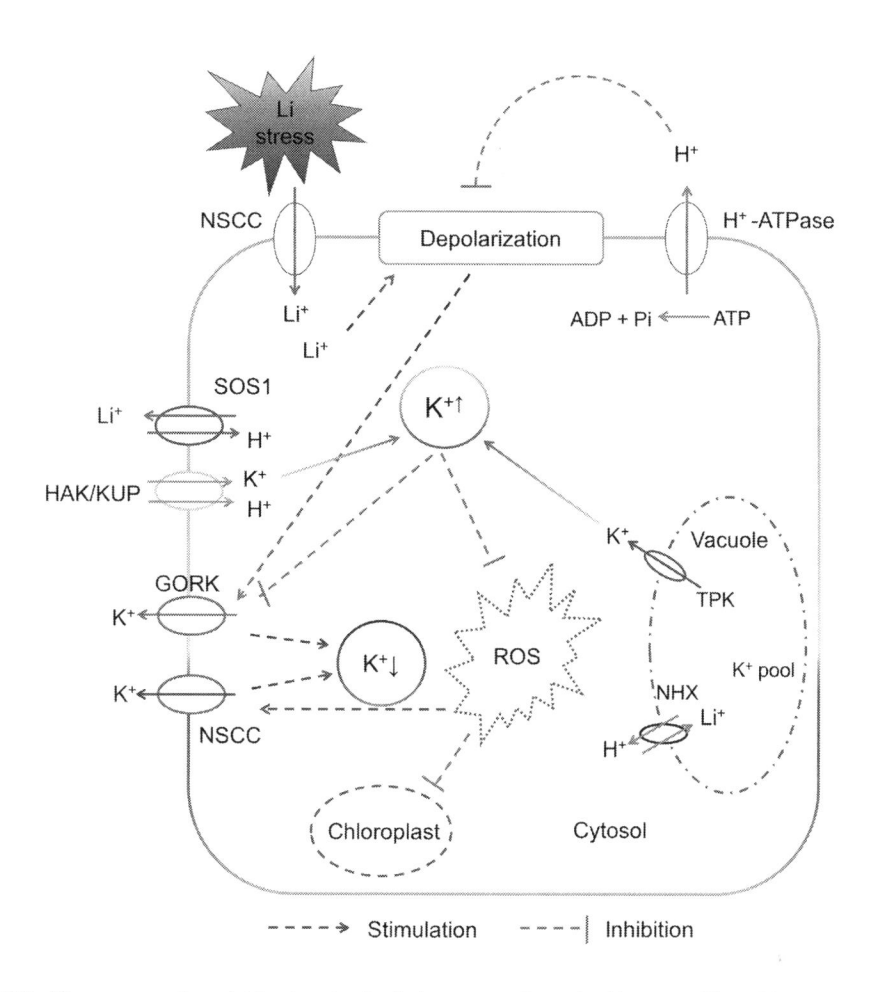

Figure 27.2 The proposed model is showing ionic homeostasis under Li stress. Upon Li stress, the plasma membrane (PM) may get depolarized due to a high influx of Li$^+$ from the external medium via a non-selective cation channel (NSCC). Increment in cytosolic Li$^+$ results in the production of ROS, and resultantly induces oxidative damage and reduces chloroplast activity. This causes K$^+$ efflux from plasma membrane wither by depolarization activated GORK channels and resultantly cytosolic K$^+$ homeostasis gets disrupted. The decline in cytosolic K$^+$ further induces the production of ROS and can trigger programmed cell death in the cell. ROS further kindle the situation by inducing ROS activated K$^+$ efflux via NSCC. In Li-tolerant plant species, a decline in cytosolic K$^+$ may compensate by releasing K from the vacuolar pool, mediated by tonoplast K$^+$- permeable channels (TPK in the model) while Li sensitive plant species may lack or inefficient in replenishing cytosolic K from the vacuole. Moreover, higher NHX exchanger activity may result in more Li$^+$ vacuolar sequestration in Li tolerant plant species.

27.4 CONCLUSION

Extensive research is required to underpin the essentiality of Li in different biological systems because the nature of Li toxicity induced alteration in plant metabolism is poorly understood. The present chapter overviewed the role of Li in plant physiology. It can be summarized from the above discussion that Li at low concentration improves plant growth by improving morphological growth, while at high concentration Li decreases plant growth by altering carbon assimilation and oxidative damage. Tolerance mechanisms in plants for Li toxicity are yet to be explored, therefore species focus should be given by considering the response of halophytes to Li toxicity.

REFERENCES

Antonkiewicz, J., Jasiewicz, C., Koncewicz-Baran, M., & Bączek-Kwinta, R. (2017). Determination of lithium bioretention by maize under hydroponic conditions. *Archives of Environmental Protection*, *43*(4), 94–104.

Aral, H., & Vecchio-Sadus, A. (2008). Toxicity of lithium to humans and the environment—A literature review. *Ecotoxicology and Environmental Safety*, *70*, 349–356.

Bakhat, H.F., Rasul, K., Farooq, A.B.U., Zia, Z., Natasha, Fahad, S., Abbas, S., Shah, G.M., Rabbani F., & Hammad, H.M. (2019). Growth and physiological response of spinach to various lithium concentrations in soil. *Environmental Science and Pollution Research*. https://doi.org/10.1007/s11356-019-06877-2

Berridge, M.J. (2009). Inositol trisphosphate and calcium signalling mechanisms. *Biochimica et Biophysica Acta (BBA)-Molecular Cell Research*, *1793*(6), 933–940.

Bingham, F.T., Bradford, G.E., & Page, A.L. (1964). Toxicity of lithium. *California Agriculture*, *18*, 6–7.

Birch, N.J. (1988). Lithium. In: Seiler, H.G. (Ed.), *Handbook on the toxicity of inorganic compounds*. Marcel Dekker, New York, pp. 382–393.

Birch, N.J. (ed.) (1991). *Lithium and the cell: Pharmacology and biochemistry*. Academic Press, San Diego.

Birch, N.J. (Ed.). (2012). *Lithium and the cell: Pharmacology and Biochemistry*. Academic Press, Cambridge, MA.

Chan, L.H., Sturchio, N.C., & Katz, A., 1997. Lithium isotope study of the Yellowstone hydrothermal system. *EOS Transactions American Geophysical Union*, *78*, F802.

Choi, H.B., Ryu, J.S., Shin, W.J., & Vigier, N. (2019). The impact of anthropogenic inputs on lithium content in river and tap water. *Nature Communications*, *10*(1), 1–7.

da Silva, R.R., de Faria, A.J.G., Alexandrino, G.D.C., Ribeiro, E.A., dos Santos, A.C.M., Deusdara, T.T., do Nascimento, I.R., & Nascimento, V.L. (2019). Enrichment of lithium in lettuce plants through agronomic biofortification. *Journal of Plant Nutrition*, *42*(17), 2102–2113.

dos Santos, A.C.M., Marques, K.R., Rodrigues, L.U., de Faria, Á.J.G., Nascimento, V.L., & Fidélis, R.R. (2019). Biofortification of soybean grains with foliar application of Li sources. *Journal of Plant Nutrition*, *42*(19), 2522–2531.

Dubey, R.S. (2005). Photosynthesis in plants under stressful conditions. In: *Handbook of photosynthesis*, Marcel Dekker, New York, pp. 859–875.

Duff, M.C., Kuhne, W.W., Halverson, N.V., Chang, C.S., Kitamura, E., Hawthorn, L., & Stieve-Caldwell, E. (2014). mRNA transcript abundance during plant growth and the influence of Li+ exposure. *Plant Science*, *229*, 262–279.

Gillaspy, G.E., Keddie, J.S., Oda, K., & Gruissem, W. (1995). Plant inositol monophosphatase is a lithium-sensitive enzyme encoded by a multigene family. *The Plant Cell*, *7*(12), 2175–2185.

Grigor'ev, N.A. (2003). Average concentrations of chemical elements in rocks of the upper continental crust. *Geochemistry International*, *41*(7), 711–718.

Hanson, B.A. (1991). The effects of lithium on the phosphoinositides and inositol phosphates of *Neurospora crassa*. *Experimental Mycology*, *15*(1), 76–90.

Harpaz-Saad, S., Azoulay, T., Arazi, T., Ben-Yaakov, E., Mett, A., Shiboleth, Y.M., Hörtensteiner, S., Gidoni D., Gal-On, A., Goldschmidt, E.E., & Eyal, Y. (2007). Chlorophyllase is a rate-limiting enzyme in chlorophyll catabolism and is posttranslationally regulated. *The Plant Cell*, *19*, 1007–1022.

Hawrylak-Nowak, B., Kalinowska, M., & Szymańska, M. (2012). A study on selected physiological parameters of plants grown under lithium supplementation. *Biological Trace Element Research*, *149*(3), 425–430.

Hull, S.L., Oty, U.V., & Mayes, W.M. (2014). Rapid recovery of benthic invertebrates downstream of hyperalkaline steel slag discharges. *Hydrobiology*, *736*, 83–97.

Jiang, L., Wang, L., Mu, S.Y., & Tian, C.Y. (2014). *Apocynum venetum*: A newly found lithium accumulator. *Flora-Morphology, Distribution, Functional Ecology of Plants*, *209*, 285–289.

Kabata-Pendias, A., & Mukherjee, A.B. (2007). *Trace elements from soil to human*. Springer, Berlin, pp. 87–93.

Kalinowska, M., Hawrylak-Nowak, B., & Szymańska, M. (2013). The influence of two lithium forms on the growth, L-ascorbic acid content and lithium accumulation in lettuce plants. *Biological Trace Element Research*, *152*(2), 251–257.

Kashin, V.K. (2019). Lithium in soils and plants of western Transbaikalia. *Eurasian Soil Science*, *52*(4), 359–369.

Kent, N.L. (1941). Absorption, translocation and ultimate fate of lithium in the wheat plant. *New Phytologist*, *40*(4), 291–298.

Klein, P.S., & Melton, D.A. (1996). A molecular mechanism for the effect of lithium on development. *Proceedings of the National Academy of Sciences*, *93*(16), 8455–8459.

Kszos, L.A., Beauchamp, J.J., & Stewart, A.J. (2003). Toxicity of lithium to three freshwater organisms and the antagonistic effect of sodium. *Ecotoxicology*, *12*, 427–437.

Kszos, L.A., & Stewart, A.J. (2003). Review of lithium in the aquatic environment: Distribution in the United States, toxicity and case example of groundwater contamination. *Ecotoxicology*, *12*, 439–447.

Li, X., Gao, P., Gjetvaj, B., Westcott, N., & Gruber, M.Y. (2009). Analysis of the metabolome and transcriptome of *Brassica carinata* seedlings after lithium chloride exposure. *Plant Science*, *177*, 68–80.

Liskow, B. (1993). Lithium and the cell: Pharmacology and biochemistry. *American Journal of Psychiatry*, *150*(11), 1748–1749.

Magalhães, J.R., Wilox, G.E., Rocha, A.N.F., & Silva, F.L.I.M. (1990). Research on lithium-phytological metabolism and recovery of hypo-lithium. *Pesquisa Agropecuária Brasileira*, *25*, 1781–1787.

Makus, D.J., Zibilske, L., & Lester, G. (2006). Effect of light intensity, soil type, and lithium addition on spinach and mustard greens leaf constituents. *Subtropical Plant Science*, *58*, 35–41.

Moore, S. (2007). Between rock and salt lake. *Industrial Minerals*, *June*, 58–69.

Mulkey, T.J. (2005). Alteration of growth and gravitropic response of maize roots by lithium. *Gravitational and Space Biology*, *18*(2), 119–120.

Naranjo, A., Romero, C., Bellés, J.M., Montesinos, C., Vicente, O., & Serrano, R. (2003). Lithium treatment induces a hypersensitive-like response in tobacco. *Planta*, *217*, 417–424.

Nciri, R., Allagui, M.S., Bourogaa, E., Saoudi, M., Murat, J.C., Croute, F., & Elfeki, A. (2012). Lipid peroxidation, antioxidant activities and stress protein (HSP72/73, GRP94) expression in kidney and liver of rats under lithium treatment. *Journal of Physiology and Biochemistry*, *68*(1), 11–18.

Oktem, F., Ozguner, F., Sulak, O., Olgar, S., Akturk, O., Yilmaz, H.R., & Altuntas, I. (2005). Lithium-induced renal toxicity in rats: Protection by a novel antioxidant caffeic acid phenethyl ester. *Molecular and Cellular Biochemistry*, *277*, 109–115.

Ribas, B. (1991). Lithium. In: Merian, E. (Ed.), *Metals and their Compounds in the Environment*, 4th edn. VCH, Weinheim, pp. 1014–1023.

Ribeiro, E.A., da Silva, L.P., da Luz, J.H.S., de Oliveira, H.P., Nunes, B.H.D.N., de Faria, A.J.G., de Freitas, G.A., Beserra, J.P.S ., Oliveira, S.D.S., de Oliveira, M., da Silva, R.R. (2019). Germination of Soybean Seeds treated with Sources and doses of Lithium for Agronomic Biofortification. *International Journal of Advanced Engineering Research and Science*, *6*(7), 670–674.

Saxena, S.C., Salvi, P., Kaur, H., Verma, P., Petla, B.P., Rao, V., Kamble, N., & Majee, M. (2013). Differentially expressed myo-inositol monophosphatase gene (CaIMP) in chickpea (*Cicer arietinum* L.) encodes a lithium-sensitive phosphatase enzyme with broad substrate specificity and improves seed germination and seedling growth under abiotic stresses. *Journal of Experimental Botany*, *64*(18), 5623–5639.

Schrauzer, G.N. (2002). Lithium: Occurrence, dietary intakes, nutritional essentiality. *Journal of the American College of Nutrition*, *21*, 14–21.

Shahzad, B., Mughal, M.N., Tanveer, M., Gupta, D., & Abbas, G. (2017). Is lithium biologically an important or toxic element to living organisms? An overview. *Environmental Science and Pollution Research*, *24*(1), 103–115.

Shahzad, B., Tanveer, M., Hassan, W., Shah, A.N., Anjum, S.A., Cheema, S.A., & Ali, I. (2016). Lithium toxicity in plants: Reasons, mechanisms and remediation possibilities–A review. *Plant Physiology and Biochemistry*, *107*, 104–115.

Singh, R., Tripathi, R.D., Dwivedi, S., Kumar, A., Trivedi, P.K., & Chakrabarty, D. (2010). Lead bioaccumulation potential of an aquatic macrophyte *Najas indica* are related to antioxidant system. Bioresource Technology, *101*, 3025–3032.

Stolarz, M., Król, E., & Dziubińska, H. (2015). Lithium distinguishes between growth and circumnutation and augments glutamate-induced excitation of Helianthus annuus seedlings. *Acta Physiologiae Plantarum*, *37*(4), 1–9.

Tanveer, M., Hasanuzzaman, M., & Wang, L. (2019). Lithium in environment and potential targets to reduce lithium toxicity in plants. *Journal of Plant Growth Regulation*, *38*(4), 1574–1586.

Tanveer, M. & Ahmed, H.A.I. (2020). ROS signalling in modulating salinity stress tolerance in plants. In: Hasanuzzaman, M. & Tanveer, M., (Eds.), *Salt and Drought Stress Tolerance in Plants: Signaling Networks and Adaptive Mechanisms*. Springer, New York, pp. 299–314.

Tanveer, M., & Shabala, S. (2018). Targeting redox regulatory mechanisms for salinity stress tolerance in crops. In: *Salinity responses and tolerance in plants, Volume 1*, Springer, Cham, pp. 213–234.

Vine, J.D., & Dooley, J.R. Jr. (1980). Where on Earth is all the lithium?, with a section on Uranium isotope studies. Fish Lake Valley, Nevada, U.S. Geological Survey Open-File Report 80-1234.

Zonia, L.E., & Tupy, J. (1995a). Lithium treatment of *Nicotiana tabacum* microspores blocks polar nuclear migration, disrupts the partitioning of membrane-associated Ca^{2+}, and induces symmetrical mitosis. *Sexual Plant Reproduction*, 8(3), 152–160.

Zonia, L., & Tupy, J. (1995b). Lithium-sensitive calcium activity in the germination of apple (Malus× domestica Borkh.), tobacco (*Nicotiana tabacum* L.), and potato (*Solanum tuberosum* L.) pollen. *Journal of Experimental Botany*, 46(8), 973–979.

CHAPTER 28

An Overview of Nickel Toxicity in Plants

Mohsin Tanveer
University of Tasmania

CONTENTS

28.1 INTRODUCTION

Nickel (Ni) is the most abundant element after iron on the earth's crust and comprises roughly 0.008% of total earth crust (Hedfi et al. 2007; Hussain et al. 2013). Approximately 10% of Ni in earth crust is being locked up in molten Fe-Ni ore (Ahmad et al. 2011). Worldwide, Ni is present in different soils ranging from 20 mg/kg in China to 420 mg/kg in the USA (Figure 28.1). Ni is a transition metal element and is widely accepted as an essential element for plants (Matraszek et al. 2016). Nonetheless excess concentration of Ni in our environment makes it a toxic element and thus release of Ni into the environment is of great concern that includes a deposition in agricultural soils (Jamil et al. 2014). Generally, optimum or sub-lethal Ni concentration should be lower than 100 ppm in soil and 0.05 ppm in surface water; however, inevitable anthropogenic activities and unnecessary industrial developments are increasing Ni concentration in soil and water (Shahzad et al. 2018). Ni is being added in our environment via different means such as smelting, burning of fossil fuels, mining, disposal of industrial waste, and effluent from Ni-based metallurgy and electroplating industry (Orlov et al. 2002). Therefore it is important to understand the role of Ni in plants. In this chapter, the essential role of Ni in plant metabolism has been discussed and Ni toxicity in plants has also been overviewed.

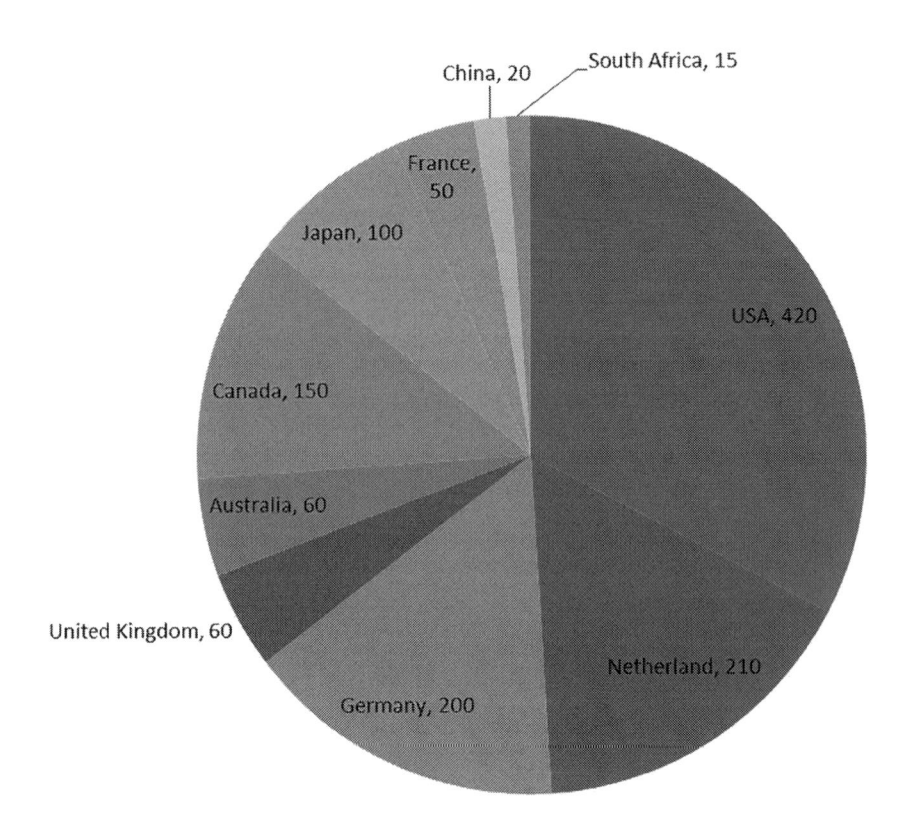

Figure 28.1 Nickel concentration (mg/kg) in the soil around the globe (Chen et al. 1999).

28.2 ESSENTIALITY OF NI AND ITS ROLE IN PLANT GROWTH AND DEVELOPMENT

An element is considered as an essential nutrient when the plant cannot perform its normal life cycle without that element. Ni was considered as toxic element ages ago, however, during 1987, Mr. Brown and his co-workers established that Ni is an essential nutrient; however, its essentiality largely depends on its concentration in the growth medium. Generally, optimal Ni concentration ranges from 0.05 to 10 mg/kg in plant tissues (Nieminen et al. 2007). Later on, researchers around the globe discovered the positive and essential role of Ni in plants. Ni has been found as a prime component of numerous enzymes, as it develops bonds with S-ligands and O-ligands (Marschner 2002). Moreover, urease is the only enzyme in plant which requires Ni as its essential component (Ojeda-Barrios et al. 2016). Urease activity is highly sensitive to Ni deficiency; even a minute change in optimal Ni concentration reduces urease activity up to 25%, which concomitantly disrupts N metabolism in plants (de Queiroz Barcelos et al. 2017; Hussain et al. 2020).

The essentiality of Ni in plants can also be observed from previous findings such as Ni deficiency results in leaf injury, necrosis, stunted root growth and nodule development in legumes, and reduced iron uptake (Chen et al. 2009; Zobiole et al. 2010; Shahzad et al. 2018). Nonetheless, the application of Ni in low concentration alleviates these symptoms and improves plant growth and is also required for hydrogenase enzyme, an important enzyme required to reduce nitrogen into ammonia (Dalton et al. 1985), and Ni deficiency significantly reduces hydrogenase activity (Ahmad and Ashraf 2012). Ni deficiency also disrupts N metabolism in plants by disturbing amino acid metabolism especially cysteine amino acid, polyamines biosynthesis pathway, and ornithine cycle intermediates (Bai et al.

2006; Sachan and Lal 2017). Moreover Ni deficiency also interferes with the TCA cycle by reducing the citrate level in plants (Bai et al. 2006). Furthermore, due to the fact that Ni plays a prime role in the activities of several enzymes, it is shown to be an important promoter of plant growth and development at low concentrations (Gajewska & Skłodowska 2005, Shahzad et al. 2018; Daneshmand et al. 2019). Ni improves disease resistance in plants by triggering disease resistance mechanism and by exerting direct antagonistic effects on pathogens (Brown 2007).

Ni at low concentration increases overall plant biomass production by improving plant growth, leaf area development, root proliferation, carbon assimilation process, and water retention in plants (Seregin and Kozhevnikova 2006; Prasad and Shivay 2019). The application of Ni at the concentration of 10 μM improved the fresh and dry weight of wheat seedlings as compared with control plants grown without Ni (Gajewska et al. 2006). Likewise, Ni improved seedling length, number of branches, and biomass accumulation in *Hibiscus sabdariffa* seedlings (Aziz et al. 2007). As mentioned above, Ni is important for urease enzyme activity, thus the presence of an adequate amount of Ni in xylem sap is important for the RNAase activity and nutrient uptake (Bai et al. 2013). Moreover several reports indicated that Ni-induced growth improvement was due to the direct and essential role of Ni in N metabolism (Gheibi et al. 2009; Khoshgoftarmanesh et al. 2011; Kutman et al. 2013, 2014; Alibakhshi and Khoshgoftarmanesh 2015). Ni at low concentration improves nitrogen reductase enzyme activity in different plant species (Gad et al. 2007; Tabatabaei 2009). Ni is also an essential component of hydrogenase (Ni-Fe) that takes part in the recycling of H_2, an obligatory product of N_2 reduction (González-Guerrero et al. 2014). Pre-treatment with Ni improved urease activity and higher N contents in grain, which was positvely correlated with increased N-remobilization or from older to younger leaves (Kutman et al. 2013, 2014; de Macedo et al. 2016). In short, Ni at low concentration is important for plant growth and development.

28.3 PLANT RESPONSES UNDER NI TOXICITY

28.3.1 Plant Growth and Development

Though Ni at low concentration improves plant growth, however, being a transition metal element, it causes tremendous yield loss by reducing plant growth and development (Shahzad et al. 2018). However mechanism underlying such growth reduction is largely unknown. Some recent studies showed that Ni toxicity induces growth reduction by altering multiple physiological mechanisms (Figure 28.2), for example, Lešková et al. (2020) found that Ni in excess reduces plant growth by inducing gravitropic defects and locally inhibits root growth by suppressing cell elongation without significantly disrupting the integrity of the stem cell niche. Moreover, Ni in excess affects the expression of many genes associated with plant cell walls (Lešková et al. 2020). Ni toxicity also reduces the width of epidermal cells and the thickness and size of vascular bundles and mesophyll cells (Kovacevic et al. 1999). Moreover, hydroxycinnamic and polygalacturonic acids in the cell wall can bind to Ni and concomitantly cell wall loses its integrity and rigidity, which can explain the strong inhibition of root growth under Ni toxicity (Meychik et al. 2014). Ni toxicity reduced root cell wall plasticity also by binding with pectin and up-regulates peroxidase activity and ROS production in the cell wall (Meychik et al. 2014). Ni-induced inhibition in cell growth is related to the inhibited cell division (Bhalerao et al. 2015). Ni also reduces lateral root growth by disorganizing the ultrastructural development of pericycle and endodermal cells during cell division and cell elongation (Samantaray et al. 1997; Hassan et al. 2020).

Seed germination and seedling establishment are the most critical stages in the plant's life cycle and Ni in excess amount inhibits seed germination and seedling growth (Shahzad et al. 2018). Several studies showed Ni induced reduction in seed germination in different plant species (Zhang et al. 2006; Bhardwaj et al. 2007; Sharma et al. 2008). Ni toxicity results in the disruption of

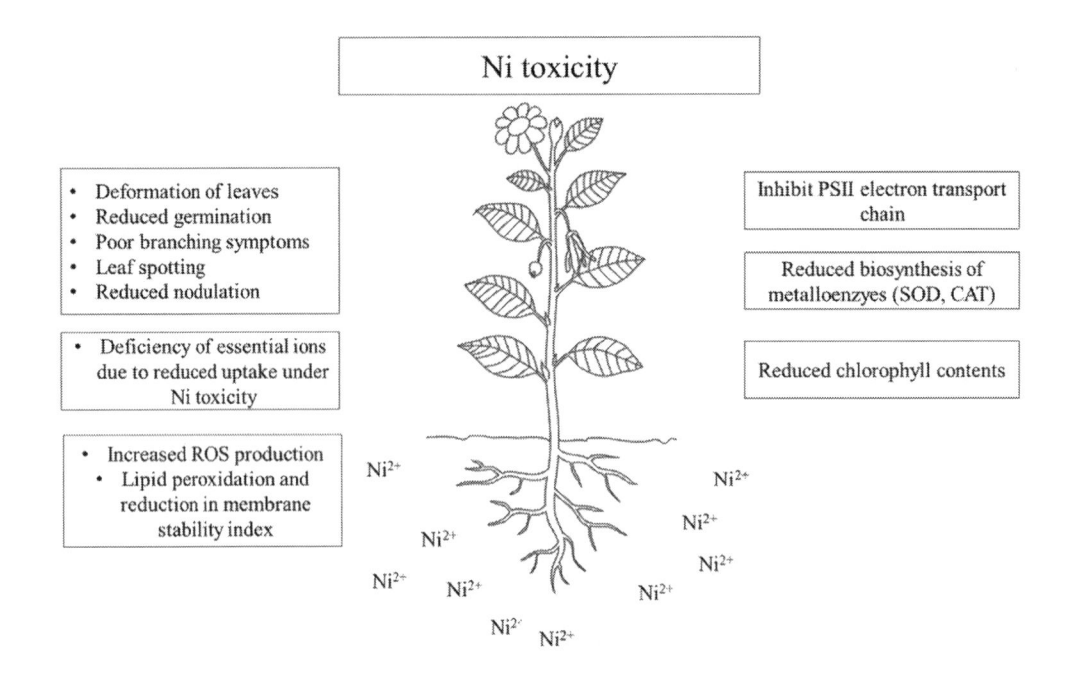

Figure 28.2 An overview of Ni toxicity in plants.

α-amylase activity, protein synthesis, and reduces the activities of different enzymes mobilizing the storage food from endosperm to developing embryo (Yadav et al. 2009). In another study it was found that Ni-induced reduction in seed germination partially due to alteration in biochemical metabolism as the availability of sugars and amino acid for the synthesis of metabolic energy and protein respectively and activation of enzymes essential for the growing embryo are generally reduced due to suppression in α-amylase and protease activities (Ashraf et al. 2011). Conclusively, Ni reduces plant growth by reducing seed germination, root development, and proliferation.

28.4 PHOTOSYNTHESIS AND PHOTOSYNTHETIC PIGMENTS

Photosynthesis is an important mechanism in plants and is highly sensitive to adverse growth conditions (Evans 2013). Ni toxicity leads to non-specific inhibition of photosynthesis through direct or indirect means (Figure 28.3, Prasad et al. 2005). Reduction in photosynthetic activity can be examined using photosynthetic-related indicators such as inhibition of chlorophyll biosynthesis, electron transport chain activity, alteration in chloroplast structure, and reduced enzymatic activities during Calvin cycle (Seregin and Ivanov 2001). Reduction in size and number of chloroplast and structural disorganization in chloroplast has been observed in response to Ni toxicity (Appenroth et al. 2010). Other factors that reduce photosynthesis include the number of thylakoid and lipid membrane composition. Ni can reduce the activity of these factors either by reducing turgor pressure in leaves or by inducing ROS-induced oxidative damage and lipid peroxidation (Wang et al. 2008; Ali et al. 2009; Hassan et al. 2020). Ni decreases chlorophyll contents in different plant species (Gajewska et al. 2006), however, mechanism underlying reduction in chlorophyll contents is still to be explored. Studies suggest that Ni stress induces Fe and Mg deficiency, which could be one reason (Kamran et al. 2016), while Ni-mediated structural alteration in chlorophyll molecule and reduction in Rubisco (ribulose-1,5-bisphosphate carboxylated oxygenase) could be the other reason (Küpper et al. 1998; Shahzad et al. 2018).

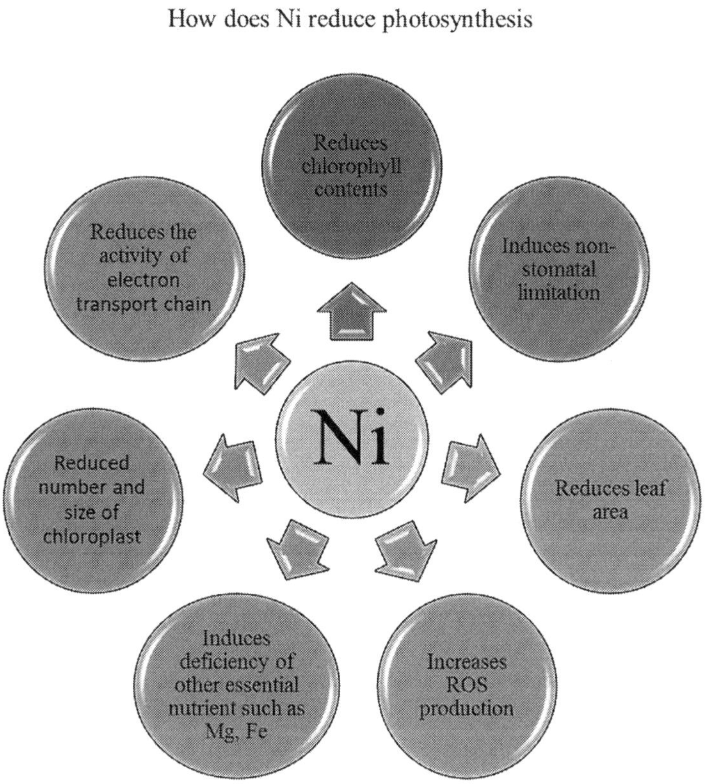

How does Ni reduce photosynthesis

Figure 28.3 An overview of Ni induced reduction in photosynthesis.

The electron transport chain is a very integral and crucial step in the photosynthesis process; Ni interacts with photosynthetic electron transport chain and intermediates the availability of cytochromes b6f and b559 in leaves (Krupa et al. 1993). Calvin cycle is also a very important component of photosynthesis in plants, and Ni toxicity significantly alters the functionality of the Calvin cycle by interfering and limiting the activities of key enzymes involved in the Calvin cycle (Shahzad et al. 2018). Important enzymes such as Rubisco, 3-phosphosglycerate kinase, fructose-1,6-bisphosphate, aldolase, and NAD- and NADP-dependent phosphoglyceraldehyde dehydrogenase. In *Populus nigra*, Ni reduces photosynthesis by reducing mesophyll conductance to CO_2 and chloroplastic CO_2 as compared with control (Velikova et al. 2011). In some cases, the accumulation of ATP and NADPH may result from the inhibition of Calvin cycle reactions, later it develops a pH gradient across the thylakoid membranes, which ultimately blocks the activity of PSII (Alam et al. 2007; Yusuf et al. 2011). Moreover Ni further limits photosynthesis by limiting stomatal conductivity and intracellular CO_2 accumulation (Shafeeq et al. 2012). These drastic effects of Ni result in a significant reduction in photosynthesis (Figure 28.3).

28.5 MEMBRANE PERMEABILITY AND ROS PRODUCTION

Being a transition metal element, Ni induces ROS production and reduces plant growth (Shahzad et al. 2018). Such Ni toxicity affects the growth and is closely related to the overproduction of ROS in different cellular compartments that causes oxidative damage to cell membranes, proteins, lipids, and DNA through lipid peroxidation (Anjum et al. 2015, 2016a), leading to developmental impairments and genetic instability in plant species (Shah et al. 2016). Malondialdehyde (MDA) is a key indicator of membrane degradation under Ni-induced oxidative damage in different plant species when exposed to Ni stress (Dubey and Pandey 2011; Hussain et al. 2013; Soares et al. 2016).

Activation of ion efflux systems in response to ROS is a subsequent effect of Ni toxicity in plants. Ni-induced ROS production stimulates the disintegration of membranes through lipid peroxidation and decreases thiolation of proteins in several plant species (Rao and Sresty 2000; Gajewska and Skłodowska 2007; Maheshwari and Dubey 2009). Ni in excess produced different ROS species such as H_2O_2, O^{2-}, and OH^- (Khaliq et al. 2015). Moreover, Ni also reduces membrane permeability and stability by causing oxidative damage to the lipo-proteins of the plasma membrane (Dat et al. 2000). Ni-induced decline in osmolyte accumulation in Ni sensitive plant speices and results in the loss of turgor pressure further degrade membrane permeability (Verma and Dubey 2003; Wang et al. 2008; Tanveer et al. 2017). Ni-induced high H_2O_2 accumulation in different plant species reduces plant growth by reducing the activities of numerous enzymes by altering their activation energy, substrate affinity, and turnover number (Boominathan and Doran 2002; Gajewska and Skłodowska 2007; Kazemi et al. 2010). Ni increased the NADPH-mediated ROS production in wheat roots, thus, reduces wheat productivity significantly under Ni stress (Hao et al. 2006). Likewise, Gajewska and Skłodowska (2007) found that Ni can increase H_2O_2 and O^{2-} production in plant leaves by 250%. In two maize hybrids, Ni toxicity increased MDA production and reduced membrane stability (Amjad et al. 2020). Ni-induced ROS production also affects other fundamental physiological processes, for instance in Ni stress, wheat changes in the contents of linolenic acid, oleic acid, palmitic acid, and palmitoleic acid, which are partly responsible for Ni-induced lipid peroxidation, oxidative stress, and increased electrolyte leakage (Gajewska et al. 2009). Kumar et al. (2015) found that tomato quality can be significantly reduced in response to high lipid peroxidation and membrane damage.

28.6 CONCLUSION

The requirements of Ni largely depend on its concentration in the growth medium. Ni at low concentration stimulates plant growth, while at high concentration causes toxicity in plants. Also, Ni at low concentration improves plant growth by improving the performance of numerous agronomic traits of plants and by increasing the net CO_2 assimilation rate. On the other hand, Ni at high concentration reduces plant growth by reducing photosynthesis and reduced activity of associated enzymes. Moreover Ni induces ROS production and causes oxidative damage to the plasma membrane. Hence based on the above discussion, Ni is an essential nutrient; however, its beneficial effects on plant growth depend on its concentration in the growth medium.

REFERENCES

Ahmad, M. S. A., & Ashraf, M. (2012). Essential roles and hazardous effects of nickel in plants. In: Whitacre, D. (ed.), *Reviews of environmental contamination and toxicology* (pp. 125–167). Springer, New York, NY.

Ahmad, M. S., Ashraf, M., & Hussain, M., (2011). Phytotoxic effects of nickel on yield and concentration of macro- and micro-nutrients in sunflower (*Helianthus annuus* L.) achenes. *Journal of Hazardous Materials, 185*(2–3), 1295–1303. doi: 10.1016/j.jhazmat.2010.10.045.

Alam, M. M., Hayat, S., Ali, B., & Ahmad, A. (2007). Effect of 28-homobrassinolide treatment on nickel toxicity in *Brassica juncea*. *Photosynthetica, 45*(1), 139–142. doi:10.1007/s11099-007-0022-4.

Ali, M. A., Ashraf, M., & Athar, H. R. (2009). Influence of nickel stress on growth and some important physiological/biochemical attributes in some diverse canola (*Brassica napus* L.) cultivars. *Journal of Hazardous Materials, 172*(2–3), 964–969.

Alibakhshi, M., & Khoshgoftarmanesh, A. H. (2015). Effects of nickel nutrition in the mineral form and com-plexed with histidine in the nitrogen metabolism of onion bulb. *Plant Growth Regulation, 75*, 733–740. doi:10.1007/s10725-014-9975-z

Amjad, M., Ameen, N., Murtaza, B., Imran, M., Shahid, M., Abbas, G., ... & Jacobsen, S. E. (2020). Comparative physiological and biochemical evaluation of salt and nickel tolerance mechanisms in two contrasting tomato genotypes. *Physiologia Plantarum, 168*(1), 27–37.

Anjum, S. A., Tanveer, M., Hussain, S., Bao, M., Wang, L., Khan, I., Ullah, E., Tung, S.A., Samad, R.A., & Shahzad, B. (2015). Cadmium toxicity in maize (*Zea mays* L.): Consequences on antioxidative systems, reactive oxygen species and cadmium accumulation. *Environmental Science and Pollution Research, 22*, 17022–17030.

Anjum, S. A., Tanveer, M., Hussain, S., Shahzad, B., Ashraf, U., Fahad, S., Hassan, W. Jan, S., Khan, I., Saleem, M.F., Bajwa, A.A., Wang, L., Mahmood, A., Samad, R.A., & Tung, S. A. (2016). Osmoregulation and antioxidant production in maize under combined cadmium and arsenic stress. *Environmental Science and Pollution Research.* doi:10.1007/s11356-016-6382-1.

Appenroth, K. J., Krech, K., Keresztes, A., Fischer, W., & Koloczek, H. (2010). Effects of nickel on the chlo-roplasts of the duckweeds Spirodela polyrhiza and Lemna minor and their possible use in biomonitoring and phytoremediation. *Chemosphere, 78*(3), 216–223.

Ashraf, M. Y., Sadiq, R., Hussain, M., Ashraf, M., & Ahmad, M. S. A. (2011). Toxic effect of nickel (Ni) on growth and metabolism in germinating seeds of sunflower (*Helianthus annuus* L.). *Biological Trace Element Research, 143*(3), 1695–1703.

Aziz, E. E., Gad, N., & Badran, N. M. (2007). Effect of cobalt and nickel on plant growth, yield and flavonoids content of *Hibiscus sabdariffa* L. *Australian Journal of Basic and Applied Sciences, 1*, 73–78.

Bai, C., Liu, L., & Wood, B. W. (2013). Nickel affects xylem Sap RNase a and converts RNase A to a urease. *BMC Plant Biology, 13*, 207. http://doi.org/10.1186/1471-2229-13-207.

Bai, C., Reilly, C. C., & Wood, B. W. (2006). Nickel deficiency disrupts metabolism of ureides, amino acids, and organic acids of young pecan foliage. *Plant Physiology, 140*(2), 433–443.

Bhalerao, S. A., Sharma, A. S., & Poojari, A. C. (2015). Toxicity of nickel in plants. *International Journal of Pure and Applied Biosciences, 3*(2), 345–355.

Bhardwaj, R., Arora, N., Sharma, P., & Arora, H. K. (2007). Effects of 28-homobrassinolide on seedling growth, lipid peroxidation and antioxidative enzyme activities under nickel stress in seedlings of *Zea mays* L. *Asian Journal of Plant Sciences, 6*(5), 765–772.

Boominathan, R., & Doran, P. M. (2002). Ni-induced oxidative stress in roots of the Ni hyperaccumulator, *Alyssum bertolonii. New Phytologist, 156*(2), 205–215. doi:10.1046/j.1469-8137.2002.00506.x.

Brown, P. H. (2007). Nickel. In A. V. Barker, & D. J. Pilbean (Eds.), *Handbook of plant nutrition* (pp. 395–402). CRC Taylor and Francis, New York, NY.

Brown, P. H., Welch, R. M., & Cary E. E. (1987). Nickel: A micronutrient essential for higher plants. *Plant Physiology, 85*, 801–803. doi:10.1104/pp.85.3.801.

Chen, J., Bradhurst, D. H., Dou, S. X., & Liu, H. K. (1999). Nickel hydroxide as an active material for the positive electrode in rechargeable alkaline batteries. *Journal of the Electrochemical Society, 146*(10), 3606.

Chen, C., Huang, D., & Liu, J. (2009). Functions and toxicity of nickel in plants: Recent advances and future prospects. *CLEAN – Soil, Air, Water, 37*(4–5), 304–313. doi:10.1002/clen.200800199.

Dalton, D. A., Evans, H. J., & Hanus, F. J. (1985). Stimulation by nickel of soil microbial urease activity and urease and hydrogenase activities in soybeans grown in a low-nickel soil. *Plant Soil, 88*, 245–258.

Daneshmand, B., Eshghi, S., Gharaghani, A., & Eshghi, H. (2019). Growth, mineral nutrient composition, and enzyme activity of strawberry as influenced by adding urea and nickel to the nutrient solution. *Journal of Berry Research, 9*(1), 27–37.

Dat, J., Vandenabeele, S., Vranova, E., Van Montagu, M., Inze, D., & Van Breusegem, F. (2000). Dual action of the active oxygen species during plant stress responses. *Cellular and Molecular Life Sciences, 57*(5), 779–795. doi:10.1007/s000180050041.

de Macedo, F. G., Bresolin, J. D., Santos, E. F., Furlan, F., Lopes da Silva, W. T., Polacco, J. C., & Junior, J. L. (2016). Nickel availability in soil as influenced by liming and its role in soybean nitrogen metabolism [original research]. *Frontiers in Plant Science, 7*(1358). doi:10.3389/fpls.2016.01358.

de Queiroz Barcelos, J. P., de Souza Osório, C. R. W., Leal, A. J. F., Alves, C. Z., Santos, E. F., Reis, H. P. G., & dos Reis, A. R. (2017). Effects of foliar nickel (Ni) application on mineral nutrition status, urease activity and physiological quality of soybean seeds. *Australian Journal of Crop Science, 11*(2), 184.

Dubey, D., & Pandey, A. (2011). Effect of nickel (Ni) on chlorophyll, lipid peroxidation and antioxidant enzymes activities in black gram (Vigna mungo) leaves. *International Journal of Science and Nature, 2*(2), 395–401.

Evans, J. R. (2013). Improving photosynthesis. *Plant Physiology, 162*(4), 1780–1793.

Gad, N., El-Sherif, M. H., & El-Gereedly, N. H. M. (2007). Influence of nickel on some physiological aspects of tomato plants. *Australian Journal of Basic and Applied Sciences, 3*, 286–293. doi:10.1002/tox.20470.

Gajewska, E., & Skłodowska, M. (2005). Antioxidative responses and proline level in leaves and roots of pea plants subjected to nickel stress. *Acta Physiologiae Plantarum, 27*(3), 329–340. doi:10.1007/s11738-005-0009-3.

Gajewska, E., & Skłodowska, M. (2007). Effect of nickel on ROS content and antioxidative enzyme activities in wheat leaves. *Biometals, 20*(1), 27–36. doi:10.1007/s10534-006-9011-5.

Gajewska, E., Skłodowska, M., Słaba, M., & Mazur, J. (2006). Effect of nickel on antioxidative enzyme activities, proline and chlorophyll contents in wheat shoots. *Biologia Plantarum, 50*(4), 653–659. doi:10.1007/s10535-006-0102-5.

Gajewska, E., Wielanek, M., Bergier, K., & Skłodowska, M. (2009). Nickel-induced depression of nitrogen assimilation in wheat roots. *Acta Physiologiae Plantarum, 31*(6), 1291. doi:10.1007/s11738-009-0370-8.

Gheibi, M., Malakouti, M., Kholdebarin, B., Ghanati, F., Teimouri, S., & Sayadi, R. (2009). Significance of nickel supply for growth and chlorophyll content of wheat supplied with urea or ammonium nitrate. *Journal of Plant Nutrition, 32*, 1440–1450. doi:10.1080/01904160903092655.

González-Guerrero, M., Matthiadis, A., Sáez, Á., & Long, T. A. (2014). Fixating on metals: New insights into the role of metals in nodulation and symbiotic nitrogen fixation. *Frontiers in Plant Science, 5*, 45. doi:10.3389/fpls.2014.00045.

Hao F, Wang X, Chen J (2006). Involvement of plasma-membrane NADPH oxidase in nickel-induced oxidative stress in roots of wheat seedlings. *Plant Science, 170*, 151–158.

Hedfi, A., Mahmoudi, E., Boufahja, F., Beyrem, H., & Aïssa, P. (2007). Effects of increasing levels of nickel contamination on structure of offshore nematode communities in experimental microcosms. *Bulletin of Environmental Contamination and Toxicology, 79*(3), 345–349. doi:10.1007/s00128-007-9261-0.

Hussain, M. B., Ali, S., Azam, A., Hina, S., Farooq, M. A., Ali, B., Bharwana, S.A., & Gill, M. B. (2013). Morphological, physiological and biochemical responses of plants to nickel stress: A review. *African Journal of Agricultural Research, 8*, 1596–1602. doi:10.5897/AJAR12.407.

Hussain, S., Khaliq, A., Noor, M. A., Tanveer, M., Hussain, H. A., Hussain, S., Shah, T., & Mehmood, T. (2020). Metal toxicity and nitrogen metabolism in plants: An overview. In: Datta R., Meena R., Pathan S., Ceccherini M. (eds.), *Carbon and nitrogen cycling in soil* (pp. 221–248). Springer, Singapore.

Jamil, M., Zeb, S., Anees, M., Roohi, A., Ahmad, I., Rehman, S., & Rha, E. S. (2014). Role of *Bacillus licheniformis* in phytoremediation of nickel contaminated soil cultivated with rice. *International Journal of Phytoremediation, 16*, 554–571.

Kamran, M. A., Eqani, S. A. M. A. S., Bibi, S., Xu, R.-K., Amna, Monis, M. F. H., Katsoyiannis, A., Bokhari, H., &, Chaudhary, H. J. (2016). Bioaccumulation of nickel by E. sativa and role of plant growth promoting rhizobacteria (PGPRs) under nickel stress. *Ecotoxicology and Environmental Safety, 126*, 256–263. doi:10.1016/j.ecoenv.2016.01.002.

Kazemi, N., Khavari-Nejad, R. A., Fahimi, H., Saadatmand, S., & Nejad-Sattari, T. (2010). Effects of exogenous salicylic acid and nitric oxide on lipid peroxidation and antioxidant enzyme activities in leaves of *Brassica napus* L. under nickel stress. *Scientia Horticulturae, 126*(3), 402–407. doi:10.1016/j.scienta.2010.07.037.

Khaliq, A., Ali, S., Hameed, A., Farooq, M. A., Farid, M., Shakoor, M. B., Mahmood, K., Ishaque, W., &, Rizwan, M. (2015). Silicon alleviates nickel toxicity in cotton seed- lings through enhancing growth, photosynthesis and suppressing Ni uptake and oxidative stress. *Archives of Agronomy and Soil Science.* doi:10.1080/03650340.2015.1073263.

Khoshgoftarmanesh, A. H., Hosseini, F., & Afyuni, M. (2011). Nickel supplementation effect on the growth, urease activity and urea and nitrate concentrations in lettuce supplied with different nitrogen sources. *Scientia Horticulturae, 130*, 381–385. doi:10.1016/j.scienta.2011.07.009.

Kovacevic G, Kastori R, Merkulov LJ (1999). Dry matter and leaf structure in young wheat plants as affected by Cd, lead, and nickel. *Biologia Plantarum, 42*, 119–123.

Krupa, Z., Siedlecka, A., Maksymiec, W., & Baszynski, T. (1993). In vitro responses of photosynthetic apparatus of *Phaseolus vulgaris* L. to nickel toxicity. *Plant Physiology, 142*, 664–668.

Kumar, P., Rouphael, Y., Cardarelli, M., & Colla, G. (2015). Effect of nickel and grafting combination on yield, fruit quality, antioxidative enzyme activities, lipid peroxidation, and mineral composition of tomato. *Journal of Plant Nutrition and Soil Science, 178*(6), 848–860.

Küpper, H., Küpper, F., & Spiller, M. (1998). In situ detection of heavy metal substituted chlorophylls in water plants. *Photosynthesis Research, 58*(2), 123–133. doi:10.1023/a:1006132608181.

Kutman, B. Y., Kutman, U. B., & Cakmak, I. (2013). Nickel-enriched seed and externally supplied nickel improve growth and alleviate foliar urea damage in soybean. *Plant and Soil, 363*(1), 61–75. doi:10.1007/s11104-012-1284-6.

Kutman, B. Y., Kutman, U. B., & Cakmak, I. (2014). Effects of seed nickel reserves or externally supplied nickel on the growth, nitrogen metabolites and nitrogen use efficiency of urea- or nitrate-fed soybean. *Plant Soil, 376*, 261–276. doi:10.1007/s11104-013-1983-7.

Lešková, A., Zvarík, M., Araya, T., & Giehl, R. F. (2020). Nickel toxicity targets cell wall-related processes and PIN2-mediated auxin transport to inhibit root elongation and gravitropic responses in Arabidopsis. *Plant and Cell Physiology, 61*(3), 519–535.

Maheshwari, R., & Dubey, R. S. (2009). Nickel-induced oxidative stress and the role of antioxidant defence in rice seedlings. *Plant Growth Regulation, 59*(1), 37–49. doi:10.1007/s10725-009-9386-8.

Marschner, H. (2002). *Mineral nutrition of higher plants* (pp. 364–369), 3rd edn. Academic Press, London.

Matraszek, R., Hawrylak-Nowak, B., Chwil, S., & Chwil, M. (2016). Macronutrient composition of nickel-treated wheat under different sulfur concentrations in the nutrient solution. *Environmental Science and Pollution Research, 23*(6), 5902–5914.

Meychik, N., Nikolaeva, Y., Kushunina, M., & Yermakov, I. (2014). Are the carboxyl groups of pectin polymers the only metal-binding sites in plant cell walls? *Plant and Soil, 381*(1–2), 25–34.

Nieminen, T. M., Ukonmaanaho, L., Rausch, N., & Shotyk, W. (2007). Biogeochemistry of nickel and its release into the environment. *Metal Ions in Life Sciences, 2*, 1–30.

Ojeda-Barrios, D. L., Sánchez-Chávez, E., Sida-Arreola, J. P., Valdez-Cepeda, R., & Balandran-Valladares, M. (2016). The impact of foliar nickel fertilization on urease activity in pecan trees. *Journal of Soil Science and Plant Nutrition, 16*(1), 237–247.

Orlov, D. S., Sadovnikova, L. K., & Lozanovskaya, I. N. (2002). *Ecology and protection of biosphere under chemical pollution*. Vysshaya Shkola, Moscow.

Prasad, R., & Shivay, Y. S. (2019). Nickel in environment and plant nutrition: A mini review. *International Journal of Plant and Environment, 5*(04), 239–242.

Prasad, S. M., Dwivedi, R., & Zeeshan, M. (2005). Growth, photosynthetic electron transport, and antioxidant responses of young soybean seedlings to simultaneous exposure of nickel and UV-B stress. *Photosynthetica, 43*(2), 177–185.

Rao, K. M., & Sresty, T. V. S. (2000). Antioxidative parameters in the seedlings of pigeonpea (Cajanus cajan (L.) Millspaugh) in response to Zn and Ni stresses. *Plant Science, 157*(1), 113–128.

Sachan, P., & Lal, N. (2017). An overview of nickel (Ni^{2+}) essentiality, toxicity and tolerance strategies in plants. *Asian Journal of Biology, 2*(4), 1–15.

Samantaray, S., Rout, G. R., & Das, P. (1997). Tolerance of rice to nickel in nutrient solution. *Biologia Plantarum, 40*(2), 295–298. doi:10.1023/a:1001085007412.

Seregin, I. V., & Ivanov, V. B. (2001). Physiological aspects of cadmium and lead toxic effects on higher plants. *Russian Journal of Plant Physiology the English-Language Translation of Fiziologiya Rastenii,48*, 523–544.

Seregin, I. V., & Kozhevnikova, A. D. (2006). Physiological role of nickel and its toxic effects on higher plants. *Russian Journal of Plant Physiology, 53*(2), 257–277. doi:10.1134/s1021443706020178.

Shafeeq, A., Butt, Z. A., & Muhammad, S. (2012). Response of nickel pollution on physiological and biochemical attributes of wheat (*Triticum aestivum* L.) var. Bhakar-02. *Pakistan Journal of Botany, 44*(1), 111–116.

Shah, A. N., Tanveer, M., Hussain, S., & Yang, G. (2016). Beryllium in the environment: Whether fatal for plant growth? *Reviews in Environmental Science and Bio/Technology, 15*(4), 549–561.

Shahzad, B., Tanveer, M., Rehman, A., Cheema, S. A., Fahad, S., Rehman, S., & Sharma, A. (2018). Nickel; whether toxic or essential for plants and environment - A review. *Plant Physiology and Biochemistry, 132*, 641–651.

Sharma, P., Bhardwaj, R., Arora, N., Arora, H. K., & Kumar, A. (2008). Effects of 28-homobrassinolide on nickel uptake, protein content and antioxidative defence system in *Brassica juncea*. *Biologia Plantarum, 52*(4), 767–770.

Soares, C., de Sousa, A., Pinto, A., Azenha, M., Teixeira, J., Azevedo, R. A., & Fidalgo, F. (2016). Effect of 24-epibrassinolide on ROS content, antioxidant system, lipid peroxidation and Ni uptake in *Solanum nigrum* L. under Ni stress. *Environmental and Experimental Botany, 122*, 115–125.

Tabatabaei, S. J. (2009). Supplements of nickel affect yield, quality, and nitrogen metabolism when urea or nitrate is the sole nitrogen source for cucumber. *Journal of Plant Nutrition, 32*(5), 713–724. doi:10.1080/01904160902787834.

Velikova, V., Tsonev, T., Loreto, F., & Centritto, M. (2011). Changes in photosynthesis, mesophyll conductance to CO_2, and isoprenoid emissions in Populus nigra plants exposed to excess nickel. *Environmental Pollution, 159*(5), 1058–1066. doi:10.1016/j.envpol.2010.10.032.

Wang, Z., Zhang, Y., Huang, Z., & Huang, L. (2008). Antioxidative response of metal-accumulator and non-accumulator plants under cadmium stress. *Plant and Soil, 310*(1), 137. doi:10.1007/s11104-008-9641-1.

Yadav, S. S., Shukla, R., & Sharma, Y. K. (2009). Nickel toxicity on seed germination and growth in radish (*Raphanus sativus*) and its recovery using copper and boron. *Journal of Environmental Biology, 30*(3), 461–466.

Yusuf, M., Fariduddin, Q., Hayat, S., & Ahmad, A. (2011). Nickel: An overview of uptake, essentiality and toxicity in plants. *Bulletin of Environmental Contamination and Toxicology, 86*(1), 1–17. doi:10.1007/s00128-010-0171-1.

Zhang, M., Zhou, C., & Huang, C. (2006). Relationship between extractable metals in acid soils and metals taken up by tea plants. *Communications in Soil Science and Plant Analysis, 37*(3–4): 347–361. doi:10.1080/00103620500440095.

Zobiole, L. H. S., Oliveira Jr, R. S., Kremer, R. J., Constantin, J., Yamada, T., Castro, C., Oliveira, F.A., & Oliveira Jr, A. (2010). Effect of glyphosate on symbiotic N2 fixation and nickel concentration in glyphosate-resistant soybeans. *Applied Soil Ecology, 44*(2), 176–180.

Commentary
A Treatise on Metal Toxicology

Debasis Bagchi
Texas Southern University
Victory Nutrition International, Inc.

Manashi Bagchi
Dr. Herbs LLC

CONTENTS

Global toxic manifestations, especially by the toxic metals and metalloids including arsenic, lead, mercury, hexavalent chromium, and cadmium, devastated and rupturing the environmental and atmospheric integrity.[1-3] In the recent past, the World Health Organization (WHO) has listed cadmium, mercury, lead, and arsenic as part of major public health concern and environmental hazards including ozone layer depletion. These heavy metals are all around us, on the ground, in the water, in both vegetarian and non-vegetarian foods that we consume on a regular basis, and in the environment. Also, chronic exposure and accumulation of these noxious metals over a long period of time cause a broad spectrum of degenerative diseases including cancer.[1-3] Also, ships and submarines regularly use anti-fouling paints containing mercury and tributyltin oxide, and these heavy metals leach out to the aquatic climate and endanger the fishes and marine creatures, as well as the entire aquatic environment.[1-3] Consider the workers (a) engaged in heavy metal industries, (b) painters regularly exposed to lead- and cadmium-based paints, (c) fisherman exposed in the area of high level of mercury, (d) user of selected herbal medicines containing high levels of mercury and other toxic minerals, and (e) canned and preserved foods that we consume on a regular basis causes chronic accumulation of these noxious heavy metals, and all these people get victim of heavy metal toxicity, which ultimately lead to oxidative stress, DNA damage and loss of genomic integrity, fatal neurodegeneration causing a number of degenerative and deadly diseases, as well as metabolic and functional dysfunctions.[1-3]

Moreover, radionuclides consisting of 12 elements including Americium-241, Cesium-137, Cobalt-60, Iodine, Plutonium, Radium, Radon, Strontium-90, Technetium-99, Thorium, Tritium, and Uranium, as a result of continued exposure to electric and electronic resources as well as environmental radiation, pose potential risks to human and animal health.[4]

On the other hand, selected heavy metals including zinc, magnesium, silicon, manganese, aluminum, copper, iron, boron, and trivalent chromium are not injurious to health provided these are consumed in the appropriate doses to keep human bodies healthy.[2,5] As quoted long back by the Eminent German Physician Paracelsus (1493–1541), The Father of Toxicology, *"What is there that is not poison? All things are poisons, and nothing is without poison. Solely the dose determines that a thing is not a poison,"* which was the exactly translation of His quoted cited in the literature.

Another heavy metal, Titanium (Ti) has been reported to be non-toxic and doesn't exert serious adverse effects.[6] It has extensive applications in paint manufacturing and dentistry.[6] However, research demonstrated that titanium, especially its overdose, may exert adverse pulmonary effects leading to pleural disease, chest pain with tightness, respiration and breathing problems, coughing, and dermal and ocular irritation.[6]

Literature exhibits that Tin (Sn) is mined in Peru, Bolivia, and Brazil, as well as "tin belt" is located extensively in China, Thailand, and Indonesia, and commercially obtained in furnace by reducing the raw ore with coal. Inorganic tin compounds are known to be non-toxic to humans because they don't exert biological adverse effects and have been demonstrated to enter and excrete out from the body very rapidly in the feces.[7] Research has demonstrated that approximately 5% is absorbed from the GI tract and distributed, however, rapidly excreted by the kidney. However, sometimes tin is deposited in lung and bone and may induce renal necrosis. No mutagenic potential has been observed for metallic tin and inorganic tin compounds. Long-term animal exposure experiments exhibited fewer malignant tumors in treated animals as compared to the control animals. However, studies exhibited that the ingestion of significantly excessive amounts of inorganic tin causes stomachaches, anemia, hepatic, and kidney problems.

Selected species of **plants** have demonstrated their ability in absorbing contaminants like lead.[7] Like other metals, Tin is extensively used for (a) high polish, (b) coating other metals including steel and can coating to resist corrosion, (c) superconducting magnets made of niobium-tin alloy, (d) several alloys of tin are used including pewter, bronze, phosphor bronze, and soft solder, (e) electrically conductive coatings are manufactured by spraying tin on glass and (f) flat surfaces of window glass are prepared by floating molten glass on molten tin. In addition, Tin (II) chloride is employed for dyeing calico and silk as a mordant, while tin (IV) oxide is employed in the manufacturing of ceramics and gas sensors. On the other hand, organic tin compounds are extremely poisonous, while tributyltin oxide is extensively used as an antifouling paint for ships, boats, submarines, and naval underwater coatings, which are polluting the aquatic environment and marine life extensively.

It has been reported that aluminum is not absorbed in the human body through the skin, lungs, and GI tract.[7] However, prolonged chronic exposure to high levels of aluminum powders and dusts may exert pulmonary diseases and disorders.[7] Scientists have shown that no evidence exists for aluminum to induce Alzheimer's and neurodegenerative diseases, however, aluminum has been banned in the European subcontinents. Medical researchers have indicated that some Alzheimer's patients can show high levels of aluminum in their neuronal and brain tissues.[7]

Now, let us discuss the noble metals Platinum, Gold, and Silver. The occurrence of platinum is extremely low in soil, water, and air. Platinum occurs naturally as the uncombined metal or as an alloy of platinum-iridium. South Africa is the major source of platinum, which provides 75% of the world's total supply of platinum, while the USSR is the second-largest producer of platinum followed by North America.[8] Platinum and/or its alloys are widely used in the chemical, electrical, glass, and aircraft industries, as well as for (a) making fine jewelry for its wear- and tarnish-resistance properties, which consumes almost 50% of the total produce, (b) manufacturing surgical tools, electrical resistance wires, laboratory utensils, and electrical contact points, (c) as a catalyst in the catalytic converter, (d) an integral constituent of the gasoline-fueled car exhaust system, (e) platinum is used for manufacturing of optical fibers and liquid crystal display glass, especially for laptops, and (f) platinum-derived cisplatin is extensively used as a chemotherapy medication for treating diverse cancers including breast, esophageal, testicular, ovarian, cervical, lung, bladder, mesothelioma and head and neck cancers, as well as brain tumors and neuroblastoma.[8] However, platinum has been reported to aggravate the toxicity of selenium.[8]

Another noble and ancient metal is gold (Au), globally accepted as the most inert of metals, which is extensively used in jewelry and in its metallic form gold has been reported to be non-toxic, and sometimes ice cream is served with gold flakes as a dessert.[9] Also, it is regularly consumed by the heavy-duty wrestlers and bodybuilders. Burned gold dust (*Sarna Vasha*) is used in Ayurvedic drugs. Also, gold is extensively used in dentistry. However, it is important to know that gold may break down in the body releasing gold ions, which may be toxic for living organisms. Very recently, it is being used in gene therapy, a recent addition to medical pharmacopoeia.[9]

Silver, in its metallic form, induce minimal or no health risks. In fact, healthy skin provides a protective barrier against the absorption of silver.[10] However, exposure to soluble silver compounds demonstrated to exert toxic manifestations as demonstrated by hepatic, renal and intestinal tract injuries, neurological, nephrotoxicity, ocular and dermal irritation, hematological injury, and adverse modulation in the blood cells.[10] Silver may cause genotoxicity, but its carcinogenic potential hasn't yet been proven.[10]

In this book, we have extensively discussed the molecular mechanistic aspects, diverse adverse effects, and toxic manifestations, as well as numerous beneficial effects of minerals and included 28 chapters from the leading experts around the world.

REFERENCES

1. Tchounwou PB, Yedjou CG, Patolla AK, Sutton DJ. Heavy Metals Toxicity and the Environment. *EXS.* 2012; 101:133–164. doi:10.1007/978-3-7643-8340-4_6.

2. Stohs SJ, Bagchi D. Oxidative Mechanisms in the Toxicity of Metal Ions. *Free Radic Biol Med.* 1995 Feb; 18(2):321–336.

3. Wu X, Cobbina SF, Mao G, Xu H, Zhang Z, Yang L. A Review of Toxicity and Mechanisms of Individual and Mixtures of Heavy Metals in the Environment. *Environ Sci Pollut Res Int.* 2016 May; 23(9):8244–8259. doi:10.1007/s11356-016-6333-x. Epub 2016 Mar 11.

4. Iryna P, Toxicity of Radionuclides in Determining Harmful Effects on Humans and Environment. *J Environ Sci Public Health.* 2017; 1(2):115–119. doi: 10.26502/JESPH.011.

5. International Labor Organization (ILO). https://www.ilo.org/legacy/english/protection/safework/cis/products/safetytm/metals.htm (accessed Apr 5, 2020).

6. Kim KT, Eo MY, Nguyen TTH, Kim SM. General Review of Titanium Toxicity. *Int J Implant Dent.* 2019 Mar 11; 5(1):10. doi:10.1186/s40729-019-0162-x.

7. Safety Assessment of Tin (IV) Oxide as Used in Cosmetics. Final Report Dec 14, 2013. http://www.cir-safety.org/sites/default/files/tinoxi122012final_faa-final%20for%20posting.pdf (accessed Apr 5, 2020).

8. Platinum Chapter 6.11. WHO Regional Office for Europe, Copenhagen, Denmark, 2000, pp. 1–13.

9. Merchant B. Gold, the noble metal and the paradoxes of its toxicology. *Biologicals.* 1998; 26(1):49–59.

10. Hadrup N, Sharma AK, Loeschner K. Toxicity of silver ions, metallic silver, and silver nanoparticle materials after in vivo dermal and mucosal surface exposure: a review. *Regul Toxicol Pharmacol.* 2018 Oct; 98:257–267. doi:10.1016/j.yrtph.2018.08.007. Epub 2018 Aug 17.

Index